High Resolution Stratigraphy

Geological Society Special Publications

Series Editor J. BROOKS

GEOLOGICAL SOCIETY SPECIAL PUBLICATION NO. 70

High Resolution Stratigraphy

EDITED BY

E. A. HAILWOOD
Department of Oceanography,
University of Southampton, UK

and

R. B. KIDD
Department of Geology,
University of Wales College of Cardiff, UK

1993

Published by

The Geological Society

London

THE GEOLOGICAL SOCIETY

The Society was founded in 1807 as the Geological Society of London and is the oldest geological society in the world. It received its Royal Charter in 1825 for the purpose of 'investigating the mineral structure of the Earth'. The Society is Britain's national learned society for geology with a Fellowship of 6965 (1991). It has countrywide coverage and approximately 1000 members reside overseas. The Society is responsible for all aspects of the geological sciences including professional matters. The Society has its own publishing house which produces the Society's international journals, books and maps, and which acts as the European distributor for publications of the American Association of Petroleum Geologists.

Fellowship is open to those holding a recognized honours degree in geology or cognate subject and who have at least two years relevant postgraduate experience, or who have not less than six years relevant experience in geology of a cognate subject. A Fellow who has not less than five years relevant postgraduate experience in the practice of geology may apply for validation and subject to approval, may be able to use the designatory letters C. Geol (Chartered Geologist).

Further information about the Society is available from the Membership Manager, The Geological Society, Burlington House, Piccadilly, London W1V 0JU, UK.

Published by The Geological Society from:
The Geological Society Publishing House
Unit 7
Brassmill Enterprise Centre
Brassmill Lane
Bath BA1 3JN
UK
(*Orders*: Tel. 0225 445046
  Fax 0225 442836)

First published 1993

British Library Cataloguing in Publication Data

A catalogue record for this book is available from the British Library

ISBN 0-903317-86-9

Typeset by Type Study, Scarborough
Printed in Great Britain by Galliard (Printers) Ltd, Great Yarmouth

Distributors

USA
  AAPG Bookstore
  PO Box 979
  Tulsa
  Oklahoma 74101–0979
  USA
  (*Orders*: Tel. (918)584–2555
    Fax (918)584–0469)

Australia
  Australian Mineral Foundation
  63 Conyngham Street
  Glenside
  South Australia 5065
  Australia
  (*Orders*: Tel. (08)379–0444
    Fax (08)379–4634)

India
  Affiliated East-West Press PVT Ltd
  G-1/16 Ansari Road
  New Delhi 110 002
  India
  (*Orders*: Tel. (11)327–9113
    Fax (11)331-2830)

Japan
  Kanda Book Trading Company
  Tanikawa Bldg
  3–2 Kanda Surugadai
  Chiyoda-ku
  Tokyo 101
  Japan
  (*Orders*: Tel. (03) 3255–3497
    Fax (03) 3255–34957

Contents

KIDD, R. B. & HAILWOOD, E. A. High resolution stratigraphy in modern and ancient marine sequences: ocean sediment cores to Palaeozoic outcrop ... 1

Techniques: Chronology and Correlations

SMITH, A. G. Methods for improving the chronometric time-scale ... 9

MUSSETT, A. E. & MCCORMACK, A. G. Magnetic polarity timescales: a new test ... 27

THOMPSON, R. & CLARK, R. M. Quantitative marine sediment core matching, using a modified sequence-slotting algorithm ... 39

Quarternary and Tertiary

SMITH, M. B., POYNTER, J. G., BRADSHAW, S. A. & EGLINGTON, G. High resolution molecular stratigraphy: analytical methodology ... 51

ROBINSON, S. G. Lithostratigraphic applications for magnetic susceptibility logging of deep-sea sediment cores: examples from ODP Leg 115 ... 65

ALI, J. R., KING, C. & HAILWOOD, E. A. Magnetostratigraphic calibration of early Eocene depositional sequences in the southern North Sea Basin ... 99

JENKINS, D. G. & GAMSON, P. The late Cenozoic *Globorotalia truncatulinoides* datum-plane in the Atlantic, Pacific and Indian oceans ... 127

RADFORD, S. S. & LI QIANYU Eocene–Miocene high latitude biostratigraphy ... 131

WEAVER, P. P. E. High resolution stratigraphy of marine Quaternary sequences ... 137

WHATLEY, R. C. Ostracoda as biostratigraphical indices in Cenozoic deep-sea sequences ... 155

KNOX, R. W. O'B. Tephra layers as precise chronostratigraphical markers ... 169

SCHWARZACHER, W. Milankovitch cycles in the pre-Pleistocene stratigraphic record: a review ... 187

Mesozoic

McARTHUR, J. M., THIRLWALL, M. F., GALE, A. S., KENNEDY, W. J., BURNETT, J. A. MATTEY, D. & LORD A. R. Strontium isotope stratigraphy for the Late Cretaceous: a new curve, based on the English Chalk ... 195

WRAY, D. S. & GALE, A. S. Geochemical correlation of marl bands in Turonian chalks of the Anglo-Paris Basin ... 211

HART, M. B. Cretaceous foraminiferal events ... 227

HANCOCK, J. M. Transatlantic correlations in the Campanian–Maastrichtian stages by eustatic changes of sea-level ... 241

COPE, J. C. W. High resolution biostratigraphy ... 257

Palaeozoic

HOUSE, M. R. & KIRCHGASSER, W. T. Devonian goniatite biostratigraphy and timing of facies movements in the Frasnian of eastern North America ... 267

BECKER, R. T., HOUSE, M. R. & KIRCHGASSER, W. T. Devonian goniatite biostratigraphy and timing of facies movements in the Frasnian of the Canning Basin, Western Australia ... 293

LOYDELL, D. K. Worldwide correlation of Telychian (Upper Llandovery) strata using graptolites   323

BRASIER, M. D. Towards a carbon isotope stratigraphy of the Cambrian System: potential of the Great Basin succession   341

High resolution stratigraphy in modern and ancient marine sequences: ocean sediment cores to Palaeozoic outcrop

ROBERT B. KIDD[1] & ERNEST A. HAILWOOD[2]

[1] *Department of Geology, University College of Wales, PO Box 914, Cardiff CF1 3YE, UK*

[2] *Department of Oceanography, University of Southampton, Southampton SO9 5NH, UK*

The Deep Sea Drilling Project (DSDP), over its initial ten years of operations, relied almost entirely on a developing biostratigraphy based on microfossil groups for its stratigraphic control. During that same period, high resolution stratigraphic methods were developed for Quaternary marine cores through combination of microfossil biostratigraphy and paleomagnetic reversal sequences. The introduction of the hydraulic piston corer in 1982 provided the means to extend the surface core studies to depth and a major impulse to paleoceanographic studies. By the end of the DSDP programme in 1984, an integrated high resolution stratigraphy for the North Atlantic had emerged, based on microbiostratigraphy, magnetostratigraphy and isotope stratigraphy which had extended the identification of Milankovitch cycles to the Pliocene.

Through the Ocean Drilling Program's operations since 1986, integrated Cenozoic stratigraphies have been extended to all the major oceans and emphasis is now placed on extending these to land sections. Resolution in pre-Cenozoic marine sequences has begun to benefit from the same integrated approach along with emerging techniques such as strontium isotope stratigraphy and 'cyclostratigraphy'.

Introduction and background

Stratigraphy: a rapidly developing field

The deciphering of geological events from the sedimentary record requires increasingly precise stratigraphic resolution if we are to advance our understanding of the Earth's system and its processes. An important development of the 1980s was the recognition that the record of orbitally-modulated late Cenozoic and Quaternary climatic fluctuations might provide a key to improving resolution further back in the stratigraphic record (e.g. Imbrie 1985). It is important to improve stratigraphic resolution on these longer timescales in order to explore the causal relationships and driving mechanisms of pro-

cesses such as sea level and climate change. Sedimentologists attempting to establish rates of erosional, transportational and depositional processes, and palaeontologists specializing in evolutionary studies require particularly fine definition of geological events.

The striking advances that have been achieved in fine-scale stratigraphic studies of ocean sediment sequences have depended upon integration of information from a range of stratigraphic techniques, in particular the combination of microfossil biostratigraphy (generally using more than one fossil group) with magnetostratigraphy and/or oxygen isotope stratigraphy (e.g. Kennett 1982). By 1989, stratigraphic resolution of better than a few thousand years was being quoted for the Pleistocene by marine stratigraphers (Ruddiman *et al.* 1989) and successes were well established in extending the integrated magneto-biostratigraphic approach to Cenozoic marine stratotypes on land (Zijderveld *et al.* 1986; Channell *et al.* 1988; Premoli-Silva *et al.* 1988).

Concept of a 'multidisciplinary' conference on high resolution in stratigraphy

The Geological Society Marine Studies Group has, for many years, been concerned with the dissemination of developments in marine geology to the wider land-based geological community, through the promotion of national and international conferences. The majority of the papers presented in this volume were originally given at a Geological Society conference in London, co-sponsored by the Marine Studies Group and the Stratigraphy Committee. The principal aims of this meeting were (i) to focus on recent advances in Quaternary and late Cenozoic stratigraphic resolution, with a particular emphasis on developments in approaches such as magnetic, isotopic, geochemical and molecular stratigraphy, (ii) to extend these considerations to aspects of stratigraphic resolution in the Mesozoic, Palaeozoic and beyond,

From HAILWOOD, E. A. & KIDD, R. B. (eds), *High Resolution Stratigraphy*
Geological Society Special Publication, No. 70, pp. 1–8.

and (iii) to address problems of intercorrelation between different stratigraphic schemes and their chronometric calibration, together with developments in sequence stratigraphy and wire-line logging stratigraphy.

The response to the call for papers was outstanding and this necessitated extending the originally-planned two-day conference to two separate, but closely-linked two-day meetings, one held in October 1990 and the other in January 1991. This tremendous response is a clear indication of the extent of interest in 'open-forum' discussions on developments in stratigraphic resolution. The subjects addressed at the two meetings were broadly separated into two parts, the first focusing mainly on Quaternary and Tertiary topics, together with developments in stratigraphic techniques, and the second focusing largely on Mesozoic and Paleozoic topics. However, some mixing of topics between the two parts was maintained and both parts of the meeting were characterised by lively discussions and debates.

What is high resolution in stratigraphy?

One clear consensus on this question, arising from the four days of presentations and discussions, was the predictable: 'it depends on where you are in the geological timescale'. Just as the meaning of 'hi-fi' is, to a large extent, in the ear of the beholder, so is the definition of 'high resolution' in stratigraphy dependant on the particular application!

An attempt at a statement of achievable resolutions for different parts of the geological timescale is shown in Table 1. This represents the authors' view of the state-of-the-art in early 1991. Some of these figures are already being improved upon, especially for the late Quaternary, where concerns about global climate change are driving rapid developments over the shorter timescales. Figures for a decade earlier are shown in brackets to emphasise the rapid improvements in stratigraphic resolution in recent years.

Table 1. *Achievable resolution for integrated stratigraphy in marine sediment sequences (circa 1991)*

Quaternary	<1 to 3 ka (1980:c.20 ka)
Late Cenozoic	5 to 10 ka (1980:c.100 ka)
Early Cenozoic	10 ka to 1 Ma
Late Cretaceous	100 ka to 1 Ma
Early Cretaceous	c. 10 Ma
Jurassic	50 ka to 150 ka*
Triassic	225 ka to 2 Ma*

* See Cope (this volume).

Contributions to the Volume

The present volume makes no attempt to represent all of the various presentations and discussions at the October 1990 and January 1991 sessions. All of the contributors were invited to submit papers for peer-review and publication in the volume. Many accepted, but others had presented review papers or contributions on developments in techniques that had been published or were planned to be published elsewhere. Nevertheless, the volume brings together much of the proceedings of the sessions and, most importantly, provides under one cover a representative cross section of the issues discussed at this multidisciplinary gathering of stratigraphers.

The purposes of this editorial are (i) to summarize some of the stratigraphic developments in marine geology that led up to the conference, (ii) to highlight some particularly significant topics addressed at the two meetings (especially where the corresponding conference presentation did not result in a contribution to this volume), and (iii) to briefly consider some of the possible ways in which stratigraphic resolution may be improved in the future.

The ocean record

At the present day, approximately 70% of the Earth's surface lies beneath a cover of 4 km or more of seawater. Open ocean pelagic sequences deposited in this environment offer the greatest potential for stratigraphic completeness available in the geological record. Sampling of this record has been largely by piston coring techniques (Kullenberg 1947; Weaver & Schultheiss 1990) or by ocean drilling (Storms 1990).

Until the early 1980s, two distinct levels of stratigraphic resolution had been established for the ocean record, relating respectively to the deep sea drilling cores recovered through the oceanwide activities of the JOIDES drilling vessel 'Glomar Challenger' and conventional piston cores taken by other research vessels. The drilling programme depended almost entirely on microfossil biostratigraphy for its age control whilst major advances were being made by combining stratigraphic techniques in the study of piston cores.

Development of an integrated stratigraphy

The limit of penetration of conventional piston corers in pelagic carbonate-rich sequences is around 20 m subbottom. A sediment sequence recovered by this method usually consists of a 6.5 cm diameter transparent plastic tube filled with the sediment, which has been cut into 1.5 m long sections. In the laboratory, these sections are split lengthwise, revealing a flat surface for analysis. One of the split halves is usually designated the 'archive' and is used for core description, photography and a range of non-destructive physical logging investigations. The other half section is designated the 'sampling section' and from this is derived the material for biostratigraphic and sedimentological analyses. The suite of stratigraphic techniques applied to such a core would generally involve:

(1) Microfossil biostratigraphy. The essence of this technique is the identification of first appearance (FAD) and extinction (LAD) data for the calcareous foraminifera and nannofossils (Bolli & Premoli-Silva 1973; Martini 1971) or for the siliceous diatoms and radiolaria (Burckle 1972; Nigrini 1970), depending on site location, water depth and other factors. Other microfossil groups may be used to enhance the biostratigraphy if they are regionally abundant. Resolution of the biostratigraphy alone depends in particular upon the characteristics and numbers of groups and species of fossil organisms present in the core and on the extent of diachroneity of the data used. (Kennett 1982; Jenkins & Gamson this volume)

(2) Magnetostratigraphy. The identification of records of geomagnetic polarity reversal sequences in oceanic sediment cores, by palaeomagnetic analyses of whole sediment cores or discrete samples, provides age-control points for the biostratigraphic data (Ryan *et al.* 1974; Berggren *et al.* 1985; Hailwood 1989). The typical duration of magnetic polarity zones is in the range 10^5 to 10^6 years. For some intervals, resolution may be enhanced by an order of magnitude or more, through the identification of short-period magnetic 'events' or geomagnetic excursions. Downcore variations in magnetic intensity and susceptibility provide further useful correlation tools. Magnetostratigraphic resolution is controlled in particular by the effectiveness of demagnetization techniques in removing magnetic overprints (Hailwood 1989) and on the extent of bioturbative mixing of the sediment.

(3) Carbonate analyses. Down-core variations in $CaCO_3$ content offer a powerful tool for inter-core correlation. They provide a record of Quaternary and older glacial cyclicity (Ruddiman & McIntyre 1976). Apart from the deepest

areas of the ocean basins, below the carbonate compensation depth (Berger & Vincent 1981), calcium carbonate is ubiquitous in the oceans. However, the proportion of carbonate present in the sediment varies as a result of changing organic productivity and ocean chemistry through time and varying input of non-carbonate material, such as ice-rafted and wind-blown material. Core correlation based on these variations provides a useful first order framework on which to 'hang' the other stratigraphic parameters. The stratigraphic resolution offered by such correlations depends in particular upon the degree of bioturbation and of redeposition of material, either as slumps or as turbidites, in the sediment sequence.

(4) Oxygen isotope stratigraphy. Downcore analyses of oxygen isotopic composition of calcareous foraminiferal tests provide detail on the past chemical state of the oceans and the nature of glacial meltwater input (Shackleton & Opdyke 1973). They also give information on glacial/Milankovich cyclicity and provide a further useful tool in core correlation. Resolution of the oxygen isotope stratigraphy depends upon the availability of suitable foraminiferal tests, on sedimentation rates and ocean turnover parameters, on the extent of dissolution and redeposition of the carbonate and on bioturbative processes. As with carbonate stratigraphy, first order age-calibration of oxygen isotope fluctuations, is achieved through use of bio- and magneto-stratigraphic information.

Where possible, stratigraphic investigations of marine cores use these four techniques together, to form the basis of an 'integrated stratigraphy'. Some workers in this field have suggested that such integrated stratigraphies should form the basis of 'stratotypes' for the marine Quaternary (e.g. Imbrie *et al.* 1984; Bowen 1984). Numerical calibration of the time-scales derived form such studies is normally achieved by referring the magnetostratigraphy to absolute ages derived from igneous rocks on land or, where suitable authigenic minerals are present, by absolute dating on the cored sediments themselves.

Certain geological 'events' effectively provide a stratigraphic control of their own and are sometimes used to enhance an integrated stratigraphy. Thus, subaerial explosive volcanism brings about regional or even global ash distributions which are deposited as tephras and form particularly useful marker beds in marine cores (e.g. Kennett 1981). Some of these events can be tied to the historical record (e.g. Ninkovich & Donn 1975). Older tephras, dated by K/Ar or

C14 single crystal dating techniques, or magnetostratigraphy of their host sediments can provide excellent control points in an integrated stratigraphy (see Knox this volume). In some marine cores from enclosed basins, organic carbon-rich sapropel layers provide marker beds of particular value in core correlation (McCoy 1974) whilst in open ocean high-productivity settings laminated 'varves' can offer the potential for annual, even seasonal, levels of resolution (Suess & Thiede 1983).

Milankovitch and CLIMAP. Probably the most comprehensive use of integrated stratigraphy on sets of piston cores was that of the CLIMAP project, a multi-disciplinary effort to provide maps of oceanwide sea-surface temperatures and other climatic parameters for the last few glacial cycles, derived from sediment and microfossil analyses (Hays *et al.* 1969; Imbrie & Kipp 1971; Ruddiman & McIntyre 1976). The CLIMAP group utilized an extensive collection of piston cores held at the Lamont-Doherty Geological Observatory, New York to successfully map the state of the oceans at the maximum of the last glacial cycle, 18 000 years BP (Cline & Hays 1976). The CLIMAP studies were instrumental in establishing the applicability of Milankovitch's (1920) concept of climatic cycles to the interpretation of fine-scale stratigraphic variations. Predictable changes in the Earth's orbital parameters, namely the eccentricity ('stretch' of the orbit around the sun) with a 100 000 year periodicity, the obliquity (tilt of the Earth's rotational axis) with a 41 000 year periodicity, and precession of the equinoxes (a 'wobble' in the longitude of the perihelion) with a 23 000 year periodicity (Imbrie *et al.* 1984), combine to produce cyclic variations in oceanic productivity, which are reflected in pelagic sediment sequences. The periodicities have not changed through the relatively short duration of Quaternary glacial history and are easily detected in the frequencies of variation in oxygen isotope parameters, carbonate content, foraminiferal abundances, etc. of sediment cores. CLIMAP's successor SPECMAP extended these studies through the entire interval of Plio-Quaternary glacial history (see Imbrie & Imbrie 1988) and the ages of the isotope stage boundaries are now known with an accuracy of about one thousand years (see Weaver this volume). Stratigraphies developed by such studies on mainly pelagic sequences have been extended to other settings, and have aided the identification of redeposition events (Weaver & Kuipers 1983; Weaver this volume).

DSDP to ODP. The rotary drilling technique with wireline coring that was the mainstay of the Deep Sea Drilling Project's drilling operations with the drillship *Glomar Challenger*, was effective in recovering sediments and rock through to the oceanic basaltic basement. However, in the upper 100 m or so of unlithified sediments the core recovery was generally poor. In most cases, the superposition of successive sedimentary units was preserved but the sequence was disturbed to varying degrees. Marine stratigraphers were left with a dichotomy of resolution, from the 'high resolution' integrated stratigraphy of the surface piston core (with the 10 000–50 000 year resolution provided by the oxygen isotope stages) to the much lower resolution (at the biozone-level) in the drill-cores recovered from deeper in the sequences.

The development by the DSDP of the Hydraulic Piston Corer (HPC) in 1982 provided the impetus for a major leap forward in marine stratigraphy. In HPC coring, a specialized core barrel is lowered down the inside of the drillpipe to the drill bit, as in standard wireline coring, but is then hydraulically 'shot' into the undisturbed sediment beyond the current base of the hole. After recovery of each core, the drill string is then lowered to the next level for hydraulic piston coring and the HPC is fired again, so that a continuous sediment section, undisturbed by the normal rotary and washing action of drilling, is recovered. This development allowed the acquisition of superior quality cores down to depths of 150 to 200 m below the sea floor, beyond which the sediment normally becomes too lithified for further penetration by the HPC tool. A second coring development by DSDP at around the same time, the extended core barrel (XCB) corer, allowed relatively undisturbed sediment cores to be recovered through to igneous basement. In this device, a wireline core barrel with its own cutting edge latches into the drillbit and extends about 15 cm beyond the roller-cones and wash nozzles that cause disturbance at the base of the hole in standard rotary drilling. By the end of the DSDP programme in 1984, it had become routine on a drilling cruise with paleoenvironmental objectives to attempt to recover complete, undisturbed, sediment sections down to oceanic basement, using combined HPC/XCB coring (Ruddiman *et al.* 1987). Stratigraphic studies now had the benefit of integrated biostratigraphy and magnetostratigraphy defined from shipboard analyses of discrete samples from the cores (Weaver & Clement 1986). These provided an important control for the drilling operations and in most cases a second or even third borehole was drilled at each location so that overlapping cores could be taken to ensure continuous recovery of the section. The most striking utilization of these techniques was in DSDP's targetting of Quaternary climate change and Milankovitch cyclicity, by extending the integrated surface core stratigraphies to much greater depth. The onset of the Northern Hemisphere glacial cycles was identified at 2.5 Ma, well into the Pliocene, in the Atlantic, and the base of the Quaternary was set at 1.65 Ma, at the start of oxygen isotope stage 22 (Ruddiman *et al*. 1986).

The ability to study long-term climate change became one of the prime reasons for the international impetus to enter a new phase of scientific ocean drilling, leading to the launch of the Ocean Drilling Program (ODP) in 1985. The new drillship '*JOIDES Resolution*' is equipped with facilities for shipboard whole-core logging of magnetic and other physical properties before the cores are split. A range of specialized downhole logging techniques, able to resolve fine-scale variations such as glacial cyclicity (Jarrard & Arthur 1989) have been developed by ODP. Shipboard scientists now have a wide range of stratigraphic and proxy-stratigraphic techniques that can be integrated together in the pursuit of high resolution stratigraphy. The linked biostratigraphic–magnetostratigraphic–oxygen isotope scale has been extended to 3.3 Ma, representing a total of 137 isotope stages (Ruddiman *et al*. 1988) and the present emphasis is on extending this fine scale stratigraphy further into the Cenozoic.

Other techniques that were developed during the early years of ODP include magnetic susceptibility core logging (Robinson this volume), P-wave and gamma ray core loggings (Weaver this volume) and molecular stratigraphy (Smith *et al*. this volume). Balanced against the striking developments in high-resolution stratigraphy arising from ODP, some of the problems and limitations inherent in the original surface core studies remain, such as the effects of diachroneity of microfossil FADs and LADs (see Jenkins & Gamson this volume), limitations of magnetostratigraphy in low latitudes, and the effects of species variation on oxygen isotope stratiography. However, refinements in resolution continue, as outlined below.

The changing face of outcrop stratigraphy

Applying integrated stratigraphy to the outcrop

At around the same time as the launch of the ODP, groups of Cenozoic stratigraphers began to turn their attention to the stratotype sections exposed in outcrop, with a view to integrating the original biostratigraphic data with new data from magnetic and other stratigraphic techniques. The position of the Miocene/Pliocene boundary at Capo Rosselo and at Capo Spartivento, Italy was refined using magnetic and planktonic foraminiferal analyses by Zijderveld *et al*. (1986) and Channell *et al*. (1988) respectively and the Eocene–Oligocene boundary at Massignano, Italy was refined using the same techniques by Premoli-Silva *et al*. (1988). Handheld devices for magnetic susceptibility logging have now become popular in the reassessment of outcrop sections (Weedon 1992) and gamma ray logging has been used by exploration teams to correlate between boreholes and outcrop. Many of these kinds of studies are currently being mounted on land sections and, whilst stratotypes for epoch boundaries are unlikely be relocated, there remains the possibility that marine 'stratotypes' may prove to be the only continuous sections available for some stage boundaries.

Developments in sequence stratigraphy

One of the major stratigraphic advances of recent years has been the application of sequence stratigraphic concepts to seismic profiles (Vail *et al*. 1977; Mitchum *et al*. 1977). Although this conference attracted presentations that addressed regional interpretations of stratigraphy using sequence nomenclature, this type of study was not central to the main theme of the meeting and none are included in this volume. Although changes in relative sea level are widely believed to be the main driving mechanism for sequence development (Vail *et al*. 1977; Haq *et al*. 1988), there is still a great deal of controversy over the role of tectonics in the development of sequence architecture (Macdonald 1992). The reader is referred to the review by Wilson (1990) for a balanced presentation of sequence stratigraphy as an interpretational tool. One important development of sequence stratigraphy has been the application of the integrated approach, involving a number of different biostratigraphic zonation schemes, along with magnetic stratigraphy, to the study of outcrop sections on a regional basis. Ali *et al*. (this volume) address the problem of age-calibrating early Eocene sequence boundaries using the magnetic time scale. Several authors have constructed integrated stratigraphies for different European and American sections, basing their interpretations on sequence stratigraphic interpretations.

Pre-Pliocene Milankovich cycles

With the recognition of Milankovich cyclicity in the Plio-Pleistocene as a powerful tool for stratigraphy (e.g. House 1985; Berger 1988; Fischer *et al.* 1990) it became clear that the potential of this approach should be realisable for much of geological time. This prospect has excited much interest among sedimentologists and stratigraphers interested in the interpretation of cyclic sequences throughout the geologic record. Schwarzacher (this volume) outlines the basis for Milankovich-related studies in the pre-Pleistocene, while House & Kirchgasser (this volume) link goniatite stratigraphy in the Devonian to Milankovich cyclicity. Clearly this is an avenue of great future potential for high resolution stratigraphic studies of the Mesozoic and Palaeozoic.

Pre-Cenozoic developments

A number of the contributions to this volume report refinements of biostratigraphic zonations or the limitations that can be placed on the use of certain fossil groups in the search for higher resolution. For the Mesozoic and Palaeozoic, perhaps the most notable new techniques to come to the fore include strontium isotope stratigraphy (see McArthur *et al.* this volume) and the use of tephras and geochemical marker horizons in regional correlations (Knox this volume; Wray & Gale this volume).

Future needs and developments

The lively discussions that were a conspicuous feature of the London conference, articulated a number of themes that have become central to subsequent stratigraphic developments and have pointed the way to future needs for high resolution stratigraphy.

There will obviously be a continued trend to utilize integrated stratigraphic techniques on outcrop investigations and further attempts to rationalise sequence stratigraphy with this approach. As the magnetic polarity timescale becomes further refined for Cretaceous and earlier periods (e.g. Cande *et al.* 1992) there will be further impetus to refine biozonations. The study of Milankovich cyclicity in the Mesozoic and Palaeozoic is likely to prove increasingly fruitful.

Whilst the land specialists continue their refinements, the marine stratigraphers are entering a new phase of ODP core analysis with the development of shipboard core-colour logging techniques (Mix *et al.* 1992) which, when integrated with other fine-scale stratigraphic techniques, offer the potential for recognition of seasonal-scale cyclicity (Leg 138 Shipboard party 1992). This is leading to exciting potential developments in the characterization and interpretation of pre-Cenozoic sedimentary cycles on the finest scales.

We are grateful to the many participants at the London meeting, who contributed to the lively discussion and debate on the current status of high resolution stratigraphy. We thank those who submitted papers to this volume, and the many diligent referees who helped us to achieve the set of high quality contributions contained herein. We are particularly grateful to the following Companies, who generously made donations towards the costs of running the conference and producing this volume: Amoco UK, BP Petroleum Development, Clyde Expro, Fina Exploration, Shell UK and Unocal UK.

References

ALI, J. R., KING, C. & HAILWOOD, E. A. 1993. Magnetostragraphic calibration of early Eocene depositional sequences in the Southern North Sea Basin. *This volume.*

BERGER, A. 1988. Milankovitch theory and climate. *Reviews of Geophysics*, **26**, 624–657.

BERGER, W. H. & VINCENT, E. 1981. Chemostratigraphy and biostratigraphic correlation: exercises in systematic stratigraphy. *Oceanologica Acta*, **4**, 115–127.

BERGGREN, W. A., KENT, D. V. & VAN COUVERING, J. A. 1985. Neogene geochronology and chronostratigraphy. *In*: SNELLING, N. J. (ed.) *The Chronology of the Geological Record.* Geological Society, London, Memoir, **10**, 211–260.

BOLLI, H. M. & PREMOLI-SILVA, I. 1973. Oligocene to Recent planktonic foraminifera and stratigraphy of Leg 15 sites in the Caribbean Sea. *In*: EDGAR, N. J., SAUNDERS, J. B. *et al.*, *Initial Reports of DSDP*, **15**. US Govt. Printing Office, Washington, 475–497.

BOWEN, D. Q. 1984. *Quaternary geology: a stratigraphic framework for multidisciplinary work.* Pergamon Press, Oxford, UK. 237pp.

BURCKLE, L. H. 1972. Late Cenozoic planktonic diatom zones from the eastern equatorial Pacific. *Beih. Zwr. Nova. Hedwegia, Helf*, **39**, 217–246.

CANDE, S. C. & KENT, D. V. 1992. A new geomagnetic polarity time-scale for the Late Cretaceous and Cenozoic. *Journal of Geophysical Research*, **97**, 13 917–13 951

CHANNELL, J. E. T., RIO, D. & HILGEN, R. C. 1988. Miocene-Pliocene boundary magnetostratigraphy at Capo Spartivento, Calabria, Italy. *Geology*, **16**, 1096–1099.

CLINE, R. M. & HAYS, J. D. (eds) 1976. *Investigations of Late Quaternary paleoceanography and paleoclimatology*; Geological Society of American memoir, **145**.

COPE, J. C. W. 1993. High resolution biostratigraphy. *This volume*.

FISCHER, A. G., DE BOER, P. L. & PREMOLI-SILVA, I. 1990. Cyclostratigraphy. *In*: GINSBURG, R. N. & BEAUDOIN, B. (eds) *Cretaceous Resources, Events and Rhythms*. Reidel, NATO ASI Series C, **304**, 139–172.

HAILWOOD, E. A. 1989. *Magnetostratigraphy*. Geological Society, London, Special Report, **19**.

HAQ, B. L., HARDENBOL, J. & VAIL, P. R. 1988. Mesozoic and Cenozoic chronostratigraphy and cycles of sea-level change. *In*: WILGUS, C. K., HASTINGS, B. S., KENDALL, G., POSAMENTIER, H. W., ROSS, C. A. & VAN WAGONER, J. C. (eds) *Sea-level Changes: an Integrated Approach*. Special Publication of the Society of Economic Paleontologists and Mineralogists, **42**, 71–108.

HAYS, J. D., SAITO, T. S., OPDYKE, N. D & BURCKLE, L. H. 1969. Pliocene-Pleistocene sediments of the eastern Pacific: their paleomagnetic, biostratigraphic, and climate record. *Geological Society of America Bulletin*, **80**, 1481–1514.

HOUSE. M. R. 1985. A new approach to an absolute timescale from measurements of orbital cycles and sedimentgary micro-rhythms. *Nature,* **315**, 712–725.

—— & KIRCHGASSER, W. J. 1993. Devonian goniatite biostratigraphy and timing of facies movement in the Frasnian of the Canning Basin, Western Australia, *This volume*.

IMBRIE, J. 1985. A theoretical framework for the Pleistocene ice ages. *Journal of Geological Society, London*, **142**, 417–432.

—— & IMBRIE, J. Z. 1980. Modelling the climatic response to orbital variations. *Science*, **207**, 943–953.

—— & KIPP, N. G. 1971. A new micropaleonotological method for quantitative paleoclimatology: application to a late Pleistocene Caribbean core. In: TUREKIAN, K. K. (ed.) *The Late Cenozoic Glacial Ages*. New Haven, Yale Univ. Press, 71–181.

—— & HAYS, J. D., MARTINSON, D. G., McINTYRE, A., MIX, A. C., MORLEY, J. J., PISIAS, N. G., PRELL, W. L. & SHACKLETON, N. J. 1984. The orbital theory of Pleistocene climate: support from a revised chronology of the marine $\delta^{18}O$ record. *In*: BERGER, A., IMBRIE, J., HAYS, J., KUKLA, G. & SALTZMAN, B. (eds) *Milankovitch and Climate*. NATO ASI Series C: Mathematical and Physical Sciences, **126**, 269–305.

JARRARD, R. D. & ARTHUR, M. A. 1989. Milankovitch paleoceanographic cycles in geophysical logs from ODP Leg 105: Labrador Sea and Baffin Bay. *In*: SRIVASTAVA, S. P., ARTHUR, M., CLEMENT, B. *et al.* (eds) *Proceedings of ODP Scientific Results*, **105**. Ocean Drilling Program, College Station, Texas, 757–772.

JENKINS, D. G. & GAMSON, P. 1993. The late Cenozoic *Globorotalia truncatuylinoides* datum-plane in the Atlantic, Pacific and Indian Oceans. *This volume*.

KENNETT, J. P. 1981. Marine tephrochronology. *In*: EMILIANI, C. (ed.) *The Sea (Vol. 7) The Oceanic*

Lithosphere. Wiley-Interscience, New York, 1373–1436.

—— 1982. *Marine Geology*. Prentice-Hall Inc, Englewood Cliffs, N.J.

KNOX, R. W. O'B. 1993. Tephra layers as precise chronostratigraphic markers. *This volume*.

KULLENBERG, B. 1947. *The piston core sampler*. Sveska Hydrogr. Biol. Komm. Series 1.

MACDONALD, D I. M. 1991. Sedimentation, tectonics and eustasy: sea-level changes at active margins. *In*: MACDONALD, D. I. M. (ed.) *Special Publication of the International Association of Sedimentologists*, **12**. Blackwell Scientific Publications, Oxford.

MARTINI, E. 1971. Standard Tertiary and Quaternary calcareous nannoplankton zonation. *In*: FARINACCI, A. (ed.) *Proceedings of the II Plankton Conference Roma, 1971*. Rome, 739–777.

McARTHUR, J. M., THIRLWALL, M. F., GALE, A. S., KENNEDY, W. S., BURNETT, J. A., MATLEY, D. & LANEL, A. R. 1993. Strontium isotope stratigraphy for the late Cretaceous: a new curve, based on the English Chalk. *This volume*.

McCOY, F. W. 1974. *Late Quaternary sedimentation in the eastern Mediterranean Sea*. PhD Thesis, Harvard University, 132pp.

MILANKOVITCH, M. 1920. *Theorie mathematique des phenomenes thermiques produits par la radiation solaire*. Gauthiers-Villars, Paris, 1941, Royal Serbian Acadamy, Special Publication, **133**.

MITCHUM, R. M., JR., VAIL, P. R. & THOMPSON, S. III. 1977. The depositional sequence as a basic unit for stratigraphic analysis. *In*: PAYTON, C. E. (ed.) *Seismic Stratigraphy – Applications to Hydrocarbon Exploration*. American Association of Petroleum Geologists Memoir **26**, 53–62.

MIX, A. C., RUGH, W., PISIAS, N. G & VIERS, S., Leg 138 Shipboard Sedimentologists, and the Leg 138 Scientific Party in press. Color Reflectance Spectroscopy: A tool for rapid characterization of deep-sea sediments. *In*: MAYER, L., PISIAS, N., JANECEK, T. *et al.* (eds) *Proceedings of ODP, Initial Reports*, **138**, Ocean Drilling Program. College Station, TX.

NIGRINI, C. 1970. Radiolarian assemblages in the North Pacific and their application to a study of Quaternary sediments in Core V20–130. *Geological Society of America Memoir*, **126**, 139–183.

NINKOVITCH, D. & DONN, W. L. 1975. Explosive Cenozoic volcanism and climatic interpretations. *Science,* **194**, 899–906.

PREMOLI-SILVA, I., COCCIONI, R., MONTONARI, A. *et al.* 1988. *The Eocene–Oligocene boundary in the Marche–Umbria Basin (Italy)*. Ancona, Italy.

ROBINSON, S. G. 1980. Lithostratigraphic applications for magnetic susceptibility logging of deep-sea sediment cores: examples from ODP Leg 115. *This volume*.

RUDDIMAN, W. F. & McINTYRE, A. 1976. Northeast Atlantic paleoclimatic changes over the past 600,000 years. *In*: CLINE, R. M. & HAYS, J. D. (eds) *Investigation of Late Quaternary Paleoceanography and Paleoclimatology*. Geological Society of American Memoir, **145**, 111–146.

——, KIDD, R. B., THOMAS, E. *et al.* 1987. *Initial Reports of DSDP,* **94**, Parts 1 and 2. US Govt. Printing Office, Washington, 615–634.

—— & MCINTYRE, A. 1986. Matuyama 41,000 year cycles: North Atlantic Ocean and northern hemisphere ice sheets. *Earth and Planetary Science Letters,* **80**, 117–129.

——, RAYMO, M. E., MARTINSON, D. G., CLEMENTS, B. M. & BACKMAN, J. 1989. Pleistocene evolution: northern hemisphere ice sheets and North Atlantic Ocean. *Paleoceanography,* **4**, 353–412.

——, SARNTHEIN, M., BALDAUF, J *et al.* 1988. *Proceedings of ODP, Initial Reports,* **108**. Ocean Drilling Program, College Station, TX.

RYAN, W. B. F. *et al.* 1974. A paleomagnetic assignment of Neogene stage boundaries and the development of isochronous datum planes between the Mediterranean, the Pacific and Indian Oceans in order to investigate the response of the world ocean to the Mediterranean "salinity crisis". *Rivista Italiana di Paleontologia,* **80**, 631–688.

SCHWARZACHER, W. 1993. Milankovitch cycles in the pre-Pleistocene stratigraphic record. *This volume.*

SHACKLETON, N.J. & OPDYKE, N. D. 1973. Oxygen isotope and paleomagnetic stratigraphy of equatorial Pacific core V28-238: oxygen isotope temperatures and ice volumes on a 10^5 year and 10^6 year scale. *Quatertarny Research,* **3**, 39–55.

SMITH, M. B., POYNTER, J. G., BRADSHAW, S. A. & EGLINGTON, G. 1993. High resolution molecular stratigraphy: analytical methodology. *This volume.*

STORMS, M. A. 1990. Ocean Drilling Program (ODP) Deep sea coring techniques. *Marine Geophysical Researches,* **12**, 109–130.

SUESS, E. & THIEDE, J. 1983. Responses of the sedimentary regime to coastal upwelling. *In:* SUESS, E. & THIEDE, J. (eds) *Coastal Upwelling: Its sediment record, Part A.* Plenum Press, New York, 000–000.

VAIL, P. R., MITCHUM, R. M. JR. & THOMPSON, S. III. 1977. Global cycles of relative changes of sea level. *In:* PAYTON, C. E. (ed.) *Seismic Stratigraphy – Applications to Hydrocarbon Exploration.* American Association of Petroleum Geologists Memoir, **26**, 83–97.

WEAVER, P. P. E. 1993. High resolution stratigraphy of marine Quaternary sequences. *This volume.*

—— & CLEMENT, B. M. 1986. Synchroneity of Pliocene planktonic foraminiferal datums in the North Atlantic. *Marine Micropaleontology,* **10**, 295–307.

—— & KUIJPERS, A. 1983. Climatic control of turbidite deposition on the Madeira Abyssal Plain. *Nature,* **306**, 360–363.

—— & SCHULTHEISS, P. J. 1990. Current methods for obtaining, logging and splitting marine sediment cores. *Marine Geophysical Researches,* **12**, 85–100.

WEEDON, G. P. 1989. The detection and illustration of regular sedimentary cycles using Walsh power spectra and filtering, with examples from the Lias of Switzerland. *Journal of the Geological Society, London,* **146**, 133–144.

WILSON, R. C. 1991. Sequence stratigraphy: an introduction. *Geoscientist,* **1**, 13–23.

WRAY, D. S. & GALE, A. S. 1993. Geochemical correlation of marl bands in Turonian chalks of the Anglo-Paris Basin. *This volume.*

ZIJDERVELD, J. D. A., ZACHARIASSE, J. W., VERHALLAN, P. J. J. & HILGEN, F. J. 1986. The age of the Miocene/Pliocene boundary. *Newsletters in Stratigraphy,* **6**, 169–181.

Methods for improving the chronometric time-scale

ALAN G. SMITH

Department of Earth Sciences, The University, Downing Street, Cambridge CB2 3EQ, UK

Abstract: The problems of improving the chronometric time-scale are discussed with reference to the chronogram method for the estimation of the Cretaceous–Tertiary boundary. The need for better assessment of stratigraphic errors is emphasized. Suggestions are made on how the position of dated horizons relative to stratigraphic boundaries may be more precisely described. The assumption that stratigraphic thickness is linearly related to time is considered in relation to the likely gaps in the stratigraphic record from a variety of environments.

A geological time-scale is made up of standard divisions, based on rock sequences, that are calibrated in years (Harland *et al.* 1990, p. 1). The chronometric scale, based on the standard second, is joined to a time-scale based on rock sequences, or a chronostratic scale. As presently defined, the chronostratic scale is based on reference points located in stratigraphic sections were fossil variations are believed to be relatively completely preserved across a stratigraphic boundary. Such sections are known as boundary stratotypes.

The magnetic polarity scale forms another chronostratic scale. For late Jurassic and later time the sequence of geomagnetic polarity changes is defined largely by fluctuations in the magnetic field profiles of ocean-floor rocks. Certain profiles have been adopted as informal reference standards. Except for details, the polarity record back to late Jurassic time is believed to be complete. Because the older record is preserved in stratigraphic sections rather than in the ocean-floor, it is discontinuous and not yet well defined. It is not customary to formally define standard reference sections for these earlier polarity changes. Other chronostratic scales are being constructed, such as those based on variations in stable isotopes, but they will not be discussed further here.

The boundary points of chronostratic scales are conventions to be agreed rather than discovered, though the positions of boundary points for some chronostratic scales, such as the magnetic polarity time-scale, are much more readily agreed than for other chronostratic scales. Their calibration in years is a matter of discovery or estimation (Harland *et al.* 1990). Because the chronostratic scales depend on international agreement, they change relatively slowly. Their calibration with the chronometric scale is subject to refinements that are dependent on advances in measurements, in turn dependent on technology. In times of rapid technical advance, as now, changes in the chronometric scale will be frequent.

In addition to the calibration of chronostratic scales with the chronometric scale, there is much interest in the calibration of one chronostratic scale with another, particularly of the biostratigraphic and magnetic polarity scales (e.g. Berggren *et al.* 1985a; Hailwood 1989). This paper discusses some of the problems encountered in calibrating parts of the Harland *et al.* (1990) time-scale and suggests how calibrations may be improved.

Present methods

Prior to about 5 Ma, the time-scale is defined chronostratically by either the magnetic polarity time-scale or by a biostratigraphic scale. These time-scales can be calibrated chronometrically using isotopic dating techniques.

(i) By dating samples from stratigraphically controlled horizons (Fig. 1) or a suite of samples that span a particular stage boundary, each of which can be correlated to the stratotypes of the stages immediately above or below the boundary (see 'Chronograms', below).

(ii) By dating magnetic chrons, such as the uppermost igneous ocean-floor known to belong to a particular chron, or dating a sample whose position within a magnetic chron is based on magnetic stratigraphy (Fig. 2).

(iii) By dating samples whose biostratigraphic position can be correlated to the magnetic polarity scale (Fig. 3).

Dating biostratigraphically controlled samples

Although a suite of samples from one section

From HAILWOOD, E. A. & KIDD, R. B. (eds), *High Resolution Stratigraphy*
Geological Society Special Publication, No. 70, pp. 9–25.

9

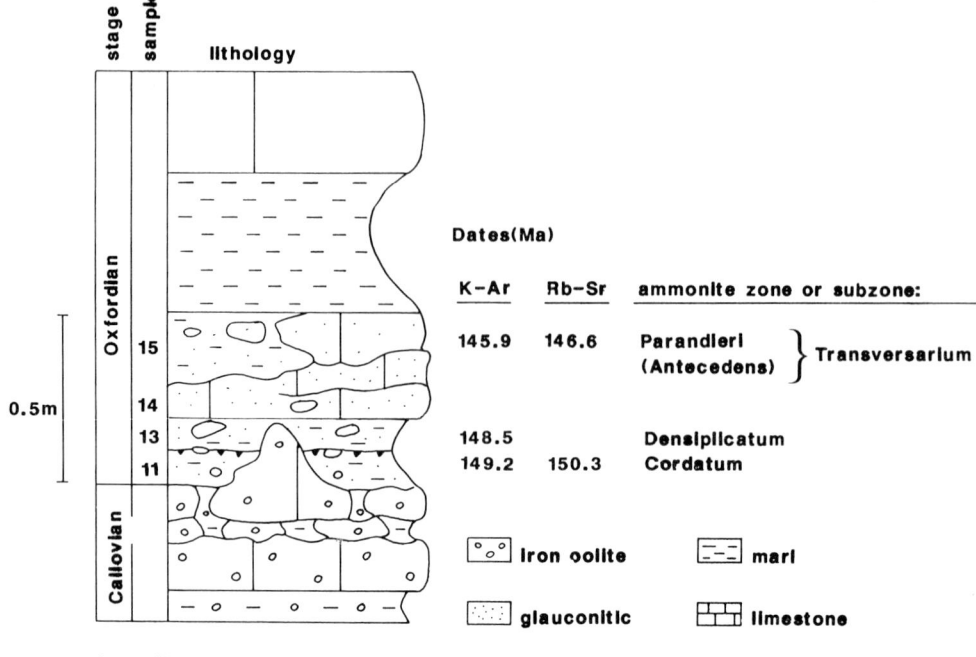

Locality: Gachlingen, Lang Ran-Jen

Fig. 1. The dated samples can be correlated directly to ammonite subzones (From Fischer & Gygi 1989). Direct correlation of dated samples to the biostratigraphic scales.

may be analysed precisely and related to biostratigraphic zones, such dates must be compared to dates obtained elsewhere. For example, Fischer & Gygi (1989) give concordant K-Ar and Rb-Sr dates from glauconites in undeformed Jurassic rocks in the Tabular Jura of Switzerland. These dates, published too late for inclusion in the Harland *et al.* (1990) database, suggest that early Oxfordian time spans the range 150–146 Ma, whereas Harland *et al.* (1990), gives *c.*157–155 Ma, a difference in duration of about 50% because the value is the difference between two numbers with relatively large errors that are relatively close to each other. The ages assigned to the base and top of the Oxfordian differ by about 5%, well outside the analytical errors.

One solution to such a discrepancy is to reject one or other set of data. However, even after the most careful selection has taken place the dates obtained at different locations from which the age of a stratigraphic boundary may be inferred will, in general, still differ significantly. The reasons for the differences will lie mostly in systematic errors in the isotopic clocks, due perhaps to diagenesis or to heating. There may also be systematic paleontological correlation

errors. However, anyone attempting to make a time-scale will not know in general what is causing the discrepancy. It is therefore important to have either a method of linking individual dates to an alternative time-scale, such as the magnetic polarity scale (Hailwood 1989), or, where such a scale is absent, as in most of Phanerozoic time, to have some systematic method of combining all dates that appear to be well determined so that a reasonably objective choice may be made of the age of a stratigraphic boundary from all of them. Chronograms attempt to do this.

Chronograms

In principle, estimating the age of a stratigraphic boundary from ideal data should be straightforward. Following Mussett & McCormack's suggestion (this volume) one could plot each date and its normal error distribution. One then selects an arbitrary age, sums that area of each error distribution that is older than the arbitrary age for samples that are stratigraphically younger than the boundary. One also sums the area of each error distribution that is younger than the arbitrary age for stratigraphically older

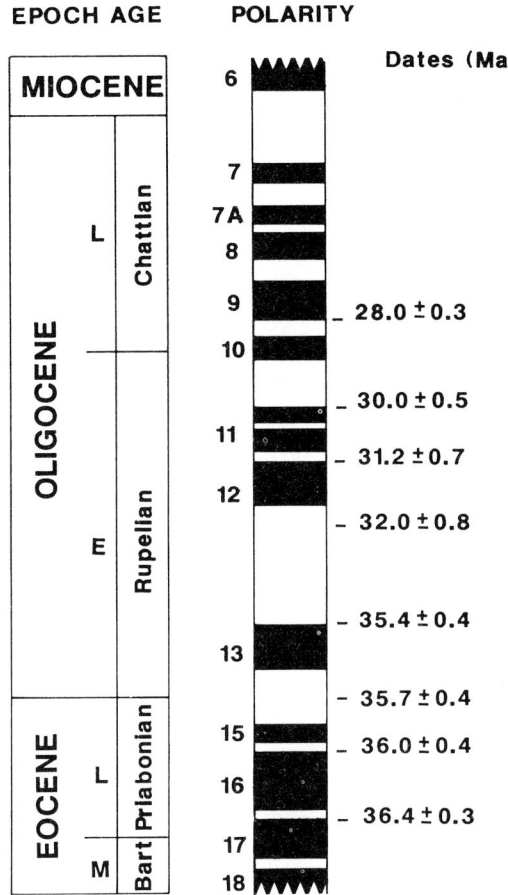

Fig. 2. Dated samples are correlated to the magnetic polarity scale and to the standard international stages (From Montanari *et al.* 1985).

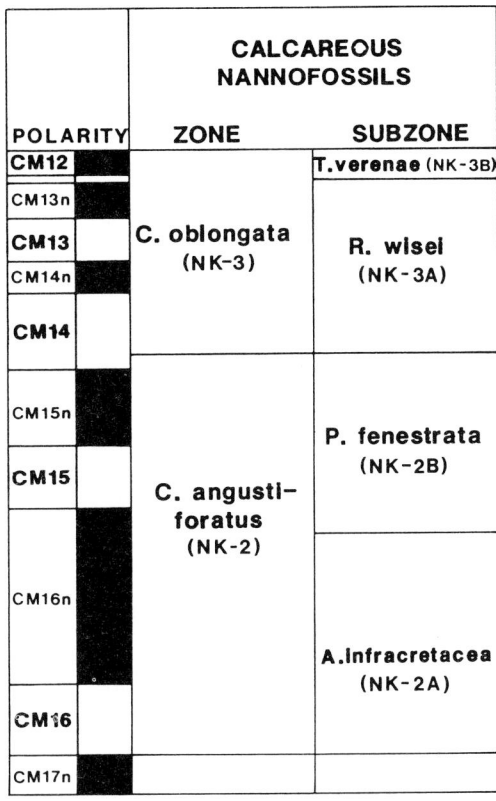

Fig. 3. By examining the magnetostratigraphy and biostratigraphy of several sections, changes in magnetic polarity can be linked directly to calcareous nannofossil zones in Berriasian and Valangian strata (From Bralower *et al.* 1989).

samples. The two areas are added. The arbitrary age is then moved to a new position and a new area calculated. The age of the boundary is then taken as that age for which the total area is a minimum.

Unfortunately the normal error distribution for each age is not well determined and the stratigraphic errors may be significant (and unknown). In practice all the relevant isotopic dates are assembled, together with their estimated chronostratic and chronometric errors. An error function is calculated for different values of the boundary. Its minimum is taken as the best estimate of the boundary age (Harland *et al.* 1990). The figure showing this function has been named a chronogram. It was initially devised to provide the best estimates of the ages of polarity transitions in cases where there were

overlapping K-Ar dates (Cox & Dalrymple 1967).

However, in the writer's view, subjective judgement is still necessary in evaluating the data for setting up a numerical time-scale for stage boundaries because one is trying to relate non-numerical data (stratigraphic boundaries from different sections) to rather scarce numerical data (the small number of isotopic dates with different errors that may be used to define a stage boundary).

The practical problems involved in applying the chronogram method to estimating the age of a stratigraphic boundary are illustrated here by reference to the Cretaceous–Tertiary boundary. There are 14 dates from the Danian stage in the Harland *et al.* (1990) database, ranging in age from 61.25 Ma to 67.14 Ma. There are also 14 dates from the Maastrichtian with an age range of 63.33 to 71.5 Ma. Clearly there is considerable overlap (Fig. 4).

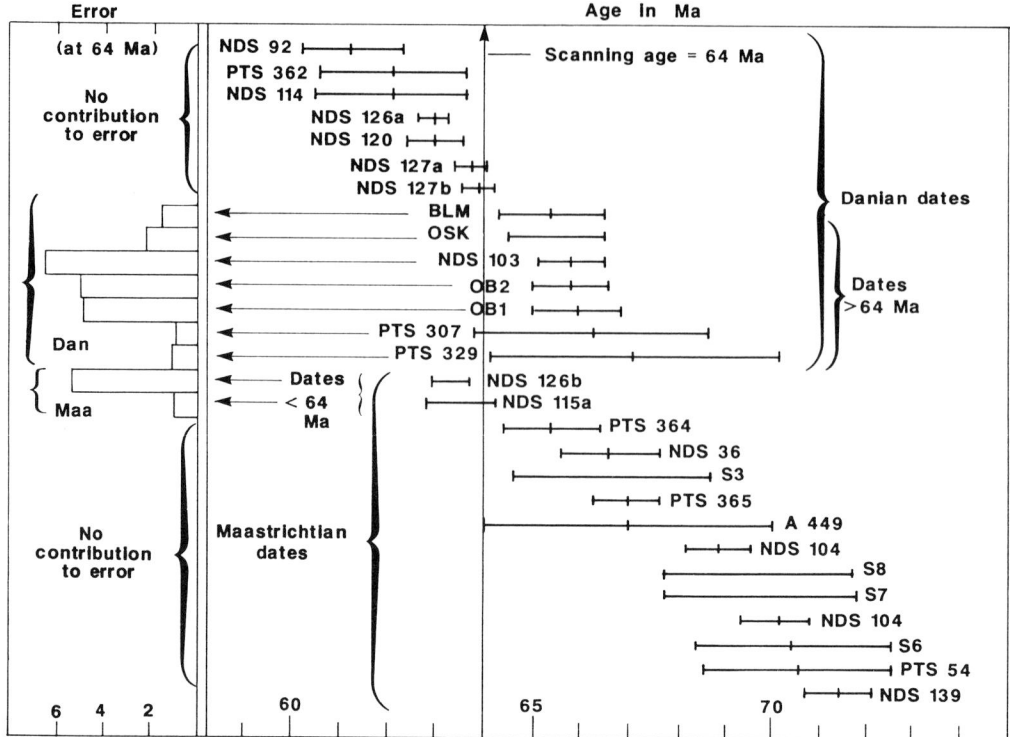

Fig. 4. Plot of all dates and their 1 sigma errors in Harland *et al.* (1990) assigned either to the Danian or the Maastrichtian stages. Those from the Danian stage are labelled on the left; those from the Maastrichtian on the right. Errors for those dates that contribute to the error function at 64 Ma are shown on the extreme left.

Visually the age of the boundary would appear to lie somewhere in the age range 64–66 Ma. To derive a more precise estimate of the boundary age a chronogram is constructed. The explanation of chronograms is given in detail in Harland *et al.* (1990, pp. 105–109). What follows is a shortened account. In essence a chronogram provides a quantitative but statistically non-rigorous method for estimating an age from a set of dates with variable errors. The chronogram displays an error function which is the sum of the individual errors for each date at a sequence of ages that straddle the likely minimum.

The error for a given isotopic date at a particular trial age is simply: (Isotopic date − trial age)^2/(1 sigma error)^2. Dates from the Danian that are younger than the trial age do not contribute to the error function; dates from the Maastrichtian that are older than the trial age also do not contribute. The oldest Danian dates and the youngest Maastrichtian dates are the ones that control the position of the minimum. Young Danian dates and old Maastrichtian dates have no effect on the minimum though

they may determine the steepness of the chronogram as one moves away from the minimum value.

In the case of the K–T boundary, a chronogram is made by scanning the data in an appropriate time interval, for example, 62–67 Ma (Fig. 5a). The contribution of data to the error function at, say, 64 Ma is found as follows. There is no contribution to the error function from any Danian dates that are younger than 64 Ma. Thus only 7 dates contribute. They are, with the 1 sigma errors in brackets: BLM 65.4 (1.1); OSK 65.5 (1.0); NDS103 65.8 (0.7); OB2 65.8 (0.8); OB1 66.0 (0.9); PTS307 66.3 (2.4); PTS329 67.1 (3.0).

Similarly, there is no contribution to the error function at 64 Ma from any Maastrichtian dates that are older than 64 Ma: only two dates contribute to the error function: NDS126b 63.3 (0.3) and NDS115a 63.5 (0.7).

The contribution from BLM at 64 Ma is (Date-scanning age)^2/(error)^2 = (65.4 − 64.0)^2/(1.1)^2 = (1.4)^2/(1.1)^2 = 1.60 (Fig. 6a); the contribution of OSK at 64 is (1.5)^2/(1.0)^2 = 2.25,

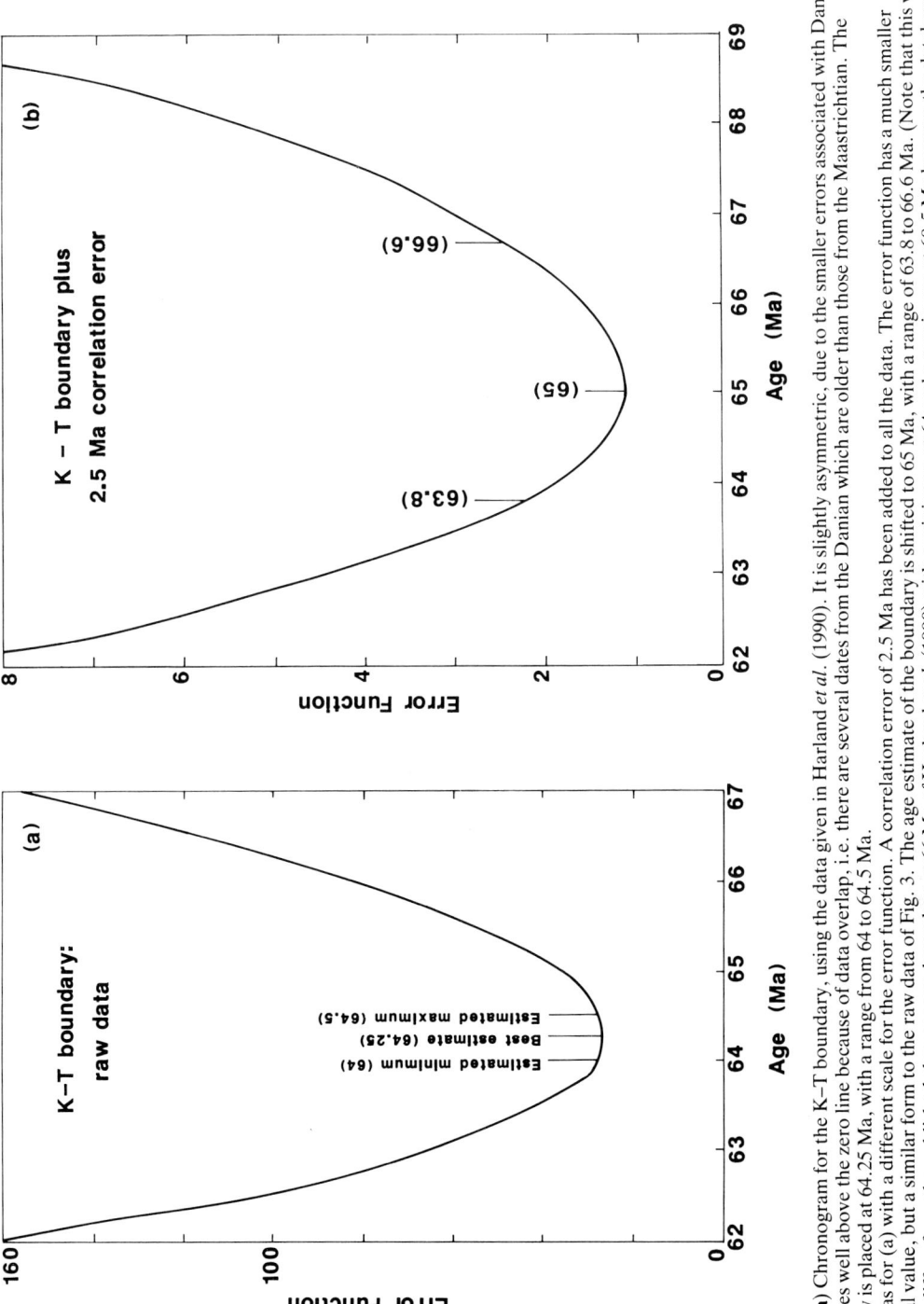

Fig. 5. (a) Chronogram for the K–T boundary, using the data given in Harland *et al.* (1990). It is slightly asymmetric, due to the smaller errors associated with Danian data. It lies well above the zero line because of data overlap, i.e. there are several dates from the Danian which are older than those from the Maastrichtian. The boundary is placed at 64.25 Ma, with a range from 64 to 64.5 Ma.
(b) Axesas for (a) with a different scale for the error function. A correlation error of 2.5 Ma has been added to all the data. The error function has a much smaller numerical value, but a similar form to the raw data of Fig. 3. The age estimate of the boundary is shifted to 65 Ma, with a range of 63.8 to 66.6 Ma. (Note that this value of 65 Ma differs from the published chronogram boundary at 66 Ma of Harland *et al.* (1990) with a minimum at 64 and a maximum at 68.5 Ma because the database was updated after all the chronograms had been drawn.)

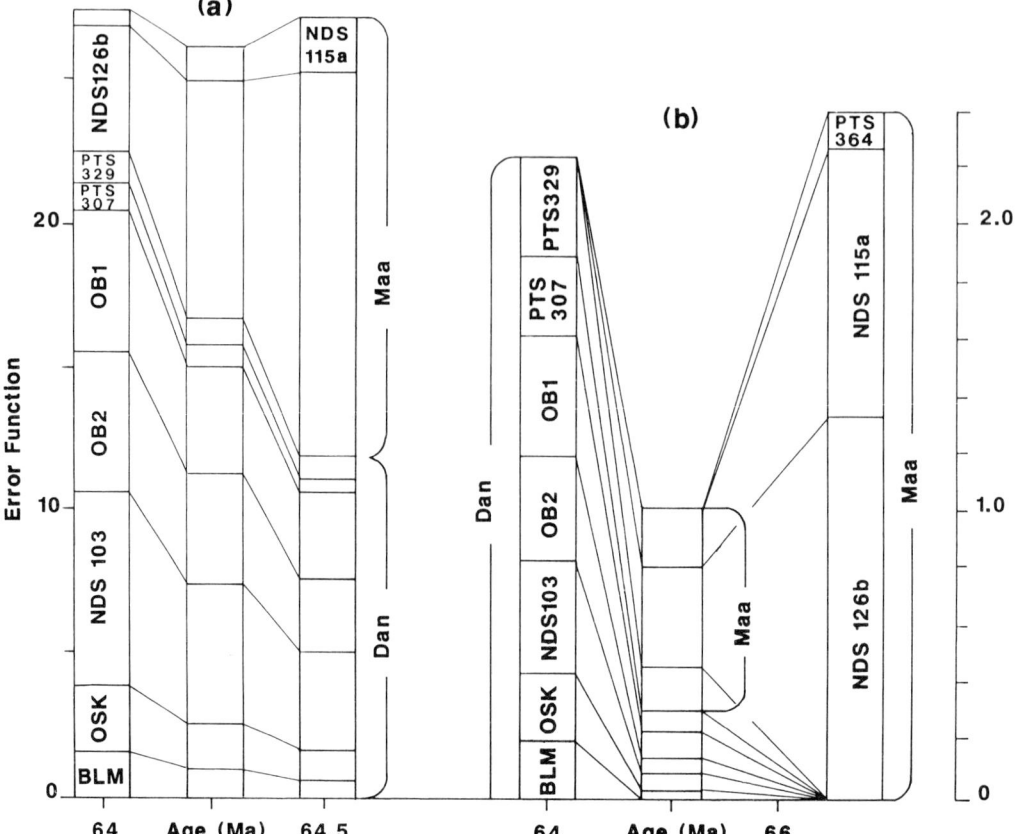

Fig. 6. (**a**) Contributions of individual dates to the chronogram for the K–T boundary. Note the dominance of NDS126b.
(**b**) Contributions of individual dates to the chronogram for the raw data plus a correlation error of 2.5 Ma. Other dates, e.g. PTS364 are beginning to influence the position of the chronogram.

and so on. The total value of the error function at 64 Ma is 1.6 + 2.25 + . . . for all relevant data, which comes to 27.3. Scanning at 0.25 Ma intervals gives a minimum at 64.25 Ma, with an error function at 26.1 (Figs 5a and 6a). The high non-zero value at the minimum is due to the overlap of dates.

In other words, the chronogram of the raw data places the K–T boundary at 64.2 Ma, which is lower than conventional estimates of 65–66 Ma. The minimum and maximum age estimates are somewhat arbitrarily taken as those ages for which the error function rises to more than 1 unit above its minimum value. In the case of the K–T boundary this yields an estimate of 64.0 as the minimum age and a maximum estimate of 64.5 Ma.

One reason for this small range of values is the small 1 sigma error of 0.32 Ma assigned to NDS126b, dated at 63.33 Ma, from Red Deer

Valley, Alberta. This has the effect of causing its contribution to the error function to increase rapidly as the scanning age is increased from 63.33 Ma. For example, at 64.25 Ma its contribution to the error function is $(63.33 - 64.25)^2/(0.32)^2 = 8.27$, or one third of the total error value (Fig. 6a). At 64.5 Ma, it contributes about one half of the total error. However, the stratigraphic position of this Red Deer Valley sample in relation to the K–T boundary is possibly contentious because that boundary is defined by the disappearance of *Aquilapollenites* (Baadsgaard & Lerbekmo 1982, p. 795). Clearly, it would be unwise to allow a single date, however precise its analysis, to dominate the position of a boundary when its correlation to the type Maastrichtian is in some doubt.

Another young Maastrichtian date, NDS115a at 63.5 Ma, comes from the *Early* Maastrichtian, or even the Campanian. As the abstractors note

(Odin & Kennedy 1982, p. 777): '. . . either there is a stratigraphic problem . . . or the geochemical history of these . . . glauconies was anomalous.' Its effect on the chronogram is less than NDS126b because its chronometric error is more than twice as large as 0.7 Ma (Fig. 6a).

To circumvent the problem of including precisely determined dates with large chronostratic errors or anomalous dates, the authors of Harland *et al.* (1990) added a 'fossil error', representing an estimate of the correlation error in Ma that might exist in correlating any stratigraphic section to its global stratotype. Because of the difficulty of systematically assessing the likely individual correlation errors without greatly delaying publication, the correlation error has been given the same value to all dates, including those cases in which its stratigraphic relation to a type section is well known. This is clearly an incorrect procedure which will overestimate the uncertainties in the ages of stratigraphic boundaries, but it is unclear what to propose as a better practical one.

In the case of correlations in Cenozoic and Late Cretaceous time, the error was taken as 2.5 Ma. The chronogram for the K–T boundary is therefore rerun from the same data as above but with the 1 sigma error replaced by (1 sigma error + 2.5 Ma). This has two effects: first, it reduces the importance of high precision dates and secondly, it increases the estimated error associated with each boundary. The chronogram now gives a value for the K–T boundary of 65.0 Ma, with a minimum age of 63.8 Ma and a maximum age of 66.6 Ma (Fig. 5b).

Because isotopic dating is currently the only way of dating events older than a few Ma, errors in chronometric scales are likely to be at least as great as 1 sigma analytical errors, estimated for current techniques at about 0.5%, except in cases where they have been reduced by a significant number of replicate analyses on the same sample. Errors will be generally greater because dated samples need to be correlated to the type sections. For Cenozoic and later Cretaceous time, the chronometric errors are unlikely to be less than 0.1 Ma for the Neogene, 0.3 Ma for the Palaeogene and 0.4 Ma for later Cretaceous time.

Limitations on chronogram method

For the time-scale of Harland *et al.* (1990), chronogram errors, which are probably overestimates due to the inclusion of high correlation errors, range up to about 4 Ma for Neogene stages; 5 Ma for Palaeogene stages and 8 Ma for later Cretaceous stages. If material suitable for dating can be found in boundary stratotypes above and below the boundary then the correlation errors would be very small, though it is not clear how to quantify them in Ma. It might seem that dating samples from such boundary stratotypes would reduce the numerical errors assigned to stage boundaries by an order of magnitude over current error estimates.

Unfortunately, there is always the possibility that dateable material in a section has been affected by a process that has systematically altered the isotopic ratios. The dates may still be internally consistent, giving a numerical sequence corresponding to the stratigraphic order, but all may have been reduced in value. Fischer & Gygi (1989) note this is as a possibility for their Oxfordian glauconite dates from the Jura. Isolated studies will therefore always be suspect. The only way of testing such a possibility is by dating materials from a range of localities, which will necessarily involve sections that do not include the stratotypes. Chronograms, or some other method of assessing the age of a boundary from a variety of data, will always be necessary; in addition, the correlation error must be quantified somehow.

A major advantage of the chronogram method is that time-scales based on it will change slowly, rather than in major jumps. This is simply a reflection of the fact that, once in the database, all data are considered relevant to the position of a stage boundary, rather than only the most recent determinations. From this viewpoint new determinations cannot be regarded as providing the definitive data for a boundary but must take their place alongside previously available dates. Clearly, as precision in isotopic and biostratigraphic dating progresses, it will be necessary to weed out imprecisely defined items.

At present, the chronogram method is limited in its application because the position each date occupies within its stage is ignored. For the K–T boundary, all Danian dates are regarded simply as Danian, and all Maastrichtian dates as Maastrichtian. The Danian stage is 4.5 Ma in duration and the Maastrichtian is 9 Ma in duration. With some chronometric errors as small as 0.3 Ma, the chronometric order of dates should correspond to their stratigraphic order within each stage. As already noted, what is required is some method of quantifying the position of samples within a stage.

Quantifying the stratigraphic position of calibration points

The need to quantify the position of samples in sections is a general one. All time-scales, chronostratic and chronometric, need to be calibrated numerically with one another. Here only three scales are discussed: the magnetic polarity time-scale, defined by the sequence of polarity transitions (m); the biostratigraphic time-scale, defined by stages (s) or fossil zones (f); and the chronometric scale, defined in time t by dates (d).

In a mathematical sense, the general problem of linking the chronostratic scales with the chronometric scale is essentially a 'mapping' problem.

$$(s,f) \rightarrow (m) \rightarrow (t)$$
[Fig. 3]   [Fig. 2]

or independently:

$$(s,f) \rightarrow (t)$$
[Fig. 1]

The correlation markers for the magnetic polarity scale are the polarity transitions. They are discontinuously distributed in time and therefore in rocks. Biostratigraphic stages and zones are defined by their fossil content. The correlation markers are the boundaries between one stage and the next stage or between one zone and the next zone. They too are discontinuously distributed in sections.

Thus the biostratigraphic and the magnetic polarity scales provide discontinuous markers for calibration purposes. Moreover, markers in one scale do not in general coincide with markers in the other. Suitable samples for dating are also discontinuously distributed in sections and therefore in time. The problem of calibrating one time-scale with another is that of mapping one set of discontinuous values onto another time-scale defined by a second discontinuous set. Mapping and calibration must therefore involve some form of interpolation. Ideally, calibrations should be numerical, including those between the non-numerical chronostratic

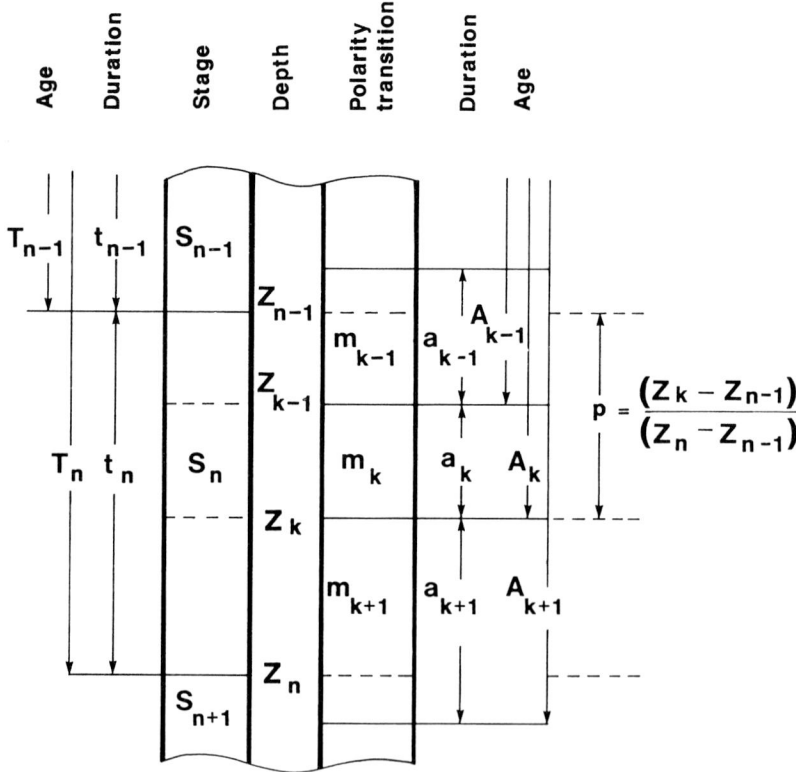

Fig. 7. A section contains the base of stage s_{n-1} and stage s_n at depths z_{n-1} and z_n with ages of T_{n-1} and T_n respectively. The duration of s_n is t_n. Polarity transitions $m_{k-1}, m_k \ldots$ take place at depths $z_{k-1}, z_k \ldots$ and are of age $A_{k-1}, A_k \ldots$ and duration $a_{k-1}, a_k \ldots$ respectively. A sample is taken from a depth z. If z is linearly related to t over the interval t_n, then the age of the sample is given by $T = T_{n-1} + t_n \times (z - z_{n-1})/(z_n - z_{n-1})$. Hence if T_{n-1} and t_n are known, then t can be expressed in terms of its relative position in the section. Similarly the age of a sample within a given polarity chron is given by $A = A_{k-1} + a_k \times (z - z_{k-1})/(z_k - z_{k-1})$.

scales. How can one unambiguously express the position of a sample in one scale in numerical terms relative to another scale?

Let $t_1 \ldots t_n$ be the time intervals (unknown) corresponding to some chronostratic scheme $s_1 \ldots s_n$, such as the stage boundaries. The base of s_1, is of age T_1, of s_2, T_2 and so on, such that the duration of s_2, t_2, is $T_2 - T_1$, and so on (Fig. 7). Let $a_1 \ldots a_n$ be the time intervals (unknown) corresponding to a second chronostratic scheme $m_1 \ldots m_n$, such as the polarity transitions. The base of m_1, is of age A_1, of m_2, A_2 and so on, such that the duration of m_2, a_2, is $A_2 - A_1$, and so on (Fig. 7).

Let a polarity transition from m_k to m_{k+1} take place at a depth z_k within a section. Let the top of stage s_n be at a depth z_{n-1} and its base at depth z_n, where $z_{n-1} < z_k < z_n$. Then, measured from the top of S_n, the relative position, p, of z_k within stage s_n is $(z_k - z_{n-1})/(z_n - z_{n-1})$. That is, when $z_k = z_{n-1}$, $p = 0$; when $z_k = z_n$, $p = 1$. p represents the fractional position of the polarity change within the section. It is equivalent to normalizing the stratigraphic position of the polarity change with respect to the thickness of stage s_n. The stratigraphic position of the polarity change can then be expressed numerically as $p \times (s_n)$.

For example, the transition from C26N to C26R might take place at 10 m below the top of the Palaeocene stage in a section, with the base of the Palaeocene lying at 15 m below the transition. The stratigraphic position of the transition is $0.4 \times$ (Palaeocene), where (Palaeocene) refers to the total thickness of the Palaeocene in the section, i.e. $p = 0.4$.

In a stratigraphic section $T = f(z)$, where $f(z)$ is some function relating depth to time. Thus $T_{n-1} = f(z_{n-1})$; $T_n = f(z_n)$. If depth is proportional to time, then $t = cz$, where $c =$ constant, z is now the thickness of the interval, i.e. $z_n - z_{n-1}$, and t is the duration of the interval. This, the simplest of all functions, can be thought of as a 'linearizing operator' (Hailwood 1989), which allows linear interpolation between points on the scales to be calibrated. The validity of this assumption is discussed below.

With it, $t = cz$ for that interval, and $p = (t - t_{n-1})/(t_n - t_{n-1})$. The expression $p \times (s_n)$ provides an estimate of the normalized position in time of the polarity transition relative to the duration of the stage concerned.

In the same way, the positions of stage boundaries can be expressed in terms of a normalized polarity scale. For example, one estimate places the K–T boundary within magnetic chron C29R (Berggren *et al.* 1985a), at 0.26

$\times$ (C29R), where (C29R) = duration of C29R. Here the fraction p represents a time estimate, because the relative stratigraphic positions of C29R(y), the K–T boundary itself and C29R(o) have been assumed to be proportional to age. [The symbols (y) and (o) mean 'younger boundary' and 'older boundary', respectively].

It would also be possible to describe the fractional position of a dated sample or of a biostratigraphic boundary measured either from the older boundary, getting younger, e.g. as C29N(o) $- p \times$ (C29N), or the younger boundary, getting older, e.g. as C29N(y) $+ p \times$ (C29N). It would also be possible to describe the position of the boundary in terms of the whole chron C29 (combined normal and reverse parts), either from the top down or the bottom up, i.e. as C29(y) $+ p \times$ (C29) or as C29(o) $+ p \times$ (C29). Using the whole chron is unsatisfactory because it would be necessary to know where the top and bottom of the entire chron is, rather than just those of the normal (or reversed) interval in which the boundary lies. On balance the author prefers describing the position in the same way as illustrated for the normalized stratigraphic position: from the top down, i.e. C29R(y) $+ 0.26 \times$ C29R as in a well log. But whatever the description, it must be unambiguous and readily understood.

Major advances have been made in the past few years in linking the biostratigraphic and magnetostratigraphic time-scales to one another. These advances allow not only stage boundaries but finer units such as chrons and biostratigraphic zones to be mapped directly onto the magnetic polarity time-scale with errors in relative values that are probably as small as 0.1 Ma. However, there is rarely a numerical description of where these boundaries are: they have to be interpolated from synthesis of the data, rather than read from the original data themselves.

As more data accumulate, it will be important for the primary data to be given. Ideally, the depths of the calibration points in a core or section should be given together with the lithologies involved. The assumption $t = cz$ cannot hold for a thick section, even if it is of the same lithology, because of differences in porosity between the top and the bottom of the section. In such cases it may be necessary to decompact the section first to make better estimates of the relative time durations. Thus it is important to give full lithologic logs if time-scale calibration is to be improved. Stratigraphic incompleteness (see below) is another factor that will disturb the $t = cz$ relationship as well as the changes in accumulation rate.

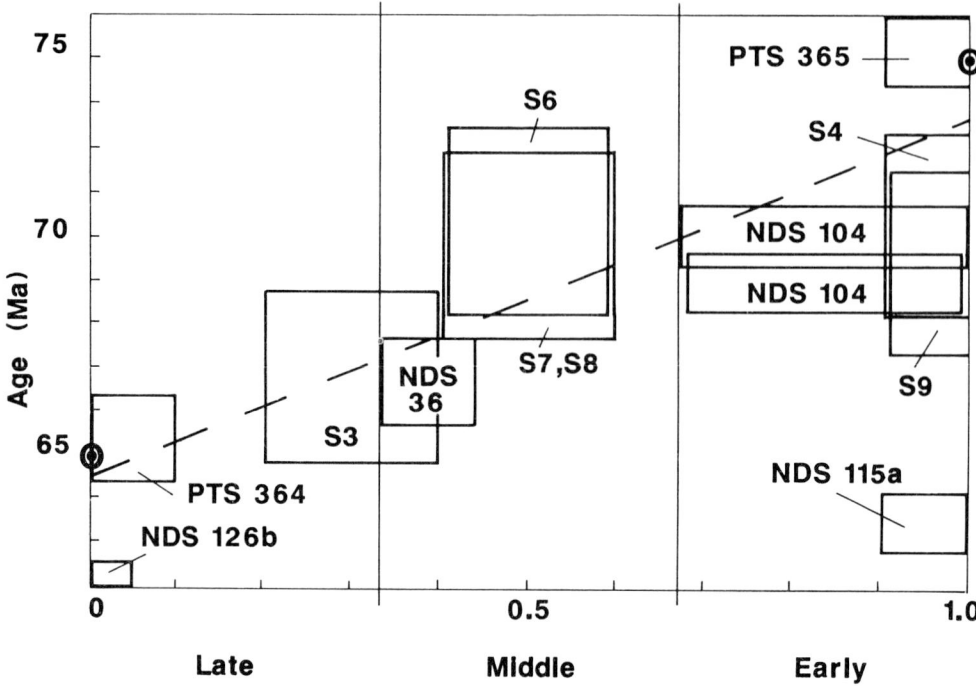

Fig. 8. The horizontal axis is the estimated normalized position of the items within the Maastrichtian stage (Table 1). Because of their range, A449 and PT554 have been omitted. NDS115a, PTS365 and S9 have been excluded from the linear regression line. A linear regression through the data which ignores the associated errors gives a value of 64.5 Ma for the top of the Maastrichtian and 72.7 Ma for its base, close to the values of 65 Ma and 74 Ma eventually adopted by Harland *et al.* (1990), which was based on chronograms for the boundaries only.

Normalization of relative stratigraphic position of isotopic dates

The Maastrichtian stage (with the abbreviation Maa not to be confused with the abbreviation Ma for millions of years) is used as an example of how normalization of stratigraphic position may eventually improve time-scale calibration. All dates within a given stage s_n can be listed as having estimated ages $p \times s_n$, where p is the appropriate decimalized position within the stage. For example, the sketch stratigraphic column provided by Shibata (1986) is a compact summary of the location, stratigraphy and biostratigraphic age of six Maastrichtian dates in the Harland *et al.* (1990) database. These are located at fractional depths of about 0.3 (S3), 0.5 (S6, S7, S8) and 1.0 × (S4, S9) below the stratigraphic top of the Maastrichtian (Fig. 8). If $t = cz$, then the estimated decimalized positions in time within the stage are 0.3 × (Maa), 0.5 × (Maa) and 1.0 (Maa) below the top of the

Maastrichtian, where (Maa) = duration of Maastrichtian stage in Ma. This is equivalent to normalizing the position of each date within the stage.

Although there is no defined middle Maastrichtian and no agreement among palaeontologists and stratigraphers about other Maastrichtian devisions (J. Kennedy, pers. comm.), for the purposes of illustration only, it is assumed that the early, middle and late Maastrichtian correspond to (1–0.67) × (Maa), (0.67–0.33) × (Maa) and (0.33–0) × (Maa), i.e. that the mapping of early, middle and late is of the form $t = c \times$ (biostratigraphic subdivisions of equal rank).

In addition to the dates of Shibata (1986), 9 other dates are available (PTS365, A449, PTS54, NDS126b, PTS364, NDS36, NDS104 [2 dates] and NDS 115a). However, of the total of 16 dates, the following six dates were omitted from the final calculation, either because their stratigraphic range was considered too large,

Table 1. *Data used to construct a Maastrichtian time-scale*

Item	p	Age	Item	p	Age
NDS126b	0.0	63.3	S7	0.5	69.8
NDS364	0.05	65.41	S8	0.5	69.8
S3	0.3	66.8	NDS104	0.84	69.0
NDS36	0.38	66.7	NDS104	0.84	70.1
S6	0.5	70.4	PTS365	1.0	74.2

The 'Item is the code and the 'Age' is the age given in Harland *et al.* (1990).

p is this author's estimate of the normalized position of the item in the Maastrichtian stage ($p = 0$ is the top; $p = 1$ is the base).

e.g. 0–1 × (Maa), or they were reported as anomalous or as minimum ages: PTS365, A449, PTS54, S4, S9 and NDS115a. The Maastrichtian dates used to construct the Maastrichtian time-scale are shown in Table 1.

As shown in Table 1, each Item in the database would then have two kinds of values associated with it: an isotopic date with a chronometric error (e.g. for S3 66.8 ± 2 Ma) and a numerical chronostratic age and error eg for S3 (0.30 ± 0.10) × (Maa). Local stages could also be used to define the precise chronostratic position, but would eventually have to be mapped onto the global scheme.

A graph of the data in Table 1, with estimated chronostratic errors, is shown in Fig. 8. A linear regression line through the data gives an age for the top of Maastrichtian of 64.5 Ma and 72.7 Ma for the base. Such estimates would be modified by the need to reconcile independent estimates of the age of the top provided by Danian data and the age of the base from Campanian data. These Maastrichtian values, which are not obtained by chronograms, are similar to the chronogram values of 66 and 71.5 Ma respectively of Harland *et al.* (1990).

Normalization could be applied throughout the entire Phanerozoic Eon. It cannot be reliably used unless a stratigraphic sections shows the locations of dated samples in relation to the top and bottom of the stage concerned. If the sample age is given merely as Early X, where X is a stage, then the additional assumption has to be made about the proportion of X that is represented in time by Early X: in practice the range might be considerable.

The Maastrichtian example suggests that linear regression of normalized stratigraphic data may provide an alternative method to chronograms for assessing the ages of stratigraphic boundaries and the fine structure of

stratigraphic stages. Alternatively, chronograms could be decimalized, so that instead of making a chronogram for the Maastrichtian/Danian boundary, one could also make chronograms for any part of the Maastrichtian, such as 0.6 × (Maa). To be able to use either method will require adoption of more informative methods for defining the stratigraphic location of dated samples than are in general use at the present time.

How good is the linear thickness versus time assumption?

The importance of the assumption that sediment thickness is proportional to time is clear from the fact that all present calibrations of the magnetic polarity and biostratigraphic scale implicitly assume that, at least locally, $t = cz$. The finest subdivisions of these scales correspond to some 10^5–10^6 years, yet even on this scale the stratigraphic record shows irregularities. What are the time and thickness intervals for the range of commonly encountered environments and lithologies for which this linear time–thickness relationship is an adequate assumption? Only a sketchy answer to this fundamental question can be attempted here.

Sadler (1981) has analysed sediment accumulation rates in relation to the completeness of the stratigraphic record. Sedimentation in general is unsteady and discontinuous. In particular, all sedimentary environments for which there are adequate data show a net decrease in accumulation rate with increasing time-span of observation, with a flattening to a steady rate for long time-spans for some environments. In other words, for short time intervals the net accumulation rate is more rapid than it is for long time periods. As Sadler notes, '. . . to interpolate directly between known ages is to apply a measured rate to a shorter time interval, and this is clearly improper', though this improper procedure is the basis for nearly all time-scale calibrations!

Sadler & Strauss (1990) define the completeness of a sedimentary section of total duration T as the fraction of time intervals of length t that have left a record. A complete section is one in which all possible intervals t have left a record. A section in which there is a hiatus of duration t, or longer, is incomplete. Expected completeness is the probability that some sediment will have been preserved during an interval of duration t. Completeness is also the ratio of the long-term rate of accumulation, S, in a section of duration, T, to the average rate of accumulation, s, at the time scale considered, t (Sadler 1981). s is rarely available for a real section and the average expected value, s_{av}, for comparable environments must be used instead:

$$S/s_{av} = (t/T)^{-m}$$

where m is the observed gradient of log(S) against log (t). From this it is possible to calculate the expected completeness for a range of time-scales in a specific setting. If m is zero, i.e. if the rate of accumulation is constant, then the sections are complete. For the range of durations for which m is constant completeness follows a self-similar law, i.e., stratigraphic completeness then forms a fractal system.

Table 2. *Gradient of time-span against rate of accumulation* (m) *and stratigraphic completeness for different time-scale ratios* (t/T)

	log(S)/ log(T) (= m)	Stratigraphic completeness t/T	
		0.1	0.01
Small basinal seas	−0.10	0.79	0.63
Lakes	−0.14	0.72	0.52
Abyssal chalks & oozes	−0.21	0.62	0.38
Rivers	−0.36	0.44	0.19
Terrigenous shelves	−0.38	0.42	0.14
Platform carbonates & reefs	−0.5	0.32	0.10

Gradients are for time-spans in the range 10^3–10^8 years and have been visually estimated from data in Sadler (1981) and Sadler & Strauss (1990).

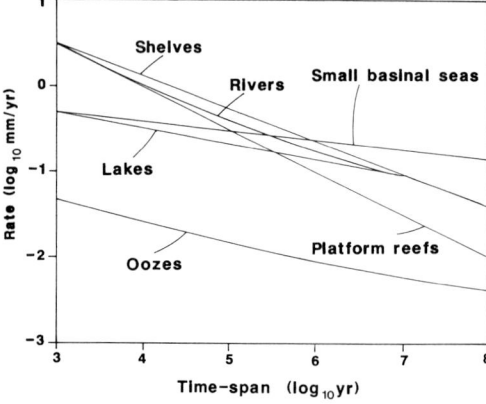

Fig. 9. Log–log graph of rate of accumulation versus time-span of stratigraphic interval for a range of depositional environments. Modified from Sadler (1981) and Sadler & Strauss (1990).

The available data for determining m are highly scattered (Sadler 1981; Sadler & Strauss 1990). The present author's visual estimates of m for time-spans in the range 10^3 to 10^8 years from data compiled by these authors are shown in Table 2 and Fig. 9. m is assumed to be linear in the time range, though m for abyssal chalks and oozes approaches that for lakes for time-scales of 10^6 years or greater.

The relative completeness will be in the order of decreasing m. This order corresponds approximately to expectations: sediments in small marginal seas, lakes and abyssal chalks and oozes would be expected to be significantly more complete than sediments laid down by rivers, on terrigenous shelves or in platform carbonates and reefal settings. DSDP/ODP cores, and pelagic sections may be expected to provide excellent material for calibrating biostratigraphic and magnetic polarity time-scales, but, apart from the occasional volcanic ash layer, will provide limited data for chronometric calibrations. Fluvial, shelf and platform environments will contain a greater proportion of hiatuses.

Completeness is readily calculated. For example, m for abyssal chalks and oozes is estimated as −0.21. Consider a section that has a time-span of T Ma. The expected completeness of the section (the probability that some sediment will be deposited in any time interval t) is $(t/T)^{0.21}$. If $t = 0.1T$, then the expected completeness is $(0.1)^{0.21}$ or 0.62 (Table 2); if t is $0.01T$ completeness is 0.38.

From the point of view of time-scale interpolation the expected completeness provides a useful guide to the likely errors involved. Consider a section of 10 Ma duration at 1 Ma resolution. With a completeness of 0.62, the expected total hiatus is 3.8 Ma. In the worst case this hiatus will be concentrated in one horizon. If it is at the top of the section, then the age range of the sediments would be 3.8 to 10 Ma, rather than 0 to 10 Ma assumed for interpolation. Thus the possible age range of the top of the preserved sediments is 0 to 3.8 Ma and this section would be complete only at the 4 Ma time-scale or greater.

Clearly, the possible errors are unacceptably large. However, what is important for interpolation purposes is the distribution of the hiatuses in the section. The most favourable case for interpolation is one in which each 1 Ma increment of time in the section is represented by 0.62 Ma of sediment and 0.38 Ma of hiatus. The section would then be complete at a 1 Ma time-scale. The maximum interpolation error would be 0.38 Ma, which would be acceptable.

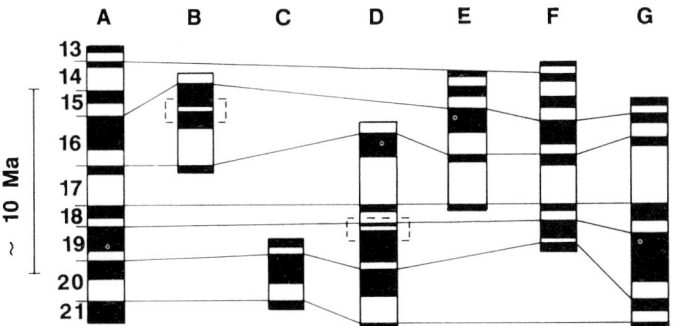

Fig. 10. The magnetostratigraphy of 6 stratigraphic sections is compared with the ocean-floor M anomaly sequence M13 to M21 (Bralower *et al.* 1989). The presence of records of all M anomalies in these sections shows them to be complete on a time-scale of about 0.5 Ma. Sections B and D contain additional magnetic data highlighted by a dashed box not present in the standard magnetic polarity scale.

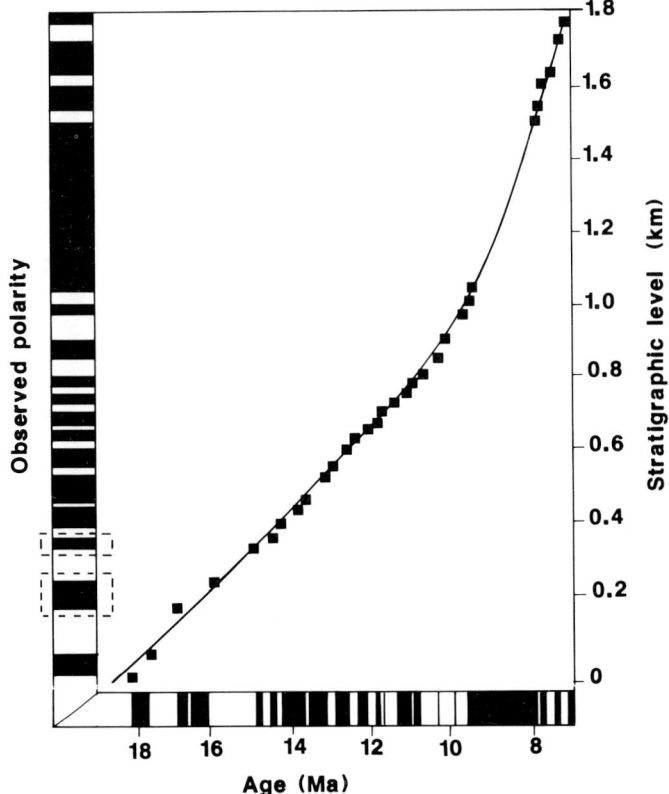

Fig. 11. Graph of net accumulation versus time for part of the fluvial Siwalik Group in Pakistan (Friend *et al.* 1988). The relatively smooth form of the accumulation curves suggests the section is complete on a time-span of about 0.5 Ma. The dashed boxes in the observed polarity scale shows that here the polarity record in the sediments is incomplete compared with the standard magnetic polarity scale.

Whether abyssal chalks and oozes approximate to the most favourable or least favourable distribution of hiatuses can be determined from magnetostratigraphy.

Bralower, Monechi & Thierstein (1989, fig. 12, Fig. 10) have identified the M13–M21 polarity changes in pelagic sections of the Jurassic–Cretaceous boundary in Spain, France, Italy and a DSDP core. The duration of each of the M13–M21 chrons is of the order of 0.5 Ma. That all the M magnetic chrons are represented in every section suggests completeness at a 0.5 Ma time-scale, in turn suggesting that hiatuses are distributed through the section on this sort of time-scale, rather than as a few large discontinuities. However, these sections may simply be of better than average completeness.

Even in fluvial environments, where large hiatuses might be expected, magnetostratigraphy shows that completeness can be preserved at a time-scale of $c.0.5$ Ma or better for durations of $c.10$ Ma. For example, Friend et al. (1988) have analysed the changes in the net accumulation rate with time in 1800 m of the fluvial Siwalik Group of Pakistan, where magnetic stratigraphy suggests that the sections are complete at $c.0.5$ Ma scale. The average rate of accumulation for the lower 1000 m of section, of duration 7 Ma, is about 0.1 mm a^{-1} (Fig. 11). Friend et al. have examined the detailed changes in a complete sedimentary cycle, 36 m of fluvial section, where the overall accumulation rate is 0.15 mm a^{-1}. As might be expected, the likely changes on a time-scale of $<10^5$ years are highly irregular, but can be plausibly fitted to the mean accumulation rate provided due attention is paid to the detailed sedimentary environment (Fig. 12).

For fluvial sediments as a whole, m is estimated at -0.36 (Table 2). The expected completeness on a time-scale of a cycle (2.5×10^5 a) for a duration of 7 Ma is $(0.25/7)^{0.36}$, or 0.30. Thus the average net accumulation rate for a cycle will be $(1/0.3) \times 0.15$ mm a^{-1}, or 0.5 mm a^{-1}. The sands and conglomerates are probably deposited at rates of 10^2–10^7 mm a^{-1} (Friend et al. 1988) and represent a negligible time on a time-scale of a cycle. The only other significant accumulations are silts and muds. Their rates of accumulation are very uncertain. Friend et al. adopted a value of 1 mm a^{-1}, which is comparable with the minimum average rate of 0.5 mm a^{-1} estimated from m, particularly when the uncertainty in the duration of hiatuses is taken into consideration.

Although terrigenous shelves have a similar m value to fluvial sequences, hiatuses in some sections are more irregularly distributed than in the example from Pakistan. Aubry et al.'s (1986) biostratigraphic and magnetostratigraphic study of the Hampshire and London basins shows at least 3 hiatuses in a section of 20 Ma duration, 2 of about 1 Ma duration and the third of 2 Ma duration (Fig. 13).

The above examples show that some fluvial and pelagic environments have existed in which

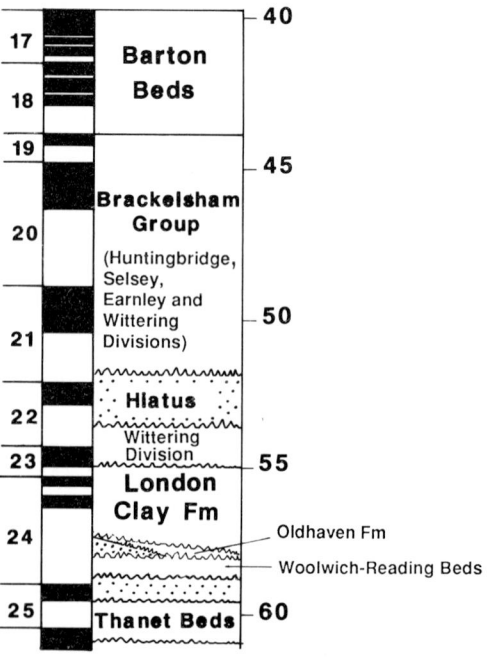

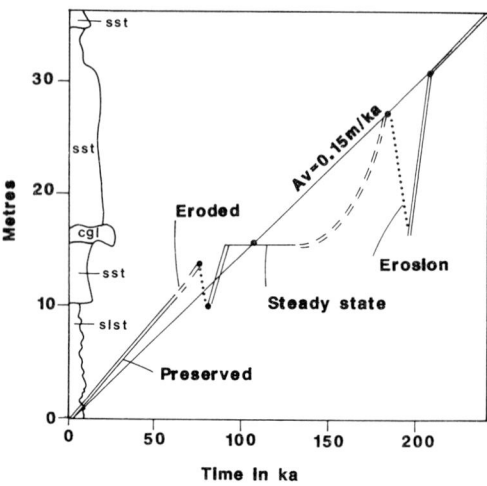

Fig. 12. Details of a stratigraphic section from the Siwalik Group showing how the rate of accumulation is incomplete on time-scales of the order of 10^3–10^4 years, but is complete on a time-scale of 2.5×10^5 years or greater (Modified from Friend et al. 1988).

Fig. 13. Comparison of the ocean-floor spreading polarity scale with observed polarity transitions in a terrigenous shelf sequence from the London and Hampshire basins (Aubry et al. 1985). The section is incomplete on a time-scale of 2 Ma or less.

sediments are essentially complete at the 0.5 Ma level, or better, whereas the terrigenous shelf section is complete only for time-spans greater than 2 Ma. Ideally, the assumption of a linear time-depth relationship for time-scale interpolation and calibration needs careful assessment for every date that is not located at a stratotype boundary. This conclusion emphasizes yet again the need for much fuller descriptions of the sections from which dated material is collected.

Over longer time periods the controlling factor will be the mechanism of subsidence. For periods of tens of Ma, the decompacted, back-stripped thicknesses of sediments deposited during the thermal cooling stages of basins formed by lithospheric stretching (McKenzie 1978; Jarvis & McKenzie 1980) necessarily follows an exponential decay law rather than a linear law, provided that sedimentation keeps pace with subsidence. For uniform lithology and short periods, the exponential decay law will approximate to linearity, but the approximation becomes more and more erroneous as the time interval increases.

Magnetic polarity scale

The primary source for the magnetic anomaly scale is the record of variations in the total magnetic field along ship's tracks in the ocean basins. The striking similarity of the relative spacing of the field variations shows that the spreading process at ridges is steady for long periods, remarkably uniform but differs in rate from ridge to ridge. For these reasons, the relative spacing of adjacent chrons represented on an ocean-floor magnetic profile is much more likely to be close to their true relative duration than is their relative thickness in stratigraphic sections, other than those in some pelagic environments. The assumption that for magnetic anomaly profiles $t = cz$, at least for a few million years, where z is the distance from the ridge crest, is therefore likely to be an excellent approximation which can be tested by comparing profiles measured across different parts of the ocean-floor.

The standard scale is derived from the Vema-20 profile in the South Atlantic at about 28°S (Heirtzler et al. 1968). Since its publication, the observed profile has been modified for calibration purposes in only a few places. In particular, anomaly 14 has been removed because it is not observed on other tracks in the oceans, and the relative spacings of a few anomalies on the track have been modified in the light of work on other anomaly sequences (see Harland et al., 1990, p. 145–149 for discussion). In a few places

then the profile used for calibration is therefore not what is actually observed, but what is believed would have been observed had the ocean-floor record been complete.

Berggren et al.'s (1985a,b) calibration of the modified Heirtzler anomaly spacing suggested that the spreading velocity in the South Atlantic had not been constant. For Cenozoic time they suggested it could be regarded as being made up of three linear segments. More recently, differences in the relative spacing of anomalies in the South Atlantic compared with the spacing on other ridges has been used to suggest additional times when the South Atlantic spreading rate has changed (Aubry et al. 1988).

Thus the function relating t to z is no longer considered to be linear throughout its length, but is regarded as being made up of linear segments, i.e. it is of the form:

$$t = (m_1z + c_1) + (m_2z + c_2) + (m_3z + c_3) + \ldots$$

where m_i and c_i are different constants applying to progressively older time intervals. Since $t = 0$ at the ridge crest, for which $z = 0$, then $c_1 = 0$. Other functions such as polynomials may eventually be found to be a better fit to the data, but in the present state of knowledge piecewise linear fits are believed to be adequate for calibration purposes.

The constants in the function can be found by tabulating all the available dates against their positions on the magnetic polarity scale and fitting regression lines to the different segments. At present, the extent of the segments has been estimated visually and includes subjective judgements.

The final calibration of Harland et al. (1990) used directly dated chron boundaries together with chronogram values that were correlated indirectly to the magnetic chron boundaries as the primary data for the Cenozoic and later Cretaceous time-scale.

High precision dating by Milankovitch cycles

The widespread interest in Milankovitch and other cycles and their expression in stratigraphic sequences (e.g. Berger 1988) potentially offers remarkable precision in chronometry. The cycles range in frequency from about 0.4 Ma for the low frequency cycles to about 0.02 Ma for the highest frequency cycles. Correlations to the nearest half-cycles can be envisaged, giving a potential resolution of ±0.01 Ma. The use of such cycles to refine the time-scale has been suggested (e.g. Fischer et al. 1990). Using oceanic cores showing probable Milankovitch cycles with a frequency of 0.0208 Ma, Herbert & D'Hondt (1990) have estimated the age of the K–T boundary either as C29R(y) + 0.229 Ma or as C29R(y) + 0.40 × (C29R). The duration of C29R is estimated at 0.580 Ma, the same value as given by Harland *et al.* (1990).

Milankovitch cycles offer the potential for an error of ±0.01 Ma, but unfortunately, for the forseeable future only relative dating with this precision can be envisaged. Even here, the method has to be used with care because of the likely incompleteness of sections. Recent data suggests that the abrupt shifts in isotopic abundances and apparently instantaneous mass extinctions in deep-sea cores straddling the K–T boundary may reflect a temporally incomplete (or extremely condensed) stratigraphic record (MacLeod & Keller 1991). In addition, cyclicity in some sediments seems to be controlled by large amplitude, irregular and unpredictable climatic variations that are not directly related to the Earth's orbit around the Sun (e.g. Weedon & Jenkyns 1990). Thus it is not clear how many future decades must pass before geologists can seriously argue whether a particular event is to be assigned to, say, orbital eccentricity cycle 1456, rather than, say, cycle 1457.

Conclusions

Will the time-scale ever be finalized? A time-scale can be regarded as final when the uncertainties in the ages of the chronometric scale are less than a particular value. For example, if the uncertainties are set at 10 Ma, then the Cenozoic biostratigraphic and magnetic polarity scale is finalized and has been for some years. But if the uncertainties are set at 1 Ma, then there is still some work to do to attain such precision for both scales in the older Cenozoic. If one considers isotopic dating for the Phanerozoic as a whole, uncertainties in Ma of ±0.5% seem achievable using presently available methods, but uncertainties of ±0.1% seem less likely to be realised for some time to come.

Order of magnitude improvements are possible to the present time-scale. This can be achieved initially by quantitatively relating all dated samples to stratigraphic, biostratigraphic and magnetic polarity logs. There is much to be said for re-examining not only boundary stratotypes but the entire stage (or stages) and recording the properties of the stratotypes by standard geophysical methods. These could provide additional local and global correlation markers. Except for the Cretaceous long normal magnetic interval, from Santonian to Aptian time, the greatest improvements in future calibrations of the Cenozoic, Cretaceous and late Jurassic time-scale are likely to come from mapping all isotopic dates, directly or indirectly, onto the magnetic polarity scale and determining the best function that makes a smooth fit to these data. For the remainder of Phanerozoic time improvements may come from normalizing the position of dates within stages and/or biostratigraphic and magnetic polarity zones. The usefulness of normalization will depend on knowing the stratigraphic completeness of the sections from which dated materials are obtained and how hiatuses are distributed within them.

P. F. Friend, E. Hailwood, J. Kennedy, D. G. Smith and G. Weedon are thanked for very helpful discussions and criticisms. The author is grateful to A. Mussett and P. Sadler for their invaluable reviews and suggestions for the improvement of the script. H. Alberti drafted the diagrams. E. Hailwood is also thanked for the invitation to speak at the time-scale meeting. This is Department of Earth Sciences Contribution 2397.

References

Aubry, M-P., Hailwood, E. A. & Townsend, H. A. 1986. Magnetic and calcareous-nannofossil stratigraphy of the lower Palaeogene formations of the Hampshire and London basins. *Journal of the Geological Society, London*, **143**, 729–735.

Aubry, M-P., Berggren, W. A., Kent, D. V., Flynn, J. J., Klitgord, K. D., Obradovich, J. D. & Prothero, D. R. 1988. Paleogene geochronology: An integrated approach. *Paleoceanography*, **3**, 707–742.

Baadsgaard, H. & Lerbekmo, J. F. 1982. The dating of bentonite beds. *In*: Odin, G. S. (ed.) *Numerical dating in Stratigraphy, Part 1*. John Wiley & Sons, New York, 423–440.

Berger, A. 1988. Milankovitch theory and climate. *Reviews of Geophysics*, **26**, 624–657.

Berggren, W. A., Kent, D. V. & Flynn, J. J. 1985*a*. Jurassic to Paleogene: Part 2, Paleogene geochronology and chronostratigraphy. *In*: Snelling, N. J. (ed.) *The Chronology of the Geological Record*. Geological Society, London, Memoir **10**, 141–195.

——, —— & Van Couvering, J. A. 1985*b*. The Neogene: Part 2, Neogene geochronology and chronostratigraphy. *In*: Snelling, N. J. (ed.) *The Chronology of the Geological Record*. Geological Society, London Memoir **10**, 211–220.

Bralower, T. J., Monechi, S. & Thierstein, H. R. 1989. Calcareous nannofossil zonation of the Jurassic-Cretaceous boundary interval and correlation with the geomagnetic polarity time scale. *Marine Micropaleontology*, **14**, 153–235.

Cox, A. V. & Dalrymple, G. B. 1967. Statistical analysis of geomagnetic reversal data and the precision of potassium-argon dating. *Journal of Geophysical Research*, **72**, 2603–2614.

Fischer, A. G. & Gygi, R. 1989. Numerical and

biochronological time scales correlated at the ammonite subzone level; K-Ar, Rb-Sr ages, and Sr, Nd, and Pb sea-water isotopes in an Oxfordian (Late Jurassic) succession of northern Switzerland. *Geological Society of America Bulletin*, **101**, 1584–1597.

——, DE BOER, P. L. & PREMOLI SILVA, I. 1990. Cyclostratigraphy. *In*: GINSBURG, R. N. & BEAUDOIN, B. (eds) *Cretaceous Resources, Events and Rhythms*. Reidel, NATO ASI Series C, **304**, 139–172.

FRIEND, P. F., JOHNSON, N. M. & McRAE, L. E. 1988. Time-level plots and accumulation patterns of sediment sequences. *Geological Magazine*, **126**, 491–498.

HAILWOOD, E. A. 1989. The Role of Magnetostratigraphy in the Development of Geological Time Scales. *Paleoceanography*, **4**, 1–18.

HARLAND, W. B., ARMSTRONG, R. L., COX, A. V., CRAIG, L. E., SMITH, A. G. & SMITH, D. G. 1990. *A geologic time scale 1989*. Cambridge University Press.

HEIRTZLER, J. R., DICKSON, G. O., HERRON, E. M., PITMAN, W. C., III & LE PICHON, X. 1968. Marine magnetic anomalies, geomagnetic field reversals, and motions of the ocean floor and continents. *Journal of Geophysical Research*, **73**, 2119–2136.

HERBERT, T. D. & D'HONDT, S. L. 1990. Precessional climate cyclicity in Late Cretaceous-Early Tertiary marine sediments: a high resolution chronometer of Cretaceous-Tertiary boundary events. *Earth and Planetary Science Letters*, **99**, 263–275.

JARVIS, G. T. & McKENZIE, D. P. 1980. Sedimentary basin formation with finite extension rates. *Earth and Planetary Science Letters*, **48**, 42–52.

McKENZIE, D. P. 1978. Some remarks on the development of sedimentary basins. *Earth and Planetary Science Letters*, **40**, 25–32.

MacLEOD, N. & KELLER, G. 1991. Hiatus distributions and mass extinctions at the Cretaceous/Tertiary boundary. *Geology*, **19**, 497–501.

MONTANARI, A., DRAKE, R., BICE, D. M., ALVAREZ, W., CURTIS, G. H., TURRIN, B. G. & DePAOLO, D. J. 1985. Radiometric time scale for the upper Eocene and Oligocene based on K/Ar and Rb/Sr dating of volcanic biotites from the pelagic sequence of Gubbio, Italy. *Geology*, **13**, 596–599.

MUSSETT, A. E. & McCORMACK, A. G. 1993. Magnetic polarity timescales: a new test. *This volume*.

ODIN, G. S. & Kennedy, W. J. 1982. *In*: ODIN, G. S. (ed.) *Numerical dating in stratigraphy*. Wiley-Interscience, Chichester, 2 volumes, 777.

SADLER, P. M. 1981. Sediment accumulation rates and the completeness of stratigraphic sections. *Journal of Geology*, **89**, 569–684.

—— & STRAUSS, D. J. 1990. Estimation of completeness of stratigraphical sections using empirical data and theoretical models. *Journal of the Geological Society, London*, **147**, 471–485.

SHIBATA, K. 1986. Isotopic ages of alkali rocks from Nemuro Group in Hokkaido, Japan: Late Cretaceous time-scale points. *Chemical Geology (Isotope Geoscience Section)*, **59**, 163–169.

WEEDON, G. P. & JENKYNS, H. C. 1990. Regular and irregular climatic cycles and the Belemnite Marls (Pliensbachian, Lower Jurassic, Wessex Basin). *Journal of the Geological Society, London*, **147**, 915–918.

Magnetic polarity timescales: a new test

ALAN E. MUSSETT, & ALAN G. McCORMACK

Department of Earth Sciences, University of Liverpool, PO Box 147, Liverpool L69 3BX, UK

Abstract: The magnetic polarity timescale, for times older than a few million years, is based upon sea-floor spreading magnetic anomalies, but the relative ages deduced from the anomalies must be calibrated to allow for non-uniform spreading rates. Calibration is carried out by using well-dated samples of known position in the magnetic timescale, but such samples are rare. This paper presents a method for testing the accuracy of a (portion of a) proposed magnetic timescale, using a body of polarity-plus-age data unconnected with sea-floor spreading anomalies. The method also suggests how a timescale could be improved by displacing or offsetting parts of it by a constant amount.

When this method is used with data for the British Tertiary Igneous Province, the timescale of Harland *et al.* (1989) is found to be superior to those of Harland *et al.* (1982) and Berggren *et al.* (1985) for part of the early Tertiary period (*c.* 62–52 Ma). Over this interval the Harland *et al.* (1989) timescale is close to optimum for this data set, but it might be improved by subtracting about 0.25 Ma from the ages of all polarity boundaries.

The magnetic polarity time-scale was first determined using K-Ar dates on extrusive rocks, but was limited to about the last 4 Ma, because, for dates earlier than this age, the errors became too large to resolve separately short polarity intervals (Cox & Dalrymple 1967). But this short time-scale was sufficient to show that the widths of the sea-floor magnetic anomalies were in proportion to the durations of the corresponding magnetic polarity intervals, a major step in establishing the reality of sea-floor spreading (Vine & Matthews 1963; Vine 1966). In turn, it was recognized that the sea-floor anomalies provided a record of geomagnetic polarity changes extending back to mid-Mesozoic times, provided the sea-floor spreading rate could be determined. Heirtzler *et al.* (1968) published a polarity time scale for the last 80 Ma, calibrated using the short polarity time-scale of Cox & Dalrymple (1967), explicitly assuming a uniform spreading rate. However, since it is now known that significant changes in spreading rate do occur, it is necessary to calibrate or adjust the magnetic time scale deduced from the sea-floor record, using radiometric dates from samples with well-defined stratigraphic positions throughout this timescale. Unfortunately, such dates are difficult to obtain because of the paucity of samples that are both reliable to date and have an unambiguous position in the magnetic polarity sequence. As a result, polarity time-scales are often based upon only a few calibration points and this has led to a succession of time-scales that differ significantly from each other. As illustration, the time-scale of Harland

et al. (1982) was claimed to have an accuracy of 1 Ma, but it placed the commencement of normal chron 24 at 53.13 Ma whereas the time-scale of Berggren *et al.* (1985) placed it at 56.14 Ma.

This paper shows that, although the analytical errors of dates older than a few million years are too large to define individual polarity boundaries and so establish a polarity time-scale, a set of polarity-plus-age data can be used to test and adjust a proposed polarity timescale.

The test

When a rock cools rapidly in the Earth's field, or otherwise acquires a magnetic remanence in a time interval short compared to the duration of geomagnetic polarity intervals, it provides a record of the magnetic polarity at an 'instant' in geological time. However, though the magnetic polarity of a rock can usually be determined unambiguously, using modern palaeomagnetic methods, when the rock is dated radiometrically the analytical error, (typically about 1%) is so large that there is usually a significant probability of the actual measured date being in a polarity interval different from that of the true age (Fig. 1). (In this paper 'age' will refer to the true age of the sample, while 'date' will be used for the result of a single dating analysis, and will generally differ from the age because of analytical errors. It is assumed that there are no other sources of error.) In principle, if the date were redetermined with a sufficient number of replications, the error on the mean could be reduced to any desired level and so obviate this

From HAILWOOD, E. A. & KIDD, R. B. (eds), *High Resolution Stratigraphy*
Geological Society Special Publication, No. 70, pp. 27–37.

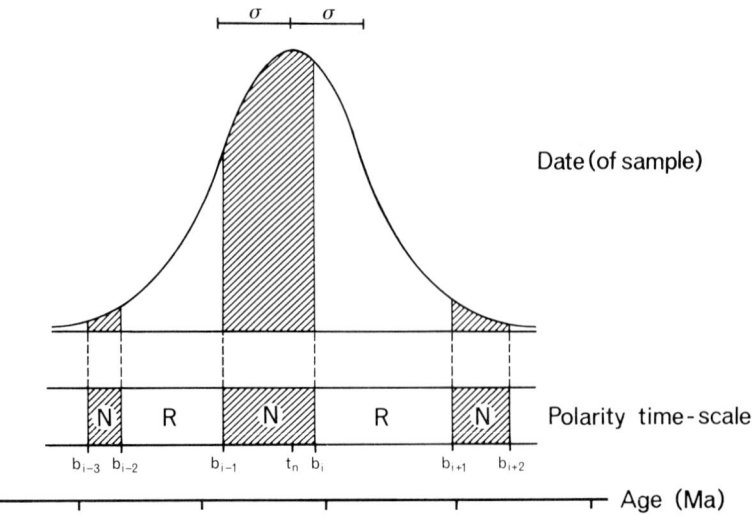

Fig. 1. Distribution of possible dates for a single datum, due to analytical error, with respect to a polarity timescale. b_i, b_{i+1} etc. represent ages of successive magnetic chron boundaries. t_n is the measured date (standard deviation σ) for a sample of given (normal) magnetic polarity.

problem but, in practice, the effort required to determine even a single reliable date is so great that this is not a feasible solution.

Instead, assume that there is a body of independent data, for each datum there being a precise polarity determination and a date having an analytical error comparable to the duration of the polarity intervals. When these data are compared with a given magnetic polarity timescale (determined in some way independent of this body of data), some data will have a polarity appropriate to their date but others, because of the analytical errors of their dates and statistical chance, will *appear* to have the wrong polarity. However, if the timescale is correct, one would expect the proportion of data with appropriate polarity to be higher than if the timescale were in error. This is the basic idea of the test presented in this paper and explained in the next section.

Figure 1 shows a single datum (in this case, with normal (N) polarity). Though its polarity is, by the assumptions of this approach, correct or *appropriate* for its true age this may not be so for the measured date, because of the analytical errors involved. In the example, there is a significant chance that the true age and the measured date do not lie in the same polarity interval. However, it is quite possible that they may lie in intervals of the same polarity; for example, the true age may lie in one particular interval of N polarity but the measured date may, by chance, fall into the succeeding N interval. In such a case the polarity would be

appropriate for the measured date. It is this uncertainty as to which polarity interval a date 'belongs' that limits the direct dating of polarity boundaries, except for those of the last few million years. Clearly, for the example shown in Fig. 1, the chance of the datum having a polarity appropriate to its date is simply the sum of the shaded areas shown, which, if the total area under the curve is normalised to unity, will be a value less than one. This value is given the symbol 'f' or, for the nth datum, f_n. The value of f can be evaluated for every datum and then the average taken, to give $\bar{f}$. The larger the value of $\bar{f}$ the better the fit of the polarity timescale to the polarity-age data. $\bar{f}$ can be evaluated using different timescales, and hence used as a measure of the fit of the timescales to the data.

What significance should be attached to any particular value of $\bar{f}$? A value of 1 would denote perfect agreement of a timescale with the data, i.e., every datum with normal polarity has a date that falls in a N polarity interval and correspondingly for R data. Apart from chance, this would require both that the timescale is correct and that the analytical error of each date is small compared to the durations of the polarity intervals, so that the measured date is close to the true age. This is the ideal case where sufficient dates spread over a given period would delineate the timescale for that period. Conversely, if the errors were very large compared to the durations of the polarity intervals a measured date could fall almost anywhere on the timescale, and so

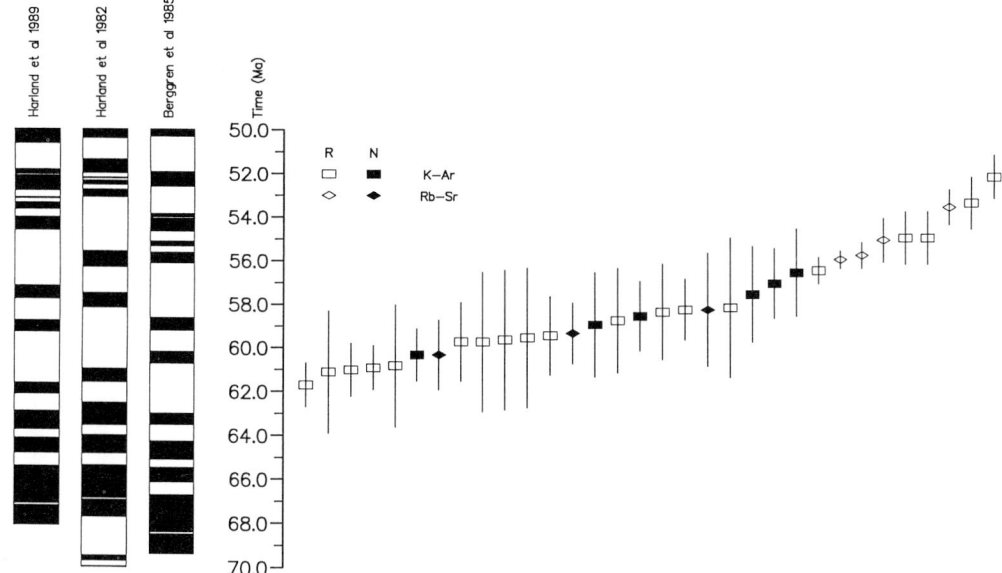

Fig. 2. Data for the British Tertiary Igneous Province, in relation to the polarity timescales of Harland *et al.* (1982) and (1989), and Berggren *et al.* (1985). Error bars represent standard deviations.

would be just as likely to have an inappropriate as appropriate polarity. In this case, $\bar{f}$ would approach 0.5. It follows that a value of less than 0.5 denotes a measure of anti-correlation (which, apart from chance, could come about if the timescale were displaced about one polarity interval, so tending to replace a N polarity by a R one and vice versa. This would be possible only if all the polarity intervals were of about the same duration). Thus the value of $\bar{f}$ is most likely to lie between 0.5 and 1. To refine further our understanding of the significance of $\bar{f}$ we must take into account the relative size of the dating errors compared to the durations of the polarity intervals.

Since the expected value of $\bar{f}$ depends not on the absolute size of the errors, σ_n, but their size *relative* to the duration of the polarity intervals, and because σ will vary from date to date, and also because the duration of the polarity intervals varies along the polarity timescale (see e.g. Fig. 2), both the range of σ and the duration of polarity intervals over the portion of the timescale covered by the dates have to be taken into account when deducing the expected value of $\bar{f}$.

In effect, we can treat the actual body of data as a sample from an infinite population, and calculate the value of $\bar{f}$ to be expected from this population. To a first approximation, we can specify this population as having the same range

of σ as the actual body of data, for all possible ages over a range which we equate with the actual range of dates. The value of $\bar{f}$ for the population will be denoted by F. Expressions for $\bar{f}$ and F are derived in the appendix.

F, the value for the population, is used to normalise the actual value calculated for the data, $\bar{f}$. The normalized value, $\bar{f}/F$, is termed FIT, and it is used to test the agreement of a timescale with the polarity-age data.

Application of the test, using data for the British Tertiary Igneous Province (BTIP)

The BTIP consists of a number of geographically separate areas of igneous activity located predominantly in NW Scotland and the northern half of Ireland. Activity consisted mainly of extensive outpourings of lavas and the development of central volcanoes, whose lower parts, now exposed by erosion, form central intrusive complexes. The stratigraphic succession in most of the areas is fairly well known (as described in the references for Table 1) and provides the basis for the radiometric and palaeomagnetic studies. In addition, there are numerous minor intrusions, particularly dykes, but as their positions in the successions are usually poorly known their contribution is correspondingly less. Results have been published in a number of

Table 1. *Data for rocks of the British Tertiary Igneous Province, selected as satisfying the criteria given in the text*

Datum	Rock unit	Dating method	Age ±1σ (Ma)	Ref.	Polarity	Ref.	Datum
1	Sgurr lava, Eigg	Ar-Ar	52.1 ± 0.5	Dickin and Jones (1983)	R	Dagley & Mussett (1986)	1
2	G4 granite, Mourne Mtns	Ar-Ar	53.3 ± 0.6	Thompson et al. (1987)	R	Dagley et al. (unpublished)	2
3	Beinn an Dubhaich Granite, Skye	Rb-Sr	53.5 ± 0.4	Dickin (1981)	R	Dagley et al. (1990)	3
4	composite sill, Skye	Ar-Ar	54.9 ± 0.6	Dagley et al. (1990)	R	Dagley et al. (1990)	4
5	G2 granite, Mourne Mtns	Ar-Ar	54.9 ± 0.6	Thompson et al. (1987)	R	Dagley et al. (unpublished)	5
6	Conochair Granite, St. Kilda	Rb-Sr	55.0 ± 0.5	Brook (1984)	R	Morgan (1984)	6
7	G3 granite, Mourne Mtns	Rb-Sr	55.7 ± 0.3	Gibson et al. (1987)	R	Dagley et al. (unpublished)	7
8	G2 granite, Mourne Mtns	Rb-Sr	55.9 ± 0.2	Gibson et al. (1987)	R	Dagley et al. (unpublished)	8
9	dyke, Lundy Island	Ar-Ar	56.4 ± 0.3	Mussett (unpublished)	R	Mussett et al. (1976)	9
10	L. Bà Felsite, Mull	Ar-Ar	56.5 ± 1.0	Mussett (1986)	N	Dagley et al. (1987)	10
11	Beinn à Ghraig Granophyre, Mull	Ar-Ar	57.0 ± 0.8	Mussett (1986)	N	Dagley et al. (1987)	11
12	Fishnish E-W dyke, Mull	Ar-Ar	57.5 ± 1.1	Mussett (1986)	N	Dagley & Mussett (1978)	12
13	L. Uisg Granophyre, Mull	Ar-Ar	58.1 ± 1.6	Mussett (1986)	R	Dagley et al. (1987)	13
14	Centre 3 granites, Mull	Rb-Sr	58.2 ± 1.3	Walsh et al. (1979)	N	Dagley et al. (1987)	14
15	Western Granophyre, Rhum	Ar-Ar	58.2 ± 0.7	Mussett (1984)	R	Dagley & Mussett (1981)	15
16	Slemish plug, Antrim	Ar-Ar	58.3 ± 1.1	Dagley et al. (unpublished)	R	Dagley et al. (unpublished)	16
17	Drumadoon Dyke, Arran	Ar-Ar	58.5 ± 0.8	Mussett et al. (1987)	N	Mussett et al. (1987)	17
18	Carlingford Complex	Ar-Ar	58.7 ± 1.2	Dagley et al. (unpublished)	R	Dagley et al. (unpublished)	18
19	dyke, Port na Feannaiche, Arran	Ar-Ar	58.9 ± 1.2	Hodgson et al. (1990)	N	Hodgson et al. (1990)	19
20	Coire Uaigneach Granite, Skye	Rb-Sr	59.3 ± 0.7	Dickin (1981)	N	Dagley et al. (1990)	20
21	Lava S007c, Mull	Ar-Ar	59.4 ± 0.9	Mussett (1986)	R	Dagley et al. (1987)	21
22	G5 granite, Mourne Mtns	Ar-Ar	59.5 ± 1.6	Evans et al. (1973)	R	Dagley et al. (unpublished)	22
23	G2 granite, Mourne Mtns	Ar-Ar	59.6 ± 1.6	Evans et al. (1973)	R	Dagley et al. (unpublished)	23
24	Lava S010, Mull	Ar-Ar	59.7 ± 1.6	Mussett (1986)	R	Dagley et al. (1987)	24
25	Ailsa Craig	Ar-Ar	59.7 ± 0.9	Hodgson et al. (1990)	N	Hodgson et al. (1990)	25
26	Northern Granite, Arran	Rb-Sr	60.3 ± 0.8	Dickin et al. (1981)	N	Hodgson et al. (1990)	26
27	Northern Granite, Arran	Ar-Ar	60.3 ± 0.6	Evans et al. (1973)	N	Hodgson et al. (1990)	27
28	Lava 78/5, Mull	Ar-Ar	60.8 ± 1.4	Mussett (1986)	R	Dagley et al. (1987)	28
29	Carlingford Complex	Ar-Ar	60.9 ± 0.5	Dagley et al. (unpublished)	R	Dagley et al. (unpublished)	29
30	Sandy Braes, Antrim	Ar-Ar	61.0 ± 0.6	Dagley et al. (unpublished)	R	Dagley et al. (unpublished)	30
31	Lava S023, Mull	Ar-Ar	61.1 ± 1.4	Mussett (1986)	R	Dagley et al. (1987)	31
32	Blind Rock Dyke, Donegal	Ar-Ar	61.7 ± 0.5	Dagley et al. (unpublished)	R	Dagley et al. (unpublished)	32
33	Lava, Muck	Ar-Ar	63.0 ± 3.4	Dagley & Mussett (1986)	R	Dagley & Mussett (1986)	33
34	Lava, Eigg	Ar-Ar	63.5 ± 2.0	Dagley & Mussett (1986)	R	Dagley & Mussett (1986)	34

papers, most of which have been considered in a summary for the whole province, in Mussett *et al.* (1988). References are given in Table 1.

Some hundreds of dates have been obtained for the BTIP, mostly by the conventional K-Ar method, but the rocks have proved hard to date accurately, possibly because of the ubiquitous hydrothermal alteration. For this reason, dates have only been accepted if they pass certain internal consistency tests:

(i) Rb-Sr dates are to be isochron dates, where the scatter of the data about the isochron is accounted for by analytical error (Brooks *et al.* 1972);
(ii) K-Ar dates are to be ^{40}Ar-^{39}Ar incremental-heating plateau dates, where the plateau satisfies the criteria of Lanphere & Dalrymple (1978) and Mussett (1986).

All dates given in this paper have been calculated, or recalculated, using the decay constants etc. recommended by Steiger & Jäger (1977), now in general use. These values of the decay constants are not definitive but have been adopted to achieve inter-laboratory standardisation. However, this is only achieved for a single decay scheme and there may be systematic errors when comparing dates determined using more than one method. For the two methods used for the dates analysed in this paper there is believed to be little disagreement (Williams *et al.* 1982).

Palaeomagnetic polarities have been assigned to sites only where several independently oriented cores have been collected and their directions agree closely ($\alpha_{95} < 10°$) after cleaning by alternating field or thermal demagnetisation. A site is considered to be normally magnetised if its mean palaeomagnetic pole is within 40° of the geomagnetic pole, and correspondingly for reversed polarity (Wilson *et al.* 1972).

The data that satisfy these requirements are set out in Table 1, and shown in Fig. 2.

As expected, the errors of the dates are so large that the data cannot be used to determine polarity boundaries, even though at this period the polarity intervals are relatively long (the average duration over the period 50 Ma to 65 Ma is about 0.7 Ma, according to the timescale of Harland *et al.* (1989)).

Before the formulae given in the appendix can be applied, the range of errors in the dates (σ_{min} to σ_{max}) and the range of dates (t_{min} to t_{max}) have to be chosen. Fig. 3 shows histograms of these quantities. The two oldest dates also have considerably the largest errors and they have been omitted because they are atypical of the body of data, and to extend t_{max} to include them would

Table 2. *Polarity timescales used*

Magnetic anomaly (Chron)	Normal polarity interval (Ma)		
	Harland *et al.* (1982)	Berggren *et al.* (1985)	Harland *et al.* (1989)
20	44.31–45.49	44.66–46.17	43.13–44.59
21	47.46–48.69	48.75–50.34	47.01–48.51
22	49.91–50.43	51.95–52.62	50.03–50.66
23	51.39–51.57	53.88–54.03	51.85–52.08
23	51.60–52.02	54.09–54.70	52.13–52.83
24	52.26–52.55	55.14–55.37	53.39–53.69
24	52.77–53.13	55.66–56.14	54.05–54.65
25	55.60–56.33	58.64–59.24	57.19–57.80
26	57.52–58.19	60.21–60.75	58.78–59.33
27	61.00–61.62	63.03–63.54	61.65–62.13
28	62.55–63.57	64.29–65.12	62.94–63.78
29	64.03–64.86	65.50–66.17	64.16–64.85
30	65.39–66.88	66.74–68.42	65.43–67.14
31	66.97–67.74	68.52–69.40	67.23–68.13

give them undue weight. The remaining data are still not uniformly distributed, as required for an ideal body of data, but the range of σ has been chosen as $\sigma_{min} = 0.15$ to $\sigma_{max} = 1.65$ Ma, and the range of ages $t_{min} = 52$ to $t_{max} = 62$ Ma.

Next, the limits of the range of timescale to be included in the integral (polarity boundaries b_1 to b_J of Expression 3 of the Appendix) have to be chosen. Though the error distribution extends to + and − infinity, the area in each tail beyond 2 σ is only 2.5%; in this paper we have set b_1 and b_J at twice the average σ beyond the range of dates, i.e. from $(t_{min} - 2\sigma)$ to $(t_{max} + 2\sigma)$.

The values of $\bar{f}$, F and FIT have then been computed for the timescales of Harland *et al.* (1982), (1989) and Berggren *et al.* (1985) (see Table 2). The results are displayed in Fig. 4. These three parameters have been calculated not only for the timescales as presented but for displacements or offsets of the timescales (when the calculation has been carried out with an offset, the condition that the range of timescale used extends to twice the average σ beyond the range of ages has been maintained). This is justified because the main difference between timescales over a limited period of time, since they all derive from sea-floor magnetic anomalies, is in the age assigned to an anomaly boundary, and only to a lesser extent to 'stretching' of the timescale, either uniformly or non-uniformly. As expected, values of F and $\bar{f}$ lie between 0.5 and considerably less than 1, and F is a smoother function than $\bar{f}$.

For the Harland *et al.* (1982) timescale, the value of FIT is near a local minimum for zero offset but significantly larger for offsets of

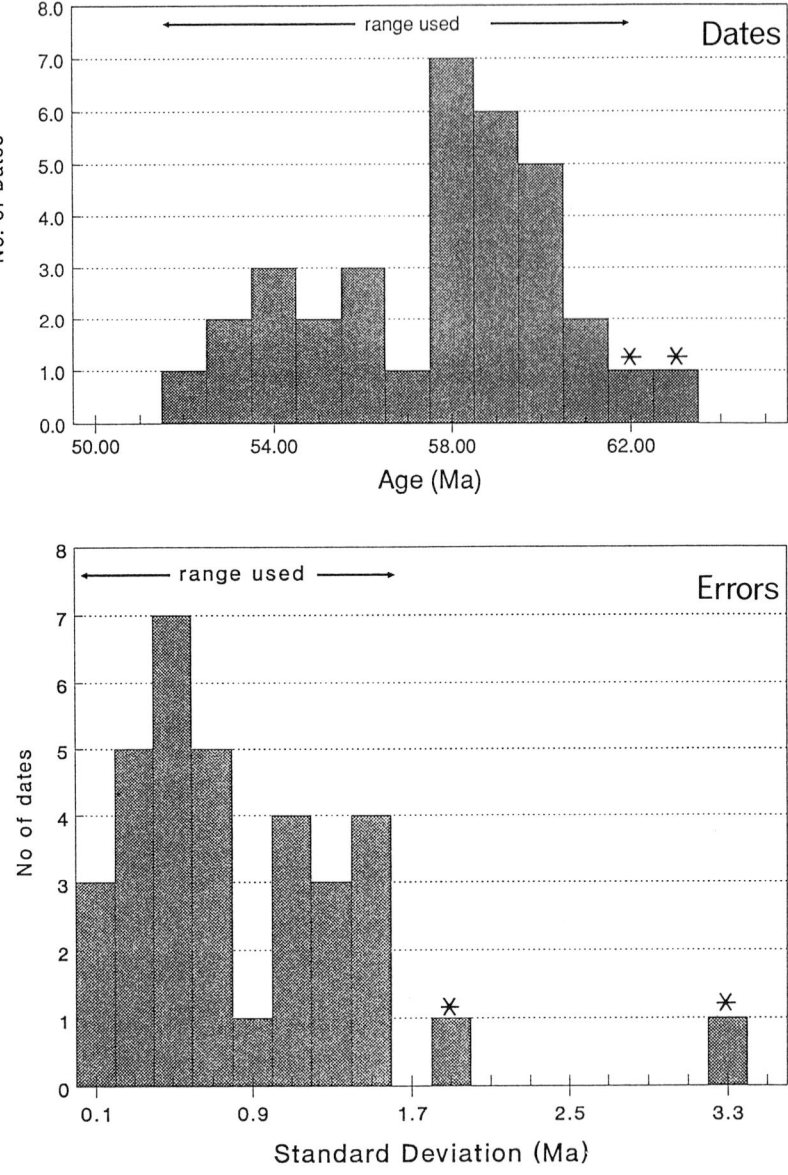

Fig. 3. Histograms of the distributions of (**a**) dates, and (**b**) standard deviations of the dates. The two data shown starred have been omitted from the evaluations of F and $\bar{f}$.

−0.75 Ma and particularly +1.25 Ma (a *positive* offset refers to *adding* a fixed amount to the ages of all polarity boundaries). The graph of FIT therefore indicates that this timescale is an inaccurate one over the interval considered and that it could be improved by adding about 1.25 Ma to, or possibly, subtracting 0.75 Ma from, the times of all polarity boundaries over the period 52–62 Ma.

As presented, the timescale of Berggren *et al.* (1985) corresponds with the bottom of a broad minimum in the FIT value, indicating poor agreement. Better agreement is achieved by offsetting the timescale over the interval considered by about −1.75 Ma or, less suitably, +1.0 Ma.

The Harland *et al.* (1989) timescale without offset corresponds to a value of FIT close to a

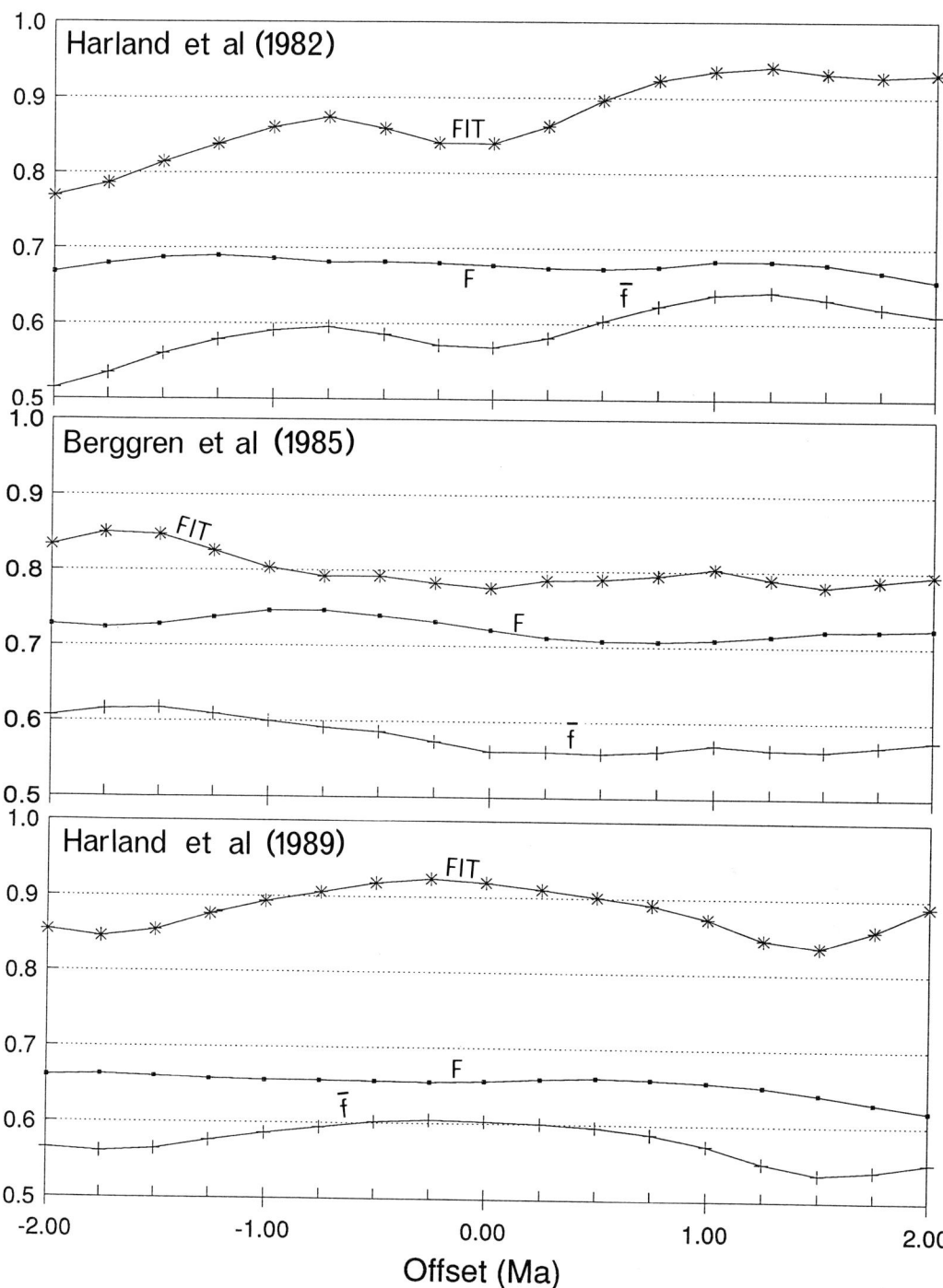

Fig. 4. Values of $\bar{f}$, F and FIT for each of the selected polarity timescales, with offsets up to ± 2 Ma (an offset is considered to be positive when it has been added to the ages of all polarity boundaries).

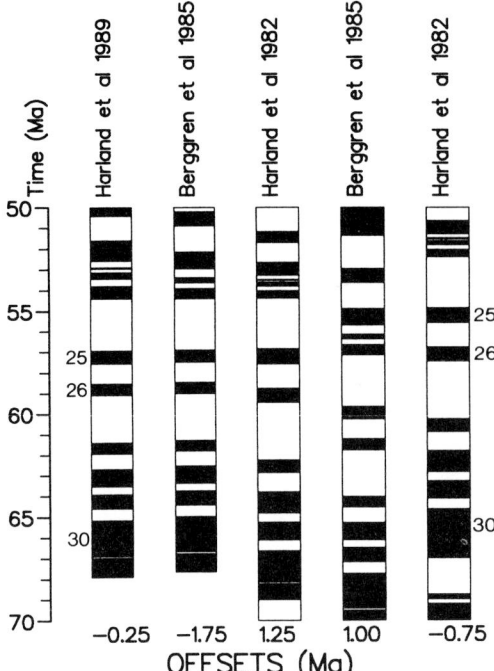

Fig. 5. The polarity timescales after being offset by the amounts corresponding to the five largest peaks of FIT shown in Fig. 4. The amounts of offsets are given at the base of each scale, + indicating a shift to older ages. Numbers at right refer to corresponding marine magnetic anomaly (chron) numbers.

broad peak, and this peak has the highest value of FIT within the range of offsets considered. Thus the body of data considered clearly supports the Harland *et al.* (1989) timescale as being the most consistent with the BTIP polarity/age data of the three timescales considered, over the period 52 to 62 Ma. However, the consistency could be improved further by subtracting about 0.25 Ma from the ages of all polarity boundaries in this interval.

Figure 5 shows how the three timescales would appear for the interval 50–70 Ma if each of the five offsets mentioned above, which give maximal values of FIT, were made to them. The three scales to the left correspond to the largest maxima, and hence the 'optimum' offset, for each of the three scales. Anomaly 25 is brought into almost perfect coincidence for these three offset scales, with other anomalies showing fair to good agreement. This indicates that the method not only can be used to test the accuracy of timescales against observed polarity/age data sets, but also can be used to suggest improvements to published scales by offsetting (though

not by differential offsetting or 'stretching') of these scales.

Discussion and conclusions

The method described in this paper is based on a number of assumptions: (i) a body of data exists, each item of which has a correct polarity and a date whose error is purely analytical; (ii) the range of ages is sufficient to encompass several polarity intervals; (iii) the number of data is large enough that it can be taken as representative of a population i.e. the range of dates and the range of their errors can be taken as being the same as that of the population. Of these assumptions (i) is probably satisfied so far as correct assignment of polarity is concerned, but it is less certain that the only sources of error are analytical, since dating systems are not always closed; it was to satisfy this condition so far as possible that dates were only accepted if they satisfied internal tests of consistency; however, non-analytical sources of error cannot be ruled out; (ii) the span of dates, 52–62 Ma, is much greater than the average duration of polarity intervals in this period (about 0.7 Ma, see above); (iii) though the number of data (32) is moderately large, the histograms of Fig. 3 show that neither the dates nor their errors are distributed uniformly over their ranges, and this is probably the most severe limitation to the application of this method; the concentration of dates in the interval 58–62 Ma will tend to put more weight on the agreement between the data and the timescales in this range. It might be possible to adjust the expressions for $\bar{f}$ and F to allow for this, but it is felt that with the present number of data and other reservations this would hardly be justified.

The final qualification is that the method can only be used to test or compare timescales already proposed and though, by using offsets, it can suggest improvements, these can only be by the addition or subtraction of a fixed interval to all polarity boundaries within the span of the data; it cannot in its present form deal with 'stretching' of a timescale, either uniformly or non-uniformly, though it might be adapted to permit uniform 'stretching'.

Subject to these reservations, the method used with data from the British Tertiary Igneous Province indicates that, for the approximate period 52–62 Ma, the earlier timescales of Harland *et al.* (1982) and Berggren *et al.* (1985) show poor consistency with the observed data, and could be improved by applying the offsets discussed in the previous section. This would bring them, particularly the latter, close to the time-scale of Harland *et al.* (1989) which is in good overall agreement with the data. However, the consistency of this last timescale with the observed data may be improved by subtracting about 0.25 Ma from the ages of all polarity boundaries in the period 52–62 Ma.

We wish to thank P. Dagley and A. G. Smith for helpful discussions, and the latter for a preview of the Harland *et al.* (1989) timescale.

Appendix: derivation of formulae for $\bar{f}$ and F

The derivation assumes that a body of polarity-age data is available, in which the polarities have been correctly determined and the errors of the dates are analytical only.

For the nth polarity-age datum, an an age t_n that has an analytical error following a normal distribution with standard deviation σ_n, the probability that a single date will have a value less than b_i, the age of polarity boundary i (see Fig. 1), is, according to the equation of the Gaussian distribution,

$$P\left(\frac{b_i - t_n}{\sigma_n}\right)$$

$$\text{where} \quad P(x) = \frac{1}{\sqrt{2\pi}} \int_{-\infty}^{x} \exp(-y^2/2)dy \quad (1)$$

Thus the probability that a single date will lie between successive polarity boundaries b_i and b_{i+1} (i.e. within a single, particular polarity interval) is

$$P\left(\frac{b_{i+1} - t_n}{\sigma_n}\right) - P\left(\frac{b_i - t_n}{\sigma_n}\right) \quad (2)$$

Hence the probability of a single date having an appropriate polarity (represented by the total shaded area of Fig. 1), over the range of polarity intervals between boundaries 1 and J, is

$$\sum_{j=1}^{j=J} \left[P\left(\frac{b_{j+1} - t_n}{\sigma_n}\right) - P\left(\frac{b_j - t_n}{\sigma_n}\right) \right] \delta_j \quad (3)$$

where δ_j is set to 1 if the polarity is appropriate, 0 if it is not. However, a correction needs to be made because, though a normal distribution tails to + and − infinity, in practice the integration is carried out only between the finite limits b_1 and b_J (the choice of these limits is discussed under Results, above), and so it is necessary to make a correction, or normalisation, by dividing the total area beneath the curve between these two limits:

$$f_n = \frac{\sum_{j=1}^{j=J} \left[P\left(\frac{b_{j+1} - t_n}{\sigma_n}\right) - P\left(\frac{b_j - t_n}{\sigma_n}\right) \right] \delta_j}{P\left(\frac{b_J - t_n}{\sigma_n}\right) - P\left(\frac{b_1 - t_n}{\sigma_n}\right)} \quad (4)$$

If f_n is averaged over all the data, all observed values of t for the J different data, we have $\bar{f}$:

$$\bar{f} = \frac{1}{J} \sum_{n=1}^{J} f_n \quad (5)$$

To find F, Expression 4 must not be summed over the observed values of t_n and σ_n but integrated over the ranges of t_n (t_{min} to t_{max}) and σ_n (σ_{min} to σ_{max}) of the population:

$$F = \frac{\int_{t=t_{min}}^{t=t_{max}} \int_{\sigma=\sigma_{min}}^{\sigma=\sigma_{max}} f_n(\sigma,t)d\sigma dt}{(\sigma_{max} - \sigma_{min})(t_{max} - t_{min})} \quad (6)$$

The ranges of t_n and σ_n of the population have to be estimated from the sample. If the sample is not too small and is reasonably uniformly distributed over the ranges of t_n and σ_n then, to a first approximation, these values can be taken to be those of the population. In practice, the number and distribution of the data probably does not justify a more sophisticated treatment.

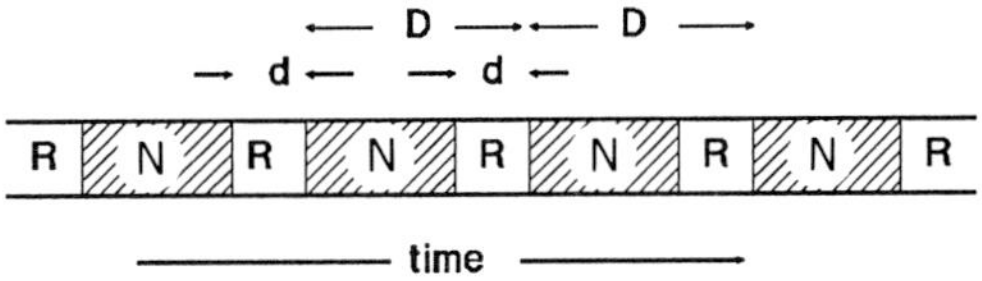

Fig. 6. Diagram showing the synthetic polarity timescale used for Test 2 (Appendix).

Tests: The computer program written to evaluate F was tested in several ways, two of which are now described. In Test 1 a synthetic timescale was created with equal N and R polarity intervals, each of 1 Ma duration. σ was varied from a value small compared to the duration of polarity intervals, to one large, so that F should decrease from nearly 1 to a little above 0.5; the results shown in Table 3 under Test 1 confirm this.

In Test 2 a synthetic timescale was used having R intervals all equal to d, and N intervals all equal to $(D - d)$ (Fig. 6). If σ is large then the dates corresponding to a single N datum will be 'smeared' almost equally across all moments of time, as confirmed by the results of the previous test, and so the proportion of dates that would be of appropriate polarity would be the fraction of the timescale that is N, viz. $(D - d)/D$. Since the chance of a date falling in a N interval is equal to the proportion of the timescale that is N, again $(D - d)/D$, the contribution to F of data with N polarity is $(D - d)^2/D^2$: similarly, for data of R polarity the contribution is d^2/D^2. Thus the value of F would be:

$$F \rightarrow \frac{(D-d)^2}{D^2} + \frac{d^2}{D^2} + 1 - \frac{2d}{D} + 2\left(\frac{d}{D}\right)^2 \quad (7)$$

The figures in Table 3, under the first column for Test 2, for $d = 0.1$ and the same range of σ as in Test 1, shows that as σ tends to very large values F tends to the value predicted by Expression 7. The second column, $d = 1.9$ Ma, is the complementary case in which the

Table 3. *Results of tests of formula for F*

| σ | | F* | | |
min (Ma)	max (Ma)	Test 1[†] d = 1 Ma	Test 2[‡] d = 0.1 Ma	d = 1.9 Ma
0.00	0.01	0.9999	0.9980	0.9980
0.10	0.11	0.9163	0.9354	0.9354
0.20	0.21	0.8365	0.9191	0.9191
0.30	0.31	0.7570	0.9130	0.9130
0.40	0.41	0.6807	0.9099	0.9099
0.50	0.51	0.6151	0.9080	0.9080
0.75	0.76	0.5240	0.9055	0.9055
0.90	0.91	0.5071	0.9051	0.9051
1.00	1.01	0.5031	0.9050	0.9050
1.10	1.11	0.5015	0.9050	0.9050
1.50	1.51	0.4996	0.9051	0.9051
1.75	1.76	0.4998	0.9049	0.9049
2.00	2.01	0.5004	0.9050	0.9050
	limit[§]:	0.5000	0.9050	0.9050

* Calculated for t_{min} = 45 Ma, t_{max} = 55 Ma
[†] The timescale has N and R intervals all equal to 1 Ma
[‡] The timescale has N + R ($=D$) = 2 Ma, d as stated
[§] Value for σ → ∞, calculated using Expression 7

durations of N and R intervals have been interchanged; from symmetry it should give the same results as for $d = 0.1$ Ma, which the table confirms is the case. (The value of F, in each of these tests, does not approach the limiting value completely monotonically. This is because, though the distribution of dates extends smoothly as σ is increased the polarity timescale with which the polarity of each datum is compared is not smooth but discrete, and as σ is increased so the spread of the dates extends first into a polarity of one sign, then into the opposite.)

References

BERGGREN, W. A., KENT, D. V., FLYNN, J. J. & VAN COUVERING, J. A. 1985. Cenozoic geochronology. *Geological Society of America Bulletin*, **96**, 1407–1418.

BROOK, M. 1984. The age of the Conochair Granite, *In*: HARDING, R. R., *et al.* (eds) St Kilda: an illustrated account of the geology. *Report of the British Geological Survey*, **16**, 7, 40–41.

BROOKS, C., HART, S. R. & WENDT, I. 1972. Realistic use of two-error regression treatments as applied to rubidium-strontium data. *Review of Geophysics and Space Physics*, **10**, 551–577.

COX, A. V & DALRYMPLE, G. B. 1967. Statistical analysis of geomagnetic reversal data and the precision of potassium-argon dates. *Journal of Geophysical Research*, **72**, 2603–2614.

DAGLEY, P. & MUSSETT, A. E. 1978. Palaeomagnetism of the Fishnish dykes, Mull. *Geophysical Journal of the Royal Astronomical Society*, **53**, p. 553–558.

—— & —— 1981. Palaeomagnetism of the British Tertiary Igneous Province: Rhum and Canna. *Geophysical Journal of the Royal Astronomical Society*, **61**, 475–491.

—— & —— 1986. Palaeomagnetism and radiometric dating of the British Tertiary Igneous Province: Muck and Eigg. *Geophysical Journal of the Royal Astronomical Society*, **85**, 221–242.

——, —— & SKELHORN, R. R. 1987. Polarity stratigraphy and duration of the Mull Tertiary igneous activity. *Journal of the Geological Society*, London, **144**, 985–996.

——, —— & —— 1990. Magnetic polarity stratigraphy of the Tertiary igneous rocks of Skye, Scotland. *Geophysical Journal International*, **101**, 395–409.

DICKIN, A. P. 1981. Isotope geochemistry of Tertiary igneous rocks from the Isle of Skye. *Journal of Petrology*, **22**, 155–189.

—— & Jones, N. W. 1983. Isotopic evidence for the age and origin of pitchstones and felsites, Isle of Eigg, NW Scotland. *Journal of the Geological Society, London*, **180**, 691–700.

——, MOORBATH, S. & WELKE, N. J. 1981. Isotope, trace element and major element geochemistry of Tertiary igneous rocks, Isle of Arran, Scotland. *Transactions of the Royal Society of Edinburgh*, **72**, 159–170.

EVANS, A. L., FITCH, F. J. & MILLER, J. A. 1973. Potassium-argon age determinations on some British Tertiary igneous rocks. *Journal of the Geological Society, London*, **129**, p. 419–443.

GIBSON, D., McCORMACK, A. G., MEIGHAN, I. G. & HALLIDAY, A. N. 1987. The British Tertiary Igneous Province: Rb-Sr ages for the Mourne Mountain granites. *Scottish Journal of Geology*, **23**, 221–225.

HARLAND, W. B., COX, A. V., LLEWELLYN, P. G., PICKTON, C. A. G., SMITH, A. G. & WALTERS, R. 1982. *A geologic time-scale*. Cambridge University Press.

——, ARMSTRONG, R. L., COX, A. V., CRAIG, L. E., SMITH, A. G. & SMITH, D. C. 1990. *A geologic time-scale 1989*. Cambridge University Press.

HEIRTZLER, J. R., DICKSON, G. O., HERRON, E. M., PITMAN, W. C. & LE PICHON, X. 1968. Marine magnetic anomalies, geomagnetic field reversals, and the motions of the ocean floor and continents. *Journal of Geophysical Research*, **73**, 2119–2136.

HODGSON, B. D., DAGLEY, P. & MUSSETT, A. E. 1990. Magnetostratigraphy of the Tertiary igneous rocks of Arran. *Scottish Journal of Geology*, **26**, 99–118.

LANPHERE, M. A. & DALRYMPLE, G. B. 1978. The use of ^{40}Ar/^{39}Ar data in evaluation of disturbed systems. *United States Geological Survey Open-file Report*, **78–701**, 241–243.

MORGAN, G. E. 1984. Palaeomagnetism. *In*: HARDING *et al.* (eds) St. Kilda: an illustrated account of the geology. *Report of the British Geological Survey*, **16**, (7), 38–39.

MUSSETT, A. E. 1984. Time and duration of Tertiary igneous activity of Rhum and adjacent areas. *Scottish Journal of Geology*, **20**, 273–279.

—— 1986. ^{40}Ar-^{39}Ar step-heating ages of the Tertiary igneous rocks of Mull, Scotland. *Journal of the Geological Society, London,* **143**, 887–896.

——, DAGLEY, P. & ECKFORD, M. 1976. The British Tertiary Igneous Province: palaeomagnetism and ages of dykes, Lundy Island, Bristol Channel. *Geophysical Journal of the Royal Astronomical Society,* **46**, 595–603.

——, —— & SKELHORN, R. R. 1988. Time and duration of activity in the British Tertiary Igneous Province *In*: MORTON, A.C. & PARSON, L. M. (eds) *Early Tertiary volcanism and the opening of the NE Atlantic,* Geological Society, London, Special Publication, **39**, 337–348.

——, ——, HODGSON, B. D. & SKELHORN, R. R. 1987. Palaeomagnetism and age of the quartz-porphyry intrusions, Isle of Arran, Scotland. *Scottish Journal of Geology,* **23**, 9–22.

STEIGER, R. H. & JÄGER, E. 1977. Subcommission on geochronology: convention on the use of decay constants in geo- and cosmochronology. *Earth and Planetary Science Letters,* **36**, 359–362.

THOMPSON, P., MUSSETT, A. E. & DAGLEY, P. 1987. Revised ^{40}Ar-^{39}Ar age for granites of the Mourne Mountains, Ireland. *Scottish Journal of Geology,* **23**, 215–220.

VINE, F. T. 1966. Spreading of the ocean sea-floor: new evidence. *Science,* **154**, 1405–1415.

—— & MATTHEWS, D. H. 1963. Magnetic anomalies over oceanic ridges. *Nature,* **199**, 947–949.

WALSH, J. N., BECKINSALE, R. D., SKELHORN, R. R. & THORPE, R. S. 1979. Geochemistry and petrogenesis of Tertiary granitic rocks from the Island of Mull, northwest Scotland. *Contributions to Mineralogy and Petrology,* **71**, 99–116.

WILLIAMS, I. S., TETLEY, N. W., COMPSTON, W. & McDOUGALL, I. 1982. A comparison of K-Ar and Rb-Sr ages of rapidly cooled igneous rocks. *Journal of the Geological Society, London,* **139**, 557–568.

WILSON, R. L., DAGLEY, P. & ADE-HALL, J. M. 1972. Palaeomagnetism of the British Tertiary Igneous Province: the Skye lavas. *Geophysical Journal of the Royal Astronomical Society,* **28**, 285–293.

Quantitative marine sediment core matching using a modified sequence-slotting algorithm

ROY THOMPSON[1] & R. M. CLARK[2]

[1] *Department of Geology & Geophysics, University of Edinburgh, West Mains Road, Edinburgh EH9 3JW, UK*

[2] *Department of Mathematics, Monash University, Victoria, Australia*

Abstract: An automated core correlation procedure has been developed in order to match sequences, such as core logs, based on a mathematical sequence-slotting approach. The programme can be applied to univariate, multivariate or compositional data. In addition to finding the optimal slotting, the programme also quantifies the quality of match. It can also be used to detect sections of the sequences which are tightly constrained, by the match, or only fit together more loosely. An important aspect of any multivariate analysis is scaling. This topic is addressed through a simplex optimization method.

We have used the sequence-slotting method to analyse high quality oxygen-isotope data from ODP Site 658 on the continental slope, west of Cap Blanc, on the northwest African coast. Excellent agreement is found with the global oxygen-18 reference curve and this allows the sediments to be dated. By and large the resulting depth-age profile is very similar to that derived in previous studies. However, small changes in the age estimates between the original stratigraphy and our sequence-slotting lead to major differences in the pattern of sediment accumulation rate.

An hiatus spanning 24 500 SPECMAP years some 50 000 years ago has been proposed in the Site 658 sediments. However, as we find no compelling mathematical evidence for the hiatus, Site 658 may contain an even better stratigraphic record of palaeoclimatic change than previously thought.

During the last two decades there has been a remarkable upsurge in environmental and sedimentological work on geologically young marine sediments. Research activity has been fuelled by the recognition that the stratigraphic resolution now attainable in the uppermost sediments is detailed enough to allow many global environmental and palaeoclimatic questions to be addressed for the first time. The aim of this paper is to explore the use of a quantitative sequence matching approach to high resolution marine stratigraphic studies.

Two types of advance stand out in creating this proliferation of research work on recent marine sediments. On the one hand there are notable technological advances. In particular, the deployment of the hydraulic piston corer, as part of the Ocean Drilling Programme, has allowed long, continuous sequences of soft sediments to be recovered with remarkably little coring disturbance. This has allowed entirely new fields of research to open up. Instrumental advances in mass spectrometry have also been extremely important in allowing the isotopic composition of marine calcareous microfossils to be measured accurately. This has created the potential for very high resolution stratigraphy. On the other hand there have been developments in theory and interpretation. Again, major changes are seen. In particular the old picture of some four Pleistocene glaciations has been swept away and replaced by a whole sequence of glacial periods. The glaciations are now recognised to have followed a cyclic pattern, determined by the periodic variations of the Earth's orbital characteristics and their effect on the distribution of the Sun's radiation over the Earth's surface. This appreciation of the periodicity of climatic change has arisen from the documentation of cyclostratigraphic variations of the oxygen isotope signature in deep sea sediment cores.

Limitations on the quality of data obtainable from the ocean depths remain however. Irregular geological processes such as bioturbation, caused by marine organisms, and major sediment disturbances, associated with changes in ocean currents or with the movements of turbidites, serve to degrade and complicate marine sediment records. The data that are now being generated by ocean core studies are of such detail and quality that they can be subjected to numerical and statistical analysis, although as noted in 1987 by the report of the Second

From HAILWOOD, E. A. & KIDD, R. B. (eds), *High Resolution Stratigraphy*
Geological Society Special Publication, No. 70, pp. 39–49.

Conference of Scientific Ocean Drilling, as yet 'quantitative techniques have been little used in deep sea . . . stratigraphy'.

A range of quantitative approaches to stratigraphic correlation have been reviewed by Agterberg & Gradstein (1988). A further approach is the inverse procedure of Martinson *et al.* (1982) which has been applied to young marine sequences by Pisias *et al.* (1984).

Sequence-slotting

Sequence-slotting (Gordon 1973), which we now describe, provides a straightforward and intuitively appealing method for matching or calibrating an ordered sequence of measurements against a given reference sequence. The method makes no *a priori* assumptions about the form of the response-time relationship in either sequence or on the sedimentation rates in the sequence being calibrated. It uses only the pair-wise differences between measurements and the known order of observations within each sequence (Fig. 1b).

The reference sequence, denoted sequence A, comprises m points (observations) A_1, A_2, . . ., A_m. The typical point A_i is made up of a $p \times 1$ vector U_i of response variables, and *known* age t_i. Sequence B, comprising n points B_1, B_2 . . ., B_n, contains n measurements $v_1, v_2 . . ., v_n$ of the same p response variables, at successive depths $d_1, d_2 . . ., d_n$ down the core. The aim is to estimate the ages, denoted $t_1^*, t_2^* . . ., t_n^*$ of each of the samples in Sequence B, noting that

$$t_1^* < t_2^* < . . . < t_n^*.$$

Sequence-slotting works by slotting together, in an optimal fashion, the two sequences into a single combined sequence, starting with either A_1 or B_1 and ending in either A_m or B_n, and

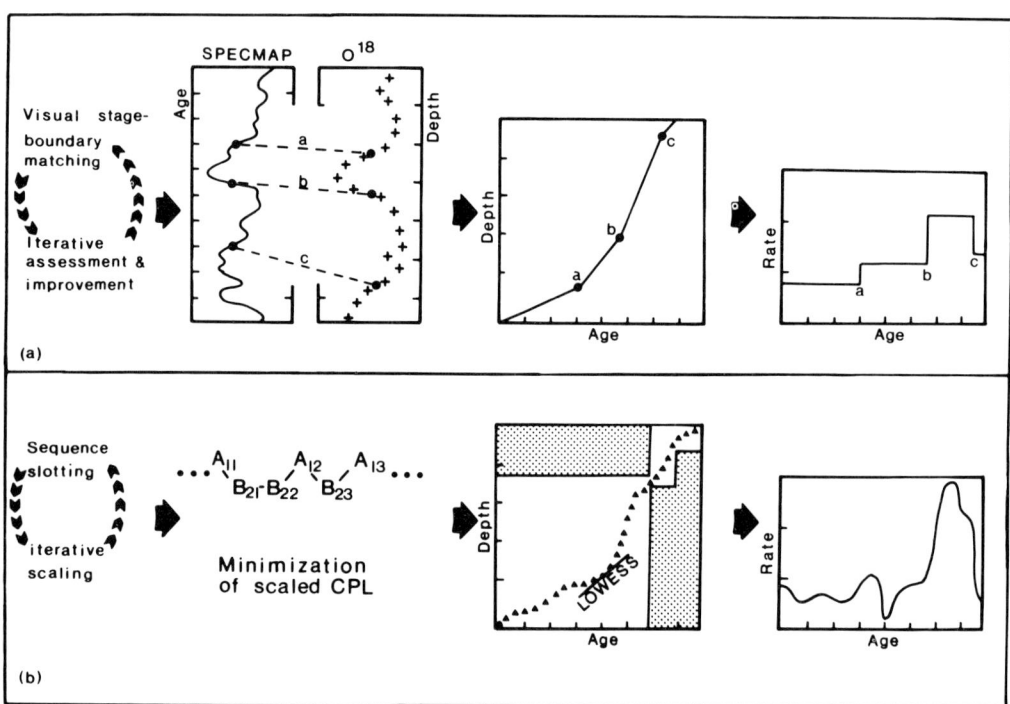

Fig. 1 (**a**) Summary of the graphical/visual stage boundary core correlation approach. New oxygen-18 measurements are used to estimate the depths of stage boundaries (solid dots) and matched (dashed lines) to the SPECMAP stage boundaries. These tie lines (a to c) can be used to construct a depth–age graph. Straight line segments on the depth–age plot yield sediment accumulation rates which vary in a step function, and (**b**) Summary of the sequence-slotting approach. A dynamic programming minimization provides an optimal matching of the sequences, while scaling is handled by an iterative search method. The resulting best match gives a detailed age–depth relationship. The matching can be constrained to avoid prescribed regions (shading) of the depth–age space. A locally weighted smoothing function (LOWESS) is used to estimate sediment accumulation rates.

preserving the natural ordering within each sequence. For any given combined sequence, the m + n points trace out a path in p-dimensional space. Hence a natural measure of the quality of the matching of sequences A and B is the total length of this combined path. This combined path length (CPL) is the sum of the distances between successive pairs of points in the combined sequence. Various measures of 'distance' may be used. In this paper, we use the Euclidean distance, i.e. the root-mean-square of the pair-wise differences in coordinates.

As a simple example, if $m = 2$, $n = 3$ the slotting

$$A_1B_1B_2A_2B_3$$

has

$$CPL = d(A_1,B_1) + d(B_1,B_2) + d(A_2,B_2) \\ + d(A_2,B_3),$$

where $d(.) = $ distance between the indicated points.

The optimal slotting is taken as that which minimizes the CPL, and this can be found by simple, highly efficient, dynamic programming techniques (Delcoigne & Hansen 1975; Gordon & Reyment 1979; Clark 1985; Gordon et al. 1988). It is also possible to minimize the CPL subject to additional order constraints (Clark 1985).

Once the optimal combined sequence has been found, the ages $\{t_j^*\}$ of the samples in sequence B may be estimated by interpolation using the successive distances making up the CPL. For example, if the optimal slotting contains the sub-sequence

$$\ldots A_{19}B_{10}A_{20} \ldots$$

this implies that the age, t_{10}^*, of sample 10 in sequence B must be somewhere between t_{19} and t_{20}, the known ages of points 19 and 20 in the reference sequence. Further, if it happens that B_{10} is twice as distant from A_{19} as from A_{20},

i.e. $d(A_{19},B_{10}) = 2 \times d(A_{20},B_{10})$

then the estimated age t_{10}^* of B10 should also be twice as far from t_{19} as t_{20},

i.e. $t_{10}^* = t_{19} + \frac{2}{3}(t_{20} - t_{19})$.

This local interpolation can be extended in an obvious fashion when there are several Bs between successive As, for example

$$\ldots A_{19}B_{10}B_{11}B_{12}B_{13}A_{20} \ldots$$

This procedure can be formalized by plotting the partial path length (PPL) = length of path through the combined sequence up to the current point, versus the age, actual or estimated, of

the current point. Notice that the estimated ages $\{t_j^*\}$ automatically satisfy the monoticity conditions.

While the minimum CPL is unique, there may be many slottings giving the same optimum CPL. This will rarely happen when p, the number of response variables, is two or more, but will almost certainly happen when $p = 1$. For example, there is an inherent ambiguity whenever an individual point of either sequence corresponds to a local extremum of the response versus time graph. In sections 4 and 5 below, we provide methods for estimating the extent and importance of multiple solutions.

Site 658

We have chosen to illustrate the application of sequence slotting to marine sediment logs by examining oxygen isotope data from the ocean drilling programme Site 658. The rapidly deposited sediment from Site 658 situated on the northwest African continental margin has been shown to be particularly suitable for detailed stratigraphic work (Ruddiman et al. 1989). Sarnthein & Teidemann (1989), in an excellent study, have obtained especially clear patterns of oxygen isotope change in Site 658 sediments as part of the general investigation of African aridity and oceanic circulation. The lithogenic component in the Site 658 sediments is taken to be an indicator of dust blowing to the site, while biological and chemical fluctuations can be related to changes in ocean current upwelling, to productivity and to sea level fluctuations. In order to interpret fully the great wealth of information obtained in studying these sediments a detailed dating scheme is needed for the hole, along with a measure of changing sediment deposition rate. Sediment accumulation rate estimates form the basis for the important process of casting the many measurements of biochemical concentration or percentage in terms of flux or accumulation rate.

Matching Site 658 to the SPECMAP reference sequence

In tackling the problem of sediment age and accumulation rate, Sarnthein & Teidemann (1989) followed the common procedure of erecting an age model by matching the new oxygen isotope record to the SPECMAP stacked oxygen isotope reference curve of Imbrie et al. (1984).

The Imbrie et al. (1984) SPECMAP reference sequence is shown in Fig. 2 (centre graph). Isotopic variations, about a mean value, were

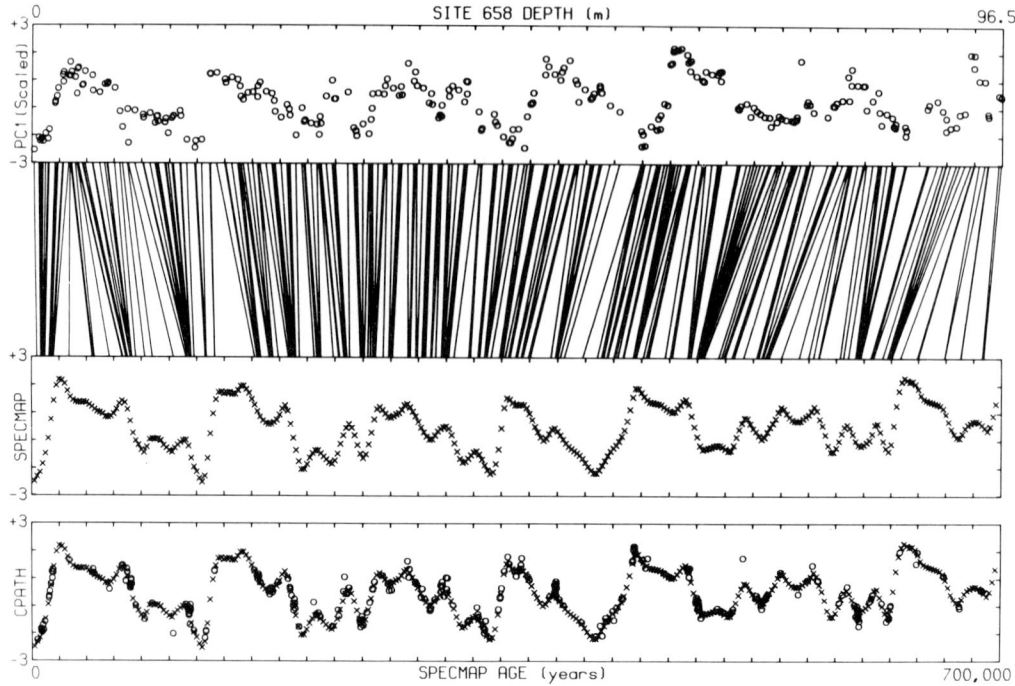

Fig. 2. Sequence-slotting match of the first principal component (PC1) of Site 658 delta oxygen-18 measurements (circles) to the SPECMAP reference sequence (diagonal crosses). The Site 658 data have been scaled to maximize the overall fit. Tie lines have been drawn for each measurement horizon in Site 658 to their sequence-slotted age in the SPECMAP reference curve. Changes in sediment deposition rate are revealed by the way the tie lines bunch together and diverge. This figure displays results of our unconstrained slotting calculations.

given by Imbrie *et al.* (1984) at 2 ka intervals. The variations have been estimated by stacking oxygen isotope data from five cores from different oceans around the world. The isotopic data were matched to astronomical functions in a detailed data processing exercise involving digital filtering and an iterative tuning process, with independent stratigraphical checking. The resulting SPECMAP curve has been widely adopted as a reference sequence by the Quaternary earth science community.

The problem at hand is how to match new isotopic data to the SPECMAP reference curve. A common approach is to note characteristic shapes in the reference curve and to try to identify them visually in the new data sequence (Fig. 1a). Isotopic stage boundaries, which are in effect places where the reference curve changes rapidly, are found to be convenient features to use in this visual correlation procedure. For the top 96.16 m of Site 658 Sarnthein & Teidemann (1989) tabulate forty stage boundary horizons that they tie to the reference curve in this way. Establishing such tie lines is essentially an itera-

tive process of building up a more and more detailed time scale. In between the tie points (i.e. stage boundaries) sediment deposition is taken to have been constant (Fig. 1a). Thus a complete, fully specified, age–depth model is finally erected for the total length of the sediment core.

In our sequence-slotting approach the dynamic programming algorithm similarly produces tie lines between the sequences. However tie lines are produced simultaneously for all data points in the two sequences. So in this example with 266 new isotope measurements and 348 reference curve horizons, each with their own positions in the combined slotting, the result is a total of 614 tie lines. For presentation purposes we have just drawn the 266 tie lines, associated with the new isotope data, as an illustration (Fig. 2) of the end result of the sequence-slotting procedure.

Recognition of hiatuses

A recurring problem in marine stratigraphy is that gaps in sediment deposition may have occurred but be very difficult to detect. Breaks in deposition may leave little or no visual or readily measurable markers in the sediment. Hence they can only be inferred by comparison with other sequences, for example through the absence of biostratigraphic indicator fossils or indicated on time-depth plots by periods of apparently very slow deposition. While the sequence-slotting approach would appear to be able automatically to take account of any hiatuses, in practice it remains extremely desirable to be able to terminate sequences at hiatuses. The sediment above and below the hiatuses can then be slotted separately.

In the Site 658 sediment cores we are considering, Sarnthein & Teidemann (1989) elected in their age model to include a hiatus at 9.1 m depth. They estimate that this hiatus covered a period of 24.5 thousand years. Further hiatuses were identified by Sarnthein & Teidemann (1989) at 96.41 m and 99.30 m depth.

The opposite of a hiatus is a period of extremely rapid deposition. In marine sequences a turbidite is a common example of exceptionally rapidly deposited sediment. Turbidites tend to be much easier to recognise than hiatuses having, for example, unusual grain size characteristics or geochemical signatures. Sarnthein & Teidemann (1989) found no need to include periods of extremely rapid deposition in their age depth model.

In sequence-slotting the mathematical representations of hiatuses and turbidites are long blocks. We address the question of hiatuses in the uppermost Site 658 sediments in 'slotting constraints' below.

We now describe some details of the specific data manipulation procedures used in our quantitative sequence-slotting of the Site 658 delta oxygen-18 data to the SPECMAP reference sequence. Then we go on to compare the results of our slotting with those of the earlier sedimentation rate investigations.

Site 658 data manipulation

Step 1: missing data

The Site 658 data comprise two records of oxygen-18 variation with depth for two different animal shells, namely *C. wuellerstorfi* and *G. inflata*. At most horizons in this core the oxygen-isotope ratio was determined for both types of animal shell. In approximately 10% of the horizons, however, measurements were available on only one type of shell. As a first step in the analysis, the missing data were estimated by

regressing the measurements for *C. wuellerstorfi* against those for *G. inflata*, and vice versa.

Step 2: principal component analysis

At each position in the SPECMAP curve, there was just one variable, whereas at each horizon in Site 658 there were two. But in both cases there is only one genuine physical quantity of interest, and the two measurements at each horizon of Site 658 are naturally highly correlated. These two features of the data can be accounted for simultaneously by performing a principal components analysis (e.g. Rock 1988 p. 314; Mardia *et al.* 1979 chapter 8). As a second step in the analysis, therefore, we calculated the first principal component (denoted PC1). This accounted for 95.4% of the variation of the oxygen-18 in the core. All subsequent analyses were performed using this principal component.

Step 3: scaling

A most important practical step involved scaling the data. Two parameters were used in a linear scaling, or transformation, by changing the mean and variance of the oxygen-18 data. The two scaling parameters were chosen by following the automated search procedure of simplex optimization (NAG Subroutine E04CCF) (Nelder & Mead 1965). Simplex is a particularly flexible and robust optimization method for applications, as here, where efficiency is not of overriding consequence. The simplex iterative procedure was here set up to minimize the normalized sequence-slotting parameter, which is in effect a scale free version of the combined path length, (see Clark 1985; Thompson & Clark 1989). During this stage of the Site 658 data processing the mean was changed from an initial estimate of -0.0837 to -0.1578 and the variance from an initial estimate of 1.0678 to 0.9341; as δ was reduced from its initial value of 0.3495 to the minimum value found of 0.3366. This low value of δ indicates an excellent degree of fit of the Site 658 oxygen-18 data to the SPECMAP reference sequence (see Thompson & Clark 1989 Table 1).

Step 4: dissimilarity and slotting

The optimum matching of Site 658 against SPECMAP was obtained by slotting the two sequences together, using Euclidean distance based on the scaled PC1 data and an extended version of the FORTRAN program of Clark (1985) (cf. Fig. 1b). As described above, the key aspect of the slotting method is the use of dynamic programming to assess all possible ways

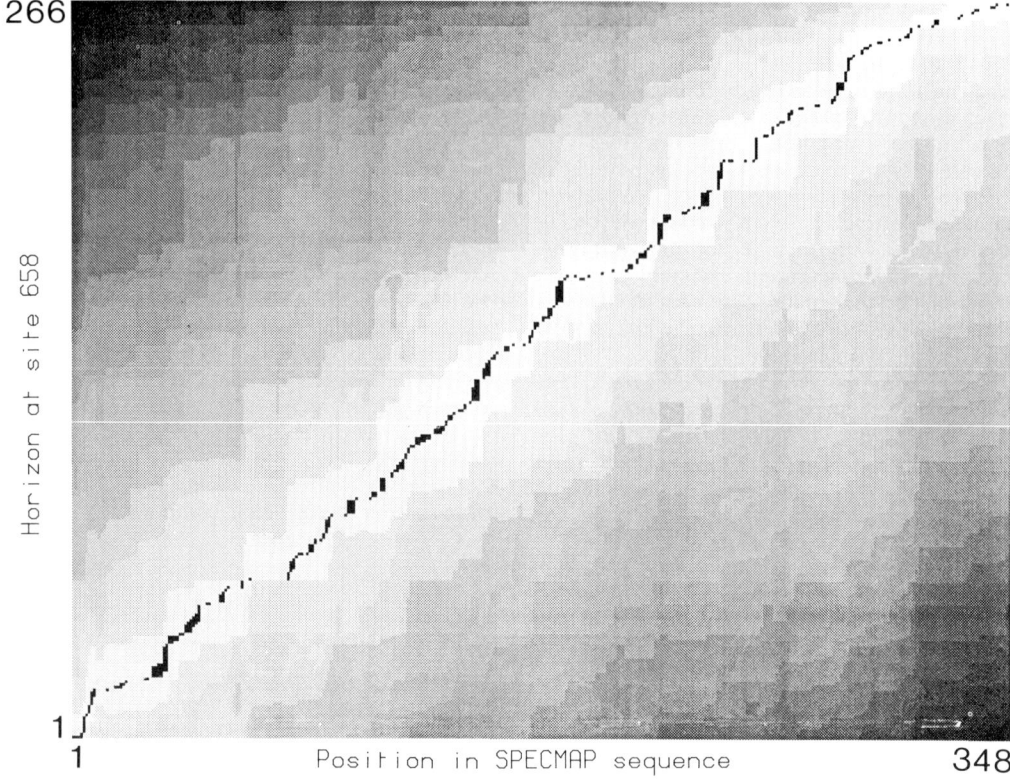

Fig. 3. Modified H-matrix associated with the match of PC1 of the Site 658 delta oxygen-18 measurements and the SPECMAP reference sequence. Successive horizons in Site 658 are associated with the vertical axis while position in the SPECMAP reference sequence is associated with the horizontal axis. The best fit depth/age sequences lie in the dark band running diagonally across the diagram. Increasing grey shade represents poorer and poorer matches of the data. As described in the text the horizontal width between the grey shades can be used as a measure (or bound) of the local tightness of slotting at each delta oxygen-18 measurement horizon.

of matching the sequences as a means of finding the minimum CPL.

Step 5: multiple solutions

Since there is only one response variable in this application, there are necessarily multiple slottings corresponding to the same minimum CPL. One of these matchings is shown in Fig. 2, while the range of alternative solutions is shown in coded form in Fig. 3, by the dark region running diagonally across the centre of the diagram.

Local tightness

There may be subsections of the combined path where the sequences are slotted together more tightly than elsewhere. We now derive a simple numerical measure of the local tightness of the

slotting of sequence B relative to the reference sequence.

Suppose, for example, that in the optimal slotting, point B_{10} (the tenth element in sequence B) lies between A_{19} and A_{20}, but not necessarily immediately adjacent to them. For convenience, we represent this subsequence of the optimal slotting by

$$\ldots A_{19}+++B_{10}+++A_{20}\ldots,$$

where '+' indicates possible additional points from sequence B and '.' elements from either sequence. To assess how tightly B_{10} is slotted relative to the As, we ask the question: how much is the CPL increased if we force B_{10} to lie between A_{20} and A_{21} (say) instead? This question can be answered easily, without repeating the slotting, by a simple modification to the H-matrix defined by Gordon *et al.* (1988).

Similarly, we can easily find how much the CPL increases if we force B_{10} to lie between A_{21} and A_{22}, A_{22} and A_{23}, and so on. In fact it is a very simple matter to compute the percentage increase in CPL due to imposing all such additional constraints

$$\ldots A_k+++B_{10}+++A_{k+1}\ldots$$

for $k = 0, 1 \ldots, m$ and moreover including the extreme cases

$$+++B_{10}+++A_1\ldots \text{ (Block of Bs at start)}$$

and

$$\ldots Am+++B_{10}+++ \text{ (Block of Bs at end)}.$$

The above sensitivity calculations result in $m + 1$ numbers, denoted $c(0)$, $c(2)$, ..., $c(m)$. These numbers measure how the minimum possible combined path length increases as point B_{10} is forced to move away from its preferred optimal slotting position.

Finally we define the local tightness of B_{10} as the number of $c(k)$s which are less than or equal to some pre-assigned tolerance.

In the same way, we can compute the local tightness of each point B_1, B_2 ..., B_j, ... in sequence B relative to the reference sequence. The relative tightness of the successive points in sequence B can be made clear by plotting the local tightness of B_j versus j, for different choices of the tolerance, e.

The starting point for the local tightness calculations for Site 658 is the modified H matrix of Fig. 3. It shows in diagrammatic form the results of the above sensitivity investigations of how much the CPL increases as any horizon, B_j in Site 658 is moved away from its optimal position in the SPECMAP reference series of sequence A. Local tightness is then found for any tolerance, by finding the width of the appropriate grey shading in Fig. 3 for each Site 658 horizon.

Clearly, the local tightness of B_j will increase as the tolerance increases. The minimum value

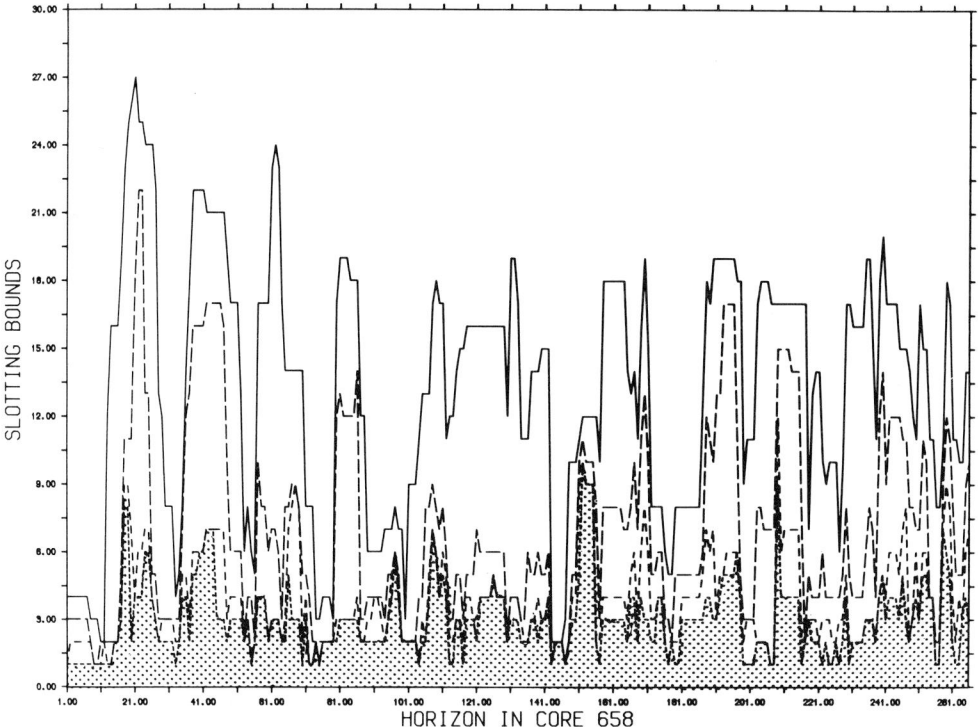

Fig. 4. Tightness of slotting bounds vs Site 658 measurements horizons for four degree of fit levels. Low slotting bounds correspond to sections of the core where the sequence-slotting is particularly well constrained (e.g. horizons 143 to 147) and high values to poorly constrained regions (e.g. horizons 152 to 156). The lower shaded area covers the variation at a goodness of fit level 1.001 times the minimum path length.

of the tolerance should be related to the intrinsic accuracy of the response variables. For example, suppose, as with Site 658, there is just one response variable, and the observed measurement for B_j were increased by 0.1 say. If B_j were located at a local maximum of the response-time graph, then the CPL would be increased by twice this measurement error. For Site 658 with a minimum CPL of 112.6 the equivalent e would be 0.2/112.6 i.e. 0.18%. Since the oxygen isotope ratios in SPECMAP units are recorded to the nearest 0.1, the preceding argument suggests that the tolerance level should be no smaller than 0.18%.

Figure 4 shows the local tightness plots for four choices of for Site 658. The tightest points are numbers 9, 10, 11, 144, 145, 146. The plots indicate rapid changes in the degree of tightness of match around horizons 15 and 150, but no overall trend along the length of the core. These horizons (15 and 150), at which local tightness changes most rapidly, coincide with particularly rapid variation in the oxygen isotope ratio, at the end of glacial periods I and IV, at depths of 3 and 51 m.

Slotting constraints

The analyses described above were all made using unconstrained slottings. A most important additional facility available in the sequence-slotting approach is that constraints can be easily included in the slotting. Thus, additional geological information can be used to guide the sequence matching. For example, all horizons of the sediment core must be older than 0 SPECMAP years. This type of information can be mathematically formulated very simply. This specific constraint has been included in our subsequent analyses by forcing the joint sequence to begin with A_1 (the first point in the SPECMAP sequence). Clark (1985) describes a very flexible array of easily implemented constraints that can be used to reflect various types of additional stratigraphic information. Other common examples of additional kinds of stratigraphic information present in deep sea core sequences, that could provide constraints on slotting, would be the depths of dated ash horizons or biostratigraphic first or last data appearance.

A second form of mathematical constraint that we have found particularly useful is the facility to impose a maximum block length on either or both sequences. In the example we are considering, a long block of As corresponds to a long period of time without any sediment accumulation, ie a hiatus in sequence B, while a long block of Bs would correspond to a long sequence of sediment forming in a short time (<2000 years), ie a turbidite in sequence B. By varying the maximum block lengths allowed we can encourage or discourage the presence of hiatuses or turbidites.

We have modified the sequence-slotting algorithm of Clark (1985) in order to allow the maximum block length in either sequence to be fixed explicitly. The maximum block lengths specified can thus be used to limit the relative deposition rates when slotting data from cores. Ideally the relationship between sequences formed by the slotting approach should not be sensitive to the block length constraints imposed. Also the minimum path length, calculated subject to the blocking constraints, should not be significantly greater than the unconstrained minimum.

Inspection of Fig. 2 reveals that unconstrained sequence-slotting produces blocks in sequence A of up to length 14 and of up to length 13 in sequence B. Imposition of moderate constraints to the A and B sequences was found, at this site, to break the long blocks into smaller units without producing any major changes to the general pattern of match or any dramatic increases in CPL. As the question of a hiatus (long block of As) around 50 thousand SPECMAP years was of some interest, slottings with a practical block length constraint on only the B sequence were also tried. It was found that the maximum block length for the B sequence could be reduced as low as four with only a small effect on the quality of match. This block length constraint reduces the number of multiple solutions, still allows the presence of hiatuses and forms the basis of our preferred slotting of Fig. 5.

We have found, with other data sets, that the overall match can be very sensitive to the block length constraints imposed. Such sensitivity is a pointer to mismatching of the sequences and is often associated with pronounced hiatuses. The only satisfactory way of handling this most unwelcome situation is to split the offending sequence into two at the hiatus. For the upper 96.41 m at Site 658 we have found no evidence of such sensitivity or mismatching.

Accumulation rate determination

In addition to sediment age another important geological quantity we would like to estimate is sediment accumulation rate. The isotopic stage boundary tie line approach yields average sediment accumulation rates within each isotopic stage (Fig. 1a). Using this method Sarnthein & Teidemann (1989) estimated stage average sedi-

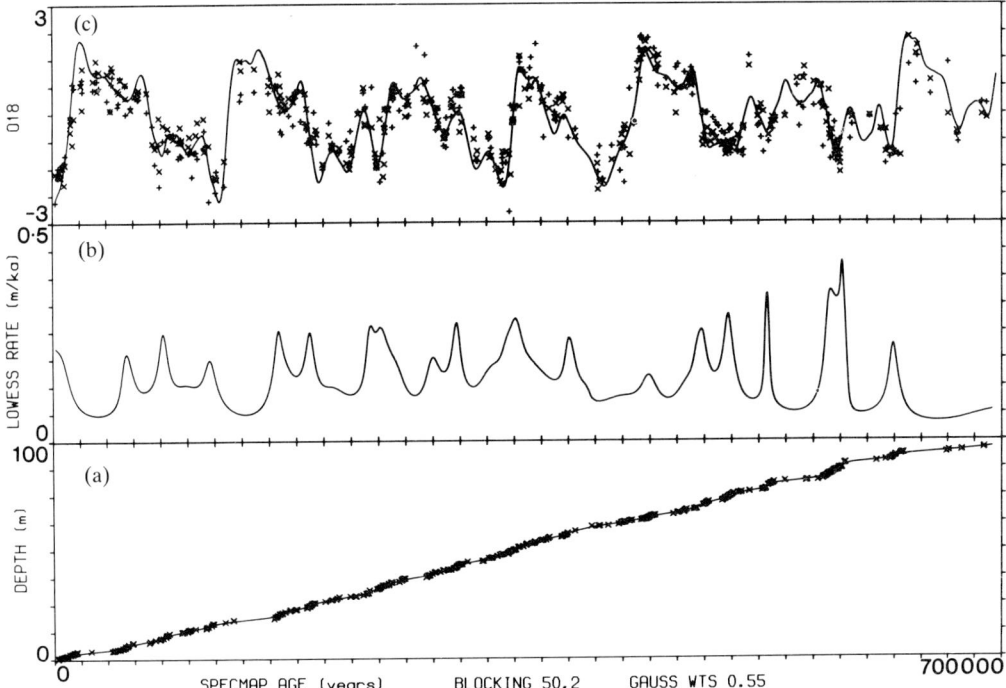

Fig. 5. Our best estimate of combined ages and accumulation rates for Site 658, as found from constrained sequence-slotting and LOWESS/Hermite calculations.
(**a**) Depth v. age diagram. Crosses are results from constrained sequence-slotting matching. Continuous curve is hermite cubic fit to LOWESS estimates.
(**b**) Accumulation rates vs. age diagram. Continuous curve is gradient of Hermite in (a).
(**c**) comparison of Site 658 original delta oxygen-18 measurements on *G. inflata* (upright crosses) and *C. wuellerstorfi* (diagonal crosses) plotted using Hermite/LOWESS ages from (a) with the SPECMAP reference sequence (continuous curve). Oxygen-18 data from Sarnthein & Tiedemann 1989.

ment accumulation rates at Site 658 to have varied between 0.0 and 26.0 cm per thousand years.

Following the sequence-slotting approach, it would appear a simple task to differentiate the resulting depth-age relationship $d_1-t_1^*$, $d_2-t_2^*$ ..., $d_n-t_n^*$ (e.g. the data of Fig. 2) and hence to obtain a detailed sediment accumulation rate history for the site. However, errors associated with individual data points, as demonstrated by the tightness calculations, are too large to allow this direct approach. It is necessary to firstly smooth the age–depth sequence.

Cleveland (1979) has described a robust locally weighted smoothing procedure (LOW-ESS) for scatterplots which deals admirably with the type of data we have. By fitting a local low-order polynomial regression our age-depth points can be smoothed by LOWESS. Cleveland (1979) notes that LOWESS is controlled by four items. These are d, the order of the polynomial

that is locally fit to each point on the age-depth plot; W, the weighting function t, the number of iterations and f, the parameter used to determine the amount of smoothing.

As Cleveland (1979) suggests, d can be most satisfactorily set to be 1 i.e. to perform a local straight line fit. A particularly helpful property of LOWESS with $d = 1$ is that it preserves the monoticity of the data. Cleveland (1979) notes that $t = 2$ is sufficient for most situations. As our data are monotonic and as we have no outliers, we have found that t can be set to 1 i.e. no iterations are necessary. We have used Gaussian weights for W, but found Tukey Biweights or Cleveland Tricubes to be equally suitable. f, the span of data points with non zero weights (or in the case of Gaussian weights, the standard deviation) in effect determines the severity of the smoothing. In situations, as in ours, where the purpose of the smoothing is to produce a smooth regression function then the smoothing parameter f is of paramount importance. A compelling demonstration of this general point is given by Hastie & Tibshirani (1990, chapter 2) in a comparison of different smoothing methods including polynomial regression, regression spline, natural spline and LOWESS, but with the same smoothing parameter. Cleveland (1979) indicates various methods for choosing f. Unfortunately these choices of f give far too little smoothing. We also experimented with a cross validation approach to determining f. Although this gave a clear, smooth cross validation minimum, as based on validation of the depth–age data, once again the degree of smoothing was far too little for our purposes. At present we have had to select f subjectively by assessment of the fit of the oxygen-18 data to the SPECMAP reference curve for different f values, i.e. different degrees of the depth–age data. Visual assessment was based on the differences (Site 658, SPECMAP) i.e. on oxygen-18 residuals. Fig. 5c shows this fit for our preferred choice of f.

At this stage in the analyses we have from LOWESS derived new estimates of the ages $t_1{}^*$, $t_2{}^* \ldots, t_n{}^*$. The final operation is to fit a smooth interpolation function to the smoothed depth–age pairs. We have used a piecewise cubic Hermite, NAG subroutine E01BEF (Fritsch 1982). The Hermite interpolant is monotonic and differentiable. It therefore is able to give us both age and sediment accumulation rates estimates at all depths between the top and bottom samples in our core. These estimates are plotted as continuous curves in Fig. 5a and b. They represent our best estimate of internally consistent ages and sediment accumulation rates for Site 658.

Sequence slotting/stage boundary the line comparisons

Although the overall age–depth relationships arising from the two approaches of sequence-slotting and stage boundary tie line matching are similar, there are many differences in the accompanying accumulation rate estimates. The main age differences concern (i) the absence of a hiatus around 9 m depth when using the sequence-slotting approach and (ii) within-stage differences which, with the sequence-slotting approach, lead to much more variable sediment accumulation rates.

Sequence-slotting has selected a more straightforward match of the Site 658 oxygen-18 data, around isotope stages 3 to 5 (around 40–90 ka) than that adopted by Sarnthein & Teidemann (1989), with no need for a hiatus.

The best estimates of accumulation rates for this period using sequence-slotting and LOWESS (Fig. 5b) are slightly above average. Furthermore the local tightness of slotting estimates near horizon 36 are also quite normal. So, for this part of the sequence, there is no great uncertainty in the slotting. All these factors demonstrate how sequence-slotting shows no tendency to place a hiatus in this section of the core.

In contrast in an analysis of another oxygen-18 record at ODP Site 722 we found that sequence-slotting selected to include marked hiatuses between 460 and 560 SPECMAP years (author's unpublished calculations), whereas investigations by Clemens & Prell (1991) using the stage boundary/tie line approach did not detect any sedimentation gaps. This practical example indicates to us that sequence-slotting is quite capable of detecting hiatuses in oxygen-18 records of deep sea sediments.

Perhaps the most evocative result of the sequence-slotting is the form of the accumulation rate variations as displayed in Fig. 5b. Here we see a pattern of pulses of sediment. For example, there is a clear pulse at around 570 to 590 thousand SPECMAP years. The sedimentation rate increases by a factor of about 5 during the period of increased sedimentation. This sediment pulse can also be seen in Fig. 5a as a steep section on the age–depth plot involving some 19 oxygen-18 measurement horizons. The alternative choices of LOWESS smoothing parameters or block length constraints, discussed above, would lead to even stronger sediment pulses than those depicted in Fig. 5 and to even greater contrasts with earlier studies.

Sarnthein & Teidemann (1989) have no such sediment pulses. Indeed this is the normal situation of many studies of deep sea sediments. The small modifications suggested to the Site 658 depth-age model by the sequence-slotting can thus be seen to have major effects on sedimentation rates. These in turn will lead to major changes in any flux calculations.

Conclusions

(i)   Sarnthein & Teidemann (1989) have produced a remarkably high resolution oxygen-18 record at ODP Site 658 which can be matched extremely well with the SPECMAP global reference sequence.

(ii)   Oxygen-18 analyses that are even more closely spaced could be expected to yield better age models, and are probably essential for constraining sediment accumulation rate estimates.

(iii) Sequence-slotting is able to indicate alternative depth-age models to those derived by the frequently used approach of stage boundary tie line construction, or to the inverse approach of Martinson *et al.* (1982).

(iv) A quantitative overall measure of core matching can be found using sequence-slotting through the statistic δ. The minimum generated of 0.3366 represents an excellent degree of fit of the Site 658 oxygen-18 data to the SPECMAP reference sequence.

(v) Through calculations of local tightness of slotting, sections of core can be delimited which are dated particularly well or particularly poorly. Although it might be expected that dating of oxygen-18 records would be most precise at stage boundaries, this situation is not found by our tightness of slotting calculations.

(vi) Marked changes in sediment accumulation rate are found as a consequence of the sequence-slotting and LOWESS calculations.

(vii) Sequence-slotting is easily applied to oxygen-18 data. Other applications to marine sequences might be, on the one hand, in helping to build up composite depth scales (e.g. using whole core susceptibility data) while on the other hand they might include improving the global oxygen-18 reference curve by formalized stacking of the master oxygen-18 records and direct matching to astronomical target functions.

References

AGTERBERG, F. P. & GRADSTEIN, F. M. 1988. Recent developments in quantitative stratigraphy. *Earth Science Reviews*, **25**, 1–73.

CLARK, R. M. 1985. A FORTRAN program for constrained sequence-slotting based on minimum combined path length. *Computers and Geosciences*, **11**, 605–617.

CLEMENS, S. C. & PRELL, W. L. 1991. One-million year record of summer-monsoon winds and continental aridity from the Owen Ridge (Site 722), northwest Arabian Sea. *In*: PRELL, W. L. *et al.* (eds) *Proceedings of ODP, Scientific Results*, **117**. College Station, TX (Ocean Drilling Program), 365–388.

CLEVELAND, W. S. 1979. Robust locally weighted regression and smoothing scatterplots. *Journal of the American Statistical Association*, **74**, 829–836.

DELCOIGNE, A. & HANSEN, P. 1975. Sequence comparison by dynamic programming. *Biometrika*, **62**, 661–664.

FRITSCH, F. N. 1982. *PCHIP Final specifications*. Lawrence Livermore National Laboratory report UCID–30194.

GORDON, A. D. 1973. A sequence-comparison statistic and algorithm. *Biometrika*, **60**, 197–200.

—— & REYMENT, R. A. 1979. Slotting of borehole sequences. *Mathematical Geology*, **11**, 309–327.

—— CLARK, R. M. & THOMPSON, R. 1988. The use of constraints in sequence-slotting. *In*: DIDAY, E. (ed.) *Data Analysis and Informatics*, **V**. North Holland, Amsterdam, 353–364.

HASTIE, T. J. & TIBSHIRANI, R. J. 1990. *Generalized Additive Models*. Chapman & Hall, London.

IMBRIE, J. *et al.* 1984. The orbital theory of Pleistocene climate: Support from a revised chronology of the marine oxygen isotopic record. *In*: BERGER, A., IMBRIE, J., HAYS, J., KUKLA, G. & SALTZMAN, B. *Milankovitch and Climate Part 1*, D. Reidel, Hingham, Mass, 269–305.

MARDIA, K. V., KENT, J. T. & BIBBY, J. M. 1979. *Multivariate Analysis* Academic Press, London.

MARTINSON, D. G., MENKE, W. & STOFFA, P. 1982. An Inverse Approach to Signal Correlation. *Journal of Geophysical Research*, **87**, 4807–4818.

NELDER, J. A. & MEAD, R. 1965. A simplex method for function minimisation. *Computer Journal*, **7**, 308–313.

PISIAS, N. G., MARTINSON, D. G., MOORE, T. C., SHACKLETON, N. J., PRELL, W., HAYS, J. & BODEN, G. 1984. High resolution stratigraphic correlation of benthic oxygen isotopic records spanning the last 300,000 years. *Marine Geology*, **56**, 119–136.

ROCK, N. M. S. 1988. *Numerical Geology*. Springer-Verlag, Berlin.

RUDDIMAN, W., SARNTHEIN, M., *et al.* 1989. *Proceedings of ODP, Scientific Results*, **108**. College Station TX (Ocean Drilling Program).

SARNTHEIN, M. & TEIDERMANN, R. 1989. Towards a high-resolution stable isotope stratigraphy of the last 3.4 million years: Sites 658 and 659 off northwest Africa. *In*: RUDDIMAN, W., SARNTHEIN, M. *et al.* (eds) *Proceedings of ODP, Scientific Results*, **108**. College Station, TX (Ocean Drilling Program), 167–185.

THOMPSON, R. & CLARK, R. M. 1989. Sequence-slotting for stratigraphic correlation between cores: theory and practice. *Journal of Palaeolimnology*, **2**, 173–184.

High resolution molecular stratigraphy: analytical methodology

MATTHEW B. SMITH, JON G. POYNTER[1], STUART A. BRADSHAW[2] &
GEOFFREY EGLINTON

*Organic Geochemistry Unit, Bristol University, School of Chemistry, Cantocks Close,
Bristol BS8 1TS, UK*

[1] *Present address: Thomson Tour Operations, Greater London House, Hampstead Road,
London NW1 7SD, UK*

[2] *Present address: Mobil Oil Company Ltd, Information Systems Department, Mobil
House, 54–60 Victoria Street, London SW1E 6QB, UK*

Abstract: In recent years Quaternary stratigraphy has become analytically intensive, given
the realisation that the sedimentary record at specific locations is sufficiently continuous and
complete for examination at 100 year resolution. This paper describes methods which
enable high resolution molecular stratigraphy to be carried out quickly and efficiently. The
primary technique for data generation is gas chromatography (GC), validated for certain
samples by gas chromatography mass spectrometry (GC-MS). GC data are reduced using a
suite of programs to provide compound abundances for entry in Lotus 1–2–3 spreadsheets.
xy plots of compound abundance versus sub-bottom depth are produced, prior to the use of
numerically intensive methods such as Fourier transform spectral analysis and principal
components analysis. The paper is illustrated with time series data for the U^k_{37} parameter
from ODP Leg 108, Sites 658 A and B.

Molecular stratigraphy: goals

Utilizing organic compounds as biogeochemical
markers ('biomarkers') in the sedimentary
record is not new (Brassell *et al.* 1986a; Poynter
& Eglinton 1991). However, it is proving to have
potential as a method by which relatively continuous stratigraphies can be produced for a
given location. In recent years sedimentary
biomarker concentrations have been used as
tools for making palaeoenvironmental reconstructions (Brassell *et al.* 1986a), based on
information gained from such stratigraphies. It
has been demonstrated that the distributions
and abundances of selected biomarker compounds in sediments and oils as determined by
GC-MS and other techniques, provide information regarding depositional environments
(Didyk *et al.* 1978), due to different biomarkers
characterising organisms contributing to the
molecular content of a sediment (Brassell *et al.*
1987). The basis for this approach lies in the
identification of the compounds present in the
sediments and correlation of their abundances
with those of biolipids present in living organisms.

Molecular stratigraphy is of use in the reconstruction of palaeoclimates and oceanic history,
for example, in relation to global climate change
(Brassell *et al.* 1986b), providing a means of
reversing the basic geological tenet, 'The

present is the key to the past', to map the past
onto the present in order that a better understanding of current climatic trends might be
derived (Marlowe 1984). In principle, molecular
stratigraphy should afford an insight into the
activity of benthic and pelagic biota and their
response to differing climatic and oceanic conditions. This information, combined with micropalaeontological data for species diversity,
evolutionary trends, and individual species
numbers might then allow modelling of palaeo
sea surface temperatures, oceanic and atmospheric circulation patterns (Sarnthein *et al.*
1981), terrestrial sediment source regions, levels
of water column oxicity, oceanic productivity
and ocean atmosphere interactions.

Molecular stratigraphy: nature of biomarkers

Biomarkers can be grouped into broad categories based on their various sources (Table 1).
Each category of biomarker provides information on various processes and events through
the temporal variation of it's member compounds. The categories are as follows.

Alkanes and alcohols. Straight chain alkanes and
alcohols (with carbon numbers > nC23) are
present in higher plant cuticular leaf waxes.

From Hailwood, E. A. & Kidd, R. B. (eds), *High Resolution Stratigraphy*
Geological Society Special Publication, No. 70, pp. 51–63.

Table 1. *Organic biomarker compounds quantitated for Hole 658 A & B extracts*

Compounds	Source
n-alkanes n-C27 (heptacosane) n-C28 (octacosane) n-C29 (nonacosane) n-C30 (triacontane) n-C31 (untriacontane) n-C32 (dotriacontane) n-C33 (tritriacontane) n-C34 (tetratriacontane)	Assigned to higher plant cuticular waxes, input as terrestrial dusts transported on wind systems (Poynter *et al.* 1990).
n-alcohols n-C24 (tetracosanol) n-C26 (hexacosanol) n-C28 (octacosanol) n-C30 (triacontanol) n-C32 (dotriacontanol)	As n-alkanes (Poynter *et al.* 1990).
Unknowns U1–U8	Provenance also unknown.
Hopanes C28 (bisnorhopane) C30 (hopane)	hopanes Assigned mainly to aerobic bacterial and possibly algal sources. Weathered ancient sediments (Robinson *et al.* 1984).
Hopanols bishomohopanol	As hopanes (Robinson *et al.* 1984).
Sterols cholest–22–en–3β–o1 cholest–5–en–3β–o1 cholestan–3β–o1 24-methyl-cholesta–5,22–dien–3β–o1 24-methyl-cholest-22-en-3β-o1 24-methyl-cholest-5-en-3β-ol 24-methyl-cholestan-3β-o1 24-ethyl-cholest-22-en-3β-ol 24-ethyl-cholest-5-en-3β-o1 24-ethyl-cholestan-3β-o1 Dinosterol Dinostanol	Phyto and Zooplankton (Yamaguchi *et al.* 1986). } Dinoflagellates (Wengovitz *et al.* 1981).
Alkanediols etc. triacontan-1,15-diol untriacontan-1,14-diol dotriacontan-1,15-diol dotriacontan-15-ket-1-o1	Assigned to Cyanobacteria or other plankton (Morris & Brassell 1988).
Alkenones heptatriacont-8,15,22-trien-2-one heptatriacont-15,22-dien-2-one methyl hexatriacont-15,22-enate octatriacont-9,16,23-trien-3-one octatriacont-9,16,23-trien-2-one octatriacont-16,23-dien-3-one octatriacont-16,23-dien-2-one nonatriacont-10,17,24-trien-3-one nonatriacont-17,24-dien-3-one	Assigned to Prymnesophycae – Coccolithophorid algae, E. Huxleyi etc. (Volkman *et al.* 1980).

These waxes coat the surface of leaves and are released into soils and terrestrial sediments as particulate material. As the sediments are eroded, they form wind born dusts which are redeposited in oceanic basins, carrying these compounds with them. Consequently, the temporal variations observed relate to the direction and strength of local winds (and consequently the pole/equator thermal gradient) with time, and the position of vegetation belts in response to global climate change, whilst the n-alkane distributions relate to species compositions within vegetation belts (Poynter & Eglinton 1991; Poynter et al. 1989a).

Alkanediols. This category has been assigned to production by cyanobacteria and can be used as an indicator of changing water depth and therefore temperature, salinity and light penetration (Morris & Brassell 1988).

Hopanes and hopanols. Produced by aerobic bacteria and algae, these compounds can be included in the sedimentary record by transport from weathered ancient sediments, or primary production in abyssal or other aerobic environments. (Robinson et al. 1984).

Sterols. Sterols are mainly the result of primary production by phyto and zooplankton in ocean surface waters (Yamaguchi et al. 1986). They indirectly reflect variations in salinity, water temperature, upwelling activity, nutrient levels, light penetration and oxicity.

Alkenones. Sourced by prymnesiophyte algae (Marlowe et al. 1990), the alkenones provide a direct correlation between ratioed amounts and changes in sea surface temperature (SST) (Poynter 1989).

Thus, a suite of long chain alkenones occurs in ocean sediments dating from the Albian, the most notable and recent producer being the coccolithophorid *E. huxleyi* (250 Ka – present) (Volkman et al. 1980; Marlowe 1984; Farrimond et al. 1986). Laboratory studies have demonstrated that biosynthesis of the di- and triunsaturated alkenones by *E. huxleyi* closely correlates growth with water temperature (Marlowe 1984; Prahl & Wakeham. 1987). Hence, the U^k_{37} ratio of these compounds can be utilized as a molecular stratigraphic parameter indicative of palaeo sea surface temperature (SST) in Quaternary marine sediments (Poynter et al. 1986b). The most recent stratigraphic implementation of this technique is referred to herein and pertains to ODP Leg 108, Site 658,

Holes A and B, drilled in the upwelling region off NW Africa (Ruddiman et al. 1988; 1989).

Focus of paper

This paper focusses on current methods and new automation processes, with illustrations of the types of data and output produced by these processes.

Sampling

Previous work

Previous studies using methods associated with high resolution molecular stratigraphy have been carried out on samples from non upwelling locations adjacent to Site 658. Leg 108 Site 659 was chosen by Ten Haven et al. (1989) as a comparison study with Site 658.

Regarding upwelling areas, samples from ODP Leg 112, Site 686B (Farrimond et al. 1990) have been analysed at Bristol. This site, like Site 658, lies under an oceanic upwelling centre. Such sites are well situated for the preservation of biomarkers and other compounds (Brassell et al. 1986a, b; Brassell et al. 1987; Prahl & Wakeham 1987; Prahl & Pinto 1987; Morris & Brassell 1988; Ten Haven et al. 1989).

The samples and sampling rationale

The U^k_{37} parameter analogues palaeo-seasurface temperatures (SST) based on the ratio of two long chain alkenones. The two alkenones considered are the 37:2 methyl ketone ($C_{37}H_{70}O$) and the 37:3 methyl ketone ($C_{37}H_{68}O$). The index is calculated using the following equation:

$$U^k_{37} = \frac{[C_{37}:2]}{[C_{37}:2 + C_{37}:3]}$$

Where: $[C_{37}:2]$ = Concentration of the 37:2 methyl ketone.
$[C_{37}:3]$ = Concentration of the 37:3 methyl ketone.

In regions of upwelling, where local SST falls, and during periods of glaciation, when global SST falls, production of $C_{37}:3$ is enhanced and so the U^k_{37} index falls, reflecting the change of SST.

Initial studies by Farrimond et al. (1990) proved the feasibility of using U^k_{37} and lipid biomarkers to reconstruct palaeoclimates, ocean productivity and glacial activity. The present dedicated molecular stratigraphic study using the U^k_{37} parameter was carried out jointly between the Bristol University (UK) and Kiel (Germany) research groups.

A series of samples were collected at positions as close as possible to those used by Sarnthein & Teidemann in the production of the $\delta^{18}O$ data (Sarnthein & Teidemann 1990). Completion of this stratigraphic study demonstrated the need for a very detailed examination of glacial termination events, therefore; an 'ultra high' resolution sample set was collected. It should be noted that the terms 'high', 'very high' and 'ultra high' are relative terms and are not directly defined by any absolute measure. Experience of working Site 686B proved invaluable in designing improved sampling strategies and analytical methods.

High and very high resolution set. This set consists of 239 samples taken at two sampling (high and very high resolution) intervals within each core section from the A and B holes. The intervals were calculated by M. Sarnthein from the SPECMAP (McIntyre *et al.* 1976), time scale for $\delta^{18}O$ and $\delta^{13}C$ studies (Sarnthein & Teidemann 1990). The nominal sampling interval from centre to centre was 100 cm for the high resolution and 20 cm for the very high resolution samples. 1 cm^3 of sample was made available by ODP/East Coast Repository (ECR). Data generated from these samples were used to produce a high resolution molecular stratigraphy based on U^k_{37}.

Ultra high resolution set. Consisting of a total of 507 samples, taken at a regular interval of 2 cm centre to centre, this set covered glacial terminations II and IV and the interglacial between cold stages 5.4 and 6 (Prell *et al.* 1986) as assessed from the $\delta^{18}O$ and $\delta^{13}C$ SPECMAP time series data (Sarnthein & Teidemann 1990). A sample volume of 0.33 cm^3 was used for this set.

Storage and sampling problems

Sediment cores at ODP/ECR are stored at a temperature of 4° C in 1.5 m long plastic storage tubes capped at both ends. The atmosphere in the storage tube is kept moist by wet sponges. The sediment itself is contained in a plastic liner as a half core within the storage tube. Sampling is achieved using either metal spatulas and glass vials or plastic insert tubes. The insert and enclosed sediment sample are thermally sealed into plastic bags. The use of plastic liners, inserts, sponges, tubing and bags is somewhat problematical as these are all potential sources of plasticizer (mainly phthalate) contamination (Fig. 1). This form of contamination can cause

the true signal to be swamped by a multitude of GC peaks which effectively obscure biomarker data. Thermal sealing of plastic bags may be a source of contamination, as the volatile constituents of the bags will condense on the sample. Additionally, the storage of cores at 4° C is not ideal for organic analysis as it is sufficiently warm for some compounds to biodegrade. Deep frozen cores such as those stored at ODP/ TAMU are preferable.

The size of samples authorized by ODP/ TAMU presented a problem. Ideally, a large sample of 5 to 10 cm^3 would minimize the effect of local inhomogeneities. Such inhomogeneities occur as a result of lateral grain size changes and sedimentation rate changes. A slice across most of the core width that only samples a small length (<0.5 cm) of the core would partially alleviate this problem, as would better information on sedimentation rate changes. Bioturbation also plays an important role in determining the selection of sampling resolutions. Bioturbation acts as a smoothing filter rather than a noise effect and therefore blurs genuine local inhomogeneities.

Analytical method

The principal method of analysis used to identify and quantify biomarkers in the present study was gas chromatography (GC). GC peaks were identified using gas chromatography mass spectrometry (GC-MS) on selected samples.

Sample extraction

The extraction used was a five stage methanol/ dichloromethane method. First, the sample was extracted using excess methanol, sonicated in a tank for ten minutes and then centrifuged. The extract was decanted into a vial and the extraction process repeated. Stage three followed the same procedure, save that the methanol was replaced by a 2:1 methanol:dichloromethane mixture. Stages four and five were replicates of stages one and two, using dichloromethane instead of methanol. The extract was then derivatized for 24 hours at a temperature of 4° C in excess bis- (trimethylsilyl)trifluoroacetamide (BSTFA) to form trimethylsilyl derivatives of the alcohols and acids.

Gas chromatography

GC analyses were performed using a Carlo Erba (Fisons) Mega® series 5300 gas chromatograph fitted with an HT-5, high temperature, capillary column (Scientific Glass Engineering). To en-

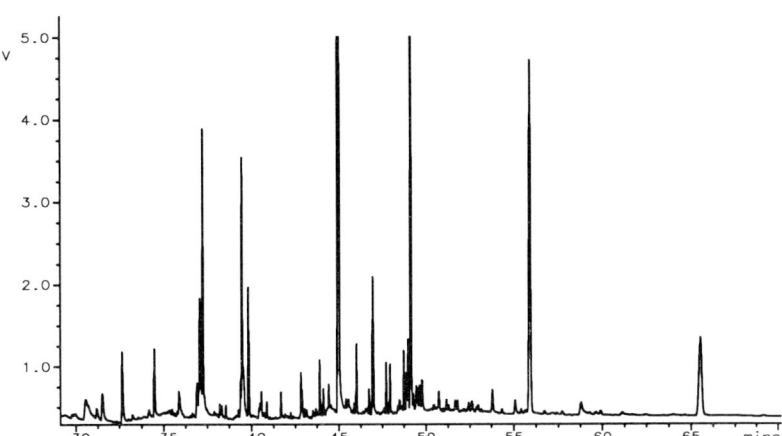

Potential Contamination from Plastics Associated with Sampling Procedures

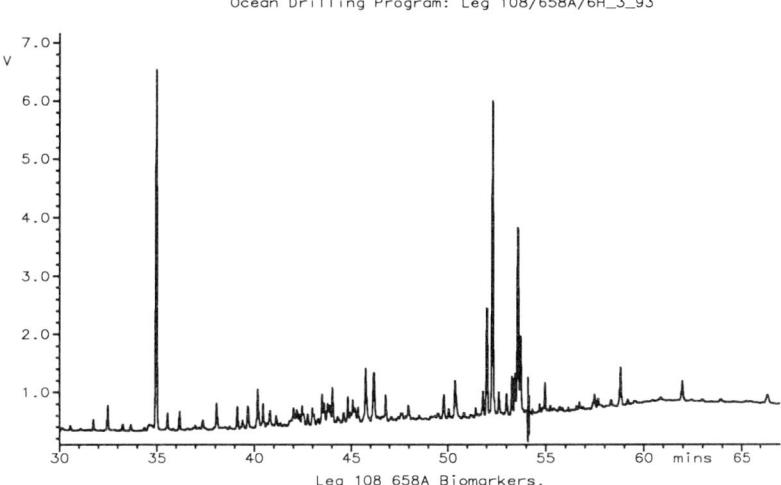

Ocean Drilling Program: Leg 108/658A/6H_3_93

Leg 108 658A Biomarkers.

Fig. 1. The problem of organic contaminants in ODP sample analysis for biomarkers. The upper of the two GC traces demonstrates a potential 'worst case' scenario whereby the operator has introduced contamination from, *inter alia* plastic inserts, foam rubber core plugs and hands. As a result, what would otherwise be a flat trace for a 'blank extract' demonstrates strong peaks due to phthalates and skin lipids. The lower trace represents a clean, contamination-free ODP sample (shown in more detail in Fig. 2.) for comparison. Extraction of quantitative, high quality biomarker data from contaminated samples is extremely difficult.

hance the throughput of samples a Varian 3400® GC supporting an ATC 200S® auto sampler was operated concurrently with the Carlo Erba 5300. The following temperature programme was devised; 50–150° C at 10° C min⁻¹, then 150–350° C at 4° C min⁻¹, isothermal hold at 350° C for 15 minutes. Total run time, 75 minutes.

Continuous 24 hour analysis on this system raised the daily turnover to 26 samples. Both GC systems were connected to a VG Data Systems Ltd Minichrom® data system. This system acquires and processes data on an IBM PC/AT® 80286 based microcomputer.

Mass spectrometry

Several GC runs were selected for compound assignment ratification and peak identification

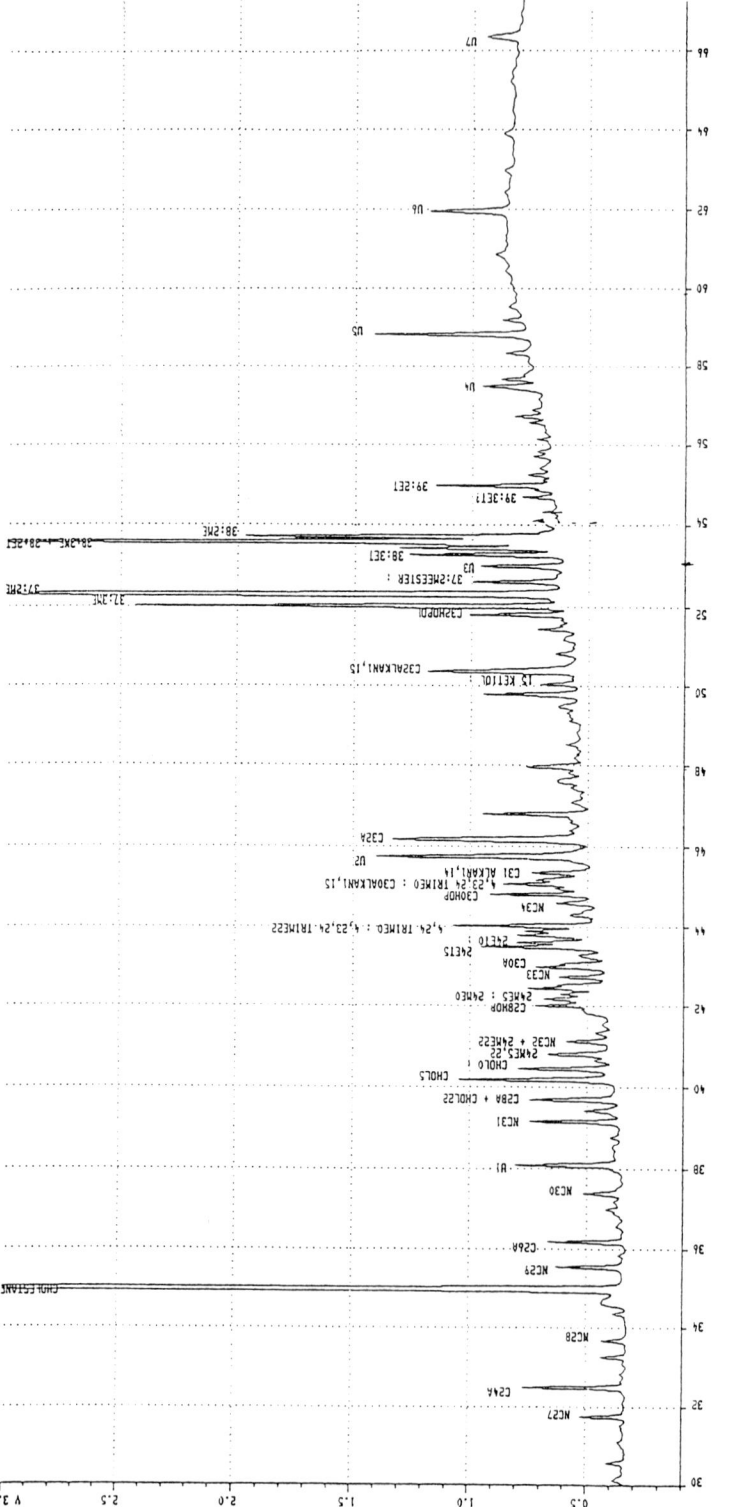

Fig. 2. Gas chromatogram (GC) trace for lipid extract of leg 108 sample (6H–3–93). Peaks have been identified using gas chromatography mass spectrometry (GC-MS). This trace shows biomarkers identified for sample 6H–3–93 from Leg 108, Site 658, Hole A. The peak at 35 minutes is the internal standard (cholestane) from which 'target' compound abundances are determined. The two prominent peaks at 52.5 minutes and 37 : 3 methyl alkenones are the 37 : 2 and 37 : 3 methyl alkenones used to calculate U^k_{37} indices.

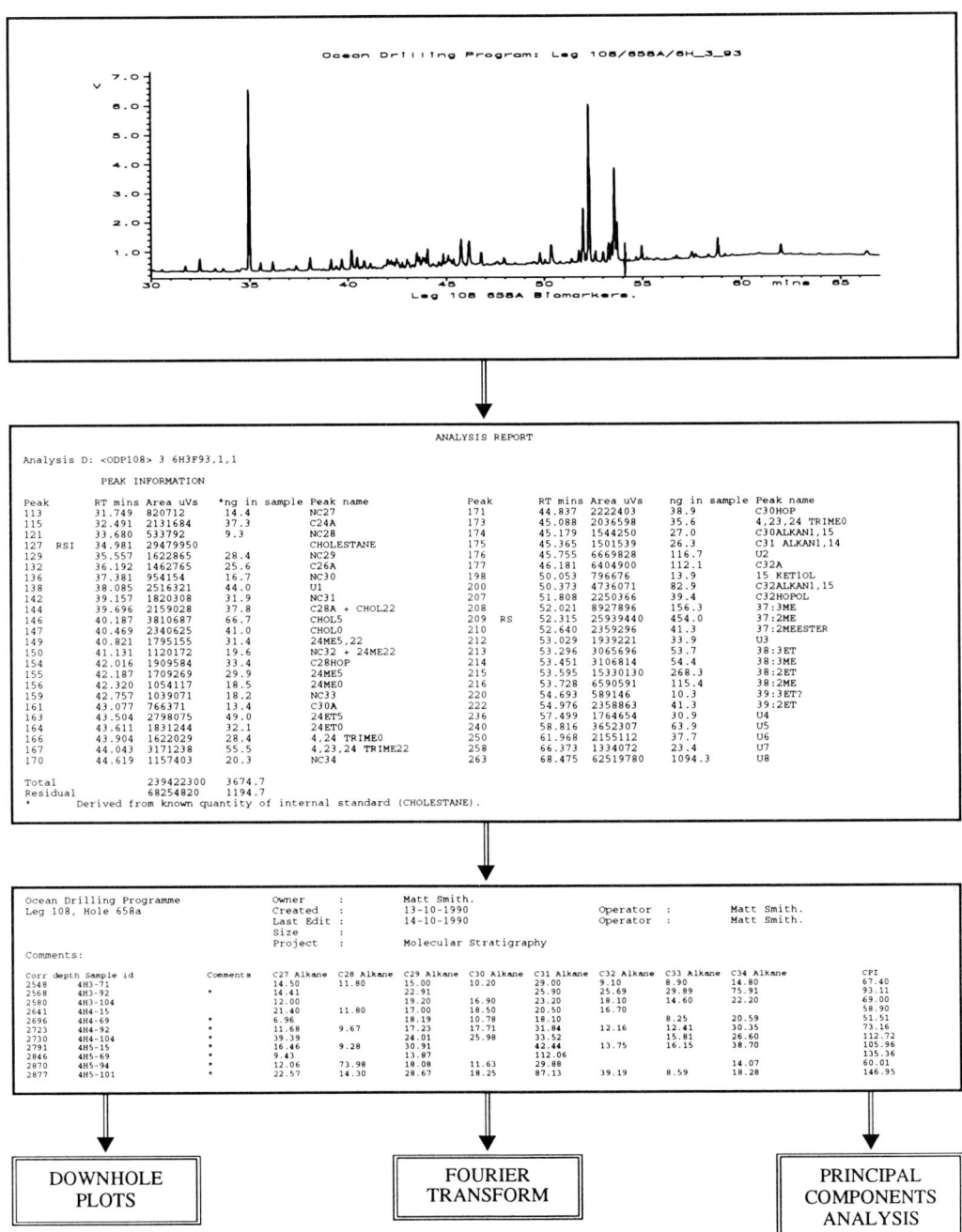

Fig. 3. Quantitative processing of gas chromatographic data. A schematic diagram, designed to illustrate the stages involved from the production of a chromatogram trace (Fig. 2.), via subsequent automated peak detection and quantitation (centre), input into Lotus 1–2–3 spreadsheets to the final numerical and graphical processes employed (bottom). The whole scheme has been designed with automation in mind, the only input from operators being at quality control and spreadsheet entry stages (bottom).

purposes using combined gas chromatography – mass spectrometry. Compounds were identified using published data as reference. The GC-MS system comprised a Mega 5160 attached to a Finnigan MAT 4500 quadrupole mass spectrometer. The GC temperature programme and column were similar to those above. The following mass spectrometer conditions were standard for that machine. Ionization Method, Electron Ionization (EI); Emission Current, 250 mA; Electron Energy, 70 eV; Ion Source Temperature, 170° C; Transfer line temperature, 300° C.

Post acquisition data handling

At this point in the procedure several hundred data files containing up to 700 peaks with an amount for each had been produced. Procedures needed to be designed and implemented to facilitate automatic data processing. The VG Minichrom® system supports a system of 'Macros' to enable batch processing and output (Figs 2 & 3) of single runs (i.e peak detection, quantification, calibration and baseline fitting as well as reporting and plotting facilities). Each 'target compound' was first identified using GC-MS, then assigned a name, retention time, retention search window, retention standard, quantitation standard and search criterion (i.e largest, widest, only or nearest). A macro was then written for each GC run, to call the peak quantification procedures, printer output routines and file dump procedures. These macros were queued to allow runs to be processed overnight. The time taken to design, write and queue forty runs was estimated at between 2.5–3 hours. Hence, a program called 'MACRO-GEN', was implemented to generate, queue and trigger the processing of a number of runs. This software proved capable of processing the forty runs in 103 seconds, a large saving in processing time. The 'top down' algorithm for MACRO-GEN is given in Appendix 1.

For the same 40 runs the system was occupied for eighteen hours, producing hardcopy and disc files for later entry into Lotus 1–2–3. Data entry proved awkward due to the size of the files produced by Minichrom®. Therefore, another piece of software called EDIP has been designed to edit these output files prior to their insertion into Lotus. The algorithm for EDIP is given in Appendix 2. EDIP uses link list dynamic structures to allow quick efficient insertion of data. In practice three lists are active at any one time, (files, contents, and lookup). Involved parsing routines and peak validity checking routines are required to construct the file contents list. Correction of peaks from the lookup table uses

conditionally recursive constructs. All software was written using Borland Turbo® Pascal® version 5.0.

Numerical analysis

Once generated, compound abundance data could then be subjected to several graphical and numeric techniques designed to highlight and recognise cyclic or periodic variations (Brassell *et al.* 1986a).

Graphical methods

Graphical methods provide the most immediate means for gaining a preliminary overview of trends and variations in the data, and allow erroneous and spurious data points to be identified and noted. Typically, the plots produced are *xy* plots of compound abundance versus sub-bottom depth. Frequently, whole compound classes may be co-plotted to give a visual representation of compound covariance down hole. A useful comparison may be made in this way, allowing numbers of variables to be reduced prior to the implementation of more quantitative technqiues. Other methods include plotting of compound abundance histograms or GC traces adjacent to sedimentary logs of the core. This allows rapid familiarization with both core logs and GC traces.

Fourier spectral analysis

As a means of deriving information relevant to frequencies and duration of events in a data set, Fourier transform (FT) spectral analysis is currently highly regarded in many areas of the geosciences and chemical sciences. By taking the Fourier transform of a time-corrected correlation matrix, it is possible to recognize and extract dominant harmonics which give information related to periodic events and the regularity and frequency of these events (Fig. 4). The most commonly used algorithm is the fast Fourier transform (FFT), (Press *et al.* 1986) although for greater accuracy a slower discrete Fourier transform (DFT) is recommended (Davis 1973). Spectral analysis is carried out at Bristol using shareware written for Cambridge University. The software, known as SPECTAX, comprised the transform software, data formatter, ager programs and a graphics kernel. Written in Ryan–McFarland FORTRAN® the routines have been ported to MSDOS®, UNIX® and proprietary operating systems, based around PCs or multi user hosts.

Spectral Density Plots: Site 658A+B

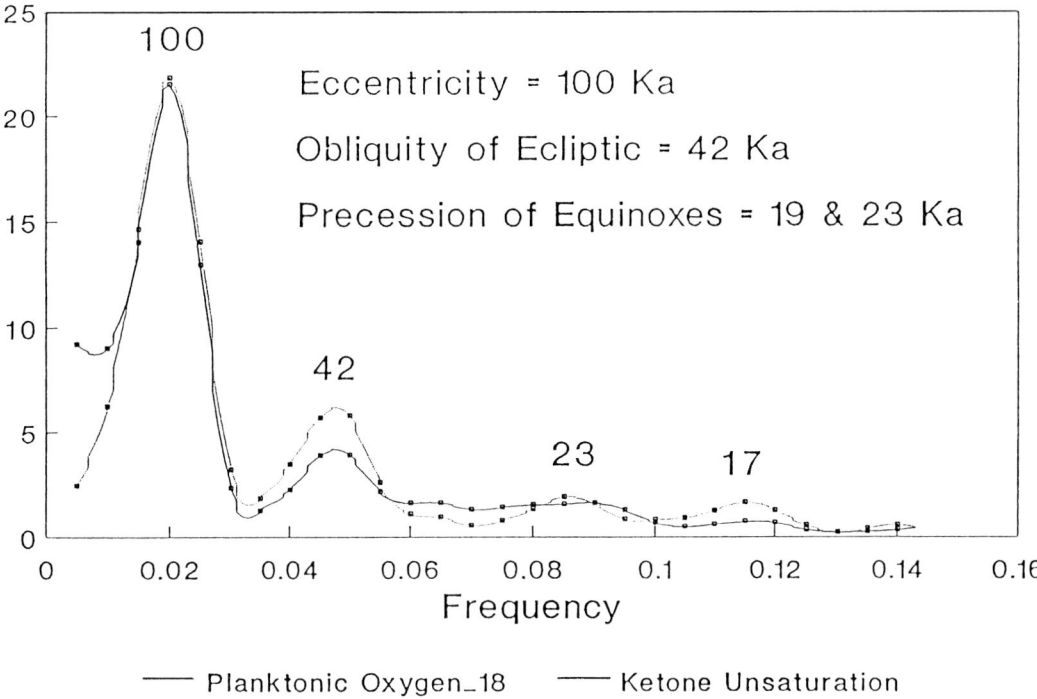

Fig. 4. Comparison of fast Fourier transform spectral plots for $\delta^{18}O$ and U^k_{37} parameters. This plot demonstrates the various cycles hidden within both data sets. This form of spectral information can usefully demonstrate in a visual form the cycles affecting particular parameters. This plot also demonstrates the similarity of response of two different parameters to the same driving mechanism.

Principal components analysis

Use of this tool has so far been limited to providing information about relative contributions of, and relations between, the large number of variables inherent in organic geochemical data sets. However, PCA has proved susceptible to the effects of valid yet exceptional points causing 'leverage' effects which serve to warp PC space due to the data having been centered and normalised prior to the PCA being run. This effect may produce heavily loaded variables which are not as significant as they appear. It is possible to alleviate this problem by using principal components regression to identify the offending points and thereby correct for the effects they cause. The use of factor, iterative factor (Yendle *et al.* in press), and cluster analysis as a means of obtaining a preliminary analysis prior to the use of PCA is currently being assessed. These techniques allow redundant data to be identified by providing an early

reckoning of covariance of the measured variables thereby eliminating those variables which do not covary with any others. Software for PCA and iterative factor analysis has been written at Bristol using Borland Turbo® Pascal® version 3.0 (Yendle 1989). It has only been made available on MSDOS® based systems and requires a Lotus 1–2–3 compatible spreadsheet to produce graphical output.

Discussion

Molecular stratigraphy: practice

Whilst relatively large sample sizes are desirable to smooth out lithological inhomogeneities, only small quantities of sample (a few hundreds of milligrams of sediment) are required for extractable lipid data to be generated. This is ideal for use on deep sea cores where sample sizes are restricted. As this extraction is achieved using small amounts of solvent, a great many samples become available for GC analysis very rapidly. The only possible means of coping with this sample volume is to use automated GC analysis. A weekly output of 100 runs per GC can be achieved in this way. Analytical errors caused by the operator (e.g. variable injection versus acquisition start times) are also a problem; however, automated GC has done much to reduce these. Planned improvements in analytical method should increase the throughput of samples and the ease with which the ensuing data mass is reduced.

Data production consequent from using automated GC systems can be substantial. The quantities of paper and magnetic media utilised generate bottle-necks in terms of interpretation, assessment and storage, a limiting factor as regards speed of data handling. Although the limitations of the method are not solely confined to data volume, this is a large consideration when producing a data set. Automated data handling and processing may enhance throughput of information. It is not desirable to fully automate data handling, as intervention is required to ensure data quality and integrity. Also, the complex nature of biomarker data means automatic detection, quantitation and reporting procedures are inherently prone to error. It is with this in mind that software such as EDIP has been written. EDIP provides both quality control and enhanced data throughput. Having organised and checked the data, treatments such as plotting and mapping are the first resort in the interpretation and numerical analysis phase, as they graphically demonstrate the features of a data set. The quantitative handling of high resolution molecular data is a developing area. For instance, Fourier transform spectral analysis is being used to give measure to observed temporal fluctuations in the data, simplifying the qualitative interpretation, whilst principal components analysis can give meaning to the covariance of known geochemical, biological and sedimentological influences on the data. The use of cluster analysis and correlative methods is under consideration as potential preprocessing methods designed to highlight redundant data.

As previously stated, the volume of data being so large, it is necessary to adopt an analysis and interpretation strategy. Currently, it is possible to study many variables for a few samples or many samples and a restricted number of variables. The ultimate goal is, of course, to study a large number of both. It is with this in mind that the previously described software and numerical methods are being developed.

Molecular stratigraphy (U^k_{37}): example, ODP Site 658

This collaborative project involved the generation of $\delta^{18}O$ data at a moderate resolution by M. Sarnthein at Keil, and corresponding U^k_{37} data at Bristol, for comparison. Examination of the two preliminary data sets encouraged the production of ultra high resolution U^k_{37} data for glacial terminations II and IV as a means of investigating potential Younger Dryas events observed in the $\delta^{18}O$ curve. U^k_{37} data gained from this exercise clearly demonstrates glacial stages I to VI and the intervening interglacials (Fig. 5). The U^k_{37} parameter, which is believed to be a measure of palaeo sea surface temperature (SST), fluctuates in response to global temperature changes during terminations II and IV (Fig. 5).

Evolutionary parameters are worth special mention in this context. The most common prymnesiophyte species today is *E. huxleyi*. It is assumed therefore, that the bulk of the production of alkenones found in present day sediments is attributable to *E. huxleyi* (Volkman *et al.* 1980; Marlowe *et al.* 1984). Species succession information demonstrates that *E. huxleyi* entered the fossil record at about 250 Ka and a clearly discernible signal can be recognized back to about 650 Ka. The inference is, that ancestors of *E. huxleyi* predominated prior to 250 Ka with similar metabolic biochemistry and responses to changing water temperatures.

Future research should explore a wide range of compounds as shown in Table 1, extending the information base to include aeolian and bacterial inputs. A more integrated strategy must develop, addressing aspects relevant to water column, depositional and diagenetic processes (Poynter 1989). Current initiatives at Bristol and Kiel will give a detailed insight into the last 30 Ka whilst providing information on sea floor diagenetic processes. Ideally these projects should integrate in such a way as to rationalise molecular data. The intention also exists to collaborate with stratigraphers working on other properties (e.g. magnetic susceptibility etc); indeed, molecular data is currently being combined with species composition data for prymnesiophytes and species of the Gephyrocapsa genus (Hine, pers. comm.) to assist in a better understanding of speciation and evolutionary effects.

Aside from U^k_{37}, a large quantity of biomarker data is contained in each analysis. Work is underway to quantify these data and generate other biomarker measures of use in molecular

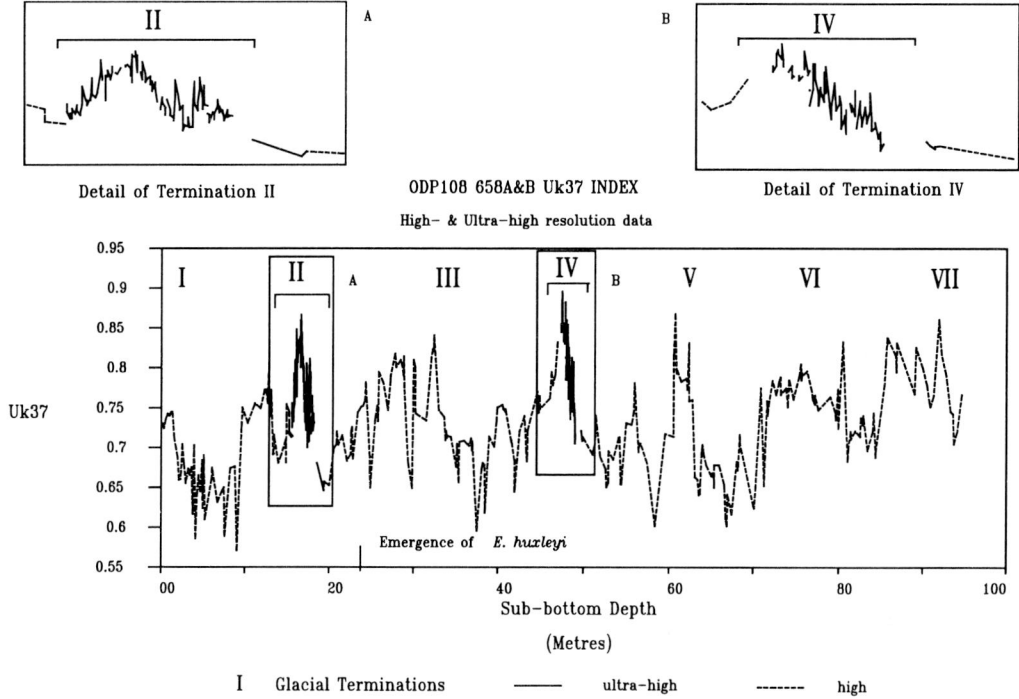

Fig. 5. High resolution molecular stratigraphy (U^k_{37} index) for the top 100 m of hole 658 A and B for ODP Leg 108. Plot showing U^k_{37} data for Hole 658, Holes A and B, ODP Leg 108. The high resolution data set (dashed line) represents data for the past 600 Ka. Insets A and B represent the ultra high resolution data set for terminations II and IV, respectively. Age control is provided by correlation with the SPECMAP time series. The roman numerals on the trace represent warming events over the last 600 Ka.

stratigraphy for Site 658 and other sediment records (Eglinton *et al.* in press).

Summary

Molecular abundance data provides the basis for the rapid generation of high and ultra high resolution stratigraphies designed for palaeoclimatic reconstructions. Further integration of molecular with other stratigraphies is a necessary future step, particularly with reference to Quaternary records.

Advances in data handling and processing methods have allowed large numbers of samples and compounds to be processed in comparatively short time-frames. Increased availability of low cost computing power and high performance software for statistical and signal processing has enhanced the quality and amount of molecular data available.

Molecular stratigraphy is one of a set of data production methods which may be integrated to maximise the amount of information produced when combined with an accurate chronostratigraphy.

The authors would like to thank M. Sarnthein and his co-workers at GPI Kiel for the provision of $\delta^{18}O$ data and NERC for financial support on grants GR3/6619 (M.B.S.), NERC/ODP research fellowship (J. G. P.), and GST/02/444 (S.A.B.).

Appendix 1. *Algorithm for computer program 'MACROGEN'*

1.   Intiialize and set defaults (flags, drives directories etc.)
2.   Check if command line parameters exist
3.   If command line parameters do exist:
3a. Parse each parameter (toggle switch or option parameter)
3b. Assign the appropriate variable with the passed value
3c. If parameter is not recognized output warning and default
4.   If command line parameters do not exist:
4a. Set up prompt screen containing the current defaults
4b. Input any changes and assign appropriate variables
5.   Get the first 'Data' file in the directory
6.   Does the equivalent 'Peak data' (Peak process) file exist
7.   If YES create an output file and write non peak detect macro, else write peak detect macro
8.   Close output file
9.   Loop until end of directory (filename = last filename)
9a. Get the next 'Data' file in the directory
9b. Does the equivalent 'Peak data' (Peak process) file exist
9c. If YES create an output file and write non peak detect macro, else write peak detect macro
9d. Close output file
10. End loop
11. End execution

Appendix 2. *Algorithm for computer program 'EDIP'*

1.   Set defaults and open system files
2.   Check for command line parameter (name of working directory)
3.   If parameter not found query user for name of working directory
4.   Set up working directory
5.   Construct linked list of files for processing ('*.PRN')
6.   Load contents of first file into heap memory (bidirectional linked list)
6a. Add each peak in turn checking peak name existence and validity
6b. If a peak name is encountered but is not the next one expected, add expected peak from the system lookup table (compound names and expected RT) and continue
7.   Write user edit screen
8.   Allow user to make changes with PC editor keys
9.   When the user signifies completion (ENTER) write named compounds to output file
10. Open next file in list
11. Continue processing until files linked list is exhausted
12. Close files
13. End execution

References

BRASSELL, S. C., EGLINTON, G., MARLOWE, I. T., PFLAUMANN, U. & SARNTHEIN, M. 1986a. Molecular Stratigraphy: a New Tool for Climatic Assessment. *Nature,* **320**, 129–133.

——, BRERETON, R. G., EGLINTON, G., GRIMALT, J., LIEBEZEIT, G., MARLOWE, I. T., PFLAUMANN, U. & SARNTHEIN, M. 1989b. Palaeoclimatic Signals Recognised By Chemometric Treatment of Molecular Stratigraphic Data. *In*: LEYTHAUSER, D. & RULLKÖTTER, J. (eds) *Advances in Organic Geochemistry*. 649–660.

BRASSELL, S. C., EGLINTON, G. & HOWELL, V. J. 1987. Palaeoenvironmental Assessment of Marine Organic Rich Sediments Using Molecular Organic Geochemistry. *In*: BROOKS, J. & FLEET, A. J. (eds) *Marine Petroleum Source Rocks*

Geological Society Special Publication, **26**, 79–98.

DAVIS, J. C. 1973. *Statistics and Data Analysis in Geology*. Wiley, New York.

DIDYK, B. M., SIMONEIT, B. R. T., BRASSELL, S. & EGLINTON, G. 1978. Organic Geochemical Indicators of Palaeoenvironmental Conditions of Sedimentation. *Nature,* **272**, 216–222.

EGLINTON, G., BRADSHAW, S. A., ROSELL, A., SARNTHEIN, M., PFLAUMANN, U. & TEIDEMANN, R. 1992. Molecular Record of Secular Sea Surface Temperature Changes on 100 Year Timescales for Glacial Terminations, I, II and IV. *Nature*, in press.

FARRIMOND, P., EGLINTON, G. & BRASSELL, S. C. 1986. Alkenones in Cretaceous Black Shales, Blake-

Bahama Basin, Western North Atlantic. *In*: LEY-THAUSER, D. & RULLKÖTTER, J. (eds) *Advances in Organic Geochemistry*. 897–904.

——, POYNTER, J. G. & EGLINTON, G. 1990. A Molecular Stratigraphic Study of Peru Margin Sediments, Hole 686B, Leg 112. *In*: SUESS, E. & VON HUENE, R. (eds) *Proceedings of the ODP 112*. 547–553.

MARLOWE, I. T. 1984. *Lipids as Palaeoclimatic Indicators*. PhD Thesis, Bristol University.

——, GREEN, J. C., NEAL, A. C., BRASSELL, S. C., EGLINTON, G. & COURSE, P. A. 1984. Long Chain (n-C_{37}-C_{39}) Alkenones in the Prymnesiophyceae. Distribution of Alkenones and other Lipids and their Taxonomic Significance. *British Phycological Journal*, **19**, 203–216.

——, BRASSELL, S. C., EGLINTON, G. & GREEN, J. C. 1990. Long Chain Alkenones and Alkylalkenoates and the Fossil Coccolith Record of Marine Sediments. *Chemical Geology*, **88**, 349–375.

McINTYRE, A., KIPP, N. G., BÈ, A. W. H., CROWLEY, T., KELLOG, T., GARDENER, J. V., PRELL, W. & RUDDIMAN, W. 1976. Glacial North Atlantic, 18,000 Years Ago: A SPECMAP Reconstruction. *In*: CLINE, R. M. & HAYES, J. D. (eds) *Investigation of Late Quaternary Palaeooceanography and Palaeoclimatology*. Geological Society of America Memoir, **145**, 000–000.

MORRIS, R. & BRASSELL, S. C. 1988. Long Chain Alkenediols: Biological Markers for Cyanobacterial Contributions to Sediments. *Lipids*, **23**, 256–258.

POYNTER, J. G. 1989. *Molecular Stratigraphy: The Recognition of Palaeoclimatic Signals in Organic Geochemical Data*. PhD Thesis, Bristol University.

—— EGLINTON, G. 1991. The Biomarker Concept – Strengths and Weaknesses. *Fresenius Journal of Analytical Chemistry*, **339**, 725–731.

——, FARRIMOND, P., ROBINSON, N. & EGLINTON, G. 1989a. Aeolian derived higher plant lipids in the marine sedimentary record: links with palaeoclimate. Palaeometeorology: Modern and Past Patterns of Global Atmospheric Transport, NATO. Advanced Workshop Series, Reidel Dordrecht.

——, ——, BRASSELL, S. C. & EGLINTON, G. 1989b. Molecular Stratigraphic Study of Sediments from Holes 658A and 660A, Leg 108 – Scientific Results, Sites 657–688 Eastern Tropical Atlantic. *Procedings of ODP*, **108**, 387–394.

PRAHL, F. G. & PINTO, L. A. 1987. A Geochemical Study of long Chain *n* – aldehydes in Washington Coastal Sediments. *Geochimica et Cosmochimica Acta*, **51**, 1573–1582.

PRAHL, F. G. & WAKEHAM, S. G. 1987. Calibration of Unsaturation Patterns in Long Chain Ketone Composition for Palaeotemperature Assessment. *Nature*, **341**, 367–369.

PRELL, W. L., IMBRIE, J., MARTINSON, D. G., MORLEY, J. J., PISIAS, N. G., SHACKLETON, N. J. & STREETER, H. J. 1986. Graphic Correlation of Oxygen Isotope Stratigraphy, Application to the late Quaternary. *Palaeoceanography*, **2**, 137–162.

PRESS, W. H., FLANNERY, B. P., TEUKOLSKY, S. A. & VETTERLING, W. T. 1986. *Numerical Recipes: The Art of Scientific Computing*. Cambridge University Press, Cambridge, Massachusetts.

ROBINSON, N., EGLINTON, G., BRASSELL, S. C. & CRANWELL, P. A. 1984. Dinoflagellate Origin for Sedimentary 4α-Methylsteroids and 5α (H)-Stanols. *Nature*, **308**, 439–441.

RUDDIMAN, W., SARNTHEIN, M. & BALDAUF, J. 1988. Site 658 – Part A – Initial Reports (section 1), Sites 657–668 Eastern Tropical Atlantic. *Proceedings of ODP*, **108**, 105–219.

——, —— & —— 1989. Site 658 – Scientific Results, Sites 657–668 Eastern Tropical Atlantic. *Proceedings of ODP*, **108**, 000–000.

SARNTHEIN, M. & TEIDEMANN, R. 1990. Younger Dryas Style Cooling Events at Glacial Terminations I–VI at ODP Site 658: Associated Benthic $\delta^{13}C$ Anomalies Constrain Meltwater Hypothesis. *Palaeoceanography*, **5**, 1041–1055.

——, TETZLAFF, G., KOOPMANN, B., WOLTER, K. & PFLAUMANN, U. 1981. Glacial and Interglacial Wind Regimes over the Eastern Subtropical Atlantic and North West Africa. *Nature*, **293**, 193–196.

TEN HAVEN, H. L., RULLKÖTTER, J. & STEIN, R. 1989. Preliminary Analysis of Extractable Lipids in Sediments from the Eastern North Atlantic (Leg 108); Comparison of a Coastal Upwelling Area (Site 658) with a Nonupwelling Area (Site 659). Leg 108 – Scientific Results, Sites 657–668 Eastern Tropical Atlantic. *Proceedings of ODP*, **108**, 351–360.

WENGOVITZ, P. S., SANDUJA, R. & ALAM, M. 1981. Dinoflagellate Sterols – 3: Sterol Composition of the Dinoflagelate *Gonyaulax monilata*. *Comp. Biochemistry and Physiology*, **69B**, 535–539.

VOLKMAN, J. K., EGLINTON, G., CORNER, E. D. S. & FORSBERG, T. E. V. 1980. Long Chain Alkenes and Alkenones in the Marine Coccolithophorid *Emiliania huxleyi*. Photochemistry, **19**, 2619–2622.

YAMAGUCHI, T., ITO, H. & HATA, M. 1986. Studies on the Sterols in some Marine Planktons. *Tohoku Journal of Agricultural Research*, **37**, 5–13.

YENDLE, P. W. 1989. *Chemometric Studies of Biochemical and Geochemical Systems*. PhD Thesis, Bristol University, 41–148.

——, POYNTER, J. G., FARRIMOND, P., MACFIE, H. J. H. & EGLINTON, G. 1991. Identification and quantitation of multiple *n* – alkane Inputs to Sediments Using Iterative Target Transformation Factor analysis. *Organic Geochemistry*, in press.

Lithostratigraphic applications for magnetic susceptibility logging of deep-sea sediment cores: examples from ODP Leg 115

SIMON G. ROBINSON

Department of Earth Sciences, University of Cambridge, Downing Street, Cambridge CB2 3EQ, UK

Abstract: Whole-core (WC) logging of low-field magnetic susceptibility (MS) is an extremely simple, rapid and non-destructive method of acquiring a quantifiable, litho-logically related signal which is ideally suited to the task of correlating between deep-sea sediment sequences recovered from related holes at ODP sites. This is particularly useful for reconstructing uninterrupted composite sequences by splicing together stratigraphically continuous subsections of offset cores from adjacent holes. Continuity of sequence is established by pattern-matching the WCMS profiles of each hole, which are based on measurements made at intervals of only a few centimetres. It is essential that interhole correlation is effected with the highest possible lithostratigraphic resolution if composite sequences are to be used, for example, in studies of Milankovitch-band cyclicity in palaeoceanographic records, involving spectral analysis of down-core data.

In addition to identifying coring irregularities such as unrecorded gaps in the recovered sequence between cores, or repetition of sequence due to repenetration of the corer, WCMS profiling can be used to identify turbidite horizons and also intervals contaminated by ferrous metal artifacts from drilling (e.g. pipe rust). The presence of contaminants often indicates that the sequence contains horizons of slumped, or mechanically reworked sediment washed-in from uphole. Thus identified, turbidites and contaminated intervals can be avoided during sampling, and eliminated from stratigraphic reconstructions or accumu-lation rate estimates. The only category of deep-sea sediment lithology which exhibits WCMS values similar to those of contaminated intervals, is that of basic volcanic ash-rich horizons. Accordingly, this makes WCMS logging a potentially valuable tool in studies of marine tephrostratigraphy.

Inter-regional correlation between the WCMS profiles for holes drilled at different ODP sites is also possible in many instances, especially when lithological variations at each site are controlled by large scale palaeoceanographic, or global (i.e. orbitally forced) palaeoclimatic changes. In such circumstances, WCMS may also be used as a proxy palaeoclimatic indicator, duly assisting more conventional microfossil and isotopic methods of climato-stratigraphical zonation of the sequences recovered.

Whole-core magnetic susceptibility (WCMS) logging of deep-sea sediments, chiefly for the purpose of high-resolution lithostratigraphic correlation, has become a routine procedure on all Ocean Drilling Program (ODP) cruises since ODP Leg 108 in 1986 (e.g. Bloemendal 1983; Bloemendal *et al.* 1988b, 1989). The present study aims to demonstrate the effectiveness of this simple, non-destructive logging tool in a broad range of marine lithostratigraphic appli-cations, using the results of shipboard WCMS measurements made entirely during ODP Leg 115 to illustrate the account. Applications described include core logging; interhole and intersite correlation; tephrostratigraphy; identification of coring irregularities such as unidentified gaps at core boundaries, recoring the same sequence, and contamination of cores by drilling artifacts; and investigations of the palaeoclimatic (orbital/Milankovitch) record of deep-sea sediments.

Early reports of the *potential* value of WCMS logging in lithostratigraphic studies of deep-sea sediments were published by Radhakrish-namurty *et al.* (1968), Amin *et al.* (1972) and Somayajulu *et al.* (1974, 1975). At the same time, however, some palaeolimnologists were already using WCMS logging for multiple cor-relation of lake sediment cores (e.g. Thompson 1973; Thompson *et al.* 1975; Oldfield *et al.* 1978, Thompson & Morton 1979), based on a tech-nique for WCMS scanning of wet sediments described by Molyneux & Thompson (1973).

The WCMS logs recorded in these early studies, however, were rather crude by compari-son with the high-resolution, high signal-to-noise ratio profiles used for detailed interhole correlation during recent Legs of the ODP (e.g. Ruddiman *et al.* 1988; Mascle, *et al.* 1988; Backman *et al.* 1988; Cochran *et al.* 1989; Prell *et al.* 1989; Peirce *et al.* 1989; Rangin *et al.* 1990; Murray & Prell 1991; DeMenocal *et al.* 1991),

From HAILWOOD, E. A. & KIDD, R. B. (eds), *High Resolution Stratigraphy*
Geological Society Special Publication, No. 70, pp. 65–98.

and for multiple, intersite correlation of deep-sea sediment cores in other studies (e.g. Robinson 1986*a*; Thompson & Oldfield 1986, p. 144). These, much higher resolution WCMS profiles were obtained using the Bartington MS system (see 'Methods' section below).

Magnetic susceptibility: definition and interpretation

Magnetic susceptibility literally means the ease with which a substance can be magnetized, i.e. the extent to which it is attracted to, or repelled by a magnetic field. Low-field volume magnetic susceptibility (κ) is defined as the ratio of induced magnetization intensity (M) per unit volume of a substance, to the strength of the applied magnetic field (H) inducing the magnetization, i.e. $\kappa = M/H$ (Thompson & Oldfield 1986, p. 25). In weak magnetic fields (e.g. <1 Oe), M is generally thought to vary with H linearly (although this assumption has been challenged recently: see Smith & Banerjee 1987); thus the magnetic moment induced in the substance when subjected to a weak magnetic field, reverts to zero when the applied field is removed (Collinson 1983, p. 14). Magnetic susceptibility is therefore not a measure of a permanent (remanent) magnetization (RM) retained by the sample on removal of an applied magnetic field, either natural (NRM) or artificial (e.g. isothermal or anhysteretic remanence (IRM, ARM)). Hence, *magnetic susceptibility is totally unrelated to the Earth's magnetic field at the deposition site, past or present.*

Magnetic susceptibility is effectively a measure of the change in the strength of a magnetic field (positive or negative) caused by placing a sample into that field, which is controlled by the concentration and composition (i.e. mineralogy and grain size/shape) of magnetizable material the sample contains (Thompson & Oldfield 1986, p. 25).

The volume magnetic susceptibility of natural materials like deep-sea sediments is largely a function of the concentration (per unit volume) of magnetizable material they contain. *Magnetizable*, as distinct from *magnetic* constituents of marine sediments include not only the ferromagnetic (*ferrimagnetic* and *imperfect antiferromagnetic*) minerals which may acquire a remanent magnetization (NRM, IRM, ARM, etc.), but also any compound containing Fe^{2+}, Fe^{3+}, or Mn^{2+} ions. These so-called *paramagnetic* substances include clay minerals, particularly chlorite, smectite and glauconite; ferromag-

nesian silicates; iron and manganese carbonates; iron disulphides like pyrite; and authigenic, ferric-oxyhydroxide mineraloids, collectively referred to as limonite, which frequently occur as colloidal complexes in the pore waters of deep-sea sediments (Burns & Burns 1981). Reviews of the most common magnetic (remanence-carrying) minerals found in deep-sea sediments (their origins, mineralogy, geochemical stability and magnetic properties) have been documented by Murray (1979), Lowrie & Heller (1982), Oldfield & Robinson (1985), Freeman (1986), and Robinson (1986*b*). Briefly, these include (1) strongly magnetic (ferrimagnetic) iron–titanium oxides like (titano)magnetite and (titano)maghemite (MS = 10^{-3} to 10^{-1} cgs) and iron monosulphides of the pyrrhotite series, also greigite (MS = 10^{-6} to 10^{-3} cgs); and (2) weakly magnetic (lattice-imperfect antiferromagnetic) iron oxides like (titano)hematite, and iron oxyhydroxides like goethite (MS = 10^{-7} to 10^{-5} cgs). Additional, but less well defined magnetic components of deep-sea sediments exist within authigenic ferromanganese concretions (nodules/micronodules) and encrustations (e.g. *see* Burns & Burns 1981 and references therein, also Nesteroff 1982).

In contrast to the strongly magnetizable (ferromagnetic) and moderately magnetizable (paramagnetic) minerals which occur mainly in the *lithogenic* fraction of deep-sea sediments, *biogenic* carbonbate and silica are not only very weakly magnetizable, but actually exhibit *diamagnetic* behaviour, that is, they are characterized by extremely weak, negative MS values (usually between -1 and -5×10^{-7} cgs). Downcore variations in the MS values of deep-sea sediments, therefore, tend to reflect changes in lithology, i.e. fluctuations in the ratio of biogenic to lithogenic components in the sediment, which is essentially the basis for lithological classification of deep-sea sediments (e.g. Backman *et al.* 1988, p. 22). Exceptions to this relationship include diamagnetic components of lithogenic origin like detrital quartz and alkali feldspar, and ferrimagnetic minerals of biogenic origin like bacterial magnetite (e.g. Petersen *et al.* 1986; Chang & Kirschvink 1987; Karlin *et al.* 1987), Chiton teeth magnetite (e.g. Kirschvink & Lowenstam 1979; Nesson & Lowenstam 1985), and possibly even bacterial iron sulphides including pyrrhotite (Farina *et al.* 1990) and greigite ($^+/_-$ paramagnetic pyrite) (Demitrack 1985; Mann *et al.* 1990).

It is important to note that variations in the MS of deep-sea sediments reflect the proportion of biogenic (carbonate and silica) and lithogenic (clay and labile mineral) components in the sequence, as distinct from the carbonate/non-carbonate ratio. Accordingly, when WCMS profiles do not reflect downcore variations in the carbonate content of deep-sea sediments, it is possible that biogenic silica (or pure, unstained quartz grains) may constitute a significant fraction of the sediment. There are, however, several exceptions to this generalization, the most important of which include (1) the disproportionately high MS values associated with volcanic-ash-rich sediments (e.g. Oldfield *et al.* 1978; Robinson 1990); (2) organic-rich sediments affected by suboxic diagenesis which often leads to the dissolution of magnetic minerals (Karlin & Levi 1983, 1985; Canfield & Berner 1987; Bloemendai *et al.* 1989; Robinson 1990); and (3) variations in the sources and/or provenance of magnetic minerals (associated with different terrigenic or volcanic components *within* the lithogenic fraction of the sediment), or changes in the relative flux-density of magnetic minerals from different sources, which are unrelated to palaeoceanographic factors controlling the relative accumulation rates of the biogenic and lithogenic fractions.

Methods

Whole-core magnetic susceptibility

On board the ODP drilling vessel, JOIDES Resolution, measurements of WCMS are made with a Bartington Instruments' MS system, which is based on a modification of the principle utilized by metal detectors (Lancaster 1966). A 1 Oe magnetic field is created by passing a 0.565 kHz alternating frequency current (AC) through a coil which forms a loop, of 80 mm inner diameter, through which whole-core sections are passed. Unlike direct, AC methods of MS measurement (see Collinson 1983, p. 21), however, the Bartington system uses a sharply-tuned oscillator circuit to detect the change in frequency of the AC waveform which results from introducing the sample into the field (i.e. passing the core through the loop). The change in frequency of the alternating current is directly proportional to the MS of the sample. The system is calibrated using a range of paramagnetic salts (e.g. copper sulphate, ferrous sulphate, magnesium carbonate) and, at the lower end of the scale, distilled water and air as standards, yielding a noise level of the order of 5×10^{-8} cgs (approx. 1×10^9 SI), and a measuring range of between 1 and 9999×10^{-6} cgs.

On ODP Leg 115, WCMS measurements were made by passing, manually, whole-core sections through an 80 mm loop-type sensor, and recording changes in MS at successive intervals downcore. Intervals of 3, 5, 6, 7, or 10 cm were used, depending on the rate of recovery, degree of disturbance, and/or lithological homogeneity of the cores measured. Since ODP Leg 125, however, the WCMS system on board JOIDES Resolution has been modified to incorporate an automated core-transport facility, and is thus comparable in efficiency to, and is measured on-line with, the GRAPE (density) and P-wave (sonic velocity) logger.

Results of WCMS measurements made in the present study are reported here in arbitrary, Bartington Meter cgs. scale units (nominally G/Oe). For each section of core measured, a correction for instrumental drift was applied to the data by initially zeroing the meter in the presence of a blank standard consisting of a section of plastic core liner filled with seawater. The MS value of the standard was then re-measured immediately on completion of each sequence of core measurements, and the difference in the two values for the standard was used to compute the drift correction (assuming uniform drift rate). The results reported here were not further calibrated, however, to take account of variations in the volume of core measured due to incompletely filled core liners (commonly occurring at the top of cores to a depth of between 0.2 and 2.0 m). The relationship between cgs units and (dimensionless) SI units of bulk susceptibility is as follows: 1 cgs unit $= 1/4\pi$ SI units. The volume susceptibility value (susceptibility per unit volume) used in this paper is obtained by dividing the bulk susceptibility value by the effective volume of the sample, specified in cm^3 for cgs units and m^3 for SI units.

Carbonate content

The bulk carbonate content of samples taken at 5 cm intervals from Core 115–711A–1H was measured using the Chittick gasometric apparatus (Dreimanis 1962), based on the method outlined by Bascomb (1974). The results obtained by this procedure were reproducible to within a range of about $^+/_-1\%$ CaCO$_3$ equivalent. All other calcimetric data presented in this study were obtained on board ship during ODP Leg 115 (see Backman *et al.* 1988, p. 36).

Results

In this section, the results of WCMS logging carried out during ODP Leg 115, which drilled 12 sites in the tropical, NW Indian Ocean (see

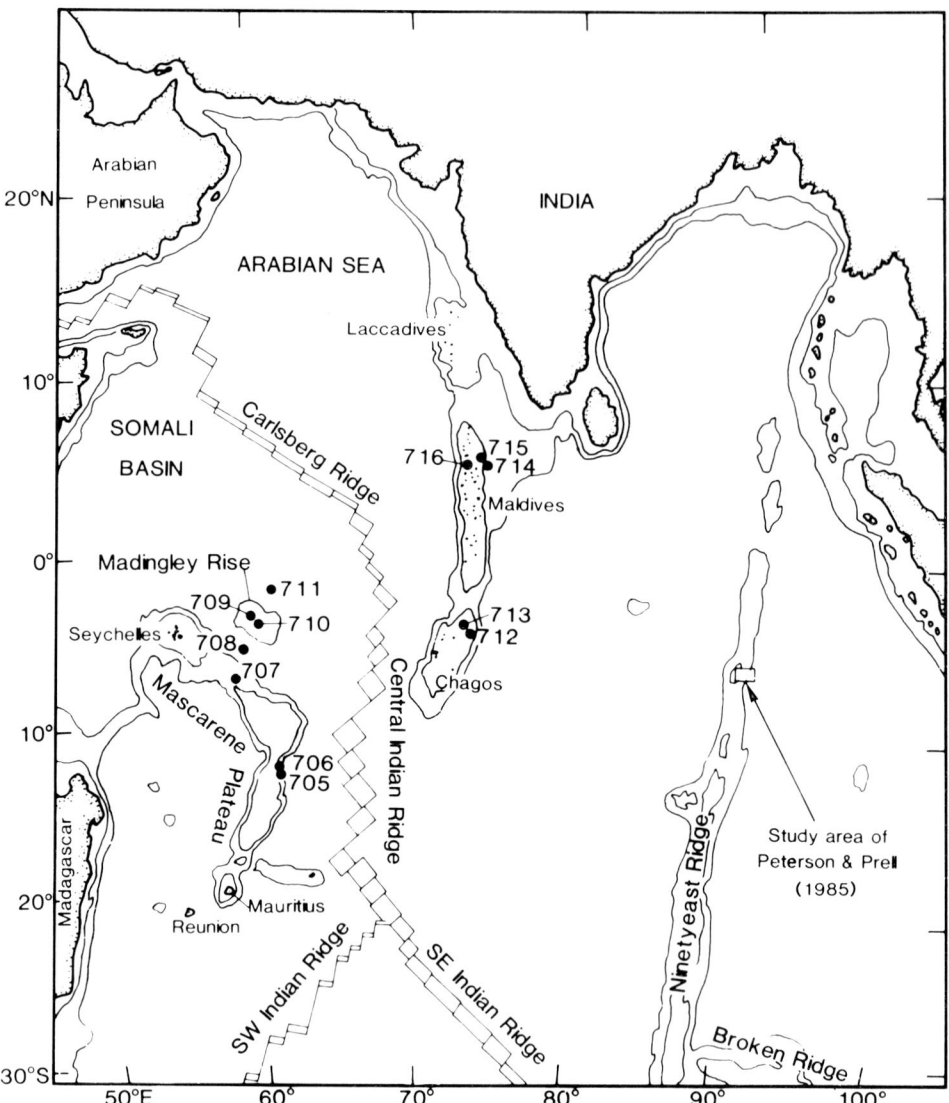

Fig. 1 Locations of ODP Leg 115 drilling sites.

Fig. 1), are used to illustrate a number of lithostratigraphic applications for this technique. Since data presented in subsequent figures may be relevant to more than one of the themes addressed in the accompanying text, some figures will be referred to in more than one of the following sections.

Core logging

There are several situations in which WCMS logging of deep-sea sediments may provide

valuable sedimentological (and sometimes palaeoclimatic) information *per se*, regardless of its implications for correlation purposes. The following examples demonstrate several aspects of this application for WCMS data, and illustrate the relationship between sediment lithology and MS at the scale of individual (9 m) cores, and entire, multicore sequences recovered from ODP holes drilled to over 200 mbsf.

In Fig. 2, the WCMS profiles of three subsections of ODP Hole 708A are plotted alongside logs of calcturbidites in the sequence taken from

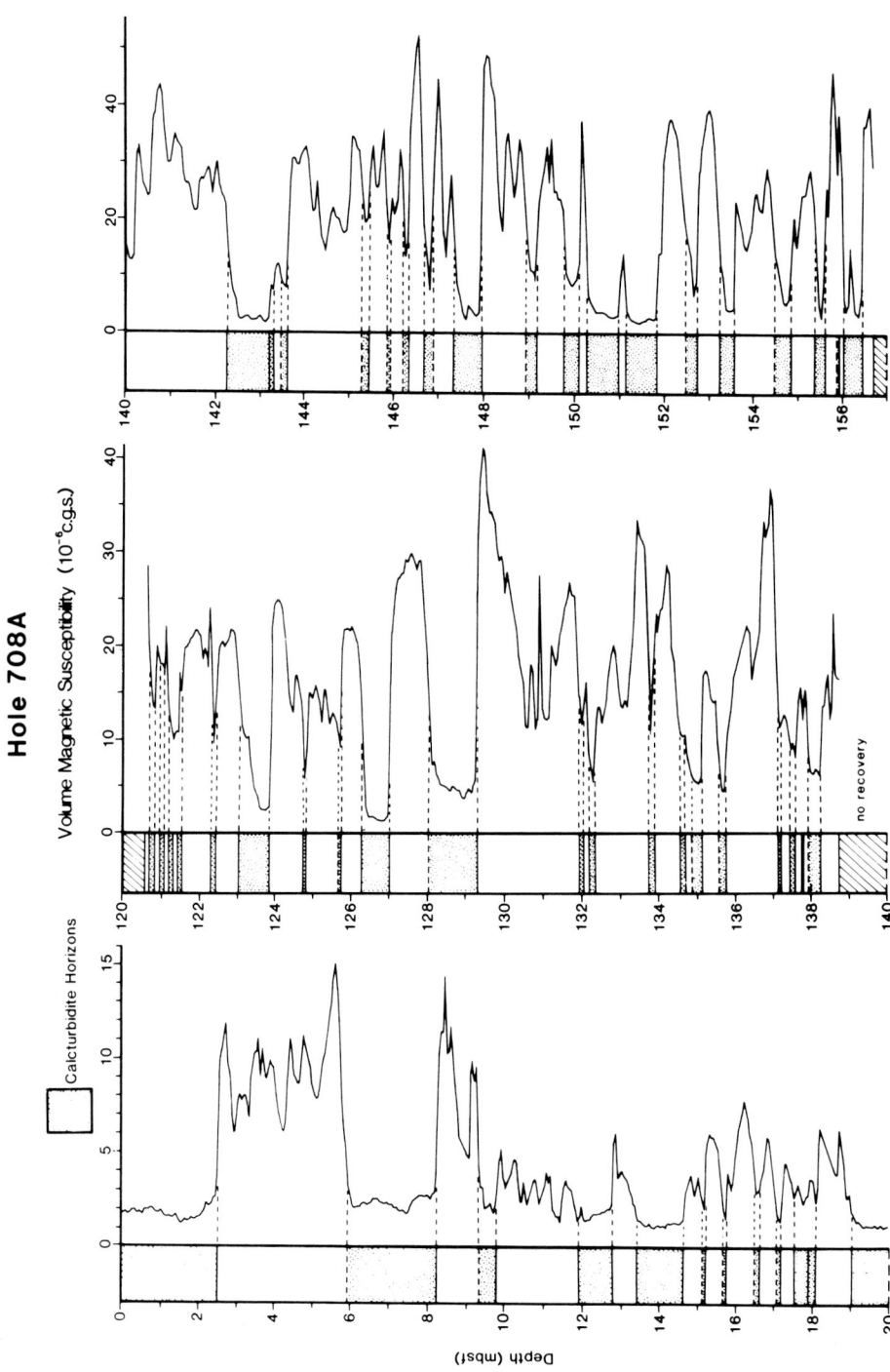

Fig. 2. WCMS profiles of the three intervals of Hole 708A, plotted alongside logs of calcturbidite horizons in the sequence (stippled intervals) taken from visual core-description data. **Left:** Quaternary interval (Cores 115-708A-1H and -2H). **Centre:** middle and late Miocene interval (Cores 115-708A-14X and -15X). **Right:** early and middle Miocene interval (Cores 115-708A-16X and -17X).

shipboard, visual core-description data. Calcturbidite horizons are clearly identified with major reductions in WCMS values in this hole, and are also characterised by generally low but erratic MS values (except for turbidites with gradational tops), reflecting the unstructured nature of these deposits. This contrasts with the highly ordered, upwardly-decreasing WCMS profiles which characterise the black mud (siliciclastic) turbidites of the distal Bengal Fan, drilled during ODP Leg 116, which exhibit classic 'upward-fining' grain size motifs (Sager & Hall 1990). Precise identification, by means of WCMS logging, of the true thickness and character of calcturbidite horizons in pelagic sequences such as that recovered from Hole 708A, considerably assists the task of estimating the

rate of pelagic sediment accumulation at such sites, thus improving the accuracy of stratigraphic sections, age/depth plots, etc., by enabling these to be expressed on a turbidite-free basis (e.g. Backman, *et al.* 1988, p. 401).

On a much smaller scale, Fig. 3 shows the WCMS (3 cm interval) and carbonate content (5 cm interval) profiles of Core 115–711A–1H, plotted alongside the lithological log of this core, taken directly from shipboard core description data (Backman *et al.* 1988, p. 697). Also shown in Fig. 3 are the palaeomagnetic and nannofossil stratigraphies for this core, from which the timescale, shown right, is derived, and upon which the approximate positions of oxygen-isotope stages are interpolated.

ODP Site 711 is located on the northern edge

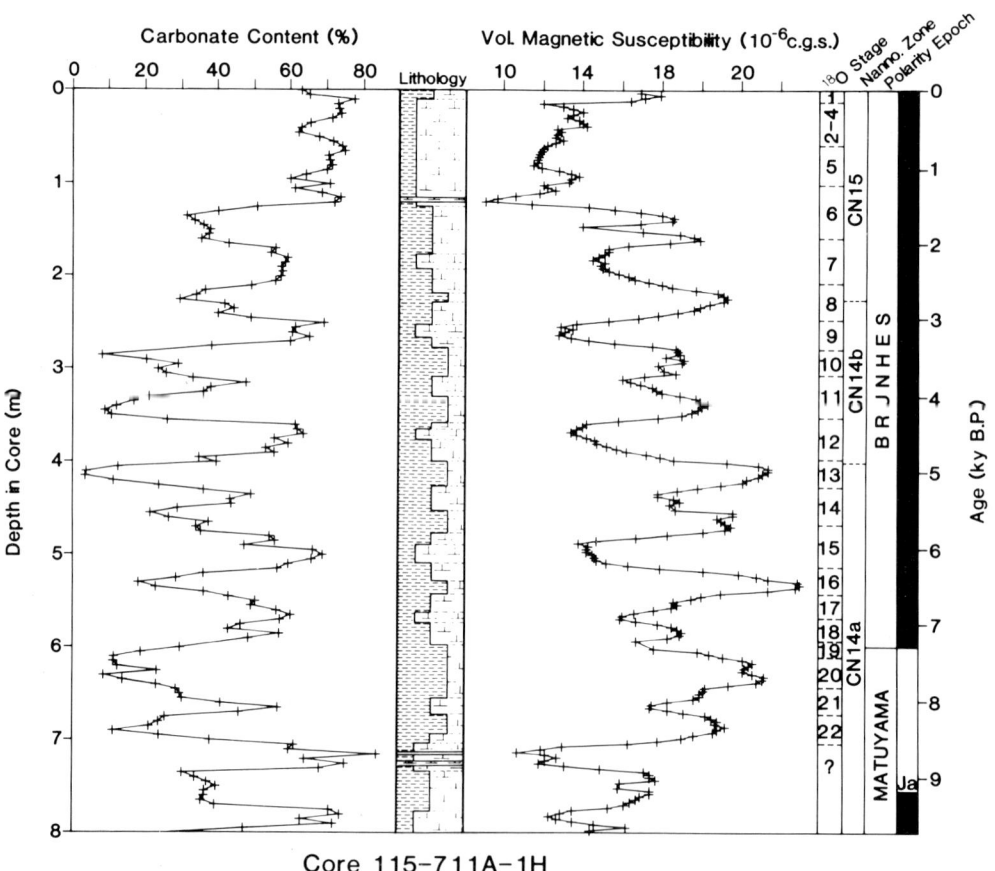

Core 115–711A–1H

Fig. 3. WCMS and percent-carbonate profiles of Core 115-711A-1H, plotted alongside a log of core lithology taken from shipboard core description data. The positions of hypothetical oxygen-isotope stage boundaries are based on the shipboard palaeomagnetic and biostratigraphic timescale for this core (shown right), using the ages for these events given by Imbrie *et al.* (1984). The percent-carbonate profile is based on measurements made for the present study on samples taken at 5 cm intervals.

of the Madingley Rise (Fig. 1), at a depth of 4428 mbsl, making it the deepest site drilled during Leg 115. This site lies beneath the present day hydrographic lysocline, and close to the carbonate compensation depth (CCD). Accordingly, with the exception of three microturbidite horizons, 5–7 cm in thickness, variations in the lithology of Core 115–711A–1H are all related to the effects of carbonate dissolution (Backman *et al.* 1988, p. 664; Peterson & Backman 1990), probably due to fluctuations in the regional CCD associated in some way with Quaternary climatic changes (cf. Peterson & Prell 1985*b*; Curry *et al.* 1990). Regular oscillations in biogenic carbonate accumulation (almost exclusively nannofossils), superimposed on a relatively constant rate of lithogenic clay input, have effected an apparently 'interbedded' marl-ooze sequence at this site, with $\%CaCO_3$ values ranging between <5% and >80%.

The sequence of regularly-spaced, dissolution-controlled cycles in the carbonate content of Core 115–711A-1H are clearly mirrored by its WCMS profile (Fig. 3). Peaks in the WCMS profile of this core correspond to clay-rich intervals, while troughs correspond to carbonate-rich intervals, both in direct proportion to the carbonate content of the sediment. The regression of MS values on carbonate percentages for this core is shown in Fig. 4. The

coefficient of correlation (Pearson's r) between these parameters is 0.971 ($p' = 99.9\%$, $n = 159$). The palaeoclimatic implications of this relationship are considered below in 'Palaeoclimatic investigations'.

Figure 5 illustrates the macroscale relationship between WCMS, carbonate content (shipboard data), and lithologic units, based on the entire sequence recovered from Hole 710A. Site 710 lies at a depth of 3812 mbsl (Fig. 1), which is close to the present-day hydrographic lysocline in the Northwest Indian Ocean (Peterson & Prell 1985*a*), but is significantly shallower than Sites 708 and 711. Dissolution of carbonate constituents in the sediment at Site 710, therefore, is rather less severe than at these deeper sites. Consequently, carbonate content of the sediment at this site is generally higher and less variable than at Site 708 and 711, thus modal WCMS values are proportionally lower and less variable, and the relationship between WCMS and carbonate content is correspondingly weaker ($r = 0.674$).

The WCMS profile of Hole 710A (Fig. 5) reveals, with comparable sensitivity, all of the lithostratigraphic information available from the percent-carbonate record of this hole. The main difference between these two profiles is the greater sensitivity with which WCMS differentiates between Lithologic Subunits IA and IB.

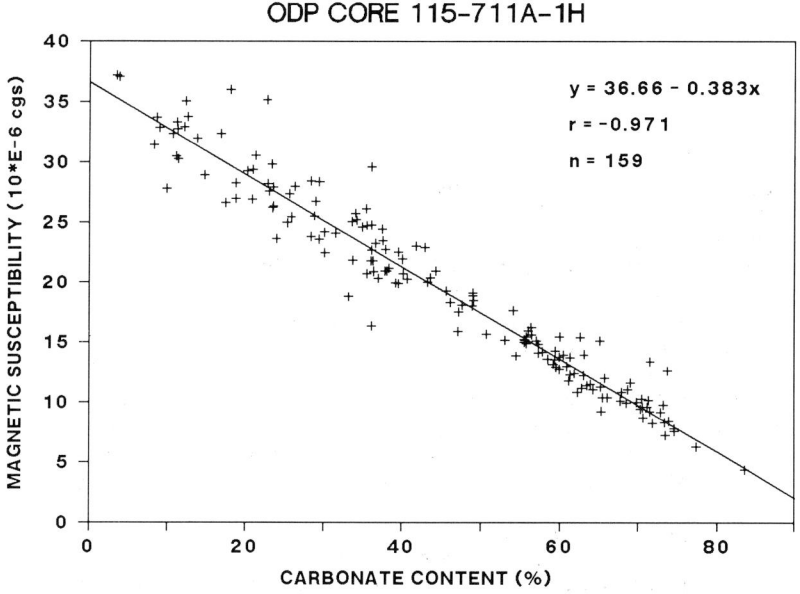

ODP CORE 115-711A-1H

$y = 36.66 - 0.383x$

$r = -0.971$

$n = 159$

Fig. 4. Relationship between the magnetic susceptibility and carbonate content of samples taken at 5 cm intervals from Core 115-711A-1H.

Hole 710A

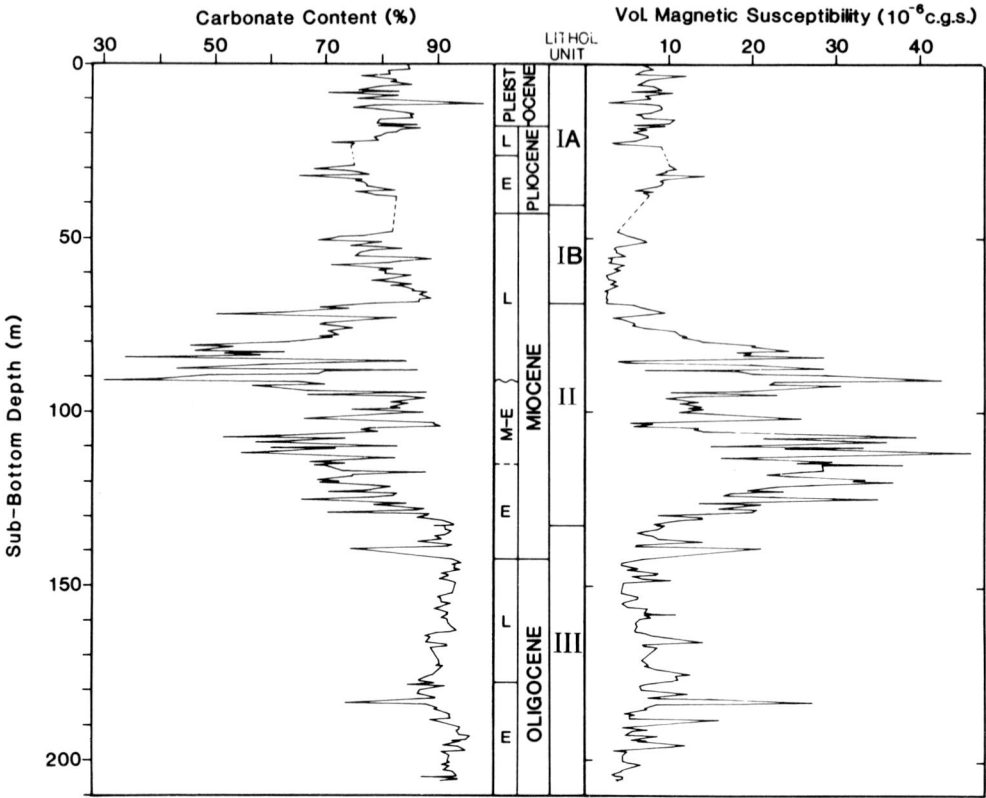

Fig. 5. WCMS profile and percent-carbonate record of Hole 710A, showing the relationship between these parameters and lithologic units at this site. The percent-carbonate profile is based on shipboard data (Backman *et al.* 1988, p. 614).

This is because Lithologic Unit I at this site is subdivided according to differences in the oxidation state of the sediment, inferred by changes in colour from brown to green, or from white to gray, which is related to a transition from Fe(III) to Fe(II) as the stable form of structural iron in clay mineral lattices, particularly smectites (Lyle 1983). Naturally, this change in the stability of iron-bearing phases with respect to Eh, may affect the magnetic ordering of iron oxide and/or oxyhydroxide minerals in the sediment, thus lowering its MS values. Hence, the oxic sediment of Unit IA exhibits generally higher WCMS values than the anoxic sediment of Unit IB, because of the magnetic disordering of iron oxides in the latter caused by suboxic diagenesis (cf. Bloemendal *et al.* 1989). A second source of disparity between the responses of WCMS and carbonate content to downhole fluctuations in

lithology at Site 710, is the inordinate sensitivity of WCMS to volcanic ash horizons in the sequence (see 'Tephrostratigraphy' section below).

The early and middle Miocene sequence of tropical Indian Ocean sediments is often condensed as a result of a period of intensified dissolution of biogenic carbonate constituents, which occurred due to a pronounced shoaling of the regional CCD during this interval (van Andel 1975; Sclater *et al.* 1977; Peterson & Backman 1990). In Hole 710A, this interval is clearly associated both with much lower average carbonate percentages and with proportionately higher WCMS values (Fig. 5). A similar pattern of variation was observed in the records of both these parameters in the early and middle Miocene interval of Holes 708, 709 and 711 (see Fig. 11, and accompanying text, below).

Core correlation

The main objective in logging MS variations of whole-core sections as they arrive on deck, is to provide a means of high-resolution correlation between adjacent holes drilled at a given ODP site. More than one hole is drilled at each ODP site because gaps in the stratigraphic record of HPC/APC-cored holes often correspond to the boundaries between cores. Accordingly, the run length of individual cored intervals in each hole, at any one site, is arranged such that the top and bottom of each core occurs at a sub-bottom depth which corresponds, more or less, to the middle of a coring run in the adjacent hole. In this way it is theoretically possible to recover a complete record of the stratigraphic sequence of each site drilled, by 'dovetailing' uninterrupted subsections of the sequence obtained by offset coring in related holes. In order to accomplish the task of reconstructing a stratigraphically uninterrupted, composite sequence from any one ODP site, therefore, a method for ultra-high-resolution, between-hole correlation is required. The aim of the following examples is to show that WCMS logging provides an ideal tool for this kind of operation, as first demonstrated during ODP Leg 108 (see Ruddiman *et al.* 1988).

Figure 6 shows a correlation between the uppermost 24 m of Holes 709A and 709C, based on their WCMS profiles. As Site 709 (Fig. 1) is significantly shallower (3040 mbsl) than any of the Leg 115 sites considered above, the sediments which accumulated here during the interval encompassed by the profiles shown in Fig. 6 (late Pliocene to Pleistocene), consist of fairly pure calcareous oozes (generally between 80% and 95% $CaCO_3$ content), with only minor amounts of lithogenic clay (Backman *et al.* 1988, p. 465). Accordingly, the relationship between MS and carbonate content of the sediment within the interval of Hole 709C shown in Fig. 6, for example, is much weaker ($r = 0.337$) than it is at the deeper Leg 115 sites. Nevertheless, the WCMS profiles of the carbonate-rich, Plio-Pleistocene sediments recovered from Holes 709A and 709C are detailed enough to provide over 100 points of lithostratigraphic correlation between these holes. However, this may be partly because WCMS variations reflect not the carbonate/non-carbonate ratio of the sediment but, rather, its biogenic/lithogenic ratio. Since the sediments of Quaternary and late Miocene age at Site 709 contain abundant, well preserved siliceous microfossils, unlike any of the sediments sampled at the deeper Leg 115 sites (Backman *et al.* 1988, p. 473), it is possible that variations in biogenic silica content may affect

significantly the WCMS signal of the sediment in these intervals.

For reference purposes, correlation points identified in Fig. 6 (and in subsequent figures) have been labelled as follows: P1, P2, P3 . . . etc., in the case of peaks in the WCMS profiles of each hole, and T1, T2, T3 . . . etc., for the intervening troughs in WCMS values. (Note that there is no implied correlation between points labelled P1, T1, P2, T2, etc., in Fig. 6, and correlation points assigned the same labels in subsequent figures.) Correlation points are identified by pattern-matching the WCMS profiles of each hole, which involves consideration of a number of criteria, including the stratigraphic position, sequence, relative amplitude, and overall configuration of WCMS variations.

As indicated by the dashed lines of correlation in Fig. 6, the section in Hole 709A between T14 and T17 is absent from the WCMS profile of Hole 709C. Similarly, the section in Hole 709C between T21 and T23 is absent from the WCMS profile of Hole 709A. In both instances, these stratigraphic discontinuities are the result of sediment lost, or simply not sampled, between coring runs, and subsequently not recorded due to inaccuracies in calculating the sub-bottom depth assignments for these holes. The use of WCMS logging to detect coring irregularities of this kind are considered in more detail below in 'Identification of coring irregularities'.

Figure 7 shows the WCMS-based correlation between Holes 709A and 709C, plotted in the form of a Shaw Diagram (Shaw 1964). This plot clearly demonstrates the high degree of lithostratigraphic resolution with which WCMS is able to correlate between these holes. In this form of correlation diagram, gaps in the stratigraphic record of each hole, resulting from loss of material between cores, are clearly revealed as segments of the correlation curve which lie parallel to one or other of the graph's axes. Gaps due to coring irregularities of this kind may, in general, be differentiated from real stratigraphic discontinuities only by the coincidence of core boundaries at the positions in the sequence where the gaps occur.

In contrast to Site 709, the sediments which accumulated at the deepest of Leg 115 sites, Site 711, exhibit generally higher, and more variable WCMS values, due to the effects of severe, though periodic, dissolution. The effectiveness of WCMS as a lithostratigraphic correlation tool at this site, therefore, is likely to be even better than it is at Site 709.

Figure 8 shows the correlation between the WCMS profiles of Holes 711A and 711B, to a sub-bottom depth of 12 m. Note the generally

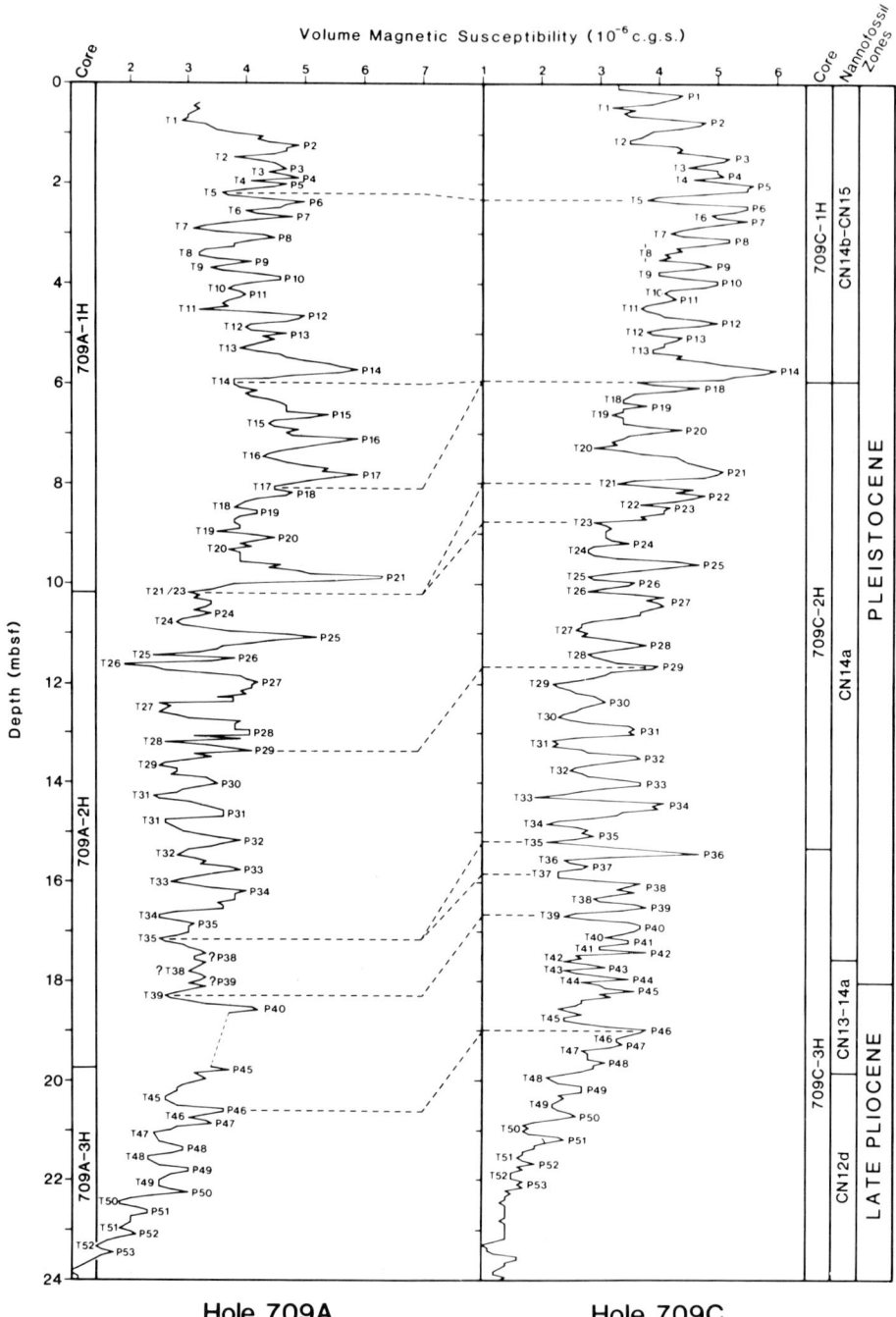

Fig. 6. Lithostratigraphic correlation between the late Pliocene and Pleistocene horizons of Holes 709A and 709C, based on their WCMS profiles. The correlation points indicated in this figure (P1, T1, P2, T2, etc.) were used to construct the Shaw Diagram shown in Fig. 7. Note the degradation of the WCMS signal in each hole below P50, due to the effects of early diagenesis.

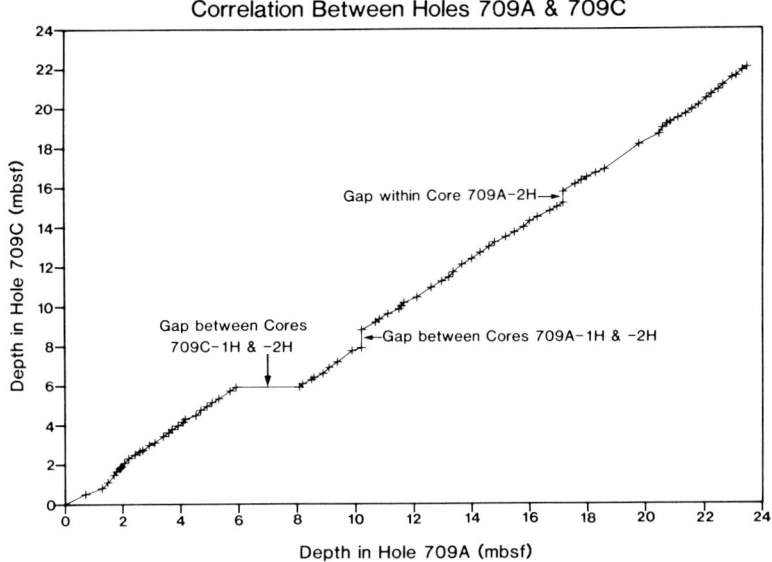

Fig. 7. Shaw Diagram correlating depth in Hole 709A with depth in Hole 709C, based on cross-matching the WCMS profiles of each hole (Fig. 6).

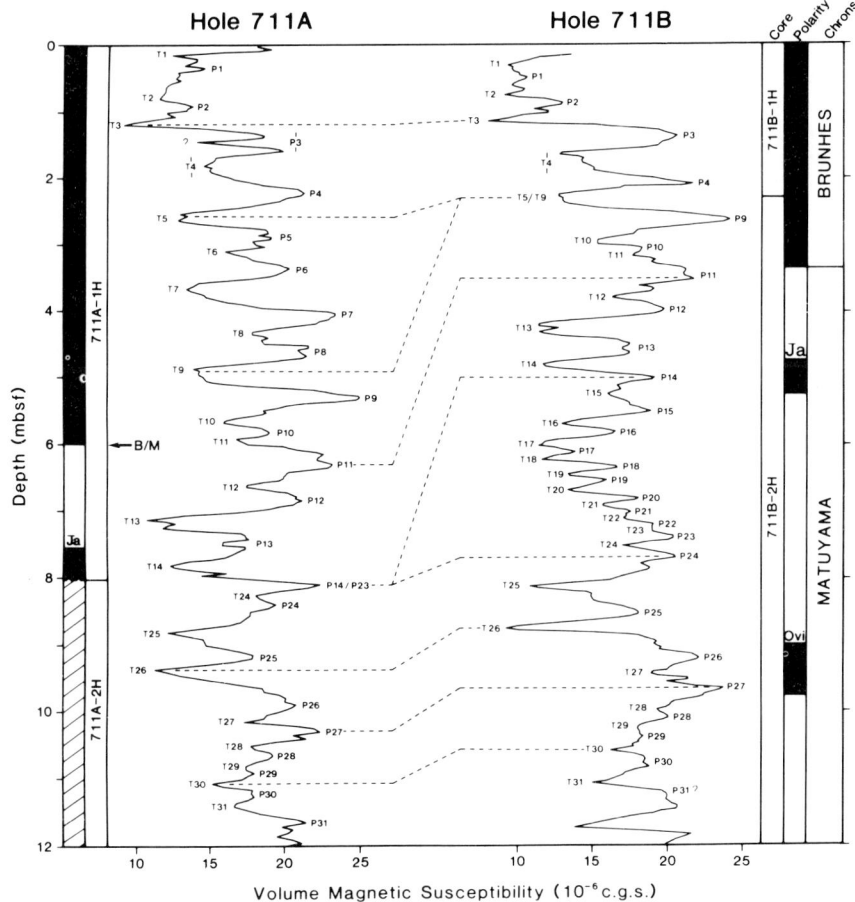

Fig. 8. Lithostratigraphic correlation between the Quaternary horizons of Holes 711A and 711B, based on cross-matching their WCMS profiles. The correlation points indicated in this figure were used to construct the Shaw Diagram shown in Fig. 9. Note the major discontinuities in the sequence of each hole at core boundaries, corresponding to 'unrecorded intervals of no recovery'.

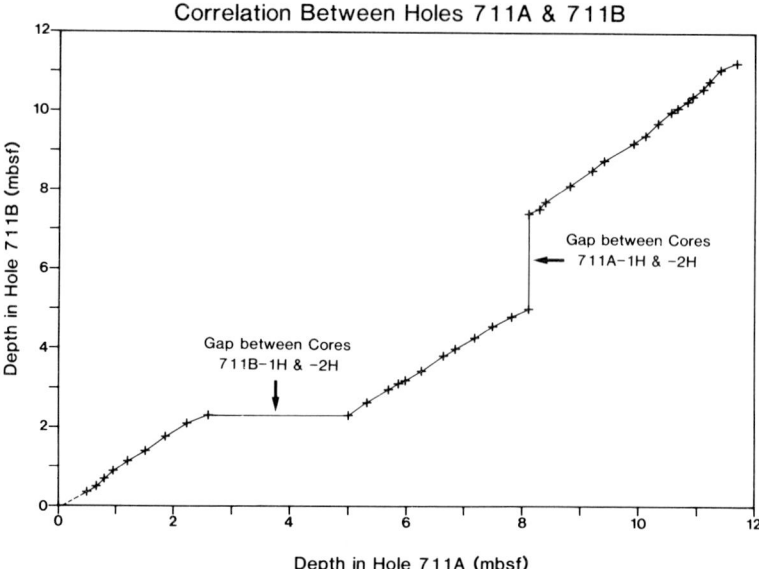

Fig. 9. Shaw Diagram correlating depth in Hole 711A with depth in Hole 711B, based on cross-matching the WCMS profiles of each hole (Fig. 8).

higher WCMS values, and also the greater amplitude of oscillations in WCMS (mirroring carbonate content, Fig. 3) at this site, relative to Site 709. The strong dependence of WCMS on carbonate content of the sediment at this site ($r = 0.971$), naturally provides a sound basis for lithostratigraphic correlation between Holes 711A and 711B, based on their WCMS profiles.

The WCMS-based correlation between Holes 711A and 711B shown in Fig. 8, is plotted in the form of a Shaw Diagram in Fig. 9. Figures 8 and 9 both clearly reveal that gaps occur in the stratigraphic record of each hole at Site 711, between individual cored intervals of the sequences recovered. In Hole 711A, an interval is missing between Cores 115–711A–1H and -2H, equivalent to the sequence in Hole 711B between P14 and P23 (= 2.4 m). In Hole 711B, an interval of the sequence is missing between Cores 115–711B–1H and -2H, equivalent to the sequence in Hole 711A between T5 and T9 (= 2.4 m). It is significant that the equivalent length of core missing from each hole at this site, between the first two cores in each sequence recovered, is exactly 2.4 m, further supporting the inference that these missing intervals are the result of coring irregularities. In Fig. 9, gaps in the sequences of Holes 711A and 711B, presumably due to coring irregularities, are clearly identified as major 'steps' in the correlation curve for these holes.

The main purposes for which the type of data

plotted in Fig. 9, are gathered, are (1) to enable bio- and magnetostratigraphic datums recognized in one hole to be extrapolated to the other, and (2) to assist the process of reconstructing a stratigraphically uninterrupted, composite sub-bottom depth model for a given site. Figure 10 shows a composite WCMS profile for Site 711, to an adjusted sub-bottom depth of 15 m, obtained by splicing the WCMS profiles of uninterrupted subsections of Holes 711A and 711B, as shown in Figure 8. Table 1 lists the depths at which a composite, WCMS-based sub-bottom depth

Table 1. *Composite sub-bottom depth model for Site 711 to approximately 40 mbsf, based on correlating between the WCMS profiles of Holes 711A and 711B (e.g., Fig. 8). This SBD model avoids stratigraphic discontinuities and intervals of disturbed sediment, and also excludes major turbidite horizons from the composite sequence*

Composite SBD (mbsf)		Depth in Hole 711A (mbsf)		Depth in Hole 711B (mbsf)	
Upper	Lower	Upper	Lower	Upper	Lower
0.00	0.15	–	–	void	void
0.15	0.99	–	–	0.15	0.99
0.99	1.92	–	–	1.17	2.10
1.92	7.65	2.28	8.00	–	–
7.65	23.36	–	–	4.80	20.51
23.36	29.58	21.08	27.30	–	–
29.58	40.98	28.53	39.93	–	–

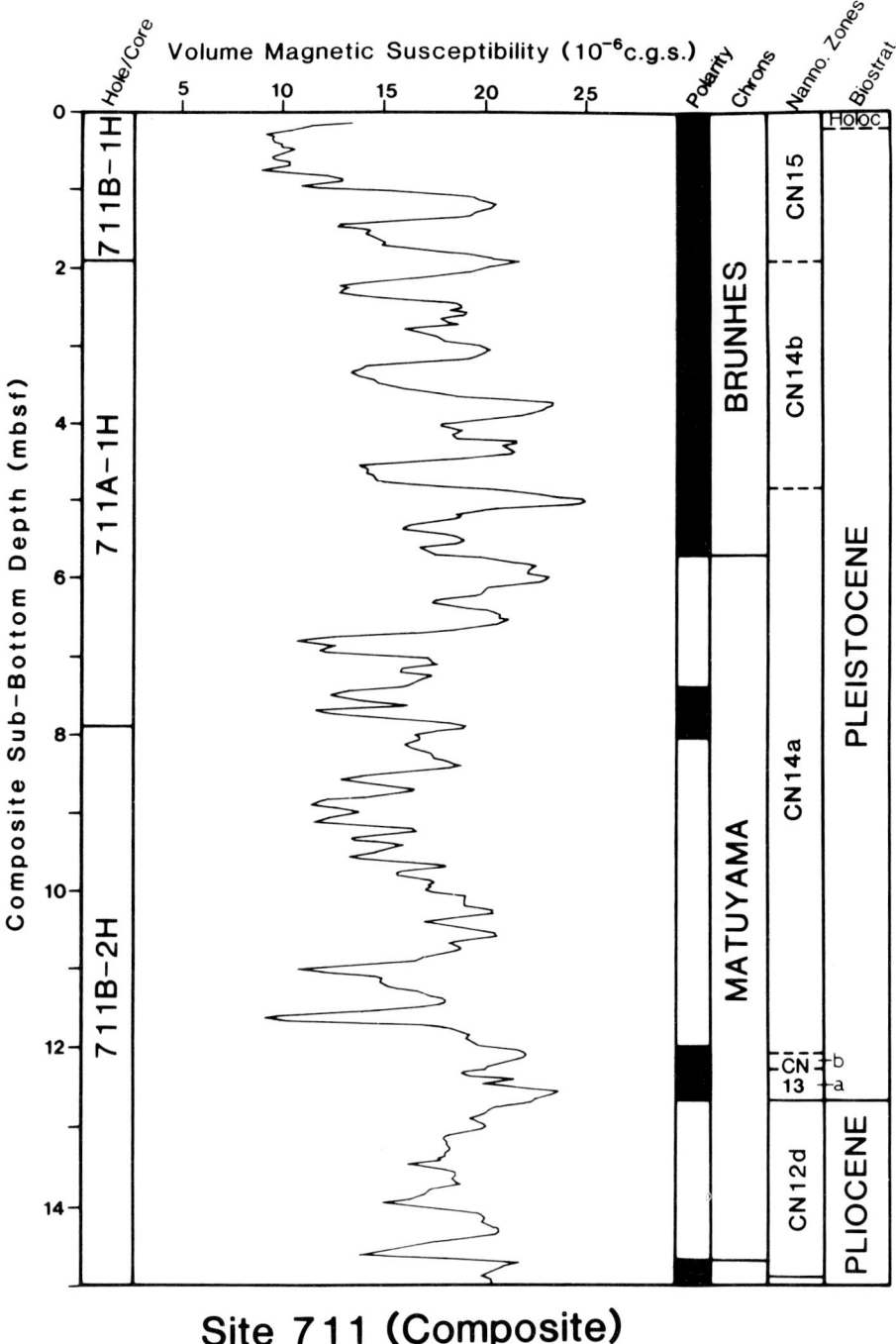

Site 711 (Composite)

Fig. 10. Composite, stratigraphically continuous WCMS profile for the Plio-Pleistocene sequence of Site 711 (0–15 mbsf). The composite sub-bottom depth model for Site 711 (0–40 mbsf), upon which this profile is based, is given in Table 1.

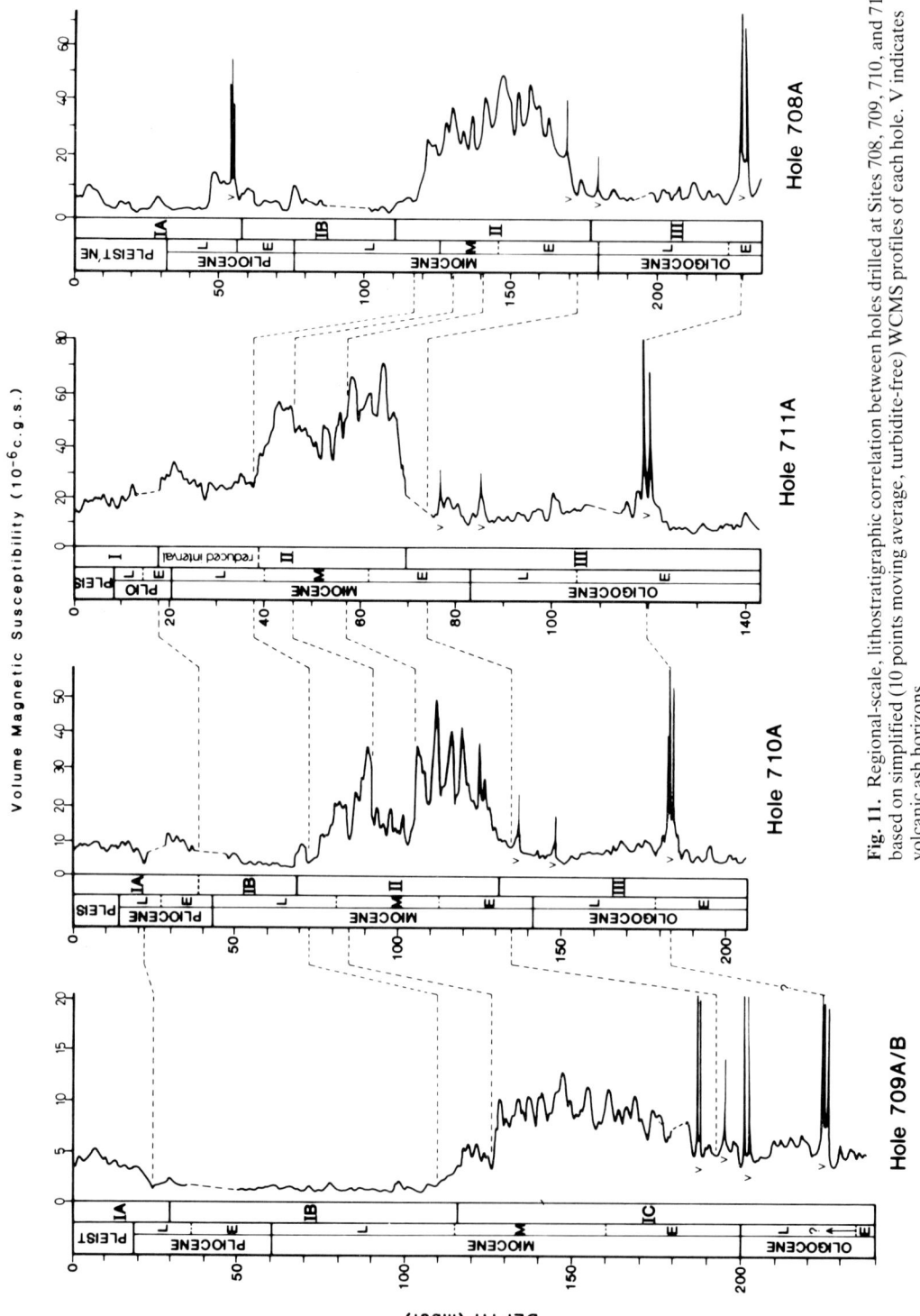

Fig. 11. Regional-scale, lithostratigraphic correlation between holes drilled at Sites 708, 709, 710, and 711, based on simplified (10 points moving average, turbidite-free) WCMS profiles of each hole. V indicates volcanic ash horizons.

model for the Site 711 (to an adjusted depth of 40 mbsf) switches between Holes 711A and 711B, in order to avoid gaps in the recovered sequences due to coring irregularities (major turbidite horizons have also been eliminated).

The last example in the present section (Fig. 11) involves a change of scale back up to that of complete, downhole WCMS profiles, several hundreds of metres long. This time, however, the correlation is between the WCMS profiles of entire holes drilled at different Leg 115 sites along a depth transect of the Madingley Rise (Fig. 1), the so-called 'Indian Ocean Carbonate Dissolution Profile' (CDP). This profile consisted of a series of closely spaced holes drilled at progressively deeper sites, for the purpose of studying the variation in pelagic carbonate accumulation as a function of water depth in the tropical Indian Ocean (Backman *et al.* 1988, p. 9; Peterson & Backman 1990; Curry *et al.* 1990).

On a scale as large as that used in Fig. 11, it is impossible to illustrate clearly the high-resolution (3–6 cm spaced) WCMS variations of each hole which provided the basis for the inter-hole correlations shown in previous figures. Hence, only a simplified (10-point moving-average, turbidite-free) version of the WCMS profiles obtained from Holes 708A, 709A/B (composite), 710A and 711A are shown in Fig. 11, in order to clarify both (1) the macroscale relationship between lithologic units (effectively equivalent to lithostratigraphic formations) and their WCMS profiles, (2) the correlation between equivalent lithologic units at different ODP sites, based on their WCMS profiles, and (3) the effect of water depth on the strength of the WCMS signal at each site, and its variation downhole, in response to carbonate fluctuations of the sediment.

Probably the most reliable between-site correlations shown in Fig. 11, are based on cross-matching the major deflections in the WCMS profiles of each hole within the early Oligocene interval, which correspond to regional volcanic ash horizons. Most notable in this figure, however, is the way in which the early and middle Miocene interval in each hole is characterized by a significant increase in WCMS values, roughly in proportion to the water depth of the drilling site. The deeper the drilling site, the more pronounced is the increase in WCMS values within the early and mid-Miocene interval of the hole, and the more 'condensed' this interval becomes relative to the remainder of the recovered sequence. This is the condensed, early through middle Miocene sequence referred to earlier in connection with Fig. 5, which, in the

western Indian Ocean, corresponds to a period of intensified carbonate-dissolution caused by a major shoaling of the regional CCD during this interval (van Andel 1975; Sclater *et al.* 1977; Peterson & Backman, 1990).

Figure 12 illustrates graphically the effect of increasing water depth on the relationship between carbonate content of the sediment and its magnetic susceptibility. This plot shows the variation in the coefficient of correlation (Pearson's *r*) for the relationship between %CaCO$_3$ and MS values at each of the sites for which data are shown in Fig. 11. The data plotted in Fig. 12 are weighted group averages of the correlation coefficients obtained for the relationships between these variables based on %CaCO$_3$ data obtained from different sources, including shipboard geochemistry and physical properties measurements, and from measurements made for the present study. Data from Site 708 are from turbidite-free intervals. Figure 12 shows clearly that the strength of the relationship between carbonate content and magnetic susceptibility of the sediment (i.e. the dependence of MS on CaCO$_3$ content) increases as a function of water depth in this region. The variation in *r*-values in Fig. 12 appears to exhibit a fairly simple, but non-linear relationship to water depth. However, no special significance is inferred from the curve fitted to the data plotted in this figure (using a 3rd-order cubic polynomial function), which simply demonstrates the non-linear trend of the relationship, the strength of which depends on water depth and thus dissolution intensity.

Also of regional (Northwest Indian Ocean) significance in Fig. 11, is the interval of reduced (greenish-grey coloured) sediment between the base of the late Miocene and uppermost Pliocene or early Pleistocene horizons of each hole (Lithological Subunit IB, or its stratigraphic equivalent at Site 711). This interval is characterized by distinctly lower WCMS values (except for volcanic ash horizons) than in the underlying and overlying units, attributable to early (suboxic) diagenesis, involving reduction of iron oxides in the sediment. This is suggested by a much weaker relationship between carbonate content and WCMS values for this reduced interval, relative to the strong correlation between these two parameters above and below this interval (e.g. Fig. 5). The effect of early diagenesis on the %CaCO$_3$ and WCMS relationship is clearly illustrated in Fig. 13, which plots the variation in WCMS as a function of carbonate content of the sediment in Hole 709A, to a depth of 140 mbsf. Samples taken from within the interval of greenish-grey coloured, anoxic

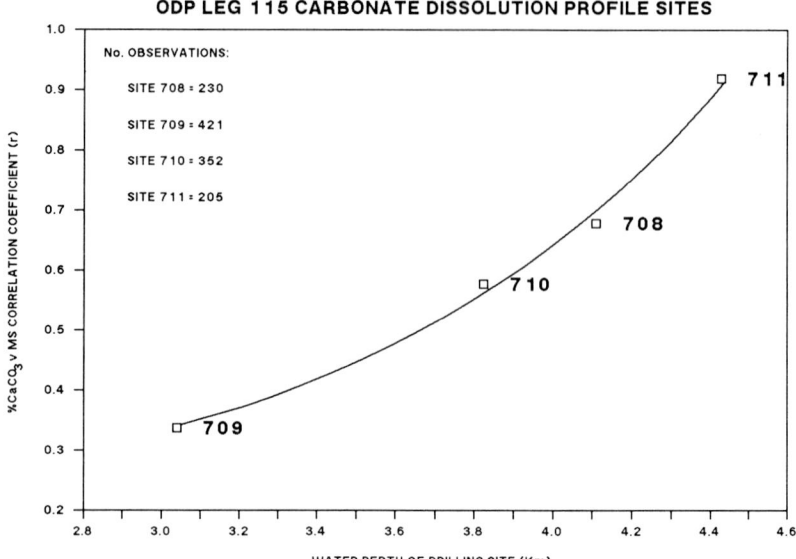

Fig. 12. Variation in the coefficient of correlation (Pearson's *r*) between WCMS and %CaCO₃, as a function of water depth, for Leg 115 'carbonate dissolution profile' Sites 708, 709, 710, and 711.

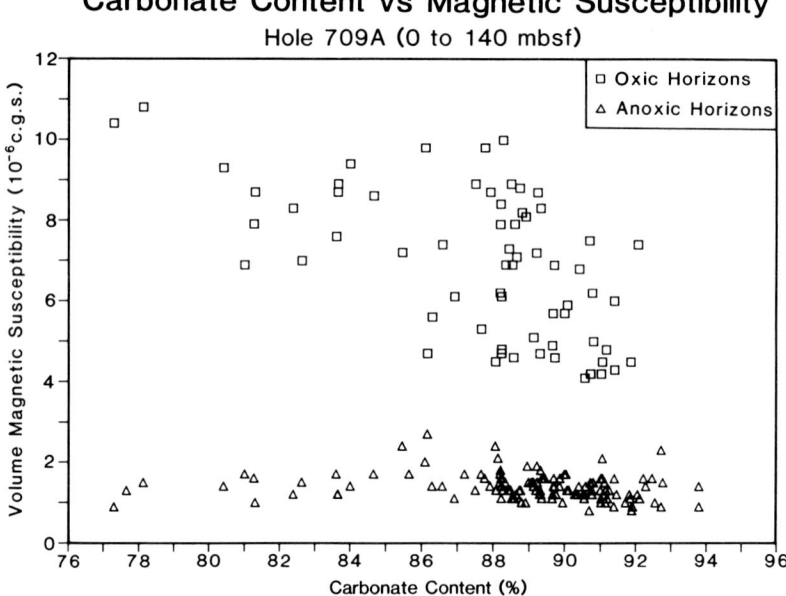

Fig. 13. Variation in WCMS with carbonate content of the sediment (shipboard 'geochemistry' data) for the upper part of Lithologic Unit I of Hole 709A (0–140 mbsf, Fig. 11), showing the effect of early (suboxic) diagenesis on this relationship. Data from oxic horizons of Hole 709A (Lithologic Units IA and IC) are plotted as squares, while data from anoxic horizons (Lithologic Unit IB) are plotted as triangles.

sediment (i.e. Lithological Subunit IB, between 118 and 22.5 mbsf) are plotted as triangles, whereas samples taken from horizons above and below the anoxic interval, belonging to different lithological units, are plotted as squares. Clearly, WCMS data from within the reduced interval exhibit an entirely different, and independent relationship to carbonate variations of the sediment than the WCMS data from elsewhere in the sequence, despite there being no significant difference in the population distribution of %CaCO$_3$ values between the two groups of samples. This suggests that the WCMS signal of sediments desposited at this site derives from a low concentration (i.e. a few ppm) of a strongly magnetic (ferrimagnetic) iron oxide, which, within the interval affected by early diagenesis, has now been replaced by a similar, or lower concentration of a non-magnetic (paramagnetic), and thus more weakly magnetizable, reduced form of iron in the sediment.

The effects of early diagenesis on the WCMS characteristics of ODP sequences are most pronounced at Site 709, the shallowest of the sites featured in Fig. 11, and become less obvious as the waterdepth of the site deepens. This suggests that the interval of anoxic sediment, presumably related to an increase in the organic matter content of the sediment (e.g. Froelich et al. 1979), may have been caused by a thickening and intensification of the oceanic oxygen-minimum layer (OML) during this period. This, in turn, was probably caused by an increase in organic productivity at this time, which could possibly have been related to the establishment of the modern equatorial water mass and associated circulation systems in the tropical Indian Ocean. This event took place at the beginning of late Miocene times, as the southern tip of the Indian subcontinent drifted north of the Equator (Vincent 1974, 1977).

Tephrostratigraphy

A clear demonstration of the usefulness of WCMS logging in studies of marine tephrostratigraphy ('tephrochronology') is provided by Fig. 14. This figure shows the WCMS profiles of Holes 706A and 706B to a sub-bottom depth of 20 m, plotted alongside lithological logs of these holes taken from shipboard core description data.

Marine tephra and volcanic ash-rich horizons of deep-sea sediments may contain as much as 10% primary (titano)magnetite when derived from a source of basic vulcanicity (e.g. Kennett 1981). Consequently, such horizons frequently exhibit WCMS values which are orders-of-

magnitude greater than those values normally associated with pelagic lithologies. Volcanic ashes associated with acid igneous activity, however, may consist dominantly of glass shards which are diamagnetic, and thus exhibit extremely weak MS values. As indicated by reference to the lithological logs shown in Fig. 14, WCMS variations at Site 706 are almost entirely controlled by the character and distribution of basic volcanic ash horizons in the sequence. The response of WCMS to variations in pelagic lithology between volcanic ash horizons at this site is totally masked by the generally high 'background' levels in WCMS, which are higher even than the maximum WCMS values attained in the most strongly dissolved intervals of Holes 709A and 710A. The high 'background' values of WCMS at Site 706 are probably related to the presence of minor amounts of volcanic ash dispersed throughout the sequence by bioturbation.

Above 'background' levels, a series of prominent peaks in the WCMS profiles of Holes 706A and 706B appear in response to discrete volcanic ash-rich horizons, and rise to values of between 1.0 and 1.5 × 10^{-4} cgs (often deflected off-scale in Fig. 14). The intensity of these WCMS peaks is probably related directly to the initial concentration, and extent of post-depositional alteration (reduction), of volcanic ash in the sediment. In many of the greenish-grey, degraded volcanic ash horizons in Holes 706A and 706B, discrete crystals of pyrite were observed which exhibited framboidal morphology, indicative of an authigenic origin involving iron and sulfate-reducing bacteria (Curtis 1983). The much lower WCMS values exhibited by degraded, as distinct from unaltered volcanic ash horizons, therefore, are probably due to 'dissolution' and pyritization of primary titanomagnetite grains which were originally present in the volcanic ash before these horizons were reduced (cf. Canfield & Berner 1987). The change in the level of 'background' WCMS values which occurs at around 7.5 mbsf in both holes at Site 706, is also related to a change in the oxidation state (indicated by a change in colour) of the sediment, and thus its magnetic iron oxide content, at this point in the sequence.

The lines of correlation drawn between the WCMS profiles in Fig. 14, highlight the gaps which occur at different depths in the sequence of each hole at Site 706, resulting both from real, stratigraphic discontinuities, and from coring irregularities. Figure 15 plots the correlation between the WCMS profiles of Holes 706A and 706B in the form of a Shaw Diagram.

A significant unconformity occurs in both

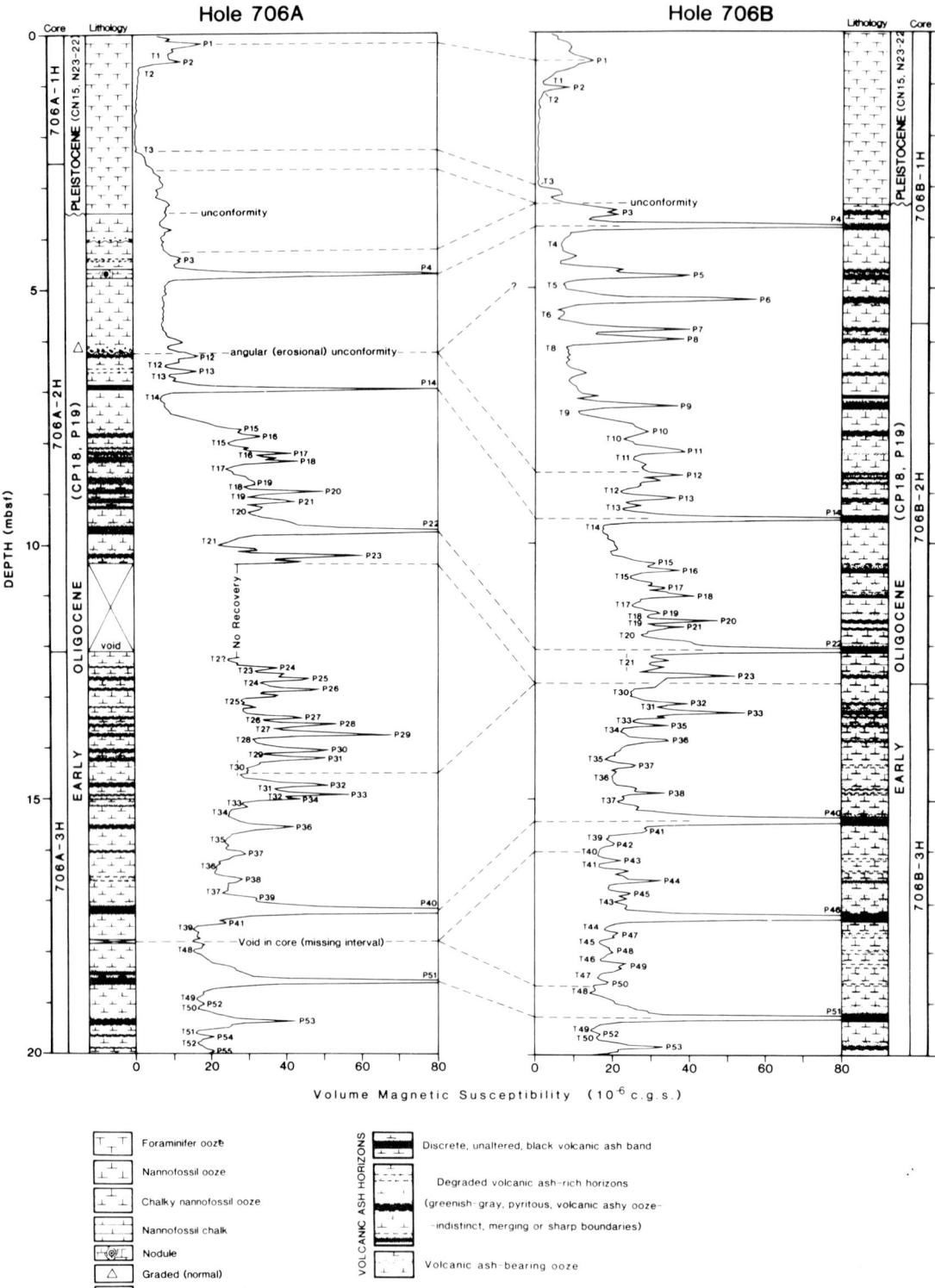

Fig. 14. Tephrostratigraphic correlation between subsidiary holes drilled at Site 706, based on their WCMS profiles. Logs of volcanic ash horizons are based on original (shipboard), visual core description data. Correlation points indicated in this figure (P1, T1, P2, T2, etc.) were used to construct the Shaw Diagram shown in Fig. 15.

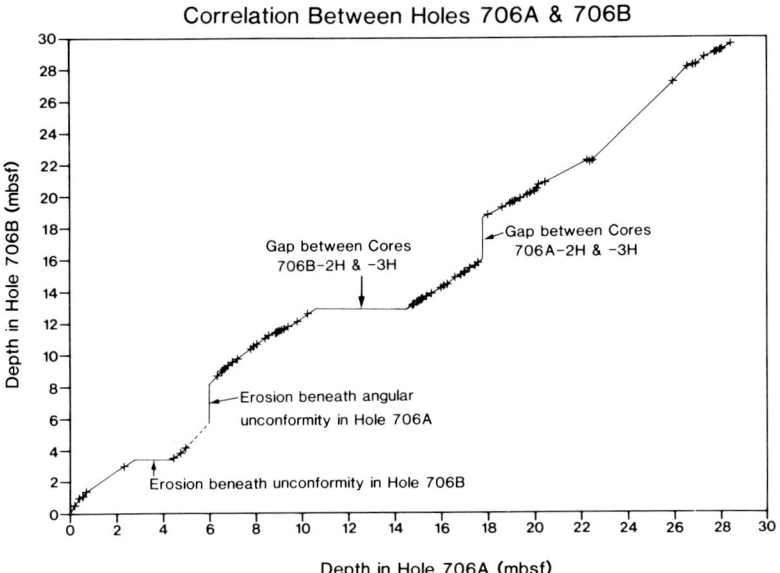

Fig. 15. Shaw Diagram correlating depth in Hole 706A with depth in hole 706B, based on cross-matching the WCMS profiles of each hole (Fig. 14).

holes at Site 706, separating the uppermost lithological unit at this site, which consists of foraminifer ooze of Pleistocene age, from the remainder of the sequence which consists of nannofossil oozes of early Oligocene age, containing numerous volcanic ash horizons. However, an interval of at least 1 m in thickness, which corresponds to a sequence in Hole 706A somewhere between P3 and T3, is missing from Hole 706B beneath the unconformity at *c.* 3.3 mbsf. Further down-sequence, an interval which is at least equivalent to that which occurs in Hole 706B between P5 and T11 (approximately 4 m in thickness), is missing from the sequence of Hole 706A due to a second, this time angular (erosional) unconformity in this hole at *c.* 6.25 mbsf.

The gaps which occur in the records of Holes 706A and 706B further down-sequence, are clearly the result of coring irregularities. In Hole 706B, an interval is missing between Cores 115–706B–2H and -3H, which is at least equal in thickness to the sequence in Hole 706A between T22 and P31 (approximately 2 m, possibly as much as 4 m if the interval of zero recovery in Hole 706A between 10.4 m and 12.1 mbsf is included). In Hole 706A, a void occurs in Core 115-706A-3H (*c.* 18.75 mbsf) which represents a gap in the recovered sequence equivalent to the interval in Hole 706B between T40 and T48

(approximately 2.75 m). Loss of material from the middle of a core like this, may have been caused by repenetration of the corer, with material lost or not sampled (i.e. washed) between successive penetrations of the corer, within a single coring run. An alternative explanation is that this gap represents a real stratigraphic discontinuity, i.e. a diastem in this instance.

Both Figs 14 and 15 clearly demonstrate the complexities involved in reconstructing the lithostratigraphic record of some ODP Sites from the sequences recovered by offset coring in adjacent holes, where both stratigraphic and coring-related discontinuities occur in close association down-sequence. To achieve the sort of centimetre-scale, stratigraphic reconstruction exemplified by Site 706 would be extremely difficult without the assistance of a practical technique for ultra-high-resolution, lithologically-related core-logging and between-hole correlation.

Identification of coring irregularities

This section considers the use of WCMS logging to detect three kinds of coring irregularity which seriously disrupt the continuity of ODP sequences. These are: (1) unidentified gaps at core boundaries; (2) repetition of sequence due to repenetration of the corer; and (3) contamination of cores by drilling artifacts, or mechanically reworked material containing such artifacts.

Unidentified gaps at core boundaries. This type of coring irregularity is most readily identified by correlating between offset cores taken from adjacent holes at a given site, by means of high-resolution WCMS logging. Several examples of this form of disturbance, and its detection by WCMS profiling, have already been illustrated above (Figs 6 to 9, 14, and 15). Further (incidental) examples are provided in Figs 16, 17 and 18 discussed below. Unrecorded gaps between successive cored intervals of a given hole may be detected when a sequence of WCMS variations in the adjacent hole at that site appears to be absent from the first hole, and the lines of correlation between the two holes, for the horizons immediately above and below the missing sequence, converge at a core boundary. Unrecorded intervals of 'no recovery' between cores result from miscalculations in the sub-bottom depth assignment for individual cored intervals, due to inadequate compensation for unwanted vertical displacements of the drill string, caused both directly by surface waves, and indirectly by the effects of subsurface currents (Ruddiman *et al.* 1987). Such uncompensated displacements may cause the drill bit to wash down further than intended (or recorded by drilling platform instruments) between successive coring runs.

Repetition of sequence due to repenetration of the corer. A thorough examination of the causes and effects of repenetration during hydraulic piston coring was conducted by Ruddiman *et al.* (1987), based on an analysis of DSDP Leg 94 cores. An earlier study of repenetration features in open-barrel, gravity cores was documented by Weaver & Schultheiss (1983). This kind of coring irregularity results in the repetition of a sequence previously cored, often at the expense of an interval underlying that which was captured during the initial penetration of the corer, and equal in thickness to that of the repeated sequence. This effect may imply that a real stratigraphic discontinuity or an 'unrecorded interval of no recovery' occurs in the adjacent hole at that site, if the repeated sequence remains undiscovered. Unless biostratigraphic datums are repeated by the recoring of a sequence, it is extremely difficult to detect the presence of repenetration features within cores without the assistance of a quantitative tech-

nique for high-resolution, lithologically-related core logging.

Figure 16 shows the WCMS profiles obtained from Holes 709A and 709B for the early Miocene interval between 153.5 and 178 mbsf. Detailed correlation between these high-resolution profiles (6 cm spaced data points) reveals the presence in Hole 709B of several stratigraphic discontinuities, resulting from loss of material between cores and repenetration of the corer. Unrecorded intervals of no recovery occur at the junction between Cores 115-709B-17H and -18H, and, more significantly, between Cores 115-709B-18H and -19H. The lengths of the sequences missing from between these cores may be gauged by reference to the WCMS profiles of the equivalent intervals in Hole 709A. These missing intervals are clearly related to the repetition of the sequence in Hole 709B which occurs within Core 115-709B-18H. The repeated sequence occurs between 163.3 and 168 mbsf in Hole 709B, and corresponds, stratigraphically, to the sequence immediately overlying it in this hole, between 158.4 and 164.5 mbsf. These stratigraphically equivalent intervals of Holes 709A and 709B are each labelled 'X' in Fig. 16.

It is a remote possibility that the uppermost of the intervals, labelled 'X_1' in Hole 709B (Fig. 16), represents a small-scale sediment slide (lateral translation), in which the dislocated material has been redeposited *en masse*, thus retaining its internal stratigraphy undeformed. It is much more likely, however, that the repetition of sequence in Hole 709B is entirely attributable to repenetration of the corer. The top of the sequence in Hole 709B labelled 'X_1' in Fig. 16, occurs at 158.3 mbsf, while the base of the sequence labelled 'X_2' occurs at 168 mbsf. This interval (158.3–168 mbsf) corresponds exactly to that of Core 115-709B-18H, and the gaps in the sequence of Hole 709B between cores, as noted above, occur at the upper and lower boundaries of this core.

A possible mechanism which may account for the repetition of sequence in Hole 709B, as shown in Fig. 16, is as follows. Following penetration of the APC to a depth of only about half its intended run length, the core barrel was temporarily withdrawn from the hole, unintentionally, possibly due to inadequate heave compensation during the passage of an abnormally large wave. It was then reinserted at a slight angle to the vertical into the side-wall of the hole, at an immediately adjacent location to the previously cored interval, thus recoring the sequence captured by the first penetration of the corer.

Figure 17 shows a further example of

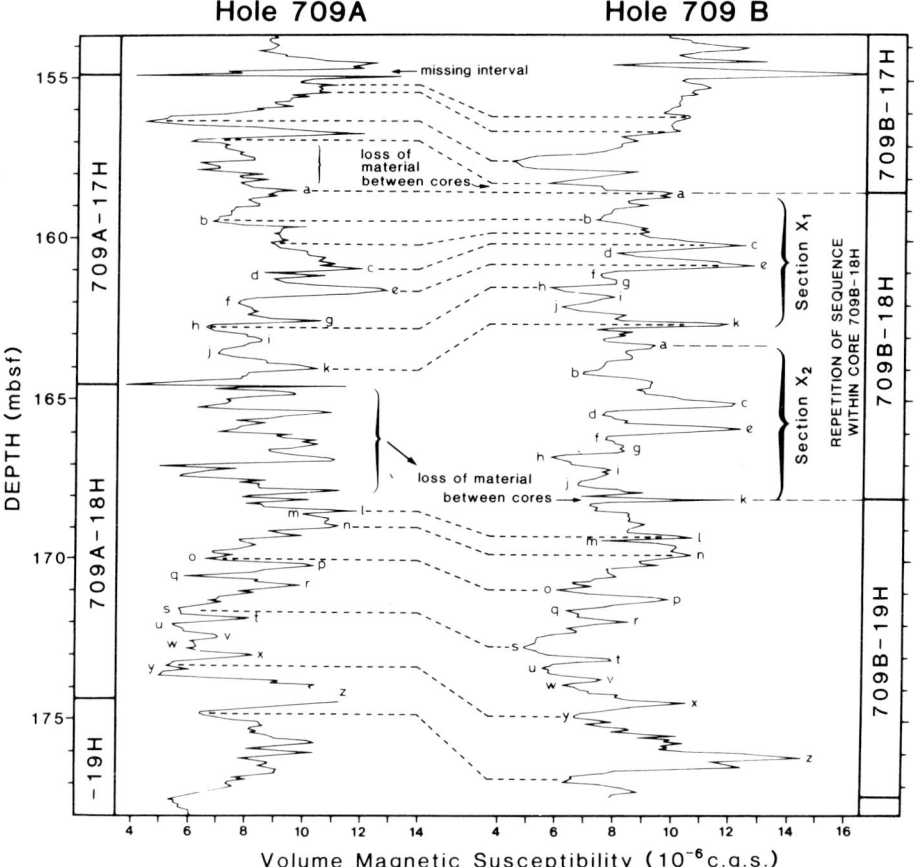

Fig. 16. Lithostratigraphic correlation between Holes 709A and 709B for the interval between 153.5 and 178 mbsf (early Miocene), based on WCMS profiles of each hole. Letter-coded WCMS peaks indicate points of auto- and cross-correlation within and between each hole.

sequence repetition caused by repenetration of the corer, this time, however, the repeated sequence corresponds to an entire core (115-710B-3H), and is stratigraphically equivalent to the sequence recovered during the previous coring run. The presence of a repeated sequence within, or corresponding to Core 115-710B-3H is also indicated by the repetition of nannofossil Subzone CN14A in this core. Figure 17 also provides a further illustration of 'unrecorded intervals of no recovery' between successive cores. Lines of correlation between selected features of the WCMS profiles shown in Figure 17 are drawn in order to demonstrate that the sequence in Hole 710A between T8 and P34, i.e. between 2.7 and 123.9 mbsf (Cores 115-710A-1H and -2H), correlates with the sequence in Hole 710B between 5.0 and 14.4 mbsf (Cores 115-710B-1H and -2H), and also with the

sequence in Hole 710B between 16.45 and 25.8 mbsf (Core 115-710B-3H). The interval in which the repeated sequence occurs, therefore, corresponds exactly to that of Core 115-710B-3H, and is stratigraphically equivalent to the sequence captured by Core 115-709-2H. A small exception to this, however, is the interval labelled 'X' in Fig. 17, which is absent from the uppermost of the two, stratigraphically equivalent sequences recovered from Hole 710B due to loss of material between Cores 115-710B-1H and -2H. The interval labelled 'X' is clearly identifiable both in the repeated sequence in Hole 710B (Core 115-710B-3H) and in Hole 710A (Core 115-710A-1H), but is missing from between Cores 115-710B-1H and -2H.

Figure 18 shows the correlation between the WCMS profiles of Holes 710A and 710B in the form of a Shaw Diagram, illustrating graphically

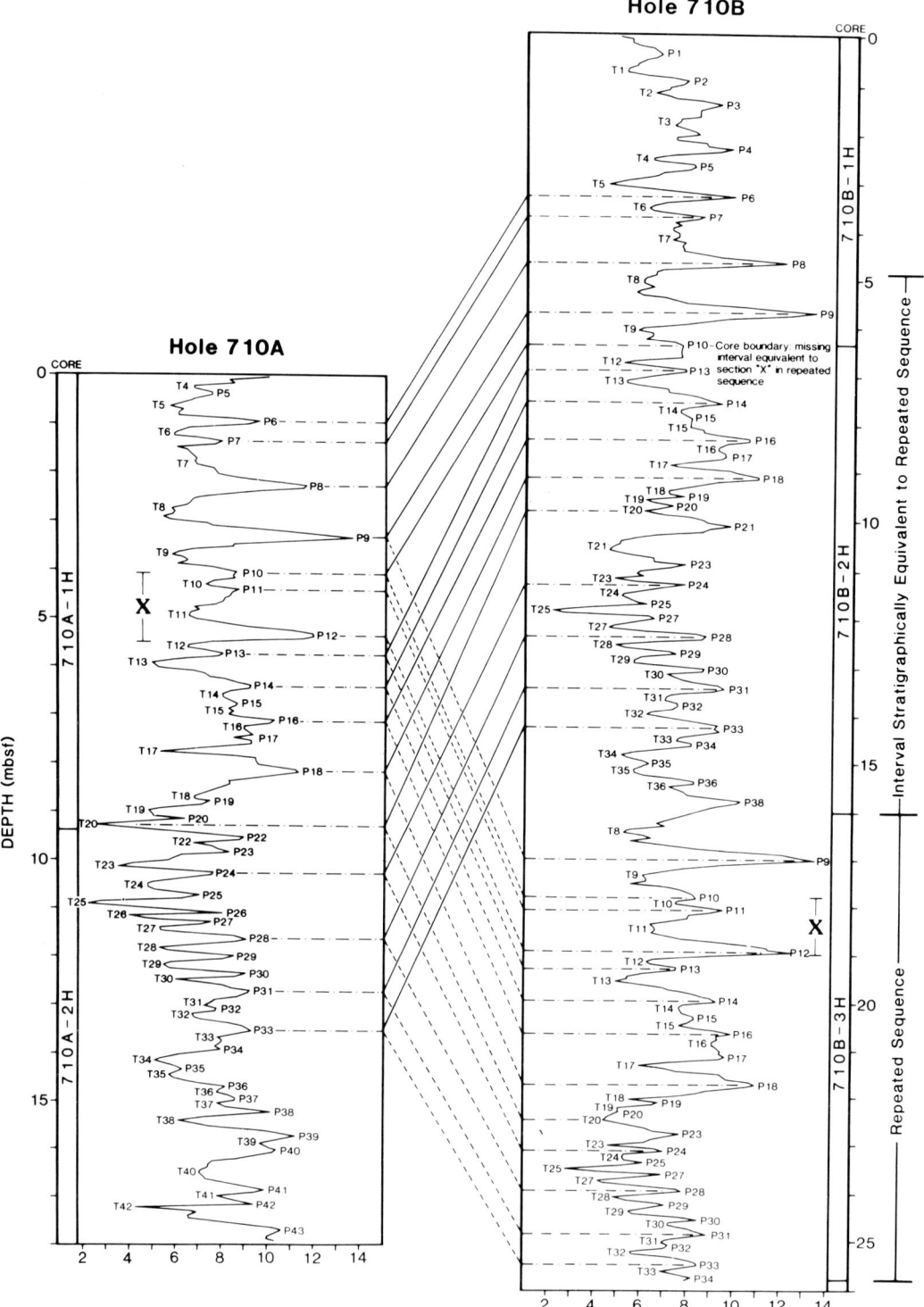

Fig. 17. WCMS-based, lithostratigraphic correlation between the Pleistocene horizons of Holes 710A and 710B, showing the repetition of sequence in Hole 710B corresponding to the entire length of Core 115-710B-2H. The correlation points indicated in this figure were used to construct the Shaw Diagram shown in Fig. 18.

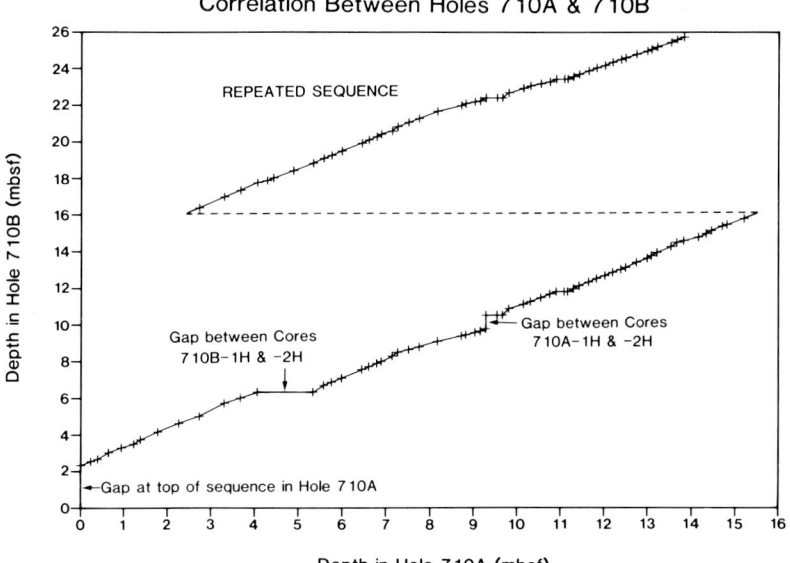

Fig. 18. Shaw Diagram correlating depth in Hole 710A with depth in Hole 710B, based on cross-matching the WCMS profiles of each hole (Fig. 17).

the effect that repenetration of the corer has had on the stratigraphic continuity of Hole 710B, relative to that of Hole 710A. With the exception of two, short, 'unrecorded intervals of no recovery', the upper and lower segments of the single correlation curve plotted in Fig. 18 lie parallel and are of equal length, indicating that the sequence in Hole 710A between 2.7 and 13.9 mbsf correlates with sequences in Hole 710B both between 5.0 and 14.4 mbsf, and between 16.45 and 25.8 mbsf, corresponding to Cores 115-710B-2H and -3H, respectively.

Although Figs 17 and 18 do not show it, the repeated sequence corresponding to Core 115-710B-3H was captured at the expense of the interval which should have occupied this core. This is indicated by the total absence of correlatable features between the WCMS profile of Hole 710A, for the interval between 15.5 and 23.6 mbsf, and that of Hole 710B below the repeated sequence (see Backman *et al.* 1988, p. 607). Therefore, the stratigraphic sequence equivalent to that of the interval between 15.5 and 23.6 mbsf in Hole 710A, is absent from the record of Hole 710B, and has been replaced by the repeated sequence. Possible explanations for the coring irregularities identified at Site 710 are discussed below.

Contamination by drilling artifacts. A recurrent problem affecting DSDP and, subsequently, ODP sequences, is that of contamination by

pipe-rust and/or similar metallic artifacts of drilling. Contamination of this kind not only precludes whole-core and, if contaminants are disseminated throughout the sediment, discrete-sample palaeomagnetic (and geochemical) measurements, it also signifies the possible presence in the sequence of mechanically reworked sediment washed-in from uphole (Sager 1986). This is because pipe-rust and drilling abrasion debris is usually derived from the outer surface of the drill string, and collects in pools of slumped material which form at the bottom of holes drilled in poorly cohesive (e.g. foraminifer-rich) sediments, upon retrieval of each interval of core. Such reworked material then becomes incorporated into the recovered sequence at the top of the next core taken. Contamination of this kind was experienced repeatedly on ODP Leg 101, prompting Sager (1986) to make a detailed study of the nature, distribution and probable sources of ferrous metal contaminants in the cores, using WCMS logging to detect their presence. Rust-flake contamination was also a problem on ODP Leg 103, seriously disrupting the palaeomagnetic records of some cores (Boillot *et al.* 1987). Naturally, the MS of ferromagnetic contaminants (e.g. fragments of pipe-rust, etc.) is orders-of-magnitude greater than that of most pelagic lithologies, with the exception of volcanic ash-rich sediments.

Figure 19 shows part of the WCMS profile

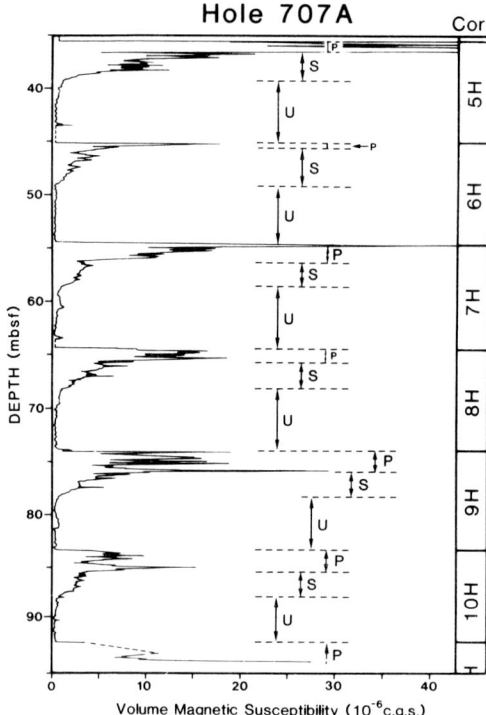

Fig. 19. WCMS profile of Hole 707A, between 40 and 100 mbsf, showing the characteristic 'sawtooth' motif associated with contamination at the top of cores by pipe-rust or similar ferrous metal artifacts of drilling, and the subsequent smearing of these contaminants around the exterior of the sediment column (inside the core liner), which occurs during penetration of the core barrel. P, pervasive contamination; S, superficial contamination (due to smearing); U, uncontaminated.

obtained from Hole 707A, between 35 and 95 mbsf. The profile consists of a series of regularly spaced, asymmetric clusters of WCMS peaks, each of which corresponds exactly to individual cored intervals of the sequence, with the culmination in WCMS peak values occurring at the top of each core. Beneath the uppermost peak, or initial cluster of peaks in the WCMS profile of each core, WCMS values generally taper off downcore, the diminution often spanning a depth in core of between 1/3 and 1/2 of its total length, before reaching 'background' values again. Background WCMS values in Hole 707A were particularly low because the sediments at this site mainly consist of carbonate-rich, sandy (thus incohesive), foraminifer-nannofossil oozes. The uppermost peak, or cluster of peaks in the WCMS profile of each

core (labelled 'P' in Fig. 19) corresponds to an interval of mechanically reworked material pervasively contaminated by metallic artifacts (pipe rust, etc.) scraped from the outer surface of the drill string. The gradual decline in WCMS values beneath this initial cluster of peaks (labelled 'S' in Fig. 19) corresponds to superficial smearing of contaminated material from the top of the core, around the exterior of the sediment column, inside the core liner. Below these two intervals in each core, the sediment remains uncontaminated (labelled 'U' in Fig. 19). In this example, therefore (see also Sager 1986), WCMS profiling is useful, not as a tool for lithostratigraphic correlation between holes, but as an indicator of contaminated intervals of the sequence, some of which may actually consist of slumped, or mechanically reworked sediment from uphole. Such intervals, once identified by WCMS logging, may subsequently be avoided when sampling the cores, and adjustments can be made to stratigraphic sections, accumulation rate estimates, etc.

Palaeoclimatic investigations

Since WCMS often reflects the ratio of biogenic to lithogenic components in deep-sea sediments, in many instances it may also reflect their palaeoclimatic record in view of the well-established relationship between fluctuations in the carbonate content of deep-sea sediments and climatic changes (e.g. Olausson 1967; Broecker 1971; Berger 1973; Ramsay 1974; Volat *et al.* 1980; Crowley 1985; Dean & Gardner 1986; Rea *et al.* 1986; Chuey *et al.* 1987). The carbonate content of deep-sea sediments may be regulated by climatic influences via one or more of the following three mechanisms. (1) Dilution of the biogenic carbonate fraction of the sediment by terrigenous detritus; (2) dissolution of carbonate constituents by seawater undersaturated with respect to $CaCO_3$ (Ca^{2+} and HCO_3^- ions); and (3) productivity variations of carbonate-secreting flora and fauna. The inter-relationship between carbonate content, magnetic mineral concentration and the palaeoclimatic record of deep-sea sediments has prompted several workers in recent years to use MS and related rock magnetic parameters like isothermal remanent magnetism (IRM) and anhysteretic remanent magnetism (ARM), as proxy palaeoclimatic indices (e.g. Kent 1982; Robinson 1982; Robinson & Bloemendal 1983; Oldfield & Robinson 1985; Mead *et al.* 1986; Robinson 1986*a,b*, 1990; Thompson & Oldfield 1986, p. 145; Bloemendal *et al.* 1988*a,b*, 1989; Hall *et al.* 1989; Clemens & Prell 1991; deMenocal *et al.* 1991).

Figures 3 and 10, above, illustrate the potential of WCMS logging as a tool for investigating the palaeoclimatic record of Quaternary deep-sea sediments, in which obviously cyclic fluctuations in carbonate content are controlled mainly by dissolution of biogenic constituents of the sediment. This contrasts with palaeoclimatic investigations involving the WCMS profiles measured during ODP Legs 108 (e.g. Bloemendal *et al.* 1989) and 117 (e.g. Bloemendal & deMenocal 1989; Clemens & Prell 1991; deMenocal *et al.* 1991), in which the basis for the cyclic WCMS signal was orbitally-forced *dilution* of the pelagic carbonate fraction of the sediment, principally by eolian terrigenous detritus.

Figure 3 shows that WCMS variations in Core 115-711A-1H, of middle to uppermost Pleistocene age, are strongly related ($r = 0.971$) to dissolution-controlled oscillations in carbonate content of the sediment. The timescale indicated in this figure is based on six nannofossil datum events identified by Okada (1990), and the shipboard magnetic reversal stratigraphy for this core (Backman *et al.* 1988, p. 669). The positions at which inferred oxygen isotope stage boundaries shown in Fig. 3 were interpolated, are based on the SPECMAP ages for these events listed in Imbrie *et al.* (1984). Direct oxygen isotope measurements on individual foraminifer species in this core were precluded by the effects of severe dissolution. Bulk oxygen and carbon isotope measurements were made on samples at 5 cm intervals from Core 115-711A-1H, but the results proved to be unusable for stratigraphic purposes due to a biasing of the signal by extremely solution-resistant forms of carbonate in the strongly dissolved intervals of the core. These results are therefore not reproduced here.

Fluctuations in the percent-carbonate and WCMS records of Core 115-711A-1H appear to exhibit a dominant, near-100 ka periodicity, corresponding to that of the Earth's orbital eccentricity cycle (e.g. Hays *et al.* 1976; Imbrie & Imbrie 1980; Imbrie *et al.* 1984). This apparent periodicity is suggested by the approximate correlation between WCMS and percent-carbonate half-cycles and most isotopic stages, and the observation that seven complete cycles in the downcore records of these parameters occur between the Brunhes-Matuyama boundary and isotopic stage 5: a feature which is characteristic of 100 ka-dominated records of palaeoclimatic indices from late Quaternary deep-sea sediments (Jenkins *et al.* 1985).

In order to test more rigorously the hypothesis that cyclicity in the WCMS and percent-carbonate records of Site 711 is orbitally forced, spectral analysis was performed on the com-

posite WCMS profile data for this site, as shown in Fig. 10, above. That this composite profile effectively mirrors carbonate dissolution cycles in the sediment, has already been demonstrated above with reference to Fig. 3. The composite WCMS profile of Site 711, therefore, represents a continuous, proxy-record of carbonate dissolution in the tropical NW Indian Ocean for the entire Quaternary, with stratigraphic and coring discontinuities avoided, and turbidite horizons removed from the sequence. The age model used for spectral analysis of this composite WCMS profile was based on shipboard palaeomagnetic data (Schneider & Kent 1990), and the nine Quaternary nannofossil events identified later by Okada (1990).

Spectral estimates were based on WCMS data linearly interpolated at 5 ka intervals, yielding Nyquist periods of between about 10 ka and 14 ka for the Site 711 composite WCMS time series, originally based on 3 cm spaced WCMS measurements. The procedure used for fast Fourier transforms of this time series are detailed by Jenkins & Watts (1968, p. 209–257). Figure 20 shows the results of this analysis in the form of two logarithmic variance-density spectra: one (Fig. 20A) for the middle and upper Pleistocene interval of the time series, i.e. between 0.92 Ma and *c.* 0.015 Ma (7.5–0.15 mbsf); and the other (Fig. 20B) for the middle and lower Pleistocene interval of the same sequence, i.e. between 0.82 Ma and 1.87 Ma (6.5–12.6 mbsf). A separate analysis was required for each interval because the dominant frequency of WCMS cycles at this site appears to change markedly within the interval between the Brunhes–Matuyama and top-Jaramillo magnetic reversal boundaries (see Fig. 10). Above this transition, relatively long wavelength (near-100 ka) cycles appear to dominate the WCMS record, whereas below the transition, WCMS variations exhibit a more complex pattern of higher frequency spectral components (containing a powerful 40 ka cycle).

Figure 20A reveals that variance in WCMS for the middle and upper Pleistocene sequence at Site 711 is concentrated at several periods approximating to dominant cycles in the Milankovitch band (e.g. Imbrie 1985), as indicated by the ETP lettering scheme in this figure (*E* represents eccentricity, *T* obliquity and *P* precession). Conspicuous peaks at the anomalous periods of 71 ka and 36 ka may represent heterodyne tones at harmonics, or possibly combination tones (non-linear interactions between pairs of the Milankovitch cycles), one of which is known to occur at a period of 69 ka (Ghil 1987). Ruddiman *et al.* (1989) also found a 70 ka period

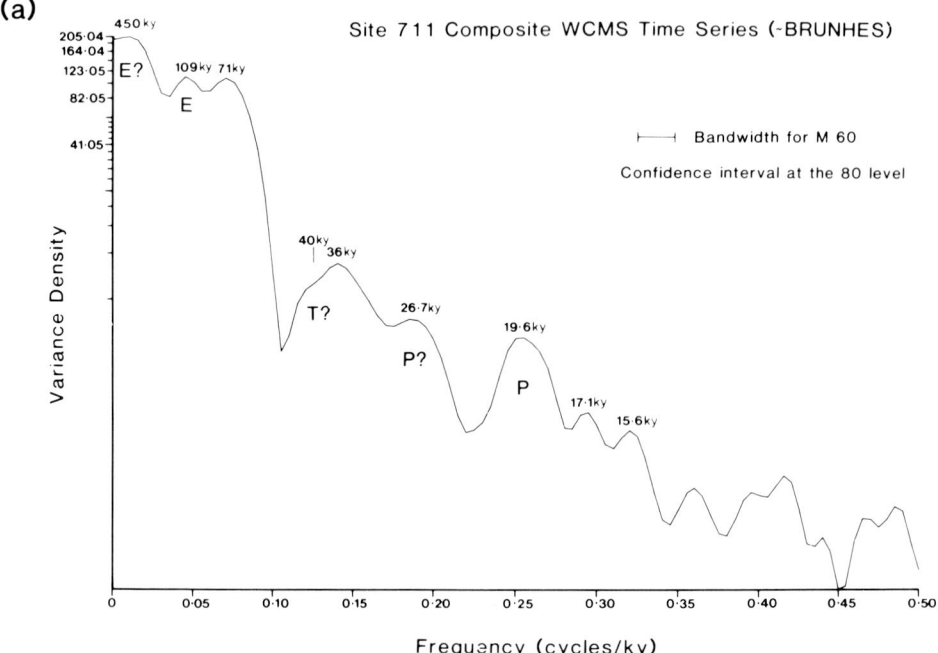

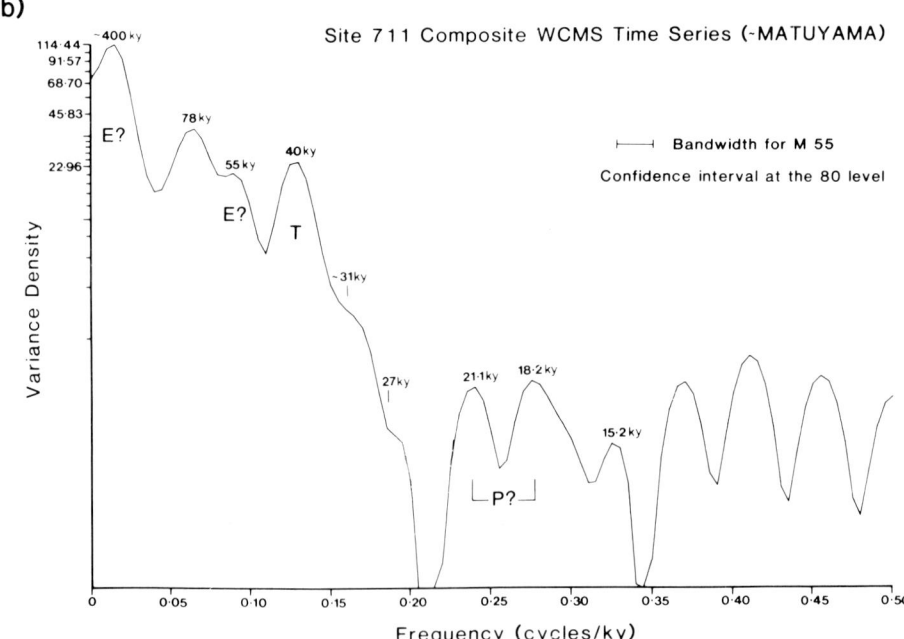

Fig. 20. Logarithmic variance-density spectra of an interpolated WCMS time series from the Pleistocene interval of Site 711, based on the composite WCMS profile for this site shown in Fig. 10. (**A**) WCMS periodogram for the middle and late Pleistocene interval at Site 711, i.e., between 0.92 and 0.015 Ma (0.15–7.5 mbsf); (**B**) WCMS periodogram for the middle and lower Pleistocene interval at Site 711, i.e., between 0.82 and 1.87 Ma (6.5–12.6 mbsf). Probable orbital forcing parameters are represented by E (eccentricity), T (obliquity), or P (precession)

in mid-Pleistocene sea-surface temperature and oxygen isotope records from the North Atlantic (DSDP Site 607), but attributed this to imperfect tuning of the SPECMAP timescale (upon which their age model was based) between c. 0.65 Ma and 0.70 Ma. With regard to the anomalous 36 ka peak, however, Morley & Hays (1978) identified a period of about 37.5 ka for variations in carbonate content of a core from the sub-tropical Atlantic.

Figure 20B shows the WCMS variance spectrum for the pre-Brunhes, middle and lower Pleistocene interval at Site 711. As noted above, the dominant periods in this record differ markedly from those of the Brunhes record at this site. In particular, the strong, near-100 ka (eccentricity) period is absent from the pre-Brunhes spectrum (Fig. 20B), although the 413 ka and 54 ka eccentricity periods (Imbrie 1985) may be represented in this spectrum by the peaks at 400 ka and 55 ka. However, the powerful spectral peaks close to the known obliquity (41 ka) and precessional frequencies (23 ka and 19 ka), strongly suggests that orbital forcing also modulates carbonate accumulation in the early, pre-Brunhes Pleistocene interval at this site. Once again, however, several significant peaks in variance density occur at periods which do no correspond to the primary Milankovitch cycles, most notably at 78 ka. The change from a WCMS spectrum dominated by obliquity-forced carbonate cycles in the Matuyama Chron, to one of dominantly eccentricity-forced cycles in the Brunhes, clearly parallels the well-documented, 'mid-Pleistocene transition' observed in Neogene oxygen isotope, SST and percent-carbonate records (e.g. Pisias & Moore 1981; Ruddiman et al. 1986, 1989; Ruddiman & Raymo 1988).

Discussion

Core-logging and correlation

As a tool for routine logging of whole-round sections as they arrive on deck, WCMS is analogous to whole-core geophysical logging tools like the GRAPE or P-wave logger, and to downhole, wireline logging systems like spontaneous potential, resistivity, sonic, gamma ray, etc. As a lithological indicator, WCMS is most directly analogous to gamma ray, spectral gamma ray, or gamma spectroscopy wireline logging tools. Accordingly, it may be useful for lithostratigraphic purposes, to routinely compare or 'cross reference' the results of WCMS measurements with data from the GRAPE and P-wave logger (e.g. Ciesieleski et al. 1988,

p. 190; Shipboard Scientific Party Leg 121 1989; Tarduno et al. 1991). The comparison could be made stratigraphically in the depth or frequency domains (the latter by cross-spectral analysis), or by cross-plotting the data as bivariate scattergrams, with the potential for lithological discrimination analogous, for example, to that of Schlumberger Log Interpretation Charts (e.g. Schlumberger 1977). One important advantage which the Bartington MS system has over the GRAPE and P-wave logging tools, however, is that of portability. The Bartington MS meter and interchangeable whole-core loop, hand-held probe, and single-sample sensors are light, compact and battery operated, fitting easily into aircraft hand-luggage for example. This facilitates MS measurements, using the same set of equipment, in the laboratory, in the field, on board ship, or in core repositories around the world.

Using the now fully-automated WCMS logging system on board JOIDES Resolution, in conjunction with a simple xy-graph plotting program, it is possible to generate WCMS profiles on a core-by-core, or even section-by-section basis, for immediate inspection by shipboard sedimentologists and stratigraphers, prior to the cores being split open for lithological examination and sampling. The potential value of this approach may be illustrated with reference to Fig. 3, above, which shows a WCMS profile of Core 115-711A-1H plotted alongside the lithological log and carbonate content record of this core. The WCMS profile shown in Fig. 3 is based on measurements made by passing whole-core sections, 3 cm at a time, through a WCMS loop-type sensor, which required approximately 40 minutes to complete. In contrast, a period of over 40 man-hours was required to measure (gasometrically) the carbonate content of samples taken at 5 cm intervals from this core, upon which its carbonate content profile (Fig. 3, left) is based.

The above comparison emphasises the great speed and simplicity with which WCMS logging can provide valuable, lithological information even before cores are split open for description and sampling. This attribute is particularly important in geochemical studies, where whole-core sections often have to remain intact and sealed, in order to prevent oxidation of the sediment consequent on its exposure to the atmosphere. It should also be remembered that, unlike most methods of carbonate content determination, WCMS measurements are totally non-destructive, and do not disturb the physical, chemical, mineralogical or magnetic properties of the sediment in any way.

The combined attributes of speed, simplicity, and non-destructiveness make WCMS logging an ideal tool for high-resolution correlation between offset cores recovered from adjacent holes at ODP sites. It is essential that interhole correlation is effected with the highest possible stratigraphic resolution if composite sequences from individual ODP sites are to be used, for example, in studies of Milankovitch-band cyclicity in palaeoceanographic records, based on spectral analysis of downcore data. In this context, the use of Shaw Diagrams, constructed by pattern-matching the WCMS profiles of subsidiary holes (e.g., Figs 7, 9, 15 and 18), provides an effective technique for reconstructing composite sequences, and for extrapolating magneto- or biostratigraphic data recognized in one hole, to the sequences recovered from adjacent holes at the same or nearby sites. For example, the Shaw Diagram for Site 710 (Fig. 18) may be used to extrapolate the nine Quaternary nannofossil datum events identified in Hole 710B (Okada 1990), in which a major repeated sequence occurs (see Fig. 17), to the relatively uninterrupted sequence recovered from Hole 710A.

Coring irregularities

Possible mechanisms which may explain how successive coring runs may recover the same sequence, thus resulting in the situation which occurred at the top of Hole 710B, are discussed in detail by Ruddiman et al. (1987). In essence, unlike repenetration during a single coring run, generally caused by inadequate heave compensation during the passage of an abnormally large wave (i.e vertical displacement of the ship), the recoring of the same sequence by successive cores is probably caused by the ship suddenly drifting off station between coring runs (i.e. lateral displacement), usually during severe weather and/or sea conditions. This causes the entire drill string and bottom-hole assembly to be lifted above the base of the hole without drilling platform instruments recording any change in the length of string advanced. As the drill bit is then 'washed down' in readiness to core the next interval in the sequence, it actually descends only to a point approximately equivalent to the top of the previously cored interval. If side-wall coring then occurs, the next interval of the sequence recovered will essentially repeat the stratigraphic record captured by the previous core. Naturally, if the ship subsequently returns, equally abruptly, to its original position above the acoustic beacon, then the drill bit may be 'washed down' further than intended (or recorded) between coring runs, thus resulting in an 'unrecorded interval of no recovery' between successive cores.

Orbital forcing of WCMS variations

Figure 20 illustrated the palaeoclimatic control, or at least the orbital forcing of WCMS variations of deep-sea sediments from the equatorial Indian Oceans (Figs 3 and 10). This example demonstrated the application for WCMS logging as a proxy palaeoclimatic indicator for Quaternary deep-sea sediments which consist of carbonate *dissolution* cycles related to orbitally forced oscillations in the regional CCD. Other recent studies (e.g. Robinson 1986a, 1990; Bloemendal et al. 1989; Clemens & Prell 1991; deMenocal et al. 1991; Tarduno et al. 1991) have also demonstrated the value of WCMS logging in revealing the palaeoclimatic record of Quaternary deep-sea sediments consisting of carbonate *dilution* cycles. All these examples also emphasise the potential value of WCMS logging as a lithologically related reconnaissance tool in studies which search for Milankovitch-type cyclicity in the geological record beyond that of the Plio-Pleistocene epochs (e.g. Weedon 1989, in press; Weedon & Jenkyns 1990; Weedon et al. in press). In this context, the high-amplitude, high-frequency oscillations in WCMS (proxy carbonate) within the condensed, early and middle Miocene sequences of several Leg 115 ODP holes (e.g. Fig. 5) suggests the possibility of orbital influences on carbonate accumulation rates in the tropical Indian Ocean during certain intervals before the onset of Northern Hemisphere glaciations in the late Pliocene.

Peterson & Prell (1985b) considered various mechanisms which have been described to explain how palaeoclimatic, and/or orbital variations could control carbonate dissolution in the tropical oceans during the Quaternary. These included: (1) global, i.e. continent–ocean exchange, organic-carbon inventory models (e.g. Shackleton 1977); (2) shelf-basin exchange, organic-carbon inventory models (e.g. Broecker 1981, 1982); (3) shelf-basin exchange, carbonate-carbon inventory models (e.g. Berger & Winterer 1974; Hay & Southam 1977); and (4) oceanic deep-water circulation models (e.g. Streeter & Shackleton, 1979; Volat et al. 1980; Curry & Lohmann 1982). Of these, Peterson & Prell favoured a deep-water circulation model to account for the orbitally-forced cycles of carbonate dissolution which they identified in their eastern equatorial Indian Ocean cores (see Fig. 1). In addition to the models considered by Peterson and Prell, carbonate accumulation rates further west in the equatorial Indian Ocean (e.g. Site 711) may possibly be influenced by orbitally-forced variations in monsoon-driven upwelling in the Somali Basin and western Arabian Sea (e.g. Prell 1984; Robinson 1990). Such variations could influence the circulation patterns and rates of carbonate productivity in surface waters of adjacent areas, leading to changes in carbonate preservation patterns at depth (i.e. the CCD and lysocline depth). Curry *et al.* (1990) also favoured a productivity-driven model to explain variations in carbonate accumulation at Site 711. They argued that orbitally-forced, global changes in surface productivity of the tropical oceans, with local terrigenous input of Ca^{2+} and CO_3^- held constant, led to variations in the level of carbonate saturation at depth. The effects of eustatic sea-level fluctuations on the area of continental shelves in the NW Indian Ocean, which affects the flux of carbonate to the deep ocean and thus its level of saturation, may also influence carbonate dissolution patterns in this region (cf. Hay & Southam 1977).

Summary

The low-field (<1 Oe), volume magnetic susceptibility (κ or MS) of a substance is a constant of proportionality which relates the intensity of magnetization (M) induced in the substance, by an applied magnetic field (H), to the strength of that field, i.e. $\kappa = M/H$. Magnetic susceptibility is therefore totally independent of the Earth's magnetic field, and is, in effect, simply a measure of the amount of magnetizable material the substance contains.

Most of the magnetizable constituents in deep-sea sediments reside in the lithogenic fraction of the sediment (with the exception of bacterial magnetofossils, if present), and not only include magnetic iron oxides (e.g. magnetite, hematite, etc.), but also all compounds containing Fe^{2+}, Fe^{3+}, and Mn^{2+} ions. These so-called paramagnetic substances include clay minerals, rock-forming ferromagnesian silicate minerals, and authigenic ferromanganese oxyhydroxide mineraloids (e.g. limonite, todorokite) often present in the pore waters of deep-sea sediments or as nodules, micronodules, etc. Biogenic carbonate and silica, in contast, are diamagnetic, which means they exhibit weak, negative MS values. Accordingly, downcore variations in the MS of deep-sea sediments often reflect changes in their lithological composition, i.e. fluctuations in the ratio of biogenic to lithogenic components in the sediment (often the same as percent-carbonate).

Whole-core (WC) measurements of volume MS therefore offer an extremely simple, rapid, and non-destructive technique for quantitatively logging variations in a lithologically-related signal in deep-sea sediment cores immediately after they arrive on deck, prior to them being split open for lithological description and sampling.

WCMS logging thus also provides a tool for ultra-high resolution lithostratigraphic correlation between subsidiary holes drilled at ODP sites. This is particularly useful when composite, stratigraphically continuous sequences for individual ODP sites are to be reconstructed by dovetailing the uninterrupted records obtained from the middle sections of offset cores recovered from subsidiary holes.

An obvious corollary to this application is the use of WCMS logging to identify stratigraphic discontinuities and repeated sequences resulting from coring irregularities, which are generally caused by abrupt, vertical or lateral displacements of the drilling vessel during severe weather or sea conditions.

Another form of disturbance which sometimes affects HPC cores, is contamination by metallic artifacts like fragments of pipe rust, etc., or mechanically reworked material containing such artifacts. Naturally, ferrous metal contaminants are strongly magnetic, and therefore may easily be identified by WCMS logging, allowing intervals contaminated in this way to be avoided during sampling, and eliminated from stratigraphic reconstructions and accumulation rate estimates.

In addition to between-hole correlations, WCMS profiles can be used to correlate between holes at different sites (i.e. between sites) when changes in the lithological composition of the sediments involved are controlled by regional, or even larger scale variations in oceanographic variables (e.g. CCD, sea-level, upwelling, productivity, bottom currents, etc.), or detrital sediment source and flux-density.

An interval of greenish-gray coloured, reduced sediment, of late Miocene to late Pliocene or early Pleistocene age, was identified in a number of holes drilled during Leg 115. The WCMS profiles of these holes indicated that within the interval of greenish-gray sediment, suboxic diagenesis had depleted the iron oxide/oxyhydroxide minerals originally present, thus degrading the WCMS signal in this interval, relative to that of 'normally' oxidised sediments in the same sequence.

The high proportion of primary titanomagnetite normally present in unaltered, basic volcanic ash, or secondary pyrite (after titanomagnetite) in degraded volcanic ash, provides the basis for

WCMS to be used as a correlation tool in studies of tephrostratigraphy ('tephrochronology'). An example of this application is provided by ODP Site 706 (0–30 mbsf), where 127 points of correlation between subsidiary holes were identified by cross-matching the WCMS profiles of each hole. Virtually all of the peaks in the WCMS profiles of these holes corresponded to volcanic ash horizons in varying stages of alteration.

It is well established that lithological variations of deep-sea sediments are frequently modulated by climatic fluctuations and/or orbital forcing (e.g. Dean & Gardner 1986) via the mechanisms of carbonate dissolution, dilution, and/or productivity variations (Volat *et al.* 1980). For sediments such as these, WCMS logging not only provides a means of long distance, inter-regional scale correlation, but may also be used as a proxy palaeoclimatic indicator, duely assisting the task of climato-stratigraphic zonation by more costly, time consuming and laborious microfossil and isotopic techniques. At ODP Site 711, Quaternary carbonate and WCMS cycles are dissolution controlled, and exhibit spectral periods corresponding to all of the primary Milankovitch responses, with a clear transition in dominance from 41 ka, obliquity-forced cycles in the Matuyama Chron, to 100 ka, eccentricity-forced cycles thereafter. Site 711 WCMS spectra also contain significant power at non-Milankovitch periods, however, e.g. at 78 ka in the spectrum for the Matuyama interval, and at 71 ka in that for the Brunhes.

I am extremely grateful to the following members of the scientific drilling party of ODP Leg 115 for sharing with me the burden of WCMS measurements: J. Backman, R. Duncan, M. Fisk, J. Greenough, R. Hargreaves, A. MacDonald, J. Tauxe, and J. Weisbruch. I am also very grateful to N. Shackleton for oxygen-isotope measurements of Core 115-711A-1H and, along with J. Backman, for giving me the opportunity to participate in ODP Leg 115. A special note of thanks is due to J. Bloemendal for his very careful reading, constructive criticism and helpful suggestions for improving the content of the original manuscript of this paper. This work was funded by the UK Natural Environment Research Council (Grant Ref. GST/02/260).

References

AMIN, B. S., LIKHITE, S. D., RADHAKRISHNAMURTY, C. & SOMAYAJULU, B. L. K. 1972. Susceptibility stratigraphy and palaeomagnetism of some deep Pacific Ocean cores. *Deep Sea Research*, **19**, 249–252.

BACKMAN, J., DUNCAN, R. A., *et al.* 1988. *Proceedings of the Ocean Drilling Program, Initial Reports*, **115**, College Station, TX (Ocean Drilling Program), 1085.

BASCOMB, C. L. 1974. Physical and chemical analyses of <2 mm samples. *In*: AVERY, B. W. & BASCOMB, C. L. (eds) *Soil Survey Laboratory Methods*. Soil Survey Technical Monograph No. 6., Rothamsted Experimental Station, Harpenden, Herts., UK, 14–41.

BERGER, W. H. 1973. Deep-sea carbonates: Pleistocene dissolution cycles. *Journal of Foraminiferal Research*, **3**, 187–195.

—— & WINTERER, E. L. 1974. Plate stratigraphy and the fluctuating carbonate line. *In*: Hsü, K. J. & JENKINS, H. (eds) *Pelagic Sediments on Land and Under the Sea*. Special Publication of the International Association of Sedimentologists, Vol. 1, Blackwell, Oxford, 11–48.

BLOEMENDAL, J. 1983. Paleoenvironmental implications of the magnetic characteristics of sediments from Deep-Sea Drilling Project Site 514, Southeast Argentine Basin. *In*: LUDWIG, W. J., KRASHENINNIKOV, V. A., *et al. Initial Reports of the Deep-Sea Drilling Project*, **71**, Washington (U.S. Government Printing Office), 1097–1108.

—— & DEMENOCAL, P. 1989. Evidence for a change in the periodicity of tropical climate cycles at 2.4 Myr from whole-core magnetic susceptibility measurements. *Nature*, **342**, 897–900.

——, LAMB, B. & KING, J. 1988a. Paleoenvironmental implications of rock-magnetic properties of late Quaternary sediment cores from the eastern Equatorial Atlantic. *Paleoceanography*, **3**, 61–87.

——, TAUXE, L. & VALET, J-P. 1988b. High resolution whole-core magnetic susceptibility logs from Leg 108. *In*: RUDDIMAN, W. F., SARNTHEIN, M., BALDAUF, J., *et al. Proceedings of the Ocean Drilling Program, Initial Reports*, **108**, College Station, TX (Ocean Drilling Program).

——, KING, J. W., TAUXE, L. & VALET, J.-P. 1989. Rock magnetic stratigraphy of Leg 108 Sites 658, 659, 661, and 665, eastern tropical Atlantic. *In*: RUDDIMAN, W., SARNTHEIN, M., *et al. Proceedings of the Ocean Drilling Program, Scientific Results*, **108**, College Station, TX (Ocean Drilling Program), 415–428.

BOILLOT, G., WINTERER, E. L., MEYER, A. W., *et al.* 1987. *Proceedings of the Ocean Drilling Program, Initial Reports*, **103**, College Station, TX (Ocean Drilling Program).

BROECKER, W. S. 1971. Calcite accumulation rates and glacial to interglacial changes in oceanic mixing. *In*: TUREKIAN, K. K. (ed.) *The Late Cenozoic Glacial Ages*. New Haven, Yale University Press, 239–265.

—— 1981. Glacial to interglacial changes in ocean and atmosphere chemistry. *In*: BERGER, A. (ed.) *Climatic Variations and Variability: Facts and Theories*. Reidel, Hingham, Mass., 109–120.

—— 1982. Glacial to interglacial changes in ocean chemistry. *Progress in Oceanography*, **11**, 151–197.

BURNS, R. G. & BURNS, V. M. 1981. Authigenic

oxides. *In:* EMILIANI, C. (ed.) *The Oceanic Lithosphere (The Sea, Vol. 7).* John Wiley and Sons, New York, 875–914.

CANFIELD, D. E. & BERNER, R. A. 1987. Dissolution and pyritization of magnetite in anoxic marine sediments. *Geochimica et Cosmochimica Acta,* **51,** 645–659.

CHANG, S-B. B. & KIRSCHVINK, J. L. 1987. Biogenic magnetite as a primary remanence carrier in limestone deposits. *Physics of the Earth and Planetary Interiors,* **46,** 289–303.

CHUEY, J. M., REA, D. K. & PISIAS, N. G. 1987. Late Pleistocene paleoclimatology of the central Equatorial Pacific: a quantitative record of eolian and carbonate deposition. *Quaternary Research,* **28,** 323–339.

CIESIELSKI, P. F., KRISTOFFERSEN, Y., *et al.* 1988. *Proceedings of the Ocean Drilling Program, Initial Reports,* **114,** College Station TX (Ocean Drilling Program).

CLEMENS, S. C. & PRELL, W. L. 1991. One million year record of summer-monsoon winds and continental aridity from the Owen Ridge (Site 722), Northwest Arabian Sea. *In:* PRELL, W. L., NIITSUMA, N., *et al. Proceedings of the Ocean Drilling, Scientific Results,* **117,** College Station, Texas (Ocean Drilling Program), 365–388.

COCHRAN, J. R., STOW, D. A. V., *et al.* 1989. *Proceedings of the Ocean Drilling Program, Initial Reports,* **116,** College Station, TX (Ocean Drilling Program).

COLLINSON, D. W. 1983. *Methods in Rock Magnetism and Palaeomagnetism: Techniques and Instrumentation,* London (Chapman and Hall).

CROWLEY, T. J. 1985. Late Quaternary carbonate changes in the North Atlantic, and Atlantic–Pacific comparisons. *In:* SUNDQUIST, E. T. & BROECKER, W. S. (eds) *The Carbon Cycle and Atmospheric CO₂: Natural Variations, Archaen to Present.* American Geophysical Union, Washington D.C., Geophysical Monographs, **32,** 271–284.

CURRY, W. B. & LOHMANN, G. P. 1982. Carbon isotopic changes in benthic foraminifera from the western South Atlantic: reconstruction of glacial abyssal circulation patterns. *Quaternary Research,* **18,** 218–235.

——, CULLEN, J. L. & BACKMAN, J. 1990. Carbonate accumulation in the Indian Ocean during the Pliocene: evidence for a change in productivity and preservation at about 2.4 Ma. *In:* DUNCAN, R. A., BACKMAN, J., PETERSON, L. C., *et al. Proceedings of the Ocean Drilling Program, Scientific Results,* **115.** College Station, TX (Ocean Drilling Program), 509–518.

CURTIS, C. D. 1983. Microorganisms and the diagenesis of sediments. *In:* KRUMBEIN, W. E. (ed.) *Microbial Geochemistry,* Blackwell, Oxford, England, 263–286.

deMENOCAL, P., BLOEMENDAL, J. & KING, J. 1991. A rock-magnetic record of monsoonal dust deposition at 2.4 Ma. *In:* PRELL, W. L., NIITSUMA, N. *et al. Proceedings of the Ocean Drilling Program, Scientific Results,* **117,** College Station, TX (Ocean Drilling Program), 389–407.

DEAN, W. E. & GARDNER, J. V. 1986. Milankovitch cycles in Neogene deep-sea sediments. *Paleoceanography,* **1,** 539–553.

DEMITRACK, A. 1985. A search for bacterial magnetite in the sediments of Eel Marsh, Woods Hole, Massachusetts. *In:* KIRSCHVINK, J. L., JONES, D. S. & McFADDEN, B. J. (eds) *Magnetite Biomineralization and Magnetoreception in Organisms. A New Biomagnetism.* Plenum Press, New York, 625–645.

DREIMANIS, A. 1962. Quantitative gasometric determination of calcite and dolomite by using the Chittick Apparatus. *Journal of Sedimentary Petrology,* **32,** 520–529.

FARINA, M., ESQUIVEL, D. M. S. LINS DE BARROS, H. G. P. 1990. Magnetic iron-sulphur crystals from a magnetotactic microorganism. *Nature,* **343,** 256–258.

FREEMAN, R. 1986. Magnetic mineralogy of pelagic limestones. *Geophysical Journal of the Royal Astronomical Society,* **85,** 433–452.

FROELICH, P. N., KLINKHAMMER, G. P., BENDER, M. L., LUEDTKE, N. A., HEATH, G. R., CULLEN, D., DAUPHIN, P., HARTMAN, B., HAMMOND, D. & MAYNARD, V. 1979. Early oxidation of organic matter in pelagic sediments of the eastern Equatorial Atlantic: suboxic diagenesis. *Geochimica et Cosmochimica Acta,* **43,** 1075–1090.

GHIL, M. 1987. Nonlinear phenomena in climate dynamics. *In:* NICOLIS, C. & NICOLIS, G. (eds) *Irreversible Phenomena and Dynamical Systems Analysis in Geosciences,* NATO ASI Series C, **194,** D. Reidel Publ. Co., 313–320.

HALL, F. R., BLOEMENDAL, J., KING, J. W., ARTHUR, M. A. & AKSU, A. E. 1989. Middle to late Quaternary sediment fluxes in the Labrador Sea, ODP Leg 105, Site 646: a synthesis of rock-magnetic, oxygen-isotopic, carbonate, and planktonic foraminiferal data. *In:* SRIVASTAVA, S. P., ARTHUR, M., CLEMENT, B., *et al. Proceedings of the Ocean Drilling Program, Scientific Results,* **105,** College Station, TX (Ocean Drilling Program), 653–665.

HAY, W. W. & SOUTHAM, J. R. 1977. Modulation of marine sedimentation by the continental shelves. *In:* ANDERSEN, N. R. & MALAHOFF, A. (eds) *The Fate of Fossil Fuel CO₂ in the Oceans.* Plenum Press, New York, 569–604.

HAYS, J. D., LOZANO, J. A., SHACKLETON, N. J. & IRVING, G. 1976. Reconstruction of the Atlantic and western Indian Ocean sectors of the 18,000 yr. B.P. Antarctic Ocean *Geological Society of American Memoir,* **145,** 337–372.

IMBRIE, J. 1985. A theoretical framework for the Pleistocene ice ages. *Journal of the Geological Society, London,* **142,** 417–432.

—— & IMBRIE, J.' Z. 1980. Modelling the climatic response to orbital variations. *Science,* **207,** 943–953.

——, HAYS, J. D., MARTINSON, D. G., McINTYRE, A., MIX, A. C., MORLEY, J. J., PISIAS, N. G., PRELL, W. L. & SHACKLETON, N. J. 1984. The orbital theory of Pleistocene climate: support from a revised chronology of the marine ¹⁸O record.

In: BERGER, A. L., IMBRIE, J., HAYS, J., KUKLA, G. & SALTZMAN, B. (eds) *Milankovitch and Part 1*. D. Reidel, Dordrecht, 269–305.

JENKINS, D. G., BOWEN, D. Q., ADAMS, C. G., SHACKLETON, N. J. & BRASSELL, S. C. 1985. The Neogene: part 1. *In*: SNELLING, N. J. (ed.) *The Chronology of the Geological Record*. Geological Society Memoir, **10**, 199–210.

JENKINS, G. M. & WATTS, D. G. 1968. *Spectral Analysis and its Applications*. Holden Day, San Francisco.

KARLIN, G. M., LYLE, M. & HEATH, G. R. 1987. Authigenic magnetite formation in suboxic marine sediments, *Nature*, **326**, 490–493.

KARLIN, R. & LEVI, S. 1983. Diagenesis of magnetic minerals in recent hemipelagic sediments. *Nature*, **303**, 327–330.

—— & —— 1985. Geochemical and sedimentological control of the magnetic properties of hemipelagic sediments. *Journal of Geophysical Research*, **90**, 10, 373–10, 392.

KENNETT, J. P. 1981. Marine tephrochronology. *In*: EMILIANI, C. (ed.) *The Oceanic Lithosphere (The Sea, Vol. 7)*. John Wiley and Sons, New York, 1373–1436.

KENT, D. V. 1982. Apparent correlation of paleomagnetic intensity and climatic records in deepsea sediments, *Nature*, **299**, 538–539.

KIRSCHVINK, J. L. & LOWENSTAM, H. A. 1979. Mineralization and magnetization of Chiton teeth: palaeomagnetic, sedimentologic and biologic implications of organic magnetite. *Earth and Planetary Science Letters*, **44**, 193–204.

LANCASTER, D. E. 1966. Electronic metal detection. *Electronics World*, (Dec.), 39–62.

LOWRIE, W. & HELLER, F. 1982. Magnetic properties of marine limestones. *Reviews of Geophysics and Space Physics*, **20**, 171–192.

LYLE, M. 1983. The brown-green colour transition in marine sediments: a marker of the FeIII-FeII redox boundary. *Limnology and Oceanography*, **28**, 1026–1033.

MANN, S., SPARKS, N. H. C., FRANKEL, R. B., BAZYLINSKI, D. A. & JANNASCH, H. W. 1990. Biomineralization of ferrimagnetic greigite (Fe_3S_4) and iron pyrite (FeS_2) in a magnetotactic bacterium, *Nature*, **343**, 258–261.

MASCLE, A., MOORE, J. C., *et al.* 1988. *Proceedings of the Ocean Drilling Program, Initial Reports*, **110**, College Station TX (Ocean Drilling Program).

MEAD, G. A., TAUXE, L. & LABRECQUE, J. L. 1986. Oligocene paleoceanography of the South Atlantic: paleoclimatic implications of sediment accumulation rates and magnetic susceptibility measurements. *Paleoceanography*, **1**, 273–284.

MOLYNEUX, L. & THOMPSON, R. 1973. Rapid measurement of the magnetic susceptibility of long cores of sediment. *Geophysical Journal of the Royal Astronomical Society*, **32**, 479–481.

MORLEY, J. J. & HAYS, J. D. 1978. Spectral analysis of climatic records from the subtropical South Atlantic. *Geological Society of America Abstracts with Programs*, **10**, 460.

MURRAY, J. W. 1979. Iron oxides. *In*: BURNS, R. G.

(ed.) *Marine Minerals: Reviews in Mineralogy*. Mineralogical Society of America Publication No. 6, 47–98.

MURRAY, W. M. & PRELL, W. L. 1991. Pliocene to Pleistocene variations in calcium carbonate, organic carbon, and opal on the Owen Ridge, northern Arabian Sea. *In*: PRELL, W. L., NIITSUMA, N., *et al. Proceedings of the Ocean Drilling Program, Scientific Results*, **117**, College Station, TX (Ocean Drilling Program), 343–355.

NESSON, M. H. & LOWENSTAM, H. A. 1985. Biomineralization processes of the radula teeth of Chitons. *In*: KIRSCHVINK, J. L., JONES, D. S. & McFADDEN, B. J. (eds) *Magnetite Biomineralization and Magnetoreception in Organisms. A New Biomagnetism*. Plenum Press, New York, 333–346.

NESTEROFF, W. D. 1982. The origin of the ferromanganese coatings of deep-sea rocks in the Atlantic Ocean. *In*: SCRUTTON, R. A. & TULWAIN, M. (ed.) *The Ocean Floor*. John Wiley and Sons, New York, 129–146.

OKADA, H. 1990. Quaternary and Paleogene calcareous nannofossils, ODP Leg 115. *In*: BACKMAN, J., DUNCAN, R. A., *et al. Proceedings of the Ocean Drilling Program, Scientific Results*, **115**, College Station, TX (Ocean Drilling Program), 129–174.

OLAUSSON, E. 1967. Climatological, geoeconomical, and paleoceanographical aspects of carbonate deposition. *Progress in Oceanography*, **4**, 245–265.

OLDFIELD, F. & ROBINSON, S. G. 1985. Geomagnetism and palaeoclimate. *In*: TOOLEY, M. J. & SHEAIL, G. M. (eds) *The Climatic Scene*. George Allen and Unwin, London, 186–205.

——, DEARING, J. A., THOMPSON, R. & GARRETJONES, S. E. 1978. Some magnetic properties of lake sediments and their possible links with erosion rates. *Polskie Archiwum Hydrobiologii*, **25**, 321–331.

PEIRCE, J., WEISSEL, J., *et al.* 1989. *Proceedings of the Ocean Drilling Program, Initial Reports*, **121**, College Station, TX (Ocean Drilling Program).

PETERSON, N., VON DOBENECK, T. & VALI, H. 1986. Fossil bacterial magnetite in deep-sea sediments from the South Atlantic Ocean. *Nature*, **320**, 611–615.

PETERSON & BACKMAN, J. 1990. Late Cenozoic carbonate accumulation and the history of the carbonate compensation depth in the western equatorial Indian Ocean. *In*: DUNCAN, R. A., BACKMAN, J., PETERSON, L. C., *et al. Proceedings of the Ocean Drilling Program, Scientific Results*, **115**, College Station, TX (Ocean Drilling Program), 467–507.

—— & PRELL, W. L. 1985a. Carbonate dissolution in recent sediments of the eastern Equatorial Indian Ocean: preservation patterns and carbonate loss above the lysocline. *Marine Geology*, **64**, 259–290.

—— & —— 1985b. Carbonate preservation and rates of climatic change: an 800 kyr record from the Indian Ocean. *In*: SUNDQUIST, E. T. & BROECKER,

W. S. (eds) *The Carbon Cycle and Atmospheric CO₂: Natural Variations, Archaen to Present.* American Geophysical Union, Washington, D.C. Geophysical Monograph **32**, 251–269.

PISIAS, N. G. & MOORE, T. C. JR. 1981. The evolution of Pleistocene climate: a time series approach. *Earth and Planetary Science Letters,* **52**, 450–458.

PRELL, W. L. 1984. Monsoonal climate of the Arabian Sea during the late Quaternary: a response to changing solar radiation. *In:* BERGER, A. L., IMBRIE, J., HAYS, J., KUKLA, G. & SALTZMAN, B. (eds) *Milankovitch and Climate.* D. Reidel, Dordrecht, 349–366.

——, IMBRIE, J., MARTINSON, D. G., MORLEY, J. J., PISIAS, N. G., SHACKLETON, N. J. & STREETER, H. F. 1989. Graphic correlation of oxygen isotope stratigraphy: application to the late Quaternary. *Paleoceanography,* **1**, 137–162.

——, NIITSUMA, N., *et al.* 1989. *Proceedings of the Ocean Drilling Program, Initial Reports,* **117**, College Station, TX (Ocean Drilling Program).

RADHAKRISHNAMURTY, C., LIKHITE, S. D., AMIN, B. S. & SOMAYAJULU, B. L. K. 1968. Magnetic susceptibility stratigraphy in ocean sediment cores. *Earth Planetary Science Letters,* **4**, 464–468.

RAMSAY, A. T. S. 1974. The distribution of calcium carbonate in deep-sea sediments. *In:* HAY, W. W. (ed.) *Studies in Paleoceanography.* Society of Economic Paleontologists and Mineralogists. Special Publication, **20**, 58–76.

RANGIN, C., SILVER, E. A., VON BREYMANN, M. T., *et al.* 1990. *Proceedings of the Ocean Drilling Program, Initial Reports,* **124**, College Station, TX (Ocean Drilling Program).

REA, D. K., CHAMBERS, L. W., CHUEY, J. M., JANECEK, T. R., LEINEN, M. & PISIAS, N. G. 1986. A 420,000-year record of cyclicity in oceanic and atmospheric processes from the eastern Equatorial Pacific. *Paleoceanography,* **1**, 577–586.

ROBINSON, S. G. 1982. Two applications of mineral-magnetic techniques to deep-sea sediment studies. *Geophysical Journal of the Royal Astronomical Society,* **69**, 294.

—— 1986a. The late Pleistocene palaeoclimatic record of North Atlantic deep-sea sediments revealed by mineral-magnetic measurements. *Physics of the Earth and Planetary Interiors,* **42**, 22–47.

—— 1986b. *Mineral magnetism of deep-sea sediments: palaeoclimatic implications.* PhD thesis, University of Liverpool, UK.

—— 1990. Applications for whole-core magnetic susceptibility measurements of deep-sea sediments: ODP Leg 115 results. *In:* DUNCAN, R. A., BACKMAN, J., PETERSON, L. C., *et al. Proceedings of the Ocean Drilling Program, Scientific Results,* **115**, College Station, TX (Ocean Drilling Program), 737–771.

—— & BLOEMENDAL, J. 1983. The application of mineral-magnetic measurements to deep-sea sediments: some preliminary results. *Transactions of the American Geophysical Union,* **64**, 241.

RUDDIMAN, W. F. & RAYMO, M. E. 1988. Northern hemisphere climate regimes during the past 3 Ma:

possible tectonic connections. *Philosophical Transactions of the Royal Society of London,* **B318**, 411–430.

——, —— & McINTYRE, A. 1986. Matuyama 41,000-year cycles: North Atlantic Ocean and northern hemisphere ice sheets. *Earth and Planetary Science Letters,* **80**, 117–129.

——, CAMERON, D. & CLEMENT, B. M. 1987. Sediment disturbance and correlation of offset holes drilled with the hydraulic piston corer: Leg 94. *In:* RUDDIMAN, W. F., KIDD, R. B., THOMAS, E., *et al. Initial Reports of the Deep-Sea Drilling Project,* **94**, Washington (U.S. Government Printing Office), 615–634.

——, SARNTHEIN, M., BALDAUF, J., *et al.* 1988. *Proceedings of the Ocean Drilling Program, Initial Reports,* **108**, College Station, TX (Ocean Drilling Program).

——, RAYMO, M. E., MARTINSON, D. G., CLEMENT, B. M. & BACKMAN, J. 1989. Pleistocene evolution: northern hemisphere ice sheets and North Atlantic Ocean. *Paleoceanography,* **4**, 353–412.

SAGER, W. W. 1986. Magnetic susceptibility measurements of metal contaminants in ODP Leg 101 cores. *In:* AUSTIN, J. A. JR., SCHLAGER, W., PALMER, A. A., *et al. Proceedings of the Ocean Drilling Program, Initial Reports,* **101**, College Station (Ocean Drilling Program), TX, 39–45.

—— & HALL, S. A. 1990. Magnetic properties of black mud turbidites from ODP Leg 116, distal Bengal Fan, Indian Ocean. *In:* COCHRAN, J. R., STOW, D. A. V., *et al. Proceedings of the Ocean Drilling Program, Scientific Results,* **116**, College Station, TX (Ocean Drilling Program), 317–336.

SCHLUMBERGER, LTD 1977. Crossplots for porosity and lithology interpretation. *In: Schlumberger Log Interpretation Charts,* Schlumberger Limited, U.S.A., 20–39.

SCHNEIDER, D. A. & KENT, D. V. 1990. Paleomagnetism of Leg 115 sediments: implications for Neogene magnetostratigraphy and paleolatitude of the Reunion hotspot. *In:* DUNCAN, R. A., BACKMAN, J., PETERSON, L. C., *et al. Proceedings of the Ocean Drilling Program, Scientific Results,* **115**, College Station, TX (Ocean Drilling Program), 717–736.

SCLATER, J. G., ABBOTT, D. & THIEDE, J. 1977. Paleobathymetry and sediments of the Indian Ocean. *In:* HEIRTZLER, J. R. *et al.* (eds) *Indian Ocean Geology and Biostratigraphy,* Washington, D.C., American Geophysical Union, 25–59.

SHACKLETON, N. J. 1977. The oxygen isotope stratigraphic record of the late Pleistocene. *Philosophical Transactions of the Royal Society of London,* **B280**, 169–182.

SHAW, A. B. 1964. *Time in Stratigraphy.* McGraw-Hill, New York.

SHIPBOARD SCIENTIFIC PARTY. 1989. Broken Ridge summary. *In:* PEIRCE, J., WEISSEL, J., *et al. Proceedings of the Ocean Drilling Program, Initial Reports,* **121**, College Station, TX (Ocean Drilling Program).

SMITH, G. M. & BANERJEE, S. K. 1987. The dependence of weak-field susceptibility on applied magnetic field. *Physics of the Earth and Planetary Interiors*, **46**, 71–76.

SOMAYAJULU, B. L. K., RADHAKRISHNAMURTY, C. & WALSH, T. J. 1974. Susceptibility as a tool for studying magnetic stratigraphy of marine sediments. *In: Proceedings of the Workshop Conference on Late Cenozoic Magnetostratigraphy – Comparison With Bio-, Climato-, and Lithostratigraphic Zones*, Tokyo.

—, WALSH, T. J. & RADHAKRISHNAMURTY, C. 1975. Magnetic susceptibility stratigraphy of Pacific Pleistocene sediments. *Nature*, **253**, 616–617.

STREETER, S. S. & SHACKLETON, N. J. 1979. Paleocirculation of the deep North Atlantic: 150,000 year record of benthic foraminifera and oxygen-18. *Science*, **203**, 168–171.

TARDUNO, J. A., MAYER, L. A., MUSGRAVE, R., *et al.* 1991. High resolution whole-core magnetic susceptibility data from Leg 130, Ontong Java Plateau. *In*: KROENKE, L. W., BERGER, W. H., JANECEK, T. R., *et al. Proceedings of the Ocean Drilling Program, Initial Reports*, **130**, College Station, TX (Ocean Drilling Program), 541–548.

THOMPSON, R. 1973. Palaeolimnology and palaeomagnetism. *Nature*, **242**, 182–184.

—— & MORTON, D. J. 1979. Magnetic susceptibility and particle-size distribution in recent sediments of the Loch Lomond drainage basin, Scotland. *Journal of Sedimentary Petrology*, **49**, 801–812.

—— & OLDFIELD, F. 1986. *Environmental Magnetism*. Allen and Unwin, London.

—, BATTARBEE, R. W., O'SULLIVAN, P. E. & OLDFIELD, F. 1975. Magnetic susceptibility of lake sediments. *Limnology and Oceanography*, **20**, 687–698.

VAN ANDEL, T. H. 1975. Mesozoic/Cenozoic calcite compensation depth and the global distribution of calcareous sediments. *Earth and Planetary Science Letters*, **26**, 187–194.

VINCENT, E. 1974. Cenozoic planktonic biostratigraphy and paleoceanography of the tropical western Indian Ocean. *In*: FISHER, R. L., BUNCE, E. T., *et al. Initial Reports of the Deep-Sea Drilling Project*, **24**, Washington (U.S. Govt. Printing Office), 1111–1150.

—— 1977. Indian Ocean Neogene planktonic foraminiferal biostratigraphy and its paleoceanographic implications. *In*: HEIRTZLER, J. R., *et al.* (eds) *Indian Ocean Geology and Biostratigraphy*. Washington (Am. Geophys. Union), 469–584.

VOLAT, J-L., PASTOURET, L. & VERGNAUD-GRAZZINI, C. 1980. Dissolution and carbonate fluctuations in Pleistocene deep-sea cores: a review. *Marine Geology*, **34**, 1–28.

WEAVER, P. P. E. & SCHULTHEISS, P. J. 1983. Detection of repenetration and sediment disturbance in open-barrel gravity cores. *Journal of Sedimentary Petrology*, **53**, 649–678.

WEEDON, G. P. 1989. The detection and illustration of regular sedimentary cycles using Walsh power spectra and filtering, with examples from the Lias of Switzerland. *Journal of the Geological Society, London*, **146**, 133–144.

—— in press. The recognition and stratigraphic implications of orbital forcing of climate and sedimentary cycles. *Sedimentology Review*.

—— & JENKYNS, H. C. 1990. Regular and irregular climatic cycles and the Belemnite Marls (Pliensbachian, Lower Jurassic, Wessex Basin). *Journal of the Geological Society, London*, **147**, 915–918.

—, ROBINSON, S. G. & JENKYNS, H. C. In press. Magnetic susceptibility as a high-resolution logging tool for Lower Jurassic mudrocks. *Sedimentology*.

Magnetostratigraphic calibration of early Eocene depositional sequences in the southern North Sea Basin

JASON R. ALI[1], CHRISTOPHER KING[2] & ERNEST A. HAILWOOD[1]

[1] *Oceanography Department, University of Southampton, Southampton SO9 5NH, UK*
[2] *Paleo Services, Unit 15, Paramount Industrial Estate, Sandown Road, Watford WD2 4XA, UK*

Abstract: We present the results of a comprehensive magnetostratigraphic study of early Eocene (Ypresian) sedimentary sequences in the London Basin, Belgium, and the Varengeville outlier of northern France. A re-investigation of part of the lower Eocene section at Whitecliff Bay (Isle of Wight), in the Hampshire Basin, has also been carried out. The palaeomagnetic results from each area are used to refine existing litho- and biostratigraphic correlations. The magnetostratigraphy of the sedimentary sequence in each area is linked to the geomagnetic polarity time-scale using biostratigraphic data from calcareous nannoplankton and dinoflagellates.

The Ypresian deposits were laid down during the interval spanning geomagnetic Chrons C24BR to C22R. In the Hampshire Basin, deposition of Lutetian (middle Eocene) sediments extended into Chron C21R, but the apparent absence of a record of Chron C22N suggests an unconformity separating the lower and middle Eocene of between 0.6 and 2.0 Ma in duration, at the basin margin.

In this paper, recently acquired Ypresian stratigraphic data are integrated with previously published data, and a high-resolution 'event-stratigraphy' is developed for the Ypresian deposits of the southern margin of the North Sea Basin. A total of ten depositional sequences ('cycles'), probably reflecting eustatic sea-level changes, are calibrated against this framework. An attempt is made to match these sequences, which represent a combination of third and fourth order cycles in the Exxon Model, with published sea-level cycle charts for the Early Eocene.

The North Sea Basin

During the Early Eocene (Ypresian), an outer neritic to upper bathyal environment existed in the southern North Sea and the north western part of mainland Europe, from the eastern London 'Basin' to Denmark (Fig. 1). Shallower shelf (inner to mid-neritic) environments existed across SE England (including the western London 'Basin' and the Hampshire 'Basin') and the Paris Basin, extending through Belgium, the Netherlands and Germany. (N.B. The London and Hampshire 'Basins', in their present configuration, are synclinal areas, largely created by mid-Cenozoic folding and subsequent denudation; originally, Eocene sediments were probably continuous across southeast England; see Fig. 1).

Throughout the late Palaeocene and Eocene the position of the shoreline fluctuated as a result of relative sea level changes. The corresponding transgressive and regressive events can be identified in the sedimentary record. An important aim of our magnetostratigraphic research has been to tie these records of sea level fluctuations to the geomagnetic polarity reversal time scale, in order to establish the extent to which they represent synchronous basin-wide events, or diachronous events which occurred at different times in different parts of the basin.

Since geomagnetic polarity reversals represent globally synchronous events, their records define a sequence of 'absolute' time planes within the sedimentary sequences. These can be used to compare the timing of the transgressive and regressive events in different areas, both within the North Sea Basin itself and also between the North Sea Basin and other areas.

The basic lithostratigraphic subdivisions of the Ypresian deposits were established during the last century. The first correlation linking the sequences in the different parts of the southern North Sea area was presented by Prestwich (1855). These correlations were based on a combination of gross lithologic characteristics and macrofossil assemblages. With increasing knowledge of the biostratigraphy and lithostratigraphy of these sediments, further correlations utilising similar criteria were proposed (Wrigley & Davis 1937; Davis & Elliott 1957; Feugueur 1963; Curry 1965) but the diversity of lithofacies and biofacies across the area prevented a generally agreed correlation from being established.

From HAILWOOD, E. A. & KIDD, R. B. (eds), *High Resolution Stratigraphy*
Geological Society Special Publication, No. 70, pp. 99–125.

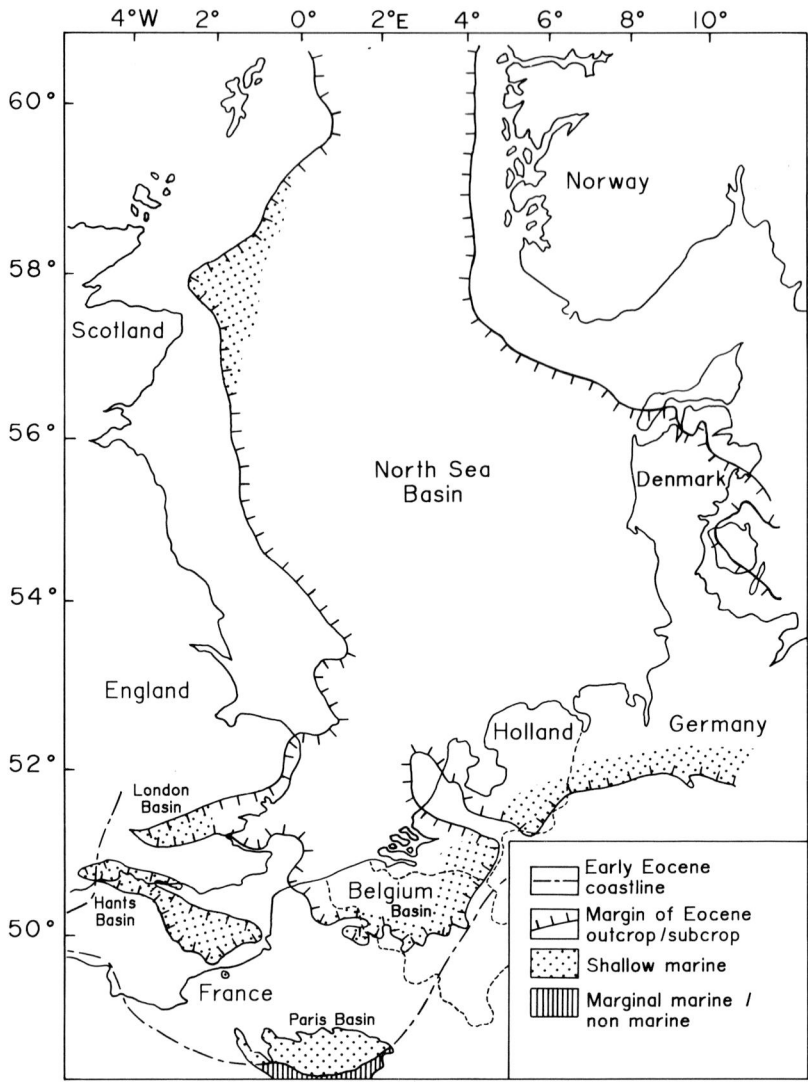

Fig. 1. Palaeogeographic map of the North Sea Basin area during early Eocene times.

Alternative methods of correlation, based on the recognition of depositional sequences, were first developed by Stamp (1921), but subsequently neglected until the 1970s. These correlations are reviewed by King (1981).

Since the mid-1960s, the use of planktonic microfossils (chiefly dinoflagellates and calcareous nannoplankton) has considerably enhanced biostratigraphical resolution in the Ypresian sediments of NW Europe. Refined biostratigraphic schemes have been developed over a period in which there has been a renewed application of sequence stratigraphy concepts to

these sediments. This has led recently to the recognition of nine principal depositional parasequences (approximately equivalent to third or fourth order parasequences in the 'eustatic' sea level curve of Haq *et al.* 1987). Each of the principal depositional sequences begins with a basal condensed unit deposited during a marine-flooding (transgressive) episode, represented by glauconitic clays and silts, with dispersed flint pebbles in the Hampshire Basin and western London Basin. This is followed by a coarsening-upwards sequence, typically beginning with silty clays and passing up to thin-bedded or laminated clays and sands. This was deposited as a result of coastal progradation during a period of relatively stable sea level. Nine such sequences were recognized within the London Clay Formation of the Hampshire and London Basins by King (1981). These were designated 'Divisions' A1 to E in ascending order (Column 5 of Fig. 2).

These 'Divisions' can now be classed as parasequences (as defined by Van Wagoner *et al.* 1990). Channelled units of cross-stratified (tidal channel?) sands which form the final unit of several 'Divisions' in the upper part of the London Clay Formation (e.g. the Whitecliff and Portsmouth Members) probably reflect episodes of relative sea level fall, with channelling of shelf sediments followed by filling of the channels during the initial phases of sea-level rise.

The coarsening-upwards trends within these 'Divisions' are fully developed only close to the palaeoshoreline, in the Hampshire Basin, the western London Basin, and (at equivalent levels) in parts of the Belgian Basin. In deeper water environments (e.g. those of the eastern London Basin and northwestern Belgium) the lithology is predominantly silty clay. Here, some of the condensed sections at the base of the 'Divisions' can be recognised lithologically, but the position of other parasequence boundaries can be identified only through biostratigraphic event-correlation.

Townsend & Hailwood (1985) and Aubry *et al.* (1986) presented the first magnetostratigraphic data from Ypresian sediments in NW Europe. The bulk of the data presented by these authors were from the well-exposed sections at Alum Bay and Whitecliff Bay, on the Isle of Wight (Hampshire Basin). In the present study, part of the Whitecliff Bay section has been

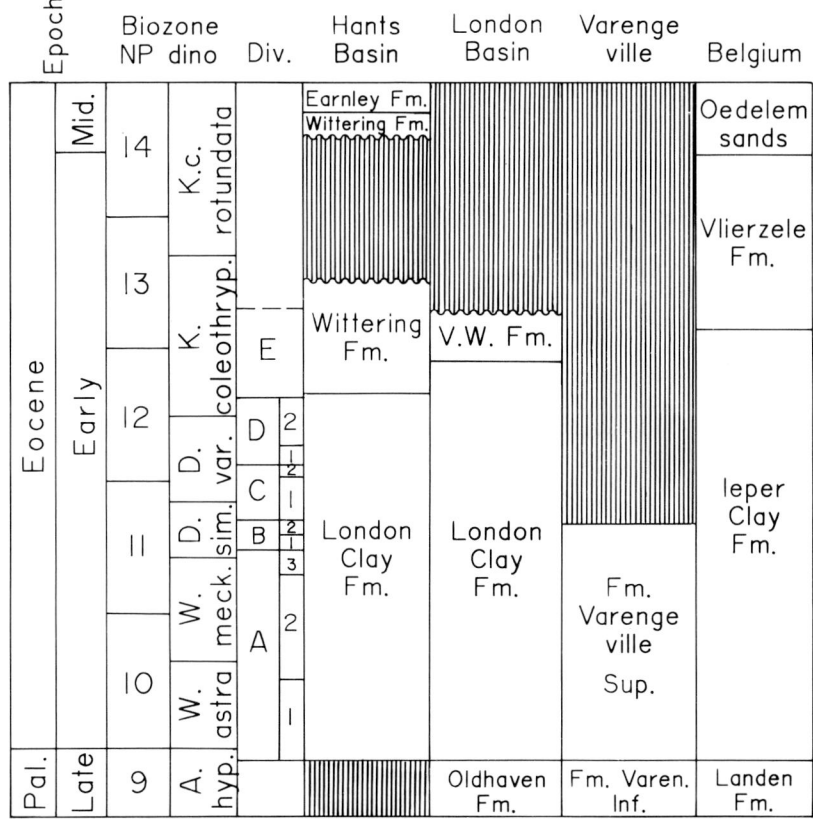

Fig. 2. Early Eocene (Ypresian) formations in the southern North Sea Basin area. Principal nannofossil (NP) and dinoflagellate cyst zones are shown. See text for discussion of position of Palaeocene/Eocene boundary relative to these zones. The successive depositional sequences within the Lonon Clay, represented by Divisions A, B, C etc of King (1981) are shown. 'V.M.Fm' is the Virginia Water Formation. Dinoflagellate cyst zonal names are abbreviated.

re-investigated, and studies have been extended to the London and Belgium Basins and to the Palaeogene outlier at Varengeville on the Channel coast of France. This has allowed the identification of both local and basinwide 'events' and, by using all available data, the placing of these events within a refined stratigraphic framework.

Stratigraphic setting

The start of the Eocene Epoch has generally been equated with the base of the Ypresian Stage. It is now generally accepted that the base of each geological stage or series be defined by a distinctive biostratigraphic marker within an uninterrupted sedimentary sequence. The base of nannoplankton zone NP10 (defined by the first appearance of *Marthasterites bramlettei*; Martini 1971) is used by most biostratigraphers (e.g. Berggren *et al*. 1985) as the biostratigraphic marker for the start of the Eocene but, unfortunately, this marker is missing from NW European sequences.

In the North Sea Basin calcareous nannoplankton and planktonic foraminifera are everywhere absent in Palaeocene/Eocene boundary sequences (reviewed by Berggren *et al*. 1985). Zone NP10 has not been identified. Correlation is based primarily on dinoflagellates, and several zonal schemes have been proposed (Costa & Downie 1976; Bujak *et al*. 1980; Knox *et al*. 1981; Powell, 1988). Costa & Müller (1978) correlated the base of zone NP10 with the first appearance of the short-ranging dinoflagellate *Wetzeliella astra* (Fig. 2) on the basis of data from DSDP Hole 117A, in the NE Atlantic. Although Morton *et al*. (1983) questioned this correlation, Berggren *et al*. (1985), in a major synthesis of Palaeogene geochronology, proposed that the base of the Eocene in the NW European sequences be calibrated to the *W. astra* marker. However, the validity of *W. astra* as a separate species has been questioned (Brown & Downie 1984), and the slightly younger event defined by the earliest occurrence of *Wetzeliella meckelfeldensis* is now regarded as more reliable for correlation.

On the basis of volcanic ash layer correlations between the North Sea Basin and the North Atlantic, Knox (1984) suggested that the base of NP10 corresponds to the base of the Sele Formation in the central North Sea, which has been correlated with the base of the *Apectodinium hyperacanthum* dinoflagellate zone (see Fig. 2).

More recent work (Powell 1988; Knox 1990) indicates that the base of nannofossil zone NP10 may in fact correspond to a level within the upper part of the *Apectodinium hyperacanthum* Zone, at the base of the *Deflandrea oebisfeldensis* Subzone (Subzone D5b of Costa & Manum 1989). In southern England, this event corresponds approximately to the base of the London Clay Formation (Jolley & Spinner 1990).

During the mid- and late Ypresian, the North Sea Basin had an open connection with the North Atlantic, as nannoplankton characteristic of zones NP11 through NP14 are present in the succession (though they are sometimes rare or of restricted occurrence). The nannoplankton biozonation complements the dinoflagellate scheme proposed by Costa *et al*. (1978) for this interval.

The stratigraphy of the localities studied in the present investigation is summarized in Fig. 2. This summary is based on Aubry (1986) for Varengeville, Edwards & Freshney (1987) and King (1981) for the Hampshire Basin, King (1981, 1984) for the London Basin and Steurbaut & Nolf (1986) and King (1991) for Belgium, together with our previously unpublished data for all these areas.

The lower Eocene of southern England

The London Clay Formation of King (1981) (Fig. 2) includes sediments previously referred to the London Clay, the Claygate Beds and (in the Hampshire Basin) several sand bodies previously referred to the 'Bagshot Sands'. The London Clay Formation is thickest in the eastern London Basin, where it attains 150 m. In the Hampshire Basin, it is about 135 m thick at Whitecliff Bay but thins to about 50 m in Dorset. It consists predominantly of silty clays, silts and very fine sands, with some units of medium to coarse cross-stratified sand. Deposition was in well-oxygenated, marine shelf to coastal environments at water depths less than 200 m. A sequence of six *Wetzeliellaceae* (dinoflagellate cyst) zones has been identified in the formation, ranging from the *Apectodinium hyperacanthum* Zone to the *Kisselovia coleothrypta* Zone (Fig. 2). The first appearance of *W. meckelfeldensis*, currently one of the more reliable earliest Eocene biostratigraphic events, occurs at about 5 m above the base of Division A2 in the eastern London Basin.

In the Hampshire Basin, the London Clay Formation is overlain by the Wittering Formation (Stinton 1975) (Fig. 2) whose stratotype section is at Whitecliff Bay on the Isle of Wight. Edwards & Freshney (1987) redefined the Wittering Formation to include Plint's (1983) beds WB2 to WB9 respectively. According to this definition, the lowest 11 m of the interval assigned by King (1981) to the Wittering Division of the Bracklesham Group at Whitecliff Bay is now included in the upper part of the London Clay Formation.

The Wittering Formation in the eastern Hampshire Basin, at Whitecliff Bay and Bracklesham Bay, can be broadly divided into three successive lithological units. The lower part consists of fine grained glauconitic sands and silty clays in thin bedded units, the middle part of greyish green glauconitic shelly sandy silts, and the upper part of brown laminated clays, including a 1 m thick lignite: the Whitecliff Bay Bed. The upper part of this middle unit is assigned to the lower part of nannoplankton zone NP13 (Aubry *et al.* 1986). Calacareous nannoplankton have not been found in the upper part of the Wittering Formation, and the next oldest nannoplankton-dated horizon (middle NP14) is in the lower part of the Earnley Formation (Aubry *et al.* 1986) (Fig. 2). Utilizing both the biostratigraphic and palaeomagnetic data from this part of the succession, Aubry *et al.* (1986) were not able to identify positively a stratigraphic interval of normal polarity magnetized sediments that might have been deposited during Chron C22N. The apparent absence of a record of Chron C22N from the succession was interpreted by these authors as indicating a hiatus, of between 0.6 and 2.0 Ma duration, within the Wittering Formation. They suggested that this hiatus lay at the level of the Whitecliff Bay Bed, at Whitecliff Bay.

The absence of any nannoplankton in the beds immediately below and above the Whitecliff Bay Bed means that the interpretation presented by Aubry *et al.* (1986) is open to question. The unconformity marked by the apparent absence of a record of Chron C22N could be positioned either at a level lower down in the Wittering Formation (in the upper part of Plint's (1983) Unit WB6, which is highly glauconitic and may represent a very condensed section), or at the parasequence boundary between the Wittering Formation and the overlying Earnley Sand Formation.

The Wittering Formation is typically 40–50 m thick in the Hampshire Basin, although individual beds can be traced only over short distances. The magnetostratigraphy of the overlying Earnley Sand Formation has been studied by Townsend & Hailwood (1985).

In the London Basin, the London Clay Formation is overlain by the Virginia Water Formation, a term introduced by King (1981), to replace the term 'Lower Bagshot Sands', originally used by Prestwich (1847). The main outcrops are in Surrey and Berkshire, with outliers in Kent and Essex. The Virginia Water Formation attains a thickness of 20–30 m in the eastern and central parts of the London Basin. The base of this formation is conformable with the top of the London Clay Formation in the London Basin. Lithologically it consists of cross-stratified sands and laminated sands with thin clay seams. Deposition was probably in an inner sub-littoral/marginal marine environment. King (1981) correlated the Virginia Water Formation with the middle part of the Wittering Formation in the Hampshire Basin (Fig. 2).

Palaeomagnetic methods

All sections were sampled for palaeomagnetic analysis at intervals typically between 0.7 and 1.0 m, with two separately orientated samples being taken from each stratigraphic level. In most cases the lithologies investigated are clays and silts, samples being obtained by the 'copper-tube' method of Townsend & Hailwood (1985).

For the majority of the specimens (>90%), the magnetization was measured using a 'Molspin' spinner magnetometer, similar to that described by Molyneux (1971). Specimens were incrementally demagnetised by the alternating field (a.f.) method, usually at steps of 5 milli Tesla, up to 30–40 mT, using either a 'Highfield' or a 'Molspin' tumbling-specimen system or a 3-axis stationary-specimen system. Most specimens with intensities less than 0.3 mA/m were analysed using a 'CCL' cryogenic magnetometer with an in-line stationary sample demagnetising system (Riddy & Hailwood, 1989). These specimens were demagnetized at 5 mT steps to a peak field of 35 mT. The purpose of the demagnetisation analyses was to investigate the stability of the remanent magnetism and to isolate stable components of magnetism (which are more likely to have been acquired at the time of deposition) from less stable components (acquired later in the rock's history).

Demagnetization characteristics and magnetic properties

The majority of the specimens have a natural remanent magnetisation (NRM) intensity of between 2 and 30 mA/m (milli Amperes per metre). Demagnetisation at the maximum applied field generally reduced the intensity to about 15 to 30% of the NRM value. Examples of the responses of typical samples from Sheppey and Belgium to a.f. demagnetization are shown in Fig. 3. Four main types of behaviour can be

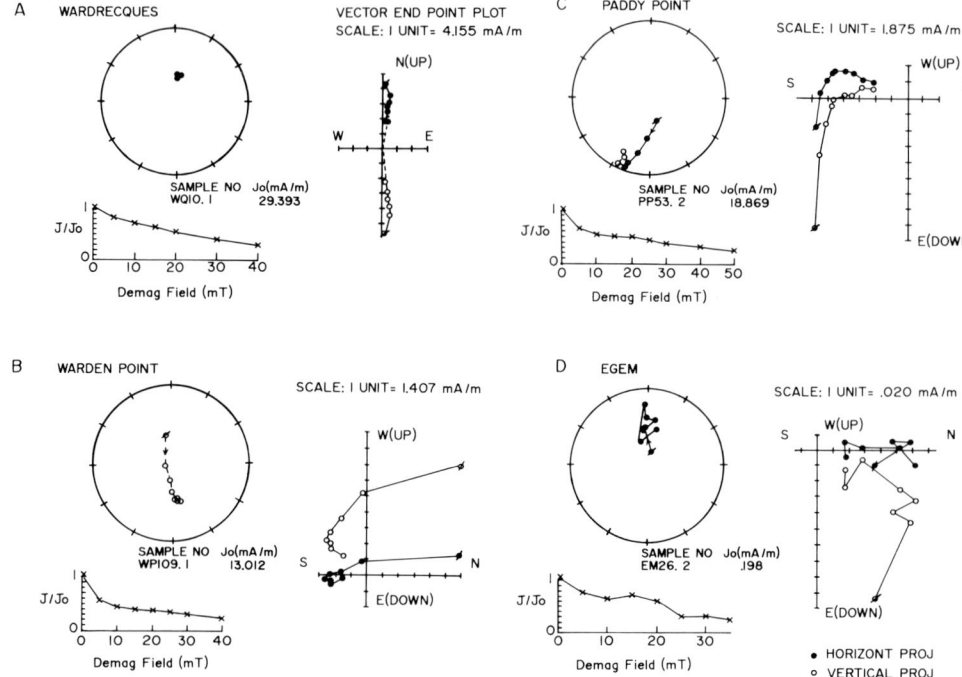

Fig. 3. Examples of responses of typical samples to alternating field (a.f.) demagnetization. In each case the magnetic vector after each demagnetization step (plotted on a stereographic projection) is shown, together with the normalized magnetic intensity (J/J_o) plot, on the left-hand side and the corresponding orthogonal plot (Zijderveld 1967) on the right-hand side. On the stereographic projections solid symbols represent positive inclinations, plotted in the lower hemisphere, and open symbols represent negative inclinations, plotted in the upper hemisphere. On the orthogonal plots solid symbols represent points on the horizontal plane and open symbols points on the vertical plane. In all cases the NRM direction is indicated by a slash through the symbol. (**A**) & (**B**) illustrate S-type (Stable End Point) behaviour. (**C**) illustrates 'T-type' behaviour (directional trend, but no SEP). (**D**) illustrates 'E-type' (erratic) behaviour.

identified. In the first, the direction of magnetization remains more-or-less unchanged while the magnetic intensity is progressively decreased to a low value (Fig. 3A) during incremental demagnetisation. After the first one or two demagnetising steps, the points on the corresponding orthogonal vector plots in both the horizontal and vertical planes fall on straight lines directed through the origin. This indicates the presence of only a single significant component of magnetisation in these samples (i.e. a demagnetization stable-end-point (SEP) is reached).

In the second type of behaviour the direction of magnetization undergoes a progressive change during demagnetization up to about 20 mT, after which a demagnetization stable-end-point is reached (Fig. 3B). Excluding the initial (NRM) vector, the points on the orthogonal vector plots commonly lie on two distinct linear segments, the first representing a low coercivity magnetic component which is re-

moved during demagnetization up to about 20 mT and the other, which is directed through the origin of the vector plot, the high coercivity component.

In samples showing these two types of behaviour the SEP usually has either a northerly declination and moderate to steep positive inclination, representing a *normal* magnetic polarity, or else a southerly declination and negative inclination, representing a *reverse* polarity. On the polarity logs (e.g. Fig. 6) samples showing this behaviour are coded 'S' (for stable-end-point). For occasional samples the SEP direction is neither clearly normal nor clearly reverse. In such cases it is coded 'I' (for intermediate). Such cases sometimes occur at polarity transition zones, between normal and reverse polarity magnetic zones.

In the third type of behaviour (Fig. 3C) the direction of magnetisation undergoes a systematic directional trend towards either a normal or a reverse polarity end-point, but no true SEP (represented by a linear segment directed through the origin on the orthogonal vector plots) is reached. In such cases the low coercivity component swamps the high coercivity component, which cannot then be completely isolated by the a.f. demagnetisation process. However, the polarity of the high stability component can usually be identified from the sense of the directional trend. Such samples are coded T (for 'trend') in the polarity logs.

In the fourth type of behaviour, directional changes during a.f. demagnetization are much more erratic and it is not possible to identify linear trends on the orthogonal vector plots. This is often a characteristic of very weakly magnetized samples, whose magnetization cannot be measured precisely. On the polarity logs such samples are coded 'E' (for erratic behaviour). In some cases the erratic directional changes are centred around a mean vector with an apparently normal or reverse polarity direction (e.g. Fig. 3D), so that a tentative polarity assignment can be made. However the reliability of such polarity determination will be much lower than for 'S' or 'T' type samples. In other cases the erratic changes are much larger, indicating a complete lack of magnetic stability, so that no polarity assignment is possible.

IRM analyses

To investigate the principal magnetic minerals carrying the magnetic remanence, representative specimens from each section were subjected to isothermal remanent magnetization (IRM) analysis. In these experiments the sample is incrementally magnetized, by the application of a progressively increased direct magnetic field, up to a maximum of 0.94 Tesla (T), the magnetic intensity being measured after each step. Two distinct types of behaviour were identified. In the first (Fig. 4A) the magnetisation saturates in an applied field of about 0.3 T. This is an indication of the presence of magnetite. In the second (Fig. 4B), the magnetic intensity continues to rise during treatment up to the maximum applied field used, i.e. saturation is not

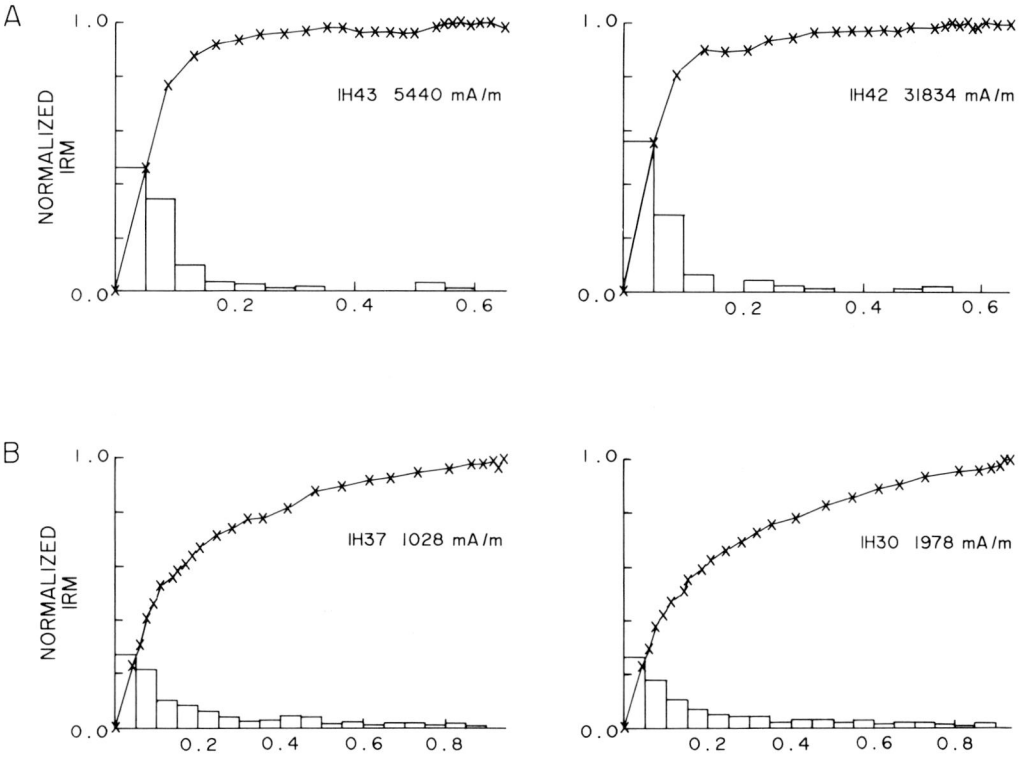

Fig. 4. Examples of IRM acquisition curves for (**A**) magnetite-bearing and (**B**) hematite-bearing sediments.

reached. This indicates the presence of iron oxide in a higher oxidation state, probably hematite. In the latter case the peak magnetic intensity is significantly weaker than in the first case (see values beneath the IRM curves in Fig. 4). A useful parameter for quantifying the shape of the IRM curve is the 'IRM ratio' (Ali 1989), which is the peak value (usually specified at 0.94 T) divided by the value at 0.3 T. For magnetite-bearing sediments (which are nearly-saturated at 0.3 T) this ratio is typically >0.9, whereas for hematite-bearing sediments it is significantly lower, typically about 0.7.

The IRM analysis indicates that the magnetization of the majority of specimens in this study is carried by magnetite. However, in several units it is carried by hematite and in some cases this can be attributed to surface weathering effects (see later discussions).

Magnetostratigraphic results from the lower Eocene of southern England

The well-exposed section of the upper London Clay Formation and Virginia Water Formation at Sheppey, and the exposures of the London Clay and Wittering Formation at Whitecliff Bay, provide the best opportunity to investigate the lower Eocene palaeomagnetic record in southern England, allowing a comparison of the

stratigraphy between the eastern part of the London Basin and the central part of the Hampshire Basin. The Sheppey sections investigated also include two boreholes which, together with the cliff section, span the entire London Clay Formation at this locality.

The palaeomagnetism of the section at Whitecliff Bay was originally studied by Townsend & Hailwood (1985). A re-examination of part of the upper London Clay Formation and the Wittering Formation has been carried out during the present study.

The Isle of Sheppey

The Isle of Sheppey is situated 70 km east of London, on the north Kent coast (Fig. 5). The world famous London Clay Formation exposures have yielded a rich and varied biota, and have been studied since the 18th Century. Important contributions to the stratigraphy of Sheppey include those of Prestwich (1847, 1854, 1888), Whitaker (1872), Davis (1936), King (1981, 1984) and Islam (1983).

The London Clay Formation at Sheppey is approximately 135 m thick, although only the upper 53.5 m is exposed. King (1984) published the first detailed account of the exposed portion. The unexposed part of the formation is known from a number of boreholes, notably the Sheer-

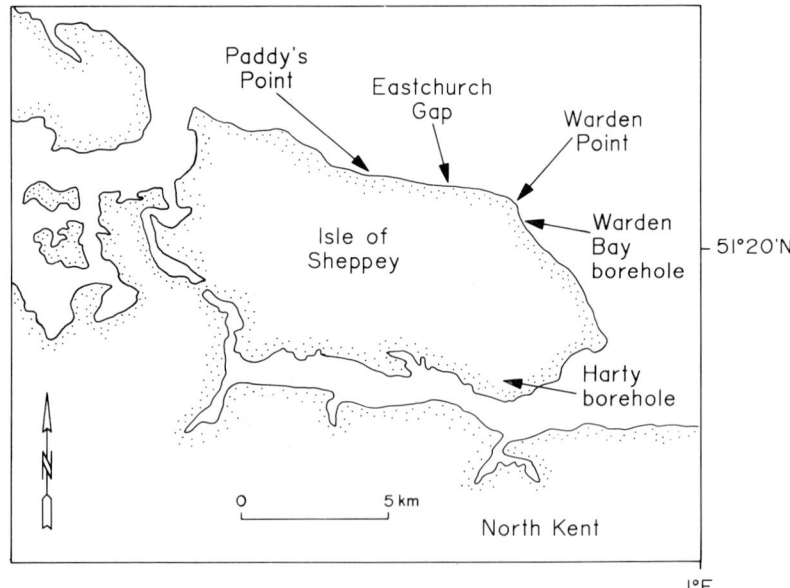

Fig. 5. Locations of coastal exposures and borehole sections sampled on the Isle of Sheppey.

ness New Town Well (Shrubsole 1878). The London Clay Formation rests uncomformably on the Oldhaven Formation and is overlain conformably by a small outlier of the Virginia Water Formation. A detailed correlation of the London Clay Formation in this area with exposures in Essex and the London area has been proposed by King (1981, 1984). According to King (1981), Division A2 of the London Clay rests on the Oldhaven Formation at Sheppey, but our present studies suggest that the basal unit of the London Clay Formation, 0.5 m of glauconitic sandy clayey silt, is probably a highly condensed representative of Division A1b, as at Herne Bay and Shelford in Kent (Ali et al. in prep.).

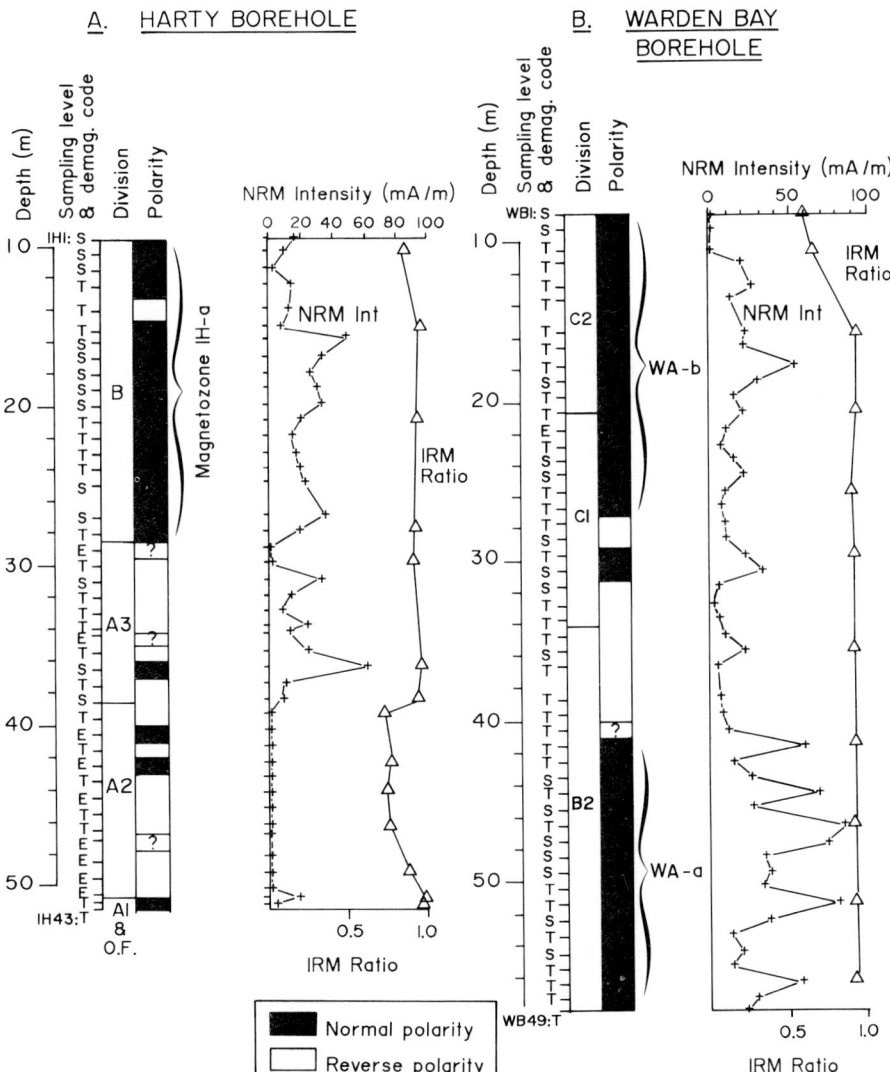

Fig. 6. Magnetostratigraphy of (**A**) the Harty borehole and (**B**) the Warden Bay borehole. Sampling levels are shown by the ticks beside the Division column. Normal polarity black, reverse polarity white. Letters S, T or E beside ticks indicate whether the corresponding polarity assignment is based on a demagnetisation stable end-point (S), directional trend (T), or whether the behaviour was erratic (E) Levels at which the polarity determination is 'intermediate' or unreliable are indicated by a query. Depths are below ground surface. Positions of Divisions A1, A2, A3, B etc. of King (1981) are indicated. O.F. = Oldhaven Formation.

The unexposed Oldhaven Formation and lower London Clay Formation were sampled in two cored boreholes. The exposed upper part of the London Clay Formation and the Virginia Water Formation were sampled at three localities along the northern coast of the island (Fig. 5). The Harty and Warden Bay boreholes (drilled in 1984 by the Kingston Polytechnic Engineering Geology Section) together penetrate Divisions A, B and the lower two-thirds of Division C (Fig. 8). There is a 12 m stratigraphic overlap between the two boreholes. The cores were sampled at approximately 1 m intervals, providing a total of 92 palaeomagnetic specimens.

Harty borehole (TR 015665). This is located at the southeastern corner of Sheppey (Fig. 5). The borehole cored the interval from the upper part of the Oldhaven Formation to the middle of London Clay Formation Division B2, a total stratigraphic thickness of 42 m.

A dominantly normal polarity magnetozone, coded IH-a, extends from the base of Division B1 to the top of the borehole (Fig. 6A). Isolated normal polarity specimens were identified also in the uppermost part of the Oldhaven Formation and Division A1, and at a number of other discrete levels within Division A (which is dominantly reverse polarity).

Warden Bay borehole (TR 024717). This cored the interval from the lower part of Division B2 to the upper part of Division C2, a total thickness of 50.3 m. Two normal polarity magnetozones, coded WA-a and WA-b, were identified (Fig. 6B). WA-a, in the lower part of the section, is 17.7 m thick. It is overlain by a reverse zone, some 15 m thick, which includes a short (1–2 m) normal polarity interval defined by two adjacent palaeomagnetic sampling levels at about 30 m below surface. WA-b commences 8 m above the base of Division C1 and continues to the top of the borehole.

With the exception of the uppermost 3 m, the sediments in the Warden Bay borehole have a relatively high magnetic intensity (typically in the range 10–50 mA/m) and IRM ratio (>0.9) (Fig. 6B) indicating that their magnetism is carried by magnetite. However, the intensity of the uppermost 3 m is significantly weaker (*c.* 2 mA/m) and the IRM ratio is lower (*c.* 0.7) indicating the presence of hematite. This appears to represent the effect of near-surface weathering and oxidation of magnetite to hematite. Thus it is possible that the normal magnetic polarity in the uppermost 3 m of this borehole is in fact an overprint in the present or recent

geomagnetic field. In the Harty borehole the sediments above 41 m depth are characterized by magnetite, whereas those below this depth (apart from the lowest two levels) are characterized by hematite (Fig. 6A) yet carry a dominantly normal polarity magnetization. In this case the presence of hematite is clearly not indicative of recent near-surface weathering, but may represent a different depositional environment or source area.

Sheppey coastal exposures

The most extensive surface outcrop of the London Clay Formation in England occurs on the northern coast of the Isle of Sheppey. Here the beds dip gently to the northwest in a 6 km long exposure, so that progressively younger beds outcrop at beach level between Warden Point and Paddy's Point (Fig. 5). The upper 53.5 m of the London Clay Formation, including the upper part of Division C2, Division D and Division E, are exposed, together with the lower 8 m of the overlying Virginia Water Formation.

Results of a detailed stratigraphic study were presented by King (1984). Fourteen lithostratigraphic units were identified (SH-1 to SH-14). Correlation was aided by tracing a number of calcareous nodule layers (coded A to P) along the exposures. The boundary between Divisions C2 and D (placed by King, 1984 at the base of the glauconitic unit SH-4) has since been re-defined at the base of the glauconitic unit SH-2. For the present study the exposures were sampled at three localities: Warden Point, Eastchurch Gap and Paddy's Point. These together provide a continuous sequence through the exposed section (Fig. 8). King's (1984) nodule layer 'A' was used as the reference datum for measurement. The base of the exposed section has a 3–4 m overlap with the top of the Warden Bay borehole.

Warden Point (TR 021724). The lowest 7.8 m of the exposed London Clay was sampled along a 500 m stretch of beach and cliff exposures close to Warden Point. The magnetostratigraphic results are presented in Fig. 7A (lower). A normal polarity magnetozone, WP-a, which is 3.5 m thick, is defined by five sampling sites in the lower part of the section. The overlying part is reverse polarity.

Eastchurch Gap (TR997730). Approximately 15 m of the London Clay Formation is exposed in the cliff section west of Hens Brook. Ten sites at stratigraphic heights ranging from 8.6 m to 19.45 m OD were sampled. Apart from an

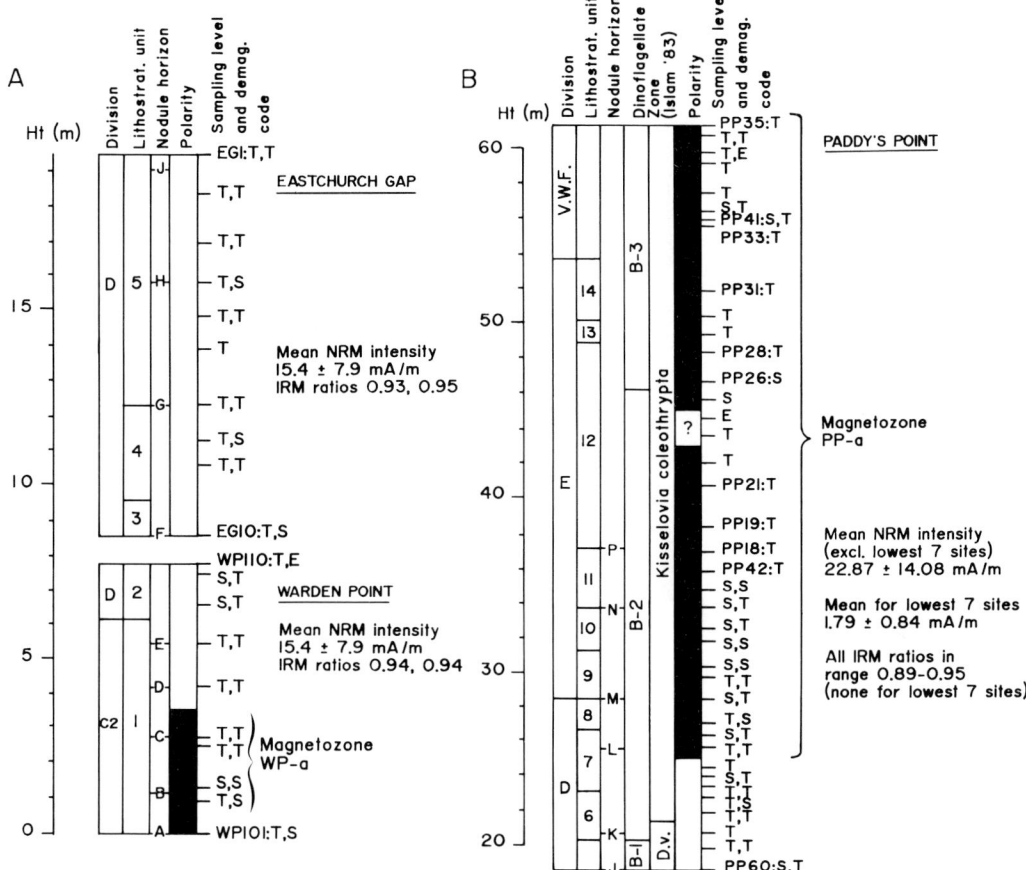

Fig. 7. Magnetostratigraphy of (**A**) the Warden Point & Eastchurch Gap sections and (**B**) the Paddy's Point section at Sheppey. Locations of Divisions C, D & E are shown, together with the lithostratigraphic units (SH1–SH14) and nodule horizons (A to P) of King (1984). The dinoflagellate cyst zonation (B–1 to B–3) of Islam (1983) is indicated for the Paddy's Point section. Demagnetization codes (S, T, E) as in Fig. 6.

'indeterminate' polarity site (EG10) at the base of the section, the succession is reverse polarity throughout (Fig. 7A upper).

Paddy's Point (TR971735). The beach, cliff and back-cliff at Paddy's Point expose a 45 m sequence extending from the upper part of the London Clay Formation to the Virginia Water Formation. Sampling commenced at nodule band 'J', at 18.45 m OD (King 1984), and extended into the lower part of the Virginia Water Formation, at 61.25 m OD. The magnetostratigraphy is summarized in Fig. 7B. A normal polarity magnetozone, coded PP-a, commences at 25.1 m OD and continues to the top of the section. The lowermost part of the section is reverse polarity.

On the basis of IRM acquisition experiments, samples from the entire surface exposure of the London Clay on the northern shore of Sheppey appear to contain magnetite as their principal magnetic constituent, with the exception of the lowest 6 m at Paddy's Point.

Synthesis of the Sheppey magnetozones

The two boreholes and three coastal exposures at Sheppey together form a composite sequence from the top of the Oldhaven Formation, through the London Clay Formation, to the lower part of the Virginia Water Formation. The magnetostratigraphic results from the five sections are combined in Fig. 8, with each section shown in its relative stratigraphic position.

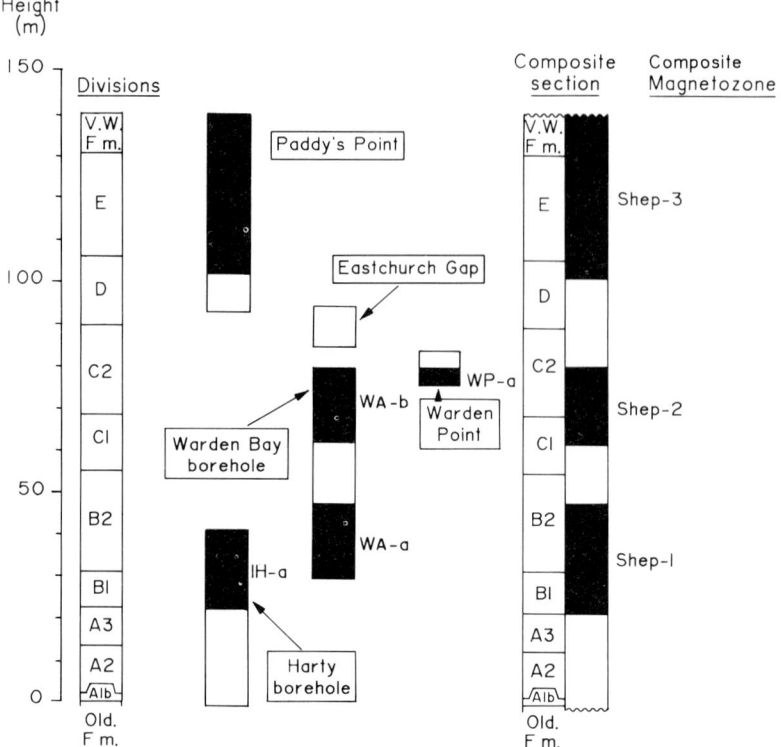

Fig. 8. Syntheses of the magnetozones defined from the Sheppey outcrop and borehole sections. Black, normal polarity; white, reverse polarity. V.W.Fm, Virginia Water Formation; Old. Fm, Oldhaven Formation.

Three discrete normal polarity magnetozones have been identified. For discussion purposes these are referred to as Shep-1 to -3 in ascending stratigraphic order.

Shep-1. This is located in the upper half of the Harty borehole (magnetozone IH-a) and the lower part of the Warden Bay borehole (zone WA-a). It commences at the base of Division B1 and terminates 7 m below the top of Division B2.

Shep-2. This appears in the upper third of the Warden Bay borehole (zone WA-b) and the lower half of the Warden Point section (zone WP-a). This magnetozone begins 8 m above the base of Division C1 and terminates 2.5 m below the top of Division C2.

Shep-3. Identified in the Paddy's Point section as zone PP-a, its base is positioned 3.35 m below the top of Division D, and it extends to the top of the sampled section (lower part of the Virginia Water Formation).

The correlation of these magnetozones with the geomagnetic polarity timescale is discussed in the final section of this paper.

The Hampshire Basin: Whitecliff Bay (SZ640860)

Earlier magnetostratigraphic studies of the Whitecliff Bay section have been reported by Townsend (1982) and Townsend & Hailwood (1985). These earlier studies were carried out with a lower sensitivity magnetometer, so that the polarity of very weakly magnetic units was less well-defined than in the present study. In the present study, initial attempts at a magneto-stratigraphic correlation of the upper part of the London Clay Formation between Sheppey and Whitecliff Bay proved difficult. After a re-examination of the earlier data, it was decided to re-investigate critical levels in the upper half of the London Clay Formation and the Wittering Formation at Whitecliff Bay. Townsend & Hailwood (1985) identified three magnetozones in this part of the succession, WB-b in the lower part of Division C, WB-c spanning the Divisions C/D boundary and WB-d in the Wittering Formation. For the present study, the 55 m thick-

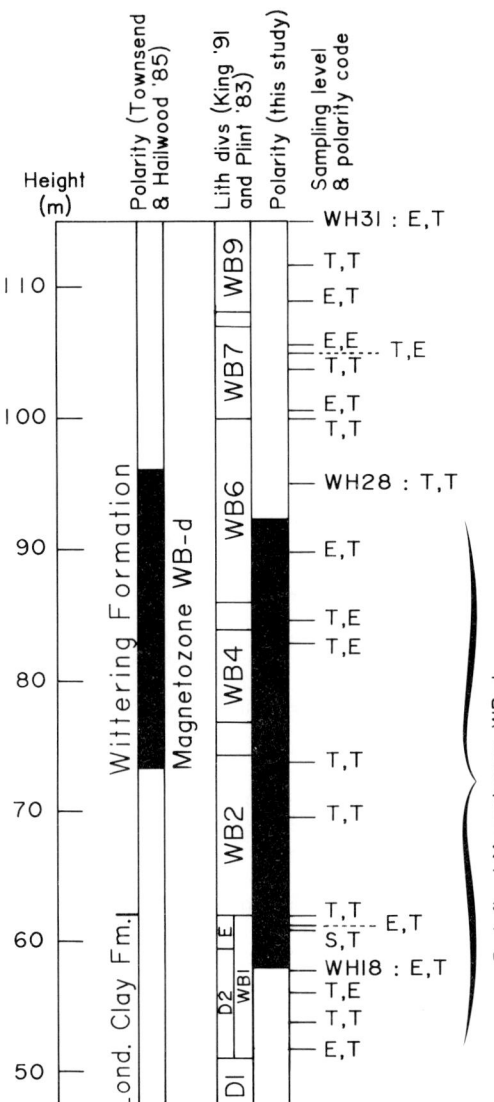

Fig. 9. Magnetostratigraphy of uppermost London Clay Formation and Wittering Formation at Whitecliff Bay. Data from present study in right hand column and for earlier study of Townsend & Hailwood (1985) in left hand column. Division boundaries (D1-E) of King (1981) and unit boundaries (WB1–WB9) of Plint (1983) are shown. Height datum is base of Division C.

dicate that the top of this magnetozone may be a few metres lower than the position originally proposed by Townsend & Hailwood (1985).

More importantly, a relatively well-defined normal polarity magnetisation has been identified in the upper part of unit WB1 and throughout unit WB2. The study of Townsend & Hailwood (1985) had suggested the presence of a poorly-defined reverse polarity magnetization in this weakly magnetic interval. The new results are considered more reliable, and it is now proposed that magnetozone WB-d extends from the middle part of Plint's unit WB6 down to the lower part of unit WB1 (including King's Division E and the top-most part of Division D2).

Divisions C1, C2 and D1 of the London Clay were also resampled at a small number of levels as part of the present study. These new palaeomagnetic results conflict with the earlier data of Townsend & Hailwood (1985). Magnetozone WB-b, which was defined by Townsend & Hailwood (1985) as extending from the base of Division C1 into the lower part of C2, could not be identified in the present study, and normal polarities were observed in the lower part of Division D1, whereas Townsend & Hailwood (1985) had observed reverse polarities in this interval. (The re-sampling did not extend over the middle part of Division D1, where Townsend & Hailwood (1985) defined magnetozone WB-c.)

The reasons for these discrepancies are uncertain at present. However this part of the section is in a dominantly sandy facies, the palaeomagnetic samples being taken from thin clay seams. It is possible that groundwater circulation through the sand has resulted in a complex and irregular remagnetization, which may have extended over several polarity reversals, giving rise to the observed wide variability of magnetic properties. The intensity of magnetization of Division C1 is anomalously high (range 2–5 mA/m) compared with that of the overlying units, but the magnetic stability is relatively low (the laboratory demagnetization field required to reduce the magnetic intensity to 50% of the initial value (i.e. the Median Destructive Field) is 7 mT for this unit, compared with a mean value of 12 mT for the rest of the section).

It is concluded that the presently available magnetostratigraphic data for Divisions C1, C2 and D1 at Whitecliff Bay are unreliable, so that the presence and positions of the magnetozones WB-b and WB-c of Townsend & Hailwood (1985) remain questionable. Consequently these data are not incorporated into the interpretations of the present study.

ness of the Wittering Formation was re-sampled at 17 levels. The palaeomagnetic results (Fig. 9) confirm the presence of the normal polarity magnetization in Units WB-3 to mid-WB6 of Plint (1983), on the basis of which Townsend and Hailwood (1985) defined normal polarity magnetozone WB-d. However, the new results in-

The lower Eocene deposits of Belgium

The Ypresian Stage was defined from the Ieper Clay Formation and associated sediments of Belgium and northern France. Willems *et al.* (1981) provide a comprehensive review of the historical background to the definition of this Stage.

The lithostratigraphy and biostratigraphy of the Ieper Clay Formation has been revised recently by Steurbaut & Nolf (1986) and King (1991). The present palaeomagnetic study, the first of its kind on these deposits, relies on the lithostratigraphic and biostratigraphic information in these two papers, together with additional unpublished information (King) to provide the basic stratigraphic framework to which the magnetostratigraphy is linked.

The Ieper Formation reaches a maximum thickness of *c.* 165 m in the north of Belgium (e.g. in the Knokke borehole, King 1990), where virtually all the succession is in a silty clay facies. To the south and east, nearer to the palaeocoast, the facies become more sandy (e.g. the Mons-en-Pevele, and Panisel Sand Members in the middle and upper part of the formation respectively).

Steurbaut & Nolf (1986) subdivided the Ieper Clay Formation into eight members, from the Orchies Clay Member at the base to the Egem Sand Member at the top. From their study of the calcareous nannoplankton, they assign the Formation to nannofossil zones NP11 and NP12, which they subdivide into nine local subzones (Fig. 10).

The lithostratigraphic scheme of King (1991) differs from that of Steurbaut & Nolf in the definition of three members in the lower part of the Ieper Formation (the Mt Heribu Member, Wardrecques Member and Bailleul Member) to represent the interval covered by the Orchies Clay Member and Roubaix Clay Member of Steurbaut & Nolf (1986). In Fig. 10, the upper Ieper Clay unit (Aalbeke Clay Member to Panisel Sand Member) are those defined by Steurbaut & Nolf (1986) while the lower units (Mt Heribu Member to Bailleul Member are those defined by King (1990). King has also extended his division scheme for the London Clay Formation (King 1981) to the Ieper Clay Formation. The scheme is based on the recognition of five major transgression–regression cycles within the formation. Glauconitic horizons are used to identify transgression surfaces and these, together with a wide variety of palaeontological data, are used to define an 'event' stratigraphy. Subsequent fieldwork and microfaunal analysis has enabled revision and

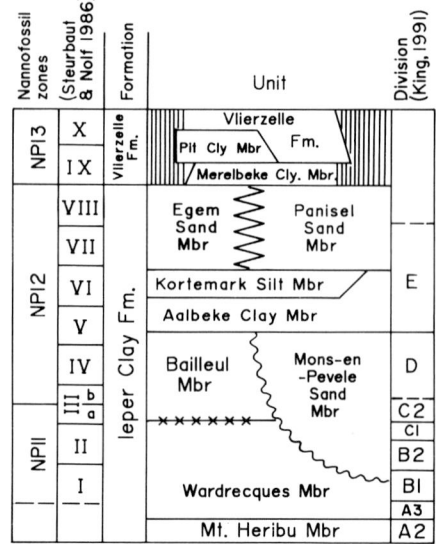

Fig. 10. The stratigraphy of the Ieper Clay Formation, as defined by Steurbaut & Nolf (1986), but incorporating the stratigraphic units defined by King (1991) beneath the Aalbeke Clay Member.

refinement of the correlations (King 1991). Divisions (parasequences) A2, A3, B1, B2, C1, C2, D and E can now be identified in Belgium, although the locations of the junctions between some of the other divisions are still uncertain. The positions of the divisions are indicated in Fig. 10 (after King 1991, and subsequent unpublished research).

Steurbaut & Nolf (1986) also revised the stratigraphy of the units immediately overlying the Ieper Clay Formation. These consist predominantly of sands and sandy clays. The Pittem Clay Member of the Vlierzele Formation was sampled for the present study at Egem. At outcrop this member is an oxidized glauconitic sandy clay and clayey silt. The lower 5 m are exposed at Egem. Although this section is barren of calcareous nannofossils, Steurbaut & Nolf (1986) assign the member elsewhere to the middle of nannoplankton zone NP13.

Magnetostratigraphic results from the lower Eocene of Belgium and NE France

The location of each of the sections investigated is shown in Fig. 11. The results from each section are discussed below in ascending stratigraphic order. When referring to lithostratigraphic units

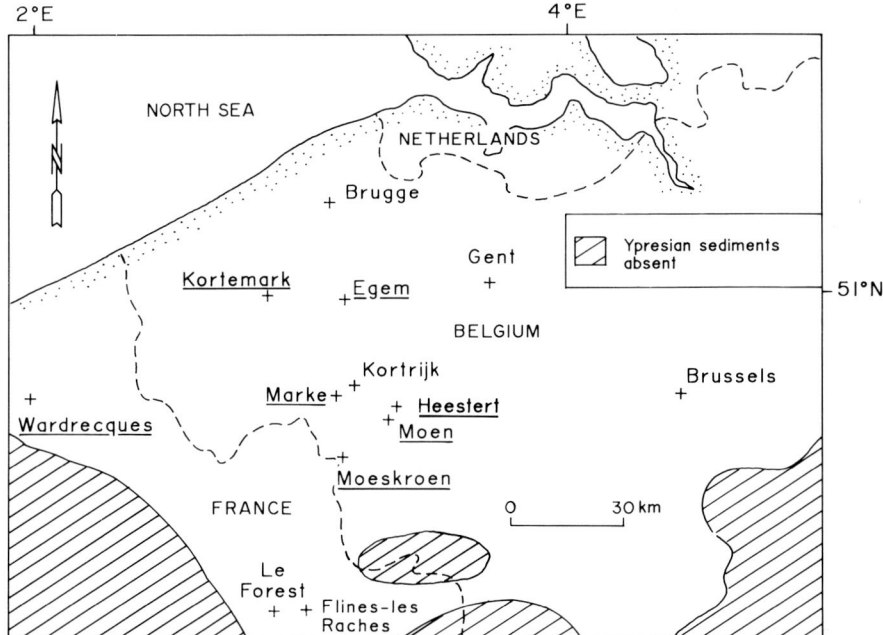

Fig. 11. Location of Ypresian sections sampled in Belgium and northeastern France.

we adopt King's scheme, unless otherwise indicated.

During the sampling program clay pits were visited in northern France at Wahagnies, Flines-les-Raches and Leforest (Steurbaut & Nolf 1986; King 1991). All three sections expose the Mt Heribu Member (equivalent to the lowest part of Steurbaut and Nolf's Orchies Clay Member). The exposures at Wahagnies were too degraded to sample. Only rather short sections (<5 m) were sampled at Flines-les-Raches and Leforest; the results from both these sections are poor and somewhat confused (Ali 1989). The detailed magnetostratigraphy of the lowest parts of the Ieper Clay Formation thus awaits the availability of new exposures, or the study of borehole material.

Wardrecques

King (1981, 1991) summarizes the stratigraphy of the section in the clay-pit at Wardrecques, 10 km SE of St Omer, in northern France. The section exposes the upper 26 m of the Wardrecques Member and the lower 3 m of the Bailleul Member. At the time of the fieldwork the Bailleul Member, exposed at the top of the pit, was rather weathered, and only the Wardrecques member was sampled (27 stratigraphic levels). The magnetic polarity results are summarized in Fig. 12. A normal polarity magnetozone, WQ-a, is defined in the interval 0.55 to 18.05 m. (The top of this magnetozone could alternatively be placed at 19.95 m, above the isolated normal polarity site WQ20) the top of WQ-a coincides with a marked upwards decrease in NRM intensity and a reduction in IRM ratio from a mean value of 0.91 below this level to 0.7 m above (Fig. 12). This suggests an upwards change at this level from magnetite to hematite as the dominant magnetic mineral. Thus it is possible that the reverse polarity interval above this level represents the effects of chemical weathering during a reverse polarity magnetochron (with the normal polarity interval at the top of the section representing a further magnetic overprint in the recent field).

However, it is also possible that the upwards change in magnetic mineralogy at about 18 m represents a change in sediment source area or depositional environment at the time of the polarity change. It has been noted earlier in this paper (Fig. 6A) that changes in magnetic mineralogy do not necessarily represent effects of chemical weathering. For our present interpretation, we assume that the top of WQ-a does represent a true record of a geomagnetic reversal, through further studies of this interval are needed to confirm this interpretation.

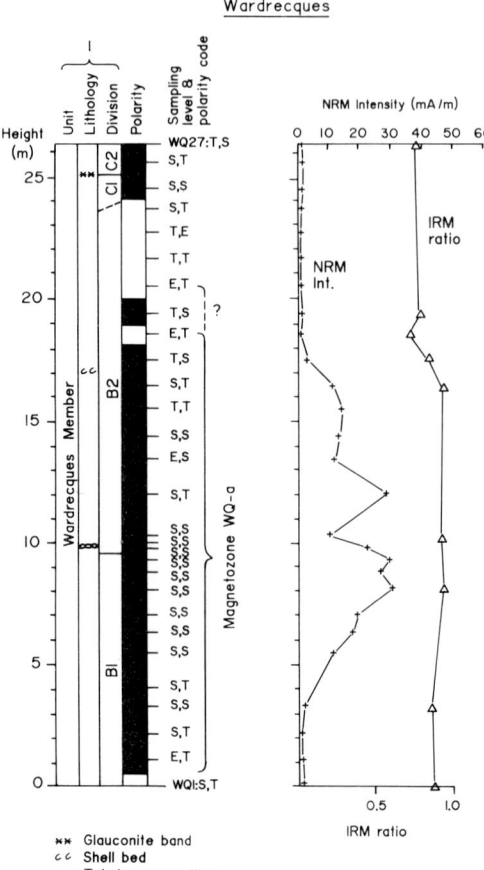

Wardrecques

Fig. 12. Magnetostratigraphic section at Wardrecques. The stratigraphic units (B1–C2) of King (1991) are shown, together with principal lithological markers. Other symbols & conventions as in Fig. 6.

The Kortrijk area

Four exposures near the town of Kortrijk provide a composite section exending from the upper part of the Wardrecques Member to the top of the Aalbeke Clay Member (Fig. 13).

Moen canal bank. The west side of the Bossuit canal cutting at Moen was chosen by Steurbaut and Nolf (1986) as the lectostratotype of their Roubaix Clay Member. King (1991) assigns the lower part of the section to the Wardrecques Member and the upper part to the Bailleul Member. A section similar to the one described by Steurbaut & Nolf was sampled on the west side of the canal bank, 50 m south of the '3500' marker painted on the towpath.

A summarized lithological log is shown together with the magnetostratigraphic sequence in Fig. 13A. Useful lithological markers include the unit of silty sand at the base of the section, a 15 cm thick glauconite-rich level at +6 m (marking the Wardrecques Member/Bailleul Member junction) and a nummulite-rich bed at +13.25 m. The upper part of the Wardrecques Member exposed here corresponds to the upper part of Division B2 and Division C1. The glauconitic bed at the base of the Bailleul Member probably marks the C1/C2 boundary. The base of Division D may be marked by a thin glauconitic bed *c.* 2–3 m above the base of the Bailleul Member.

A normal polarity magnetozone, MO-a, extends from the base of the section to 8.45 m above datum (the top is defined at site MO11 which has an intermediate polarity).

Steurbaut & Nolf (1986) identified the base of the Aalbeke Clay Member at the top of the Moen section. The exposure continues some distance above our Site MO18, but since this part of the section appeared weathered, no palaeomagnetic samples were taken from it.

Marke. The section at Marke is in the Koekelberg clay pit (map sheet 29/15; co-ords: $x = 69.0$, $y = 166.8$). The uppermost 1 m of the Wardrecques Member, the whole of the Bailleul Member (14.2 m), and several metres of the Aalbeke Clay Member are exposed.

The magnetostratigraphic results are presented in Fig. 13B. The base of the Aalbeke Clay Member is taken as the site datum. Important lithostratigraphic markers are present at −12.7 m (the glauconitic horizon which defines the base of the Bailleul Member) and at −10.8 m and −8.3 m (shell beds). According to Steurbaut & Nolf (1986, p. 131), the NP11/NP12 nannoplankton zonal boundary occurs at about −9 m in the sampled section. A normal polarity magnetozone, MK-a, was identified at the base of the section. The top of MK-a is very precisely located between sites MK20 and MK2, at 10.62 m below datum.

Heestert. The section at Heestert is in the Kwadestraat pit (map sheet 29/6; co-ords: $x = 80.55$, $y = 165.55$). The pit exposes the uppermost 0.5 m of the Bailleul Member, the full thickness of the Aalbeke Clay Member (12 m) and the lower 3 m of Steurbaut & Nolf's (1986) Panisel Sand Member.

Only the lower 4.2 m of the Aalbeke Clay Member was sampled (six sites), as the upper part of this unit is weathered to a soft yellow clay. The magnetic polarity results are presented

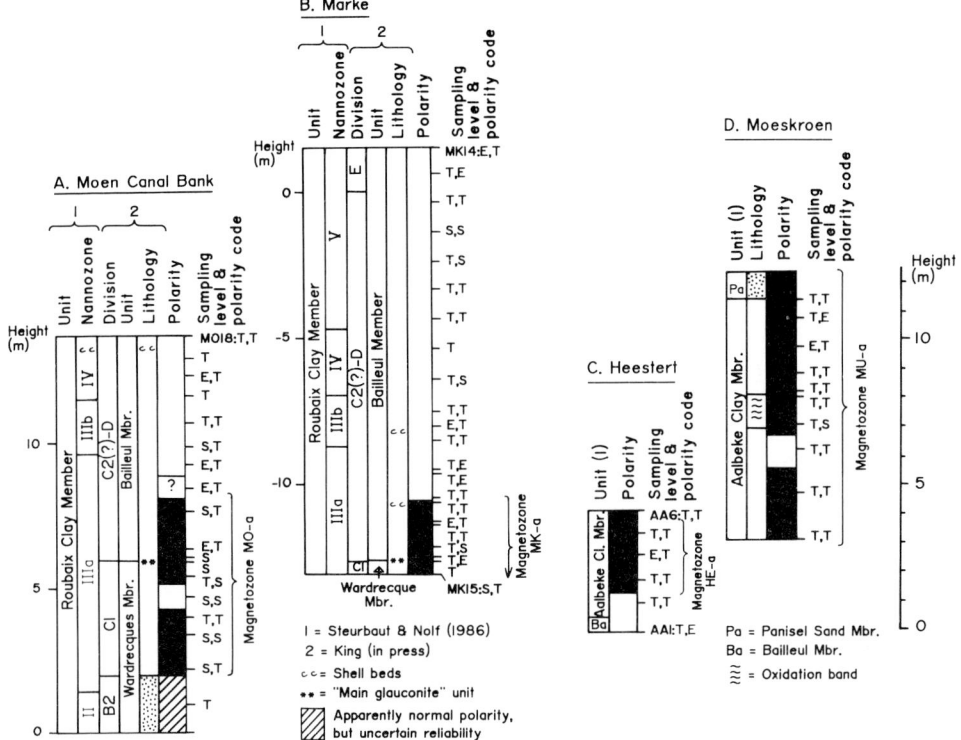

Fig. 13. Magnetostratigraphy of sections at (**A**) Moen Canal Bank, (**B**) Marke, (**C**) Heestert & (**D**) Moeskroen. Symbols and conventions as in Figs 6 & 12.

in Fig. 13C. A normal polarity magnetozone, HE-a, commences at 1.4 m above the base of the Aalbeke Clay Member and continues to the top of the sampled section.

Moeskroen. The section at Moeskroen is in the Bois Fichau pit (map sheet 29/6; co-ords: $x = 79.775$, $y = 164.725$). The upper 7 m of the Aalbeke Member and the lower 3 m of the Panisel Sand Member (Steurbaut & Nolf 1986) were sampled at a total of ten sites. The magnetostratigraphic results are presented in Fig. 13D. The silty red oxidation band in the middle of the Aalbeke Clay Member, which can be identified throughout the Kortrijk area and provides a useful lithostratigraphic marker, extends from +3.68 to +4.92 m in the sampled section. A dominantly normal polarity magnetozone MU-a is present from the base to the top of the section.

Kortemark

This section is at the Desimpel clay pit (mapsheet 20/3; co-ords: $x = 58.05$, $y = 190.4$).

Steurbaut & Nolf (1986) describe the pit as exposing the top of the Aalbeke Clay Member, the Kortemark Silt Member (stratotype), and the lower half of the Egem Sand Member. The pit is worked on two levels; the lower level exposes 12 m of the succession whilst the upper one exposes 20 m. Sampling sites spanned a total thickness of 27.7 m (Aalbeke Clay, 4 sites; Kortemark Silt, 25 sites: Egem Sand, 2 sites).

The magnetostratigraphic results are presented in Figure 14A. The section carries a normal polarity magnetisation throughout (magnetozone KM-a).

Egem

This section is in the Ampe clay and sand pit (mapsheet 21/1; co-ords: $X = 70.15$, $Y = 190.15$). The upper third of the Kortemark Silt Member, the Egem Sand Member stratotype, and the lower part of the Pittem Clay Member (Steurbaut & Nolf 1986) are exposed. The base of King's (1991) Egem Member is here taken as the base of Steurbaut's (1987, p. 348) unit 5, which marks the base of the main sand unit; the

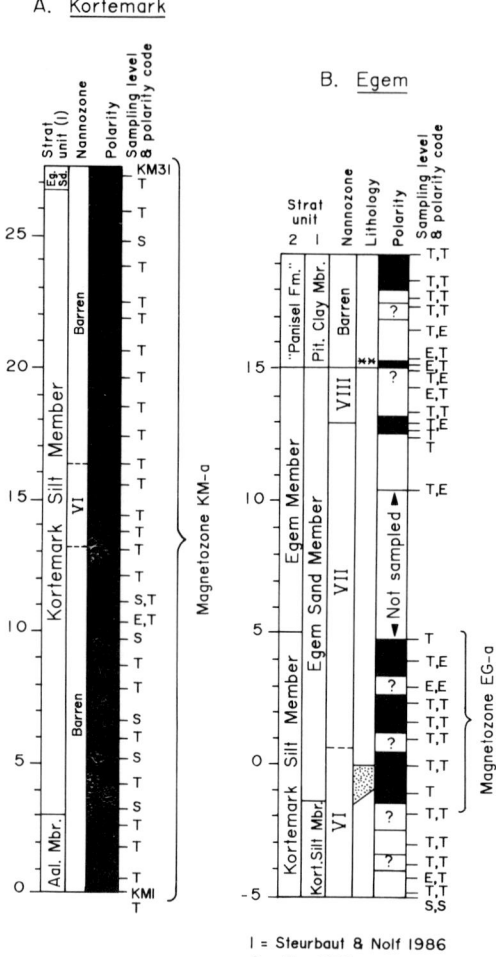

I = Steurbaut & Nolf 1986
2 = King 1991

Fig. 14. Magnetostratigraphy of (**A**) Kortemark and (**B**) Egem sections. Symbols and conventions as in Figs. 6 & 12. The nannofossil zonation of Steurbautt & Nolfe (1986) is shown.

thin lenticular sand unit 5 m lower, which is taken by Steurbaut & Nolf (1986) to define the base of their Egem Sand Member, is here regarded as a sand lens within the Kortemark Silt Member.

A total of 28 sites was sampled, including an oriented block from the shelly sandstone at the base of the Pittem Clay Member. The lowest 5 m of the Egem Member (*sensu* King 1991), which is totally unlithified, was not sampled. The majority of sample sites above this level in the Egem Member were located in thin clay seams, except for a few sites at the top of the member, which were collected from sandier horizons.

The magnetic polarity results are presented in Fig. 14B. It must be noted that the majority of specimens from this section are weakly magnetic (0.1 to 0.4 mA/m); consequently the polarity results are somewhat less reliable than those obtained from the other Belgian sections.

A mixture of normal, reverse and indeterminate polarities was found in the Kortemark Silt Member and the lower half of the Egem Sand Member (Steurbaut & Nolf's units). The dominantly normal polarity magnetic zone associated with the lower part of the Egem Sand Member is coded EM-a and its top positioned at about 5 m above the base of this member. The remainder of the section above is dominantly reverse polarity. The two normal polarity determinations (EM26 and EM27) at the top of the Pittem Clay Member should be treated with caution, as this part of the section appears rather weathered.

Synthesis of the Ieper Clay Formation magnetozones

The magnetostratigraphic results from the Belgium sections form a continuous sequence extending from the middle of the Wardrecques Member up to the middle part of the Pittem Clay Member of the Panisel Formation. The polarity results from the different sections are shown in Fig. 15, with each drawn in its relative stratigraphic position. Three normal polarity magnetozones have been identified. For discussion purposes these are referred to as Ieper-1 to Ieper-3, in ascending stratigraphic order.

Ieper-1

The local magnetozone WQ-a in the Wardrecques section defines regional magnetozone Ieper-1. This magnetozone lies in the upper part of the Wardrecques Member which King (1991) correlates with Divisions B1 and B2 of the London Clay Formation. The top of Ieper-1 is provisionally defined by the upwards transition from normal to reverse polarity at about 18 m OD in the Wardrecques section.

Ieper-2

The normal polarity magnetozone Ieper-2 is defined from the magnetostratigraphic results for the sections at Marke and Moen (local magnetozones MK-a and MO-a respectively). The base of Ieper-2 may lie at the upwards transition from reverse to normal polarity at about 24 m OD in the Wardrecques section, which corresponds with the probable base of

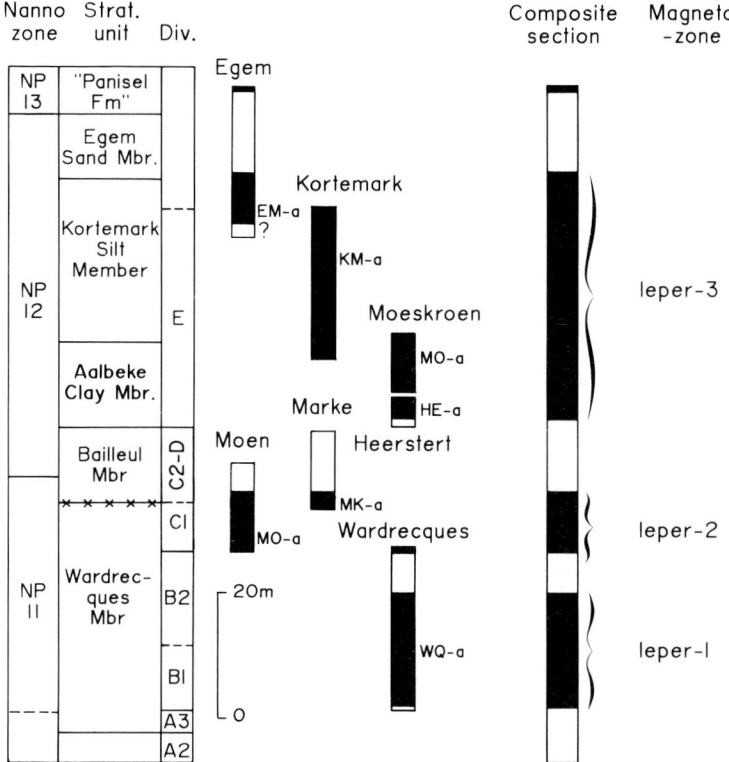

Fig. 15. Synthesis of the Ieper Clay Formation magnetozones.

Division C1 (Fig. 12). However, it is possible that this near-surface polarity transition may represent the effects of chemical weathering. In the Moen Canal Bank section the corresponding normal polarity magnetozone (MO-a) extends through Divisions C1 and C2. The single sample from the underlying sandy upper part of Division B2 also has a normal polarity (Fig. 13). However the Division B2/C1 boundary is likely to represent an interval of condensation or non-deposition so that this normal polarity sample could represent an older magnetozone. Consequently, the base of Ieper-2 is provisionally placed at the B2/C1 boundary in the Moen section. The top of Ieper-2 is located in the Marke and Moen sections at 2.13 ± 0.03 m and 2.43 ± 0.40 m respectively above the base of the Bailleul Member. This level is in the upper part of Steurbaut & Nolf's (1986) Roubaix Clay Member, and lies close to the Division C/ Division D boundary, as currently identified.

Ieper-3

The base of Ieper-3 is defined by the base of the magnetozone HE-a at Heerstert, being pos-

itioned 1.4 m above the base of the Aalbeke Member. Magnetozone Ieper-3 extends through the entire Aalbeke Member and the lower part of the Panisel Sand Member (terminology of Steurbaut & Nolf 1986) in the section at Moeskroen (MO-a). It continues up through to the top of the Kortemark Silt Member at Kortemark (KM-a), and terminates at about 5 m above the base of the Egem Sand Member at Egem (where it is coded EM-a). It is possible that this magnetozone may include a short reverse polarity interval, represented by the poorly-defined results from the uppermost 3.6 m of the Kortemark silt at Egem (Fig. 14B).

Varengeville, northern France

The importance of the outlier of Palaeogene sediments at Varengeville, near Dieppe, on the north coast of France for linking the Late Palaeocene/Early Eocene successions of the Anglo-Paris–Belgium Basin has long been recognized (e.g. Prestwich 1855). The more important stratigraphic studies of this section include those of Bignot (1962), Gamble (*in* Destombes *et al.* 1977), Chateauneuf & Gruas-Cavagnetto (1978) and Aubry (1986). The lower part of the section is assigned lithologically to the Sparnacian (a locally applied French stage name for latest Palaeocene sediments), the upper two-thirds being assigned to the Formation de Varengeville. Gamble (in Destombes *et al.* 1977), suggested that the Formation de Varengeville be informally divided into a lower (inferieur) and upper (superieur) unit to facilitate comparisons with the equivalent sequence in England. The 'Formation de Varengeville inferieur' overlies the Sparnacian, and consists of well-sorted glauconitic sands, very similar to the Oldhaven Formation of southern England, as noted already by Whitaker (1871). This unit is 5.1 m thick and is assigned to the *A. hyperacanthum* dinoflagellate zone (Costa & Muller 1978), making it late Palaeocene in age. The 'Formation de Varengeville superieur' is equivalent (and similar in lithology) to the lower and middle parts of the London Clay Formation (King 1981). The base of this unit is marked by a 1.4 m oxidized glauconitic sandy, clayey silt, equivalent to Division A1b of the London Clay Formation. Above this bed, units equivalent to Divisions A2 and A3 (spanning 16 m) can be recognised lithologically. Division A3 is overlain by an intensely glauconitic silty clay unit about 2.5 m thick. This unit contains a calcareous microfauna which indicates that it is a complex and highly condensed interval corresponding to Division B1 and the lower part of Division B2. The overlying silts and clays (2 m) contain a calcareous fauna, which in the lower part includes the bivalve *Pecten duplicatus*. This suggests correlation with the middle or upper part of Division B2. This is the highest unit preserved in the outlier. Aubry (1986) and Steurbaut (1988) assign the glauconitic interval of this formation to the upper part of NP11 (Subzone IIIa of Steurbaut 1988).

Palaeomagnetic results from Varengeville

The Palaeogene outlier at Varengeville was sampled at the Phare d'Ailly, in the cliff section near the lighthouse. The sampled part of the 'Formation de Varengeville superieur' totals 17 m. The magnetostratigraphic results are presented in Fig. 16. The section is almost entirely of reverse polarity, except for two normal polarity sites in Division A1b. It is emphasized that the glauconitic clay bed towards the top of the section, believed to represent Division B1 and the lower part of Division B2, has a reverse polarity.

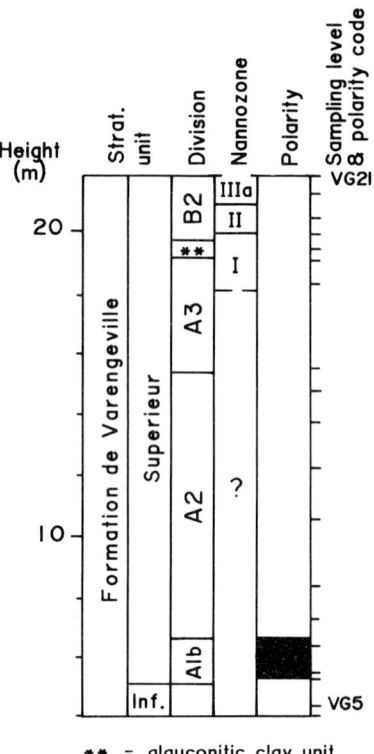

Fig. 16. Magnetostratigraphy of the Formation de Varengeville superieur at Phare d'Ailly. Symbols and conventions as in Figs 6 & 12. Divisions are from King (1991) and nannofossil zonation from Steurbautt & Nolf (1986).

Overall correlation and summary

We have presented above the magnetic polarity stratigraphy for the lower Eocene, as recorded in the London, Hampshire and Belgium Basins and at Varengeville, in northern France. In this final section the magnetostratigraphic data from each of these areas are synthesized with the corresponding lithological data, to investigate the history of sedimentation for this period across the southern margin of the North Sea Basin. As a standard, we use the magnetobiostratigraphic scale of Berggren *et al.* (1985) for the period 58 to 51 Ma. (The scale presented by Berggren *et al.* (1985) is based primarily on a synthesis of deep sea data obtained from DSDP/ODP investigations together with the results from the numerous studies of exposed pelagic sequences.) We will present magnetostratigraphic data for the Palaeocene–Eocene boundary beds of southern England in a separate paper (Ali *et al.* in prep.).

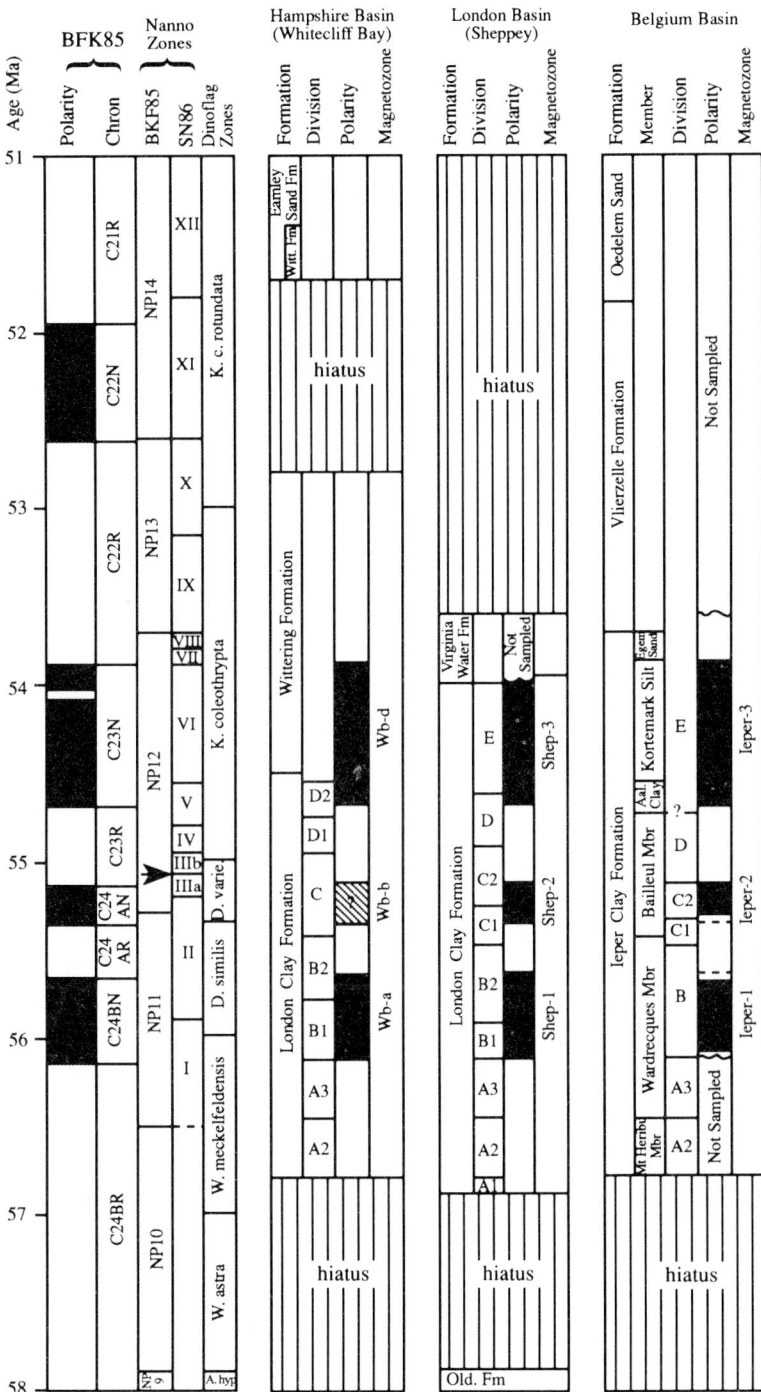

Fig. 17. Magnetostratigraphic correlation of Hampshire Basin, London Basin and Belgium Basin Ypresian sequences with geomagnetic polarity time scale of Berggren *et al.* (1985). Arrow in Nanno Zone column indicates revised position of NP11–12 boundary of Steurbaut & Nolf (1986).

Chron C24BR. This spans the Palaeocene–Eocene boundary and extends from the upper part of nannofossil zone NP9 to the lower part of zone NP11 (Berggren *et al.* 1985; Fig. 17). In the present study, the latter part of this chron has been identified in King's Divisions A1, A2 and A3 at Sheppey (London Basin), Whitecliff Bay (Hampshire Basin) and Varengeville. This is confirmed by data from the sections at Alum Bay and Herne Bay (Townsend & Hailwood 1985), and at Walton-on-the-Naze (Ali 1989).

Chron C24BN. Aubry *et al.* (1986) correlated the normal polarity magnetozone WB-a at Whitecliff Bay (defined by Townsend & Hailwood 1985) with Chron C24BN. This was based on the association of this unit with nannoplankton zone NP11 at Lower Swanwick, Hampshire. In the present study, normal polarity magnetozones have been identified in Divisions B1 and the lower part of B2 at Sheppey (Shep-1) and Wardrecques (WQ-a, which defines zone Ieper-1), suggesting a correlation with Chron C24BN (Fig. 17). The magnetozones WB-a and Shep-1 both commence at the base of Division B1, and terminate at 28 m and 7 m respectively below the top of Division B2. At Wardrecques, neither the base nor top of WQ-a (Ieper-1) has been positively identified, and so the precise stratigraphic position of the reversals bounding the Chron C24BN record cannot yet be fixed in the Belgium Basin.

At Varengeville, a reverse polarity magnetization is associated with the glauconitic unit and the overlying silty clay in the 'Formation de Varengeville superieur' (Fig. 16). King (1981) correlates the sediments above the glauconitic unit with part of Division B2. Aubry (1986) and Steurbaut (1988) correlate these levels with the upper part of nannoplankton zone NP11. The reverse polarity magnetisation associated with the part of the succession above the glauconitic unit probably represents a record of Chron C24AR. The glauconitic bed marks a period of condensation, which may account for the absence of a record of Chron C24BN.

Chron C24AN. Aubry *et al.* (1986) defined a composite normal polarity magnetozone (WB-b) which extended upwards from the base of Division C to the lower part of Division D at Whitecliff Bay. They correlated this with Chron C24AN. Reinvestigation of this part of the Whitecliff Bay section during the present study indicates that the magnetization of this interval is complex, possibly representing the effects of a prolonged remagnetization history of this dominantly sandy interval. Consequently, the earlier

results of Townsend & Hailwood (1985) and Aubry *et al.* (1986) are now regarded as unreliable. Further studies of this part of the section (or, preferably of a borehole section spanning this interval in the Hampshire Basin) are required before these uncertainties can be resolved.

The data for the corresponding stratigraphic interval at Sheppey and in Belgium are more reliable, due to the stronger magnetic signal in these sediments. At Sheppey, the base of magnetozone Shep-2 (defined in the Warden Bay borehole) is positioned at 8 m above the base of Division C1 and its top (defined at Warden Point) is placed 8 m below the Division C/D junction. Using a second order correlation of Division C relative to the NP11/12 nannoplankton zone boundary (Aubry 1986), Shep-2 can be seen to represent a record of Chron C24AN.

In Belgium, the normal polarity magnetozone Ieper-2 extends from the lower part of Division C1 (which here is highly condensed) to the upper part of Division C2 (Fig. 17). This magnetozone terminates about 2.5 m above the glauconite-rich horizon marking the Division C1/C2 boundary. These levels are associated with the uppermost part of NP11, suggesting that the Ieper-2 magnetozone represents Chron C24AN.

Chron C23N. Aubry *et al.* (1986) correlated the Wittering magnetozone of Townsend & Hailwood (1985) in the Hampshire Basin with Chron C23N. This magnetochron was identified at Bracklesham Bay, Alum Bay and Whitecliff Bay (as WB-d). Re-investigation of the Whitecliff Bay section suggests that the base of WB-d lies 3.5 m (±0.85 m) below the base of the Wittering Formation, close to the Division D/E junction. (Figs 9 and 17). The top of this magnetozone lies in the middle part of Plint's (1983) Unit WB6, at about 30 m above the base of the Wittering Formation.

At Paddy's Point, Sheppey, the base of magnetozone Shep-3 is positioned 3.35 m below the Division D/E boundary. This is just above the level indicated by Islam for the base of the *Kisselovia coleothrypta* dinoflagellate zone. This magnetozone continues up to the top of the section (into the lower part of the Virginia Water Formation). As Shep-3 occupies a similar stratigraphic position to the Wittering Magnetozone, a correlation with Chron C23N is proposed.

Magnetozone Ieper-3 can also be correlated with Chron C23N. It commences 1.4 m above the base of the Aalbeke Clay Member and terminates in the lower part of the Egem Sand Member. The base of the Aalbeke Member is tentatively equated with the base of Division E;

the Aalbeke Clay Member is assigned to the lower part of nannoplankton zone NP12 (Steurbaut & Nolf 1986).

Chron C22R. The upper part of the Wittering Formation at Whitecliff Bay has a reverse polarity. Aubry *et al.* (1986) proposed a hiatus at the level of the Whitecliff Bay (lignite) Bed (Plint's (1983) Unit WB8) in the upper part of the Wittering Formation, due to the apparent absence of Chron C22N from the succession there. This interpretation implies that Chron C22R is recorded in the sediments below the Whitecliff Bay Bed, and Chron C21R is recorded in the sediments above. However, these levels are barren of nannoplankton, and the unconformity could be positioned at two alternative levels. The first is in the topmost part of Plint's Unit WB6, which is an intensely glauconitic, highly bioturbated sand, indicating major condensation or a break in deposition. It is overlain very sharply by the base of Unit WB7, represented by a blocky silty clay with weak silt laminae. The second is between the Wittering Formation and the Earnley Sand Formation. This contact is a discontinuity (parasequence) boundary, marked by an interburrowed junction on which rests a thin pebbly glauconitic sand.

The magnetostratigraphic results presented here support the earlier studies of Hailwood and Townsend (1985) and Aubry *et al.* (1986). The apparent absence of Chron C22N from the sequence suggests a minimum duration for this hiatus of 0.6 Ma. The true position of this unconformity is difficult to establish, but the two alternatives outlined above appear equally plausible to that proposed initially by Aubry *et al.* (1986).

Magnetostratigraphic dating of Ypresian depositional sequences

A major aim of the magnetostratigraphic studies reported in this paper is to provide a high resolution time framework for dating the sea level changes represented by the depositional sequence boundaries in the Ypresian sequences of NW Europe and the southern North Sea Basin. The ages of the depositional sequence (division) boundaries in each of the areas studied (London, Hampshire and Belgium Basins) can be estimated by interpolation between adjacent magnetic chron boundaries. These ages are shown in Fig. 18. Before addressing their possible significance, the reliability of certain age assignments is first discussed.

The base of Division A1 is correlated with the early part of Chron C24BR (Fig. 17). However,

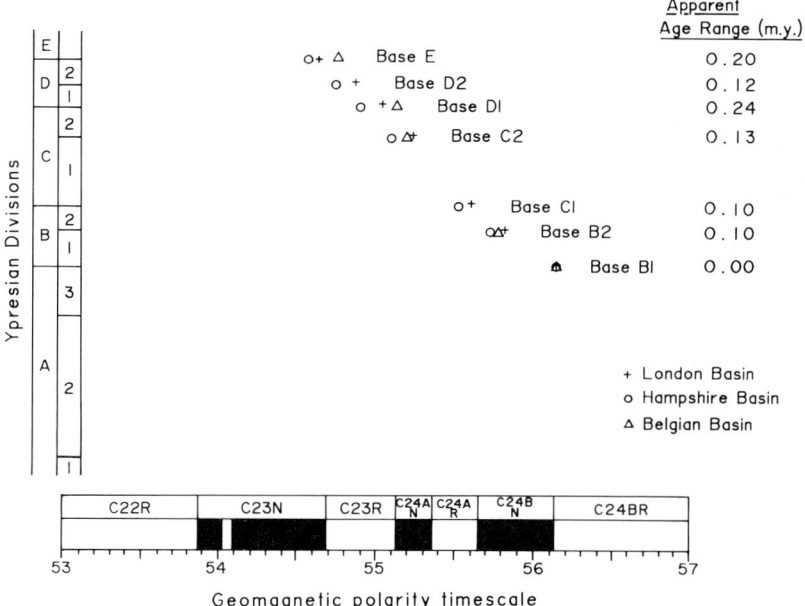

Fig. 18. Relative ages of Ypresian depositional sequences (represented by Division boundaries) in London, Hampshire and Belgium Basins, calibrated against geomagnetic polarity time scale of Berggren *et al.* (1985). The relative thicknesses of the Divisions shown on the vertical axis refer to those in the Hampshire Basin.

no record of the preceding Chron C25N has been observed in these sequences, so that it is not possible to interpolate the age of the base of Division A1 from its position between Chrons C25N and C24BN.

The base of Division B2 can be interpolated in all three areas from its position between the older and younger boundaries of Chron C24BN. However, since a record of Chron C24AN has not yet been identified reliably in the Hampshire Basin, it is necessary to interpolate between the end of C24BN (magnetozone WB-a) and the start of C23N (magnetozone WB-d), in order to date the bases of divisions C1, C2 and D in this region. Because of this relatively coarse interpolation, the ages of these division boundaries in the Hampshire Basin are less reliable than in the London Basin.

The end of Chron C23N (magnetozone Shep-3) was not identified in the London Basin, so the base of Division E in this region has been dated by extrapolation from the start of Chron C23N.

In the Belgian Basin the principal source of uncertainty is the base of magnetozone Ieper-2 (representing Chron C24AN), which lies in the highly condensed interval between the upper part of Division B2 and the lower part of C1. Consequently, in this region the base of C1 cannot be defined reliably and the base of C2 is dated by interpolation between the end of Chron C24BN (top of magnetozone Ieper-1) and the end of C24AN (top of Ieper-2).

In all three regions the base of Division B1 corresponds to the start of Chron C24BN (Fig. 18). Although, by their nature, division boundaries may correspond with sedimentary discontinuities or hiatuses, the apparently synchronous nature of the base of Division B1 across the three regions suggests that the durations of such hiatuses are probably short in this case. In contrast to the base of B1, the bases of each of the younger divisions (with the exception of the base of C1, which corresponds with a condensed interval) appear to be systematically younger in the Hampshire Basin than in the London and Belgium Basins (Fig. 18). This might be taken to indicate a possible systematic diachroneity of the division boundaries across the region, with each division base (transgression to deeper water sediment) appearing progressively younger from the Belgium area to the shallower near-shore environment of the Hampshire Basin. This could be interpreted as representing a slow progressive flooding of the shelf area from the centre of the basin towards the basin margins during each transgression. The apparent age difference of the division bases between these

areas is typically in the range 0.1 to 0.2 Ma and this could be taken as a measure of the time taken for the transition to advance from Belgium to southern England.

However, a more probable explanation for the apparent diachroneity of the division boundaries is the effect of the different sedimentary facies between Belgium and southern England. Thus, in the deeper water environment of western Belgium the division boundaries are marked mainly by thin glauconite beds; the regressive units are not well-developed, whereas in the shallower water environment of southern England the regressive phases are commonly marked by conspicuous changes to more sandy units. The sandy regressive units in southern England are commonly thicker than their finer-grained counterparts in Belgium. The effect is to cause the base of the subsequent transgressive phase to appear relatively closer to a succeeding chron boundary in southern England than in Belgium. Consequently the interpolated ages of the division boundaries will tend to appear somewhat younger in southern England.

Further information from other areas is necessary before these mechanisms can be fully understood. In the meantime, within the resolution of the available magnetostratigraphic dating, the division boundaries can be taken as representing effectively synchronous transgressive events across the southcentral and southwestern sectors of the North Sea Basin.

The magnetostratigraphic data provide an important set of 'absolute' time planes which when coupled with complementary biostratigraphic data, ultimately will allow a direct comparison of the timing of Ypresian depositional sequence boundaries in different parts of the world, in order to establish the extent to which these boundaries reflect either 'global' eustatic sea level changes or more local tectonic uplift or downwarp. We are currently extending these studies to the Ypresian sediments of the Gulf Coast of the USA, in order to carry out such a test.

It is instructive to attempt a direct comparison of the principal Ypresian depositional sequence boundaries in the southern North Sea area with the eustatic sea level curve proposed by Haq *et al.* (1987). This comparison may be achieved through matching the magnetozones defined in the southern North Sea Basin sequences with the positions of geomagnetic chron boundaries which Haq *et al.* correlated with their eustatic curve (Fig. 19). Unfortunately, these authors provide no information on the method used for linking their eustatic curve to the polarity time scale, but it is likely to have involved a second-

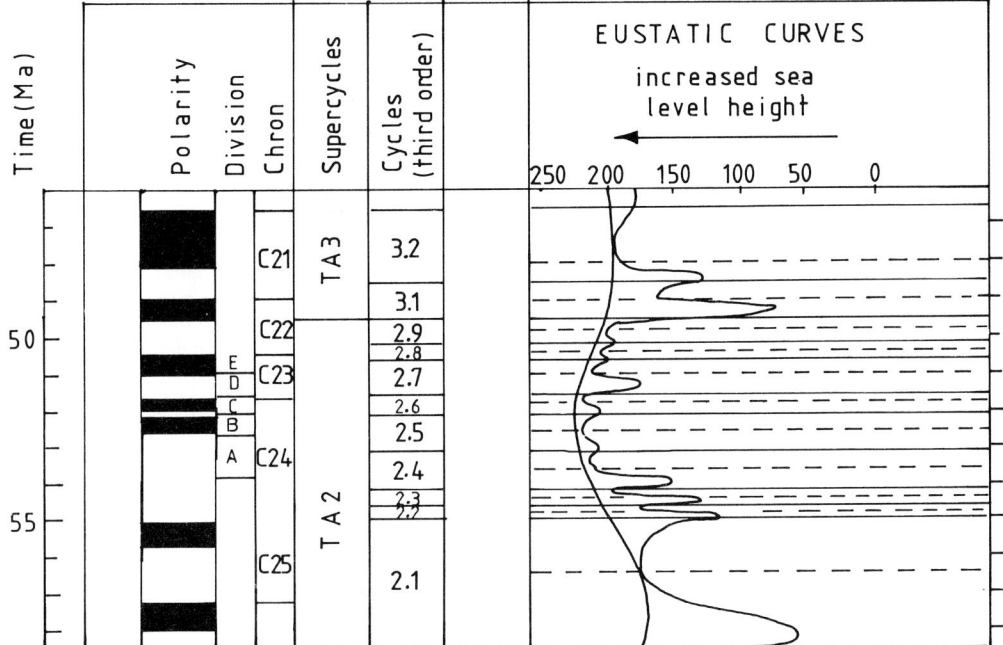

Fig. 19. Comparison of principal depositional sequence (division) boundaries in the southern North Sea area with the eustatic sea level curve of Haq *et al.* (1987), based on magnetostratigraphic correlation.

order biostratigraphic correlation. Thus there are likely to be significant uncertainties in their positioning of individual third-order sea level cycles in relation to chron boundaries. Despite this limitation, several observations (see below) can be made on the comparison of the Haq *et al.* eustatic curve with the Hampshire Basin record (where the depositional sequences are best understood) based on magnetostratigraphic correlation between the two (Fig. 19).

(1) The major marine flooding event which is reflected by the basal London Clay transgression (base of division A1) can be correlated with the rapid sea level rise in the middle part of cycle TA 2.4.

(2) The generally high sea levels through the upper part of cycle TA 2.4 to cycle TA 2.6 correspond to the relatively high sea levels indicated by the prevalence of marine silty clay facies through Divisions A3 to C2 (King 1981).

(3) The rapid fall in sea level defining the base of Cycle TA 2.7 can be correlated with the fall in sea level considered to have initiated deposition of medium to coarse cross-stratified sands (Portsmouth Member and Whitecliff Member) in shallow marine ti-

dally influenced environments in the upper part of Division C2 and Division D1 in the Hampshire Basin.

(4) The prominent rise in sea level within the middle of cycle TA 2.7 can be correlated with the base of Division E, which marks the return (briefly) of marine silty clay facies to the Hampshire Basin.

The close correspondence of coastal onlap changes in the Hampshire Basin to the 'global' sea-level curve strongly suggests that the depositional sequences seen in the London Clay Formation reflect largely eustatic sea level changes. The greater complexity of the record in the Hampshire Basin, with a larger number of coastal onlap cycles, probably reflects superimposition of lower-order cyclicity on the third-order cycles recognised by Haq *et al.* (1987). There is evidently potential for further analysis of these patterns, now that reliable geomagnetic calibration has been achieved.

Conclusions

This study has demonstrated the potential of magnetostratigraphy for the detailed correlation of early Cenozoic sedimentary sequences in the

southern North Sea Basin. The magnetic polarity history of the Ypresian deposits is now
well-defined for the Anglo-Belgium Basin. The
combined litho-, bio- and magnetostratigraphic
data which are used to define an "event stratigraphy" for these deposits highlight the complexities in applying the global sea-level cycle charts
to a relatively small portion of a passive continental margin. Only when a variety of techniques is applied in an integrated manner is it
possible to confidentially identify and date third
and fourth-order cycles in epicontinental sequences.

Work on the northern European deposits has
enabled a high resolution correlation of the
standard biozone boundaries in these deposits
with the magneto-biostratigraphic time-scale.
The results make a significant contribution to
further refinement of the Palaeogene time-scale.

The Natural Environment Research Council is gratefully acknowledged for financial support to J. R. A,
during his PhD studies. We would like to thank N.
Johnston, K. Padley and L. Ali for invaluable help in
the laboratory and field, and especially J. Watson for
her care and patience in typing the manuscript and K.
Saull for drafting the figures. We are grateful to
various colleagues for helpful and stimulating discussions during the development of the research,
particularly R. Knox, E. Steurbaut and N. Hamilton.
We thank M.-P. Aubry, G. Jenkins, J. Kennett and R.
Knox for helpful and constructive reviews of the
paper.

References

ALI, J. R. 1989. *Magnetostratigraphy of early Palaeo-
gene sediments from N.W. Europe.* PhD thesis,
University of Southampton.
AUBRY, M.-P. 1986. Paleogene calcareous nanno-
plankton biostratigraphy of northwestern
Europe. *Palaeogeography, Palaeoclimatology,
Palaeoecology,* **55**, 267–334.
——, HAILWOOD, E. A. & TOWNSEND, H. A. 1986.
Magnetic and calcareous-nannofossil stratigraphy
of the lower Palaeogene formations of the Hamp-
shire and London Basins. *Journal of the Geologi-
cal Society, London,* **143**, 729–735.
BERGGREN, W. A., KENT, D. V. & FLYNN, J. J. 1985.
Palaeogene geochronology and chronostratigra-
phy. *In*: SNELLING, N. J. (ed.) *Geochronology of
the geological record.* Memoir of the Geological
Society, London, **10**, 141–195.
BIGNOT, G. 1962. Étude micropaléontologique de la
Formation de Varengeville du gisement éocene
du Cap d'ailly (Seine-Maritime). *Revue de Micro-
paléontologie* **5(3)**, 161–184.
BROWN, S. & DOWNIE, C. 1984. Dinoflagellate cyst
biostratigraphy of Late Paleocene and Early
Eocene sediments from Holes 551, 553A and 555,
Leg 81, Deep Sea Drilling Project (Rockall

Plateau). *In*: ROBERTS, D. G. *et al. Initial Reports
of the Deep Sea Drilling Project,* **81**, 565–579.
BUJAK, J. P., DOWNIE, C., EATON, G. L. & WILLIAMS,
G. L. 1980. *Dinoflagellate cysts and acritarchs
from the Eocene of southern England.* Special
Papers in Palaeontology. No. 24.
CHATEAUNEUF, J. J. & GRUAS-CAVAGNETTO, C. 1978.
Les zones de Wetzeliellaceae (Dinophyceae) du
Bassin de Paris. Comparison et correlations avec
les zones du Paleogene des bassins du Nord-Ouest
de l'Europe. *Bulletin du Bureau de Recherches
Geologique et Minieres Series 2, Sect. IV, No. 2,*
59–93.
COSTA, L. I. & DOWNIE, C. 1976. The distribution of
the dinoflagellate *Wetzeliella* in the Palaeogene of
North-Western Europe. *Palaeontology,* **19**, 591–
614.
—— & MANUM, S. 1988. The description of the
inter-regional zonation of the Palaeogene and the
Miocene. *In*: VINKEN, R. (ed.) *The Northwest
European Tertiary Basin. Geologische Jahrbuch,*
100, 321–330.
—— & MULLER, C. 1978. Correlation of Cenozoic
dinoflagellate zones from the NE Atlantic and
NW Europe. *Newsletters on Stratigraphy,* **77**,
65–72.
——, DENISON, C. N. & DOWNIE, C. 1978. The
Palaeocene/Eocene Boundary in the Anglo-Paris
Basin. *Journal of the Geological Society, London,*
135, 261–644.
CURRY, D. 1965. The Palaeogene Beds of SE England.
Proceedings of the Geologists' Association, **77**,
437–467.
DAVIS, A. G. 1936. The London Clay of Sheppey and
the location of its fossils. *Proceedings of the
Geologists' Association,* **47**, 328–345.
—— & ELLIOT, G. F. 1957. The palaeogeography of
the London Clay Sea. *Proceedings of the Geolo-
gists' Association,* **68**, 255–277.
DESTOMBES, P., GAMBLE, H. H., JUIGNET, P. & OWEN,
H. G. 1977. Cretaceous and lower Tertiary of
Seine-Maritime, France: a guide to key localities.
Proceedings of the Geologists' Association, **83**,
471–478.
EDWARDS, R. A. & FRESHNEY, E. C. 1987. Lithostrati-
graphical classification of the Hampshire Basin
Palaeogene deposits (Reading Formation to
Headon Formation). *Tertiary Research,* **8(2)**,
43–73.
FEUGUEUR, L. 1963. *L'Ypresien du Bassin de Paris.*
Memoire du Service de la Carte Geologique de
France Paris.
HAQ, B. U., HARDENBOL, J. & VAIL, P. R. 1987.
Chronology of fluctuating sea levels since the
Triassic. *Science,* **235**, 1156–1167.
ISLAM, M. A. 1983. Dinoflagellate cysts from the
Eocene cliff sections of the Isle of Sheppey,
southeast England. *Revue de Micropaleontologie,*
25, 231–250.
JOLLEY, D. W. & SPINNER, E. 1990. Some dinoflagell-
ate cysts from the London Clay (Palaeocene-
Eocene) near Ipswich, Suffolk, England. *Review
of Palaeobotany and Palynology,* **60**, 361–370.
KING, C. 1981. *The stratigraphy of the London Clay*

and associated deposits. Tertiary Research Special Paper. No. 6. W. Backhuys, Rotterdam.

—— 1984. The stratigraphy of the London Clay Formation and Virginia Water Formation in the coastal sections of the Isle of Sheppey (Kent, England). *Tertiary Research,* **5(3)**, 121–160.

—— 1990. Eocene stratigraphy of the Knokke borehole (Belgium). *Toelichtende Verhandelinger Geologische en Mijjnkaarten van Belgie,* **29**, 67–102.

—— 1991. Stratigraphy of the Ieper Formation and Argile de Flandres (Early Eocene) in western Belgium and northern France. *Bulletin de la Société Belge de Geologie,* **97**, 349–372.

KNOX, R. W. 1984. Nannoplankton zonation and the Palaeocene/Eocene boundary beds of NW Europe: an indirect correlation by means of volanic ash layers. *Journal of the Geological Society, London,* **141**, 993–999.

—— 1990. Thanetian and Early Ypresian chronostratigraphy in south-east England. *Tertiary Research,* **11**, 57–64.

——, MORTON, A. C. & HARLAND, R. 1981. Stratigraphical relationships of Palaeocene sands in the UK sector of the Central North Sea. *In*: ILLING, L. V. & HOBSON, G. D. (eds) *Petroleum Geology of the Continental Shelf of North West Europe.* Heyden & Sons, London, 14.

MARTINI, E. 1971. Standard Tertiary and Quaternary calcareous nannoplankton zonation. *In*: FARINACCI, A. (ed.) *Proceedings of the II Planktonik Conference, Roma 1970.* Edizioni Tecnoscienza, Rome, 739–785.

MOLYNEUX, L. 1971. A complete results magnetometer for measuring the remanent magnetisation of rocks. *Geophysical Journal of the Royal Astronomical Society,* **24**, 429–433.

MORTON, A. C., BACKMAN, J. & HARLAND, R. 1983. A reassessment of the stratigraphy of DSDP Hole 117A, Rockall Plateau: implications for the Palaeocene-Eocene boundary in NW Europe. *Newsletters in Stratigraphy,* **12(2)**, 104–111.

PLINT, A. G. 1983. Facies, environments and sedimentary cycles in the Middle Eocene Bracklesham Formation of the Hampshire Basin: evidence for global sea level changes? *Sedimentology,* **30**, 625–653.

POWELL, A. J. 1988. A modified dinoflagellate cyst biozonation for latest Palaeocene and earliest Eocene sediments from the central North Sea. *Review of Palaeobotany & Palynology,* **56**, 327–334.

PRESTWICH, J. 1847. On the main points of structure and the probable age of the Bagshot Sands. *Quarterly Journal of the Geological Society of London,* **3**, 378–409.

—— 1854. On the thickness of the London Clay. *Quarterly Journal of the Geological Society of London,* **10**, 401–419.

—— 1855. On the correlation of the Eocene Tertiaries of England, France and Belgium. *Quarterly Journal of the Geological Society of London,* **11**, 206–246.

—— 1888. Further observations on the correlation of the Eocene strata in England, Belgium, and the north of France. *Quarterly Journal of the Geological Society of London.* **44**, 88–111.

RIDDY, P. J. & HAILWOOD, E. A. 1989. Stationary-sample A. F. demagnetisation using a coupled cryogenic magnetometer/demagnetiser system (Abst.). *Geophysical Journal of the Royal Astronomical Society,* **88**, 597.

SHRUBSOLE, W. H. 1878. On the New Town Well at Sheerness. *Proceedings of the Geologists' Association,* **5**, 355–362.

STAMP, L. D. 1921. On cycles of sedimentation in the Eocene strata of the Anglo-Franco-Belgian Basin. *Geological Magazine,* **58**, 108–114, 146–157, 194–200.

STEURBAUT, E. 1987. The Ypresian in the Belgian Basin. *Bulletin de la Societé belge de geologie,* **96**, 339–351.

—— 1988. New Early and Middle Eocene calcareous nannoplankton events and correlation in middle to high latitudes of the Northern Hemisphere. *Newsletters in Stratigraphy,* **18(2)**, 99–115.

—— & NOLF, D. 1986. Revision of Ypresian stratigraphy of Belgium and northwestern France. *Mededelingen van Werkgroep voor Tertiaire en Kwartaire Geologie,* **23(4)**, 115–172.

STINTON, F. C. 1975. Fish otoliths from the English Eocene. *Monographs of the Palaeontographical Society,* **1**, 1–56.

TOWNSEND, H. A. 1982. *Magnetostratigraphy of Early Palaeogene sediments from southern England.* PhD Thesis, University of Southampton.

—— & HAILWOOD, E. A. 1985. Magnetostratigraphic correlation of Palaeogene sediments in the Hampshire and London Basins, southern UK. *Journal of the Geological Society, London,* **142**, 1–27.

VAN WAGONER, J. C., MITCHUM, R. M., CAMPION, K. M. & RACHMANIAN, V. D. 1990. *Siliciclastic sequence stratigraphy in well-logs, cores, and outcrops.* American Association of Petroleum Geologists. Methods in Exploration Series No. 7.

WHITAKER, W. 1871. On the cliff-sections of the Tertiary beds west of Dieppe in Normandy, and at Newhaven in Sussex. *Quarterly Journal of the Geological Society, London,* **27**, 263–268.

—— 1872. *The geology of the London Basin: Part 1.* Memoir of the Geological Survey, UK.

WILLEMS, W., BIGNOT, G. & MOORKENS, T. 1981. Ypresian. *In*: POMEROL, C. (ed.) *Stratotypes of Paleogene Stages. Bulletin d'Information des Geologues du Bassin de Paris, Memoire h.s.,* **2**, 267–299.

WRIGLEY, A. G. & DAVIS, A. G. 1937. The occurrence of Nummulites planulatus in England, with a revised correlation of the strata containing it. *Proceedings of the Geologists' Association,* **48**, 203–228.

ZIJDERVELD, J. D. A. 1967. AC demagnetization of rocks: analysis of results. *In*: COLLINSON, D. W., CREER, K. M. & RUNCORN, S. K. (eds) *Methods in Palaeomagnetism.* Elsevier, New York, 254–286.

The late Cenozoic *Globorotalia truncatulinoides* datum-plane in the Atlantic, Pacific and Indian oceans

D. GRAHAM JENKINS[1] & PAUL GAMSON[2]

[1] *Department of Geology, National Museum of Wales, Cardiff CF1 3NP, UK*
[2] *Department of Earth Sciences, Open University, Milton Keynes MK7 6AA, UK*

Abstract: Since 1967 it has been thought that the planktonic foraminiferal taxon *Globorotalia truncatulinoides* (d'Orbigny) evolved in the oceans at about 1.9 Ma, with an anomalous date of 2.5 Ma for its first appearance in the Southwest Pacific. A new model is presented which demonstrates that *G. truncatulinoides* evolved in the Southwest Pacific at 2.5 Ma and later spread, while still evolving, into the South Atlantic via Drake Passage and then into the North Atlantic arriving there at 1.9 Ma; it also migrated and arrived in the Central and North Pacific at this time. The diachronous nature of the first appearances of both *G. truncatulinoides*, and *Globorotalia inflata* (d'Orbigny) which also evolved in the Southwest Pacific and later spread to other oceans, has implications for first appearances of species in the Cenozoic that have been labelled 'datum-planes'.

Accurate biostratigraphic correlations of sedimentary sequences are dependent on a number of hypotheses. The evolutionary initial appearance of a species has been held to be one of the most reliable markers for correlation of Cenozoic marine sequences (Jenkins 1965, 1973).

Berggren *et al.* (1967) recorded the first evolutionary appearance of *Globorotalia truncatulinoides* (d'Orbigny) during the Olduvai normal polarity magnetic subchron, in core CH61-171 at lat. 26°N in the subtropical North Atlantic at 1.85 Ma (Fig. 1). This evolutionary event was subsequently used, up to 1986 (when the International Geological Congress decided on another boundary), as a marker for the Pliocene/Pleistocene boundary; later Berggren *et al.* (1985) revised the *G. truncatulinoides* datum to 'just below the Olduvai event' at 1.9 Ma.

In New Zealand it has been known for some time that *G. truncatulinoides* first appeared at *c.* 2.5 Ma (Jenkins 1971), and this was confirmed by plotting DSDP Leg 90 results against the palaeomagnetic record at four Sites (587, 588, 590, 592) (Dowsett 1988; Jenkins 1992).

A model is proposed which explains the evolution of *G. truncatulinoides* in the Southwest Pacific and its later apparent evolutionary appearance in the North Atlantic.

Oceanographic distribution

The modern distribution of *G. truncatulinoides* has an important bearing on its evolution and migration in the Late Pliocene. *G. truncatulinoides* lives below 100 m and although it has a widespread distribution between Latitude 60°S and 60°N, its greatest numbers occur within mid-latitude water-mass gyres in the North and South Atlantic, and in the Indian and South Pacific oceans (Bé 1977). In the Indian Ocean it lives in waters with a relatively high salinity (35–55‰) and low phosphate concentrations (Bé 1977).

Palaeogeographic distribution

The initial discovery by Berggren *et al.* (1967) that *G. truncatulinoides* had evolved from *Globorotalia tosaensis* Takayanagi and Saito in the subtropical Atlantic at 1.85 Ma was later followed by intensive research in the North Atlantic. Backman & Shackleton (1983) re-examined core CH61–171 and re-calibrated the first appearance of *G. truncatulinoides* to be slightly older than 2.0 Ma. Weaver (1986) and Weaver & Clement (1986) were able to date the first appearance of *G. truncatulinoides* at DSDP Sites 606, 607, 609 and 610 between latitudes 37°–53°N in the North Atlantic as 1.80–1.85 Ma; all these sites had good palaeomagnetic records. Weaver (1986) recorded rare *G. tosaensis* at each of the sites, but only at the southernmost Site 606 was the evolutionary transition documented.

Further evidence of the first appearance of *G. truncatulinoides* at *c* 1.85 Ma came from (1) the central Pacific (Hays *et al.* 1969; Saito *et al.* 1975; Haq *et al.* 1977), (2) the South Atlantic (Haq *et al.* 1977; Berggren *et al.* 1983; Pujol 1983).

Lamb (1969) placed the first appearance of *G. truncatulinoides* at 2.4 Ma in the Gulf of Mexico but this stratigraphic position was qualified thus: 'Age approximate; based solely on stratigraphic position'. Lamb considered the age of the

From HAILWOOD, E. A. & KIDD, R. B. (eds), *High Resolution Stratigraphy*
Geological Society Special Publication, No. 70, pp. 127–130.

127

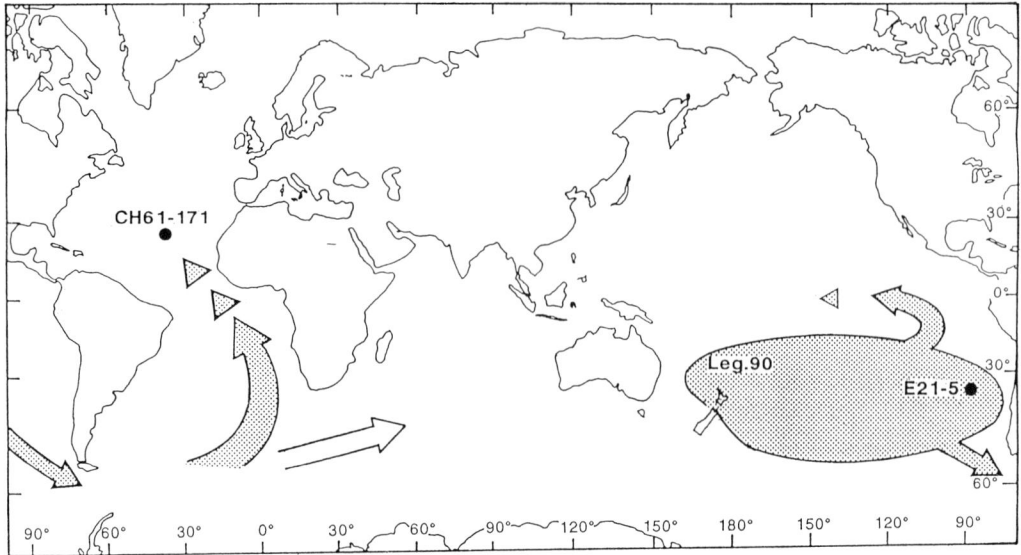

Fig. 1. Locations of cores CH61–171, E21–5 and DSDP Leg 90. Shaded area in the South Pacific is where *G. truncatulinoides* evolved prior to its migration at *c.* 1.9 Ma (shaded arrows); its time of migration into the Indian Ocean is not known (unshaded arrow).

Pliocene/Pleistocene boundary to be 2.8 Ma and this hypothesis was further fostered by Smith & Beard (1973) in their work on DSDP Leg 10 in the Gulf of Mexico; the age of the boundary was predicted to coincide with the first major cooling in the northern hemisphere. A more modern view was advanced by Saunders *et al.* (1973) from their work on DSDP Leg 15 in the Caribbean Sea when they stated: 'The lowest occurrence of *Globorotalia truncatulinoides truncatulinoides* is widely accepted as the base of the Pleistocene and is considered to have an age of 1.8 million years.' Since there is a strong oceanic connection between the Caribbean Sea and the Gulf of Mexico, it is here assumed that *G. truncatulinoides* appeared in both areas at *c.* 1.8 Ma.

The first indication that the first appearance of *G. truncatulinoides* in the South Pacific was different came from Kennett & Geitzenauer (1969) who analysed Eltanin Core 21–5 from the Southeast Pacific (Lat. 36° 41′S; Long. 93° 38′W); after identifying the Pliocene/Pleistocene boundary at the extinction of some discoasters (*D. surculus* and *D. pentaradiatus*) they stated that *G. truncatulinoides* appeared earlier than this event and cautioned its use as a marker. Kennett & Geitzenauer (1969) also pointed out that the appearance of *G. truncatulinoides* in different areas was diachronous. Kennett (1973) was of the opinion that the evolution

of *G. truncatulinoides* from *G. tosaensis* was complex and that *G. truncatulinoides* appeared at different times in various water masses. In the Southwest Pacific area Jenkins (1971) recorded the first appearance of *G. truncatulinoides* before 2.5 Ma in New Zealand, and Dowsett (1988) and Jenkins (1992) recorded its first evolutionary appearance at the DSDP sites 587, 588, 590, 592 to be at *c.* 2.5 Ma. Scott *et al.* (1990) recorded the first appearance of *G. truncatulinoides* at *c.* 2.0 Ma in the lower Nukumaruan in New Zealand which, on the modified Berggren *et al.* (1985) time-scale of Hornibrook *et al.* (1989), is at *c.* 2.5 Ma. The stratigraphic ranges of *G. tosaensis* and *G. truncatulinoides* in the North Atlantic, Southwest Pacific and in New Zealand are shown in Fig. 2.

Evolution and migration

Cifelli & Scott (1986), after reviewing the data on the evolution of *G. truncatulinoides* in the world's oceans, concluded that 'there was not necessarily a uniform phyletic rate of development'. If this interpretation was correct, it meant that *G. truncatulinoides* had evolved from *G. tosaensis* at different times in different oceans. The consequence of this interpretation was that similar developments of keels in Cenozoic planktonic foraminifera could also be varyingly

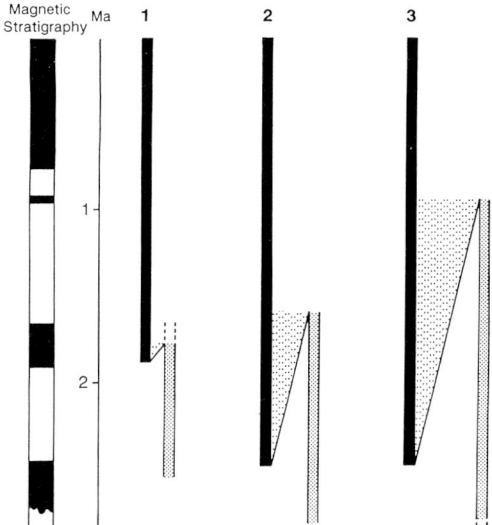

Magnetic Stratigraphy

Fig. 2. Stratigraphic ranges of *G. truncatulinoides* (black) and *G. tosaensis* (dense stippling) plotted against the paleomagnetic and radiometric time scales; the lightly stippled areas show the evolutionary transition between the two species. 1, North Atlantic (Lats 37°–53°N); 2, Southwest Pacific (Lats 26°–40°S); 3, New Zealand (Lats 39°–40°S). Magnetic polarity timescale is shown (black, normal polarity; white, reverse polarity).

diachronous, and the use of such morpho-characters as stratigraphic markers for accurate correlations would be open to question.

An alternative model is presented here. *G. truncatulinoides* evolved in the South Pacific gyre at *c.* 2.5 Ma and between 2.0–1.9 Ma it spread northwards to the Central and North Pacific and also eastwards into the North Atlantic via Drake Passage and the South Atlantic (Fig. 1). *G. truncatulinoides* reached the southern part of the North Atlantic by 1.9 Ma, and later spread northeastwards reaching DSDP Site 611 at 1.35 Ma (Weaver & Clement 1986) and the Goban Spur Site 550 at 0.95 Ma (Pujol & Duprat 1985). In the Southern Ocean, the much later arrival of *G. truncatulinoides* in the northern Subantarctic waters, at 0.30 Ma, was recorded by Kennett (1970).

There is some evidence that *G. truncatulinoides* also spread into the North Pacific at *c.* 1.8 Ma (Keller 1978) but there are no data on the Indian Ocean.

Support for the hypothesis that *G. truncatulinoides* evolved in the South Pacific and later spread into other oceans comes from the study of the *Globorotalia inflata* (d'Orbigny) evolutionary lineage. It has long been established

that *Globorotalia puncticulata* (Deshays) evolved into *G. inflata* in the southwest Pacific (Walters 1965; McInnes 1965; Kennett 1973; Jenkins 1975; Scott 1980; Malmgren & Kennett 1981). *G. inflata* first appeared in the Southwest Pacific at 2.9 Ma (Malmgren & Kennett 1981) but did not appear in the North Atlantic until 2.00–2.26 Ma at DSDP sites 606, 607, 609, 610 and 611 (Weaver & Clement 1986). Weaver (1986) rightly concluded that the appearance of *G. inflata* in the North Atlantic was a migrational event and this was confirmed by Jenkins *et al.* (1988) in their study of the Coralline Crag and DSDP Leg 94, Site 609B.

Therefore *G. inflata* evolved in the Southwest Pacific and migrated to the North Atlantic via the Drake Passage and South Atlantic. The same sequence of events seems to have been repeated in the *G. truncatulinoides* lineage about 0.1–0.2 Ma later.

Conclusions

The study of the *G. truncatulinoides* and *G. inflata* lineages clearly demonstrates that some Cenozoic planktonic foraminifera evolved in the South Pacific and later spread into other oceans and water masses. The implications are that some of the well documented evolutionary lineages in the Cenozoic may show similar patterns of evolution being limited to discrete oceanic water masses followed later by migration into other oceans (Jenkins 1992). If this is true, then some of these so-called 'datum planes' are diachronous. The South Pacific appears to have been a fertile area for evolution and speciation.

We would like to thank J. P. Kennett for his valuable suggestions for improving the manuscript.

References

BACKMAN, J. & SHACKLETON, N. J. 1983. Quantitative biochronology of Pliocene and early Pleistocene calcareous nannoplankton from the Atlantic, Indian and Pacific Oceans. *Marine Micropaleontology,* **8,** 141–170.

BÉ, A. W. H. 1977. An ecological, zoogeographic and taxonomic review of Recent planktonic foraminifera. *In*: RAMSAY, A. T. S. (ed.) *Oceanic Micropalaeontology,* **1,** 1–100, Academic Press.

BERGGREN, W. A., PHILLIPS, J. D., BERTELS, A. & WALL, D. 1967. Late Pliocene-Pleistocene stratigraphy in deep sea cores from the South-Central North Atlantic. *Nature,* **216**(5112), 253–255.

——, AUBRY, M. P. & HAMILTON, N. 1983. Neogene magnetostratigraphy of Deep Sea Drilling Project Site 516 (Rio Grande Rise, South Atlantic). *In*: BARKER, P. F., CARLSON, R. L. *et al. Initial Reports of the Deep Sea Drilling Project,* **72,**

675–713, Washington (U.S. Govt. Printing Office).

——, KENT, D. V. & VAN COUVERING, J. A. 1985. Neogene geochronology and chronostratigraphy. In: SNELLING, N. J. (ed.) The Chronology of the Geological Record. Geological Society of London, Memoir, 10, 211–260.

CIFELLI, R. & SCOTT, G. 1986. Stratigraphic record of the Neogene globorotalid radiation (Planktonic Foraminiferida). Smithsonian Contributions to Paleobiology, 58, 1–101.

DOWSETT, H. J. 1988. Diachrony of late Neogene microfossils in the southwest Pacific Ocean: application of the graphic correlation method. Paleoceanography, 3(2), 209–222.

HAYS, J. D., SAITO, T., OPDYKE, N. D. & BURCKLE, L. H. 1969. Pliocene-Pleistocene sediments of the equatorial Pacific: their paleomagnetic, biostratigraphic and climatic record. Geological Society of America, Bulletin, 80, 1481–1514.

HAQ, B. U., BERGGREN, W. A. & VAN COUVERING, J. A. 1977. Corrected age of the Pliocene/Pleistocene boundary. Nature, 269, 483–488.

HORNIBROOK, N. DE B., BRAZIER, R. C. & STRONG, C. P. 1989. Manual of New Zealand Permian to Pleistocene Foraminiferal Biostratigraphy. New Zealand Geological Survey Bulletin, 56, 1–175.

JENKINS, D. G. 1965. Planktonic foraminifera and Tertiary intercontinental correlations. Micropalaeontology, 11(3), 265–277.

—— 1971. The reliability of some Cenozoic planktonic foraminiferal "datum-planes" used in biostratigraphic correlation. Journal of Foraminiferal Research, 1(2), 82–86.

—— 1973. The present status and future progress in the study of Cenozoic planktonic foraminifera. Revista Espanola Micropaleontologia, 5, 133–146.

—— 1975. Cenozoic planktonic foraminiferal biostratigraphy of the southwestern Pacific and Tasman Sea – DSDP Leg 29. In: Initial Reports of the Deep Sea Drilling Project, 29, 449–467, Washington (U.S. Govt. Printing Office).

—— 1992. The paleogeography, evolution and extinction of Late Miocene-Pleistocene planktonic foraminifera from the southwest Pacific. In: ISHIZAKI, K. & SAITO, T. (eds) Centenary of Micropaleontology in Japan, Terra Scientific Publ. Co. Tokyo, 27–35.

—— 1988. Planktonic foraminifera from the Pliocene Coralline Crag of Suffolk, Eastern England. Journal of Micropalaeontology, 7(1), 1–10.

KELLER, G. 1978. Late Neogene planktonic foraminiferal biostratigraphy and palaeoceanography of the northern Pacific: Evidence from DSDP Sites 173 and 310 at the North Pacific Front. Journal of Foraminiferal Research, 8(4), 332–349.

KENNETT, J. P. 1970. Pleistocene paleoclimates and foraminiferal biostratigraphy in subantarctic deep-sea cores. Deep-Sea Research, 17, 125–140.

—— 1973. Middle and late Cenozoic planktonic foraminiferal biostratigraphy of the southwest Pacific – DSDP Leg 21. In: Initial Reports of the Deep Sea Drilling Project, 21, 575–640.

—— & GEITZENAUER, K. R. 1969. Pliocene-Pleistocene Boundary in a South Pacific deep-sea core. Nature, 224, 899–901.

LAMB, J. L. 1969. Planktonic foraminiferal datums and late Neogene Epoch boundaries in the Mediterranean, Caribbean, and Gulf of Mexico. Transactions of the Gulf Coast Association of Geological Societies, 19, 559–578.

McINNES, B. A. 1965. Globorotalia miozea Finlay as an ancestor of Globorotalia inflata (d'Orbigny). New Zealand Journal of Geology and Geophysics, 8(1), 104–108.

MALMGREN, B. A. & KENNETT, J. P. 1981. Phyletic gradualism in a Late Cenozoic planktonic foraminiferal lineage, DSDP Site 284: southeast Pacific. Paleobiology, 7, 230–240.

PUJOL, C. 1983. Cenozoic planktonic foraminiferal biostratigraphy of the southwest Atlantic (Rio Grande Rise). In: BARKER, P. F., CARLSON, R. L. et al. Initial Reports of the Deep Sea Drilling Project, 72, 645–656, Washington (U.S. Govt. Printing Office).

—— & DUPRAT, J. 1985. Quaternary and Pliocene planktonic foraminifers of the northeastern Atlantic (Goban Spur), DSDP Leg 80. In: DE GRACIANSKY, P. C., POAG, C. W. et al. Initial Reports of the Deep Sea Drilling Project, 80, 683–723, Washington (U.S. Govt. Printing Office).

SAITO, T., BURCKLE, L. H. & HAYS, J. D. 1975. Late Miocene to Pleistocene biostratigraphy of equatorial Pacific sediments. In: SAITO, T. & BURCKLE, L. H. (eds) Late Neogene Epoch Boundaries. Micropaleontology Special Paper, 1, 226–244, New York.

SAUNDERS, J. B., EDGAR, N.T., DONNELLY, T. W. & HAY, W. W. 1973. Cruise Synthesis. Initial Reports of the Deep Sea Drilling Project, 15, 1077–1111.

SCOTT, G. H. 1980. Globorotalia inflata lineage and G. crassaformis from Blind River, New Zealand: recognition, relationship and use in latest Miocene–Early Pliocene biostratigraphy. New Zealand Journal of Geology and Geophysics, 23, 665–677.

——, BISHOP, S. & BURT, B. J. 1990. Guide to some Neogene globorotalids (Foraminiferida) from New Zealand. New Zealand Geological Survey Bulletin, 61, 1–135.

SMITH, L. A. & BEARD, T. H. 1973. The late Neogene of the Gulf of Mexico. Initial Reports of the Deep Sea Drilling Project, 10, 643–677.

WALTERS, R. 1965. The Globorotalia zealandica and G. miozea lineages. New Zealand Journal of Geology and Geophysics, 8(1), 109–127.

WEAVER, P. P. E. 1986. Late Miocene to Recent planktonic foraminifera from the North Atlantic: DSDP Leg 94. In: Initial Reports of the Deep Sea Drilling Project, 94, 815–829, Washington (U.S. Govt. Printing Office).

—— & CLEMENT, B. M. 1986. Synchroneity of Pliocene planktonic foraminiferal datums in the North Atlantic. Marine Micropaleontology, 10, (1986), 295–307.

Eocene–Miocene high latitude biostratigraphy

SALLY S. RADFORD & LI QIANYU

Department of Geology, Royal School of Mines, Imperial College London,
Prince Consort Road, London SW7 2BP, UK

Abstract: Studies of high-latitude planktonic foraminiferal faunas indicate that certain biostratigraphic problems may be resolved by the application of bioevents among the three microperforate families, Guembelitriidae, Chiloguembelinidae and Tenuitellidae. With the aid of the scanning electron microscope, these small tests appear to be useful stratigraphic markers in the condensed sequences of deep-sea cores, where their evolution can be correlated with magnetostratigraphy.

In the early–mid-Eocene, the *Jenkinsina–Cassigerinelloita* lineage assists the recognition of zones based on the larger, muricate *Acarinina*. Four zones are proposed in a revised lineage zonation of the tenuitellids, presented as a complementary biostratigraphic scheme for differentiating the Eocene–Oligocene boundary interval in high latitudes: (1) the Praetenuitella insolita Zone in the late Eocene; (2) the Tenuitella gemma Zone in the early Oligocene; (3) the Tenuitella munda Zone in the mid-Oligocene; (4) the Tenuitellinata angustiumbilicata Zone in the late Oligocene.

The foundations of high-latitude biostratigraphy

Early studies of high latitude planktonic fora-minifera in the Northern Hemisphere were centred around the USSR (Subbotina 1953) and Scandinavia (Troelsen 1957; Berggren 1960; Hofker 1960). Early work in the Southern Hemisphere was pioneered in New Zealand by Finlay (1939), and Hornibrook (1958). In a series of papers, Jenkins (1963, 1964, 1965, 1966, 1985) laid the foundations of modern high-latitude biostratigraphy. Recent studies based on DSDP and ODP Sites have greatly increased the amount of data on high latitude planktonic foraminifera (Kennett 1973) and an Austral biostratigraphic zonation scheme has recently been compiled for the Palaeogene (Stott & Kennett 1990).

Austral biostratigraphy

Berggren (1991) described rich planktonic for-aminiferal faunas from ODP Leg 120 Sites 747, 748 and 749 south of the Polar Front in the Indian Ocean. The stratigraphic distribution of small species of the microperforate families Chiloguembelinidae, Guembilitriidae and Tenuitellidae was investigated by Li *et al.* (1991). Bioevents in these taxa complement the Austral biostratigraphic scheme of Stott & Kennett (1990) and assist in the identification of biozones based on larger species. Originations and extinctions of certain microperforate species are correlated with the magnetic stratigraphy

and provide a high resolution zonation for the Eocene–Oligocene boundary interval.

Magnetobiostratigraphy and Microperforate Planktonic Foraminifera

The Eocene Lineage of Jenkinsina and Cassigerinelloita

Figure 1 shows the stratigraphic distribution of the microperforate *Jenkinsina triseriata* and its descendant *Cassigerinelloita amekiensis*. No palaeomagnetic data exist for the mid-Eocene interval in Site 749 but an indirect correlation is possible with these species and the magnetic stratigraphy of Austral Zones AP8 to AP10 of Stott & Kennett (1990). Both species occur in ODP Leg 113 Site 689 and are illustrated by Stott & Kennett (1990, plate 5). Their fig. 9 shows *J. triseriata*, as *Chiloguembelitria* sp. and figs 11–14 show *C. amekiensis* as *Globigerina* sp. B. In Site 689, *C. amekiensis* appeared in the Early Eocene in the *Acarinina wilcoxensis berggreni* Zone AP6 and extended to Zone AP8 of the mid-Eocene. Its origin is thus correlated with the series of short, normal polarity intervals from Chron 24N to Chron 23N. These Antarctic records confirm the stratigraphic range in the type locality of Nigeria (Stolk 1965). An isotopic temperature of 14° C was registered by *C. amekiensis* in Site 749 and this explains its tropical to polar distribution in the mild oceans of the Eocene.

In Site 689 *J. triseriata* is recorded from a higher level, in zone AP9 but its first appearance

From HAILWOOD, E. A. & KIDD, R. B. (eds), *High Resolution Stratigraphy*
Geological Society Special Publication, No. 70, pp. 131–136.

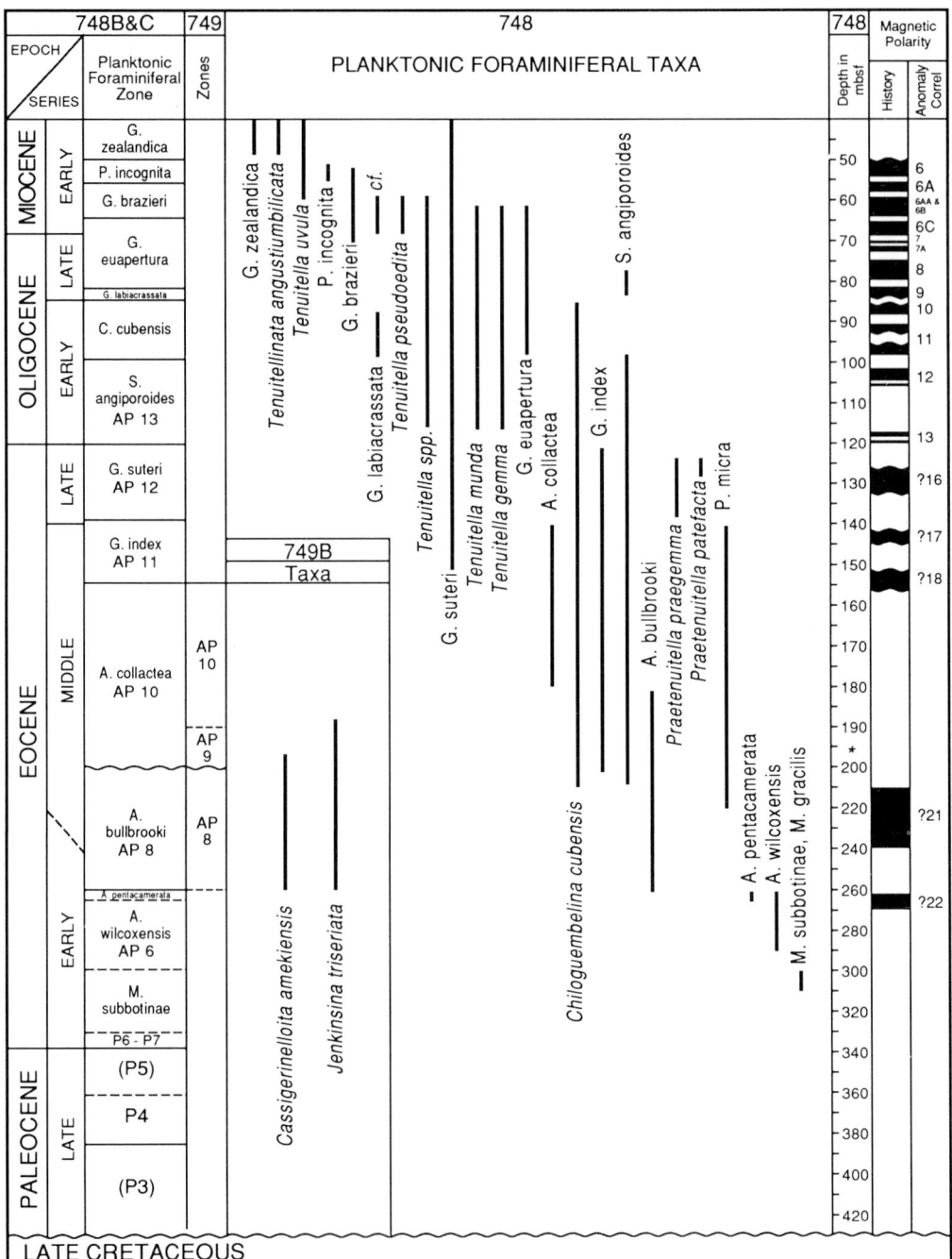

Fig. 1. Stratigraphic range of significant planktonic foraminifera in ODP Holes 748B & C (Berggren 1990) and 749B (inset; Li & Radford 1990). The distribution of the microperforate tenuitellids (in italics) is shown with that of the major zonal indices (in roman type). Chrons 21 and 22 and Zones AP6–AP13 from Stott & Kennett (1990). * Note scale change.

is not given. Its origin thus cannot be correlated with magnetostratigraphy.

Both species coexisted in Site 749 in the Middle Eocene Zone AP8 of *A. bullbrooki* (Li & Radford 1991). The acme of *C. amekiensis* precedes its rapid decline and extinction, and is followed by a similar pattern in *J. triseriata* at the top of Zone AP9. The decline and extinction of this distinctive lineage of microperforate species occurs within a reversed polarity interval correlated with magnetic Chron C20R (between normal polarity Chrons C21N and C20N).

The Late Eocene–Oligocene lineage of Praetenuitella and Tenuitella

Figure 1 shows the distribution of the microperforate genera, *Praetenuitella* and *Tenuitella* in Site 748, along with muricate, medioperforate and macroperforate zonal indices.

Praetenuitella evolved in the reversed polarity interval between Chron 17N and Chron 16N, in the *G. suteri* zone AP12 of Stott & Kennett (1990), defined at Sites 748 and 749. The entire lineage, *P. insolita*, *P. praegemma* and *P. patefacta* was recorded in these Sites (Li *et al.* 1991; Berggren 1991). *P. insolita* was also re-

corded in Antarctica (Stott & Kennett 1990) and in New Zealand (Jenkins 1966).

Tenuitella evolved from *Praetenuitella* in the reversed polarity interval above Chron 13N (Fig. 1). The plexus originates with *T. gemma*, after the global extinction of the late Eocene marker, *Globigerinatheka index* (AP 11). Most records confirm an early Oligocene age for typical *T. gemma*, but primitive *T. gemma* appeared in the late Eocene in the Eastern Indian Ocean (Gamson 1990) and in Site 689 (Stott & Kennett, 1990). Although rare, *T. gemma* is widely distributed in high and middle lattitudes (Jenkins 1985). In the Austral Oceans the plexus evolved and proliferated through *T. munda* and *Tenuitellinata juvenilis* in the mid-Oligocene. The earliest Miocene extinctions of *Tenuitella gemma* and *T. munda* correlate with Chron 6AAN in Site 748 (fig. 1) and with the evolution of *Tenuitellinata selleyi*, followed by *Tenuitella jamesi*. (Li *et al.* 1991). The lineage radiated with the establishment of typical *Tenuitellinata angustiumbilicata* in the late Oligocene in Chron 9N and with *T. pseudoedita*, the origin of which correlates with Chron 6CN, at the Oligocene–Miocene boundary. Finally, the evolution and radiation of *T. uvula* correlates with Chron 6AN, in the Early Miocene.

Epoch		N/P zones	AP zones	Sites 747A and 749B zonation		Lineage zonation of microperforate species	Datum
Miocene	middle			*N. nympha*		(unzoned)	< LAD *T. pseudoedita/T. minutissima*
	early	N8 - N10 (-?N12)		*G. woodi - G. praescitula*		*T. pseudoedita*	
							< FAD *T. jamesi*
		N7		*G. zealandica/ G. pseudomiozea*		*T. clemenciae*	
							< FAD *T. clemenciae/T. selleyi*
		N6				*T. minutissima*	
							< LAD *T. angustiumbilicata*
		N4 - N5		*P. incognita/ P. semivera*		*T. juvenilis*	
							< FAD *G. glutinata* s.s.
Oligocene	late	P22	?	*S. euapertura*	*S. labiacrassata*	*T. angustiumbilicata*	< FAD *T. uvula*
		P21b	AP14b			*T. munda*	
							< LAD *C. cubensis*
	early	P21a	AP14a		*C. cubensis*	*T. gemma*	
		P18 - P20	AP13	*S. angiporoides*			
							< FAD *T. gemma*
late Eocene		P16 - P17	AP12	*P. insolita*		*P. insolita*	
							< FAD *P. insolita*
		P15		*G. suteri*		(unzoned)	

Fig. 2. Correlation of revised lineage zones of *Praetenuitella* and *Tenuitella* with the Austral Zonation. (Modified from Li *et al.* 1991).

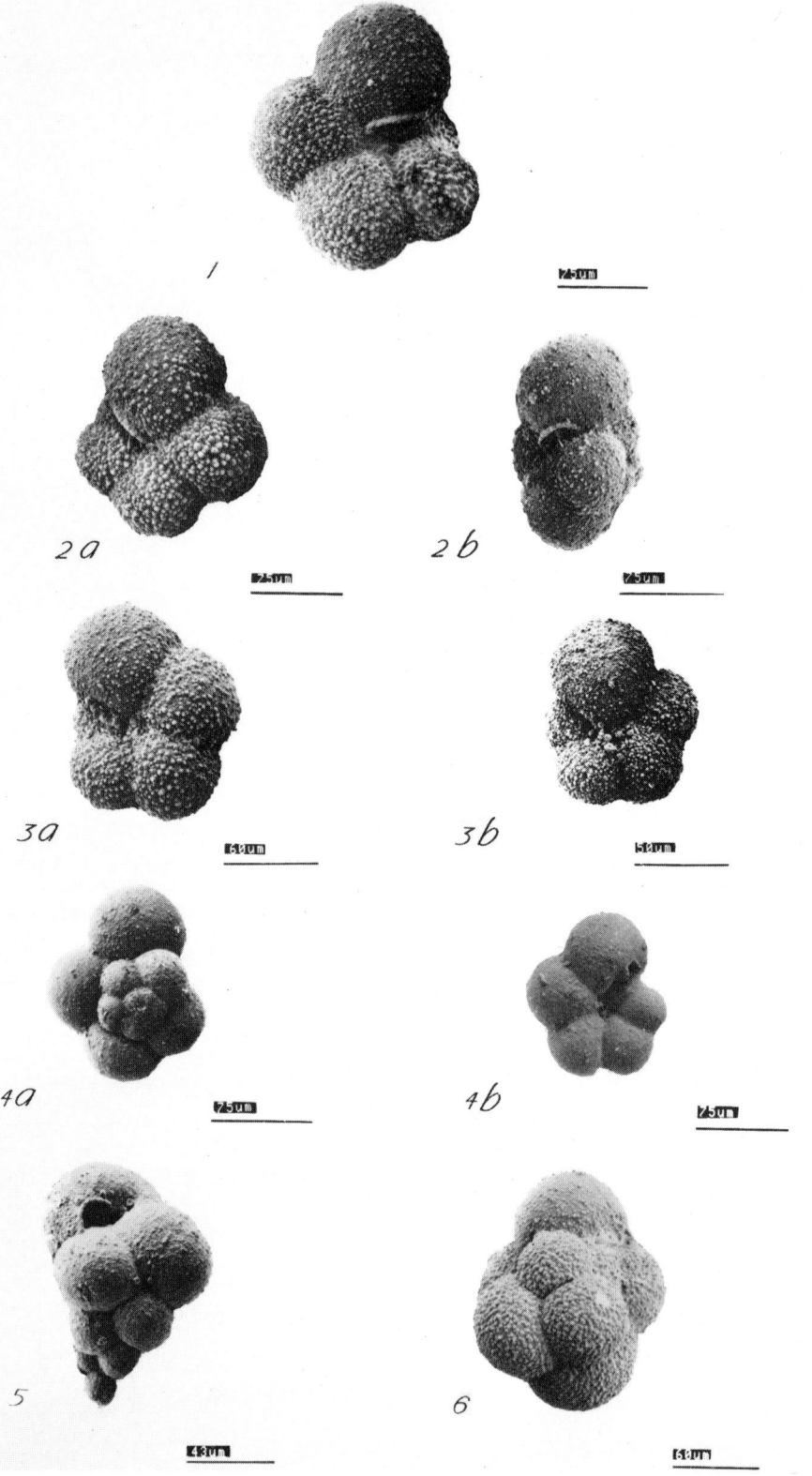

Fig. 3. Illustrations of microperforate foraminiferids discussed in this paper. (**1**) *Tenuitella angustiumbilicata* (Bolli), Late Oligocene; (**2**) *Tenuitella munda* (Jenkins), Mid-Oligocene; (**3**) *Tenuitella gemma* (Jenkins), Early Oligocene; (**4**) *Praetenuitella insolita* (Jenkins), Late Eocene; (**5**) *Jenkinsina triseriata* (Terquem), Mid-Eocene; (**6**) *Cassigerinelloita amekiensis* Stolk, Mid-Eocene. All from the Kerguelen Plateau, Southern Indian Ocean, ODP Sites 120–749B and ODP 120–747a (4b).

Revision of lineage zones of microperforate taxa

A revision of the lineage zones published by Li *et al.* (1991) is presented in Fig. 2. The species are illustrated in Fig. 3.

The Praetenuitella insolita Zone

This zone is present in several localities in the southern hemisphere. It was first established at DSDP Site 77 in the eastern Equatorial Pacific Ocean (Jenkins & Orr 1972). The top was defined by the extinction of *P. insolita* and the base was undefined.

It was recognized in ODP Sites 747 and 749 by the total range of the zonal marker, *P. insolita* and other *Praetenuitella* species (Li *et al.* 1991). It is thus a taxon range zone. However, since it is a segment of the tenuitellid lineage, the top is here re-defined by the first appearance of the genus *Tenuitella* which is correlated with magnetic Chron 13N. This event, represented by typical *T. gemma*, is more widespread and easily recognisable than the extinction of the rarer *P. insolita*. The base is defined by the first appearance of *Praetenuitella insolita*, in the reversed polarity interval between Chron 17N and Chron 16N. Other species present in the zone include *P. praegemma* and *P. patefacta*.

Figure 1 shows the stratigraphic distribution of *Praetenuitella*.

The Tenuitella gemma Zone

Jenkins (1985, fig. 4) correlated *Tenuitella gemma*, used in Australia as a zonal marker, along with *G. angiporoides* and *G. brevis*, with the *G. brevis* Zone at the Eocene–Oligocene boundary, the base of which correlates with the top of the *P. insolita* Zone. The first appearance of *T. gemma* was not clearly differentiated at ODP Sites 747, 748 and 749 where the Eocene–Oligocene boundary sequence was either condensed or missing. In view of its global distribution and particularly its presence in high latitudes, however, *T. gemma* is a useful index for the early Oligocene.

The *T. gemma* zone is here proposed as a segment of the tenuitellid lineage zonation (Fig. 2).
Category: Concurrent range zone
Age: Early Oligocene
Definition: Base defined by initial appearance of typical *Tenuitella gemma*, the nominate taxon, in Chron 13N; top defined by extinction of *Chiloguembelina cubensis*, a global datum, correlated with magnetic Chrons 9–10N.

Characteristic species: In addition to the zonal index, other tenuitellids appear: *Tenuitella munda*, *Tenuitellinata angustiumbilicata*, *T. juvenilis*. The base correlates with the *Subbotina angiporoides* Zone AP13 and the top with the *Chiloguembelina cubensis* Zone AP14, in the Austral scheme.

The Tenuitella munda Zone

The age of this zone is here modified to mid-Oligocene. The base is defined by the extinction of *Chiloguembelina* and the top by the appearance of *Tenuitellinata uvula*. It is characterized by the appearance of abundant, typical, four-chambered *T. munda*. It is equivalent to Austral zone AP14b (AP15 in Berggren 1991) and Tropical zone P21b.

Tenuitellinata angustiumbilicata Zone

The age of this zone is maintained as late Oligocene. The base is defined by the appearance of *Tenuitellinata uvula* and the top by the appearance of the bullate *Globigerinita glutinata* and *G. naparimaensis*. The zone is characterised by the abundance of typical *T. angustiumbilicata*, *T. juvenilis* and *G. praestainforthi*. It is equivalent to the *G. euapertura* Zone, AP16, (Berggren 1991) and Tropical zone P22.

This study was initiated by W. Berggren, shipboard micropalaeontologist on ODP Leg 120. We are grateful to the ODP for providing samples. R. Corfield measured the isotopes at Oxford University. Financial support was provided by the Hui Fellowship at Imperial College and we thank the staff for assistance, A. Brown for drafting the figures and A. Cash for photography. This paper is dedicated to F. Banner of University College London for guidance and encouragement.

References

BERGGREN, W. A. 1960. Some planktonic foraminifera from the Lower Eocene (Ypresian) of Denmark and northwestern Germany. *Stockholm Contributions in Geology*, **5**, 41–108.
—— 1991. Paleogene planktonic foraminiferal magneto-biostratigraphy of the Southern Kerguelen Plateau: ODP 120 Sites 747–749. *Proceedings of the ODP, Scientific Results*, **120**, College Station, TX (Ocean Drilling Program).
FINLAY, H. J. 1939. New Zealand foraminifera: key species in stratigraphy. No. 2. *Transactions and Proceedings of the Royal Society of New Zealand*, **69**, 89–128.
GAMSON, P. 1990. *Late Eocene to Early Miocene Planktonic Foraminifera of the Indian and Atlantic Oceans* PhD thesis, Open University.

HOFKER, J. 1960. Planktonic foraminifera in the Danian of Denmark. *Contributions of the Cushman Foundation on Foraminteral Research*, **11**, 73–86.

HORNIBROOK, N. DE B. 1958. New Zealand foraminifera: Key species in stratigraphy – No. 6. *New Zealand Journal of Geology and Geophysics*, **4**, 653–76.

JENKINS, D. G. 1963. New Zealand mid-Tertiary stratigraphical correlation. *Nature*, **200**, 1087.

—— 1964. A new planktonic foraminiferal subspecies from the Australasian Lower Miocene. *Micropaleontology*, **10**, 72.

—— 1965. The genus *Hantkenina* in New Zealand. *New Zealand Journal of Geology and Geophysics*, **8**, 518–526.

—— 1966. Planktonic foraminiferal zones and new taxa from the Danian to Lower Miocene of New Zealand. *New Zealand Journal of Geology and Geophysics*, **8**, 1088–1126.

—— 1985. Southern mid-latitude Paleocene to Holocene planktic foraminifera. *In*: BOLLI, H. M., SAUNDERS, J. B. & PERCH-NIELSEN, K. (eds) *Plankton Stratigraphy*. Cambridge, 263–282.

—— & ORR, W. N. 1972. Planktonic foraminiferal biostratigraphy of the east equatorial Pacific – DSDP Leg 9. *In*: HAYS, J. D. *et al. Initial Reports of Deep Sea Drilling Project*, Washington, 1059–1193.

KENNETT, J. P. 1973. Middle and Late Cenozoic planktonic foraminiferal biostratigraphy of the Southwest Pacific – DSDP Leg 21. *Initial Reports of Deep Sea Drilling Project*, **21**, 575–640.

LI QIANYU & RADFORD, S. S. 1991. Evolution and biogeography of Paleogene microperforate planktonic foraminifera. *Palaeogeography, Palaeoclimatology, Palaeoecology*, **83**, 87–115.

——, —— & BANNER, F. T. 1991. Distribution of microperforate tenuitellid planktonic foraminifers in ODP Leg 120 Holes 747A and 749B, Kerguelen Plateau: *Proceedings of ODP, Scientific Results*, **120**, College Station, TX (Ocean Drilling Program).

STOLK, J. 1965. Contribution a l'etude des correlations microfauniques du Tertiare inferieur de la Nigeria Meridional. *Memoire du Bureau de Recherches Geologique et Minieres, Orléans*, **32**, 247–278.

STOTT, L. D. & KENNETT, J. P. 1990. Antarctic Paleogene planktonic foraminifer biostratigraphy: ODP Leg 113 Sites 689 and 690. *Proceedings ODP, Scientific Results*, **113**, College Station, TX (Ocean Drilling Program).

SUBBOTINA, N. N. 1953. *Fossil Foraminifera of the USSR: Globigerinidae, Hantkeninidae, and Globorotaliidae.* Trudy Vnigri, new series, **76**. (In Russian). [English Translation by E. Lees. Collett's Ltd.]

TROELSEN, J. C. 1957. Some planktonic foraminifera of the type Danian and their stratigraphic importance. *In*: LOEBLICH, A. R. *et al.* (eds) *Studies in Foraminifera. Bulletin of U.S.N.M.*, **215**, 125–131.

High resolution stratigraphy of marine Quaternary sequences

P. P. E. WEAVER

*Institute of Oceanographic Sciences, Brook Road, Wormley, Godalming, Surrey
GU8 5UB, UK*

Abstract: The highest resolution stratigraphy of marine sequences can be obtained in the
Quaternary. Oxygen isotope stratigraphy can be used to divide the Quaternary (0–1.66 Ma)
into 63 isotope stages representing alternating glacials and interglacials. These climate
changes are controlled by astronomical forcing involving eccentricity of the earth's orbit
(100 000 year periodicity); obliquity of the earth's axis (41 000 year periodicity); and
precession of the equinoxes (23 000 year periodicity). These factors can be combined to
produce a time scale which has been used to tune the isotope records, providing ages for all
the isotope stage boundaries. The percentage of calcium carbonate also varies in many
marine sediments with similar frequencies to the isotope record and has been used in similar
ways. Both these methods are unreliable in areas with reworking or sediment disturbances.
They work best in combination with independent stratigraphic zonations such as those
derived from calcareous nannofossils. Planktonic foraminifera provide palaeoclimate data
but they are less well suited to detailed biostratigraphy in the Quaternary. Other microfossil
groups including diatoms and radiolaria provide useful biostratigraphic information in some
areas but they are often poorly preserved due to dissolution and when present they have
rarely been used to provide high resolution stratigraphy.

The Quaternary provides the opportunity for
much higher resolution stratigraphy than elsewhere in the geological record, due to accessi-
bility of material and the rapidly oscillating
climates which have influenced the composition
of marine sequences, both in terms of lithology
and fossil content. In the oceans, Quaternary
sediments are by far the most accessible because
they are easily sampled by simple devices such as
piston and box corers. Large quantities of Quaternary sediments have been collected to study
palaeoceanographic and sedimentological problems and much emphasis has been placed on
high resolution stratigraphy. The availability of
an oxygen isotope stratigraphy (Shackleton &
Opdyke 1973) has permitted much finer
resolution than with other conventional micropalaeontological stratigraphies such as the
nannofossil stratigraphy of Martini (1971) or the
planktonic foraminiferal stratigraphy of Bolli &
Premoli-Silva (1973). This is because interglacial
and glacial stages of a few tens of thousands of
years duration can be separated by the isotope
stratigraphy and these intervals are globally
synchronous. The oxygen isotope stratigraphy
has now been developed to a considerable extent
(Ruddiman *et al.* 1986; Sarnthein & Tiedemann
1989; Shackleton & Hall 1989) and a total of 137
isotope stages have been defined for the last
3.3 Ma with 63 for the Quaternary, the base of
which is placed at the top of the Olduvai
palaeomagnetic subchron at 1.66 Ma.

The problem with isotope stratigraphies is
that missing or repeated intervals can rarely be
detected, since isotope stages are largely based
on sequences of high and low values of a
particular isotope. Unless these sequences show
intervals with distinct patterns (e.g. oxygen
isotope stage 5; Shackleton 1969) the stratigraphy is based on counting back from independently identified datum levels such as
palaeomagnetic chron boundaries. Palaeomagnetic boundaries are used because they are
globally synchronous, but micropalaeontological datum levels can also be used to identify
particular isotope stages if they can be shown to
be equally reliable. Data have been collected on
the age and synchroneity of first and last appearances (FADs and LADs) of a few calcareous
nannofossils (e.g. Thierstein *et al.* 1977) and this
shows that calcareous nannofossils can give very
reliable information. Apart from the conventional stratigraphy based on FADs and LADs,
coccoliths also show a sequence of acme zones
through the Quaternary which can be used to
give extra information leading to a higher resolution stratigraphy (e.g. Pujos-Lamy 1977;
Weaver & Hine in press). These high resolution
zonations provide a very powerful stratigraphic
tool when combined with the oxygen isotope
stratigraphy or other methods of determining
the Quaternary climate changes.

Oxygen isotope stratigraphy

Oxygen isotope stratigraphy can be used to
distinguish between glacial and interglacial intervals throughout the Quaternary by measuring

From HAILWOOD, E. A. & KIDD, R. B. (eds), *High Resolution Stratigraphy*
Geological Society Special Publication, No. 70, pp. 137–153.

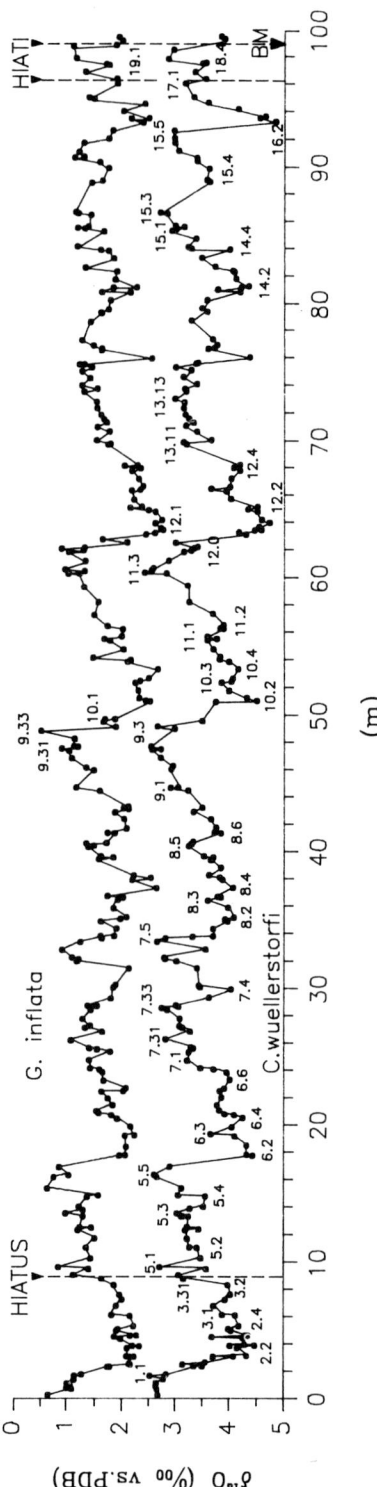

the ratio of heavy to light isotopes of oxygen, usually in planktonic or benthic foraminifera (Fig. 1). The early work was concentrated on piston cores and limited to the upper part of the Quaternary, but the advent of hydraulic piston coring by DSDP and its later use by ODP led to the recovery of continuously cored Quaternary sequences. The numbering of isotope stages has consequently been continued throughout the whole Quaternary (Ruddiman *et al.* 1986; Shackleton & Hall 1989; Fig. 2). The base of the Quaternary at 1.66 Ma lies in isotope stage 63 according to Ruddiman *et al.* (1986). The isotope record does not simply divide into intervals representing glacial and interglacial time but is more 'saw toothed' with sharp changes between glacial and interglacial periods (Terminations) and irregular transitions from interglacial to fully glacial periods. In high accumulation rate sites individual isotope stages can often be subdivided (Shackleton 1969) suggesting climatic fluctuation at higher frequencies.

The boundaries between individual isotope stages can be dated with great accuracy. This was originally attempted by linear interpolation, assuming constant sediment accumulation rates, between known datums such as palaeomagnetic chron boundaries (Ruddiman *et al.* 1987). This method, however, can be subject to error because the ages of the palaeomagnetic boundaries themselves are dated by linear interpolation between a small number of radiometrically dated boundaries (Berggren *et al.* 1985). A much better method of dating isotope stage boundaries is to use the frequencies of climate change predicted by Milankovitch (1920) to tune the isotope signal. The three primary factors are eccentricity ('stretch' of the earth's orbit around the sun) with a 100 000 year periodicity; obliquity of the earth's axis (tilt of the earth's rotational axis) with a 41 000 year periodicity; and precession of the equinoxes (a wobble causing variation in the longitude of the perihelion) with a 23 000 year periodicity (Imbrie *et al.* 1984). These periodicities are controlled by

Fig. 1. High resolution oxygen isotope curve for the Brunhes normal polarity chron from ODP Site 658 off West Africa. Upper curve represents oxygen isotope measurements on a planktonic species and the lower curve measurements on a benthic species. Note subdivisions of the isotope stages and the sawtooth pattern caused by sharp changes at the end of glacial periods. Numbers on graphs refer to isotopic events. (After Sarnthein & Tiedemann 1989).

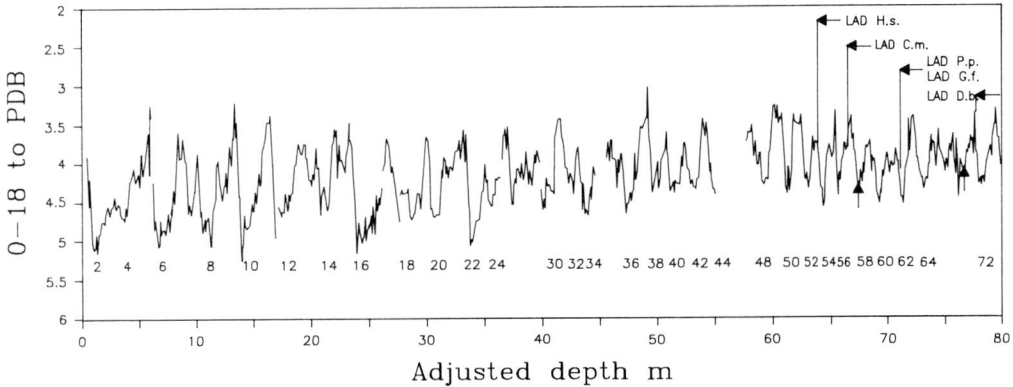

Fig. 2. Oxygen isotope record for the whole Quaternary from ODP Site 677 in the Equatorial Pacific. Numbers refer to even numbered oxygen isotope stages. Biostratigraphic levels: H.s., *Hellicosphaera selli*; C.m., *Calcidiscus macintyrei*; P.p., *Pterocanium prismatium*; G.f., *Globigerinoides fistulosus*; D.b., *Discoaster brouweri*. Upward pointing arrows indicate possible sediment gaps. (After Shackleton & Hall, 1989).

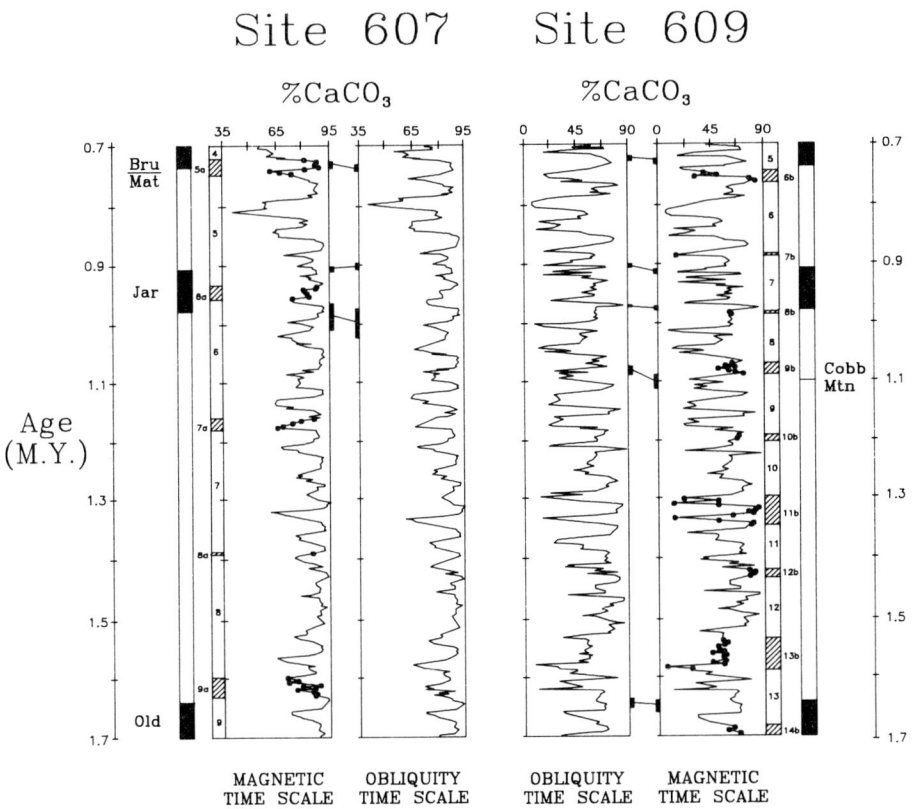

Fig. 3. Early Pleistocene records of % CaCO₃ from DSDP Site 607 and 609 in the North Atlantic. Each record is referred to a magnetic time scale and to an orbitally turned obliquity timescale. (After Ruddiman *et al.* 1986). Intervals of normal magnetic polarity are shown in black; corresponding points on the carbonate curves are shown by black dots.

Table 1. *Oxygen isotope stages in the Quaternary and the ages of the base of each stage*

$\delta^{18}O$ stage	Age (Ma BP)	$\delta^{18}O$ stage	Age (Ma BP)
1	0.012	33	1.033
2	0.024	34	1.055
3	0.059	35	1.076
4	0.074	36	1.096
5	0.130	37	1.116
6	0.190	38	1.138
7	0.244	39	1.160
8	0.303	40	1.169
9	0.339	41	1.198
10	0.362	42	1.218
11	0.423	43	1.239
12	0.478	44	1.260
13	0.524	45	1.281
14	0.565	46	1.301
15	0.620	47	1.321
16	0.659	48	1.341
17	0.689	49	1.362
18	0.726	50	1.382
19	0.746	51	1.402
20	0.767	52	1.421
21	0.789	53	1.442
22	0.809	54	1.465
23	0.828	55	1.486
24	0.848	56	1.505
25	0.869	57	1.523
26	0.890	58	1.542
27	0.910	59	1.564
28	0.931	60	1.585
29	0.954	61	1.605
30	0.975	62	1.626
31	0.995	63	1.646
32	1.014		

Ages of stages 1 to 8 after Martinson *et al.* (1987); stages 9–18 after Imbrie *et al.* (1984); stages 19 to 63 after Ruddiman *et al.* (1986).

Table 2. *Comparison of ages of palaeomagnetic boundaries based on different criteria. The base of the Quaternary is usually taken at the base of the Olduvai*

Isotope stage	Correlative reversal	Age[1]	Age[2]	Age[3]
base 19	base Brunhes	0.73	0.73	0.78
mid 27	top Jaramillo	0.92	0.90	0.99
mid 31	base Jaramillo	0.98	0.97	1.07
base 35	Cobb Mountain	1.10	1.10	1.19
base 63	top Olduvai	1.66	1.65	1.77
base 71	base Olduvai	1.88	1.82	1.95

1 From Berggren *et al.* (1985) and Maniken & Gromme (1982) based on K/Ar dating.
2 From Ruddiman *et al.* (1989): Raymo *et al.* (1989) based on an obliquity timescale.
3 From Shackleton *et al.* (1990) based on a combined obliquity/precessional timescale.

planetary physics and should not have changed over the relatively short duration of the Quaternary. Such an orbitally tuned timescale was produced by Imbrie *et al.* (1984) for isotope stages 1 to 22 representing the last 892 000 years, and the 0 to 300 000 year part of this record was retuned by higher resolution methods by Martinson *et al.* (1987). This tuning and dating of the isotope records was continued through the rest of the Quaternary by Ruddiman *et al.* (1986) so that the ages of all Quaternary isotope stage boundaries are now known to a quoted accuracy of about one thousand years (Fig. 3; Table 1). More recently Shackleton *et al.* (1990) have suggested a recalibration of the Imbrie *et al.* timescale below isotope stage 16 (620 000 years) based on ODP Site 677 which shows a stronger precessional signal. They did not recalculate the ages of all isotope stage boundaries but they did recalculate the ages of the palaeomagnetic reversals in the Quaternary (Table 2).

Calcium carbonate stratigraphy

In many areas of the North Atlantic the oxygen isotope record is mirrored by records of calcium carbonate in marine sediments which fluctuates with the same frequency and at the same time. These changes in calcium carbonate content are brought about due to fluctuations in supply of carbonate or non-carbonate material, or due to fluctuations in dissolution. There may be an increased input of non-carbonate material to the ocean during glacials, caused by the addition of ice rafted and wind blown material (Ruddiman & McIntyre 1976). In the high latitudes this is often accentuated by a decrease in productivity of calcium carbonate-producing coccoliths that do not thrive in cold waters. Other changes in the ocean system associated with changing climate such as changes in wind speed and direction, changes in runoff patterns and reduction in areas of submerged shelf will also lead to calcium carbonate variations in some areas of the ocean. Downcore plots of calcium carbonate percentage can therefore be used rather like an isotope curve (Ruddiman *et al.* 1987) and provide a powerful stratigraphic tool (Fig. 3).

The calcium carbonate content of marine sediments also varies in the deeper parts of the ocean due to dissolution, whereby it is dissolved on the seabed before it has time to be buried by further accumulation. The strength of dissolution is pressure-dependent and so increases with depth from 0 to 100% over an interval known as the lysocline, which may be several hundred metres thick (Farrell & Prell 1989). The depth of the base of the lysocline (CCD) is controlled by the abyssal carbonate ion concentration ($[CO_3^=]$) and these levels vary considerably between different water masses. Since bottom water masses may change between glacials and interglacials, in some areas the depth of the CCD may also fluctuate in sequence with climate change (Crowley, 1983). In the Atlantic these fluctuations due to dissolution accentuate the signal caused by varying inputs of carbonate and non-carbonate material and produce a distinct carbonate stratigraphy which is very similar to the oxygen isotope stratigraphy (Jansen *et al.* 1986). The dissolution cycles can be used to provide a chemostratigraphy as discussed by Berger & Vincent (1981).

The recognition of individual isotope stages (through isotope or calcium carbonate measurements) is most easily achieved in sequences which show high accumulation rates and no hiatuses. The technique depends heavily on curve matching (see Thompson & Clark, this volume) and there are few intervals which show strong characteristic patterns (an exception being oxygen isotope stage 5). It is therefore advisable to obtain independent age control such as from palaeomagnetic data or from micropalaeontological zonations; used in combination an extremely powerful stratigraphic tool can be developed.

Palaeoclimate analysis

The pioneering work of Imbrie & Kipp (1971) showed that palaeotemperature signals could be derived from detailed statistical analysis of planktonic foraminiferal faunas. Although the full procedure to determine a palaeotemperature by statistical methods would not normally be taken to determine stratigraphic ages for sediments, simple ratios of cold to warm tolerant species have been used. Weaver (1983) showed that abundance ratios of selected species of

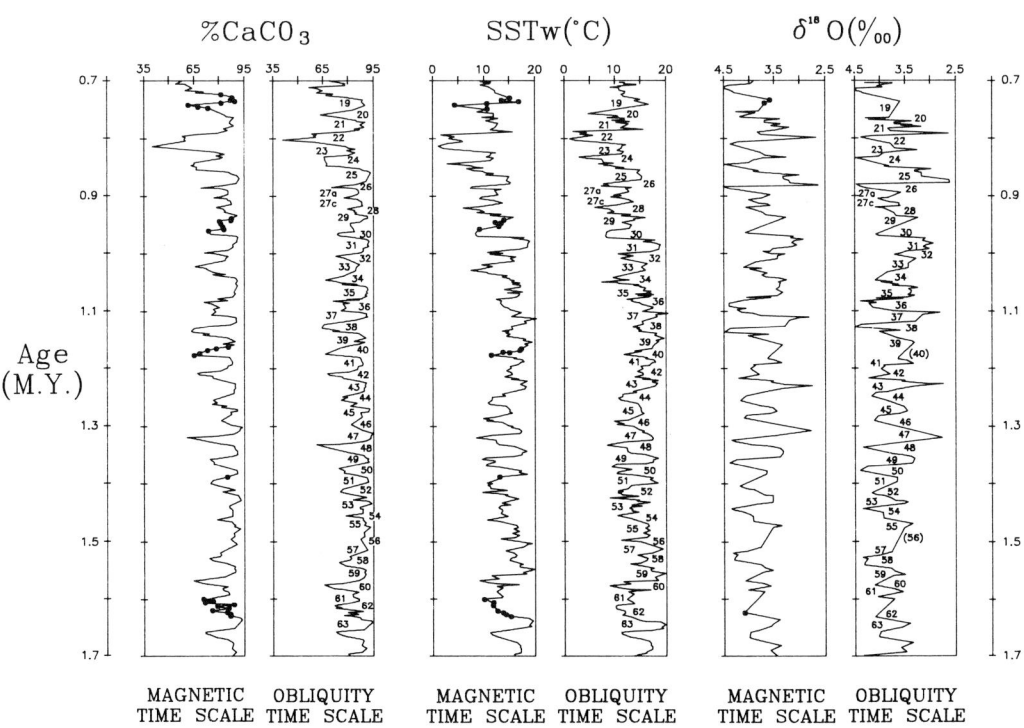

Fig. 4. Early Pleistocene records of % CaCO$_3$, estimated sea surface temperature (SST) and δ^{18}O from DSDP Site 607 in the North Atlantic. Each record is referred to a magnetic time scale and to an orbitally tuned obliquity timescale. Small numbers refer to oxygen isotope stages. Points shown by black dots refer to normal magnetic polarity intervals (After Ruddiman *et al.* 1986).

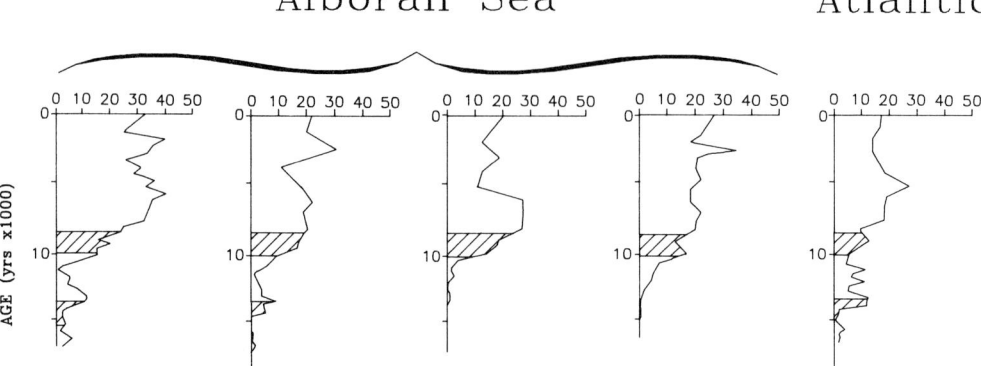

Fig. 5. Variation in abundance of a warm water species of calcareous nannofossil (*Gephyrocapsa oceanica*) and a cool water species (*Gephyrocapsa muellerae*) plotted as a ratio *G.o*/(*G.m*. + *G.o*.) through the termination of the last ice age and the Holocene. Cores from the Alboran Sea, (W. Mediterranean) and the adjacent Atlantic are illustrated. The shaded areas represent the two steps of the deglaciation separated by the cold Younger Dryas period. Note very low ratio of *G. oceanica* during the ice age and during the Younger Dryas cool event (between the two shaded intervals). (After Weaver & Pujol 1988).

planktonic foraminifera could be used to distinguish warm and cold intervals in Quaternary sediments. More detailed studies were carried out by Ruddiman *et al*. (1986) who determined a sequence of isotope stages using variations in sea surface temperature derived from planktonic foraminiferal analyses (Fig. 4). They were able to show strong correlations between the temperature curves, oxygen isotope variations and calcium carbonate curves for the same site in the North Atlantic. All of these data, however, need to be corroborated by independent dating methods.

The use of nannofossils to determine palaeotemperature change is not as easy, because the flora through the Quaternary is dominated by species that are relatively insensitive to temperature change, with the more sensitive species having lesser abundances. Because of these, and other factors, early attempts to use nannofossils in the same way as planktonic foraminifera were not very successful (Geitzenauer *et al.* 1976). Weaver & Pujol (1988), however, showed that a simple ratio of two nannofossil species could be used to show relative temperature change (Fig. 5). This work has been followed up by Hine (1990) who has shown large oscillations in abundance of warm and cold species with similar frequencies to those of the isotope curves. The exact relationship of these nannofossil ratios with the isotope records has not yet been established, but a correlation between the two

AGE	M.Y.	Planktic Foraminiferal Zones (Blow 1969)	DIATOM ZONES					
			Low Latitudes (Burkle 1972 Barron 1985)	Indian Ocean (Schrader, 1974)	North Pacific (Koizumi 1973, Barron 1980, 1985)	N.E.Pacific (Schrader 1973, Barron 1976)	Norwegian Sea (Schrader & Fenner 1976)	Southern Ocean (Weaver & Gombos 1981)
QUATERNARY 0		N22	P. doliolus	1	D.seminae	I	T. oestrupii	C. letigenosus
				2	Rh.curvirostris $\frac{b}{a}$	II	Rh. barboi	C. elliptopora- A. ingens
	1.5		N. reinholdi B	3	A. oculatus	III & IV		
				4				
			A	5				R. barobi N.kerguelensis

Fig. 6. Correlation of various diatom zonations for the Quaternary. Based on similar table in Barron (1985).

would lead to a method of determining isotope stages simply by analysing nannofossil assemblages.

Micropalaeontological zonations of the Quaternary

Diatoms and radiolaria

Siliceous microfossils have a restricted distribution in marine sediments due to most oceanic waters being undersaturated with silica, which causes severe dissolution. These microfossils have therefore been used less than calcareous fossils for biostratigraphy, except in a few areas where their preservation exceeds that of planktonic foraminifera and calcareous nannoplankton. There are three main areas where diatom and radiolarian biostratigraphy has been concentrated; these are the low-latitude Pacific and Indian Oceans, the mid- to high-latitude North Pacific and the Southern Ocean. Even in these areas they rarely provide the high resolution stratigraphy available from other groups, but they may often be the only fossil group represented.

Separate diatom zonations have been developed for the Southern Ocean, North Pacific and low latitudes because each area has its own endemic population (see review by Barron 1985). Each zonation scheme generally defines only two or three Quaternary zones (Fig. 6) and thus the stratigraphy is not of very high resolution. Diatom abundances, however, vary with changing climate and relative abundance data can be used to infer Quaternary climatic fluctuations, particularly in the high latitudes (Burckle & Cooke 1983; Jordan & Pudsey in press). Radiolarian biostratigraphy has been developed in similar areas and is discussed by Sanfilippo et al. 1985.

An extensive review of siliceous fossil biostratigraphy is not included here but the reader is referred to papers in Bolli et al. (1985) and recent DSDP and ODP drilling volumes.

Planktonic foraminifera

These organisms provide limited stratigraphic information in the Quaternary due to a lack of correlatable FADs and LADs, although they do show important variations in species abundance related to climate change. The interval is normally placed entirely in the *Globorotalia truncatulinoides* Zone. This species has its FAD at 2.5 Ma in the Southwest Pacific but did not radiate to other oceans until about 1.9 Ma

(Jenkins & Gamson, this volume). In either case its FAD is older than the base of the Quaternary at 1.66 Ma. In northern latitudes the FAD of sinistrally coiled heavily encrusted *Neogloboquadrina pachyderma* occurs at 1.66 Ma (Weaver & Clement 1986; Raymo et al. 1989; Pujol & Duprat 1984) and it is thus a useful marker for the base of the Quaternary.

Early attempts to subdivide the Quaternary on the basis of planktonic foraminifera were made by Ericson et al. (1954) who used coiling

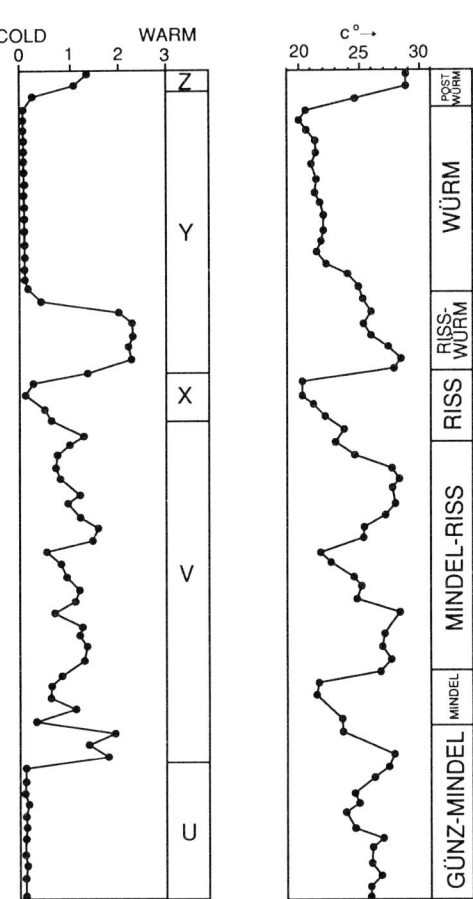

Fig. 7. Variations in abundance of *Globorotalia menardii* compared to oxygen isotope data in a core from the Caribbean. Würm, Riss and Mindel represent oxygen isotope stages 2 to 4, 6 and 8 respectively. *G. menardii* curves are based on the ratios of number of *G. menardii* to weight of material coarser than 74 μm. The *G. menardii* Zones u to z can be correlated across large areas of the Atlantic Ocean. (After Ericson et al. 1961).

right coiling changed downcore. Strong strati-direction changes in *Globorotalia truncatuli-noides*. They found distinct provinces of domi-nantly left and dominantly right coiled individuals in surface sediments from the North Atlantic and they found that the ratios of left to graphic correlations are found between adjacent cores. Pujol and Caralp (1974) showed that other species of *Globorotalia*, including *G. hirsuta* and *G. crassaformis*, could be used in the same way, each having its own pattern of coiling directions spatially and with time. The reasons for these changes in coiling direction are un-clear, but in other species such as *Neogloboqua-drina pachyderma* water temperature appears to be the controlling factor.

Ericson *et al.* (1961) defined 6 zones in the late Quaternary of the Atlantic Ocean based on the presence or absence of *Globorotalia menardii*. They labelled the zones (z) to (u) with (z) at the top, and inferred that *G. menardii* was common only during warm intervals (Fig. 7).

Bolli & Premoli-Silva (1973) proposed five subzones for the Quaternary based on the FAD of *Globorotalia crassaformis hessi*, the FAD of

Globigerina calida calida, the LAD of *Globo-rotalia tumida flexuosa* and the FAD of *Globo-rotalia fimbriata* (Fig. 8). This zonation has proved useful in the Caribbean region where it was first determined and it has been used in other areas with limited success (Rogl 1974; Pflaumann & Krasheninnikov 1978) but in other tropical areas it has not proved reliable (e.g. Weaver & Raymo 1989). Kennett (1973) pro-posed a subdivision of the Quaternary into a *G. truncatulinoides tosaensis* subzone below and *G. truncatulinoides* subzone above, based on the LAD of the former species. Thompson & Sciar-rillo (1978) had earlier dated this LAD at 0.6 Ma, but this datum has not been widely used, due to sporadic occurrences of *G. tosaensis*. Other recorded ages of planktonic foraminiferal events include the LAD of *Globoquadrina pseudofoliata* at 0.22 Ma (Thompson & Saito 1974), the FAD of *Pulleniatina finalis* at 1.3 Ma (Saito *et al.* 1975) and the LAD of *Globigeri-noides fistulosus* at 1.6 Ma (Saito *et al.* 1975). Whilst these may have local value, none has been used in global stratigraphic zonations. Perhaps the most reliable foraminiferal datum in

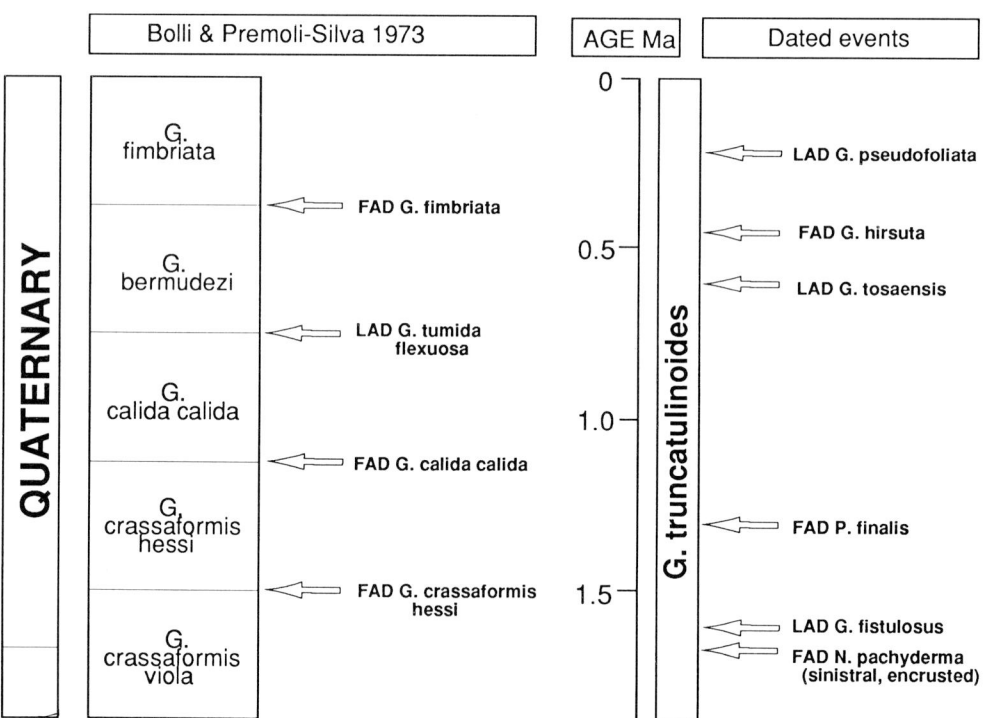

Fig. 8. Planktonic foraminiferal stratigraphy of the Quaternary. The principal dated foraminiferal events are shown on the right; those used in a zonation by Bolli & Premoli-Silva (1973) have not been dated.

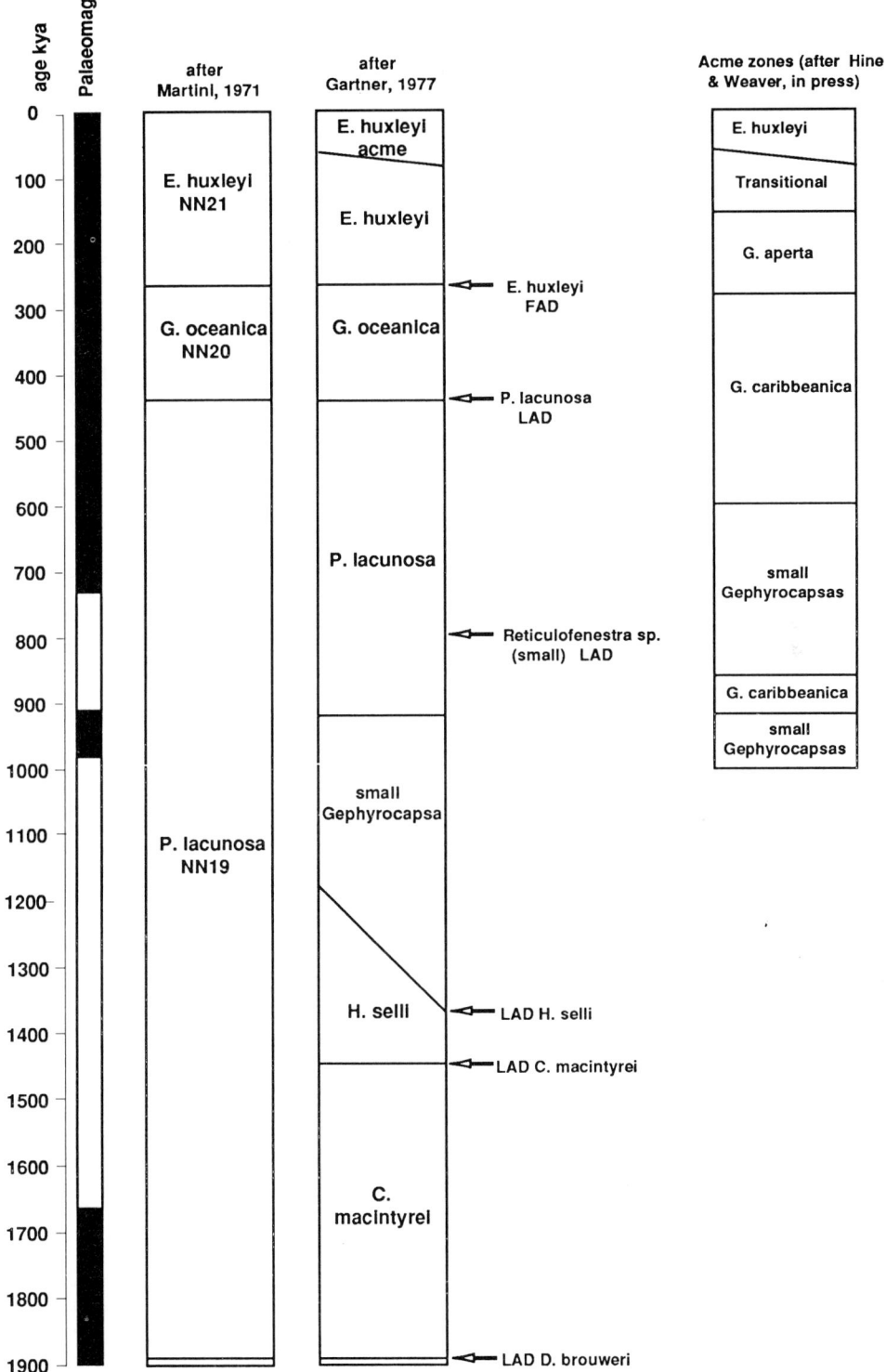

Fig. 9. Quaternary nannofossil stratigraphy incorporating FAD and LAD data with acme zones for the latter part of the sequence. (After Weaver & Hine in press)

the Quaternary of the North Atlantic is the FAD of *Globorotalia hirsuta*, which Pujol & Duprat (1984) suggested occurred in oxygen isotope stage 12 about 450 000 years ago. Weaver (1987) found similar ages for its FAD across a range of latitudes in the North Atlantic.

Calcareous nannofossil stratigraphy

Standard biozonations

Martini's (1971) 'Standard Calcareous Nanno-fossil Zonation' subdivides the Quaternary into 3 zones based on the LAD of *Pseudoemiliania lacunosa* and the FAD of *Emiliania huxleyi* (Fig. 9). Gartner (1977) added more datum levels, subdividing the *E. huxleyi* Zone at the onset of an acme of *E. huxleyi* and using the LADs of *Hellicosphaera selli* and *Calcidiscus macintyrei* to subdivide the *P. lacunosa* Zone (Fig. 9). Thierstein *et al.* (1977) examined the distribution of *Pseudoemiliania lacunosa* and *Emiliania huxleyi* in a global array of cores, all of which had been independently dated by oxygen isotope stratigraphy. They were able to correlate the LAD of *P. lacunosa* with oxygen isotope stage 13 and the FAD of *E. huxleyi* with isotope stage 8, showing that these were truly globally synchronous events. Further data from Weaver & Hine (in press) backs up the work of Thierstein *et al.* with ages of 440 ka for the LAD of *P. lacunosa* and 242–268 kya for the FAD of *E. huxleyi* for cores from the North Atlantic. The onset of the *E. huxleyi* acme was found by Thierstein *et al.* to be slightly diachronous between oxygen isotope stage 5b in the tropics and subtropics to stage 4 in mid-latitudes. These data show that individual isotope stages can be clearly identified by coccolith distributions in marine sediments.

The LAD of *Calcidiscus macintyrei* was dated at 1.45 Ma by Backman & Shackleton (1983) and the LAD of *Helicosphaera selli* at 1.37 Ma at the equator and 1.22 Ma in higher latitudes. One other useful datum is the LAD of a small *Reticulofenestra* sp., identified by Hine (1990) with an age of 0.8 Ma (isotope stage 22) occurring throughout the North Atlantic.

Acme zones and their synchroneity

The potential for using intervals of single species dominance of the flora (acme zones) for Quaternary nannofossil stratigraphy was first established by Gartner & Emiliani (1976) who erected the *Emiliania huxleyi* Acme Zone in the upper part of the *E. huxleyi* Zone, NN21, of Martini (1971). Other acme zones, based mainly on species of the genus *Gephyrocapsa* have been suggested since then, and these were built into a formal sequence by Weaver & Hine (in press). The sequence is not yet established for the whole Quaternary, but 7 acme zones (Fig. 9) can be identified through the last 1.2 million years as follows.

(1) *Emiliania huxleyi* Acme Zone. This zone is defined at its base by the dominance reversal between *E. huxleyi* and large *Gephyrocapsa* spp and continues to the present day. The dominance reversal between *E. huxleyi* and large Gephyrocapsas is considered a time transgressive event, and was dated by Thierstein *et al.* (1977) as occurring in oxygen isotope stage 5b in the low latitudes, and stage 4 in transitional waters. Gard (1989) dates this dominance reversal in Stage 4 in the high latitude North Atlantic.

(2) Transitional Zone. An interval dominanted by *Gephyrocapsa muellerae*. The base of the interval is the dominance reversal between *G. muellerae* and *Gephyrocapsa aperta*, and the top of the interval is the base of the *E. huxleyi* acme. It occupies oxygen isotope stages 4, 5 and most of 6. Although *G. muellerae* is very common throughout this interval it does not dominate the flora for the whole time, being frequently outnumbered by *G. aperta* around the oxygen isotope stage 5–6 boundary. We therefore do not elevate this interval to acme zone status.

(3) *Gephyrocapsa aperta* Acme Zone. An interval dominated by *G. aperta*. The base is defined by the reversal in dominance between *G. caribbeanica* and *G. aperta*, and the top by a reduction in the abundance of *G. aperta* to less than 90% of the Gephyrocapsas. It occupies the lower part of oxygen isotope stage 6, all of stage 7 and the top part of stage 8.

(4) *Gephyrocapsa caribbeanica* Acme Zone. An interval dominated by *G. caribbeanica*. The base is defined by the reduction in abundance of small Gephyrocapsas, and the top by the base of the *G. aperta* acme. The zone spans the lower part of oxygen isotope stage 8, and includes stages 9 to 15 (Hine 1990).

(5) Small *Gephyrocapsas* Zone. An interval dominated by small Gephyrocapsas, mainly *G. aperta* and small specimens of *G. caribbeanica*. This interval can be distinguished from interval 3, also with small Gephyrocapsas, due to the occurrence of *P. lacunosa* which became extinct during interval 4. The top of interval 5 is defined by the reversal of

dominance with *G. caribbeanica* and the base is defined by the top of another interval of dominant *G. caribbeanica*. This interval includes the lower part of oxygen isotope stage 15 to stage 25.

(6) An interval dominated by *G. caribbeanica*. This interval can be distinguished from interval 4, which also has dominant *G. caribbeanica*, by the presence of *Reticulofenestra* spp ($> 6.5 \mu$m). These large *Reticulofenestra* specimens have an LAD in the lower part of interval 5. This interval spans isotope stages 25 to 28.

(7) Small *Gephyrocapsas* Zone of Gartner (1977). The top of this zone is defined at the decline of small Gephyrocapsas and the base at the LAD of *Helicosphaera sellii* at 1.2 Ma. This zone spans isotope stages 28 to 41.

The evidence presented above indicates that there was an almost continuous turnover of dominant species through the last 1.2 Ma with individual acmes lasting an average of 170 000 years, although some of these, e.g. interval 3, may have been much shorter. These acmes are easily recognisable, and in combination with FAD and LAD data they can be clearly distinguished from each other. The base of the *E. huxleyi* interval is slightly transgressive but the other intervals give reliable correlation throughout the NE Atlantic.

An application of high resolution stratigraphy

In 1979 the Seabed Working Group of the Nuclear Energy Agency set up a project to investigate the feasibility of disposal of High Level Radioactive Waste in the deep-sea (Sheppard *et al*. 1988). As part of this project intensive investigations were made of late Quaternary sediments from the Madeira Abyssal Plain in the NE Atlantic and these included sequences of thick turbidites (individually up to 5 m thick) separated by thin (1–20 cm) pelagic units. It was necessary to show detailed core correlation and extent of any erosion across the whole plain. The plain lies at a water depth of 5400 m and suffered moderate calcium carbonate dissolution during interglacials but severe dissolution during glacials when bottom water masses in the area were more corrosive. The carbonate record can therefore be used to given an indication of glacial or interglacial conditions but it does not identify particular isotope stages. Oxygen isotope measurements could not be made directly because foraminifera were not present in the

dissolved glacial intervals, and so a more intricate stratigraphic scheme was devised (Weaver & Kuijpers 1983).

The detailed stratigraphy involved analysing the calcareous nannofossil flora from sediments adjacent to the plain, where a pelagic record unaffected by turbidite input could be determined, and then determining isotope stages by proxy using a combined carbonate/nannofossil stratigraphy. The nannofossils can be used to identify some of the isotope stages (e.g. Stages 5, 7 and 12) and the stage boundaries and intervening isotope stages are inferred from the carbonate record. In this way isotope stages 1 to 13 were recognized (Weaver & Kuijpers 1983), (Fig. 10). The pelagic intervals from the abyssal plain sequence were then reassembled without the intervening turbidities and compared to the truly pelagic sequence both in terms of lithology and nannofossil stratigraphy. This showed that there was no appreciable erosion of the pelagic sediments by incoming turbidites, since the thicknesses of each pelagic layer were similar in the turbidite and non-turbidite cores. This high resolution stratigraphy also showed that the turbidites were deposited preferentially at the lithological boundaries during periods of climate/sealevel change (Fig. 10).

Individual turbidites on the Madeira Abyssal Plain contain abundant nannofossils and although these represent reworked assemblages, the ratios of species in the reworked mixtures were found to be fairly constant for each turbidite but different for different turbidites (Fig. 11) (Weaver & Kuijpers 1983; Weaver *et al*. 1986). These mixtures were then used to correlate individual turbidite flows across the whole plain and correlation was found to agree with other correlations such as from geochemical determinations (Pearce & Jarvis in press). One intriguing aspect of the nannofossil mixtures in these turbidites is the age range of material they represent. In most cases it can be shown that the turbidites represent a time interval of only 200 000 to 400 000 years. This can be achieved by comparing the turbidite mixture with synthetic mixtures derived from adding the average nannofossil abundance ratios for each isotope stage in a normal pelagic sequence. Ratios for successive isotope stages are added together until the synthetic mixture approximates the turbidite mixture (Fig. 12). The average nannofossil abundance ratios for pelagic sequences are known for the late Quaternary (Weaver 1983) and appear to be constant over wide latitudinal ranges. The time duration of each isotope stage varies but is accounted for by multiplying the ratios for each stage by the

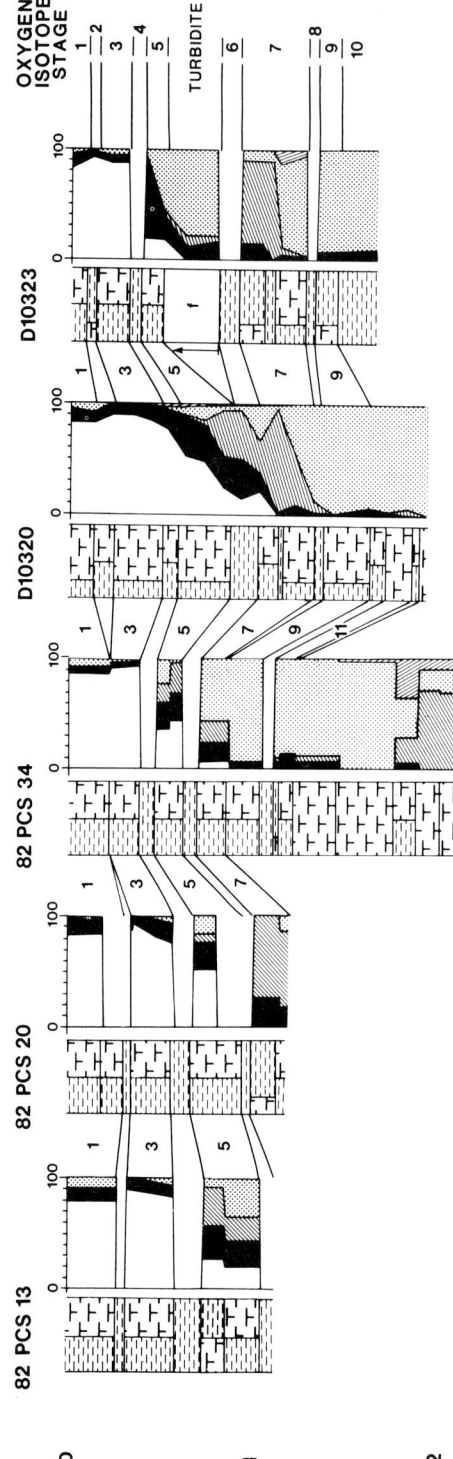

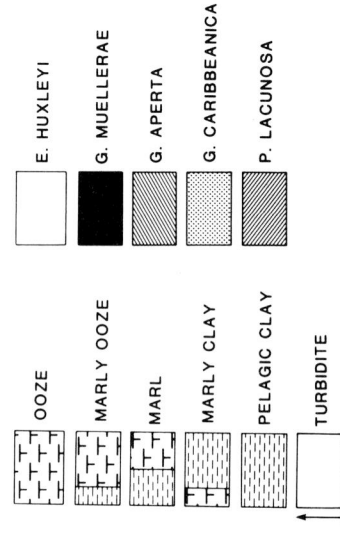

Fig. 10. Pelagic intervals of turbidite cores 82PCS13, 82PCS20 and 82PCS34 from the Madeira Abyssal plain (North Atlantic) plotted without intervening turbidites and compared to pelagic cores D10320 and D10323 from nearby hills. Coccolith stratigraphy plotted as cumulative percent of five species. Blank sections due to complete dissolution. Oxygen isotope stage boundaries placed within the constraints of the coccolith stratigraphy at lithological changes. Note similar thicknesses of pelagic layers in core suggesting incoming turbidites caused no erosion of the underlying sediment surface.

82 PCS 13   82 PCS 20   82 PCS 34

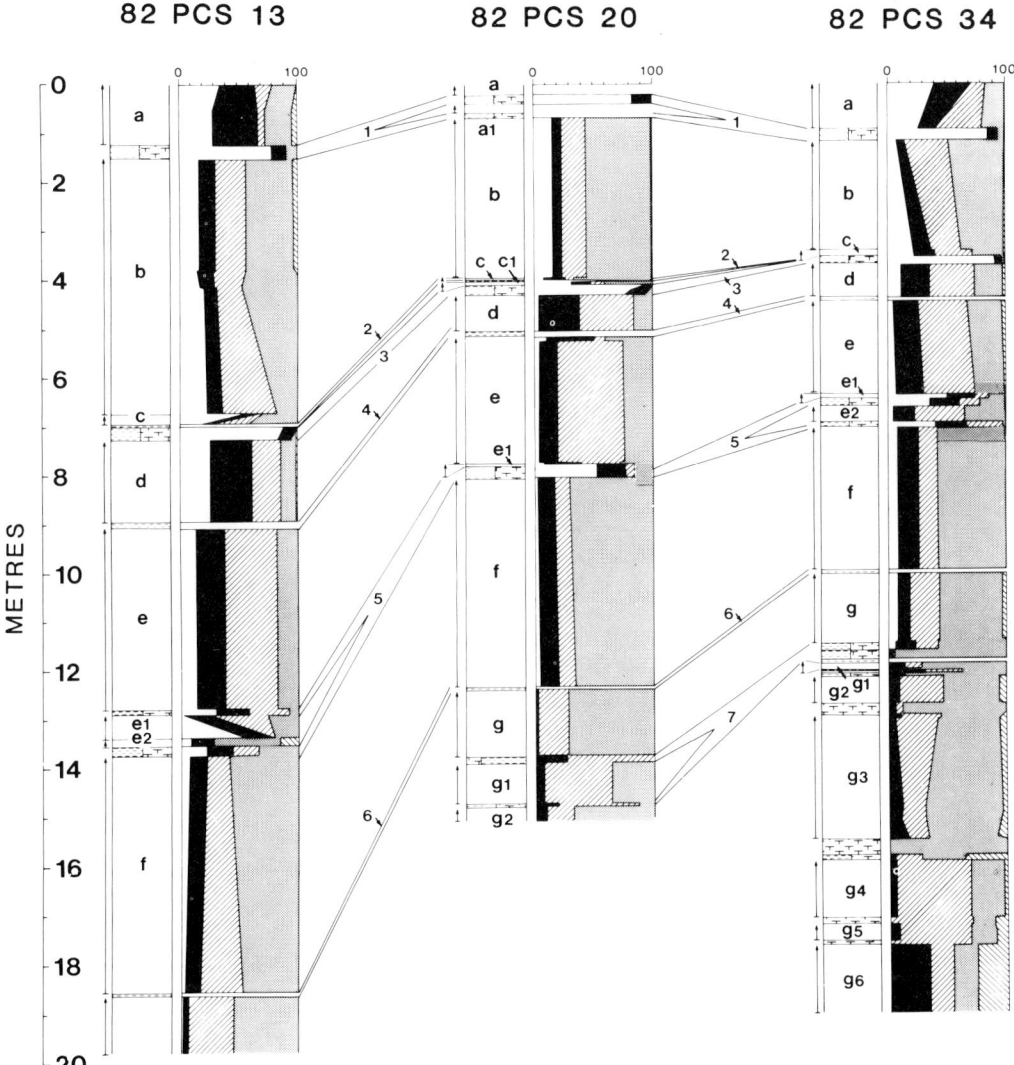

Fig. 11. Lithology of cores 82PCS13, 82PCS20 and 82PCS34 including both turbidites and pelagic intervals. Correlation lines derived from Fig. 10. Turbidite units present in all three cores lettered consecutively from the top. Note different lithologies at top and base of each turbidite suggesting deposition during periods of climate/sealevel change. (After Weaver & Kuijpers 1983. Key as for Fig. 10.)

duration of the stage in thousands of years and assuming constant accumulation rates.

This type of high resolution stratigraphy can therefore provide a detailed temporal framework for core correlation and interpretation of events. It also allows other sedimentological questions to be addressed such as age range of reworked sediments, from which depths of erosion can be calculated.

Conclusion

High resolution stratigraphy with an accuracy of a few thousand years is attainable throughout the Quaternary in marine sequences. The highest resolution methods involve measuring oxygen isotopes or calcium carbonate to identify flucutations related to Quaternary climate changes. These methods, however, do not

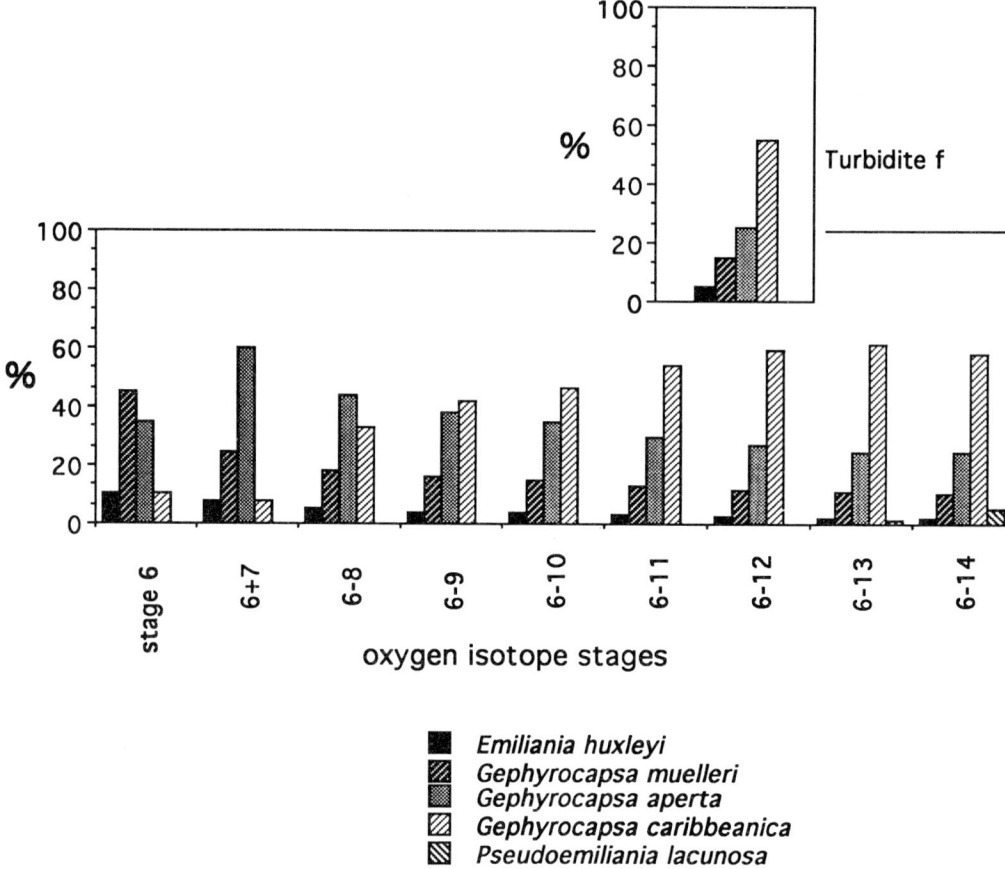

Fig. 12. Comparison of coccolith mixture in turbidite f (upper graph) with synthetic coccolith mixtures created by adding coccolith abundances from successive oxygen isotope stages, starting with stage 6, which immediately underlies tubidite f. Account has been taken of the length of time represented by each isotope stage in creating the synthetic mixtures. Each of the graphs therefore represents the expected mixture of coccoliths if that number of isotope stages was eroded to form the turbidite. Note the strong comparison between the actual mixture in the upper graph with the synthetic mixture for isotope stages 6 to 12, suggesting erosion of this age range of material. The species *P. lacunosa* became extinct in isotope stage 12 and is not found in turbidite f; It is therefore, unlikely that this turbidite contains material older than stage 12/13. Based on cores from the Madeira Abyssal Plain.

provide definitive ages and the stratigraphy must be interpreted assuming no erosion or reworking so that the cycles can be counted back in time. Obviously, the method is more accurate if independent age control is available and this can be provided by palaeomagnetic determinations on cores, and through biostratigraphy. Planktonic foraminifera provide poor stratigraphic information in the Quaternary but calcareous nannofossils provide several excellent datum levels, some of which have been shown to be globally synchronous. Nannofossils provide traditional first and last occurrence data and also a stratigraphy based on a sequence of acme zones.

Used in combination, these can subdivide the Quaternary into units of about 100 000 year duration. When combined with the oxygen isotope record a very powerful stratigraphic scheme is produced.

All of the stratigraphic methods discussed are at their optimum in open ocean sites showing high accumulation rates and minimal reworking. Measurements of oxygen isotopes can be severely limited in sites showing strong dissolution, but unless there is near complete removal of calcium carbonate, it is often possible to find some calcareous nannoplankton and determine a lower resolution stratigraphy. The type of high resolution stratigraphy outlined here can be applied to older sequences, and it has already been shown to work well in the Late Pliocene. Apart from its obvious use in correlating marine sequences, the high resolution can be used to test the accuracy of other biostratigraphic datum planes, providing information for modelling biostratigraphic accuracy.

I thank J. Murray, C. Pujol, B. Whitmarsh and two anonymous reviewers for their comments. The diagrams were drawn by H. Marlow and J. Taylor. This is IOSDL contribution number 92034.

References

BACKMAN, J. & SHACKLETON, N. J. 1983. Quantitative biochronology of Pliocene and early Pleistocene calcareous nannofossils from the Atlantic, Indian and Pacific Oceans. *Marine Micropalaeontology*, **8**, 141–170.

BARRON, J. A. 1985. Miocene to Holocene planktic diatoms. *In*: BOLLI, H. M., SAUNDERS, J. B. & PERCH-NIELSEN, K. (eds) *Plankton Stratigraphy*. Cambridge University Press, Cambridge, 763–809.

BERGER, W. H. & VINCENT, E. 1981. Chemostratigraphy and biostratigraphic correlation: exercises in systematic stratigraphy. *Oceanologica Acta*, **4**, 115–127.

BERGGREN, W. A., KENT, D. V. & VAN COUVERING, J. A. 1985. Neogene geochronology and chronostratigraphy. *In*: SNELLING, N.J. (ed.) *The Chronology of the Geological Record*. Geological Society, London, Memoir, **10**, 211–260.

BOLLI, H. M. & PREMOLI-SILVA, I. 1973. Oligocene to Recent planktonic foraminifera and stratigraphy of Leg 15 sites in the Caribbean Sea. *In*: EDGAR, N. T., SAUNDERS, J. B. *et al. Initial Report of DSDP*, **15** US Govt. Printing Office, Washington, 475–497.

——, SAUNDERS, J. B. & PERCH-NIELSEN, K. (eds) 1985. *Plankton Stratigraphy*. Cambridge University Press, Cambridge.

BURCKLE, L. H. & COOKE, D. W. 1983. Late Pleistocene *Eucampia antarctica* abundance stratigraphy in the Atlantic sector of the Indian Ocean. *Micropalaeontology*, **29**, 6–10.

CROWLEY, T. J. 1983. Calcium-carbonate preservation patterns in the central North Atlantic during the last 150,000 years. *Marine Geology*, **51**, 1–14.

ERICSON, D. B., EWING, M., WOLLIN, G. & HEEZEN, B. C. 1961. Atlantic Deep-Sea sediment cores. *Geological Society of America Bulletin*, **72**, 193–286.

——, WOLLIN, G. & WOLLIN, J. 1954. Coiling direction of *Globorotalia truncatulinoides* in deep-sea cores. *Deep-Sea Research*, **2**, 152–158.

FARRELL, J. W. & PRELL, W. L. 1989. Climatic change and CaCO$_3$ preservation: an 800,000 year bathymetric reconstruction from the central equatorial Pacific Ocean. *Paleoceanography*, **4**, 447–466.

biostratigraphy: the eastern Arctic Ocean Record. *In*: HERMAN, Y.(ed.) *The Arctic Seas: Climatology, Oceanography, Biology and Geology*. Van Nostrand Reinhold, New York, 445–449.

GARTNER, S. 1977. Calcareous nannofossil biostratigraphy and revised zonation of the Pliocene. *Marine Micropaleontology*, **2**, 1–25.

——& EMILIANI, C. 1976. Nannofossil biostratigraphy and climate stages of the Brunhes epoch. *Bulletin of the American Association of Petroleum Geologists*, **60**, 1562–1564.

GEITZENAUER, K. R., ROCHE, M. B. & McINTYRE, A. 1976. Modern Pacific coccolith assemblages: derivation and application to Late Pleistocene paleotemperature analysis. *In*: CLINE, R. M. & HAYS, J. D. (eds) *Investigation of Late Quaternary Paleoceanography and Paleoclimatology*. Geological Society of America Memoir, **145**, 423–448.

HINE, N. 1990. *Late Cenozoic Calcareous Nannoplankton from the Northeast Atlantic*. PhD thesis, University of East Anglia.

IMBRIE, J. & KIPP, N. G. 1971. A new micropaleontological method for quantitative paleoclimatology: application to a late Pleistocene Caribbean core. *In*: TUREKIAN, K. K. (ed.) *The Late Cenozoic Glacial Ages*. New Haven, Yale University Press, 71–181.

——, HAYS, J. D., MARTINSON, D. G., McINTYRE, A., MIX, A. C., MORLEY, J. J., PISIAS, N. G., PRELL, W. L. & SHACKLETON, N. J. 1984. The orbital theory of Pleistocene climate: support from a revised chronology of the marine δ^{18}O record. *In*: BERGER, A., IMBRIE, J., HAYS, J., KUKLA, G. & SALTZMAN, B. (eds) *Milankovitch and Climate*. NATO ASI Series C: Mathematical and Physical Sciences, **126**, 269–305.

JANSEN, J. H. F., KUIJPERS, A. & TROELSTRA, S. R. 1986. A mid-Brunhes climatic event: long term changes in global atmosphere and ocean circulation. *Science*, **232**, 619–622.

JENKINS, D. G. & GAMSON, P. 1993. The late Cenozoic *Globorotalia truncatulinoides* datum-plane in the Atlantic, Pacific and Indian Oceans. *This volume*.

JORDAN, R. W. & PUDSEY, C. J. in press. High-resolution diatom stratigraphy of Quaternary sediments from the Scotia Sea. *Marine Micropalaeontology*.

KENNETT, J. P. 1973. Middle and Late Cenozoic planktonic foraminifer biostratigraphy of the southwest Pacific- DSDP Leg 21. *In*: BURNS, R. E., ANDREWS, J. E. *et al. Initial Reports of DSDP*, **21**, US Govt. Printing Office, Washington, 575–639.

MANKINEN, E. A. & GROMME, C. S. 1982. Paleomagnetic data from the Coso Range, California and current status of the Cobb Mountain normal Geomagnetic Polarity Event. *Geophysical Research Letters*, **9**, 1279–82.

MARTINI, E. 1971. Standard Tertiary and Quaternary calcareous nannoplankton zonation. *In*: FARINACCI, A. (ed.) *Proceedings of the 2nd Plankton Conference, Roma, 1971*: Rome, 739–777.

MARTINSON, D. G., PISIAS, N., HAYS, J. D., IMBRIE, J.,

MOORE, T. C. & SHACKLETON, N. J. 1987. Age dating and the orbital theory of the ice ages: development of a high resolution 0 to 300,000-year chronostratigraphy. *Quaternary Research*, **27**, 1–30.

MILANKOVITCH, M. 1920. *'Théorie mathématique des phénomènes thermiques produits par la radiation solarie'*, Royal Serbian Academy, Special Publication, **133**, Gauthiers-Villars, Paris 1941.

PEARCE, T. J. & JARVIS, I. in press. Applications of geochemical data to modelling sediment dispersal patterns in distal turbidites: Late Quaternary of the Madeira Abyssal plain. *Journal of Sedimentary Petrology*.

PFLAUMANN, U. & KRASHENINNIKOV, V. A. 1978. Quaternary stratigraphy and planktonic foraminifers of the eastern Atlantic, Deep-Sea Drilling project, Leg 41. *In:* LANCELOT, Y., SEIBOLD, E. *et al. Initial Reports of DSDP*. Supplement of vols. **38**, **39**, **40** and **41**, US Govt. Printing Office, Washington, 883–911.

PUJOL, C. & CARALP, M. 1974. Variations du sens d'enroulement des Foraminifères planctoniques dans l'interprétation stratigraphique du Quaternaire terminal de l'ocean Atlantique Nord. *Bulletin de Institute de Géologie de Bassin Aquitaine*, **16**, 31–50.

—— & DUPRAT, J. 1984. Quaternary and Pliocene planktonic foraminifers of the Northeastern Atlantic (Goban Spur), DSDP Leg 80. *In:* GRACIANSKY, P. C., POAG, C. W. *et al. Initial Reports of DSDP, 80*, US Govt. Printing Office, Washington, 683–723.

PUJOS-LAMY, A. 1977. *Emiliania et Gephyrocapsa* (nannoplankton calcaire): biométrie et intéret biostratigraphique dans le Pléistocène supérieur marin des Acores. *Revista Espanola de Micropaleontologia*, **9**, 69–84.

RAYMO, M. E., RUDDIMAN, W. F., BACKMAN, J., CLEMENT, B. M. & MARTINSON, D. G. 1989. Late Pliocene variation in Northern Hemisphere ice sheets and North Atlantic deep water circulation. *Paleoceanography*, **4**, 413–446.

ROGL, F. 1974. The evolution of the *Globorotalia truncatulinoides* and *Globorotalia crassaformis* group in the Pliocene and Pleistocene of the Timor Trough, DSDP Leg 27, Site 262. *In:* VEEVERS, J. J., HEIRTZLER, J. R. *et al. Initial Reports of DSDP, 27*, US Govt. Printing Office, Washington, 743–767.

RUDDIMAN, W. F. & McINTYRE, A. 1976. Northeast Atlantic paleoclimatic changes over the past 600,000 years. *In:* CLINE, R. M. & HAYS, J. D. (eds) *Investigation of Late Quaternary Paleoceanography and Paleoclimatology*. Geological Society of America Memoir, **145**, 111–146.

——, —— & RAYMO, M. E. 1987. Paleoenvironmental results from North Atlantic sites 607 and 609. *In:* RUDDIMAN, W. F., KIDD, R. B. *et al. Initial Reports of DSDP, 94*. US Govt. Printing Office, 855–878.

——, RAYMO, M. E., MARTINSON, D. G., CLEMENT, B. M. & BACKMAN, J. 1989. Pleistocene evolution: northern Hemisphere ice sheets and North Atlantic Ocean. *Paleoceanography*, **4**, 353–412.

——, —— & McINTYRE, A. 1986. Matuyama 41,000 year cycles: North Atlantic Ocean and northern hemisphere ice sheets. *Earth and Planetary Science Letters*, **80**, 117–129.

SAITO, T., BURCKLE, L. H. & HAYS, J. D. 1975. Late Miocene to Pleistocene biostratigraphy of equatorial Pacific sediments. *In:* SAITO, T. & BURCKLE, L. (eds) *Late Neogene Epoch Boundaries*. Micropaleontology Special Paper, **1**, New York, 226–244.

SANFILIPPO, A., WESTBERG-SMITH, M. J. & REIDEL, W. R. 1985. Cenozoic radiolaria. *In:* BOLLI, H. M., SAUNDERS, J. B. & PERCH-NIELSEN, K. (eds) *Plankton Stratigraphy*. Cambridge University Press, Cambridge, 631–712.

SARNTHEIN, M. & TIEDEMANN, R. 1989. Toward a high resolution stable isotope stratigraphy of the last 3.4 million years: Sites 658 and 659 off northwest Africa. *In:* RUDDIMAN, W. F., SARNTHEIN, M. *et al. Proceedings of the Ocean Drilling Program Scientific Results*, **108**, College Station, TX, 167–185.

SHACKLETON, N. J. 1969. The last interglacial in the marine and terrestrial records. Proc. Roy. Soc. Lond. B 174, 135–154.

SHACKLETON, N. J., BERGER, A. & PELTIER, W. R. 1990. An alternative calibration of the lower Pleistocene timescale based on ODP Site 677. Trans. Roy. Soc. Edinburgh: Earth Sciences, 81, 251–261.

SHACKLETON, N. J. & HALL, M. A. 1989. Stable isotope history of the Pleistocene at ODP Site 677. *In:* BECKER, H., SAKAI, H. *et al.* Proc. ODP, *Sci. Results*, 111: College Station, TX (Ocean Drilling Program) 295–316.

SHACKLETON, N. J. & OPDYKE, N. D. 1973. Oxygen isotope and paleomagnetic stratigraphy of equatorial Pacific core V28–238: oxygen isotope temperatures and ice volumes on a 10^5 year and 10^6 year scale. Quat. Res. **3**, 39–55.

SHEPPARD, L. E., AUFFRET, G. A., BUCKLEY, D. E., SCHUTTENHELM, R. T. E. & SEARLE, R. C. 1988. Feasibility of disposal of High-Level Radioactive Wastes into the Seabed: Geoscience characterisation studies. SANDIA Report 1913.UC-70, 303pp.

THIERSTEIN, H. R., GEITZENAUER, K. R., MOLFINO, B. & SHACKLETON, N. J. 1977. Global synchroneity of Late Quaternary coccolith datum levels: validation by oxygen isotopes. *Geology*, **5**, 400–404.

THOMPSON, P. R. & SAITO, T. 1974. Pacific Pleistocene sediments: planktonic Foraminifera dissolution cycles and geochronology. *Geology*, **2**, 333–335.

—— & SCIARRILLO, J. R. 1978. Planktonic foraminiferal biostratigraphy in the Equatorial Pacific. *Nature*, **276**, 29–33.

THOMPSON, R. & CLARK, R. M. 1993. Quantitative marine sediment core matching, using a modified sequence-slotting algorithm. *This volume*.

WEAVER, P. P. E. 1983. An integrated stratigraphy of the upper Quaternary of the King's Trough Flank area, N. E. Atlantic. *Oceanologica Acta,* **6**, 451–456.

—— 1987. Late Miocene to Recent planktonic foraminifers from the North Atlantic: DSDP leg 94. *In*: RUDDIMAN, W. F., KIDD, R. B., THOMAS, E., *et al. Initial Reports of DSDP,* **94**, US Govt. Printing Office, Washington, 703–727.

—— & CLEMENT, B. M. 1986. Synchroneity of Pliocene planktonic foraminiferal datums in the North Atlantic. *Marine Micropaleontology,* **10**, 295–307.

—— & HINE, N. in press. Calcareous nannofossils from the Quaternary of the Northeastern Atlantic. *In*: LORD, A. R. (ed.) *Stratigraphic Index of Calcareous Nannofossils.* British Micropalaeontological Society Special Publication.

—— & KUIJPERS, A. 1983. Climatic control of turbidite deposition on the Madeira Abyssal Plain. *Nature,* **306**, 360–363.

—— & PUJOL, C. 1988. History of the last deglaciation in the Alboran Sea (Western Mediterranean) and adjacent North Atlantic as revealed by coccolith floras. *Palaeogeography, Palaeoclimatology, Palaeoecology,* **64**, 35–42.

—— & RAYMO, M. E. 1989. Late Miocene to Recent planktonic foraminifers from the equatorial Atlantic, Leg 108. *In*: RUDDIMAN, W. & SARNTHEIN, M. *et al. Proceedings of the Ocean Drilling Program, Scientific Results,* **108**, College Station TX, 71–91.

——, SEARLE, R. C. & KUIJPERS, A. 1986. Turbidite deposition and the origin of the Madeira Abyssal Plain. *In*: SUMMERHAYES, C. P. & SHACKLETON, N. J. (eds) *North Atlantic Paleoceanography.* Geological Society, London, Special Publication, **21**, 131–143.

Ostracoda as biostratigraphical indices in Cenozoic deep-sea sequences

R. C. WHATLEY

Micropalaeontology and Palynology Research Group, Institute of Earth Studies, University College of Wales, Aberystwyth, Dyfed, UK

Abstract: Some of the disadvantages inherent in the use of benthonic deep-sea Ostracoda such as their lack of pelagic larval stages and the extent to which relative paucity in deeper abyssal environments is prejudicial to diversity, are balanced against the advantages of being pandemic in distribution and independent of latitudinal and surface current temperature controls. The application of Ostracoda to the biostratigraphy of the deep water Cenozoic of the North Atlantic is outlined in some detail and it is shown that podocopid Ostracoda can be used to determine the various nannofossil zones.

Ostracoda have never been considered as one of the leading fossil groups in terms of their utility as biostratigraphical indices. Their great forte in applied aspects of micropalaeontology is with respect to palaeoenvironmental analysis (Whatley 1983*a, b,* 1988 *a* and *b*). The main reason for this is that most fossil ostracods belong to orders that are either exclusively or largely benthonic in habit. They also lack pelagic larval stages.

Shallow benthonic marine organisms, such as the podocopid and palaeocopid ostracods, because of the extent to which they are substrate and depth controlled, are of limited geographical distribution and this must reduce their efficacy in biostratigraphy. Temperature, which is the primary factor in depth and latitudinal distribution, is another great limiting factor to shelf species as are land masses and oceanic depths. It is these very controls, of course, that render benthonic ostracods such valuable environmental and palaeoenvironmental indicators.

Planktonic organisms are also controlled by temperature, ocean currents and water mass, but this has less constraint on their biogeographical distribution than do the limiting factors on the shelf benthos. After all, oceans are much larger than their adjacent shelves. It is the wider distribution of planktonic animals which renders them more suitable biostratigraphical indices than benthos, although many plankton specialists (and most students) seem to believe that by virtue of being plankton *per se* they evolve faster than benthos. That this is not true has been shown by Whatley & Stephens (1976) who, in their study of the evolution of Mesozoic Ostracoda, demonstrate that at times speciation among such benthonic families as the

Cythereideidae, Protocytheridae, Progonocythereidae and Trachylberididae, was at least as rapid as in any group yet studied. Whatley (1986, 1988*a, b,* 1990*a, b*) has further demonstrated the rapidity of evolution of the group at certain times in the Mesozoic. Notwithstanding this, benthonic organisms are relatively rarely used in biostratigraphy and usually only when planktonic fossils are not available; such is the case with rugose corals in the British Lower Carboniferous.

This is not to say that such groups as benthonic ostracods in shelf or in non-marine environments are not of biostratigraphical importance. Bate & Robinson (1978) have shown the extent to which ostracods can be used biostratigraphically from the Ordovician to the Quaternary in the context of the British Isles and the present author is well aware of their potential in the Callovian and Oxfordian of the UK (Whatley 1964, 1965, 1970).

The late Jurassic and Cretaceous non-marine ostracod genus *Cypridea* (and its subgenera), is certainly a strong contender to be the most valuable biostratigraphical taxon of all time (Anderson 1971, 1973, 1985). It owes its virtually ubiquitous distribution in the Lower Cretaceous 'Wealden' deposits, from southern Argentina to Manchuria and from North America to West Africa, to the fact that its immediate ancestors at some time in the Late Jurassic evolved the ability to reproduce parthenogenetically and to produce a desiccation and freezing resistant egg. Whatley (1990*a, b,* 1991, 1992) has shown that it was these changes in the reproductive and dispersion strategies of this group of cyprid ostracods which allowed them, with extraordinary rapidity, to conquer

From HAILWOOD, E. A. & KIDD, R. B. (eds), *High Resolution Stratigraphy*
Geological Society Special Publication, No. 70, pp. 155–167.

155

the continental waters of the planet. He also demonstrated that it was probably their increased ability as swimmers which enabled them to dominate the pre-existing limnocytherid and darwinulid ostracods in these environments.

Ostracoda in deep-sea environments

The deep sea is a harsh environment of extreme cold, no light (except bioluminescence), high pressure, and problematical availability of such essentials as calcium carbonate at certain depths. Reliance on primary production, which is almost entirely in the photic zone, renders the benthonic ecosystem of the slope and abyssal plain completely dependent upon events influencing the surface layers of the water column. The consequences of kenoxic (dysaerobic) events in the worlds oceans during the past with respect to Ostracoda are increasingly well known and have been documented by Jarvis *et al.* (1988), Horne *et al.* (1990), Boomer (1991), Whatley (1990*a,b*, 1991), Whatley & Arias (in press). This evidence demonstrates both the vulnerability of the deep-sea ostracod assemblage and its resilience in its ability to reconstitute itself with time from virtual catastrophic extinction.

The deep sea does possess, however, distinct advantages not present in shallow waters. There are, for example, no diurnal nor seasonal fluctuations in temperature or light intensity and, in low latitudes, there is probably a regular supply of nutrients from the photic zone. In higher latitudes, despite the filtering effect of differential sinking rates and feeding patterns *en route* to the bottom, there is probably a reflection of spring and autumn plankton blooms (Angel 1990). Despite this, however, the deep-sea environment, if harsh, is a very stable one.

There is another factor which is unique to the deep sea. There is an enormous continuity of shelf environments around continents which, if depth were the sole criterion governing distribution, would allow eastern Atlantic organisms access to the Pacific *via* a northern or southern route, and Asian, Indian, Arabian, African and European organisms to consort together at will. In practice, such associations are denied by a much more fundamental constraint, temperature. In the deep sea, however, both slope and abyssal-plain species are able to migrate, interoceanically for the former and globally for the latter, within the same parameters of temperature and depth, without let or hindrance.

It is this factor of pandemism which makes the zoogeographical distribution of the deep-sea benthos uniquely widespread and entirely independent of the planet's temperature and climatic zones. While all the world's oceans and areas of these oceans have their endemic ostracod faunas, most of the deep-sea Ostracoda are pan-abyssal (Whatley & Ayress, 1988; Coles *et al.* 1990). Even such specialized ostracod communities as those inhabiting sunken wood on the deep-sea floor seem to be world-wide (Maddocks & Steineck 1987; Steineck *et al.* 1990). This near ubiquity of many deep-sea ostracod genera and species is now an accepted fact and runs counter to earlier (and some later) suggestions that ostracods differed from other deep-sea benthos in not being widely distributed (Benson 1975, 1979; Hartmann & Hartmann-Schroeder 1988).

The biostratigraphical potential of deep-sea Ostracoda

Given that many species and genera of ostracods occur in the same abyssal and bathyal environments in more than one ocean basin (Whatley & Ayress 1988; Coles *et al.* 1990), they should potentially be of value in correlating oceanic deposits between such basins, provided that they evolved rapidly enough and were also distributed rapidly enough from one basin to another as to appear as though this happened 'synchronously' in the geological record.

Benson (1975, 1979), Benson & Peypouquet (1983), Whatley (1985), Whatley *et al.* (1983), Whatley & Coles (1987, 1991), Coles (1990) and a number of unpublished doctoral theses completed at Aberystwyth (Downing, 1985; Millson, 1988; Ayress, 1988; Coles, 1990) have all shown in various ways that elements of the deep-sea Cenozoic ostracod fauna have at times evolved very rapidly. Whatley & Ayress (1988), however, have shown that (although Coles *et al.* 1990 dispute this to a certain extent), there is a considerable time-lag between the origination of a species in one ocean and its migration to adjacent and neighbouring oceans. This lag is not surprising, given the fact that without pelagic larval stages, these benthonic ostracods can only migrate as far as they can walk or be passively dispersed by currents during their lifetimes. There is, of course, a lag in migration times of planktonic foraminifera and calcareous nannoplankton, but it is so short in geological terms as to appear synchronous or, as we are increasingly beginning to understand, nearly synchronous.

While migration time counts against the correlation potential of deep-sea ostracods, this is not as serious a problem as is their relative

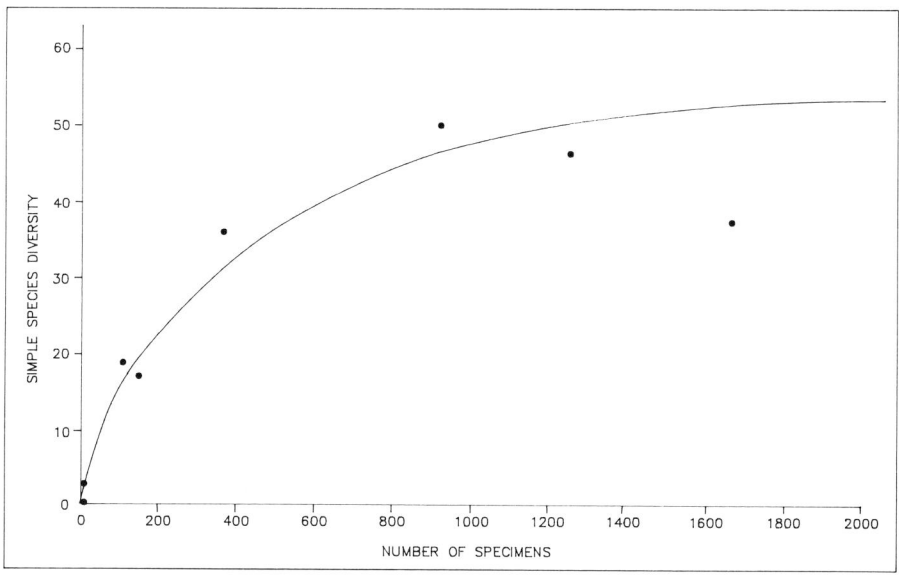

Fig. 1. Number of specimens plotted against number of species (abundance against simple species diversity) for all Palaeogene samples from DSDP Site 207, Lord Howe Rise, South West Pacific.

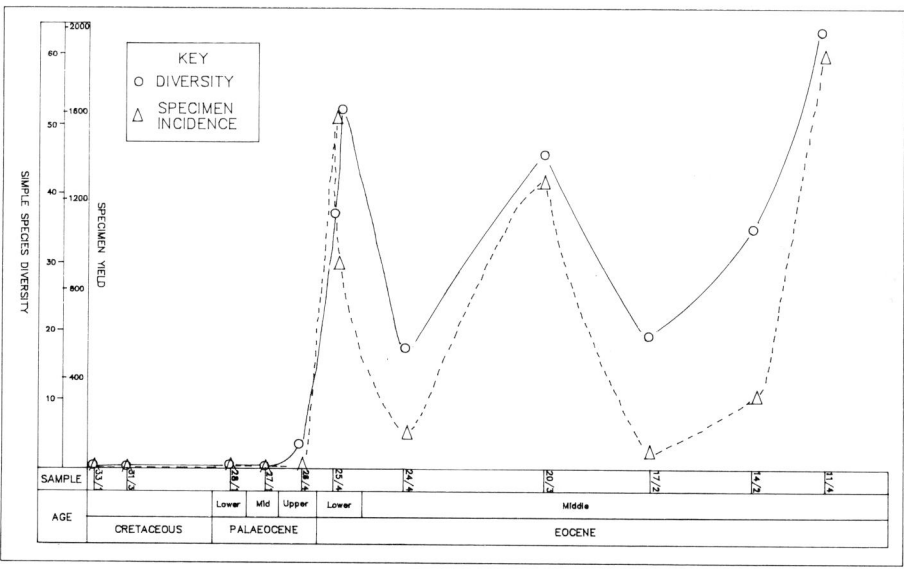

Fig. 2. Graphs showing simple species diversity (number of species) per sample and incidence (number of specimens) per sample for the Cretaceous and Palaeogene of DSDP Site 207, Lord Howe Rise, South West Pacific.

paucity in deep-sea sediments when compared to the often overwhelming numbers of planktonic organisms, whose dead remains make up so much of deep-sea biogenic sediments. What percentage of a typical nanno- or Globigerina ooze is made up by Ostracoda can only be guessed, but is likely to be much less than 1%.

Millson (1988) showed by the use of the

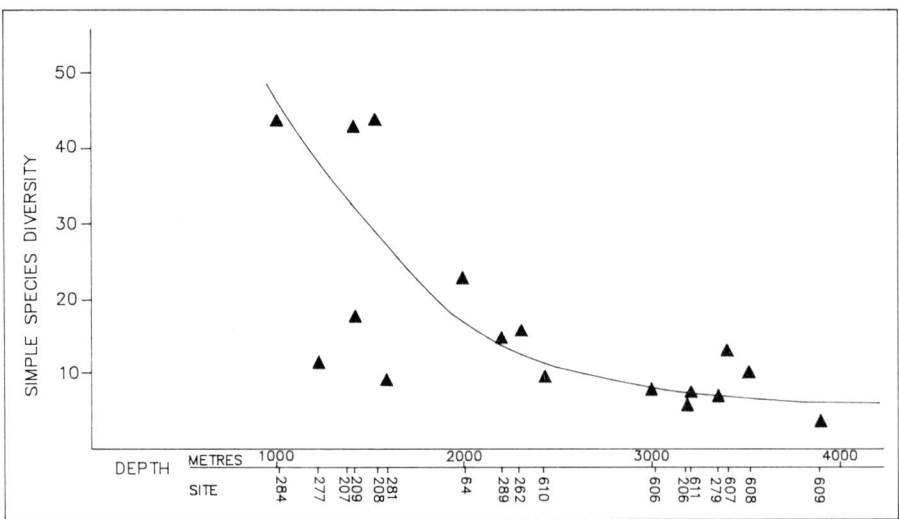

Fig. 3. Mean simple species diversity of Neogene samples from a number of DSDP sites in the South West Pacific plotted against supposed depth of deposition.

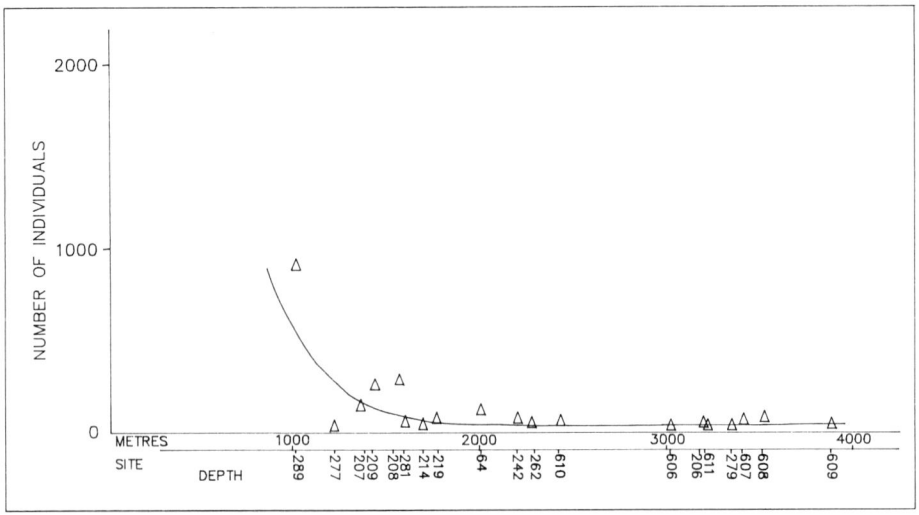

Fig. 4. Mean specimen abundance of Neogene samples from a number of DSDP sites in the South West Pacific plotted against supposed depth of deposition.

Sanders (1968) method of rarefaction, that some samples in her study of the Palaeogene Ostracoda of the southwestern Pacific, contained too few specimens for the fauna to be representative. These samples, where ostracods were rare were, of necessity, excluded from the formulation of biostratigraphical and palaeoenvironmental conclusions. A major controlling factor in DSDP/ODP studies is that the maximum available sample volume is 50 cc and this,

particularly from deeper water sites, is often inadequate to allow of the presence of a representative number of ostracods.

Benson (1975) has argued that ostracod diversity decreases with depth in the deep sea. To an extent his data are flawed since he often did not consider the fauna from the fine fractions of the sediment and, thereby, overlooked many species (Whatley, 1983c; Van Harten, 1990). In the main, however, his observations are correct

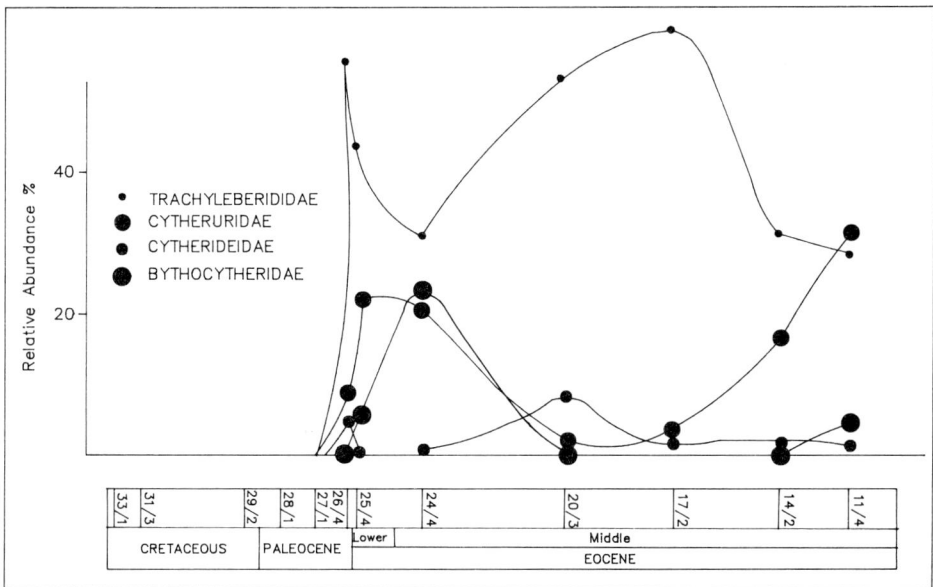

Fig. 5. Graphs showing the relative abundance of the principal families of the ostracod superfamily Cytheracea in samples from the Cretaceous and Palaeogene of DSDP Site 207, Lord Howe Rise, South West Pacific.

but as Millson (1988) has shown, it is the all important relationship between incidence and diversity which is probably the prime reason why diversity, as measured against area or volume, decreases in a general way with depth in the deep sea (Figs 1 and 2). Actually, of course, it is not the case that diversity *per se* decreases with depth, only our perception of it per unit area or volume; what we are actually measuring in the deep sea is apparent diversity. Figure 2 in particular shows the relationship between incidence and diversity and also demonstrates how, if the diversity peaks were plotted alone, they could be interpreted as reflecting major palaeoceanographic events. They may, of course, be related to major global oceanographic changes which took place in the early Palaeogene, but given their relationship to incidence, this would be difficult to demonstrate unless the same peaks were coincident at a number of different sites.

Figures 3 and 4, also taken from Millson (1988), illustrate in a different way the relationship between depth, diversity and incidence. Figure 3 shows how simple species diversity in the Neogene of the SW Pacific falls off with depth as demonstrated by Benson (1975). Fig. 4 demonstrates how this diversity decline is, in part at least, an artefact of the rarity of specimens, which increases with depth.

The author has seen a number of samples from very deep water, for example one from 5400 m

from west of the Azores, which were exceptionally rich in ostracods and which yielded a very high diversity. Such samples are usually collected by epibenthic sledge but a number of exceptionally rich samples are from standard DSDP cores from the Tertiary of the Philippine Sea.,

Figure 5 illustrates yet another difficulty which may be encountered when using deep-sea ostracods (or indeed any other group of organisms, benthonic or planktonic) in biostratigraphy. The diagram plots the fortunes, in terms of their relative abundances, of four families of Cytheracea throughout the Palaeocene–Mid-Eocene, as recorded from DSDP Site 207, on the Lord Howe Rise off New Zealand. None of the four families could have been used in the Cretaceous or early Palaeocene, simply because they were not present. The Trachyleberididae, almost throughout the entire section the dominant family, nonetheless is subject to large fluctuations in its relative abundance and, if this group were chosen to provide the bulk of the index taxa for the zonation of this section, at times these would be 50% more likely to be recovered than at other times. These fluctuations, caused by a complex interaction of evolutionary and environmental parameters, would affect the utility of the other three families even more profoundly.

What might have been

While the discussion so far has been concerned entirely with ostracods belonging to the Order Podocopida and mainly concerned with the Suborder Podocopina, all of which are benthonic in marine environments, other ostracods also occur in the oceans. In the surface layers and deeper down in the water column of modern oceans are to be found very large numbers of ostracods belonging to the Order Myodocopida. Of these, the Suborder Cladocopina is poorly and patchily represented in deep sediments, mainly in the form of *Polycope* which is a benthonic swimmer. The Suborder Myodocopina, however, despite the large number of individuals belonging to the Halocypridacea and Cypridinacea which inhabit especially the surface waters of the oceans, are almost entirely absent from Recent deep-sea sediments and entirely so from fossil oozes. The author and his students have examined many hundreds of mainly Tertiary DSDP and other deep-sea cores and have never seen even a fragment of a myodocopinid ostracod. Also, in the large number of Recent oozes we have examined, halocyprids are entirely absent and cypridinids are so rare as to almost not register. There are very few records of fossils of the former group and many of those records are of taxa quite dissimilar to those of modern oceans (Colin & Andreau, 1990). Records of the former group, although probably not from deep-sea sediments, are much more common in the Palaeozoic than subsequently.

The structure of modern myodocopid populations and particularly the halocyprids, by analogy with the Globigerinacea, is suggestive of their having evolved rapidly during the Tertiary and that if they had been preserved fossil, they would have rivalled the planktonic Foraminifera as biostratigraphical indices.

The reason why myodocopid ostracods are not preserved fossil is due mainly to the fact that, during evolution and presumably as a means to improve their buoyancy, they have radically reduced the amount of calcium carbonate used to build their carapaces; in some taxa this is entirely lost and the valves are 'chitinous'. An additional factor militating against their preservation in the fossil record is that, without a heavy calcareous carapace they will, on death, sink more slowly and be more likely to be consumed *en route*.

The application of Ostracoda to biostratigraphy in the deep sea

Notwithstanding the paucity of ostracods in the deeper parts of the abyss, they can be quite satisfactorily used biostratigraphically as demonstrated by Whatley & Coles (1987) for the late Miocene to Recent interval in the North Atlantic and, subsequently (1991) for the entire Tertiary of the same region.

Millson (1988) produced a good ostracod zonation for each of the DSDP sites she studied in the Palaeogene of the Australasian sector of the SW Pacific, while Downing (1985) has done the same for the Pliocene of the same area. Ayress (1988) had produced a Quaternary ostracod biostratigraphy for a wider area of the SW Pacific and Indian Ocean. Whatley & Coles (1991) have published a detailed and comprehensive ostracod biostratigraphy for the entire Cenozoic of the North Atlantic. Previous studies were by Brady (1880), Brady & Norman (1889), Tressler (1941), Davies (1981) and Van Harten (1990) on the Recent faunas and, for the Cenozoic, Ducasse & Peypouquet (1979) on DSDP Leg 48 cores from the Rockall Plateau and the Bay of Biscay, Guernet (1982) on the Palaeogene Faunas of DSDP Site 390 in the Bahama Basin, Cronin & Compton-Gooding (1987) on Eocene, Oligocene and Plio-Pleistocene faunas of DSDP Leg 95 from offshore New Jersey. Porter (1984) and Harpur (1985) studied Quaternary ostracods from the northern part of the area and off the entrance to the Mediterranean respectively. Coles and Whatley (1989) described many new taxa from the North Atlantic.

The ostracod range charts for each of the sites and holes of Leg 94 (Whatley & Coles 1987) yielded great detail on the biostratigraphical distribution of ostracods over the late Miocene to late Quaternary (NN9-21) interval in the central North Atlantic; data from Legs 80 (Sites 549, 550) and 82 (Sites 558, 563) were subsequently incorporated and the whole was published by the same authors (Whatley & Coles 1991).

Although 230 species were recorded from the entire Cenozoic of the North Atlantic, in the compilation of the range chart (Whatley & Coles 1991) the number was reduced to 184 species by the elimination of all shallow water contaminants, unique or exclusively juvenile occurrences and Recent species which lack a fossil history. These range charts are incorporated into table 1 of Whatley & Coles (1991) and demonstrate the facility with which all the classic divisions of the Cainozoic are recognizable on the basis of ostracod distributions. This is also illustrated by Table 1 of the present paper. Both tables show how easy it would be to erect a system of zones based on ostracods for the

Table 1. *First and last appearances of North Atlantic Cenozoic deep water Ostracoda by NP and NN zones*

Stratigraphical division/ nannofossil zone	First appearance	Last appearance
Lower Palaeocene NP3	*Bairdoppilata* cf. *cassida*	NIL
	Cytherella harmoniensis	
	Cytheropt. cf. *paucipunctatum*	
	Krithe sp.4.	
	Phacorhabdotus sp.	
	Pterygocythere mucronalatum	
	Xestoleberis cf. *profundis*	
Upper Palaeocene NP4	*Eucythere circumcostata*	NIL
	Trachyleberidea pisinensis	
NP5	NIL	NIL
NP6	*Abyssocythere cenozoica*	NIL
NP7	*Agrenocythere ordinata*	NIL
	Argilloecia sp.4.	
	Aversovalva alveiformis	
	A. sp.1.	
	Eucythere paralaevis	
	Krithe sp.C.	
	Krithe sp.G.	
NP8	NIL	NIL
NP9	*Aratrocypris maddocksae*	*Abyssocythere cenozoica*
	Argilloecia angulata	*Krithe* sp.G.
	Cardobairdia ovata	
	Eucythere multipunctata	
	Krithe cf. *parvula*	
	Palmoconcha parastriata	
Lower Eocene NP10	*Bairdia* gr. *subcircinata*	*Krithe* sp.C.
	Cytherella sp.1.	* *Krithe* sp.D.
	Eucytherura calabra	*Krithe* sp.K.
	Hemiparacythere forteornatum	
	Krithe dolichodeira	
	Krithe morkhoveni	
	* *Krithe* sp.D.	
	Legitimocythere presequenta	
	Paraceratina sp.	
	Pennyella fortedimorphica	
	Saida micropunctata	
NP11	*Australoecia* cf. *micra*	NIL
	Chejudocythere sp.	
	Cytherella gamardensis	
	Cytheropteron sp.2.	
	Heinia sp.	
	Krithe cf. *hiwanneensis*	
	Propontocypris sp.	
	Xestoleberis planoventralis	
NP12	*Aversovalva pinarense*	*Heinia* sp.
	Eucytherura mediopunctata	*Palmoconcha parastriata*
	Profundobythere multipunctata	*Xestoleberis* cf. *profundis*
	Rimacytherop. rotundapunctata	*Phacorhabdotus* sp.
NP13	*Argilloecia* sp.3.	*Bairdoppilata* cf. *cassida*
	Bathypterocythereis bathypteron	*Cytheropteron* cf. *paucipunctatum*
	Bythoceratina cf. *umbonata*	*Argilloecia* sp.4.
	Eucytherura pseudoantipodum	*Aversovalva* sp.1.
	Krithe sp.5	*Hemiparacytheridea forteornatum*
	Pedicythere sp.	

Table 1. *cont.*

Stratigraphical division/ nannofossil zone	First appearance	Last appearance
Middle Eocene NP14	*Argilloecia* gr. *hiwanneensis* *Argilloecia* sp.8 *Bairdoppilata cassida*	NIL
NP15	*Abyssobairdia anisovalva* *Abyssocythere trinidadensis* *Agrenocythere bensoni* *Argilloecia* sp.6. *Australoecia posteroacuta* *Bairdia* sp.1. *Bythocypris aturica* *Bythocypris* sp.1. *Cardobairdia asymmetrica* *Cytheropteron branchium* *Cytheropteron* sp.4. *Dutoitella eocenica* *Eucytherura* sp.4. *Heinia* sp.2. *Henryhowella* gr. *asperrima* *Krithe pernoides* *Krithe* sp.A. *Macrocypris* cf. *rhodana* *Phacorhabdotus anteronudus* *Profundobythere bathytatos* *Profundobythere splendida* *Pseudocythere caudata* *Xestoleberis profundis* *Xestoleberis moriahensis*	*Cardobairdia ovata* *Krithe* cf. *parvula* *Bythocypris* sp.1.
NP16	*Anchistrocheles antemacella* *Argilloecia* sp.7. *Krithe trinidadensis* *Parakrithe vermunti* (normal overlap)	*Krithe* sp.4. *Krithe* sp.A. *Profundobythere multipunctata* *Argilloecia* sp.3.
NP17	NIL	NIL
NP18	*Buntonia textilis* *Eucythere laevis* *Eucytherura* sp.1. *Krithe crassicaudata* *Phacorhab. posteropunctissima*	*Pariceratina* sp. *Bairdia* sp.1. *Cytheropteron* sp.4.
Upper Eocene NP19	*Argilloecia* sp.10. *Australoecia micra* *Bairdia* sp.2. *Cytherella* gr. *serratula* *Cytheropteron lineoporosa* *Parahemingwayella downingae* *Pelecocythere sylvesterbradleyi* *Pennyella praedorsoserrata*	*Eucytherura* sp.4.
NP20	*Eucythere* sp. *Parakrithe vermunti* (normal overlap) *Peleocythere trinidadensis*	*Agrenocythere ordinata* *Eucythere* sp.
Eocene–Oligocene Boundary NP21	*Agrenocythere hazelae* *Aversovalva formosa* *Aversovalva hydrodynamica* * *Bairdoppilata* sp.2. *Bythocypris* sp.2. *Cytheropteron paucipunctatum* *Cytheropteron* sp.1.	*Cytheropteron* sp.2. * *Bairdoppilata* sp.2.

Table 1. *cont.*

Stratigraphical division/ nannofossil zone	First appearance	Last appearance
Lower Oligocene NP22	*Echinocythereis echinata* *Eucythere parapubera* *Poseidonamicus pseudorobustus* *Ambocythere* cf. *elongata* *Aratrocypris* cf. *prealta* *Cytheropteron garganicum* *Cytheropteron massoni* *Cytheropteron pherozigzag* *Cytheropteron syntomalatum* *Mayburya pulchra*	*Aratrocypris maddocksae* *Macrocypris* cf. *rhodana*
Upper Oligocene NP22	*Bradleya dictyon* *Eucytherura* sp.2. *Krithe* sp.F. *Xestoleberis abyssoris*	*Chejudocythere* sp. *Cytherella gamardensis* *Bathypterocythereis bathypteron* *Bythoceratina* cf. *umbonata* *Dutoitella eocenica* *Krithe* sp.3. *Phacorhabdotus anteronudus* *Krithe crassicaudata* *Phacorhabdotus posteropunctissima* *Bythocypris* sp.2.
NP24	*Bairdoppilata* sp.3. *Cytheropteron* sp.6. *Eucythere concinna* *Legitimocythere acanthoderma* *Pennyella dorsoserrata* *Poseidonamicus* cf. *pintoi* *Semicytherura coeca* * *Xylocythere* sp.1.	*Cytherella harmoniensis* *Cytherella* sp.1. *Saida micropunctata* *Australoecia* cf. *micra* *Propontocypris* sp. *Bairdoppilata cassida* *Eucytherura pseudoantipodum* *Xestoleberis moriahensis* *Argilloecia* sp.10. *Aversovalva formosa* *Aratrocypris prealta* *Krithe* sp.F. *Xylocythere* sp.1.
NP25	NIL	*Trachyleberidea pisinensis* *Aversovalva alveiformis* *Eucythere paralaevis* *Xestoleberis planoventralis* *Aversovalva pinarense* *Eucytherura mediopunctata* *Krithe* sp.5 *Agrenocythere bensoni* *Australoecia posteroacuta* *Argilloecia* sp.7. *Bairdia* sp.2. *Parahemingwayella downingae* *Cytheropteron* sp.1. *Poseidonamicus pseudorobustus* *Ambocythere* cf. *elongata* *Cytheropteron* sp.6. *Eucythere concinna* *Semicytherura coeca*
Lower Miocene NN1	*Aversovalva atlantica* *Dutoitella parasuhmi* *Heinia dryppa* *Krithe* sp.1.	*Abyssobairdia anisovalva*

Table 1. *cont.*

Stratigraphical division/ nannofossil zone	First appearance	Last appearance
NN2	NIL	*Argilloecia* sp.8.
		Pennyella praedorsoserrata
NN3	*Henryhowella dasyderma*	
	Parakrithe cf. *elongata*	
NN4	NIL	*Bythocypris aturica*
Lower–Middle	*Cytherella* sp.2.	*Krithe* cf. *hiwanneensis*
Miocene Boundary	*Krithe* sp.2.	*Argilloecia* sp.6.
NN5		*Eucythere laevis*
		Bairdoppilata sp.3.
Middle Miocene	*Krithe reversa*	NIL
NN6		
NN7	NIL	*Dutoitella praesuhmi*
NN8	NIL	NIL
Upper Miocene	*Abyssocythere sulcatoperforata*	*Rimacytheropteron rotundapuncta*
NN9	*Ambocythere caudata*	*Abyssocythere trinidadensis*
	Eucythere hyboma	*Cytherella* sp.2.
	Pelecocythere foramena	
	Poseidonamicus minor	
	Poseidonamicus praenudus	
NN10	*Krithe* sp.6.	*Pennyella fortedimorphica*
	Rockallia enigmatica	
NN11	*Ambocythere* cf. *caudata*	*Argilloecia* gr. *hiwanneensis*
	Buntonia pyriformis	*Poseidonamicus minor*
	Cytheropteron porterae	
	Cytheropteron testudo	
	Dutoitella suhmi	
	Pennyella horrida	
	Poseidonamicus cf. *major*	
NN12	NIL	*Pelecocythere trinidadensis*
		Krithe sp.6.
Lower Pliocene	NIL	NIL
NN13		
NN14	*Argilloecia* sp.1.	NIL
	Argilloecia sp.2.	
	Argilloecia sp.5.	
	Bythocypris cf. *mozambiquensis*	
	Eucythere triangula	
	Cytheropteron tenuialatum	
NN15	*Abyssocythere atlantica*	*Heinia* sp.2.
	Ambocythere ramosa	*Aratrocypris rectoporrectus*
	* *Aratrocypris rectoporrectus*	
	Bathycythere audax	
	Rimacytheropteron longipunctata	
Lower–Upper	*Eucythere pubera*	*Xestoleberis abyssoris*
Pliocene boundary	*Krithe* sp.9.	*Ambocythere* cf. *caudata*
NN16		
Upper Pliocene	*Bythoceratina scaberrima*	NIL
NN17	*Cytheropteron carolinae*	
	Cytheropteron tressleri	
	Pedicythere phryne	
NN18	NIL	*Parakrithella vermunti*
		(reversed overlap)
		Krithe sp.9.

Table 1. *cont.*

Stratigraphical division/ nannofossil zone	First appearance	Last appearance
NN19	* *Cytheropteron circumuralla*	*Pedicythere* sp.
	Cytheropteron trifossata	*Abyssocythere sulcatoperforata*
	Pedicythere polita	*Pelecocythere foramena*
		Poseidonamicus praenudus
		* *Cytheropteron circumuralla*
Quaternary	NIL	NIL
NN20		
NN21	* *Bythocythere reflexa*	*Eucythere circumcostata*
	Bythocypris reinformis	*Xestoleberis profundis*
	Bradleya normani	*Eucythere* sp.1.
	* *Cytheropteron pinnatum*	*Eucythere hyboma*
	Cytheropteron retrosulcatum	*Bythocypris* cf. *mozambiquensis*
	* *Eopaijenborchella cymbula*	*Cytheropteron tenuialatum*
	Krithe keiji	*Rimacytheropteron longipunctata*
	* *Loxoconchidea minima*	*Cytheropteron carolinae*
	* *Macrocypris tenuicauda*	*Cytheropteron tressleri*
	Monoceratina insignis	*Pedicythere phryne*
	Parahemingwayella tetrapteron	*Cytheropteron trifossata*
	Pedicythere mirabilis	* *Bythocypris reflexa*
	Rectobuntonia inflata	*Cytheropteron pinnatum*
	Rectobunbtonia miranda	* *Eopaijenborchella cymbula*
		* *Loxoconchidea minima*
		* *Macrocypris tenuicauda*
		Cardobairdia asymmetrica
		Cytheropteron branchium
		Anchistrocheles antemacella
		Krithe sp.1.

Cainozoic of the North Atlantic. For the present however, it is thought that their utility in identifying the position of a sample with respect to existing nannofossil zones is sufficient.

As (Whatley & Coles 1991, table 1) and Table 1 of the present paper show, many of the species were of short duration while others ranged through long intervals. A complication is the fact that some species temporarily disappear from the record, often several times, only to reappear higher in the section. These Lazarus taxa complicate the interpretation of table 1 of Whatley & Coles (1991), because, at present, it is not possible to determine whether these temporary absences are brought about by mutual exclusion, result from environmental changes, or are the product of inadequate sample size.

Table 1 represents a synthesis of the biostratigraphical range charts published in Whatley & Coles (1991). This is an important data set upon which an appreciation of North Atlantic ostracod biostratigraphy can be based. The table shows that through the early part of the Palaeogene, there was zero extinction (shown as NIL in the right hand last appearances column). This

absence of extinction was a function of the lack of competition in a newly created oceanic environment and the level of extinction remains low until nannofossil Zone NP12, in the late Lower Eocene. Originations (first appearances) continued to outstrip extinctions until the late Oligocene when there was a major faunal crash.

Throughout the Cenozoic of the North Atlantic, there are a sufficiency of first and last appearances of ostracod species per nannofossil zone to allow of the recognition of these zones with considerable confidence. This allows ostracod workers to assign relative ages to their samples without the necessity for assistance from other workers or having to become a specialist in nannofossils.

Conclusions

While one does not expect ostracods to rival other, mainly planktonic, microfossils in biostratigraphy, despite all their drawbacks as index fossils in a general sense, in the deep sea their independence of global climatic belts and surface currents is very much in their favour, since it allows many species to be pan-abyssal in distribution. The current biostratigraphy is being constantly improved and a scheme of inter-oceanic correlation based on Cenozoic podocopid ostracods is not far off.

The author acknowledges K. Millson, who conceived Figs 1–5. C. Maybury is thanked for greatly improving the MS and for nursing a dying word processor so successfully as to allow the author to complete this paper almost on time.

References

ANDERSON, F. W. 1971. The Ostracoda. *In*: ANDERSON, F. W. & BAZLEY, R. A. B. (eds) *The Purbeck Beds of the Weald (England)*. Bulletin Geological Survey Great Britain, London. **34**, 1–173.

—— 1973. The Jurassic–Cretaceous transition: the non-marine faunas. *In*: CASEY, R & RAWSON, P. F. (eds) *The Boreal Lower Cretaceous, Geological Journal*, Special Issue 5, Liverpool, 101–110.

—— 1985. Ostracod faunas in the Purbeck and Wealden of England. *Journal of Micropalaeontology*, **4**, 1–68.

ANGEL, M. V. 1990. Food in the deep Ocean. *In*: WHATLEY, R. C. & MAYBURY, C. (eds) *Ostracoda and Global Events*, Chapman & Hall, 274–285.

AYRESS, M. A. 1988. *Late Pliocene to Quaternary deep-sea Ostracoda from the eastern Indian and south western Pacific Oceans*. PhD thesis, University of Wales.

BATE, R. H. & ROBINSON, E. 1978. (eds) *A stratigraphical index of British Ostracoda*, Seel House Press.

BENSON, R. H. 1975. The origin of the psychrosphere as recorded in changes of deep-sea ostracod assemblages. *Lethaia*, **8**, 69–83.

—— 1979. In search of lost oceans: a paradox in discovery. *In*: GRAY, J. & BOUCOT, A. J. (eds) *Historical biogeography, plate tectonics, and the changing environment*. Oregon State University Press, 379–389.

—— & PEYPOUQUET, J-P. 1983. The upper and mid-bathyal Cenozoic ostracod faunas of the Rio Grande Rise found on Leg 72 of the Deep Sea Drilling Project. *In*: BARKER, P. F. *et al.* (eds) *Initial Reports DSDP*, **72**, 805–818, U.S. Govt. Printing Office, Washington.

BOOMER, I. 1991. Lower Jurassic Ostracod Biozonation of the Mochras borehole. *Journal of Micropalaeontology* **9**, (2), 205–218.

BRADY, G. S. 1880. Report on the Ostracods dredged by *H.M.S. Challenger* during the years 1873–1876. *Report on the voyage of H.M.S. Challenger. Zoology*, **1**, 1–184.

—— & NORMAN, A. M. 1889. A Monograph of the marine and freshwater Ostracoda of the North Atlantic and of N.W. Europe. Section 1: Podocopa. *Scientific Transactions of the Royal Dublin Society*, **4**(2), 63–270.

COLES, G. P. 1989. *Cainozoic evolution of Ostracoda from deep waters of the North Atlantic and adjacent shallow water regions*. PhD thesis, University of Wales.

—— 1990. A comparison of the evolution, diversity, and composition of the Cainozoic Ostracoda in the deep water North Atlantic and shallow water environments of North America and Europe. *In*: WHATLEY R. C. & MAYBURY, C. (eds), *Ostracoda and Global Events*, Chapman & Hall 71–86.

—— & WHATLEY, R. C. 1989. New Palaeogene to Miocene genera and species of Ostracoda from DSDP sites in the North Atlantic. *Revista Espanola Micropaleontologia*. **21**(1), 81–124.

——, AYRESS, M. & WHATLEY, R. C. 1990. A comparison of North Atlantic and Pacific Cainozoic deep-sea Ostracoda. *In*: WHATLEY, R. C. & MAYBURY, C. (eds) *Ostracoda and Global Event*, Chapman & Hall 287–304.

COLIN, J-P. & ANDREU, B. 1990. Cretaceous halocyprid Ostracoda. *In*: WHATLEY, R. C. & MAYBURY, C. (eds) *Ostracoda and Global Events*. Chapman & Hall 515–526.

CRONIN, T. M. & COMPTON-GOODING, E. E. 1987. Cainozoic Ostracoda from Deep Sea Drilling Project Leg 95 off New Jersey (Sites 612 & 613). *In*: POAG, C. W. *et al.* *Initial Reports DSDP*, **95**, 439–451, US Government Printing Office, Washington.

DAVIES, H. C. 1981. *The areal and depth distribution of N.E. Atlantic Ostracoda*. MSc thesis, University of Wales.

DOWNING, S. E. 1985. *The taxonomy, palaeoecology, biostratigraphy and evolution of Pliocene Ostracoda from the West Pacific*. PhD thesis, University of Wales.

DUCASSE, O. & PEYPOUQUET, J-P. 1979. Cenozoic ostracods: their importance for bathymetry, hydrology, and biogeography. *In*: MONTADERT, L. & ROBERTS, D. G. (eds) *Initial Reports DSDP*, **48**, 343–363, US Government Printing Office, Washington.

GUERNET, C. 1982. Contribution a l'etude des faunes abyssales: Les ostracodes Paleogenes du Bassin des Bahamas, Atlantique Nord (DSDP Leg 44). *Revue de Micropaleontologie*, **25**, 279–295.

HARPUR, W. K. 1985. *Late Quaternary deep-sea Ostracoda from the extra-Iberian Portal region*. Msc thesis, University of Wales.

HARTMANN, G. F. & HARTMANN-SCHROEDER, G. 1988. Deep-sea Ostracoda, taxonomy, distribution and morphology. *In*: HANAI, T. *et al.* (eds) *Evolutionary biology of Ostracoda. Its fundamentals and applications*. Kodansha-Elsevier, 699–708.

HORNE, D., JARVIS, I. & ROSENFELD, A. 1990. Recovering from the effects of an Oceanic Anoxic Event: Turonian Ostracoda from S.E. England. *In*: WHATLEY, R. C. & MAYBURY, C. (eds) *Ostracoda and Global Events*, Chapman & Hall 123–138.

JARVIS, I., CARSON, G. A., COOPER, M. K. E., HART, M. B., LEARY, P. N., TOCHER, B. A., HORNE, D. & ROSENFELD, A. 1988. Microfossil Assemblages and the Cenomanian–Turonian (Late Cretaceous) Oceanic Anoxic Event. *Cretaceous Research*, **9**, 3–103.

MADDOCKS, R. F. & STEINECK, P. L. 1987. Ostracoda from experimental wood-island habitats in the deep sea. *Micropalaeontology,* **33**, 318–355.

MILLSON, K. J. 1988. *The palaeobiology of Palaeogene Ostracoda from Deep Sea Drilling Project cores in the South West Pacific,* PhD thesis, University of Wales.

PORTER, C. 1984. *Late Quaternary Ostracoda from the N.E. Atlantic.* MSc thesis, University of Wales.

SANDERS, H. L. 1968. Marine benthic diversity: A comparative study. *The American Naturalist,* **102**, 243–282.

STEINECK, P. L., MADDOCKS, R. F., TURNER, R. D., COLES, G. & WHATLEY, R. 1990. Xylophile Ostracoda in the deep sea. *In*: WHATLEY, R. C. & MAYBURY, C. (eds) *Ostracoda and Global Events,* Chapman & Hall, 307–319.

TRESSLER, W. L. 1941. Geology and biology of North Atlantic deep-sea cores between Newfoundland and Ireland. *United States Geological Survey, Professional Paper,* **196**(4), 95–106.

VAN HARTEN, D. 1990. Modern abyssal ostracod faunas of the eastern Mid-Atlantic Ridge area in the North Atlantic and a comparison with the Mediterranean. *In*: WHATLEY, R. C. & MAYBURY, C. (eds) *Ostracoda and Global Events,* Chapman & Hall, 321–328.

WHATLEY, R. C. 1964. The ostracod genus *Progonocythere* in the English Oxfordian. *Revue Micropaleontologie,* **7**, 188–194.

—— 1965. *Callovian and Oxfordian Ostracoda of Great Britain.* PhD thesis, University of Hull.

—— 1970. Scottish Callovian and Oxfordian Ostracoda. *Bulletin British Museum (Natural History) Geology,* **19**, 6, 297–358.

—— 1983a. The application of Ostracoda to palaeoenvironmental analysis. *In*: MADDOCKS, R. F. (ed.) *Applications of Ostracoda,* University of Houston, Geoscience, Houston, 51–77.

—— 1983b. Some simple procedures for enhancing the use of Ostracoda in palaeoenvironmental analysis. *Norwegian Petroleum Directorate, Bulletin,* **2**, 129–146.

—— 1983c. Evolution of the ostracods *Bradleya* and *Poseidonamicus* in the deep-sea Cainozoic of the S.W. Pacific. *Special Papers in Micropalaeontology,* **33**, 103–116.

—— 1985. Some aspects of the palaeobiology of Tertiary deep-sea Ostracoda from the S.W. Pacific. *Journal of Micropalaeontology,* **2**, 83–104.

—— 1986. Biological events in the evolution of Mesozoic Ostracoda. *Lecture Notes in Earth Sciences,* **8**, *Global Bio-Events.* Springer-Verlag, 257–265.

—— 1988a. Ostracoda and Palaeogeography. *In*: DE DECKKER, P. *et al.* (eds) *Ostracoda in the Earth Sciences.* Elsevier, 104–123.

—— 1988b. Population structure of ostracods: some general principles for the recognition of palaeoenvironments. *In*: DE DECKKER, P. *et al.* (eds) *Ostracoda in the Earth Sciences,* Elsevier, 104–123, 245–256.

—— 1988c. Patterns and rates of evolution in Mesozoic Ostracoda. *In*: HANAI, T. *et al.* (eds) *Evolutionary biology of Ostracoda. Its fundamentals and applications.* Kodansha-Elsevier, 1021–1040.

—— 1990a. The relationship between extrinsic and intrinsic events in the evolution of Mesozoic non-marine Ostracoda. *In*: KAUFFMAN, E. G. & WALLISER, O. H. (eds) *Extinction Events in Earth History.* Lecture Notes in Earth Sciences, **30**, Springer-Verlag, 253–264.

—— 1990b. Ostracoda and Global Events. *In*: WHATLEY, R. C. & MAYBURY, C. (eds) *Ostracoda and Global Events,* Chapman & Hall, 3–24.

—— 1991. The platycopid signal: a means of detecting kenoxic events using Ostracoda. *Journal of Micropalaeontology,* **10**, 181–185.

—— 1992. The reproductive and dispersal strategies of Cretaceous non marine Ostracoda: The key to pandemism. *In*: MATEER, N. J. & CHEN, P. J. (eds) *Aspects of non-marine Cretaceous geology.* China Ocean Press, Beijing, 177–192.

—— & ARIAS, C. (in press). The use of Ostrocoda to detect kenoxic events: a case history from the Spanish Toarcian. *Geobios.*

—— & AYRESS, M. 1988. Pandemic and endemic distribution patterns in Quaternary bathyal and abyssal ostracods. *In*: HANAI, T. *et al.* (eds) *Evolutionary biology of Ostracoda. Its fundamentals and applications.* Kodansha-Elsevier 739–755.

—— & COLES, G. 1987. The late Miocene to Quaternary Ostracoda of Leg 94, Deep Sea Drilling Project. *Revista Espanola Micropaleontologia,* **19**, 33–97.

—— & —— 1991. Global change and the biostratigraphy of deep water Ostracoda from the North Atlantic. *Journal of Micropalaeontology,* **9**(2), 119–132.

—— & STEPHENS, J. M. 1976. The Mesozoic explosion of the Cytheracea. *Abhandlungen und Verhandlungen des Naturwissenschaftlichen Vereins in Hamburg* (NF 18/19 (Supplement): 63–76.

——, HARLOW, C. J., DOWNING, S. E. & KESLER, K. J. 1983. Some observations on the origin, evolution, dispersion and ecology of the genera *Poseidonamicus* Benson and *Bradleya* Hornibrook. *In*: MADDOCKS, R. F. (ed.) *Applications of Ostracoda.* University of Houston, Geoscience, Houston 492–509.

Tephra layers as precise chronostratigraphical markers

R. W. O'B. KNOX

British Geological Survey, Keyworth, Nottingham NG12 5GG, UK

Abstract: Tephra layers provide the stratigrapher with very precise, tangible chronostrati-graphical marker horizons. These can be used not only for correlation between specific sequences, but also as a means of assessing the accuracy and resolution of correlations based on more conventional stratigraphical criteria. Such tephrostratigraphical applications are, however, dependent on the recognition of individual tephra layers, to which end a variety of mineralogical and chemical fingerprinting techniques have been developed.

Tephra layers also provide a unique opportunity for direct and precise dating of sedimentary sequences. Recent developments in grain-specific dating techniques have greatly increased the precision and applicability of tephra dating, which is now seen as playing an essential role in the refinement of the Phanerozoic time scale. Also, as the resolution of dating techniques approaches that of standard biozones, it is becoming apparent that tephra dates may in themselves provide a means of direct correlation on a global basis; such correlations should be of particular value in resolving problems associated with climatically or geographically controlled faunal provinciality.

Tephra layers may be defined as discrete beds, composed originally of volcanic ash, but more commonly preserved in lithified form as tuff, or in argillized form as bentonite (smectite-rich) or tonstein (kaolinite-rich). Their value as precise lithologial and chronostratigraphical markers has long been appreciated. Early studies were carried out mostly in areas of active volcanism, where tephra layers could be related directly to their source volcanoes. An additional stimulus to these early studies was the possibility of relating tephra layers to specific eruptions chronicled in early historical records, such as the account by Pliny the Younger of the eruption of Vesuvius in A.D.79, and records of Icelandic eruptions dating back to that of Hekla in 1104 (Thorarinsson 1981). Tephra studies have also played a part in elucidating prehistoric events, such as the eruption of Santorini that is thought to have been involved in the destruction of the Minoan civilization on Crete in about 1500 BC (see Watkins *et al.* 1978). Following on from early work in the Mediterranean and Icelandic regions, Quaternary tephras have now been studied in volcanically active areas throughout the world, especially in North America and Japan. An added dimension has been provided by the recovery of tephra layers in marine drilling programmes, an early study being that of Bramlette & Bradley (1941). As a result of several decades of combined onshore and offshore studies, it is now possible to build up a detailed eruption history for many of the world's volcanically active regions (see Kennett 1981).

The stratigraphical use of tephra layers in pre-Quaternary rocks was pioneered by Danish geologists in the late nineteenth century, in their investigation of tuffs in the early Eocene Mo-clay diatomites of northwest Denmark. These studies culminated in the comprehensive publication of Bøggild (1918). However, the altered state of many pre-Quaternary tephra layers caused them to be overlooked by early workers, and it was not until the volcanic origin of bentonites became generally accepted that the potential of tephra correlation in pre-Quaternary sequences became widely appreciated, especially in North America (see, for example, Ross 1925). A volcanic origin for the majority of tonsteins (see Burger 1990) did not become generally accepted until much later, following studies carried out in Germany (e.g. Stach 1950), Great Britain (Francis 1961), and France (e.g. Bouroz 1966; Chalard 1967).

This review is concerned primarily with recent developments in the identification and correlation of tephra layers and with their dating. Most workers on Quaternary sequences use the term 'tephrochronology' for both the correlation and the dating of tephra layers, following the original definition of Thorarinsson (1974) as 'a chronology based on measurement, connection and dating of tephra layers'. It should, however, be borne in mind that this definition was related to the study of Recent tephra, which, through historical records, could be used to provide precise dating of glaciological, geomorphological and archaeological features (see Thorarinsson 1981, p. 1). In older sequences, the correlation and absolute dating of tephra layers are essentially separate procedures.

Westgate & Fulton (1975) suggested that the

From HAILWOOD, E. A. & KIDD, R. B. (eds), *High Resolution Stratigraphy*
Geological Society Special Publication, No. 70, pp. 169–186.

term 'tephrostratigraphy' should be used for the 'definition, description and age of tephra layers' and that 'tephrochronology' should be limited to the 'application of the results of tephrostratigraphical studies to geochronological problems'. The two terms are used here in a broadly similar fashion, but with the concept of age more clearly divided into absolute age, included under 'tephrochronology', and relative age, included under 'tephrostratigraphy'. On this basis, *tephrostratigraphy* is concerned with the characterization and correlation of tephra layers, and hence of their associated sequences; *tephrochronology* is concerned with the absolute dating of tephra layers, and hence of their associated sequences.

Case histories are used to illustrate the various aspects of tephra correlation and dating, including the contributions that tephra layers have made to high-resolution integrated stratigraphical analysis. Some of these case histories have already featured in earlier reviews of Quaternary tephra studies (Kennett 1981; Self & Sparks 1981), but are repeated here to emphasise the importance of such studies as a stimulus to the wider use of tephra layers in the stratigraphical analysis of older sequences.

Tephrostratigraphy

Generation of widespread tephra layers

The tephra layers that are of most value in stratigraphical correlation are the thin, widespread tuffs, bentonites and tonsteins that represent the distal deposits of ash eruptions. An understanding of the mechanisms of ash dispersal and accumulation is therefore important in appreciating the geographical and sedimentological constraints on the distribution of tephra layers in the geological record.

Studies of recent volcanoes have shown that only Plinian eruptions have the power to produce an eruption column of sufficient height to allow widespread dispersal by high-altitude winds (Walker 1981; Francis 1985). It has, however, been argued that equally extensive ash deposits may be generated by exceptionally violent ignimbrite eruptions, which may generate bubble-wall shards with unusually low settling velocities (Rose & Chesner 1987). Ash deposits produced by Surtseyan activity are normally less extensive, but may be of use in local, intrabasinal correlation (Francis 1985).

The intensity and dispersive power of eruptions (see Walker 1981 for definitions) are normally related to composition, with most Plinian eruptions being associated with silicic magmas and most Surtseyan eruptions with mafic magmas. Exceptionally, however, basaltic magmas appear to have been involved in Plinian eruptions or at least in eruptions of Plinian proportions (Williams 1983; Walker *et al.* 1984; Morton & Knox 1990).

Although studies of recent eruptions provide valuable information on the dispersal and fallout of volcanic ash, they are largely concerned with terrestrial settings, in which ash accumulations suffer almost immediate redistribution and alteration. Detailed records of the accumulation of individual ash deposits in marine environments are few, but are of considerable importance in stratigraphical studies, since the great majority of tephra layers in the pre-Quaternary record occur in marine sediments, with the remainder mostly occurring in deposits in lacustrine or paludal facies. General accounts of marine tephrostratigraphy in several volcanically active regions include those of Ninkovich *et al.* (1966) for the North Pacific, Bowles *et al.* (1973) for the equatorial Pacific, Keller (1981) and Paterne *et al.* (1988) for the Mediterranean, Ledbetter (1981) and Sigurdsson & Carey (1981) for the Caribbean, and Sigurdsson & Loebner (1981) for the North Atlantic.

Ash-fall deposits from several Recent Plinean eruptions have exceeded $1 \times 10^6 \, km^2$ in extent and tens of thousands of cubic kilometres in volume (see Izett *et al.* 1978, table 8). For example, ash from the Toba eruption covered an area of more than $10 \times 10^6 \, km^2$ (Ninkovich 1979; Rose & Chesner 1987). Along the axis of such an ash lobe, deposits 5 cm thick or more may be encountered 1000 km from the site of eruption. Ancient tephra layers of comparable areal extent and volume have been described by Slaughter & Earley (1965, p. 48), Bouroz *et al.* (1983), Francis (1985) and Knox & Morton (1988).

Such widespread tephra layers are of particular interest to the stratigrapher since they may extend across regions of contrasting climatic and tectonic settings, and therefore provide a means of correlation between sequences of different facies and belonging to different faunal or floral provinces.

Tephra correlation techniques

Although tephra layers clearly have the potential to provide the ultimate in high-resolution chronostratigraphical correlation, this potential can be realized only if an individual layer or group of layers can be reliably correlated from one section to another. In early studies, correlation was based on distribution pattern, physical appearance, and petrological analysis of the groundmass and phenocryst phases. Because of physical modification during transport and deposition, together with chemical modification during diagenesis, such correlation is not always straightforward. A considerable effort has therefore been made to develop those methods of tephra identification that are least affected by depositional and diagenetic processes, as discussed below.

Physical correlation. In favourable circumstances, physical distribution alone can produce reliable and detailed correlations. For example, Bøggild (1918) established a numbered series of over 170 tuff layers that could be traced in sporadic outcrops over an area of 2000 km² in the Lower Eocene of Denmark. Most of these tuffs are of uniform, basaltic composition, so that layers were largely identified on the basis of thickness and vertical distribution pattern (Fig. 1).

Other instances in which correlation of individual tephra layers has been achieved solely

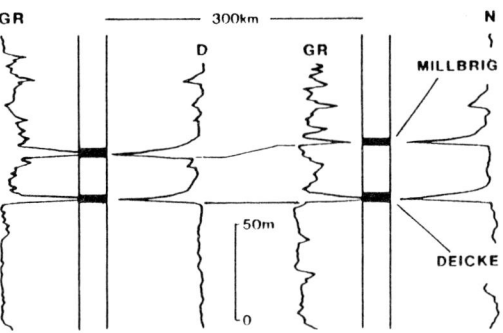

Fig. 2. Wireline-log correlation of K-bentonites associated with Ordovician limestones, USA. GR, gamma-ray log; D, density log; N, neutron log. After Huff & Kolata 1990 (fig.2).

on the basis of physical stratigraphy are provided by Slaughter & Earley (1965) for Cretaceous bentonites of Wyoming, and Kimpe (1967, figs 1 and 3) for Carboniferous tonsteins of northern Europe, even though in the latter case the author was not then aware of their volcanic origin. More extensive correlation of the northern European Carboniferous tonsteins has subsequently been proposed by Spears & Kanaris-Sotiriou (1979) and by Burger (1985) although, because of the wide separation of the coal basins, few individual tonsteins can be traced throughout with certainty.

Because of the distinctive physical and mineralogical constitution of tephra layers, they can sometimes be identified and correlated on the basis of geophysical criteria. In boreholes, tephra layers may be recognisable on wireline logs, so that correlation can be established even in uncored sections. Anomalously high gamma-ray responses are characteristic of evolved tephras, which typically contain relatively high concentrations of radioactive minerals. Distinctive signatures may also be recognized on electric and density logs (Fig. 2; Huff & Kolata 1990). In the field of marine studies, Bowles *et al.* (1973) have shown how tephra layers, identified as such through drilling, may be traced for considerable distances on shallow seismic records (Fig. 3).

In most sequences however, long-distance correlation of individual tephra layers cannot be achieved purely on the basis of physical characteristics. Sections may be too distantly spaced, layers may be discontinuous, and the physical and chemical signature of the tephra may vary between different facies as a result of detrital contamination or contrasting diagenetic modifications. The resulting problems of correlation

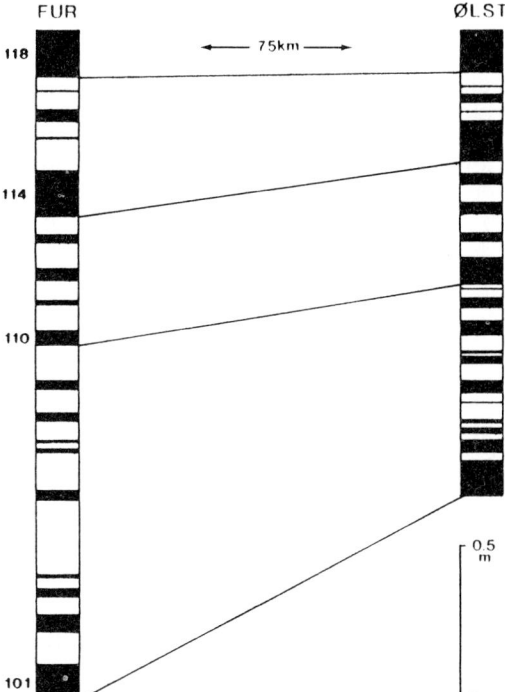

Fig. 1. Correlation of early Eocene tuffs (black) in Denmark, using vertical distribution pattern coupled with regionally extensive, thick marker tuffs (numbered). Data from Bøggild (1918) and Andersen (1937).

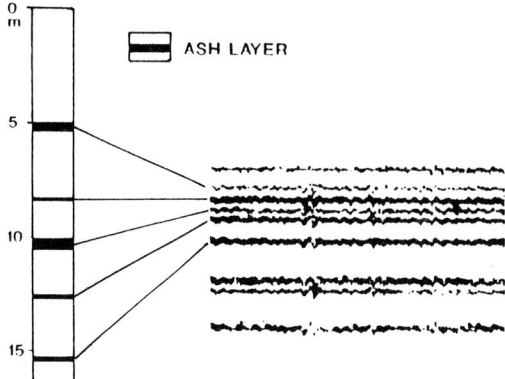

Fig. 3. Correlation of tephra layers with reflectors on a 3.5 kHz seismic profile from the Pacific coast of central America. After Bowles *et al.* (1973, fig.4).

may be compounded by the presence of successive layers of broadly similar bulk composition. In such cases, the unequivocal identification of individual tephra layers requires recognition of distinctive petrographical, mineralogical, or chemical characteristics. Many of the techniques used to achieve such 'fingerprinting' of tephra layers have evolved through the intensive study of Quaternary tephras over the last two decades, and are the subject of reviews by Westgate & Gorton (1981) and Kennett (1981). In this review, therefore, greater emphasis is placed on the more recent developments and on the application of fingerprinting techniques to older, and generally more altered, tephra layers.

Petrographical and mineralogical correlation. For more-or-less unaltered tephra layers, the preservation of glass phases greatly enhances discrimination on the basis of petrographical observations. Tephra of contrasting composition may often be differentiated by glass colour (related to variation in iron content) and particle shape (reflecting magma viscosity). The refractive index of fresh glass has been shown to be a useful discriminator in some Quaternary tephra sequences (e.g. Westgate & Evans 1978, fig. 3). The presence of more than one glass phase, resulting from liquid immiscibility, is another possible discriminator.

In older tephras, correlation on the basis of overall petrological character may still be possible. In the relatively unaltered tuffs of the Danish Lower Eocene, Bøggild (1918) was able to use features such as colour, refractive index, microcrystalline texture and grain shape to assist in a detailed layer-by-layer correlation. With

increasing alteration, however, many of the distinctive primary petrographical characteristics are lost. Nevertheless, through detailed petrographical analysis, Bouroz (1967, 1972) was able to distinguish between tonsteins in the coalfields of northern France. Despite extensive kaolinitization, he was able to recognize consistent variations in the proportions of residual shards and amygdales, feldspars, embayed quartz, quartz splinters, and accessory minerals (apatite, zircon, and altered biotite). On this basis he was able to correlate sections over distances of several hundred kilometres.

Correlation can also be achieved on the basis of phenocryst phases. In some sequences, individual tephra layers may be recognized through the presence of a specific diagnostic mineral. For example, Juvigné (1990) was able to distinguish between two widespread Quaternary tephras layers in northwest Europe on the presence or absence of enstatite. Tephras may also be recognized by the distinctive textural characteristics of a particular phenocryst phase, such as the distinctive 'spongy-textured', inclusion-rich ilmenites described by Westgate & Gorton (1981, fig. 4).

Where a rich and varied suite of minerals is present, successful discrimination of tephras may be achievable on the basis of contrasting bulk phenocryst assemblages. Suitably varied assemblages are most commonly found in Quaternary tephra layers (e.g. Keller *et al.* 1978; Kohn 1979; Juvigné 1990); they also occur in some older tephras and were used, for example, by Bøggild (1918) to identify specific non-basaltic tuff layers in the Danish early Eocene tephra sequence. A rigorous quantitative approach to such correlation is, however, rarely possible because in most instances mineral proportions are significantly modified by the differential settling of minerals as a result of contrasting density and shape (see, for example, Juvigné 1983, but contrast Kittleman 1973, p. 2963). Indeed, some tephra layers display near total loss of phenocrysts in their distal portions (Sarna-Wojcicki *et al.* 1987, p. 208).

The problems of correlation by means of phenocryst minerals are not restricted to primary differentiation during transport, since post-depositional alteration can also lead to radical alteration of the mineral assemblages through the selective removal of the less stable minerals. With some minerals, such as olivine, their intrinsic instability is such that they commonly undergo dissolution or alteration, irrespective of their facies association or subsequent burial history. With others, dissolution or alteration is dependent more on the diagenetic environment. Early diagenetic depletion of the phenocryst assemblage is usually least marked in marine successions, where tephra layers may be represented by more or less unaltered tuffs or by bentonites. Despite the alteration of the glass phase to smectite or illite–smectite derivatives, the phenocryst minerals in marine tephras generally survive with little or no modification (Weaver 1963), the most common being quartz, sanidine, anorthoclase, plagioclase, hornblende, biotite, zircon, apatite, magnetite, ilmenite, and titanite (sphene). In sandy, reworked bentonites, the degree of dissolution is often greater because of the increased permeability. In such cases minerals such as amphibole and the intermediate to calcic plagioclases may also be removed. Selective dissolution of minerals may also take place as a result of deep burial. Burial stability follows trends established for detrital heavy mineral suites (see Morton 1984), with hornblende and titanite being the most susceptible to dissolution.

Alteration of tephra under terrestrial conditions generally leads to more substantial modification of the phenocryst assemblage. Under semi-arid conditions, alteration may be restricted to the oxidation of biotite and amphibole; under humid conditions, alteration may be more extensive, with kaolinitization of the glassy constituents being accompanied by the dissolution or alteration of many of the phenocryst phases. Commonly, the phenocryst phases are reduced to an ultrastable quartz and zircon assemblage. The abundance of sanidine in some German Carboniferous tonsteins (Hess & Lippolt 1986) is an unusual feature that probably relates to the exceptionally K-rich nature of the sanidine (64–68% Or, author's unpublished data).

A surprising feature of some tonsteins, considering the abundance of authigenic kaolinite, is the preservation of apatite (Spears 1971, p. 505). Since apatite is known to be particularly susceptible to dissolution by acid meteoric waters, it may be that the environment of kaolinite formation is not so strongly acidic as generally supposed. High phosphate concentrations in the groundwater could be another contributing factor. Amphibole and biotite are not normally found in tonsteins, though interstratified kaolinite-mica ('levierrite') crystals are believed to be pseudomorphs after biotite (Spears 1971, p. 504).

From the above discussion, it is clear that although bulk phenocryst assemblages may be of value in correlating unaltered tephra layers, the effects of transport and diagenesis can be such that the mineralogical signature of the primary ash is effectively obscured. For a less ambiguous means of correlation, therefore, it is necessary to seek diagnostic features that are more-or-less unaffected by depositional and post-depositional processes. The most precise fingerprinting techniques are based on chemical analysis, either of the bulk rock or of selected primary constituents, and are described below.

Correlation by chemical fingerprinting

Bulk chemical analysis. With young, unaltered, and uncontaminated tephra, the chemical and mineralogical composition of the parent magma can be assessed with some certainty, and a combination of major-element and trace-element analysis will often suffice to distinguish one layer from another. As discussed by Amajor & Lerbekmo (1980), there appears to be no single criterion for the differentiation of tephras on the basis of bulk chemical analysis. Selection of major-element discriminators will depend on the gross composition of the parent magma, whereas that of trace-element discriminators will depend additionally on the origin of the magma. Even in unaltered tephras, however, chemical fingerprinting is not necessarily straightforward. In particular, the bulk composition of a tephra layer may change significantly with distance from the source as a result of the preferential settling of the less buoyant constituents, particularly the denser and more equant phenocrysts. Under such circumstances, trace-element composition may be most strongly affected, since many trace elements are concentrated in the phenocryst phases (Randle *et al.* 1971; Westgate *et al.* 1977).

With altered tephra layers, the problems of bulk chemical analysis are even greater, since many major and trace elements display substantial mobility, and relative mobility may vary with the specific conditions under which alteration takes place. In a study of tonsteins, Spears & Rice (1973) have shown how some elements (Zr, U, Y) are retained in the phenocryst minerals; others are concentrated in authigenic minerals, such as kaolinite (Ga, Th) and sulphides (Pb, Zn). Even those elements that are considered to be more-or-less immobile have been shown to display significant mobility during the argillization of tephra; for example Zielinski (1982) demonstrated the loss of uranium, Spears & Kanaris-Sotiriou (1979) the probable loss of nickel, and Morton & Knox (1990, p. 431) the loss of yttrium. Nevertheless, bulk major and trace element chemistry will often allow discrimination between altered tephra layers where these possessed sufficient contrast in their

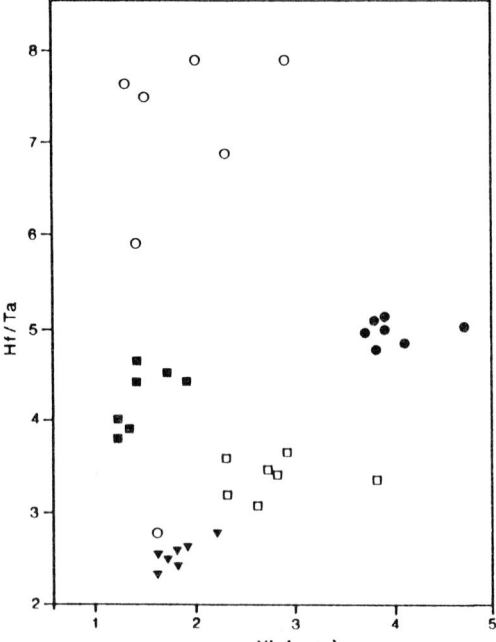

Fig. 4. Bivariate discrimination plot of trace element ratios, demonstrating the ability to differentiate between five K-bentonites (indicated by differing symbols) from the Ordovician of central Pennsylvania. After Cullen-Lollis & Huff (1986, fig.4).

sediment surface affected by current or wave action. Even in relatively quiet water environments contamination can take place through bioturbation, in which detrital sediment is introduced via sediment-filled burrows. Clearly, either of these forms of contamination can lead to a marked change in bulk composition that will affect both major and trace element analyses.

Many of the problems associated with the fingerprinting of tephra layers by bulk petrological, mineralogical, or chemical analysis can be overcome by studying selected constituents, either as bulk separates or as single grains. Provided, of course, that some primary constituents have remained unaltered, such techniques will generally overcome the problems associated with contamination and bulk alteration.

Glass analysis. Analysis of the glassy fraction is a widely used technique in the chemical fingerprinting of Quaternary tephras (see reviews by Kennett 1981 and Westgate & Gorton 1981), but is rarely possible in older tephras because of the inherent instability of volcanic glass. Although analyses can be carried out on bulk separates, most workers follow the technique of Smith & Westgate (1969), in which microprobe analysis is carried out on single grains (Fig. 5).

Phenocryst analysis. The concept of fingerprinting tephras by mineralogical analysis is based on the premise that contrasts in the bulk composition of tephra layers are often reflected in

primary composition and where alteration has taken place under broadly comparable conditions. In this way, Cullen-Lollis & Huff (1986) were able to correlate five Ordovician K-bentonites in central Pennsylvania through chemical fingerprinting (Fig. 4), using discriminant analysis to confirm that differences in chemical composition were significantly greater between the bentonites than within them. A similar statistical approach was taken by Kolata *et al.* (1987) in their study of Ordovician K-bentonites of the Mississippi Valley, and a more specific account of statistical methods in chemical tephrostratigraphy is given by Huff & Kolata (1989). Bulk major-element and trace-element chemistry has also been used successfully to establish magmatic and tectonic affinities; this aspect of tephra analysis is, however, outside the scope of this review.

The bulk chemical composition of a tephra can also be modified through contamination by detrital material. The most common form of contamination is by silt and sand, and is especially marked where the ash has settled on to a

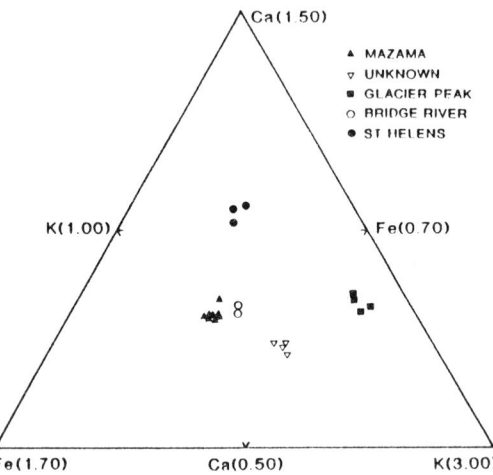

Fig. 5. Discrimination of tephra layers on the basis of elemental composition, as determined by electron microprobe analysis of glass particles. Quaternary, North America. After Smith & Westgate (1969, fig.2).

the physical and chemical characteristics of the phenocryst minerals.

Typical physical varietal characteristics are related to crystal form and colour, and to the presence and nature of inclusions. Chemical varietal characteristics involve major, minor or trace elements, which may be analysed using bulk mineral separates or single crystals. Single-crystal analysis, using an electron microprobe, is most effective in detecting major element variations, although with the increasing sophistication of the analytical equipment, minor and even trace elements can be analysed in this way. For accurate quantitative assessment of rare-earth and other trace elements, however, it is still necessary to use bulk techniques such as X-ray fluorescence, induction coupled plasma, and instrumental neutron activation analysis.

The advantage of single-crystal microprobe analysis over bulk analysis is that it involves a number of discrete analyses, and thus allows detection of detrital contaminants. In bulk analyses these would result in an undetectable skewing of the mean composition. Also, the much greater number of individual analyses allows information to be gained on the range in composition of phenocryst phases, the resulting distribution pattern often being more distinctive than the mean composition alone. A range in composition within a single mineral phase commonly reflects zonation, but a strongly bimodal or polymodal pattern may indicate the presence of two or more distinct populations. Such polymodal distributions may result from the presence of co-magnetic phases or from the presence of xenocrystic or detrital contaminant populations (see Fig. 6). Another advantage of numerous single-grain analyses is that they are well suited to statistical analysis.

To date, mineral-specific fingerprinting has been carried out mostly by major-element microprobe analysis. Where preserved, the ferromagnesian minerals are particularly suited to such analysis; for example, Juvigné (1990) used pyroxenes to discriminate between Quaternary tuffs in the Eifel region of NW Europe. Such analyses are rarely possible in older tephra layers, however, because of the inherent instability of ferromagnesian minerals. Nevertheless, Yen & Goodwin (1976) were able to correlate Eocene tuffs using biotite compositions.

Titanomagnetite and feldspar have great potential for fingerprinting, since they occur widely in fresh volcanic ash and are relatively resistant to diagenetic alteration. Feldspar displays a substantial range in composition within the Or–Ab–An series, while titanomagnetite shows significant variations in the proportion of iron,

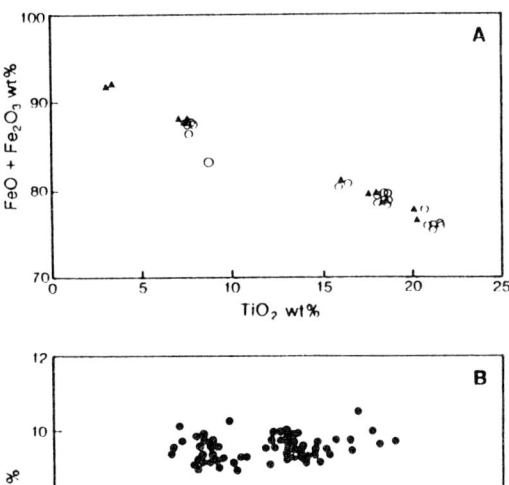

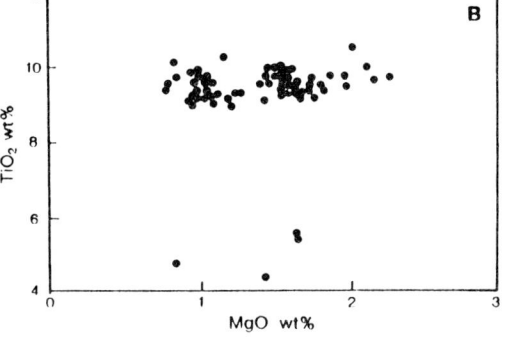

Fig. 6. Chemical analysis of titanomagnetites from North American Quaternary tephra. (**A**) Correlation of Wascana Creek (open circles) and Pearlette (triangles) tephras on the basis of the close similarity in composition of titanomagnetite phases. The three Ti-rich phases are believed to be xenocrystic. After Westgate *et al.* (1977, fig.9). (**B**) Titanomagnetite analyses for the Old Crow tephra. The Ti-poor crystals are believed to be contaminants, possibly detrital. After Westgate *et al.* (1985, fig.8).

titanium, magnesium and aluminium. Both minerals show variations in trace element proportions, though these have not been much used for fingerprinting ash. Titanomagnetite analysis has been carried out on Quaternary tephra in both North America and New Zealand (e.g., Westgate & Fulton 1975; Westgate *et al.* 1977; Kohn 1979), using a variety of cross-plots to identify separate phases within the magnetite populations. The studies by Westgate *et al.* (1977) and Kohn (1979) are of particular interest in that they revealed the presence of probable xenocrystic titanomagnetite components (Fig. 6A). The secondary population detected by Westgate *et al.* (1985) (see Fig. 6B) most probably represents detrital contamination, since it occurs only in distal samples. Knox (1984) used the composition of feldspar phenocryst populations to fingerprint and correlate individual

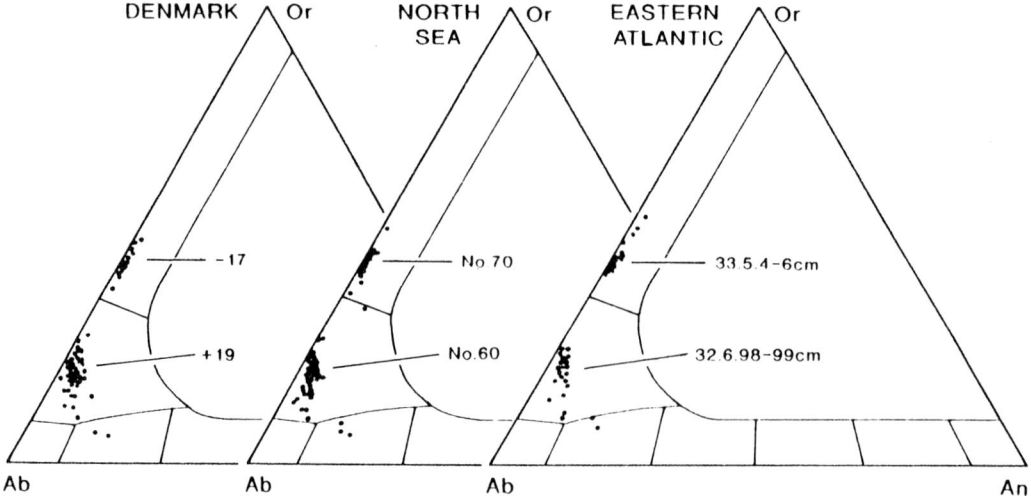

Fig. 7. Correlation of early Eocene tephra layers between Denmark, the North Sea and the eastern Atlantic (DSDP Site 550) by means of feldspar composition. +19 and −17 refer to layers in the numbered sequence of Bøggild (1918); see also Fig. 10. After Knox (1984, fig.3).

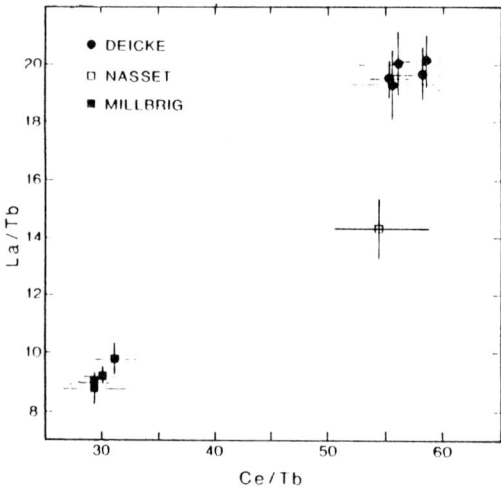

Fig. 8. Discrimination between three North American Ordovician K-bentonites by means of trace-element compositions of apatite phenocrysts. After Samson *et al.* (1988, fig. 3).

tephra layers in the Lower Eocene of the northwest Europe (Fig. 7).

Quartz, zircon, apatite, and titanite display little or no major-element variation, but, along with the other minerals mentioned, have potential for minor and trace element analysis. In one of the few studies of this kind, Samson *et al.* (1988, 1989) used the rare-earth and trace-element composition of apatite phenocrysts (Fig. 8) to discriminate between Ordovician bentonites in North America (but see also

discussion by Haynes & Huff 1990 and Samson *et al.* 1990).

In all bulk and mineral-specific fingerprinting studies, it must be borne in mind that similar compositions may be displayed by two or more tephra layers, especially where these are genetically related. Wherever possible, therefore, it is advisable to investigate correlation of individual layers within the context of the full tephra sequence, since in some cases it may be the sequence of compositions that provides the firmest basis for correlation.

Applications

The ability to establish precise chronostratigraphical correlation by means of tephra layers

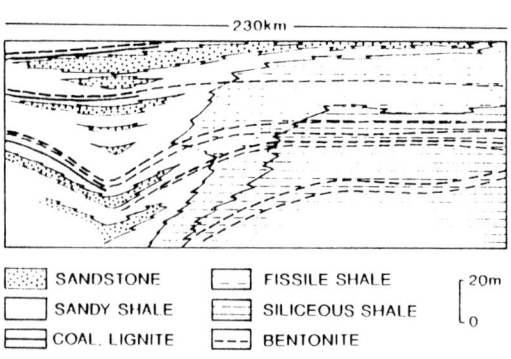

Fig. 9. Tephra correlation in the Cretaceous upper Mowry/lower Frontier formations, Wyoming. After Slaughter & Early (1965, fig.12).

is important not only in geological studies, but also in archaeology and in the study of recent climatic change.

The primary geological objective is, of course, to establish precise chronostratigraphical correlation of specific horizons from one sequence to another. Once such correlations have been established, however, a wide range of other possibilities present themselves, many of which have not yet been fully explored.

Tephrostratigraphical correlation allows sequences to be viewed in terms of a series of time planes that may be used to reconstruct the geographical distribution of lithofacies and of faunal and floral populations. Terrestrial and paralic sedimentation, in particular, can result in highly complex facies relationships which the presence of tephra layers may help to resolve (e.g. Westgate *et al.* 1987). Tephra layers may also be traced from non-marine into marine facies, thus allowing correlation between marine and terrestrial biozones. Because of the limited long-term preservation potential of tephra layers in terrestrial environments, however, such correlations are largely restricted to Quaternary sequences. Examples include those of Ninkovich & Heezen (1967) for tephra from Thera Island in the eastern Mediterranean, Drexler *et al.* (1980) for Quaternary tephra in central America and the adjacent oceans, Mangerud *et al.* (1984) for a Quaternary tephra layer in western Norway, and Sarna-Wojcicki *et al.* (1987) for tephra layers in the upper Cenozoic of the western United States and eastern Pacific. Tephra correlation from paralic to marine facies in the Lower Cretaceous Mowry Formation of Wyoming (Slaughter & Earley 1965) provides an example from older strata (Fig. 9).

Tephra correlation can provide a link between incompatible biostratigraphical schemes that reflect the provincial nature of faunas in adjacent basins (Fig. 10). It can also reveal erroneous correlations based on facies-controlled faunal events, as demonstrated by Bouroz (1967) in his study of Upper Carboniferous tonsteins of

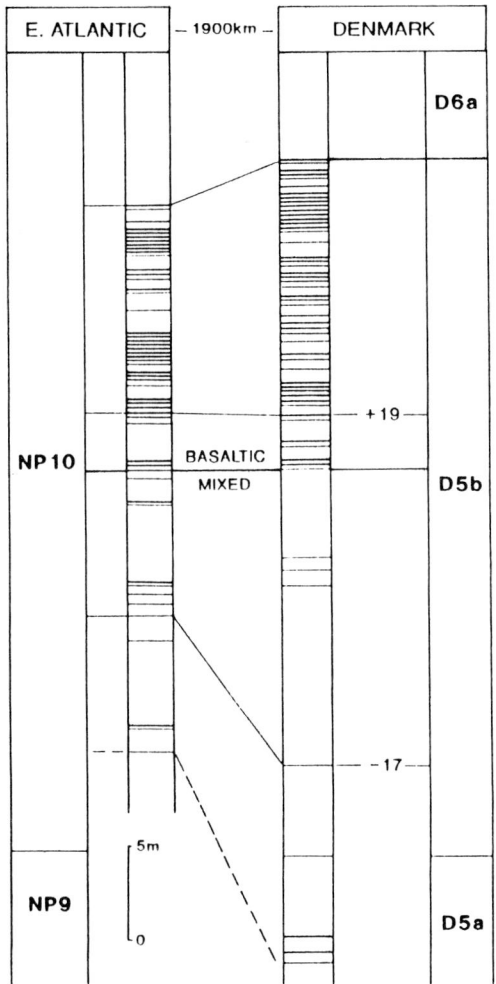

Fig. 10. Correlation between early Eocene open marine calcareous nannoplankton (NP) zones of the North Atlantic and restricted marine dinoflagellate cyst (D) zones of the North Sea Basin. Correlation is based on the overall distribution pattern, a gross upward change in tephra composition (from mixed to almost exclusively basaltic) and on the occurrence of two mineralogically distinctive marker layers. Modified from Knox (1984, fig.4).

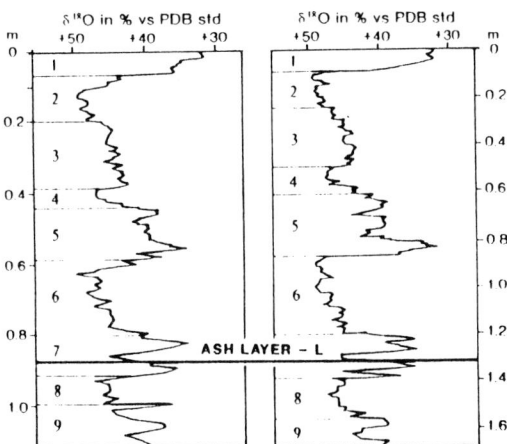

Fig. 11. Tephrostratigraphical confirmation of the correlation of oxygen isotope profiles in deep-sea sediments of the Panama region. After Ninkovich & Shackleton (1975, fig.12).

northern France. The full potential of tephra layers in assessing the time constancy of individual biostratigraphical events has not been fully realised, but the opportunity clearly exists to use individual tephra layers to explore geographical variation in faunal and floral assemblages along specific time planes. The chronostratigraphical validity of other correlation techniques may similarly be checked using tephrostratigraphy. For example, tephra correlation has been used to support correlation of oxygen isotope profiles (Fig. 11; Ninkovich & Shackleton 1975) and magnetic polarity sequences, especially the short-lived magnetic excursions recorded in Quaternary sequences (Fig. 12; Ninkovich *et al.* 1966; Westgate *et al.* 1990).

Tephrostratigraphy has also been used to demonstrate the correlation and time-constancy

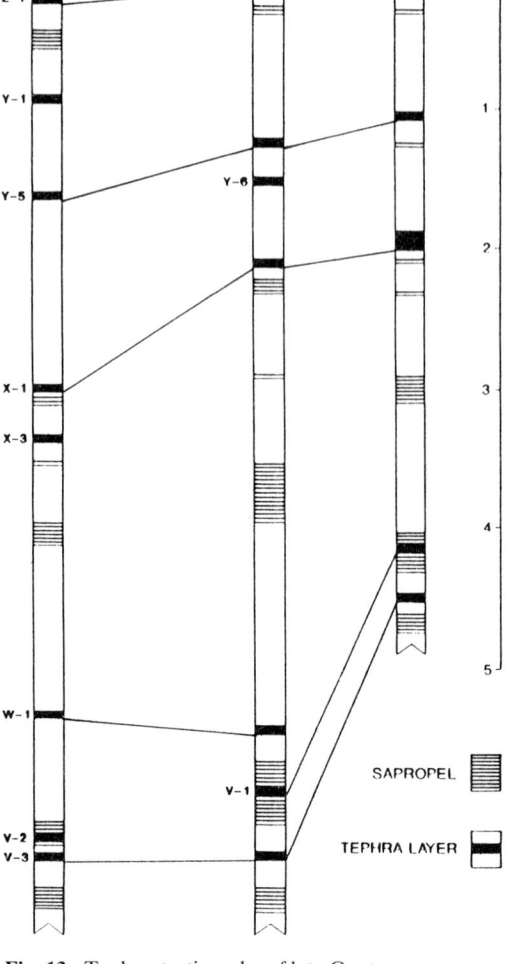

Fig. 13. Tephrostratigraphy of late Quaternary cores from the eastern Mediterranean, demonstrating the more or less synchronous development of anoxic 'sapropel' units. After Keller *et al.* (1978, fig.3).

of certain types of lithological unit, for example sapropel units with marine Quaternary sequences of the Mediterranean (Fig. 13; Keller *et al.* 1978). Correlation of tonsteins in coal-bearing sequences in NW Europe has confirmed the isochronous nature of the marine bands (Fig. 14; Spears & Kanatis-Sotiriou 1979). Similarly, widespread coal seams can be shown to be broadly isochronous (Francis 1961, pl. 6), though detailed local studies often reveal minor diachronism (Burger 1985, figs 6–8). Particularly detailed chronostratigraphical analysis of coal seams is possible where individual seams include more than one tonstein layer (Fig. 15; Ryer *et al.* 1980).

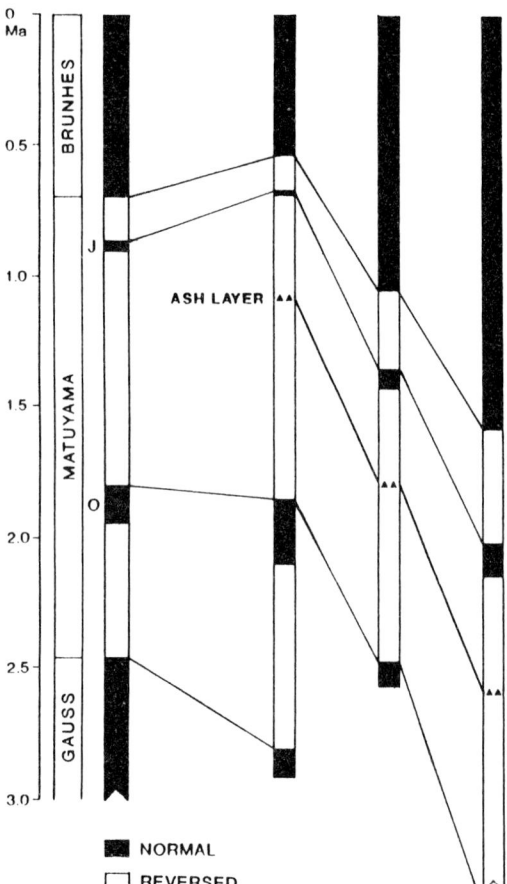

Fig. 12. Tephrostratigraphical confirmation of the correlation of magnetic polarity sequences in North Pacific deep-sea cores. After Ninkovich *et al.* (1966, fig.5).

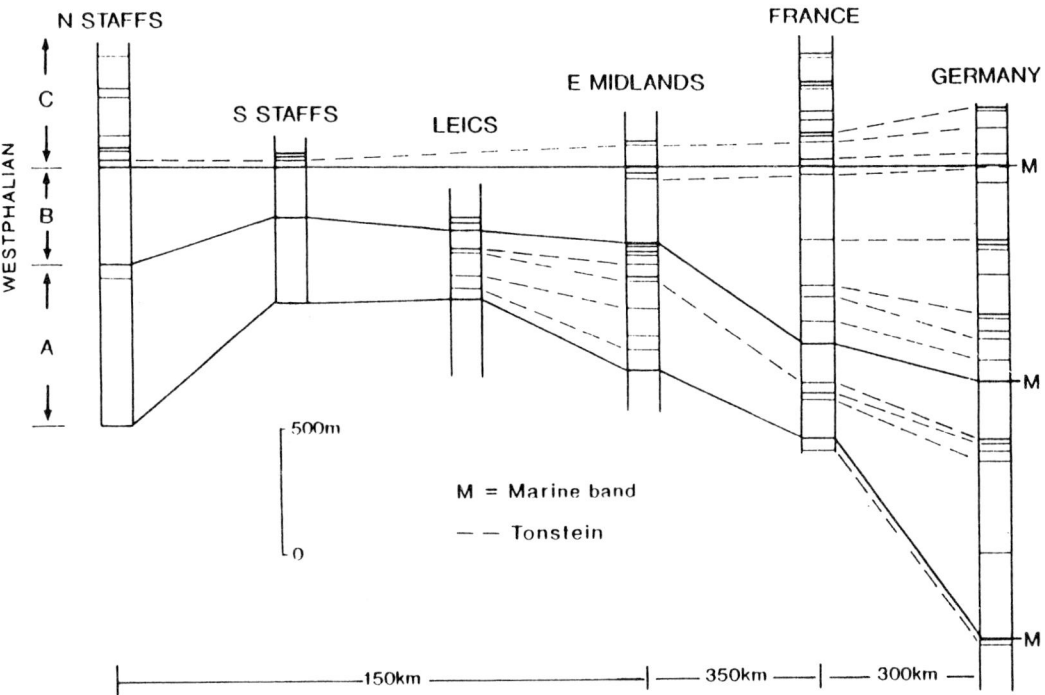

Fig. 14. Correlation of tonsteins in the Westphalian of Northwest Europe, demonstrating the isochronous character of the widespread marine bands. After Spears & Kanaris-Sotiriou (1979, fig.2).

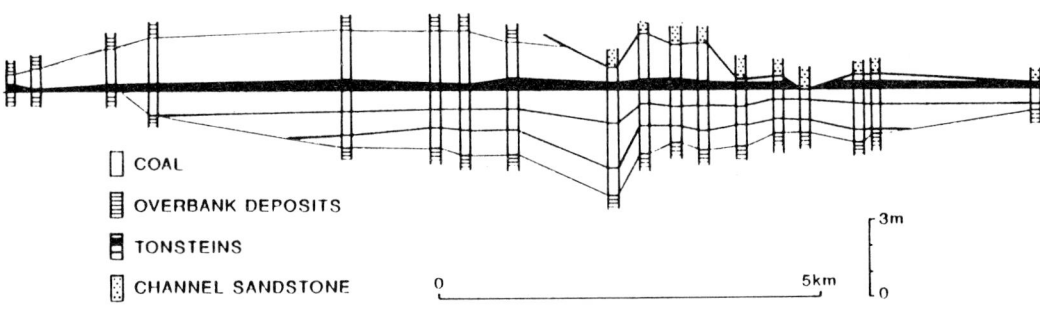

Fig. 15. Stratigraphical subdivision of a coal seam into isochronous units by means of intercalated bentonites. After Ryer *et al.* (1980, fig.5).

Tephrostratigraphical analysis of late Quaternary sequences is related as much to geoanthropology, archaeology and climatology as to geology. Terrestrial tephra deposits are of particular significance, being directly associated, for example, with hominid remains in the Olduvai Gorge (Hay 1976) or with archaeological remains in Iceland (Thorarinsson 1981). Vitaliano *et al.* (1981) and Juvigné & Gewelt (1988) have shown how tephra in cave deposits can be correlated with specific regional tephra layers, thereby allowing them to be placed in a conventional stratigraphical context. Climatological applications stem in particular from the occurrence of volcanic ash within ice sheets. Gow & Williamson (1971), in a study of ice cores from Antarctica, recorded more than 2000 tephra layers, representing volcanic activity over the past 43000 years. Such tephra sequences have great potential for correlating widely separated ice cores (Palais *et al*. 1987, 1989). They may also yield valuable information on the possible relationship between global volcanicity and climatic change (Gow & Williamson 1971), a theme that can be extended into the possible influence of volcanic eruptions on the development of tree-rings (Vogel *et al*. 1990). Tephrostratigraphy thus has an important application to anthropological and archaeological studies, both through the direct association of artefacts with tephra deposits and through the possible influence of volcanism on climatic and hence ethnographical change.

Tephrochronology

Because tephra layers provide a unique opportunity to carry out direct dating of sedimentary sequences by means of high-temperature volcanic minerals, they are critical to the establishment of an accurate geological time-scale. Discussion of individual dating techniques is beyond the scope of this paper, and the reader is referred to Baadsgaard & Lerbekmo (1982) for a general account of the radioisotopic methods of tephra dating. This section is limited to a brief résumé of recent developments in the relevant techniques and to a discussion of the significance of tephrochronological studies in refining the geological time scale.

Radiometric dating of tephra layers has been achieved largely by the K-Ar, ^{40}Ar/^{39}Ar, Rb-Sr, U-Pb and fission-track techniques. Where these are carried out on whole-rock samples or bulk mineral separates, they are subject to problems of alteration and contamination comparable to those encountered in mineralogical or chemical fingerprinting. Consequently, dating of tephra layers has increasingly focused on single-grain analytical techniques. Reliable whole-rock dates are most likely to be obtained from fresh volcanic ash, in which the glass has undergone little or no devitrification or argillization. Since such material is largely restricted to Quaternary tephra, dating has been mostly by the K-Ar and ^{40}Ar/^{39}Ar methods, which are suitable for material older than about 500 Ka. Whole-rock dating of older rocks is possible (e.g., Rundle 1986) but is likely to be less precise than more selective methods and to be more open to problems of interpretation.

The uncertainties inherent in whole-rock dating are substantially reduced by analysis of selected glassy groundmass or phenocryst constituents (see Table 1). A widely used technique is *fission-track dating* (Westgate & Briggs 1980). For Quaternary tephra, analysis is commonly carried out on the glassy groundmass (Naeser *et*

Table 1. *Commonly occurring phenocryst minerals used for dating of tephra layers, showing the appropriate techniques for each mineral*

	K/Ar	Ar/Ar	Rb/Sr	U-Pb	F-T
Sanidine	*	*			
Anorthoclase		*			
Plagioclase		*			
Biotite/ phlogopite	*	*	*		
Hornblende	*	*			
Zircon				*	*
Apatite				*	*
Titanite (sphene)					*

F-T, fission-track.

al. 1980), although acceptably precise dates are difficult to obtain from glass younger than about 100 Ka (Naeser & Naeser 1988, p. 4). Following early studies such as that of Seward (1974), fission-track dating has been much used in Quaternary tephrochronology, especially in North America (e.g., Westgate *et al*. 1977). Fission-track analysis of glass was initially hampered by the effects of hydration and incipient alteration, which lead to track fading. However, Westgate (1989) has presented a refinement of the isothermal plateau fission-track technique that yields precise ages from hydrated silicic glass, provided that the glass has had a simple thermal history.

Fission-track dating of phenocrysts involves the uranium-bearing minerals zircon, apatite, and titanite (sphene), and was the first single-crystal'dating technique to be developed. The greater accuracy of mineral dating over glass dating was demonstrated by Seward (1979), and zircon and apatite dating has been widely used on Quaternary tephra (e.g., Easterbrook *et al.* 1988). Fission-track analysis may also be applied to much older tephra, provided that they have not been subjected to excessive heating. For example, Ross *et al.* (1982) used fission-track analysis of zircons and apatites to date tephra layers in British Ordovicin and Silurian stratotypes (see Tucker *et al.* 1990 for a later assessment).

Thermoluminescence dating is another technique that can be applied to glassy constituents (e.g., Berger 1987). It can be applied to whole-rock samples, but more reliable results are obtained from separated glass grains (Berger 1985). A particularly useful aspect of this technique is that it may be used to date tephra between 40 ka and 300 ka in age, thereby spanning the gap between radiocarbon dating and fission-track dating. Though not carried out directly on tephra material, *radiocarbon dating* can provide accurate ages for individual tephra layers where these are closely associated with organic-rich sediments (see Briggs, N.D. in Naeser *et al.* 1981, pp. 26–30).

For isotopic dating of bulk separates of tephra minerals, the choice of technique depends largely on phenocryst mineralogy (see Table 1), although the age of the tephra is also a consideration. *K-Ar* and $^{40}Ar/^{39}Ar$ *dating* are the most widely applicable, since they can be carried out on a range of minerals (sanidine, anorthoclase, plagioclase, biotite, phlogopite and hornblende), and over a wide range of ages. The ability to date two or more co-magnetic phases from a single tephra deposit is particularly advantageous, as it provides a basis for assessment of the relative reliability of the dates obtained (e.g. Turner *et al.* 1980; Baadsgaard *et al.* 1988, table 7; Izett *et al.* 1988).

The $^{40}Ar/^{39}Ar$ technique has several advantages over the K-Ar technique, including relatively small sample size, much greater analytical precision, and the ability (through step heating) to detect disturbed age spectra (see, for example Hess & Lippolt 1986). The development of *single-crystal laser-fusion $^{40}Ar/^{39}Ar$ dating* has greatly increased the potential of the $^{40}Ar/^{39}Ar$ method (e.g., van den Bogaard *et al.* 1987). Added advantages are the very small sample size required, the ability to detect and eliminate detrital and xenocrystic components (e.g., Lo Bello *et al.* 1987), and high analytical precision (e.g., Swisher & Prothero 1990). Both $^{40}Ar/^{39}Ar$ techniques can be used to date material ranging in age from Precambrian (Layer *et al.* 1987) to Quaternary (e.g., van den Bogaard *et al.* 1989).

U-Pb dating is carried out on zircon and, less commonly, on apatite. The technique is particularly useful for dating tephra in which the less stable minerals, such as feldspar and biotite, have been altered or removed by dissolution (e.g., Bralower *et al.* 1990; Tucker *et al.* 1990). A commonly encountered problem with zircons, however, is the presence of inherited cores or later overgrowths, which yield anomalously old and young ages, respectively. Both problems are overcome by the recently developed technique of *single-crystal ion-probe $^{206}Pb/^{238}U$ zircon dating* (Compston *et al.* 1984), in which the ion probe can be directed to specific locations within the zircon crystal.

In bentoites, biotite is the only mineral suitable for *Rb-Sr dating* (Baadsgaard & Lerbekmo 1982); in younger tephra, additional analysis of coexisting sanidine is needed in order to establish initial $^{87}Sr/^{86}Sr$ ratios. Apart from these

limitations, the method is useful for dating tephra of evolved composition, particularly those of early Tertiary or older age (e.g. Baadsgaard & Lerbekmo 1983; Yanagi *et al.* 1988). *Sm-Nd dating* has not yet been attempted on distal tephra layers, but the dating by Thirlwall & Fitton (1983) of garnets in English Ordovician lavas and ignimbrites raises the possibility of applying the technique to the garnet-bearing Ordovician bentonites of North America (see Samson *et al.* 1989).

Although each of the above techniques can be used in isolation, a greater measure of confidence is obtained by application of two or more techniques to the same tephra layer or by the application of one technique to more than one of the coexisting mineral phases. Tephra older than about 500 ka provide an opportunity to compare results obtained by different dating techniques. That different techniques can yield closely similar dates has been clearly demonstrated by Hurford & Hammerschmidt (1985) and by Westgate (1989, table 3), in his summary of data from selected Quaternary North American tephra deposits. Discrepancies may result from a variety of causes, depending on the technique used; for example, Turner *et al.* (1980, figs 2 and 3), in a study of early Tertiary tephra layers, suggested that detrital contamination probably accounted for the anomalously old K-Ar dates obtained on some feldspar and hornblende separates compared with fission-track dates obtained from associated zircons. A combination of U-Pb, Rb-Sr, and K-Ar analysis has been successfully used in several studies, for example those of Baadsgaard & Lerbekmo (1983), Baadsgaard *et al.* (1988), and Yanagi *et al.* (1988) on tuffs and bentonites ranging from Ordovician to Tertiary in age.

Pre-Quaternary zircon-bearing tephra provide an opportunity to carry out different techniques on a single mineral phase; for example Tucker *et al.* (1990) showed that U-Pb and fission-track analysis can yield comparable dates provided that the fission-tracks are not affected by thermal resetting. The ability of a single technique to obtain comparable dates from different mineral phases is well illustrated by the single-crystal $^{40}Ar/^{39}Ar$ dating of biotite, anorthoclase, and plagioclase from late Eocene and early Oligocene tephra layers in North America (Swisher & Prothero 1990).

As the techniques for dating tephra become increasingly sophisticated, the full potential of tephrochronology in refining the geological time scale is becoming more widely appreciated. Notable tephrochronological contributions to the current geological time-scale include those of Tucker *et al.* (1990) for the Ordovician (zircon U-Pb dating), Hess & Lippolt (1986) for the late Carboniferous (sanidine $^{40}Ar/^{39}Ar$ dating), Bralower *et al.* (1990) for the early Cretaceous (zircon U-Pb dating), Obradovich & Cobban (1976) for the late Cretaceous (sanidine and biotite K-Ar dating), Swisher and Prothero (1990) for the Eocene/Oligocene boundary transition (single-crystal biotite, anorthoclase, and plagioclase $^{40}Ar/^{39}Ar$ dating), and Izett *et al.* (1988) for the Plio-Pleistocene (biotite, sanidine, and plagioclase K-Ar dating).

A range of tephra dating methods will be needed in the refinement of the existing time scale, reflecting the range in age and mineralogy of the available tephra layers. There seems little doubt, however, that increasing emphasis will be placed on single-crystal dating techniques. As discussed above, the $^{40}Ar/^{39}Ar$ technique is particularly attractive because of its precision and its applicability to several commonly occurring volcanic minerals. The principal limitation of the $^{40}Ar/^{39}Ar$ technique is that it cannot be applied to tephra in which the feldspars and ferromagnesian minerals have been dissolved or altered or as a result of diagenesis or low-grade metamorphism, this being the case with most pre-Mesozoic tephra and with those associated with freshwater deposits. Fortunately, the equally promising technique of single-crystal ion-probe $^{206}Pb/^{238}U$ zircon dating can be applied in most of these cases, since zircon is extremely resistant to diagenetic and metamorphic alteration. The potential of the technique in geochronological studies is clearly demonstrated by recent studies on Carboniferous and Permian-Triassic boundary strata (Claoué-Long *et al.* 1991, in press). Its applicability to late Mesozoic and younger strata has yet to be adequately tested, but an important proviso is that the dating of such younger tephras requires that the zircons possess greater than average uranium contents.

With the continued development of new techniques and the refinement of existing techniques, the precision of radiometric dates is approaching that of biostratigraphical zones in Mesozoic and younger strata. As confidence in radiometric dates increases, we may therefore expect to see tephrochronology used directly as a chronostratigraphical tool. Although not approaching the precision of individual tephrostratigraphical correlations, correlation by tephra dating could be of considerable importance in establishing the temporal relationships of sequences belonging to different faunal provinces or of sequences in which biostratigraphical control is relatively poor. Another potential

application of tephrochronology, though limited to suitably tephra-rich sequences, lies in the direct assessment of subsidence and sedimentation rates, a procedure that is at present carried out indirectly via chronologically calibrated biozones. Similarly, accurately dated tephra layers should allow a detailed reconstruction of the volcanic history of certain regions, which in turn should throw new light on associated tectonic events and crustal history.

Conclusions

Although the unique status of tephra layers in stratigraphical studies has long been appreciated, it is only in the last decade or so, with the development of advanced chemical analysis techniques, that their full potential has begun to be realised in the fields of tephrostratigraphy and tephrochronology.

In tephrostratigraphical studies, microchemical analysis of individual phenocryst minerals has been shown to provide the most reliable means of fingerprinting individual tephra layers, especially those in which the composition of the glassy groundmass has been radically altered by argillization.

In tephrochronological studies, dating techniques based on single-crystal or even within-crystal analysis have been shown to hold the greatest promise for accurate dating. Wide application of these techniques should lead to further significant refinement of the geological time-scale, and at the same time should help resolve long-standing problems of global correlation by means of high-resolution dating of widely separated tephra-bearing sequences.

The helpful comments of R. J. Merriman, A. C. Morton, D. A. Spears and an unknown reviewer are gratefully acknowledged. The paper is published with the approval of the Director, British Geological Survey (NERC).

References

AMAJOR, L. C. & LERBEKMO, J. F. 1980. Subsurface correlation of bentonite beds in the Lower Cretaceous Viking Formation of south-central Alberta. *Bulletin of Canadian Petroleum Geology,* **28,** 149–172.

ANDERSON, S. A. 1937. *De vulkanske Askelag i Vejgennemskæringen ved Ølst og deres Udbredelse i Danmark.* Danmarks Geologiske Undersøgelse, række 2, **59.**

BAADSGAARD, H. & LERBEKMO, J. F. 1982. The dating of bentonite beds. *In:* ODIN, G. S. (ed.) *Numerical dating in stratigraphy.* Wiley & Sons, 423–440.

—— & LERBEKMO, J. F. 1983. Rb-Sr and U-Pb dating

of bentonites. *Canadian Journal of Earth Sciences*, **20**, 1282–1290.

——, LERBEKMO, J. F. & McDOUGALL, I. 1988. A radiometric age for the Cretaceous-Tertiary boundary based upon K-Ar, Rb-Sr, and U-Pb ages of bentonites from Alberta, Saskatchewan, and Montana. *Canadian Journal of Earth Sciences*, **25**, 1088–1097.

BERGER, G. W. 1985. Thermoluminescence dating of volcanic ash. *Journal of Volcanology and Geothermal Research*, **25**, 333–347.

—— 1987. Thermoluminescence dating of the Pleistocene Old Crow tephra and adjacent loess, near Fairbanks, Alaska. *Canadian Journal of Earth Sciences*, **23**, 1975–1984.

BØGGILD, O. B. 1918. *Den vulkanske Aske i Moleret*. Danmarks Geologiske Undersøgelse, række 2, **33**.

BOUROZ, A. 1966. Fréquence des manifestations volcaniques au Carbonifère supérieure en France. *Comptes Rendus Hébdomadaires des Séances de l'Académie des Sciences, Série D*, **263**, 1025–1028.

—— 1967. Corrélations des tonsteins d'origine volcanique entre les bassin houillers de Sarre-Lorraine et du Nord – Pas-de-Calais. *Comptes Rendus Hébdomadaires des Séances de l'Académie des Sciences, Série D*, **264**, 2729–2732.

—— 1972. Utilisation des marqueurs d'origine volcanique en stratigraphie. Exemples d'application dans les gisements houillers. *Mémoires du BRGM*, **77**, 473–492.

——, SPEARS, D. A. & ARBEY, F. 1983. *Review of the formation and evolution of petrographic markers in coal basins*. Mémoires de la Société Géologique du Nord, **16**.

BOWLES, F. A., JACK, R. N. & CARMICHAEL, I. S. E. 1973. Investigation of deep-sea volcanic ash layers from equatorial Pacific cores. *Geological Society of America Bulletin*, **84**, 2371–2388.

BRALOWER, T. J., LUDWIG, K. R., OBRADOVICH, J. D. & JONES, D. L. 1990. Berriasian (Early Cretaceous) radiometric ages from the Grindstone Creek Section, Sacramento Valley, California. *Earth and Planetary Science Letters*, **98**, 62–73.

BRAMLETTE, M. N. & BRADLEY, W. H. 1941. Lithology and geologic interpretations. *In*: BRADLEY, W. H. *et al.* (eds) (1942) *Geology and biology of North Atlantic deep-sea cores between Newfoundland and Ireland*. US Geological Survey Professional Paper, **196**, 1–34.

BURGER, K. 1985. Kohlentonsteine im Oberkarbon NW-Europas: ein Beitrag zur Geochronologie. *In*: *Compte Rendu Dixième Congrès de Stratigraphie et de Géologie du Carbonifère, Madrid 1983*, **4**, 433–447.

—— 1990. Vulkanogene Glasscherben-Relikte in Kohlentonsteinen des Saar-Lothringer Oberkarbons sowie Herkunft und Menge der Pyroklastika. *Geologische Rundschau*, **79**, 659–691.

CHALARD, J. 1967. Réflections sur la définition et la genèse des tonstein. (Application possible au calcul des temps de dépôt relatifs du charbon et des stériles). *Annales de la Société Geologique du Nord*, **87**, 87–93.

CLAOUÉ-LONG, J. C., JONES, P. J., ROBERTS, J. & MAXWELL, S. In press. The age of the Devonian-Carboniferous boundary. *Earth and Planetary Science Letters*.

——, ZHANG, Z., MA, G. & DU, S. 1991. The age of the Permian-Triassic boundary. *Earth and Planetary Science Letters*, **105**, 182–190.

COMPSTON, W., WILLIAMS, I. S. & MEYER, C. E. 1984. U-Pb geochronology of zircons from lunar breccia 73217 using a sensitive high mass-resolution ion microprobe. *In*: BOYNTON, W. V. (ed.) *Proceedings of the Fourth Lunar and Planetary Science Conference. Journal of Geophysical Research*, Section B, **89**, Supplement, 525–534.

CULLEN-LOLLIS, J. & HUFF, W. D. 1986. Correlation of Champlainian (Middle Ordovician) K-bentonite beds in central Pennsylvania based on chemical fingerprinting. *Journal of Geology* **94**, 865–874.

DREXLER, J. W., ROSE, W. I., JR, SPARKS, R. S. J. & LEDBETTER, M. T. 1980. The Los Chocoyos ash, Guatemala: a major stratigraphic marker in middle America and in three ocean basins. *Quaternary Research*, **13**, 327–345.

EASTERBROOK, D. J., ROLAND, J. L., CARSON, R. J. & NAESER, N. D. 1988. Application of paleomagnetism, fission-track dating, and tephra correlation to Lower Pleistocene sediments in the Puget Lowland, Washington. *In*: EASTERBROOK, D. J. (ed.) *Dating Quaternary sediments*. Geological Society of America Special Paper, **227**, 139–165.

FRANCIS, E. H. 1961. Thin beds of graded kaolinized tuff and tuffaceous siltstone in the Carboniferous of Fife. *Bulletin of the Geological Survey of Great Britain*, **17**, 191–215.

—— 1985. Recent ash-fall: a guide to tonstein distribution. *In*: *Compte Rendu, Dixième Congrès International de Stratigraphie et de Géologie du Carbonifère, Madrid 1983*, **4**, 189–195.

GOW, A. J. & WILLIAMSON, T. 1971. Volcanic ash in the Antarctic ice sheet and its possible climatic implications. *Earth and Planetary Science Letters*, **13**, 210–218.

HAY, R. L. 1976. *The geology of Olduvai Gorge: a study of sedimentation in a semiarid basin*. University of California Press, Berkeley, Ca.

HAYNES, J. T. & HUFF, W. D. 1990. Origin and tectonic setting of Ordovician bentonites in North America: isotopic and age constraints. Discussion. *Geological Society of America Bulletin*, **102**, 1439–1440.

HESS, J. C. & LIPPOLT, H. J. 1986. $^{40}Ar/^{39}Ar$ ages of tonstein and tuff sanidines: new calibration points for the improvement of the upper Carboniferous time-scale. *Chemical Geology (Isotope Geoscience Section)*, **59**, 143–154.

HUFF, W. D. & KOLATA, D. R. 1989. Correlation of K-bentonite beds by chemical fingerprinting using multivariate statistics. *In*: CROSS, T. A. (ed.) *Quantitative dynamic stratigraphy*. Prentice Hall, Englewood Cliffs, NJ, 567–577.

—— & KOLATA, D. R. 1990. Correlation of the Ordovician Deicke and Millbrig K-bentonites between the Mississippi Valley and the southern

Appalachians. *American Association of Petroleum Geologists Bulletin*, **74**, 1736–1747.

HURFORD, A. J. & HAMMERSCHMIDT, K. 1985. ^{40}Ar/^{39}Ar and K/Ar dating of the Bishop and Fish Canyon tuffs: calibration ages for fission-track dating standards. *Chemical Geology (Isotope Geosciences Section*, **58**, 23–32.

IZETT, G. A., OBRADOVICH, J. D. & MEHNERT, H. H. 1988. *The Bishop ash bed (Middle Pleistocene) and some older (Pliocene and Pleistocene) chemically and mineralogically similar ash beds in California, Nevada, and Utah*. United States Geological Survey Bulletin, **1675**.

JUVIGNÉ, E. 1983. Les variations minéralogiques dans les retombées de 1982 du volcan El Chichon (Chiapas, Mexique) et leur intérêt pour la tephrostratigraphie. *Annales de la Société Géologique de Belgique*, **106**, 311–325.

—— 1990. About some widespread Late Pleistocene tephra horizons in Middle Europe. *Neues Jahrbuch der Geologie und Paläontologie Monatshefte* **1990**, 215–232.

—— & GEWELT, M. 1988. Tephra et depots de grottes: intérêt stratigraphique réciproque. *Annales de la Société Géologique de Belgique*, **111**, 135–140.

KELLER, J. 1981. Quaternary tephrochronology in the Mediterranean region. *In*: SELF, S. & SPARKS, R. S. J. *q.v.*, 227–244.

——, RYAN, W. B. F., NINKOVICH, O. & ALTHERR, R. 1978. Explosive volcanic activity in the Mediterranean over the past 200,000 yr as recorded in deep-sea sediments. *Geological Society of America Bulletin*, **89**, 591–604.

KENNET, J. P. 1981. Marine tephrochronology, *In*: EMILIANI, C. (ed.) *The sea. Volume 7: The ocean lithosphere*. Wiley-Interscience, 1373–1436.

KIMPE, W. F. M. 1967. Occurrence, development and distribution of Upper Carboniferous tonsteins in the paralic West German and Dutch coalfields and their use as stratigraphic markers. *Mededelingen van de Geologische Stichting, Niewe Serie*, **18**, 31–38.

KITTLEMAN, L. R. 1973. Mineralogy, correlation, and grain-size distributions of Mazama tephras and other Postglacial pyroclastic layers, Pacific Northwest. *Geological Society of America Bulletin*, **84**, 2957–2980.

KNOX, R. W. O'B. 1984. Nannoplankton zonation and the Palaeocene/Eocene boundary beds of NW Europe: an indirect correlation by means of volcanic ash layers. *Journal of the Geological Society, London*, **141**, 993–999.

—— & MORTON, A. C. 1988. The record of early Tertiary N Atlantic volcanism in sediments of the North Sea Basin. *In*: MORTON, A. C. & PARSON, L. M. (eds) *Early Tertiary volcanism and the opening of the NE Atlantic*. Geological Society, London, Special Publication, **39**, 407–419.

KOHN, B. P. 1979. Identification and significance of a Late Pleistocene tephra in Canterbury District, South Island, New Zealand. *Quaternary Research*, **11**, 78–92.

KOLATA, D. R., FROST, J. K. & HUFF, W. D. 1987. Chemical correlation applied to K-bentonite beds in the Middle ordovician Decorah Subgroup,

upper Mississippi Valley. *Geology*, **15**, 208–211.

LAYER, P. W., HALL, C. M. & YORK, D. 1987. The derivation of ^{40}Ar/^{39}Ar age spectra of single grains of hornblende and biotite by laser step-heating. *Geophysical Research Letters*, **14**, 757–760.

LEDBETTER, M. T. 1981. Tephrochronology at DSDP site 502 in the western Caribbean. *In*: SELF, S. & SPARKS, R. S. J. *q.v.*, 281–288.

LO BELLO, PH., FÉRAUD, G., HALL, C. M., YORK, D. LAVINA, P. & BERNAT, M. 1987. ^{40}Ar/^{39}Ar step-heating and laser fusion dating of a Quaternary pumice from Neschers, Massif Central, France: the defeat of xenocrystic contamination. *Chemical Geology (Isotope Geosciences Section)*, **66**, 61–71.

MANGERUD, J., LIE, S. E., FURNES, H., KRISTIANSEN, I. L. & LØMO, L. 1984. A younger Dryas ash bed in western Norway, and its possible correlations with tephra in cores from the Norwegian Sea and the North Atlantic. *Quaternary Research*, **21**, 85–104.

MORTON, A. C. 1984. Stability of detrital heavy minerals in Tertiary sandstones of the North Sea Basin. *Clay Minerals*, **19**, 287–308.

—— & KNOX, R. W. O'B. 1990. Geochemistry of late Palaeocene and early Eocene tephras from the North Sea Basin. *Journal of the Geological Society, London*, **147**, 425–437.

NAESER, C. W. & NAESER, N. D. 1988. Fission-track dating of Quaternary events. *In*: EASTERBROOK, D. J. (ed.) *Dating Quaternary sediments*. Geological Society of America Special Paper, **227**, 1–11.

——, Izett, G. A. & OBRADOVICH, J. D. 1980. *Fission-track and K-Ar ages of natural glasses*. United States Geological Survey Bulletin, **1489**.

——, BRIGGS, N. D., OBRADOVICH, J. D. & IZETT, G. A. 1981. Geochronology of Quaternary tephra deposits. *In*: SELF, S. & SPARKS, R. S. J. *q.v.*, 13–47.

NINKQVICH, D. 1979. Distribution, age and chemical composition of tephra layers in deep-sea sediments off western Indonesia. *Journal of Volcanology and Geothermal Research*, **5**, 67–86.

—— & HEEZEN, B. C. 1967. Physical and chemical properties of volcanic glass shards from Pozzuolana ash, Thera Island, and from upper and lower ash layers in Eastern Mediterranean deep-sea sediments. *Nature*, **213**, 582–584.

—— & SHACKLETON, N. J. 1975. Distribution, stratigraphic position and age of ash layer "L", in the Panama Basin region. *Earth and Planetary Science Letters*, **27**, 20–34.

——, OPDYKE, N., HEEZEN, B. C. & FOSTER, J. H. 1966. Paleomagnetic stratigraphy, rates of deposition and tephrachronology in North Pacific deep-sea sediments. *Earth and Planetary Science Letters*, **1**, 476–492.

OBRADOVICH, J. D. & COBBAN, W. A. 1976. A time-scale for the late Cretaceous of the Western Interior of North America. *Geological Association of Canada Special Paper*, **13**, 31–54.

PALAIS, J. M., KYLE, P. R., MOSLEY-THOMPSON, E. & THOMAS, E. 1987. Correlation of a 3,200 year old tephra in ice cores from Vostok and South Pole stations, Antarctica. *Geophysical Research Letters,* **14**, 804–807.

——, KIRCHNER, S. & DELMAS, R. 1989. Identification and correlation of volcanic eruption horizons in a 1,000-year ice-core record from the South Pole. *Antarctic Journal of the United States: 1989 Review,* 101–104.

PATERNE, M., GUICHARD, F. & LABEYRIE, J. 1988. Explosive activity of the South Italian volcanoes during the past 80,000 years as determined by marine tephrochronology. *Journal of Volcanology and Geothermal Research,* **34**, 153–172.

RANDLE, K., GOLES, G. G. & KITTLEMAN, L. 1971. Geochemical and petrological characterization of ash samples from Cascade Range volcanoes. *Quaternary Research,* **1**, 261–282.

ROSE, W. I. & CHESNER, C. A. 1987. Dispersal of ash in the great Toba eruption, 75 ka. *Geology,* **15**, 913–917.

ROSS, C. S. 1925. Beds of volcanic material as key horizons. *Bulletin of the American Association of Petroleum Geologists,* **9**, 341–343.

ROSS, R. J., JR, NAESER, C. W., IZETT, G. A., OBRADOVICH, J. D., BASSETT, M. G., HUGHES, C. P., COCKS, L. R. M., DEAN, W. T., INGHAM, J. K., JENKINS, C. J., RICKARDS, R. B., SHELDON, P. R., TOGHILL, P., WHITTINGTON, H. B. & ZALASIEWICZ, J. 1982. Fission-track dating of British Ordovician and Silurian stratotypes. *Geological Magazine,* **119**, 135–153.

RUNDLE, C. C. 1986. Radiometric dating of a Caradocian tuff horizon. *Chemical Geology (Isotope Geosciences Section),* **59**, 111–115.

RYER, T. A., PHILLIPS, R. E., BOHOR, B. F. & POLLASTRO, R. M. 1980. Use of altered volcanic ash falls in stratigraphic studies of coal-bearing sequences: an example from the Upper Cretaceous Ferron Sandstone Member of the Mancos Shale in central Utah. *Geological Society of America Bulletin,* **91**, 579–586.

SAMSON, S. D., KYLE, P. R. & ALEXANDER, E. C. 1988. Correlation of North American Ordovician bentonites by using apatite chemistry. *Geology,* **16**, 444–447.

——, PATCHETT, P. J., RODDICK, J. C. & PARRISH, R. R. 1989. Origin and tectonic setting of Ordovician bentonites in North America: Isotopic and age constraints. *Geological Society of America Bulletin,* **101**, 1175–1181.

——, PATCHETT, P. J., RODDICK, J. C. & PARRISH, R. R. 1990. Origin and tectonic setting of ordovician bentonites in North America: Isotopic and age constraints. Reply to discussion. *Geological Society of America Bulletin,* **102**, 1441.

SARNA-WOJCICKI, A. M., MORRISON, S. D., MEYER, C. E. & HILLHOUSE, J. W. 1987. Correlation of upper Cenozoic tephra layers between sediments of the western United States and eastern Pacific Ocean and comparison with biostratigraphic and magnetostratigraphic age data. *Geological Society of America Bulletin,* **98**, 207–223.

SELF, S. & SPARKS, R. S. J. 1981. *Tephra studies.* Riedel.

SEWARD, D. 1974. Age of New Zealand Pleistocene substages by fission-track dating of glass shards

from tephra horizons. *Earth Science and Planetary Letters,* **24**, 242–248.

—— 1979. Comparison of zircon and glass fission-track ages from tephra horizons. *Geology,* **7**, 479–482.

SIGURDSSON, H. & CAREY, S. N. 1981. Marine tephrochronology and Quaternary explosive volcanism in the Lesser Antilles arc. *In*: SELF, S. & SPARKS, R. S. J. *q.v.*, 255–280.

—— & LOEBNER, B. 1981. Deep-sea record of Cenozoic explosive volcanism in the North Atlantic. *In*: SELF, S. & SPARKS, R. S. J. *q.v.*, 289–316.

SLAUGHTER, M. & EARLEY, J. E. 1965. *Mineralogy and geological significance of the Mowry bentonites, Wyoming.* Geological Society of America Special Paper, **83**.

SMITH, D. G. W. & WESTGATE, J. A. 1969. Electron probe technique for characterising pyroclastic deposits. *Earth and Planetary Science Letters,* **5**, 313–319.

SPEARS, D. A. 1971. The mineralogy of the Stafford tonstein. *Proceedings of the Yorkshire Geological Society,* **38**, 497–516.

—— & KANARIS-SOTIRIOU, R. 1979. A geochemical and mineralogical investigation of some British and other European tonsteins. *Sedimentology,* **26**, 407–205.

—— & RICE, C. M. 1973. An Upper Carboniferous tonstein of volcanic origin. *Sedimentology,* **20**, 281–294.

STACH, E. 1950. Vulkanische Aschenregen über dem Steinkohlenmoor. *Glückauf,* **86**, 41–50.

SWISHER, C. C. III & PROTHERO, D. R. 1990. Single-crystal ^{40}Ar/^{39}Ar dating of the Eocene-Oligocene transition in North America. *Science,* **249**, 760–762

THIRLWALL, W. F. & FITTON, J. G. 1983. Sm-Nd garnet age for the Ordovician Borrowdale Volcanic Group, English Lake District. *Journal of the Geological Society, London,* **140**, 511–518.

THORARINSSON, S. 1974. The terms *tephra* and *tephrochronology*. *In*: WESTGATE, J. A. & GOLD, C. M. (eds) *World bibliography and index of Quaternary tephrochronology.* University of Alberta, Edmonton, xvii–xviii.

—— 1981. Tephra studies and tephrochronology: an historical review with special reference to Iceland. *In*: SELF, S. & SPARKS, R. S. J. *q.v.*, 1–12.

TUCKER, R. D., KROGH, T. E., ROSS, R. J. & WILLIAMS, S. H. 1990. Time-scale calibration by high-precision U-Pb dating of interstratified volcanic ashes in the Ordovician and Lower Silurian stratotypes of Britain. *Earth and Planetary Science Letters,* **100**, 51–58.

TURNER, D. L., TRIPLEHORN, D. M., NAESER, C. W. & WOLFE, J. A. 1980. Radiometric dating of ash partings in Alaskan coal beds and upper Tertiary paleobotanical stages. *Geology,* **8**, 92–96.

VAN DEN BOGAARD, P., HALL, C. M., SCHMINKE, H.-U. & YORK, D. 1987. ^{40}Ar/^{39}Ar laser dating of single grains: ages of Quaternary tephra from the East Eifel volcanic field, FRG. *Geophysical Research Letters* **14**, 1211–1214.

——, HALL, C. M., SCHMINKE, H.-U. & YORK, D. 1989. Precise single-grain ^{40}Ar/^{39}Ar dating of a cold to warm climate transition in Central Europe. *Nature,* **342**, 523–525.

VITIALANO, C. J., TAYLOR, S. R., FARRAND, W. R. & JACOBSEN, T. W. 1981. Tephra layer in Franchthi Cave, Peleponnesos, Greece. *In*: SELF, S. & SPARKS, R. S. J., *q.v.*, 373–379.

VOGEL, J. S., CORNELL, W., NELSON, D. E. & SOUTHON, J. R. 1990. Vesuvius/Avellino, one possible source of seventeenth century BC climatic disturbances. *Nature,* **344**, 534–537.

WALKER, G. P. L. 1981. Generation and dispersal of fine ash and dust by volcanic eruption. *Journal of Volcanology and Geothermal Research,* **11**, 81–92.

——, SELF, S. & WILSON, L. 1984. Tarawera 1886, New Zealand – a basaltic plinian fissure eruption. *Journal of Volcanology and Geothermal Research,* **21**, 61–78.

WATKINS, N. D., SPARKS, R. S. J., SIGURDSSON, H., HUANG, T. C., FEDERMAN, A., CAREY, S. & NINKOVICH, D. 1978. Volume and extent of the Minoan tephra from Santorini Volcano: new evidence from deep-sea sediment cores. *Nature,* **271**, 122–126.

WEAVER, C. E. 1963. Interpretative value of heavy minerals from bentonites. *Journal of Sedimentary Petrology,* **33**, 343–349.

WESTGATE, J. A. 1989. Isothermal plateau fission-track ages of hydrated glass shards from silicic tephra beds. *Earth and Planetary Science Letters,* **95**, 226–234.

—— & BRIGGS, N. D. 1980. Tephrochronology and fission-track dating. *Geoscience Canada,* **7**, 3–10.

—— & EVANS, M. E. 1978. Compositional variability of Glacier Peak tephra and its stratigraphic significance. *Canadian Journal of Earth Sciences,* **15**, 1544–1567.

—— & FULTON, R. J. 1975. Tephrostratigraphy of Olympia interglacial sediments in south-central British Columbia, Canada. *Canadian Journal of Earth Sciences,* **12**, 489–502.

—— & GORTON, 1981. Correlation techniques in tephra studies. *In*: SELF, S. & SPARKS, R. S. J. *q.v.*, 73–94.

——, CHRISTIANSEN, E. A. & BOELLSTORFF, J. D. 1977. Wascana Creek Ash (Middle Pleistocene) in southern Saskatchewan: characterization, source, fission track age, palaeomagnetism and stratigraphic significance. *Canadian Journal of Earth Sciences,* **14**, 357–374.

——, WALTER, R. C., PEARCE, G. W. & GORTON, M. P. 1985. Distribution, stratigraphy, petrochemistry, and palaeomagnetism of the late Pleistocene Old Crow tephra in Alaska and the Yukon. *Canadian Journal of Earth Sciences,* **22**, 893–906.

——, EASTERBROOK, D. J. NAESER, N. D. & CARSON, R. J. 1987. Lake Tapps tephra: an early Pleistocene stratigraphic marker in the Puget Lowland, Washington. *Quaternary Research,* **28**, 340–355.

——, STEMPER, B. A. & PÉWÉ, T. L. 1990. A 3m.y. record of Pliocene-Pleistocene loess in interior Alaska. *Geology,* **18**, 858–861.

WILLIAMS, S. N. 1983. Plinean ash-fall deposits of basaltic composition. *Geology,* **11**, 211–214.

YANAGI, T., BAADSGAARD, H., STELCK, C. R. & McDOUGALL, I. 1988. Radiometric dating of a tuff bed in the middle Albian Hulcross Formation at Hudson's Hope, British Columbia. *Canadian Journal of Earth Sciences,* **25**, 1123–1127.

YEN, F.-S. & GOODWIN, J. H. 1976. Correlation of tuff layers in the Green River Formation, Utah using biotite compositions. *Journal of Sedimentary Petrology,* **46**, 345–354.

ZIELINSKI, R. A. 1982. The mobility of uranium and other elements during alteration of rhyolite ash to montmorillonite: a case study in the Troublesome Formation, Colorado, U.S.A. *Chemical Geology,* **35**, 185–204.

Milankovitch cycles in the pre-Pleistocene stratigraphic record: a review

WALTHER SCHWARZACHER

School of Geosciences, The Queen's University of Belfast, Belfast BT7 1NN, UK

Sedimentary cycles

Sedimentary cycles have been widely used for detailed correlations and as an aid for mapping. Milankovitch cycles, which can be regarded as constant time markers, open up a wide field of stratigraphic and sedimentological research. Such cyclicity cannot be used for the absolute dating of the more remote geological past but it can be used for relatively accurate interval timing. For example, it is possible to time, and compare the duration of different fossil-zones, provided that the Milankovitch cycles can be correctly identified. The cycles therefore can be used to construct a high resolution stratigraphy of intervals which cannot be subdivided otherwise.

Before using cyclicity as a stratigraphic tool, one would ideally like to know the mechanism of cycle formation. Three features of cyclic sedimentation in particular must be examined in detail. These are: the duration of cycles, their regularity, and the physical size or domain of the cycle producing system.

Timing sedimentary cycles, with few exceptions, can only be approximate. In most situations, timing depends on average accumulation rates which are calculated for time intervals which are much longer than the cycles themselves. The accuracy of this procedure depends very much on the steadiness of the sedimentation rates. Apart from any sedimentological examination which may locate discontinuities such as non-depositional or erosional intervals, the examination of the cycle regularity and complexity is very important for establishing a time-thickness relationship in a sequence. If a sequence contains a regular repetition of similar sediments, for example a constant thickness of beds, one either has to accept that equal thickness intervals represent equal time intervals or one has to find a mechanism which explains the formation of cycles which are independent of time. Various possibilities exist for the latter, for example, the repetition of similar events, such as storms, or earthquakes produces similar sedimentary deposits at irregular time intervals. Such explanations however, become increasingly improbable with any increasing complexity of the cycle. The complexity may increase either because the individual beds contain distinct sedimentation events, or there may be an combination of several shorter cycles into a larger cycle. The formation of stratification cycles (Schwarzacher 1987) is particularly common. In such cycles a more or less constant number of beds is repeated in groups which are delimited by some different lithology or a bed or exceptional thickness (see Fig. 1). Such repeated complex groups are very unlikely to be a random phenomenon (Schwarzacher 1947), and it is therefore likely that they are time related. Stratification cycles were originally referred to as 'bundles' (Schwarzacher 1952) but the term 'stratification cycle' is more appopriate since the 'bundle' can degenerate into a single cycle by losing its subdivisions.

The importance of the size or domain of the cycle-producing system was recognised by Brinkmann (1932) who, in discussing the origin of sedimentary bedding, introduced the terms 'autonomic' and 'exogenic' stratification. An autonomic sequence is generated within a strictly defined area. An exogenic series is the result of outside influences.

The concept of autonomic stratification is similar to that of 'autocyclic' sedimentation, a term introduced by Beerbower (1964) and used to describe fluviatile cycles in the 'depositional prism'. Unfortunately, Beerbower does not define what he means by this expression, but it clearly includes not only the area of deposition but also processes on the alluvial plains and possibly in the source area. It should be noted that the terms 'auto' and 'allocyclic' are completely meaningless if they are used without specifying the area or domain to which they refer. The terms cannot be used to explain cycles since the explanation of the cycles has to be known before the terms can be sensibly applied. Any cycle is autocyclic if the domain chosen is large enough. All cycles become allocyclic on a small scale.

It is in the nature of linear and non-linear oscillating systems that their size is related to the frequency with which they operate. Small local systems, for example seiches in lakes or predator–prey relationships in defined areas, can

From Hailwood, E. A. & Kidd, R. B. (eds), *High Resolution Stratigraphy*
Geological Society Special Publication, No. 70, pp. 187–194.

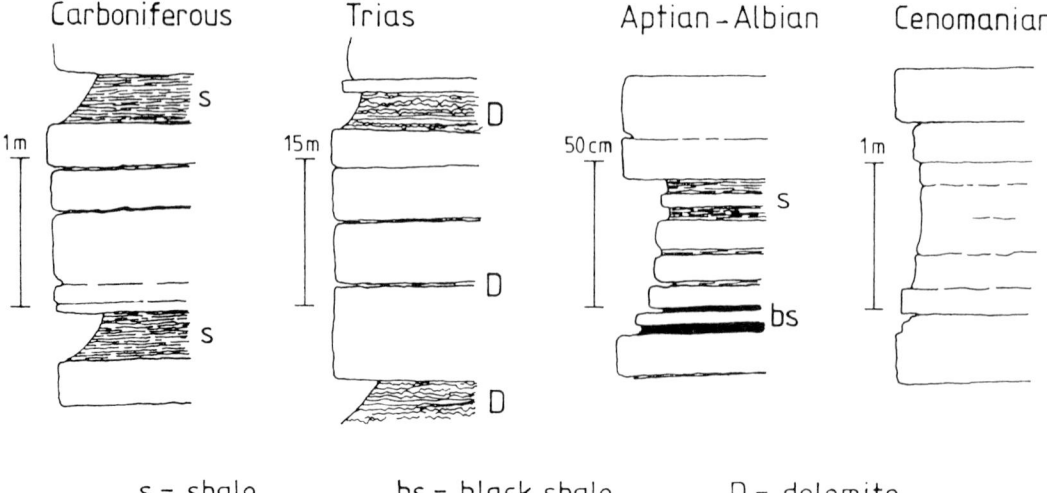

Fig. 1. Four typical limestone stratification cycles from: Carboniferous limestones of NW Ireland, Upper Triassic platform deposits of the calcareous Alps, Aptian–Albian Scisti a Fucoidi of central Italy, and Cenomanian pelagic limestones of central Italy. Note the differences in scale.

only produce relatively short cycles because they have no way of generating long time lags. Large systems on a world-wide scale can generate much longer cycles, but again there are limits to the low frequencies. For example, it is estimated that the total mixing of all the oceans would take about 2000 years and longer periods could not be generated by models using ocean circulation exclusively. Various stochastic feedback models which have been proposed to explain glacial periods by strictly terrestrial climatic processes, are capable of generating cycles of the order of tens of thousands of years. The terrestrial carbon cycle could possibly generate very long periods also. Although no quantitative information is available, the carbon cycle involves a very complex interaction of vegetation, deep-sea and shelf sedimentation and carbon dioxide accumulation in the upper atmosphere.

Considerably less is known about possible tectonic cycles but, in theory, long time lags could be generated by slow mass transports in the mantle for example. When considering theoretical systems of this type, foremost importance must be given to the question of damping. Regular oscillations can only develop in systems which are relatively free of damping and random interferences. Damping will develop particularly in cycles of long duration where the likelihood of interference increases. It may often be very questionable as to whether regular oscillations can occur at all. Indeed, this

has never been demonstrated for many of the tectonic models which have been proposed.

Astronomically controlled cycles are a special case. The oscillating system in this case is the solar system, which is naturally of a completely different magnitude from terrestrial systems and therefore considerably more stable. Although changes have occurred (see later), the system can be regarded as being practically free of damping for long intervals and it is the only system known which can definitely produce cycles of 100 000 years and longer.

The Milankovitch theory

The astronomical theory of planetary movement goes back to Laplace, who in his celestial mechanics developed the differential equations which describe the interaction and movement of several planets around the sun. A first numerical solution for the movement of the eight principal planets was given by Stockwell in 1873. An excellent modern introduction to the theory is given by Berger (1988).

The orbital elements of the earth are the eccentricity, the inclination of the earth's axis or obliquity and the longitude of the perihelion which is called the precession. A considerable number of scientists realised from an early stage that if the orbital elements change, the amount of solar radiation received by the various parts of the globe will vary. Clearly this must have an effect on the climate and Sir John Herschel (1832) was possibly one of the first to realise that records of such changes ought to be found in the geological record. In particular, it was suggested by Adhemar (1842) and Croll (1875) that the ice ages may have been caused by astronomically-controlled climatic changes. Lyell in his 'Principles' devotes a whole chapter to the relationship between climate and astronomical changes. However, it was Milankovitch (1941) who first calculated the solar radiation received by the Earth's surface for the last 650 thousand years and who used the data to explain and date the Pleistocene ice ages.

The most modern and also the most accurate calculations are due to Berger (1976). His calculations for the last 5 Ma. give the quasiperiod of the eccentricity as 95 000 years and the wavelength of the obliquity as 41 000 years. The revolution of the vernal point relative to the perihelion has a length of 21 700 years. The periodicities are referred to as quasi-periods because they are the product of several harmonics with phase differences, so that the resulting 'cycle' becomes variable. For example, the length of the eccentricity period has a mean of 103 000 years with a standard deviation of 2400 years.

The orbital perturbations and the consequent changes in solar radiation, produce changes in the relative intensity of the seasons and changes in the geographical distribution of the climatic belts. A low obliquity implies a relatively small difference between winter and summer but strong differences between insolation in equatorial and polar regions. If the orbit is exactly circular with zero eccentricity, then the precession has no climatic effect. Coupled with an elliptical orbit, the precession increases seasonality and affects the distribution of climatic belts. For this reason precession is combined with the eccentricity to give the precessional index (see Fig.2).

With the methods developed by Berger it is possible to calculate the insolation for any latitude and time up to several million years before the present. For earlier times, the calculations become more uncertain. Conditions in the solar system may have changed within the geological time span. Berger (1989) calculated the lengths of the precession and obliquity cycles based on estimates of earlier day lengths and moon–earth distances for the last 450 Ma. According to his calculations, the obliquity cycles show the largest change and have increased by about 10 000 years since Cambrian times.

Even if it is possible to calculate the intensity and distribution of insolation on the atmos-

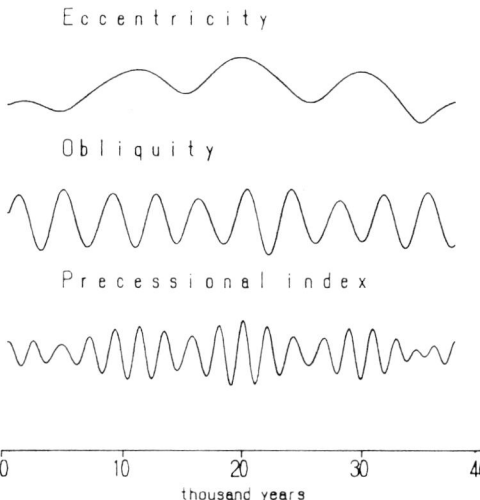

Fig. 2. Eccentricity, obliquity and precessional index for the last 400 thousand years.

phere-free earth surface, it is not possible to construct a realistic climatic model from such data. Additional basic information such as the land/sea distribution, topography, vegetation cover, composition of the atmosphere, distribution of ocean currents and so on, is essential. Such factors interact with each other to produce a very complex system, incorporating various thresholds, feedbacks and non linear relationships. The result of this is only the climate. There is yet another system of possibly equal complexity which links the climate to the sediment that will be observed eventually. Indeed it is as yet impossible to trace every step which leads from the sedimentary record to the astronomical causes. Models which attempt this are of necessity very simple if not simplistic.

Sedimentary cycles and the Milankovitch theory

Within the last two decades, great progress has been made in developing Pleistocene stratigraphy. This progress was made possible by the analysis of deep sea sediments using new techniques such as paleomagnetic dating and isotope studies. Oxygen isotopes can be used to estimate temperature and the relative amount of ice stored in the polar regions at the time a sediment formed. It was first demonstrated by Emiliani & Geiss (1957) that the $^{18}O/^{16}O$ ratio fluctuates with an approximate cycle of 100 ka (see Fig. 3). Power spectral analysis showed (Hays et al. 1976) that there is good evidence that all the

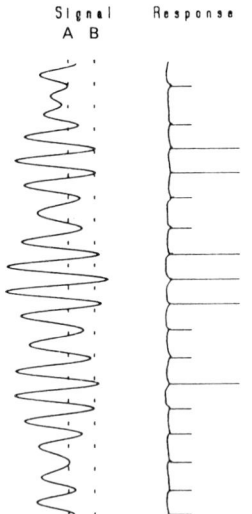

Signal   Response
A B

Fig. 3. Diagrammatic explanation of the formation of stratification 'bundles'. Two threshold levels are assumed. The lower level leads to bed formation. The higher level leads to the formation of cycle boundaries.

frequencies which the Milankovitch theory postulated are present. Based on this evidence, the stratigraphic record of deep sea sediments can be taken as definite proof that orbital variation causes changes in the environment.

The study of cycles in older sediments differs in many respects from that of the Pleistocene marine records. Numerical dating of the section becomes increasingly less accurate with increasing age. The sediments in general are considerably altered by diagenesis and their interpretation in paleoclimatic terms is much more difficult. On the positive side, the pre Pleistocene sediments permit the study of much longer time series and accumulations of uniform cyclic sediment representing millions of years are not rare.

The collection of data from long sections can present problems. Ideally, one would like to have quantitative information on the fauna, composition, grain size and any other important parameter of the sediment in as much detail as possible. This, however, cannot be done frequently, simply because the work involved in obtaining literally thousands of analyses is too expensive. As a substitute, simplified methods can be used. Variously defined features such as the type of bed, composition or colour can be coded in a semi quantitative way and recorded, together with the measured thicknesses. Such data can be collected from surface exposures or

cores. Continuous records can be obtained by measuring the colour or transmissivity of light from photographs or peels (Fischer *et al.* 1985). High resolution wire line logs promise to be a further good source of data (Fischer 1986; Melnyk & Smith 1989). Yet another promising logging technique is measuring the magnetic susceptibility which can be characteristic of certain lithologies (Robinson 1990, this volume).

The direct causes of cycles

Stratification cycles of various complexity are known from the Precambrian to the present. They have been recorded from lacustrine and marine environments and in theory, could be preserved in any sequence which is long enough and which was formed under essentially constant conditions of sedimentation. Limestone and evaporite sequences are particularly suitable, but cycles are also found in clastic sediments.

A very common cycle is the limestone–marl or limestone–shale alternation. Theoretically, there are three types (Arthur *et al.* 1984): the 'production cycle', where there is a constant influx of non-carbonate material and variable carbonate production; the 'dilution cycle', where constant carbonate production is diluted by variable terrigenous influx and the 'dissolution cycle', where variable dissolution of carbonates in deep water produces the lithological variation. Of course all three causes can operate in various combinations but the mode of formation can sometimes be found by regional facies studies of cycles (Hattin 1986).

Limestone–shale cycles are often combined with changes in redox conditions. The limestone is deposited during the well oxygenated intervals, but during the anoxic stages, shale with a relatively high carbon content is formed. The shales are often well laminated which indicates a restricted benthic fauna. Various explanations have been given for the formation of redox cycles.

Apart from overall changes in the oceanic circulation, anoxic bottom environments could develop through over-production of plankton, with oxygen being used up in the breakdown of organic matter (de Boer 1983). Alternatively, the influx of low salinity surface currents can cause a reduction in plankton development and lead to an isolation of the deep waters (Arthur *et al.* 1984). The latter mechanism could also explain the coinciding shale production as an influx of terrigenous matter.

Strong variations in bottom currents can lead to scour cycles (Bottjer *et al.* 1986) in which winnowing occurs and which could be caused by sea level fluctuations or possibly even by storm activity.

A special cycle is found in carbonate platforms which develop in extremely shallow water and often represent changes from subtidal or supratidal environments. The classical examples are found in the Trias of the calcareous Alps (Schwarzacher 1954; Fischer 1964). In such environments, both the subsidence and deposition rates were exceptionally high and it is estimated that the relative sea level fluctuations were of the order of 5–20 m. Whether this can be associated with a glacio-eustatic movement is not known.

The analysis of some Milankovitch cycles does suggest that they can be directly linked to climatic changes. This is particularly the case in lacustrine environments where it is clearly the rainfall and evaporation conditions which determined environmental changes. The distribution of Milankovitch cycles in the stratigraphic system strongly suggests that they can manifest themselves even during periods without major glaciations. Tectonic activity, which was the favoured explanation of cycles by most geologists (Duff *et al.* 1967) can produce effects which can be very similar to climatic effects, but there is not theory which can explain the type of cyclicity discussed. At the same time, the palaeoclimatology of the pre-Pleistocene is hardly known and it seems premature to construct theoretical climatic models based on the few available data.

The recognition of Milankovitch cycles

The recognition of Milankovitch cycles involves several stages. Timing the cycle to establish whether it falls into the Milankovitch range of roughly 10 to 500 ka is essential.

Very accurate timing is possible by actually counting individual varves. Anderson (1984) counted and measured over 250 000 varves in the Permian Castile evaporites and could clearly show the existence of the precession and eccentricity cycles. Varves have also been used to obtain average sedimentation rates in the Tertiary Green River formation (Bradley 1929) and the Triassic Lockatong formation (van Houten 1964; Olsen 1984). Olsen calculated the duration of cycles from radiometric dates as well as from varve counts and found that the latter were consistently shorter by about 15%. In general, time estimates calculated from long sections will give better average rates but the longer the sequence, the more likely is there to be a trend which also makes the series non-stationary.

An alternative approach to timing is to accept the existence of Milankovitch array of 20, 41, 100, and 400 ka. If this is achieved, then the Milankovitch hypothesis is accepted. However, the quality of the data must be good enough for the high as well as the low frequencies to be recognised. Failure to recognize high frequencies is often due to inadequate sampling or to the use of a parameter which is unsuitable for the description of high frequencies. For example, a bed thickness index cannot describe cycles of the same thickness or less than the average bed. Not only are high frequencies very sensitive to sampling, but they can also be destroyed by fluctuations in sedimentation rates, by bioturbation or by diagenesis.

Very low frequencies are also difficult to recognise. Clearly, for seeing long cycles, one needs long records. As already said, the longer the record, the more likely are trend-like variations to exist, the power of which may swamp the maxima of the low frequencies. Since trend often affects sedimentation rates, it cannot be removed easily without interfering with the time space relationship of the stratigraphic sequence.

Sequences in which a large number of different Milankovitch cycles can be recognised are rare. The very detailed faunal studies in the Aptian to Albian of the Piobbico core in Central Italy (Premoli-Silva *et al.* 1989; Herbert & Fischer 1986) have yielded maxima corresponding to the 100 kyr eccentricity, the approximately 40 ka obliquity and the 21 ka precession

Table 1. *Suggested examples of Milankovitch cyclicity*

	E1	E2	O	P	E2/P
Pliocene Trubi marls[1]	2016	545		115	4.7
Cenomanian Scaglia bianca[1]		99.8		16.3	6.1
Aptian–Albian Scisti a Fucoidi[2]		55.2	22	14.5	3.8
Berriasian Miravetes Form:[3]		646		133.6	4.8
Upper Trias Lockatong Form.[3]	9976	2494		498.8	5.0
Upper Trias Dachstein Kalk[4]		1600		397	4.0
Lower Carboniferous Dartry Limestone[1]	1280	320		39	8.2

[1] Schwarzacher (1988)
[2] Premoli-Silva *et al.* (1989)
[3] ten Kate & Sprenger (1989)
[4] Schwarzacher & Haas (1986)
   For explanation, see text.

cycles. It was even possible to show that the precession maximum was in fact a double peak. This must be regarded as very convincing evidence for Milankovitch cyclicity.

In many other examples one has to combine estimates of sedimentation rates with the result of spectral analysis to obtain supportive evidence. A number of suggested Milankovitch cycles are given in Table 1. The table lists the thicknesses in centimetres of cycles believed to represent the 400 ka eccentricity cycle (E1) the 100 ka eccentricity cycle (E2), and the obliquity (O) and precession cycles (P). The sedimentation rates in mm ka^{-1} are found by dividing column E2 by ten. Such rates are in reasonable agreement with the rates expected from the environment of their formation.

The examples in Table 1 are typical stratification cycles in which the 'bundle' boundaries have been interpreted as the 100 ka eccentricity cycle and the individual beds as the precession cycles. If this is true, then the ratio of cycle thickness divided by bed thickness should be approximately 5. The actual number of precession cycles which fall between two eccentricity maxima can vary between 3 and 7 with an average of 4.7 for the last 10 Ma. The ratio eccentricity/precession (see Table 1) is in reasonable agreement with the theory. The thickness ratios, however, are average values, and actual counts of beds per bundle can be very variable. It is believed that this is, in many cases, due to the bad definition of 'bedding'.

Figure 3 shows how a signal like the precessional index can be related to the formation of stratification cycles containing an average of five beds. The hypothesis assumes that an observable bedding plane can be formed only when the signal has reached a certain level. Different threshold levels can produce a variety of stratification patterns. In the diagram, only two levels A and B are assumed, whereby B is assumed to be a level leading to the formation of stratification cycle boundaries.

If it is assumed that the levels for the formation of bedding planes undergo random fluctuations, some very realistic sequences can be simulated. The statistical model explains the unreliability of bed counts, since bed formation is not strictly defined by the signal and subsidiary bedding planes can be formed whenever the signal is high. The variability of bed counts in stratification cycles (see Table 1) can be partly explained by this. The model becomes much more complicated when different sedimentation rates for different signal strengths or lithologies are assumed.

In examining the number of beds in a stratification cycle, one has to consider also the alternative, that the bundle boundaries are irregular. The decision between these alternatives depends on which is the more regular one in *time*, and this will be difficult to establish.

Methods of analysis

Two basic methods for examining cyclicity in sediments are available. They are power-spectral analysis by Fourier transforms and power-spectral analysis by Walsh transforms. The former is ideally suited for continuous functions such as the Milankovitch signal which is composed of sinusoids. The Walsh transforms are ideally suited for the analysis of square waves, for example coded data which record the presence or absence of a sediment property. Walsh spectral analysis is, however, also suitable for any stepwise function and it is often faster to calculate.

The choice of analysis depends very much on the quality of data. The type of variable describing the sediment as well as the way in which such a variable is recorded by the sediment has important bearings on choosing the method. It is generally understood that the relationship between sediment thickness and time includes random variations and that sedimentation gaps can introduce discontinuities. However, it is often neglected that many if not most stratigraphic data are discontinuous by nature. Most reactions of a sediment to the environment are controlled by thresholds which lead to stepwise time sequences. The same is true for faunal responses. Diagenesis usually enhances the discontinuous nature of the stratigraphic sequence. Incomplete sampling can hide this discontinuity or it can be artificially removed by smoothing.

Normally both methods can be used and if the data-sets are not too extensive it is interesting to apply both Walsh and Fourier transforms. Any waveform which departs from the sine-wave in the Fourier analysis and from the square wave in the Walsh analysis generates harmonics which can be difficult to interpret. It is particularly important to realise that it is impossible to 'fine tune' to any extent, without knowing the physical process which is supposed to be analysed.

Like all methods of data analysis, spectral analysis can be misused. Without a clear understanding of the method as well as the geological problem, some very misleading results can be obtained.

Summary and conclusions

There is no doubt that bedding and stratification

cycles can be used for very detailed stratigraphic correlation. If such cycles are to be used for absolute interval timing then orbital periodicities must be identified. It is not possible to recognize Milankovitch cycles simply by looking at cycles and the following criteria have to be examined. The time interval of the cycle must fall into the Milankovitch frequency band and at least two and preferably more frequency maxima of the sedimentary record, should correspond to Milankovitch cycles.

Additional criteria include foremost a regularity of cycles based on a time periodicity of the controlling processes. Suggestive of climatic control is cyclic behaviour, stretching over a large geographical and time domain. This implies that such cycles can transgress facies belts and persist through different facies in vertical sections.

Complex regularly repeated patterns like stratification cycles consisting of five or any other constant number of beds, are likely to be controlled by periodic processes. Experience so far seems to suggest that the most common stratification cycles can be related to the 100 ka eccentricity cycle with the individual beds being caused by the 21 ka precession. It would be very wrong however, to assume automatically that all bundled cycles are of this type.

References

ADHEMAR, J. 1842. *Revolution de la Mer*. Deluges Periodique. Privat publ., Paris.

ANDERSON, R. Y. 1984. Orbital forcing of evaporite sedimentation. *In*: BERGER, A. L. *et al.* (eds) *Milankovitch and climate*. 1, Reidel, 147–162.

ARTHUR, M. A., DEAN, W. E., BOTTJER, D. & SCHOLLE, P. A. 1984. Rhythmic bedding in Mesozoic-Cenozoic pelagic carbonate sequences: the primary and diagenetic origin of Milankovitch like cycles. *In*: BERGER, A. L. *et al.* (eds) *Milankovitch and climate*, 1, Reidel, 191–222.

BEERBOWER, I. R. 1964. Cyclothems and cyclic depositional mechanisms in alluvial plain sedimentation. *Kansas Geological Survey, Bulletin*, 169, 31–42.

BERGER, A. L. 1976. Obliquity and precession for the last 5,000,000 years. *Astronomy and Astrophysics*, 51, 127–135.

—— 1988. Milankovitch Theory and Climate. *Reviews in Geophysics*, 26, 624–657.

—— 1989. Pre-Quaternary Milankovitch frequencies. *Nature*, 342, 133.

BOTTJER, D. J., ARTHUR, M. A., DEAN, W. E., HATTIN, D. E. & SAVRDA, C. E. 1986. Rhythmic bedding produced in Cretaceous pelagic carbonate environments: sensitive recorders of climatic cycles. *Paleoceanography*, 1, 467–481.

BRADLEY, W. H. 1929. The varves and climate of the Green River epoch. *US Geological Survey Professional Paper*, 158-E, 87–110.

BRINKMAN, R. 1932. Uber Schichtung und ihre Bedingungen. *Fortschritte in der Geologie und Palaeontologie*, 11, 187–202.

CROLL, J. 1875. *Climate and time in their geological relations*. Appleton, New York.

DE BOER, P. L. 1983. *Aspects of Middle Cretaceous pelagic sedimentation in southern Europe*. Geologica Ultraictina, 31.

DUFF, P., McL. D. HALLAM, A. & WALTON, E. K. 1967. *Cyclic sedimentation*. Elsevier, Amsterdam.

EMILIANI, C. & GEISS, J. 1957. On glaciations and their causes. *Geologisch Rundschau*, 46, 576–601.

FISCHER, A. G. 1964. Lofer cyclothems of the alpine Trias. *Kansas Geological Survey Bulletin*, 169, 107–148.

—— 1986. Climatic Rhythms recorded in strata. *Annual Reviews of Earth and Planetary Science*, 14, 351–376.

——, HERBERT, T. D. & PREMOLI-SILVA, I. 1985. Carbonate bedding cycles in Cretaceous pelagic and hemipelagic sequences. *In*: PRATT, L. M. *et al.* (eds) *SEPM field trip guidebook No 4*. Midyear meeting Golden, Colorado, 1–10.

HATTIN, P. E. 1986. Interregional model for deposition of Upper Cretaceous rhythmites, U.W. Western Interior. *Paleoceanography*, 1, 483–495.

HAYS, J. D., IMBRIE, J. & SHACKELTON, N. J. 1976. Variations in the earth's orbit: pacemaker of the Ice Ages. *Science*, 194, 1121–1132.

HERBERT, T. D. & FISCHER, A. G. 1986. Milankovitch climatic origin mid-Cretaceous black shale rhythms in central Italy. *Nature*, 321, 739–743.

HERSCHEL, J. F. W. 1832. On the Geological Causes which may influence Geological Phaenomena. *Geological Transactions*, 3, 293–299.

MELNYK, D. H. & SMITH, D. G. 1989. Outcrop to subsurface cycle correlation in the Milankovitch frequency band: middle Cretaceous, central Italy. *Terra Nova*, 1, 432–436.

MILANKOVITCH, M. 1941. Kanon der Erdbestrahlung und seine Anwendung auf das Eiszeitproblem. *Akadamie Royale Serbe*, 133, 1–633.

PREMOLI SILVA, I., RIPEPE, M. & TORNAGHI, M. E. 1989. Planktonic foraminiferal distribution record productivity cycles: evidence from the Aptian-Albian Piobbico core (central Italy). *Terra Nova*, 1, 443–445.

OLSEN, P. E. 1984. Periodicity of lake-level cycles in the late Triassic Lockatong formation of the Newark basin. *In*: BERGER, A. *et al.* (eds) *Milankovitch and Climate*, 1, 127–146.

ROBINSON, S. G. 1990. Applications for whole core magnetic susceptibility measurements of deep sea sediments: Leg 115 results. *Proceedings of the Ocean Drilling Program*, 115, 737–771.

—— 1993. Lithostratigraphical applications for magnetic susceptibility logging of deep-sea sediment cores: examples from ODP Leg 115. *This volume*.

SCHWARZACHER, W. 1947. Uber die sedimentare Rhvtmik des Dachsteinkalkes von Lofer. *Verhandlungen der Geologischen Bundesanst.*, 175–188.

—— 1952. Zum kartieren mit sedimentaren Rhythmen. *Verhandlungen der Geologischen Bundesanst.* Heft 4.

—— 1954. Die Grossrhytmik des Dachsteinkalkes. Mineralogisch Petrographishen Mitteilungen, **4**, 44–54.

—— 1987. The analysis and interpretation of stratification cycles. *Paleoceanography.* **2**, 79–94.

—— 1988. On the possible causes of complex cycles in carbonate sequences. *In*: AGTERBERG, F. P. & RAO, C. N. (eds) *Recent advances in quantitative stratigraphic correlation.* Hindustani Publ. Corp. Dehli, 155–164.

—— & HAAS, J. 1986. Comparative statistical analysis of some Hungarian and Austrian Upper Triassic peritidal carbonate sequences. *Acta Geologica Hungarica,* **29**, 175–196.

TEN KATE, W. G. & SPRENGER, A. 1989. On the periodicity in a calcilutite-marl succession (S.E. Spain). *Cretaceous Research,* **10**, 1–31.

VAN HOUTEN, F. 1964. Cyclic lacustrine sedimentation, Upper Triassic Lockatong formation, central New Jersey and adjacent Pennsylvania. *Kansas Geological Survey Bulletin,* **169**, 497–532.

Strontium isotope stratigraphy for the Late Cretaceous: a new curve, based on the English Chalk

J.M. McARTHUR[1], M.F. THIRLWALL[2], A.S. GALE[3], W.J. KENNEDY[4],
J.A. BURNETT[1], D. MATTEY[2] & A.R. LORD[1]

[1] *Department of Geological Sciences, University College, London, Gower Street, London WC1E 6BT, UK*

[2] *Department of Geology, Royal Holloway and Bedford New College, Egham Hill, Egham, Surrey TW20 0EX, UK*

[3] *Department of Geology, Royal School of Mines, Imperial College, Prince Consort Road, South Kensington, London SW7 2BP and Palaeontology Department, Natural History Museum, Exhibition Road, South Kensington, London SW7 5BD, UK*

[4] *Department of Earth Sciences, Parks Road, Oxford OX1 3PR, UK*

Abstract: Marine $^{87}Sr/^{86}Sr$ decreases from 0.70775 in the Cenomanian to 0.70730 in the middle Turonian before increasing in a near-linear manner to >0.70775 in the early Maastrichtian. This variation has been defined using samples from the English Chalk that are closely integrated with the macrofossil and microfossil biostratigraphy of northwestern Europe. With this new isotope curve a stratigraphic resolution is attainable in correlation that is typically ±0.8 Ma for the Santonian and Campanian stages. Isotopic and biostratigraphic correlations between Dorset and Norfolk, in the UK, agree within the limit of analytical error in $^{87}Sr/^{86}Sr$.

Strontium isotope stratigraphy is used increasingly for correlation and dating of marine carbonates, evaporites and phosphates (Wickman 1948; Elderfield 1986; Veizer 1989; McArthur *et al.* 1990; Hodell *et al.* 1991; DePaolo & Finger 1991). The method works well for those periods for which isotopic calibration curves exist that are both accurate and calibrated against well-documented biostratigraphy and magneto-stratigraphy (eg. Miller *et al.* 1988). Neither condition currently applies to the Cretaceous. Extant data are poorly constrained biostratigraphically and either scatter too much to permit definition of a good curve (Burke *et al.* 1982; Koepnick *et al.* 1985) or are sparse and derive from recrystalized samples (Hess *et al.* 1986).

As a first step towards providing a standard curve for the Late Cretaceous we have determined the $^{87}Sr/^{86}Sr$ of nannofossil chalk and macrofossils through a section of the English Chalk, at Trunch in Norfolk, UK (Fig. 1), where the British Geological Survey (then the Institute of Geological Sciences) cored 469 m of Cenomanian–Maastrichtian chalk. In order to provide a degree of objective assessment of the effect of diagenesis on $^{87}Sr/^{86}Sr$ we have compared $^{87}Sr/^{86}Sr$ in macrofossils and their adhering nannofossil chalk. Furthermore, we have tested the quality of the isotope curve by comparing

isotopic and biostratigraphic correlations of the chalk strata of Norfolk to chalk strata at Studland Bay, Dorset, 350 km southwest of Trunch (Fig. 1).

Our isotope curve for the Coniacian–early Maastrichtian is a good record of marine $^{87}Sr/^{86}Sr$ and provides a quantitative isotopic template with which to correlate Late Cretaceous strata of northwestern Europe with strata worldwide, and against which to match and integrate diverse stratigraphic schemes via common isotopic signatures.

Previous publications of Sr-isotopic curves have universally plotted $^{87}Sr/^{86}Sr$ against a *linear* numerical age scale. This practise gives a specious impression of accuracy to such curves. It does so in two ways. Firstly, it obscures the considerable problems inherent in assigning numerical ages to biostratigraphic schemes, a process that requires, amongst other things, interpolation, extrapolation, and assumptions about sedimentation rate, all of which introduce unquantifiable error. Secondly, primary data (the physical position of a sample in a section) is made subordinate to derived secondary data (numerical age) without making the connection between them very clear (with, perhaps, the exception of Miller *et al.* 1988). To avoid these problems we plot our isotopic data against depth

From HAILWOOD, E. A. & KIDD, R. B. (eds), *High Resolution Stratigraphy*
Geological Society Special Publication, No. 70, pp. 195–209.

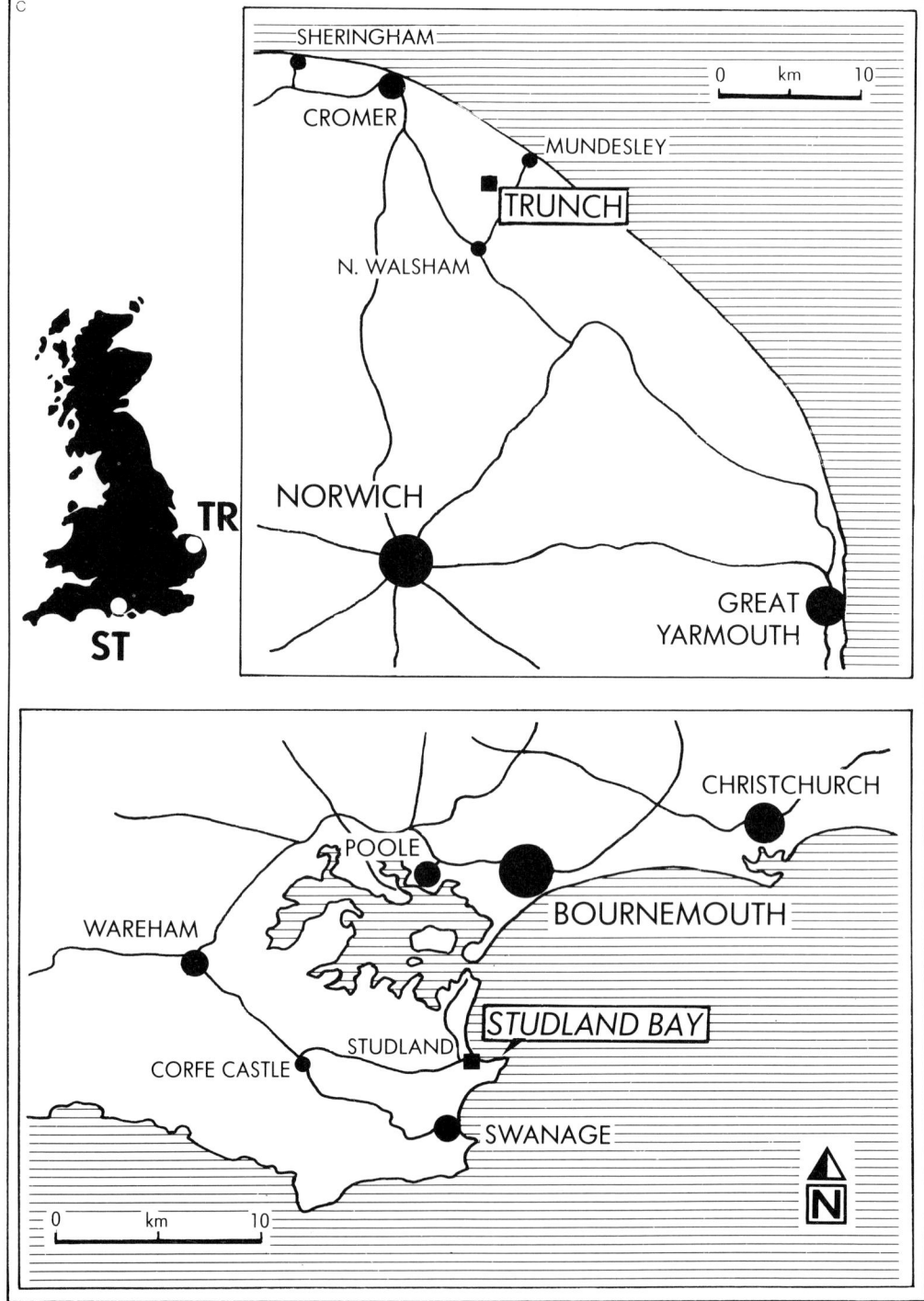

Fig. 1. Location of UK sample sites. Arrow at Studland Bay marks sample point.

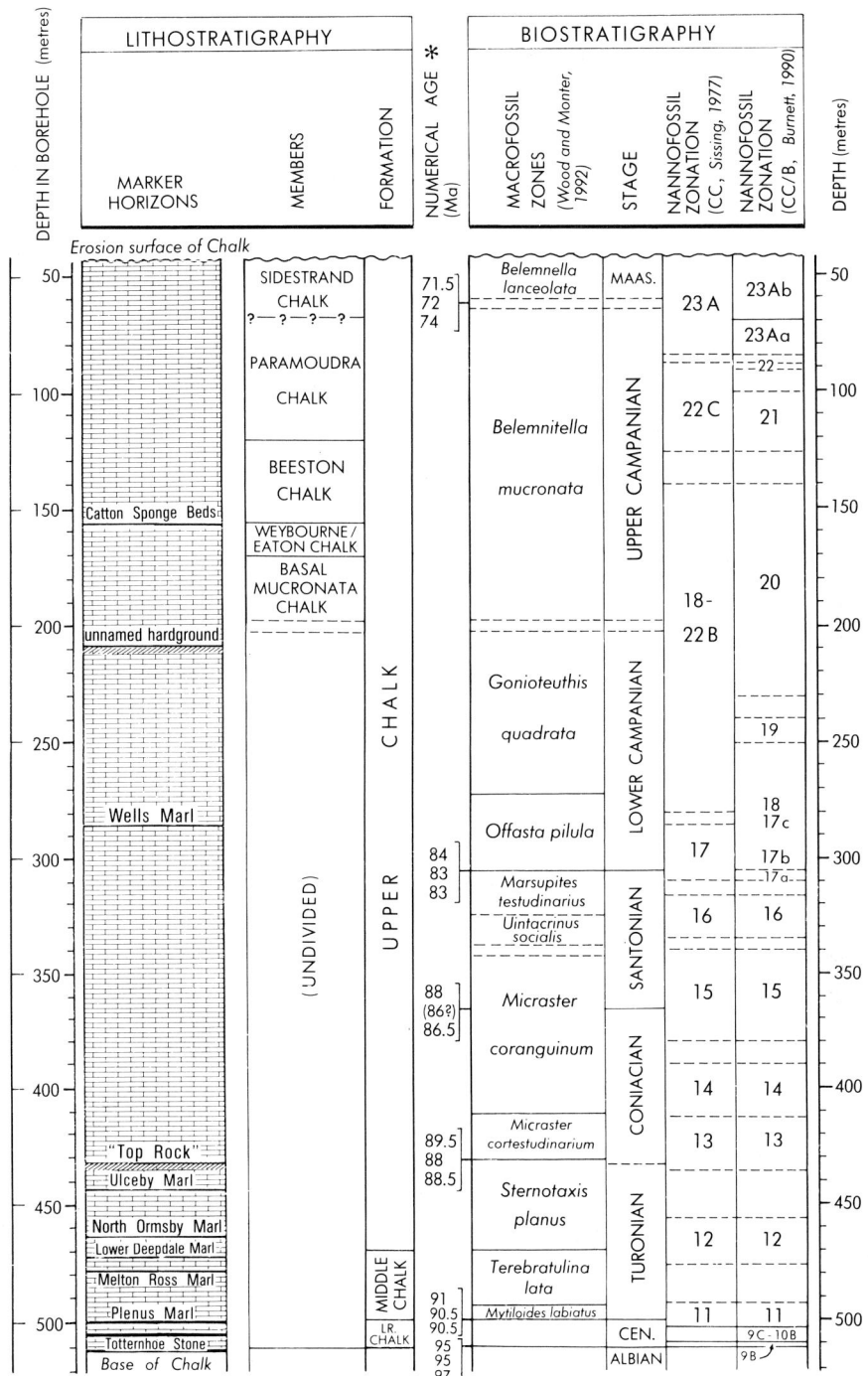

* Top to bottom: Hallam et al. (1985), Odin (1985), Harland et al. (1989)

Fig. 2. Stratigraphy of the Trunch Borehole, Norfolk, UK.

in the borehole. We have added numerical ages as a non-linear calibration using numerical ages assigned to stage boundaries by Hallam *et al.* (1985), Odin (1985) and Harland *et al.* (1989). Our preferred dates are those of Hallam *et al.* (1985) and we use these for all calculations and interpolations made in this paper. The uncertainty on all of these dates is at least ±2 Ma, and may be more; the good agreement between authors does not mean that their dates are accurate, only that they are precise. For example, these authors agree to within 0.5 Ma on the date of about 90.5 Ma for the Cenomanian/Turonian boundary, yet laser-probe ^{39}Ar/^{40}Ar dates of 94.7 ± 0.4 Ma and 94.5 ± 0.2 Ma have recently been reported by Kowallis *et al.* (1989) for uppermost Cenomanian bentonites from the US Western Interior. Some of the uncertainty in assigning numerical ages to boundaries stems from problems of boundary definition and recognition (see, for example, Lillegraven & Ostresh 1990; Lillegraven 1991, and cited papers).

The English Chalk at Trunch

The British Geological Survey borehole at Trunch (Figs 1 and 2) was drilled through the most complete development of Chalk in the UK. It penetrated 469 m of Cenomanian – lower Maastrichtian Chalk. The litho-, nannofossil, and macrofossil stratigraphy are well documented (Gallois and Morter, 1975; Morter, 1984; Burnett, 1988, 1990; Wood and Morter, in press). The English Chalk was deposited in an epicontinental sea and contains a diverse biota of fully marine aspect.

Lithology

Lithological division of the section follows Wood & Morter (in press), Rawson *et al.* (1978) and Peake & Hancock (1961). The Chalk comprises mostly calcareous nannofossils with subsidiary quantities of foraminifera, calcispheres and macrofossil debris, mostly bivalves, echinoderms and bryozoa (Hancock 1975; Bromley 1979). Non-carbonate material, predominantly smectite, illite and quartz, constitutes 30% to 40% of the Lower Chalk, but only 1% to 3% of the Middle and Upper Chalk (Hancock 1975; Bath & Edmunds 1981). In eastern England the Upper Chalk generally is poorly-cemented and retains up to 40% porosity, having 'largely resisted recrystalization and lithification' (Bath & Edmunds 1981). Diagenesis locally has resulted in the formation of nodular chalks and

hardgrounds, which are frequent in the Lower and Middle Chalk of the Trunch borehole. In the Upper Chalk the proportion of cement, present as outgrowth, is extremely low and obscures only the fine detail of coccolith structure (Burnett 1990). This thin overgrowth acts as a weak cement by bridging coccolith contacts. A small amount of microspar in voids is randomly present (Hancock 1975). In the Middle and Upper Chalk nodular flints are frequent, commonly occurring in discrete layers. Celestite has been noted at 314, 332 and 370 m (Morter 1984) and may occur at other levels; below 100 m concentrations of Sr in pore-water are in equilibrium with celestite (25 ± 5 mg l^{-1}, Bath & Edmunds 1981), which suggests this mineral may be ubiquitous below 100 m, although largely unnoticed.

Lithostratigraphy

The lithostratigraphy of the Trunch borehole follows Wood & Morter (in press). The succession shows many similarities to outcrops in eastern and northern England (Norfolk and Lincolnshire), and within it can be recognized many regional marker beds (marls, hardgrounds, fossil beds), the most important of which are shown on Fig. 2.

The Lower Chalk (12.5 m) rests non-sequentially on Albian Red Chalk and comprises 12.5 m of condensed marly nodular chalks which contain abundant inoceramus debris and many hardgrounds and omission surfaces. The Totternhoe Stone, and a condensed equivalent of the Plenus Marl, are recognised at 506.30 m (base) and 501.10 m (base), respectively. The Middle Chalk (31 m) is nodular and contains abundant inoceramid debris in its lower part; the upper part contains distinctive regional marker marls (North Ormesby, Ulceby, Wells, Melton Ross, Lower Deepdale), and nodular and tabular flints. The Upper Chalk (425 m) contains flints throughout and two well-developed hardgrounds, at 209 m and 432 m. The lower represents the Top Rock, or Navigation Hardground, of southern England. Stylolytic marly partings and omission surfaces are common between 400 and 470 m. The section 432 m to 470 m contains several minor hardgrounds as well as numerous omission surfaces. Within the Upper Chalk the lithological divisions of Sidestrand, Paramoudra, Beeston and Weybourne/Eaton and Basal Mucronata Chalk can be recognized (Fig. 2).

Excluding condensed sections of the Cenomanian and Turonian, some 370 m of Chalk were deposited between the Coniacian/

Turonian boundary (88 ± 2 Ma) and the Maastrichtian/Campanian boundary (72 ± 2 Ma). This represents a sedimentation rate between 19 and 31 m/Ma (see section on numerical ages).

Macrofossil biostratigraphy

The macrofossil zonation shown in Fig. 2 is based on Wood & Morter (in press) and follows Peake & Hancock (1961) and Rawson et al. (1978). Stage boundaries were recognized by the following criteria, following Birkelund et al. (1984).

Albian/Cenomanian, 512.5 m. Unconformity. The basal Paradoxica Bed elsewhere yields Cenomanian fossils.

Cenomanian/Turonian, 500.1 m. First occurrence of inoceramid bivalve *Mytiloides*.

Turonian/Coniacian, 432.5 m. Base of the Top Rock, or Navigation Hardground. In southeastern England this unit had yielded *Forresteria petrocoriense* (Gale & Woodroof 1981).

Coniacian/Santonian, 365.8 m. First occurrence of *Cladoceramus undulatoplicatus* (Roemer).

Santonian/Campanian, 307.4 m. Last occurrence of *Marsupites testudinarius* (Schlotheim).

Campanian/Maastrichtian, 62.5 m. First occurrence of *Belemnella* sp. in correlative beds on the Norfolk coast.

Our uncertainty in placing macrofossil boundaries within the core is estimated to be <±3 m, which is equivalent in time to a period of <±0.12 Ma. As the core represents a random small sample of the Chalk the real uncertainty in true boundary positions is unquantifiable; it is likely to be greater, although constrained by recognition of distinctive lithologies and marker beds.

Nannofossil biostratigraphy

The calcareous nannofossil zonation of the borehole (Fig. 2; Burnett 1988) is based on the standard calcareous nannofossil zonation scheme (CC zones) of Sissingh (1977), as amended by Perch-Nielsen (1979, 1985). The nannofossil zonation has been further refined by the definition of Boreal subzones (Fig. 2; Burnett, 1990) which are also recognised in the Late Cretaceous standard sections in northern Germany (Lägerdorf and Kronsmoor). Nannofossil boundaries have been located with variable

accuracy of between ±11 m (CC12/CC13) and ±1.5 m (CC13/CC14). These uncertainties are determined mostly by sampling interval and are shown in Fig. 2.

The English Chalk at Studland Bay, Dorset

The uppermost part of the Chalk succession at Studland Bay (the uppermost 5 m of the section) lies within the B. mucronata zone and is considered on biostratigraphic criteria to be equivalent to the lower part of the Eaton/Weybourne Chalk of Norfolk. The Chalk at Studland Bay is exceptionally soft and devoid of cement. It has been buried to effective depths of less than 500 m and apparently is devoid of the effects of diagenesis; in this respect it is better preserved than the Chalk at Trunch.

Samples analysed

We have analysed nannofossils and macrofossils through the 469 m of chalk recovered from the Trunch borehole. The nannofossil samples were white chalks above 500 m, and grey marls below 500 m, with varying contents of $CaCO_3$ (Tables 1 and 2). Samples from depths less than 400 m were soft to moderately cemented; those below were noticeably harder, particularly below 430 m. The macrofossils were fragments of inoceramids, echinoids, belemnites and brachipods. Some were partially silicified and therefore contain less than 56% CaO (Table 2).

We have also analysed three macrofossils and their adhering nannofossil chalk collected from a single stratigraphic level within the uppermost 5 m of the Chalk section exposed as low cliffs (3–5 m) on the north edge of The Foreland at Studland Bay, Dorset (Fig. 1, Table 3). These samples were not silicified and the macrofossils preservation appeared excellent; in particular the inoceramid showed excellent preservation of the finest prismatic shell-structure.

Table 1. *$^{87}Sr/^{86}Sr$ values obtained using different dissolution methods for nannofossil chalk sample 38 from 360 m in the Trunch Borehole, Norfolk*

6 Molar HCl	0.707499 ± 7
2.5 Molar HCL	0.707490 ± 9
Acetic acid	0.707448 ± 9
	0.707456 ± 8
	0.707457 ± 17
	0.707461 ± 8
Weak acid,	0.707462 ± 7
Ion exchanger	0.707464 ± 8

Table 2. *Sample numbers and depths, and isotopic and chemical data for nannofossil chalks and macrofossils from the Trunch Borehole, Norfolk*

Sample	Depth m	$^{87}Sr/^{86}Sr$	$\delta^{13}C$	$\delta^{18}O$	CaO%	Sr ppm
1	47	0.707785 ± 8	+1.85	−1.70	53.8	580
		0.707785 ± 9				
Belemnite	49	0.707731 ± 7	+0.79	−0.14	45.0	840
3	50	0.707773 ± 9	+1.81	−1.53	52.1	740
		0.707785 ± 7				
4	58	0.707787 ± 7	+1.83	−2.04	51.5	480
5	60	0.707773 ± 9	+1.85	−1.91		
Cretirhynchia sp.	64	0.707749 ± 8			56.0	855
8	68	0.707751 ± 8	+1.77	−2.17	55.9	640
9	71	0.707760 ± 8				
10	81	0.707750 ± 8	+1.96	−2.05	53.1	610
Belemnite	85	0.707698 ± 8			52.6	1320
Inoceramid	90	0.707694 ± 8			54.2	360
11	91	0.707732 ± 12	+1.83	−1.84	53.2	835
Belemnite	100	0.707685 ± 10	+2.48	−0.19	53.5	1174
12	101		+1.88	−1.84		
T(M)3	101	0.707702 ± 8	+1.98	−1.89	54.2	830
13	111	0.707692 ± 7	+1.95	−1.75	51.8	765
Carneithyris sp.	126	0.707637 ± 12	+2.27	−1.19	50.3	755
16	140	0.707663 ± 14	+1.97	−2.35	53.3	760
Belemnite	148	0.707622 ± 8			55.5	1180
18	160	0.707646 ± 8	+1.95	−2.20	54.5	590
Irreg. echinoid	175	0.707592 ± 9			55.3	385
20	180	0.707602 ± 9	+1.96	−2.23	54.5	775
21	191	0.707582 ± 10	+2.40	−1.68	54.7	825
Brachiopod	200	0.707557 ± 12			55.5	805
Belemnite	210	0.707555 ± 8			55.8	1310
23	211	0.707568 ± 7	+2.34	−2.04	55.2	830
24	221	0.707550 ± 8	+2.36	−2.06	52.2	730
Brachiopod	226	0.707544 ± 9				
4002 M	226	0.707584 ± 8			52.3	875
Irreg. echinoid	226	0.707569 ± 9	+2.32	−1.17	52.9	435
Bourgueticrinus	229	0.707601 ± 11			53.4	295
25	231	0.707571 ± 11	+2.29	−2.36	52.6	805
26	240	0.707581 ± 11	+2.28	−2.54	53.0	800
Belemnite	249	0.707528 ± 8			35.4	890
27	250	0.707536 ± 7	+2.16	−2.58	52.8	770
28	260	0.707542 ± 7	+2.31	−2.52		
Oyster	275	0.707522 ± 9			54.3	730
5777 M	275	0.707540 ± 11			54.3	880
30	280	0.707544 ± 9	+2.22	−3.02	53.1	785
		0.707519 ± 8				
Inoceramid	300	0.707472 ± 9	+2.66	−0.90	13.4	6950
32	300	0.707506 ± 13	+2.62	−2.55	52.2	890
33	310	0.707527 ± 8	+2.27	−3.29	55.3	1040
34	320	0.707486 ± 10			55.3	960
		0.707504 ± 9				
35	330	0.707497 ± 9	+2.36	−2.74	54.7	860
		0.707476 ± 9				
37	350	0.707471 ± 7	+2.49	−2.59	54.6	715
Inoceramid	352	0.707451 ± 8			14.7	110
38	360	0.707448 ± 9			55.4	800
		0.707456 ± 8				
		0.707457 ± 17				
		0.707461 ± 8				
		0.707462 ± 7				
		0.707464 ± 8				

Table 2. *Cont.*

Sample	Depth m	$^{87}Sr/^{86}Sr$	$\delta^{13}C$	$\delta^{18}O$	CaO%	Sr ppm
39	370	0.707444 ± 9	+1.86	−2.45	55.6	730
		0.707417 ± 11				
40	380	0.707427 ± 8	+2.56	−2.67	55.5	800
		0.707410 ± 17				
		0.707390 ± 10				
41	390	0.707394 ± 11	+2.38	−2.64	54.9	705
42	396		+2.52	−2.70		
43	397	0.707395 ± 7	+2.71	−2.88	54.8	690
44	399	0.707390 ± 13	+2.68	−3.44	55.4	560
Inoceramid	402	0.707370 ± 7			16.4	75
47	405	0.707395 ± 8	+2.41	−2.42	55.5	590
48	407	0.707399 ± 8	+2.16	−2.35	55.7	590
50	411	0.707382 ± 9	+2.11	−2.78	55.3	700
		0.707379 ± 9				
54	419	0.707368 ± 9	+2.18	−2.79	55.4	650
58	427	0.707351 ± 8				
60	431		+1.86	−2.89		
61	433	0.707398 ± 7	+1.92	−2.81	55.5	580
63	443	0.707402 ± 9	+1.78	−3.85	54.5	650
		0.707401 ± 11				
		0.707401 ± 8				
Brachiopod	449	0.707430 ± 9			55.5	460
65	453	0.707397 ± 9	+1.82	−3.81	54.8	545
67	463	0.707408 ± 16	+2.05	−3.17	53.9	520
68	468	0.707383 ± 9	+1.82	−2.53	54.1	810
69	473	0.707364 ± 7	+2.00	−3.15	54.8	630
		0.707392 ± 7				
Inoceramid	474	0.707326 ± 15			53.1	646
70	478	0.707405 ± 9	+1.93	−3.67	54.4	690
71	482	0.707399 ± 9	+2.24	−2.51	55.1	600
72	490	0.707402 ± 7	+2.24	−3.64	54.4	500
		0.707406 ± 10				
73	495	0.707472 ± 8	+2.78	−3.49	54.4	315
74	499	0.707457 ± 9	+3.32	−2.53	55.2	325
		0.707471 ± 10				
77	504	0.707493 ± 8	+2.76	−3.31	51.7	670
79	506	0.707514 ± 8	+2.23	−2.66	49.3	700
		0.707489 ± 9				
81	508	0.707504 ± 16	+1.97	−2.13	52.5	650
83	510	0.707515 ± 9	+1.79	−3.05	53.2	655
84	511	0.707495 ± 9	+2.01	−2.71	49.1	630
87	513	0.707525 ± 9	+1.64	−3.34	51.4	735

Table 3. $^{87}Sr/^{86}Sr$ *in matrix/macrofossil pairs from Dorset*

Sample	$^{87}Sr/^{86}Sr$	Stage
The English Chalk, Studland Bay, UK		
Pycnodonte sp.	0.707593 ± 8	
Adhering chalk	0.707583 ± 7	
Belemnite	0.707583 ± 11	lowermost
Adhering chalk	0.707583 ± 7	Upper
Inoceramid	0.707598 ± 8	Campanian
Adhering chalk	0.707587 ± 8	

Analytical methods

Macrofossil fragments were prepared for analysis by (1) washing off adhering chalk under the tap, (2) coarsely fragmenting them with a pestle and mortar, (3) ultrasonically removing remaining chalk in ultra-pure water, (4) briefly cleaning them in 0.5M HCl, to remove the surface layer of calcite, (5) rinsing in ultra-pure water and drying. Macrofossils were then dissolved in 2.5M HCl. After centrifugation 2 ml aliquots were evaporated to dryness with 6M HCl.

Preparation of nannofossil samples involved two stages: (1) a cleaning stage, designed to remove adsorbed and exchangeable Sr from mineral surfaces and exchange sites and to dissolve calcite overgrowth and cement, and (2) dissolution of the sample and separation of Sr. For stage 1, 100 mg of sample were ultrasonically disaggregated in ultra-pure water; even the hardest was easily disaggregated in distilled water with the aid of ultrasonic agitation under gentle pressure from a glass rod. After disaggregation 20 μl of sub-boiled 100% acetic acid were added. After reaction for 30 minutes under ultrasonic agitation, and then centrifugation, the supernatant was discarded (this step dissolves 15% of the sample, which at this point consists of separated coccolith plates so this step must dissolve any cement that coats the coccolith surfaces. It also provides a solution in which abundant Sr and Ca can displace any radiogenic Sr on exchangeable sites, e.g. clays, and thereby minimizes this source of contamination). A further 3 ml of ultra-pure water and 20 μl of sub-boiled 100% acetic acid were then added; this dissolves a further 15% of the sample and ensures that the final pH of the solution is high, thus avoiding continued acid attack on non-carbonate material. After reaction and centrifugation, the supernatant was removed and re-centrifuged and 2 ml evaporated to dryness with 6M HCl.

Following preparation, Sr was separated by standard methods of ion-exchange chromatography and ^{87}Sr/^{86}Sr was determined with a VG-354 five-collector mass spectrometer using a peak-jumping, five-collector routine (Thirlwall in press). All data were normalized to an ^{86}Sr/^{88}Sr ratio of 0.1194. Data were collected during three periods of analysis, during which SRM 987 gave ^{87}Sr/^{86}Sr of 0.710249 ± 19 ($n = 36$), 0.710259 ± 13 ($n = 25$) and 0.710247 ± 18 ($n = 55$), where errors are at 2σ and n is the number of standards run in each period. We adopt an analytical error of ±18 × 10^{-6} in this work. Two SRM-987 standards were run per turret of 14 samples. Data collected within each period have been adjusted to a value of 0.710248 for SRM 987 by addition of the value [0.710248 − Period Mean]; i.e. −1, −11, and +1, respectively. Replicates of 13 samples were run during the period of analysis and the data are shown in Table 1. Sr/Rb ratios, determined on 6 samples by isotope-dilution mass spectrometry, were >5000 so no corrections for radiogenic Sr have been made.

In order to assess the maximum disturbance of carbonate-^{87}Sr/^{86}Sr that dissolution of non-carbonate material could cause, should any be attacked by our dissolution procedure, sample 38 was also prepared by dissolution in 2.5M and 6M HCl (Table 1). Our acetic-acid preparation procedure yields the same ^{87}Sr/^{86}Sr for sample 38 as does the dissolution of carbonate in a weak-acid, ion-exchange resin at pH 5 (Table 1; Kralik 1984), a procedure that is unlikely to affect non-carbonate minerals.

Stable isotopic data were obtained with a VG Prism three-collector gas-sourced mass spectrometer running an on-line, automated, H$_3$PO$_4$ acid-bath operating at 90° C. During the analysis NBS-18, run as sample, gave δ^{13}C = −5.06 ± 0.04‰ and δ^{18}O = −3.12 ± 0.08‰; NBS-19 gave δ^{13}C = +1.97 ± 0.08‰ and

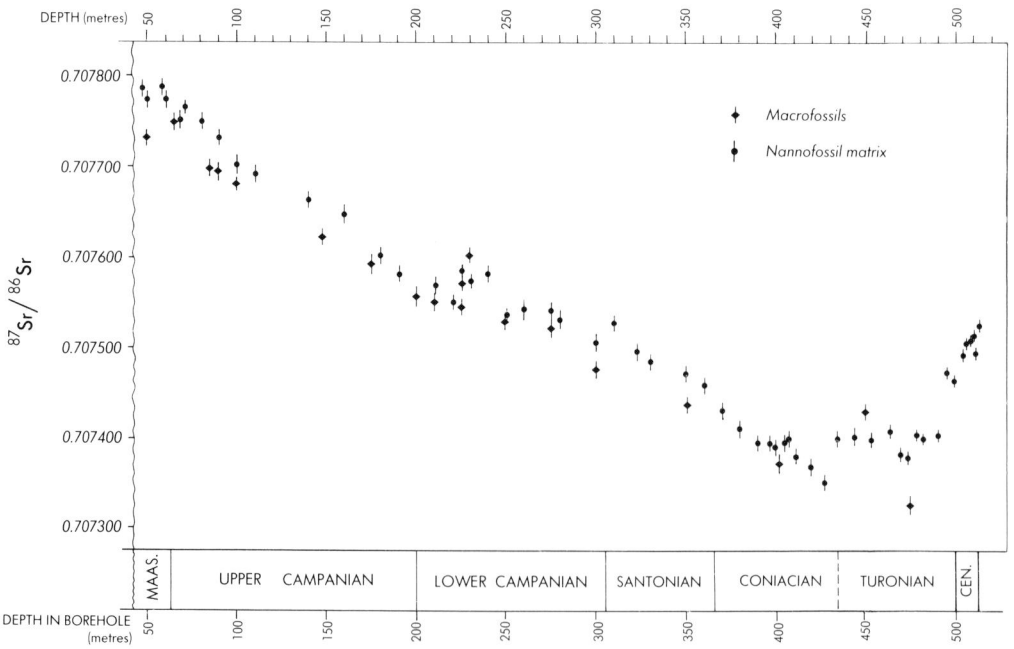

Fig. 3. ^{87}Sr/^{86}Sr with depth for the Trunch Borehole, Norfolk, UK.

$\delta^{18}O = -2.64 \pm 0.09\permil$. Analysis for Sr and Ca concentrations were done by flame AAS after dissolving samples in 1M HCl. Precision for both is <5%.

Results and discussion

The analytical results are given in Tables 1–3. Figure 2 shows the stratigraphy of the borehole and Fig. 3 shows $^{87}Sr/^{86}Sr$ against depth in the borehole. All Sr-isotopic data are normalized to an $^{86}Sr/^{88}Sr$ ratio of 0.1194, and adjusted to an SRM 987 of 0.710248. All errors in $^{87}Sr/^{86}Sr$ quoted in tables are $\pm 2 \times$ s.e. of mass spectrometer counting statistics.

We use our isotopic data for correlation, which involves matching a sample $^{87}Sr/^{86}Sr$ to an $^{87}Sr/^{86}Sr$ calibration line. Adding errors at two standard deviation ($\pm 18 \times 10^{-6}$) from both sample measurement and calibration line, yielding a total error of $\pm 36 \times 10^{-6}$, gives an unduly pessimistic measure of the combined real error. In most instances the error in our $^{87}Sr/^{86}Sr$ results from analytical error, not natural variability, and so represents a normal distribution of error about the mean value. In such instances error distributions overlapped at one standard deviation define a common area that contains the sample result in 95% of cases. We therefore plot

data using errors on $^{87}Sr/^{86}Sr$ of one standard deviation of the period mean.

Stable isotope and major element profiles

Values of $\delta^{18}O$ scatter considerably, especially in the lower part of the section. On this scatter is superimposed a trend from values of around $-3\permil$ in the Cenomanian and Turonian to around $-2\permil$ in the basal Maastrichtian (Fig. 4), a trend very similar to that seen in Cretaceous calcareous nannoplankton and pelagic foraminifera from DSDP sites in the NW Pacific (Haq 1984) and in a Campanian/Maastrichtian sequence at Bidart, southwestern France (Clauser 1987). Values of $\delta^{13}C$ show heavy-carbon excursions at 500 m: (the Cenomanian/Turonian boundary; cf. Schlanger et al. 1987) and possibly in the middle Coniacian, around 400 m. From the middle Coniacian to the lower Maastrichtian, values of $\delta^{13}C$ decrease from around $+2.7\permil$ to $+1.8\permil$.

The Sr and Mg content of the borehole are discussed by Bath & Edmunds (1981), so we refrain from duplicating that discussion. We merely note that Sr concentrations in the Chalk are highest around 300 m depth and do not correlate with lithology or $^{87}Sr/^{86}Sr$. There are no

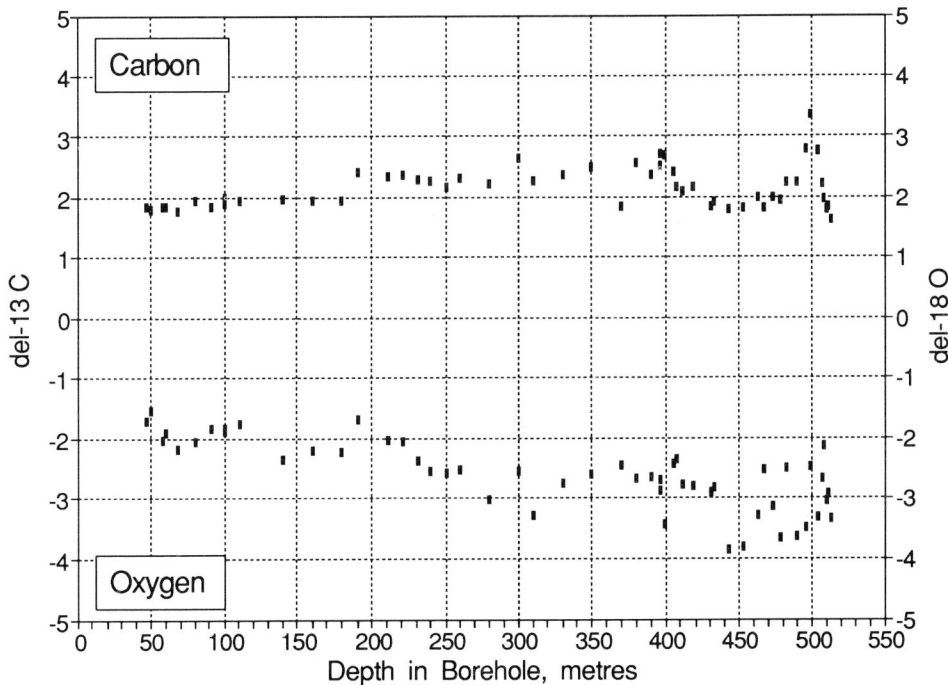

Fig. 4. Stable isotopic variation with depth of nannofossil chalk from the Trunch Borehole, Norfolk, UK.

anomalous concentrations of Sr around 225 m, where the Sr-isotope curve shows a pronounced inflexion.

Preservational quality of Trunch samples

The value of our isotope curve (Fig. 3), and the quality of isotopic correlations and dates obtainable from it, are dependent on the faithfulness with which it reflects the marine $^{87}Sr/^{86}Sr$ of the time. We believe that in at least one, and maybe two, parts of the section $^{87}Sr/^{86}Sr$ has been severely disturbed by diagenesis. A major hardground in the section (the Top Rock) occurs at 432 m. At this level a pronounced discontinuity exists in the $^{87}Sr/^{86}Sr$ of the nannofossil matrix, and $^{87}Sr/^{86}Sr$ below this level, to 470 m, appears anomalously high. Between 470 m and 432 m several minor hardgrounds and many omission surfaces also are present. The diagenesis involved in their formation has almost certainly increased $^{87}Sr/^{86}Sr$, so the flat portion of the curve is almost certainly not a true reflection of the real trend in isotopic evolution during this period. A second inflexion, at 226–240 m, occurs some way below the only other strong hardground in the section, at 209 m. It may represent a diagenetic effect and be connected to hardground formation, but its first appearance some 10 m below this hardground makes this explanation equivocal. Full explanation of this part of the curve awaits analysis of this time interval in other Cretaceous sections. We therefore propose that the elevated $^{87}Sr/^{86}Sr$ around 432–470 m, and possibly for the section around 225 m, reflects diagenetic effects consequent on hardground formation, and we ignore the data when drawing our best fit curve to the $^{87}Sr/^{86}Sr$ line. We speculate that hardground formation may produce a permeability barrier that prevents the escape of fluids during compaction, and allows localized precipitation of isotopically-anomalous cement some distance below the hardground itself.

We regard the remainder of our section as providing reasonable preservation of original $^{87}Sr/^{86}Sr$. Values of $^{87}Sr/^{86}Sr$ for macrofossils are lower by $20–30 \times 10^{-6}$ than the ratio for nannofossil chalk from the same level (Table 2, Fig. 3). This difference persists throughout the section we consider to be least diagenetically altered (above 430 m). Conflicting criteria are available for deciding which data set most closely reflect the original $^{87}Sr/^{86}Sr$.

Two criteria suggest that most of the nannofossil chalk is preserving an original signal. Firstly, nannofossils appear a little overgrown under the SEM, but this diagenetic calcite does

nothing more than obscure the finer coccolith structure. Our sample preparation method incorporates a pre-clean in acetic acid which dissolves 15% of the sample, thus removing this surface overgrowth and so minimising the effect of cementation on original $^{87}Sr/^{86}Sr$. Secondly, the diagenetic model of Richter & DePaolo (1987, 1988) predicts a lowering of matrix-$^{87}Sr/^{86}Sr$ during diagenesis in sections where $^{87}Sr/^{86}Sr$ decreases with depth in the section. This predicted effect is the opposite of that seen in our data, where the matrix is higher than the associated macrofossils.

Other evidence suggests that the macrofossil data are more faithfully preserving an original isotope signature. Firstly, preservation of macrofossils is variable. Some samples, for example the inoceramid at 474 m and the *Cretirhynchia* sp. at 64 m, showed excellent preservation of the fine shell-structure, whilst others have been replaced to varying degrees by silica, for example, the inoceramids at 300 and 353 m (Table 1). Despite this variable replacement, macrofossils are uniformly $20–30 \times 10^{-6}$ lower in $^{87}Sr/^{86}Sr$ than the corresponding nannofossil matrix, (apart from one sample at 450 m). The inoceramid specimen at 300 m, which contains only 24% calcite also contains 6950 ppm Sr, probably mostly from trace celestite. Nevertheless, it plots no further below the matrix curve than do well-preserved macrofossils. These observations are consistent with good preservation of the unsilicified portion of the shells, and very local derivation of Sr during celesite precipitation. Secondly, tests of our data against a well-defined biostratigraphic correlation, given later, shows that good isotopic correlation is possible using either matrix data or macrofossil data, but that macrofossil data seems to give marginally better correlation, suggesting the macrofossils preserve a less altered $^{87}Sr/^{86}Sr$ than do the nannofossils.

We believe that, on a balance of probability, the macrofossil data are preserving a $^{87}Sr/^{86}Sr$ that is less altered than that preserved in the nannofossil matrix, and that it may be a faithful record of the Cretaceous sea. As macrofossils suitable for analysis are not common in the borehole material, the macrofossil trend should be constrained by the shape of a curve drawn through the more abundant nannofossil data.

The Late Cretaceous Sr-isotope curve

Our $^{87}Sr/^{86}Sr$ curve (Fig. 3) shows three distinct sections. Between 512 m, the base of the Chalk, and 470 m $^{87}Sr/^{86}Sr$ decreases upward through the condensed Cenomanian and lower Turonian section. Between 470 m and 432 m five matrix samples plot close to 0.707400, and one fossil plots above this trend. These samples are below the 'Top Rock' which is at 432 m, and their departure from the general trend of the data is ascribed to diagenetic effects associated with the formation of the 'Top Rock'. If these six samples are excluded from consideration the curve appears to decrease to a projected minimum of about 0.70730 in the middle of the Turonian section before increasing upward through the rest of the section. Below 432 m the incompleteness of the sequence, and probable diagenetic alteration, limits the value of the curve, which may show more structure if examined over a more expanded, and less diagenetically-altered, section.

From 432 m upwards the sequence is more complete; hardgrounds are rare, but minor omission surfaces are common between 400 m and 432 m. Above 432 m, through most of the Coniacian and younger strata, the $^{87}Sr/^{86}Sr$ increases steadily to the top of the chalk section just above the Campanian/Maastrichtian boundary. Although the general rate of increase approximates to linearity, the rate of increase per metre of sediment is slightly lower in the Lower Campanian than either above or below this level. This is ascribed to a faster rate of sedimentation in the Lower Campanian. Above 432 m variations of $^{87}Sr/^{86}Sr$ in nannofossil chalk and macrofossils show very similar trends, with $^{87}Sr/^{86}Sr$ of the macrofossils being consistently some $20–30 \times 10^{-6}$ lower than $^{87}Sr/^{86}Sr$ of the nannofossil matrix. The inflexion at 226–240 m, and the flat part of the isotope curve between 432 m and 470 m are discussed above.

Where the biostratigraphy is poorest, the $^{87}Sr/^{86}Sr$ stratigraphy is best developed i.e. has the steepest gradient. Some of the standard Tethyan nannofossil marker species are, unfortunately, absent from UK strata for a substantial part of this interval (sub-zones CC18-22B, Fig. 2). Nevertheless, Boreal subdivisions of the standard calcareous nannofossil zones have recently been recognised (Burnett 1990) and these are shown on Fig. 2. Above 432 m eight macrofossil zones are recognized, although finer subdivision is provided in the wide B. mucronata zone by lithological subdivisions, which are held to equate closely to biostratigraphic subdivisions (Fig. 2).

Numerical ages for Trunch

We have added numerical ages to stage boundaries (Fig. 2) according to Hallam et al. (1985), Odin (1985) and Harland et al. (1989). The approximately linear increase in $^{87}Sr/^{86}Sr$ that occurs through the part of our section that is diagenetically least altered (above 432 m, from the early Coniacian to the early Maastrichtian) suggests that neither the temporal evolution of marine $^{87}Sr/^{86}Sr$, nor the sedimentation rate of the Chalk, varied greatly during this period. Near-perfect linearity in $^{87}Sr/^{86}Sr$ with thickness of section is shown by other Late Cretaceous sections (McArthur et al. in press). In view of this, numerical ages may be assigned to other biostratigraphic boundaries by interpolation, assuming a constant $\triangle^{87}Sr/^{86}Sr/Ma$. We have not made such interpolations owing to the large uncertainties on the ages of stage boundaries. Instead, we have estimated the relative duration of the macrofossil zones by assuming a linear $\triangle^{87}Sr/^{86}Sr/Ma$ from the early Coniancian onwards and adjusting zonal widths so that they fit a linear $\triangle^{87}Sr/^{86}Sr/metre$ of section. The relative adjusted thickness of sediment for each zone therefore represents the relative duration of the zones. The relative durations are given in Table 4.

Between the Campanian/Maastrichtian boundary, at 72 ± 2 Ma, and the Turonian/Coniacian at 88 ± 2 Ma (Hallam et al. 1985; Odin 1985) $^{87}Sr/^{86}Sr$ increase by 400×10^{-6}. Thus, during this interval the rate of increase in $^{87}Sr/^{86}Sr$ is between 20 and 33×10^{-6} Ma^{-1}, the range deriving from the errors on the numerical dates. The mean rate is 25×10^{-6} Ma^{-1}. For the purpose of correlation, the error in analysis is 9×10^{-6} (see methods section) at one standard deviation. Combining errors from sample and isotope curve gives a combined error of 18×10^{-6}, giving a possible time resolution in correlation of ± 0.8 Ma.

Table 4. *Relative durations of macrofossil zones and stages, based on a linear fit of sediment thickness to* $^{87}Sr/^{86}Sr$

Zone		Stage	
B. mucronate	100	Campanian	100
G. quadrata	14	(Upper)	(83)
O. pilula	7	(Lower)	(17)
M. testudinarius	13	Santonian	31
U. socialis	5	Coniacian	24
M. coranguinum	48		
M. cortestudinarium	21		

Depths less than 432 m only. Macrofossil zones relative to an arbitrary value of 100 for the B. mucronata Zone. Stage durations relative to 100 for the whole Campanian.

Correlation in the Late Cretaceous

For much of the Upper Cretaceous Chalk of the UK refined bio- and lithostratigraphy (Bailey 1984; Mortimore 1986; Robinson 1986; Gale *et al*. 1988) provides a stratigraphic resolution that is at least equal to that theoretically possible with ^{87}Sr/^{86}Sr. We have therefore tested the quality of our data for correlation in the Late Cretaceous by comparing biostratigraphic and isotope correlations between Norfolk and Studland Bay in Dorset, 350 km to the southwest.

Preservational quality of samples from Studland Bay

Samples from Studland Bay are not silicified and the macrofossils preservation appeared excellent; in particular the inoceramid showed excellent preservation of the finest prismatic shell-structure. This observation is commensurate with the extreme softness of the strata at Studland Bay, which have been buried to an effective depth of less than 500 m. Diagenesis at Studland Bay has affected samples less than it has at Trunch, and in our view had had a negligible affect on ^{87}Sr/^{86}Sr. This view is supported by the high degree of coherence for the isotopic data for samples from Studland Bay, particularly the agreement between nannofossil/ macrofossil pairs. It does not seem reasonable to suppose that diagenetic alteration could affect equally the ^{87}Sr/^{86}Sr in the solid calcite guard of a belemnite and the soft adhering nannofossil chalk, whilst leaving the prismatic inoceramid structure excellently preserved in its delicate, fine-detail, and the chalk uncemented.

Correlation between Trunch and Studland Bay

The samples (macrofossil and nannofossil) from a single stratigraphic level at Studland Bay have ^{87}Sr/^{86}Sr of 0.707588 ± 10 (Table 3). The coherence of this data confirms our view that macrofossils and nannofossils (the matrix) at Studland Bay preserve the original ^{87}Sr/^{86}Sr value and have not undergone re-equilibration of ^{87}Sr/^{86}Sr during diagenesis. From Fig. 5 an ^{87}Sr/^{86}Sr value of 0.707588 ± 10 (Table 3) correlates to the Trunch borehole at two levels. For macrofossil data the level is 175 ± 18 m, which correlates to an interval that includes the Weybourne/Eaton Chalk and the upper half of the Basal Mucronata Chalk, all within the *Belemnitella mucronata* Zone (Fig. 1). Using nannofossil data the Studland Bay strata correlate to the Trunch borehole at 195 ± 18 m,

which equates to an interval that includes the lower two thirds of the Basal Mucronata Chalk and about 8 m of underlying Upper Chalk. Both correlations agree reasonably well with the biostratigraphic correlation, but the macrofossil correlation appears to agree better, the central correlative level of 175 m being only 5 m below the base of the Weybourne/Eaton Chalk. Our data suggest that stratigraphic correlation with Sr isotopes in a well-studied basin can approach the quality attainable by the best biostatigraphic criteria.

Modelling the Sr-isotope curve

Variations in the ^{87}Sr/^{86}Sr of marine Sr reflect mainly the dual effects of changing fluxes, and changing ^{87}Sr/^{86}Sr, of hydrothermal volcanism at mid-ocean ridges and riverine supply (Elderfield 1986; Veizer 1989). The predominant influence is commonly held to be riverine supply (Palmer & Elderfield 1985), and, as a consequence, both climate (Capo & DePaolo 1990) and tectonics (Raymo *et al*. 1988; Hodell *et al*. 1989; Hodell 1991; Palmer & Edmond 1989), which affect riverine supply, have been hypothesized to be the dominant control on the rate and direction of change of ^{87}Sr/^{86}Sr for marine Sr. The specific nature of these postulated controls make them difficult to document in the geological record, particularly one as old as the Cretaceous, so we refrain from speculating on controls on the shape of the Late Cretaceous curve. We merely point out that there appears not to be a correlation between the published variations in sea level (Haq *et al*. 1987) and the slope, or points of inflexion, shown by our curve.

The increase of 400×10^{-6} in ^{87}Sr/^{86}Sr between the Campanian/Maastrichtian boundary (72 ± 2 Ma) and the Turonian/Coniacian boundary (88 ± 2 Ma) gives a temporal rate of increase in ^{87}Sr/^{86}Sr of between 20 and 33×10^{-6} Ma^{-1}. We speculate that if these rates of increase were to persist through the Maastrichtian, the ^{87}Sr/^{86}Sr at the Cretaceous/Tertiary boundary would be between 0.70840 and 0.707920, if this boundary is assigned an age of 66 ± 2 Ma. With mean values of 66 Ma and $\triangle^{87}$Sr/^{86}Sr/Ma of 25×10^{-6} the value would be 0.707870.

Summary

Isotopic analysis of the English Chalk and contained macrofossils shows that marine ^{87}Sr/^{86}Sr decreases from 0.70775 in the Cenomanian to 0.70730 in the middle Turonian before increasing in a near-linear manner to >0.70775 in the early Early Maastrichtian. Hardground for-

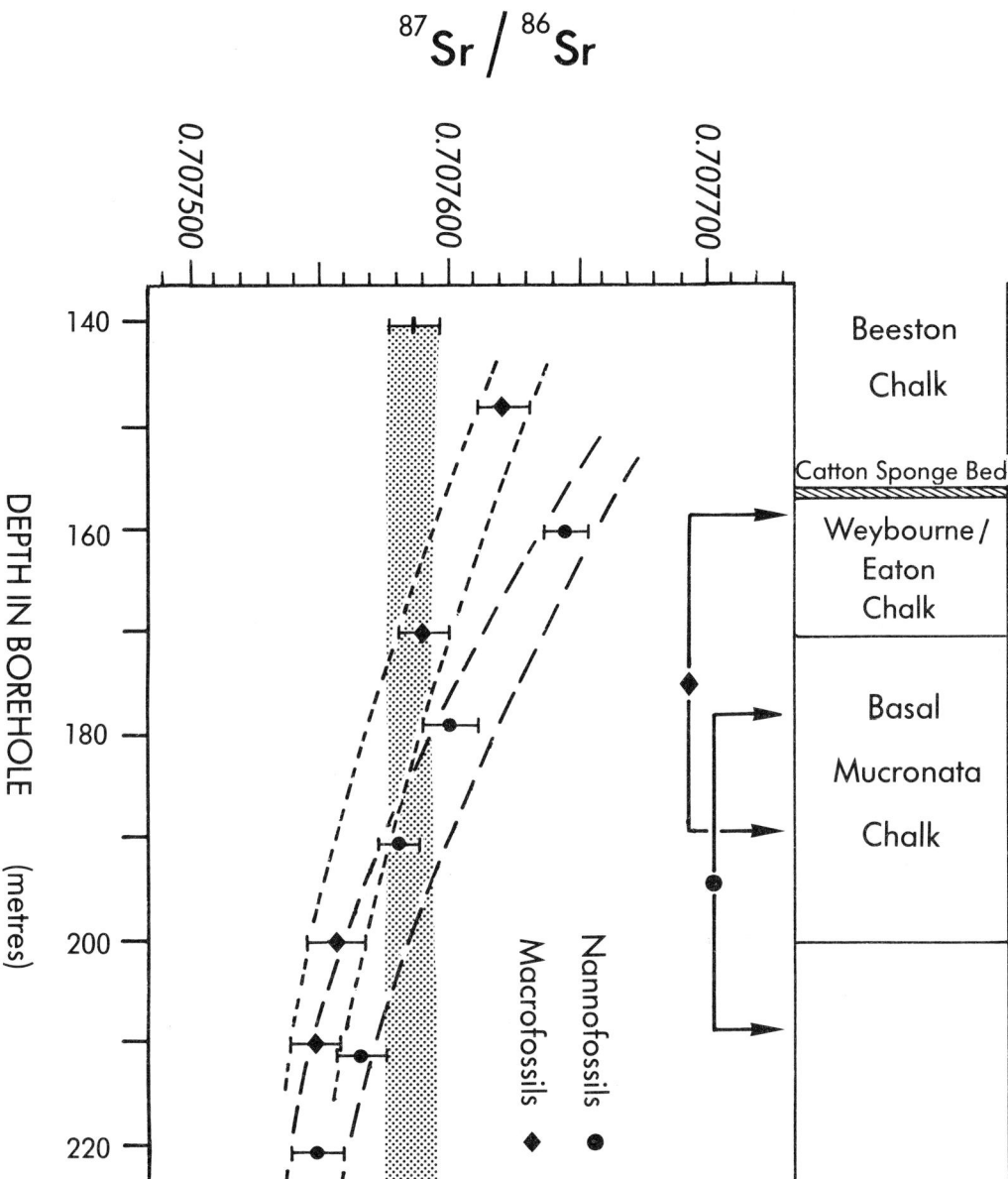

Fig. 5. Sr-isotopic correlation between Norfolk and Studland Bay, Dorset.

mation disturbs the $^{87}Sr/^{86}Sr$ values between 432 m and 470 m, and possibly also between 210 m and 250 m. During the Coniacian–Maastrichtian the rate of increase is between 20 and $33 \times 10^{-6} \, Ma^{-1}$. Macrofossils yield $^{87}Sr/^{86}Sr$ that are $20–30 \times 10^{-6}$ lower than nannofossil data. Within the limits of analytical error the

isotopic correlation between Chalk strata in Dorset and Norfolk agrees with the biostratigraphic correlation. Given the quality of the correlations we demonstrate that the English Chalk, and its macrofossils in particular, appear, for the most part, to be good preservers of Late Cretaceous marine $^{87}Sr/^{86}Sr$.

This work was supported by NERC Research Grant GR3/7716. The British Geological Survey are thanked for allowing us to sample the Trunch Borehole and for providing us with access to stratigraphic information. The macrofossil biostratigraphy of the Trunch Borehole by A. A. Morter is published with permission of the Director of BGS. Particular thanks are due to A. Whittaker, R. G. Thurrell, B. Owens and C. Wheatley of BGS for their collaboration and assistance. Micropalaeontological work was supported by a NERC Research Studentship to J. Burnett and NERC Research Grant GR3/6767 to J. M. Hancock, W. J. Kennedy and A. R. Lord. The Radiogenic and Stable Isotope Laboratories at RHBNC are supported, in part, by the University of London as Intercollegiate Facilities. We thank G. Ingram for assistance with the isotopic determinations and A. O. Osborn for the chemical data on Ca and Sr. C. Stuart and J. Baker drew the diagrams and M. Gray provided photographic assistance. K. G. Miller, P. Sugarman, H. Elderfield and one anonymous reviewer are thanked for helpful reviews.

References

BAILEY, J. W. 1984. Biostratigraphic criteria for the recognition of the Coniacian to Maastrichtian Stage boundaries in the Chalk of North West Europe, with particular reference to southern England. *Bulletin of the Geological Society of Denmark*, **33**, 31–39.

BIRKELUND, T., HANCOCK, J. M., HART, M. B., RAWSON, P. F., REMANE, J., ROBASZYNSKI, F., SCHMID, F. & SURLYKE, F. 1984. Cretaceous stage boundaries; proposals. *Bulletin of the Geological Society of Denmark*, **33**, 3–20.

BATH, A. H. & EDMUNDS, W. M. 1981. Identification of connate water in interstitial solution of chalk sediment. *Geochimica et Cosmochimica Acta*, **45**, 1449–1461.

BROMLEY, R. G. 1979. Chalk and Bryozoan limestone facies, sediments and depositional environments. *In*: BIRKLUND, T. & BROMLEY, R. G. (eds) *Cretaceous–Tertiary boundary events. 1. The Maastrichtian and Danian of Denmark.* Univ. Copenhagen, 16–35.

BURNETT, J. A. 1988. *North-West European Late Cretaceous calcareous nannofossils: biostratigraphy and selected evolutionary lineages.* PhD thesis, University College London.

—— 1990. A new nannofossil zonation scheme for the Boreal Campanian. *INA Newsletter*, **12/3**, 67–70.

BURKE, W. H., DENISON, R. E., HETHERINGTON, E. A., KOEPNICK, R. B., NELSON, H. F. & OTTO, J. B. 1982. Variation of $^{87}Sr/^{86}Sr$ throughout Phanerozoic time. *Geology*, **10**, 516–519.

CAPO, R. J. & DEPAOLO, D. J. 1988. Seawater strontium isotopic variations from 2.5 million years ago to the present. *Science*, **249**, 51–55.

CLAUSER, S. 1987. Évolution de la composition isotopique de l'oxygène des carbonates durant le Campanian–Maastrichtian. Données préliminaires de la série de Bidart (Pyrénées-Atlantiques). *Comptes Rendus de l'Acadmie des Sciences, Paris*, **304**, Serie II, 11, 579–584.

DEPAOLO, D. J. & FINGER, K. L. 1991. High resolution strontium-isotope stratigraphy and biostratigraphy of the Miocene Monterey Formation, central California. *Geological Society of America Bulletin*, **103**, 112–124.

ELDERFIELD, H. 1986. Strontium isotope stratigraphy. *Palaeogeography, Palaeoclimatology Palaeoecology*, **57**, 71–90.

GALE, A. S. & WOODROOF, P. B. 1981. A Coniacian ammonite from the 'Top Rock' in the Chalk of Kent. *Geological Magazine*, **118**, 557–560.

——, WOOD, C. J. & BROMLEY, R. G. 1988. Lithostratigraphy and marker bed correlation of the White Chalk (Late Cenomanian-Campanian), southern England. *Mesozoic Research*, **1**, 107–118.

GALLOIS, R. W. & MORTER, A. A. 1975. *East Anglia and South-East England District. Mundesley (132) sheet. Trunch Borehole (TG 2933 3455).* Institute of Geological Sciences Report 76/10, 8–10.

HANCOCK, J. M. 1975. The petrology of the Chalk. *Proceedings of the Geologists' Association*, **86**, 499–535.

HALLAM, A., HANCOCK, J. M., LaBRECQUE, J. L., LOWRIE, W. & CHANNELL, J. E. T. 1985. Jurassic to Paleogene: Part 1 Jurassic and Cretaceous geochronology and Jurassic to Paleogene magnetostratigraphy. *In*: SNELLING, N. J. (ed.) *The Chronology of the Geological Record*. Geological Society, London, Memoir, **10**, 118–140.

HAQ, B. U. 1984. Paleoceanography: A synoptic overview of 200 million years of Ocean history. *In*: HAQ, B. U. & MILLIMAN, J. (eds) *Marine Geology of the Arabian Sea and coastal Pakistan.* van Nostrand Reinhold, NY, 201–231.

——, HARDENBOL, J. & VAIL, P. R. 1987. Chronology of fluctuating sea levels since the Triassic. *Science*, **235**, 1156–1167.

HARLAND, W. B., ARMSTRONG, R. L., COX, A. V., CRAIG, L. E., SMITH, A. G. & SMITH, D. G. 1989. *A Geologic Timescale 1989*. Cambridge University Press.

HESS, J., BENDER, M. L. & SCHILLING, J-G. 1986. Evolution of the ratio of strontium-87 to strontium-86 in seawater from Cretaceous to Present. *Science*, **231**, 979–984.

HODELL, D. A. 1991. Variations in the strontium isotopic composition of seawater during the Neogene. *Geology*, **19**, 24–27.

——, MUELLER, P. A., McKENZIE, J. A. & MEAD, G. A. 1989. Strontium isotope stratigraphy and geochemistry of the late Neogene ocean. *Earth and Planetary Science Letters*, **92**, 165–178.

KOEPNICK, R. B., BURKE, W. H., DENISON, R. E., HETHERINGTON, E. A., NELSON, H. F., OTTO, J. B. & WAITE, L. E. 1985. Construction of the seawater $^{87}Sr/^{86}Sr$ curve for the Cenozoic and Cretaceous: supporting data. *Chemical Geology (Isotope Geosciences Section)*, **58**, 55–81.

KOWALLIS, B. J., CHRISTIANSEN, E. H. & DEINO, A. 1989. Multi-characteristic correlation of Upper Cretaceous volcanic ash beds from southwestern

Utah to central Colorado. *Utah Department Natural Resources (Utah Geological & Mining Survey) Miscellaneous Publication*, 89–5, 22pp.

KRALIK, K. 1984. Effects of cation-exchange treatment and acid leaching on the Rb-Sr system of illite from Fithian, Illinois. *Geochimica et Cosmochimica Acta, 48*, 527–533.

LILLIGRAVEN, J. A. 1991. Stratigraphic placement of the Santonian-Campanian boundary (Upper Cretaceous) in the North American Gulf Coastal Plain and Western Interior, with implications to global geochronology. *Cretaceous Research, 12*, 115–136.

—— & OSTRESH, L. M. JNR. 1990. Late Cretaceous (earliest Campanian/Maastrichtian) evolution of western shorelines of the North American Western Interior Seaway in relation to known mammalian faunas. *In*: BOWN, T. M. & ROSE, K. D. (eds) *Dawn of age of mammals in the northern part of the Rocky Mountain interior, North America*. Geological Society of America Special Paper, **243**, 1–30.

MCARTHUR, J. M., SAHAMI, A. R., THIRLWALL, M. F., OSBORN, A. O. & HAMILTON, P. J. 1990. Dating phosphogenesis with Sr isotopes. *Geochmica of Cosmochimica Acta, 54*, 1343–1351.

——, GALE, A. S., KENNEDY, W. J., CHEN, M. & THIRLWALL, M. F. 1992. Strontium–isotope stratigraphy in the Late Cretaceous: international correlation of the Campanian/Maastrichtian boundary. *Terra Nova, 4*, 332–345.

MILLER, K. G., FEIGENSON, M. D., KENT, D. V. & OLSON, R. K. 1988. Upper Eocene to Oligocene isotope ($^{87}Sr/^{86}Sr$, $\delta^{18}O$, $\delta^{13}C$) standard section, Deep Sea Drilling Project Site 522. *Paleoceanography, 3*, 223–233.

MORTER, A. A. 1984. Unpublished Report British Geological Survey.

MORTIMORE, R. N. 1986. Stratigraphy of the Upper Cretaceous White Chalk of Sussex. *Proceedings of the Geologists' Association*, **97**, 97–139.

ODIN, G. S. 1985. Concerning the numerical ages proposed for the Jurassic and Cretaceous geochronology. *In*: SNELLING, N. J. (ed.) *The Chronology of the Geological Record*. Geological Society, London, Memoir, **10**, 196–198.

PALMER, M. R. & ELDERFIELD, H. 1985. Sr isotope composition of sea water over the past 75 Myr. *Nature, 314*, 526–528.

—— & —— EDMOND, J. M. 1989. The strontium isotopic budget of the modern ocean. *Earth and Planetary Science Letters, 92*, 11–26.

PEAKE, N. B. & HANCOCK, J. 1961. The Upper Cretaceous of Norfolk. *Transactions of the Norfolk and Norwich Naturalists' Society, 19*, p293–339.

PERCH-NIELSON, K. 1979. Calcareous nannofossils from the Cretaceous between the North Sea and the Mediterranean. *I.U.G.S. Series A, 6*, 223–272.

—— 1985. Mesozoic calcareous nannofossils. *In*: BOLLI, H. M., SAUNDERS, J. B. & PERCH-NIELSEN, K. (eds) *Plankton stratigraphy* (1985)

RAWSON, P. F., CURRY, D., DILLEY, F. C., HANCOCK, J. M., KENNEDY, W. J., NEALE, J. W., WOOD, C. J. & WORSSAM, B. C. 1978. *A correlation of Cretaceous rock in the British Isles*. Geological Society, London, Special Report, **9**.

RAYMO, M. E., RUDDIMAN, W. F. & FROELICH, P. N. 1988. Influence of late Cenozoic mountain building on ocean geochemical cycles. *Geology, 16*, 649–653.

RICHTER, F. M. & DEPAOLO, D. J. 1987. Numerical models for diagenesis and the Neogene Sr isotopic evolution of seawater from DSDP Site 590B. *Earth and Planetary Science Letters, 83*, 27–38.

—— & —— 1988. Diagenesis and strontium isotope evolution of sea water using data from DSDP 590B and 575. *Earth and Planetary Science Letters, 90*, 382–394.

ROBINSON, N. D. 1986. Lithostratigraphy of the Chalk Group of the North Downs, southeast England. *Proceedings of the Geologists' Association, 97*, 141–170.

SCHLANGER, S. O., ARTHUR, M. A., JENKYNS, H. C. & SCHOLLE, P. A. 1987. The Cenomanian-Turonian oceanic anoxic event, I. Stratigraphy and distribution of organic carbon-rich beds and the marine ^{13}C excursion. *In*: BROOKS, J. & FLEET, A. (eds) *Marine Petroleum Source Rocks*. Geological Society, London, Special Publication, **26**, 371–399.

SISSINGH, W. 1977. Biostratigraphy of Cretaceous Nannoplankton. *Geologie en Mijnbouw, 56*, 37–65.

THIRLWALL, M. F. 1991. Long term reproducibility of multi-collector Sr and Nd isotope ratio analysis. *Chemical Geology (Isotope Geoscience Section)*, **94**, 85–104.

VEIZER, J. 1989. Strontium isotopes in seawater through time. *Annual Review of Earth and Planetary Science, 17*, 141–167.

WICKMAN, F. E. 1948. Isotope ratios: a clue to the age of certain marine sediments. *Journal of Geology, 56*, 61–66.

WOOD, C. J. & MORTER, A. A. In press. Biostratigraphy of the Trunch Borehole. *In*: ARTHURTON, R. S., BOOTH, S. J., MORIGI, A. N. & ABBOTT, M. A. W. (ed) *Geology of the Country around Great Yarmouth*. Memoir BGS, Sheet 162 (England and Wales).

Geochemical correlation of marl bands in Turonian chalks of the Anglo-Paris Basin

DAVID S. WRAY[1] & A. S. GALE[2]

[1] Geochemistry Unit, School of Earth Sciences, University of Greenwich, Walburgh House, Bigland Street, London E1 2NG, UK

[2] Department of Geology, Imperial College, Prince Consort Road, London, SW7 2BP and Department of Palaeontology, Natural History Museum, Cromwell Road, London SW7

Abstract: Geochemical analysis of a series of laterally correlative Middle and Upper Turonian marls collected from exposures in Kent, Sussex and on the French coast reveal that it is possible to chemically fingerprint individual marls. The fingerprint is derived from four elements (scandium, titanium, vanadium and yttrium) which are believed to be located within detrital clay minerals. The true potential of this method is demonstrated by applying it to a series of Turonian marls within and below the condensed succession overlying the London Platform and western edge of the Wessex Basin (the Chalk Rock and Spurious Chalk Rock). Previously, correlation of these marls with those in the main part of the Basin was difficult and tentative. Using the geochemical fingerprinting method it is possible to make accurate marl band correlations which demonstrate that, on the London Platform, much of the Middle and Upper Turonian is condensed into a series of closely spaced hardgrounds.

Marl bands in Turonian and younger chalks of the Anglo-Paris Basin are laterally continuous units of clay-rich (marly) chalk which can be correlated lithostratigraphically over thousands of square kilometres. They do not occur continuously throughout the Chalk, and are concentrated in the Turonian to Lower Coniacian, Upper Santonian to Lower Campanian, and Upper Campanian (see Fig. 1). This paper concentrates on marls in the Middle and Upper Turonian because of their excellent exposure in coastal sections on either side of the English Channel.

The marly chalk is either concentrated within a discrete (3 to 20 cm thick) bed termed a marl seam, or in a series of interwoven flaser-like wisps, individually less than 5 cm in thickness, and collectively termed a flaser marl (Figs 2–4). It must be stressed that this paper is not concerned with the origin and correlation of flaser chalks as described by Garrison and Kennedy (1977), since these are secondary features formed as a result of diagenetic compaction and dissolution.

Marls often show evidence of bioturbation: clay-rich chalk is commonly found infilling burrows below and above a seam, and sporadic white chalk filled burrows can be found cross-cutting them. The marly chalk filled burrows above and below a seam often show evidence of compaction and dissolution, making them similar in appearance to solution flasers (*sensu* Garrison & Kennedy 1977). There is little field evidence of dissolution within thick marl seams, and the absence of secondary lithification of white chalk above and below marl seams implies that carbonate solution during diagenesis, followed by local precipitation, was not an important process. Recent micropalaeontological studies (Leary & Wray 1989) have demonstrated that marls were not formed as a result of carbonate dissolution on the sea floor as had previously been proposed (Curry 1982; Ernst 1982), but rather that they reflect a sudden environmental change associated with an influx of non-carbonate material. Mineralogically they contain the same clay minerals as are found in white chalk above and below the seam (usually a combination of smectite and illite), although the relative proportions of the clays may vary. Their irregular stratigraphic distribution suggests that they are not directly related to Milankovitch-type orbital influences.

Over the last few years a series of papers has shown that marls are of great value in lithostratigraphic correlation (Jefferies 1963; Wood & Smith 1978; Bromley & Gale 1982; Bailey *et al.* 1983, 1984; Mortimore 1983, 1986; Mortimore & Wood 1986; Robinson 1986), and that individual marls can be correlated between sections in

From Hailwood, E. A. & Kidd, R. B. (eds), *High Resolution Stratigraphy*
Geological Society Special Publication, No. 70, pp. 211–226.

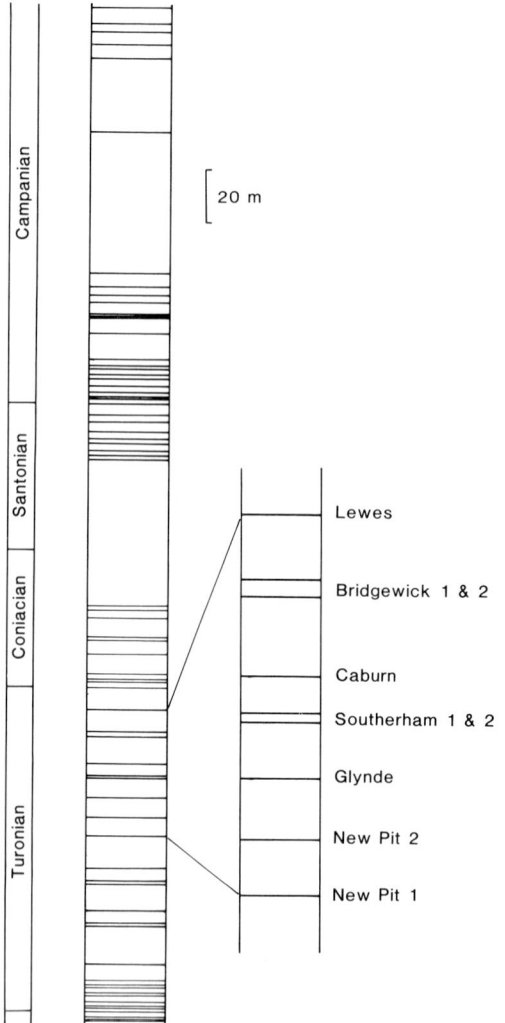

Fig. 1. Simplified log, based on sections in Sussex and the Isle of Wight, showing the stratigraphic distribution of marls within the Anglo-Paris Basin, and an expanded view of the Middle and Upper Turonian Marls discussed in this work (nomenclature from Mortimore 1986).

Fig. 2. Bridgewick Marls 1 and 2 (numbered) on the coast to the east of Dover. Note the presence of nodular chalks below and between the two bands; prominent flint bands occur below, between and above these marls. Dog is 68 cm high.

Kent, Sussex and the French coast (Mortimore & Pomerol 1987). More long-ranging correlations using marls have also been proposed, with attempts being made to correlate the Chalk of Lincolnshire and Yorkshire with that in the Anglo-Paris Basin (Mortimore & Wood 1986), and to correlate German and English chalks (Wood et al. 1984).

The aim of this paper is to demonstrate that marl bands can be recognized geochemically and to show that, using this technique, it is possible to extend the established lithostratigraphic correlation into the poorly correlated, condensed succession found on the London Platform and in Dorset.

Assumptions

As the carbonate composition of chalks remains reasonably constant throughout the Middle and Upper Chalk (Hancock 1975, table 1), work has concentrated on trying to characterise marls using elements found within the clay fraction. Table 1 shows that many of the major elements present in the clays are also present in other mineral phases (for example: silica is also present in quartz, iron is also present in limonite), complicating their use. Work therefore concentrated on trace elements which were likely to occur predominantly within clay minerals (Ba, Cu, Li, Na, Sc, Ti, V, Y, Zn and Zr).

Because the percentage of clay varies between marls, direct comparison of chemical data is

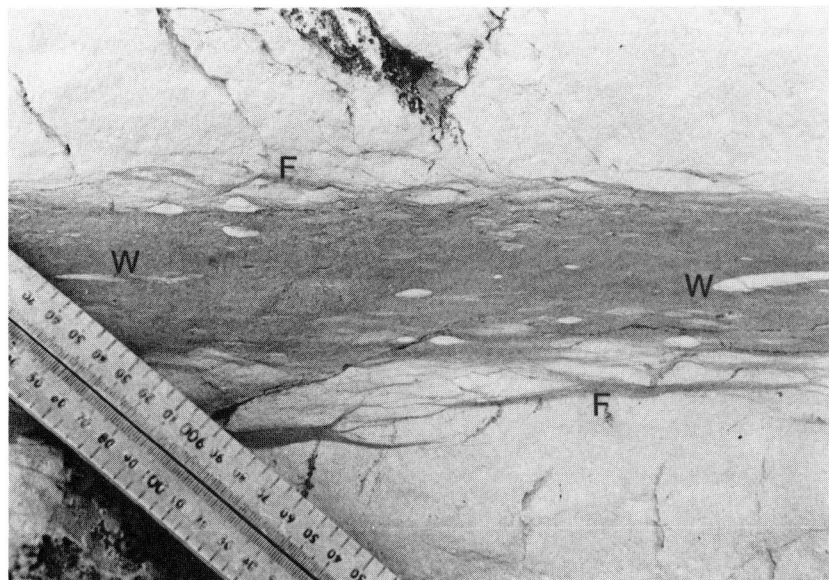

Fig. 3. A marl band containing a marl seam (New Pit Marl 2) on the French coast to the east of Dieppe. Note the flaser-like compacted burrows above and below the seam (labelled F) which are infilled with marly chalk, and the white chalk filled burrows within the seam (labelled W). Scale is in millimetres.

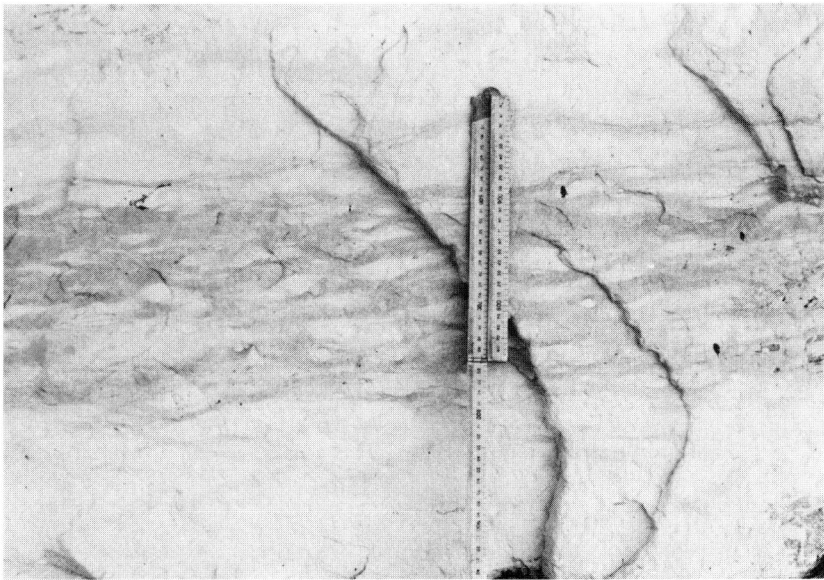

Fig. 4. A flaser marl (New Pit Marl 1) on the French coast to the east of Dieppe. Scale is in millimetres.

difficult. Examination of Table 1 shows that aluminium is concentrated in the clay minerals; a small percentage also occurs within glauconite, potassium feldspar and clinoptinolite, but all three minerals are only present in trace amounts, if at all (Weir & Catt 1965; Hancock

Table 1. *Chemical composition of the major and trace minerals found in marly chalks*

Major minerals in marly chalk
Calcite	$CaCO_3$
Carbonate-apatite	$Ca_5(PO_4,CO_3OH)_3(F,OH)$
Illite	$K_{1-1.5}Al_4[Si_{7-6.5}Al_{1-1.5}O_{20}](OH)_4$
Quartz	SiO_2
Smectite	$(\frac{1}{2}Ca,Na)_{0.7}(Al,Mg,Fe)_4(Si,Al)_8$ $O_{20}(OH)_4.nH_2O$

Trace minerals in marly chalk
Clinoptilolite	$(Ca,Na_2)[Al_2Si_7O_{18}].6H_2O$
Glauconite	$(K,Na,Ca)_{1.2-2.0}(Fe,Al,Mg)_4$ $[Si_{7-7.6}Al_{1-0.4}O_{20}](OH)_4.nH_2O$
Kaolinite	$Al_4[Si_4O_{10}](OH)_8$
Limonite	$FeO.OH.nH_2O$
Potassium feldspar	$(K,Na)[AlSi_3O_8]$
Pyrite	FeS_2

(Chemical formulae from Deer *et al.* 1966.)

1975; Wray 1990). Accepting that aluminium occurs predominantly within the clays, dividing the results of other clay-associated elements by the aluminium value of a sample will normalize the data and allow direct comparison between marls.

Geochemical correlation within the Anglo-Paris Basin

In order to evaluate the potential of marl band fingerprinting, a number of laterally correlative Middle and Upper Turonian marls were sampled at Akers Steps and Langdon Stairs, Kent (see sections in Robinson 1986 and Mortimore & Pomerol 1987), Beachy Head and pits surrounding Mount Caburn, Sussex (Lake *et al.* 1987), and the cliff section east of Dieppe on the French coast (between Puys and Belleville; see Mortimore & Pomerol 1987, figs 11, 12 and 13). Locations are shown in Fig 5. All samples were soaked in frequently changed, de-ionized water for two weeks to remove any seawater contamination, before being dried, powdered and dissolved using a HF/HClO$_4$ acid digestion similar to that described by Thompson & Walsh (1983). Analysis of the samples was carried out using an ARL 3510 Inductively Coupled Plasma – Atomic Emission Spectrometer based at the School of Earth Sciences, University of Greenwich (for details of operating conditions see Appendix).

Examination of the results showed that four of the trace elements (scandium, titanium, vanadium and yttrium) have similar intra-marl but different inter-marl aluminium normalized values. Plots of the results from Kent, France and Sussex can be seen in Fig. 6a & b, and a summary of the mean values and standard deviations is given in Table 2. Results for some of the marls are similar, but providing the stratigraphic order of the samples is remembered, many of the overlaps can be reconciled. There is no predictable stratigraphic pattern to the results, probably best exemplified by the markedly different results from Southerham Marls 1 and 2 which are separated by only two metres of sediment. These results demonstrate that it is possible to geochemically distinguish Middle and Upper Turonian marls within the Anglo-Paris Basin, and thereby add support to the lithostratigraphic correlations of Mortimore & Pomerol (1987).

Correlation of the Glynde Marl complex

It is often possible to recognise a series of thin flaser marls above the widely correlative Glynde Marl 1 *sensu* Mortimore & Pomerol (1987).

Table 2. *Mean values and standard deviations of the four elements found to be laterally constant within marls of the Anglo-Paris Basin*

Marls	Sc/Al_2O_3 $\bar{x}$	σ	TiO_2/Al_2O_3 $\bar{x}$	σ	V/Al_2O_3 $\bar{x}$	σ	Y/Al_2O_3 $\bar{x}$	σ
Lewes	0.40	0.13	0.032	0.002	6.3	1.09	8.0	0.62
Bridgewick 2	0.61	0.0	0.057	0.0	7.5	0.24	6.6	1.0
Bridgewick 1	0.47	0.07	0.030	0.002	5.3	0.58	6.4	1.0
Caburn	0.5	0.03	0.033	0.002	6.4	0.16	5.0	0.3
Southerham 2	0.79	0.002	0.107	0.002	13.1	1.25	3.9	0.56
Southerham 1	0.42	0.07	0.035	0.000	4.26	0.07	3.3	0.38
Glynde 1	0.33	0.09	0.033	0.002	4.6	1.24	2.3	1.05
New Pit 2	0.89	0.03	0.039	0.001	6.7	1.05	4.4	0.56
New Pit 1	1.00	0.05	0.051	0.001	8.0	0.41	5.6	0.49

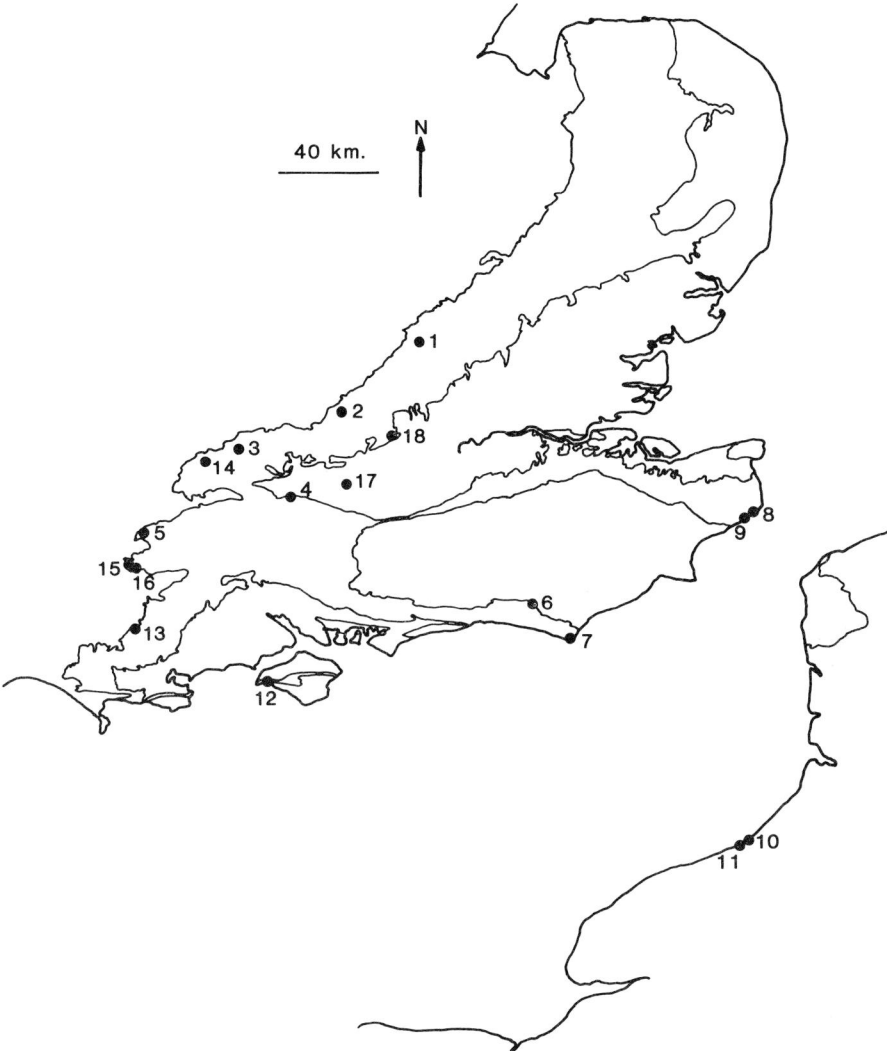

Fig. 5. Location map. 1, Kensworth, TL 017196; 2, Ewelme, SU 655893; 3, Fognam Farm, SU 296800; 4, Burghclere, SU 457589; 5, Beggars Knoll, ST 890506; 6, Mount Caburn, TQ 447089; 7, Beachy Head, TV 576953; 8, Langdon Stairs, TR 345425; 9, Akers Steps, TR 297394; 10, Section east of Dieppe; Belleville, 1.29.,55,51; 11, Section east of Dieppe; Puys, 1.36.,55,48; 12, Military Road and Compton Bay, Isle of Wight, SZ 362856; 13, Shillingstone Hill, SU 823098; 14, Ogbourne Maizey, SU 180716; 15, Mere, ST 808342; 16, Charnage Down, ST 836328; 17, Faircross Borehole; 18, Taplow Borehole.

Robinson (1986) noted a total of six on Akers Steps, and six can also be seen on the coast east of Dieppe. Samples of these marls from Kent and France have been collected and analysed (see Fig. 7a & b). The results obtained show that the correlative marls (Marl 1 at Puys and Marl 2 at Dover) aluminium-normalized values are quite distinct from the others, and that results from Marls higher in the complex are very variable. The plotted data show a broad scattering, making it difficult to propose suitable correlations for the higher marls between the two localities and, more importantly, many of the marls have scandium and titanium values similar

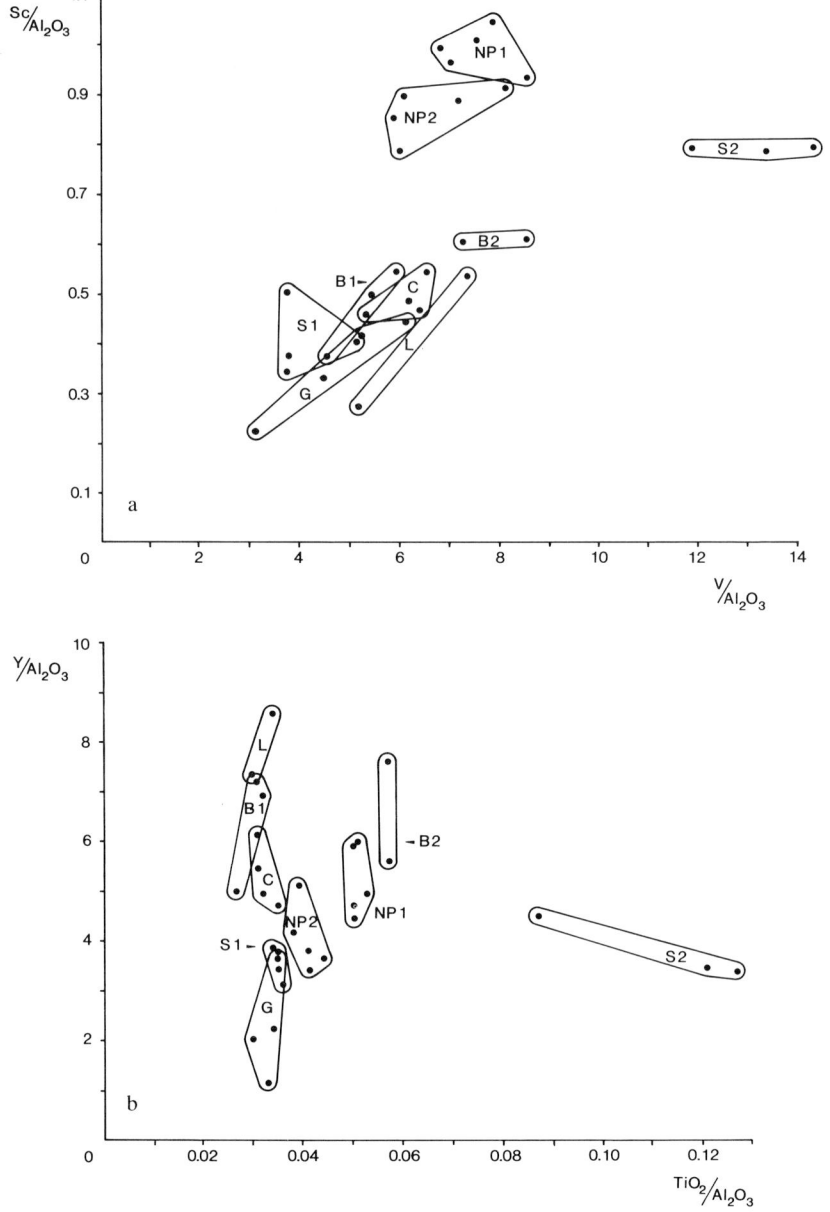

Fig. 6. Aluminium-normalized plots of the laterally constant elements from the main part of the Anglo-Paris Basin (Kent, France and Sussex), showing intra-marl similarities and inter-marl differences. (**a**) Scandium versus vanadium; (**b**) yttrium versus titanium. NP1, New Pit Marl 1; NP2, New Pit Marl 2; G, Glynde Marl 1 (*sensu* Mortimore & Pomerol 1987); S1, Southerham Marl 1; S2, Southerham Marl 2; B1, Bridgewick Marl 1; B2, Bridgewick Marl 2; L, Lewes Marl.

to those of the New Pit Marls, whilst yttrium and vanadium values are only slightly higher. Because of these difficulties, some caution must be exercised when proposing that an 'unknown' marl is a lateral equivalent of one of the New Pit Marls (see below).

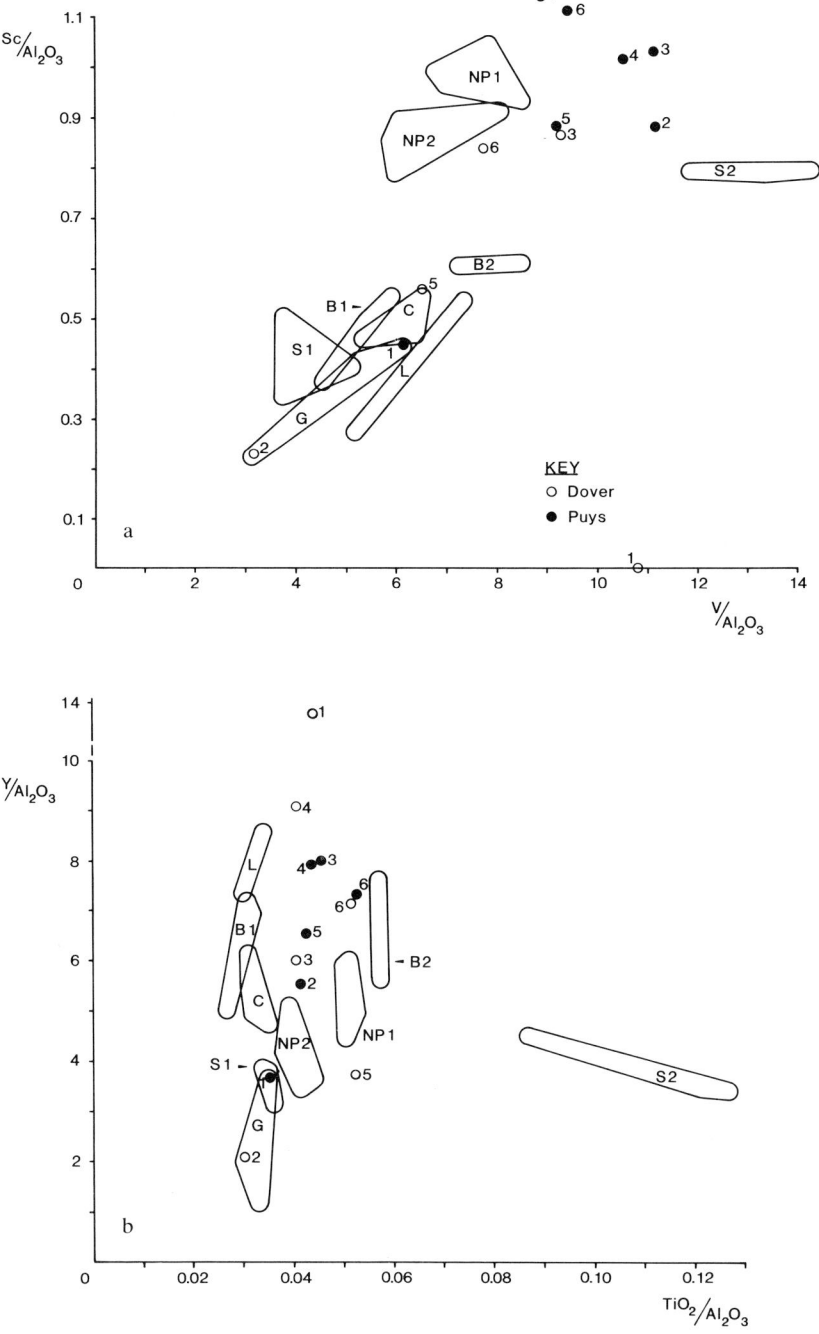

Fig. 7. Aluminium-normalized plots of marls from the Glynde Complex of Puys and Dover, demonstrating the marked difference between the correlative marl (Puys 1, Dover 2) and the higher marls (with the exception of Dover 5). The similarity between the higher marls and the New Pit Marls can also be seen, although yttrium values tend to be higher. Dover 1 has a default scandium value of 0 because it was below the detection limit of the analytical technique used. Key to the abbreviations can be found in the explanation to Fig. 6. (**a**) Scandium versus vanadium; (**b**) yttrium versus titanium.

Correlation of the Chalk Rock and Spurious Chalk Rock with the Anglo-Paris Basin succession

The ability to geochemically correlate individual Middle and Upper Turonian marl bands from geographically widely spaced localities is a useful tool in confirming the already established lithostratigraphic correlations. Potentially, its most important use is in the identification of marls which, at the present time, cannot easily be correlated with the main part of the Basin. Two such ares are the Chilterns and Dorset, where part of the Turonian is condensed into a series of lithified and mineralized hardgrounds known as the Chalk Rock and Spurious Chalk Rock.

The Chalk Rock

The Chalk Rock Member (*sensu* Gale *et al.* 1987) is a condensed unit consisting of a number of mineralized and phosphatized hardgrounds

(Fig. 8) which extend in a SW–NE direction from Dorset to Hertfordshire. Bromley & Gale (1982) described in detail the internal stratigraphy and correlation of individual hardgrounds, but did not attempt to demonstrate how the Member correlated with the main part of the Anglo-Paris Basin. They subdivided the Chalk Rock into three hardground suites, and noted

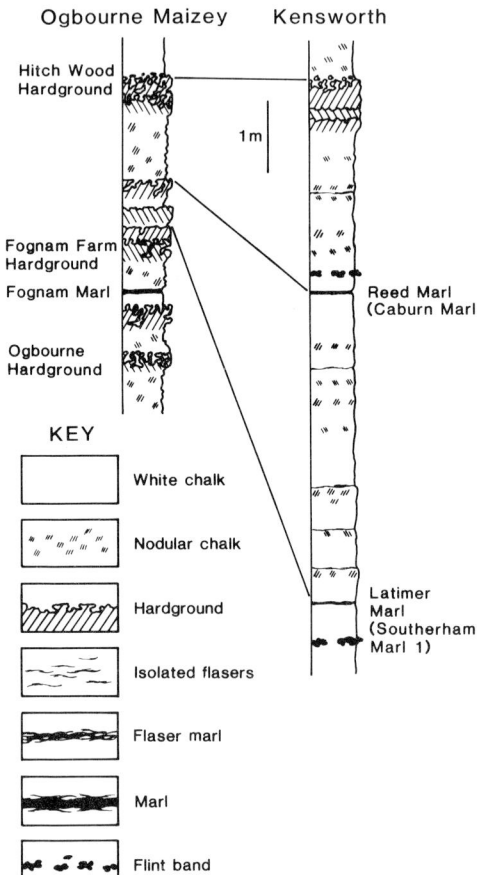

Fig. 9. Position of the Fognam, Latimer and Reed Marls within the Chalk Rock as illustrated by the sections at Kensworth (Hertfordshire) and Ogbourne Maizey (Wiltshire) according to Bromley & Gale (1982). The Reed and Latimer Marls were only described by Bromley and Gale in the more expanded successions found in Hertfordshire; the Fognam Marl extends throughout the Chalk Rock area except in west Wiltshire. Subsequent work by Mortimore & Wood (1986) and data contained within this paper indicates that the Latimer and Reed Marls correlate with the southern England Southerham Marl 1 and Caburn Marl.

Fig. 8. Chalk Rock hardgrounds at Fognam Farm. FF, Fognam Farm Hardground; LH, Leigh Hill Hardground; HW, Hitch Wood Hardground. Notebook at base is 10 cm × 8 cm.

the presence of three marls within it which they named the Fognam, Latimer and Reed Marls. The Fognam Marl occurs between the lower and middle hardground suites and is present over much of the area covered by the Chalk Rock. The Reed and Latimer Marls occur within the middle suite, but are only developed in expanded successions found in Hertfordshire (Fig. 9).

Previous attempts at correlation. Although Bromley & Gale (1982) made no attempt to

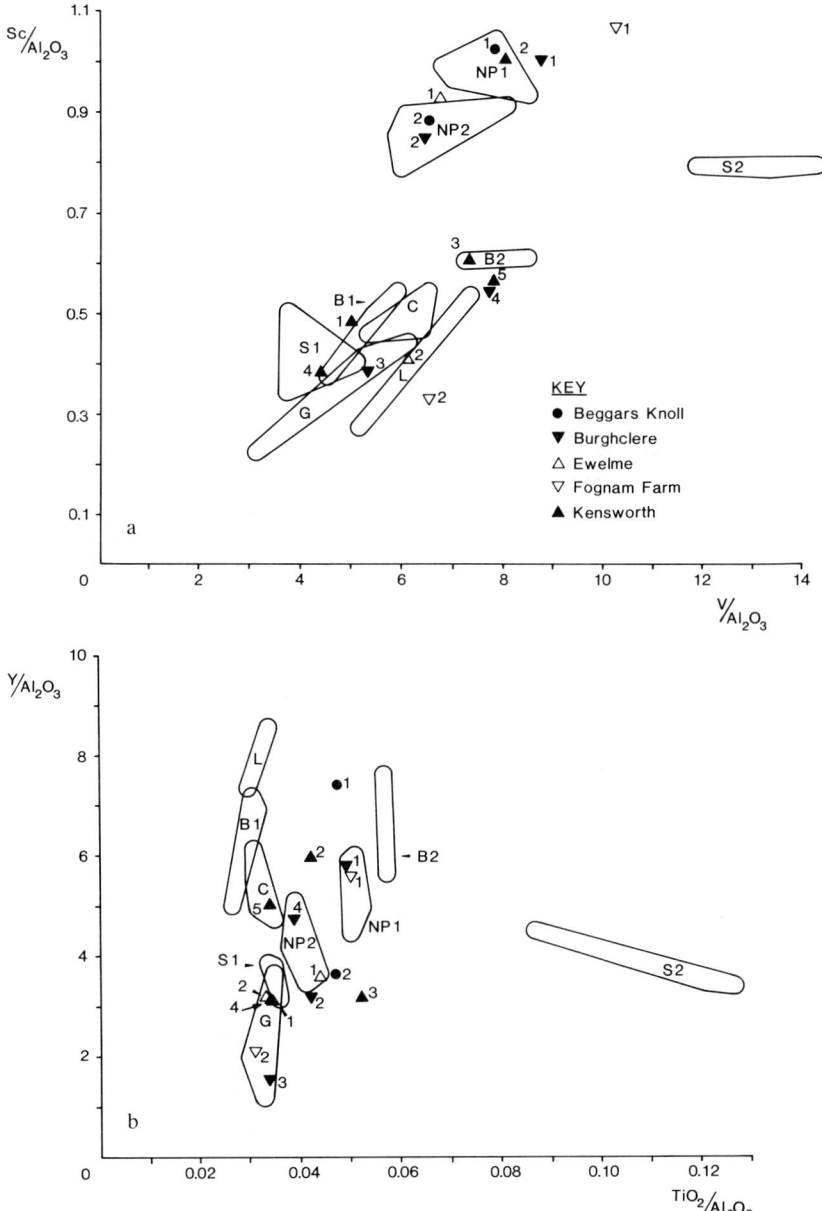

Fig. 10. Aluminium-normalized plots of marls collected within and below the Chalk Rock, see Fig. 6 for the key. (**a**) Scandium versus vanadium; (**b**) yttrium versus titanium.

correlate the Chalk Rock with expanded successions in the main part of the Basin, several
papers have attempted partial correlations:
Mortimore (1983, fig. 4a) proposed that at Mere
and Charnage Down (Witshire) Glynde Marl 1
occurred beneath the Chalk Rock, the Souther-
ham, Caburn, Bridgewick and Lewes Marls all
being condensed within it. Mortimore &
Pomerol (1987, fig. 6) proposed that in the
Faircross Borehole succession south of Reading
a Southerham Marl occurred below the Chalk
Rock, the Caburn Marl within it, and the
Bridgewick Marls just above it; they also sug-
gested that in the Taplow Borehole succession a
Southerham Marl occurred below the Chalk
Rock, and that the Caburn, Bridgewick and
Lewes Marls were all condensed within it.
Mortimore & Wood (1986) proposed that at
Kensworth the Latimer Marl was equivalent to
Southerham Marl 1, while the Reed Marl cor-
related with the Caburn Marl. Mortimore (1987,
Fig. 5) proposed that at Beggars Knoll one of the
Southerham Marls occurred beneath the Chalk
Rock, whilst at Fognam Farm the Fognam Marl
was the equivalent of the Caburn Marl.

Localities studied and results

In an attempt to clarify the correlation of the
Chalk Rock with the Anglo-Paris Basin Turonian succession, marl seams from many of the
above sections and from a temporary exposure
near Burghclere were collected and analysed
using an identical method to that used on the
samples from the main part of the Anglo-Paris
Basin (see Fig. 12 for logs of the sections). The
aluminium-normalized Sc, Ti, V and Y results
for all the marls have been plotted on graphs
containing outlines of the results from the main
part of the Anglo-Paris Basin (Fig. 10a & b).

Burghclere (SU 457589). A temporary exposure
created during modifications of the A34 road
south of Burghclere presented a superb section
through much of the Middle and Upper Turonian, including the Chalk Rock. Four marls,
three from below the Chalk Rock and one from
within it, were collected and analysed. The
lower three exhibited distinctive features in the
field, allowing them to be tentatively identified as
(in ascending order) New Pit Marls 1 and 2 and
Glynde Marl 1. The geochemical signatures
obtained confirmed the field identifications (see
graphs; points 1 to 3 in ascending order). The
highest marl (4 on the graphs), predicted in the
field to be a Southerham Marl, plots some

distance away from the field of either of the
Southerham Marls, and appears far more likely
to be the lateral equivalent of the Caburn Marl.
This proposition is supported by the results of
analysis of the Reed Marl from Kensworth (see
below), which has near identical scandium and
vanadium values.

Fognam Farm (SU 296800). Geochemical analysis of the Fognam Marl from its type locality
(point 2 on the graphs) produces a fingerprint
similar to that of Glynde Marl 1, although the
vanadium value is a little high. A thick marl
occurring approximately 1 m below the Ogbourne Hardground was also sampled (point 1
on the graphs), producing a signature similar to
that of New Pit Marl 1, although once again the
vanadium result is slightly high.

Ewelme (SU 655893). A sample of the Fognam
Marl from this locality produces a fingerprint
reminiscent of Glynde Marl 1 (point 2 on
graphs), agreeing with the Fognam Farm result.
A second, lower marl was also sampled and gives
a New Pit Marl 2 signature (point 1 on graphs).
The exact position of this marl below the Chalk
Rock is difficult to ascertain due to poor exposure, but it seems unlikely that the distance
between the two marls is more than 3.5 m.

Beggars Knoll (ST 890506). Samples were collected from two marl seams occurring below the
Chalk Rock. The results obtained imply that the
higher marl (point 2 on graphs) is the equivalent
of New Pit Marl 2, and the lower (point 1 on
graphs) the equivalent of New Pit Marl 1
(although the yttrium result is higher than
expected). It is unlikely that these marls are part
of the Glynde Marl complex (see above) because
overlying them is the Ogbourne Hardground
(Bromley & Gale 1982), and at Fognam Farm
the lateral equivalent of Glynde Marl 1 (the
Fognam Marl) occurs above Ogbourne.

Kensworth (TL 017196). Samples of both the
Latimer (point 4 on graphs) and Reed (point 5
on graphs) Marls have been collected and
analysed. When plotted, the results agree with
the proposal made by Mortimore & Wood
(1986), i.e. that Latimer is the lateral equivalent
of Southerham Marl 1 and Reed is the equivalent of the Caburn Marl. Samples were also
collected from three marls further down the
succession: the lowest marl collected (1) produces a fingerprint similar to Glynde Marl 1, the
middle one (2) plots as a New Pit Marl, and the
highest (3) gives no clear signal. The most likely
interpretation of this data is that marl 1 is Glynde

Marl 1, and marls 2 and 3 are representatives of higher Glynde Marls.

The Spurious Chalk Rock

The approximately E–W-trending Spurious Chalk Rock crops out around Shillingstone and Piddletrenthide in inland Dorset, on the Dorset coast, and on the Isle of Wight. Bromley & Gale (1982) proposed that it was a lateral extension of one of the lower Chalk Rock hardgrounds (probably Ogbourne but possibly Fognam Farm) and that the second marl above the Spurious Chalk Rock at Shillingstone Hill

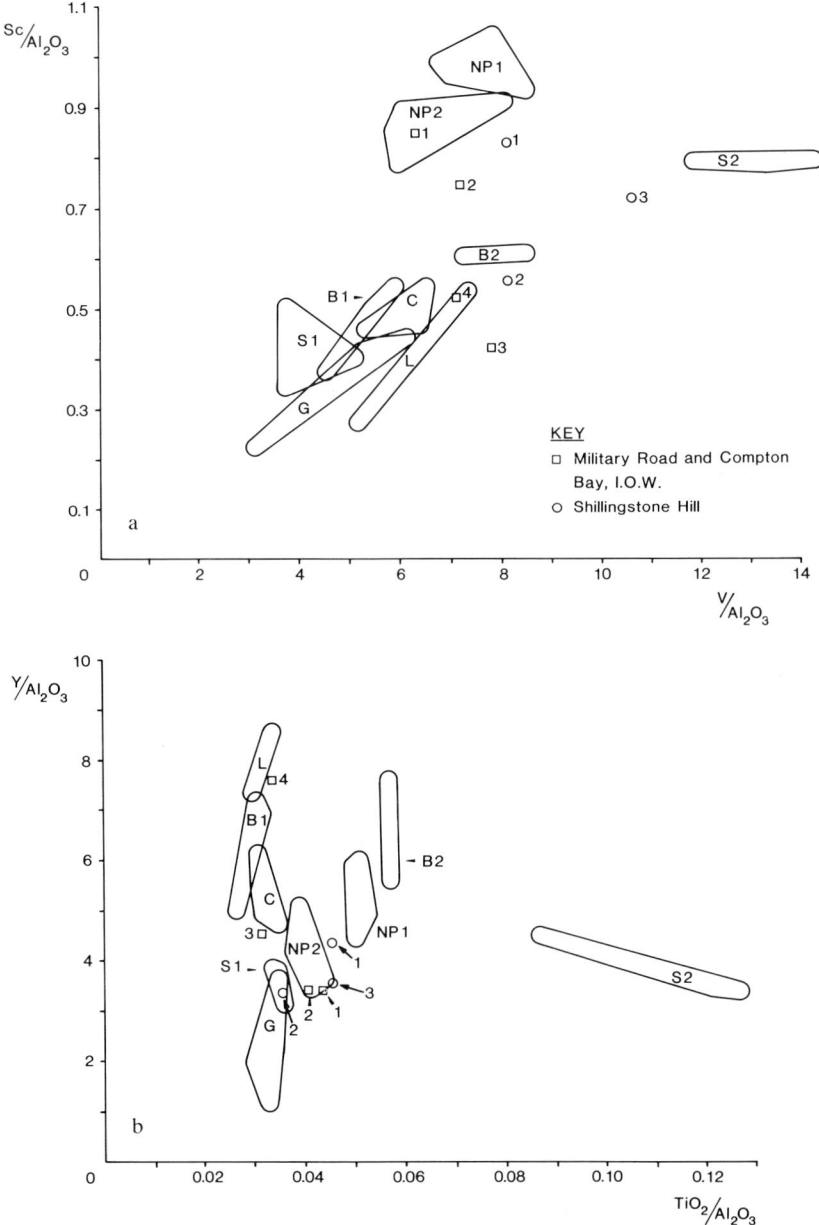

Fig. 11. Aluminium-normalized plots of marls collected within and below the Spurious Chalk Rock; see Fig. 5 for key. (**a**) Scandium versus vanadium; (**b**) yttrium versus titanium.

correlated with the Fognam Marl. Correlations with the main Anglo-Paris Basin successions have been made for marls above and below the Spurious Chalk Rock by several authors. Mortimore (1983) and Mortimore & Wood (1986, fig. 2.3) argued that on the Isle of Wight, a marl found a short distance above the Spurious Chalk Rock (the black marl-band of Rowe 1908) correlated with Southerham Marl 1 and the subsequent marl (the grey marl-band of Rowe 1908) was equivalent to the Caburn Marl. Below the Spurious Chalk Rock they claimed to find Glynde Marl 1. Their correlations should be treated with some caution as they are based on a numeric correlation of marl bands found in Sussex, coupled with the occurrence above the grey marl-band of *Bicavea rotaformis*, a bryozoan which is uncommon above the Caburn Marl in Sussex and apparently not present at that horizon at Dover (C. J. Wood pers. comm.). Indeed, on purely lithological criteria the black marl-band is more likely to correlate with Glynde Marl 1, as it is thin, dark grey, and relatively plastic, a combination of features which only Glynde Marl 1 exhibits in Turonian chalks of southern England. At Shillingstone Hill, Mortimore (1987) again proposed that Glynde Marl 1 occurred below the Spurious Chalk Rock, while the two marls above the hardground were inferred to correlate with the Southerham Marls.

Localities studied and results

Two localities were sampled: Shillingstone Hill and the Compton Bay–Military Road sections on the Isle of Wight. The results obtained have again been plotted on graphs which have outlines of the results from the main part of the Anglo-Paris Basin (Fig. 11a & b).

Shillingstone Hill (SU 823098). Three marls were collected which straddle the Spurious Chalk Rock. Marl 1, from 3.2 m below the hardground produces a fingerprint which suggests that it is one of the New Pit Marls, probably Marl 2. A problem with this interpretation is that lithologically the marl is very different from either of the New Pit Marls, being thin and flasered (characteristically both New Pit Marls are thick and well developed). The possibility therefore exists that it is a higher marl from within the Glynde Complex (see above). Both these possibilities differ from the proposal made by Mortimore (1987), further work to solve this problem was hampered by a 2–3 m high slope of scree which obscured the section below.

The two marls collected and analysed from above the hardground also produce less than ideal results. The lower of the two (point 2 on graphs) produces values which suggest that it is probably Southerham Marl 1, although the vanadium value is markedly high. Marl 3 gives rise to a plot reminiscent of Southerham Marl 2, the only unfortunate exception being titanium which has a much lower value than might be expected. Overall, although not perfect, the geochemical results from these marls seem to imply that they could be the lateral equivalents of the Southerham Marls, agreeing with Mortimore (1987).

Military Road and Compton Bay, Isle of Wight (SZ 362856). Four samples were collected across the Spurious Chalk Rock; two from below, and the black marl-band and grey marl-band from above.

The two marls from below (points 1 and 2 on graphs) both give a fingerprint reminiscent of the New Pit Marls, although it is difficult to differentiate them further. As with the sample from below the Spurious Chalk Rock at Shillingstone, it is also possible that the marls are from the higher part of the Glynde Complex.

The black marl-band (point 3 on graphs) also produces a confusing signature with both vanadium and yttrium values being higher than expected if it were either Glynde Marl 1 or Southerham Marl 1. The grey marl-band (point 4 on graphs) again has vanadium and yttrium values which are higher than usual for the Caburn Marl. Treating the marls as a pair, the consistently high results for two of the elements may be a reflection of the extensive weathering suffered by the Military Road section. This does not resolve the problem regarding the correlation of the black marl-band because of the similar geochemical fingerprint of Southerham Marl 1 and Glynde Marl 1.

Summary of Marl Band correlations within the Chalk Rock and Spurious Chalk Rock

Most samples plot in or close to the fields produced by unequivocally identified marls found in the main part of the Basin.

The three marls within and below the Chalk Rock named by Bromley & Gale (1982) can be correlated with the main part of the Anglo-Paris Basin as follows:

>   Reed Marl = Caburn Marl
>   Latimer Marl = Southerham Marl 1
>   Fognam Marl = Glynde Marl 1

The above correlation agrees with the published correlation of Kensworth (Mortimore & Wood

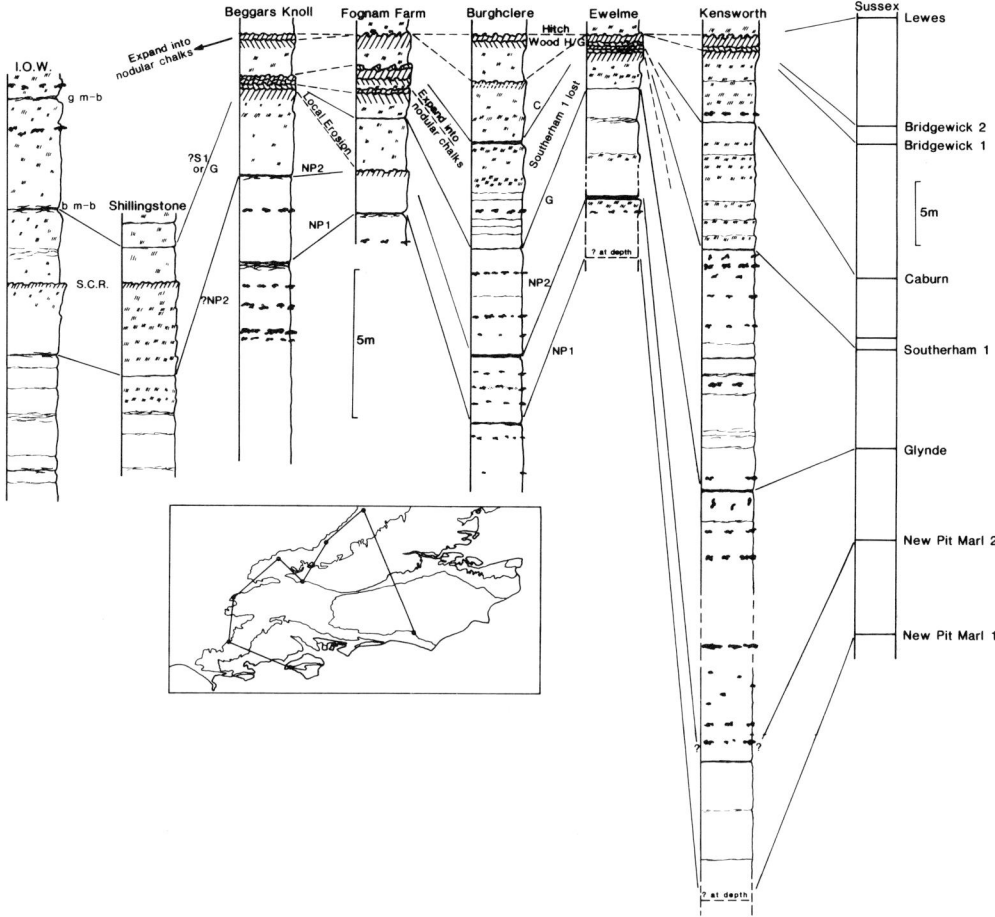

Fig. 12. Summary diagram of the correlation between marls associated with the Chalk Rock and Spurious Chalk Rock, and those found in the main part of the Anglo-Paris Basin as illustrated by a sketch section from Sussex. Note the difference in scale between the Sussex section and the remainder. Solid lines show marl band correlations, where they stop short of a section it is implied that they have been 'cut out' on a hardground. Dashed lines show hardground correlations based on Bromley & Gale (1982). At Ewelme only New Pit Marl 2 has been found, but it is likely that New Pit Marl 1 is present at depth. A marl is present between Glynde Marl 1 at Kensworth which probably correlates with New Pit Marl 2 (C.J. Wood pers. comm.), although it was not analysed during the course of the project (New Pit Marl 1 is assumed to be present at depth). The correlation of the Spurious Chalk Rock with the Chalk Rock is tentative (see text). The Isle of Wight section is a combination of those found alongside the Military Road and the cliff section at Compton Bay (b m-b, black marl-band; g m-b, grey marl-band of Rowe 1908). For a key to the logs see Fig. 9.

1986) for both the Reed and Latimer Marls. It does not agree with the proposals made by Mortimore (1987) regarding the Fognam Marl, which he believed to be of significantly younger age.

Care must be taken when attempting to interpret the marls occurring beneath the Chalk Rock because of the great similarity between the New Pit and higher Glynde Marls. At Ewelme and Fognam Farm, the correlation of the lower marls with the New Pit Marls (Fig. 12) seems likely as they are lithologically similar and are overlain by the Fognam Marl (correlated with Glynde Marl 1). At Beggars Knoll the marls below the Chalk Rock are more likely to be the New Pit Marls because of Bromley & Gale's

(1982) proposal that the lowest hardground is Ogbourne; at Fognam Farm the latter is overlain by the Glynde Marls.

The results from the Isle of Wight and Shillingstone Hill partially agree with the proposals of Mortimore & Wood (1986) and Mortimore (1987), in that the marl above the Spurious Chalk Rock at both localities may be Southerham Marl 1. On the Isle of Wight it is probably succeeded by the Caburn Marl, while at Shillingstone Hill the next marl in the sequence is most likely to be Southerham Marl 2. On solely lithological grounds it is possible to argue that the black marl-band is the lateral equivalent of Glynde Marl 1, unfortunately the present data are unable to resolve this problem. Mortimore & Wood's proposal that the marl beneath the Spurious Chalk Rock correlates with Glynde Marl 1 is doubtful; it seems more likely that it is either one of the higher Glynde Marls or that it is one of the New Pit Marls (see discussion above).

Middle and Upper Turonian subsidence and structural activity: evidence from the correlation

Chadwick (1985a,b) has proposed that to the south of the London Platform, subsidence during the Late Cretaceous was primarily due to compaction of underlying sediments, with relatively little tectonic subsidence. On the London Platform itself, Chadwick argued that tectonic subsidence was more important because of the smaller amount of underlying compactible sediment.

Recently published data (Whittaker 1985) has revealed much concerning the strata and structure underlying the Chalk. Included in this work are the positions of a large number of E–W-trending pre-Permian faults which underlie the Chalk to the south of the London Platform (Whittaker 1985, map 3), and which have been recognised as acting as a control on Jurassic sedimentation (Hallam 1958). The correlations presented above indicate that these faults also influenced sedimentation during the Turonian.

The great thickness of older Mesozoic sediments in southern Dorset and the Isle of Wight should have allowed continuous subsidence during the Turonian due to sediment compaction. The occurrence of the approximately E–W-trending Spurious Chalk Rock in southern Dorset and on the Isle of Wight is therefore somewhat unexpected. Importantly, it overlies a concentration of E–W-trending pre-Permian faults. Because an early hardground (probably Ogbourne but possibily Fognam Farm) is de-

veloped, while later ones are not, it seems likely that tectonic activity prevented subsidence and may even have caused slight uplift during the Middle Turonian. The absence of later hardgrounds implies that subsidence commenced again during the upper Middle Turonian, at a rate faster than the late Turonian sea level fall. The presence of a 'complete' Chalk Rock succession across the northwestern end of the Wessex Basin, south of the London Platform, implies regional non-subsidence or uplift for much of the Middle and Upper Turonian, with similar depositional conditions to those over the London Platform. The thinning of the Chalk Rock in the Warminster–Mere region, noted by Bromley & Gale (1982), corresponds to a horst-like structure termed the Bruton High (Smith 1985), implying that localized fault activity may also have influenced sedimentation rates.

The position of a basement fault to the north of Beggars Knoll (downthrowing to the south) may explain the preservation of New Pit Marl 2 beneath the Ogbourne Hardground, in contrast to the situation found at Fognam Farm. A difference in the subsidence rate across the fault prior to, or during the formation of the Chalk Rock may have either allowed a greater initial accumulation of sediment above New Pit Marl 2, or retarded scouring of already deposited material Likewise, the slightly expanded succession at Burghclere occurs on the downthrow side of one of the faults defining the edge of the London Platform.

The thinning of the Chalk Rock from the Chilterns into Hertfordshire (Bromley & Gale 1982) cannot at this stage be explained using the idea of basement fault influence as none have been reported in this region. There are however significant changes in the age of the basement in this area (Whittaker 1985, map 2), possibly implying some form of tectonic structure at depth.

Conclusions regarding correlation of the Chalk Rock and Spurious Chalk Rock

The geochemical fingerprinting of a series of marl bands from within, above and below the Chalk Rock and Spurious Chalk Rock provides a means of correlating these units with the main part of the Anglo-Paris Basin sequence. A correlation of marls within the Chalk Rock and Spurious Chalk Rock with the expanded succession in Sussex is shown in Fig. 12, the choice of hardground on which each of the marls is 'cut out' being based on data from Bromley & Gale (1982). These results support the proposal by Bromley & Gale (1982) that, in places, the

Chalk Rock represents condensation of much of the Middle and Upper Turonian.

Considering the degree of correlation now achieved between the successions at Kensworth and in southern England, and the less than ideal nature of the Latimer Marl stratotype section (the Hitch Wood Hardground is not exposed), we propose that the use of Latimer Marl and Reed Marl can be superceded by the southern England names Southerham Marl 1 and Caburn Marl.

Not all samples analysed produced a reliable fingerprint, most notably those from beneath the Spurious Chalk Rock; further work, including the examination of additional localities, is required to resolve this problem.

In addition to the periods of non-deposition required for the formation of hardgrounds, there is also evidence of erosion of underlying sediments. At Fognam Farm, the lowest (Ogbourne) hardground appears to have 'cut out' New Pit Marl 2. In contrast, at Beggars Knoll the apparent correlation of New Pit Marl 2 is found some 3 metres below the Ogbourne Hardground. Accepting that both the hardground correlation of Bromley & Gale (1982) and the marl band correlation introduced in this work are correct, it must be concluded that at Fognam Farm erosion of underlying sediments occurred during the formation of the Ogbourne Hardground, probably removing at least 3 m of sediment, including New Pit Marl 2.

Much of this work was undertaken whilst D. Wray was in receipt of an ILEA Research Assistantship. Both authors would like to thank A. Baxter for permission to use the laboratories at University of Greenwich, and B. Saunderson for his expert technical assistance. I. Jarvis and C. Wood offered much helpful advice during the course of the project.

Appendix: Analytical conditions

Method

Analyses have been carried out on an ARL 3510 Inductively Coupled Plasma Atomic Emission Spectrometer based in the School of Earth Sciences at University of Greenwich. Matched synthetic standards were used for calibration, rock standards being analysed as a check on accuracy. The analytical lines used were those recommended by Thompson & Walsh (1983, pp. 89 & 96). To minimize effects caused by differences in the matrixes of the synthetic standards and the 'unknown' rocks, background correction was applied during analysis. In order to correct for machine drift during analysis a monitor solution containing all the elements was analysed after every five samples, any drift being corrected on a spreadsheet at a later date.

ICP-AES operating conditions

Operation: Sequential
Nebulizer type: Meinhard (pumped using peristaltic pump)
Injector gas flow: 11 per minute
Sample uptake rate: 2 ml per minute
Power input: 1.5 kW

Wavelengths used and detection limits

Element	Wavelength	Detection limits as sprayed (μg g^{-1})
Al	308.215	0.15
Sc	361.384	0.005
Ti	337.28	0.05
V	311.07	0.02
Y	371.03	0.01

Limits are based on 10 times the standard deviation of the background emission at the analytes wavelength. The precision at LQD is ±10%

References

BAILEY, H. W., GALE, A. S., MORTIMORE, R. N., SWIECICKI, A. & WOOD, C. J. 1983. The Coniacian – Maastrichtian Stages of the United Kingdom, with particular reference to southern England. *Newsletters on Stratigraphy*, **12**, 19–42.

——, ——, ——, & —— 1984. Biostratigraphical criteria for the recognition of the Coniancian to Maastrichtian Stage boundaries in the Chalk of North West Europe, with particular reference to southern England. *Bulletin of the Geological Society of Denmark*, **33**, 31–39.

BROMLEY, R. G. & GALE, A. S. 1982. The lithostratigraphy of the English Chalk Rock. *Cretaceous Research*, **3**, 273–306.

CHADWICK, R. A. 1985a. Cretaceous sedimentation and subsidence (Cenomanain to Maastrichtian). *In*: WHITTAKER, A. (ed.) *Atlas of onshore sedimentary basins in England and Wales: post-Carboniferous tectonics and stratigraphy*. British Geological Survey, Blackie, 59–61.

—— 1985b. Permian, Mesozoic and Cenozoic structural evolution of England and Wales in relation to the principles of extension and inversion tectonics. *In*: WHITTAKER, A. (ed.) *Atlas of onshore sedimentary basins in England and Wales: post-Carboniferous tectonics and stratigraphy*. British Geological Survey, Blackie, 9–26.

CURRY, D. 1982. Differential preservation of foraminiferids in the English Upper Cretaceous – consequential observations. *In*: BANNER, F. T. & LORD, A. R. (eds) *Aspects of micropalaeontology*. George Allen and Unwin, 240–261.

DEER, W. A., HOWIE, R. A. & ZUSSMAN, J. 1966. *An introduction to the rock forming minerals*. Longmans, London.

ERNST, H. 1982. The marl layer M100 in the Maastrichtian of Hemmoor – an example of selective

CaCO$_3$ dissolution, *Geologisches Jahrbuch,* **61**, 109–127.

GALE, A. S., WOOD, C. J. & BROMLEY, R. G. 1987. The lithostratigraphy and marker bed correlation of the White Chalk (late Cenomanian-Campanian) in southern England. *Mesozoic Research,* **1**, 107–118.

GARRISON, R. E. & KENNEDY, W. J. 1977. Origin of solution seams and flaser structures in Upper Cretaceous chalks of southern England. *Sedimentary Geology,* **19**, 107–137.

HALLAM, A. 1958. The concept of Jurassic axes of uplift. *Science Progress (London),* **46**, 441–488.

HANCOCK, J. M. 1975. The petrology of the Chalk. *Proceedings of the Geologists' Association,* **86**, 499–535.

JEFFERIES, R. P. S. 1963. The stratigraphy of the *Actinocamax plenus* Subzone (lowest Turonian) in the Anglo-Paris Basin. *Proceedings of the Geologists' Association,* **74**, 1–33.

LAKE, R. D., YOUNG, B., WOOD, C. J. & MORTIMORE, R. N. 1987. *Geology of the country around Lewes.* Memoir of the British Geological Survey, Sheet 319 (England and Wales).

LEARY, P. N. & WRAY, D. S. 1989. The Foraminiferal assemblages across three middle Turonian marl bands and a note on their genesis. *Journal of Micropalaeontology,* **8**, 143–148.

MORTIMORE, R. N. 1983. The stratigraphy and sedimentation of the Turonian – Campanian in the Southern Province of England. *Zitteliana,* **10**, 27–41.

—— 1986. Stratigraphy of the Upper Cretaceous White Chalk of Sussex. *Proceedings of the Geologists' Association,* **97**, 97–140.

—— 1987. Upper Cretaceous Chalk in the North and South Downs, England: a correlation. *Proceedings of the Geologists' Association,* **98**, 77–86.

—— & POMEROL, B. 1987. Correlation of the Upper Cretaceous White Chalk (Turonian to Campanian) in the Anglo-Paris Basin. *Proceedings of the Geologists' Association,* **98**, 97–143.

—— & WOOD, C. J. 1986. The distribution of flint in

the English Chalk with particular reference to the "Brandon Flint Series" and the high Turonian flint maximum. *In*: SIEVEKING, G. DE G. & HART, M. B. (eds) *The scientific study of flint and chert. Papers from the 4th. International flint symposium.* Cambridge University Press, 7–20.

ROBINSON, N. D. 1986. Lithostratigraphy of the Chalk Group of the North Downs, southeast England. *Proceedings of the Geologists' Association,* **97**, 141–170.

ROWE, A. W. 1908. The zones of the white Chalk of the English coast. V. The Isle of Wight. *Proceedings of the Geologists' Association,* **20**, 209–352.

SMITH, N. J. P. 1985. The pre-Permian subcrop map. *In*: WHITTAKER, A. (ed.) *Atlas of onshore sedimentary basins in England and Wales: post-Carboniferous tectonics and stratigraphy.* British Geological Survey, Blackie, 6–9.

THOMPSON, M. & WALSH, N. J. 1983. *A handbook of inductively coupled plasma spectrometry.* Blackie, 85.

WEIR, A. H. & CATT, J. A. 1965. The mineralogy of some Upper Chalk samples from the Arundel area, Sussex. *Clay Minerals,* **6**, 97–110.

WHITTAKER, A. (ed.) 1985. *Atlas of onshore sedimentary basins in England and Wales: post-Carboniferous tectonics and stratigraphy.* British Geological Survey, Blackie.

WOOD, C. J. & SMITH, E. G. 1978. Lithostratigraphical classification of the Chalk in North Yorkshire, Humberside and Lincolnshire. *Proceedings of the Yorkshire Geological Society,* **42**, 263–287.

WOOD, C. J., ERNST, G. & RASEMANN, G. 1984. The Turonian-Coniacian stage boundary in Lower Saxony (Germany) and adjacent areas: the Saltzgitter-Sadler Quarry as a proposed international standard section. *Bulletin of the Geological Society of Denmark,* **33**, 225–238.

WRAY, D. S. 1990. *The petrology of clay-rich beds in Turonian (Upper Cretaceous) Chalks of the Anglo-Paris Basin.* PhD Thesis (CNAA) City of London Polytechnic.

Cretaceous foraminiferal events

MALCOLM B. HART

*Department of Geological Sciences, University of Plymouth, Drake Circus, Plymouth
PL4 8AA, Devon, UK*

Abstract: Bioevents range in importance from the first/last appearance of a single taxon, through zonal boundaries of local/regional/international significance, major non-sequences, to (at the highest level) global bioevents. The UK mid-Upper Cretaceous succession is now sampled at better than a 1 m spacing from many localities providing over 2000 samples for detailed analysis and development of high resolution stratigraphy. Examples of detailed correlation are provided for the Albian/Cenomanian boundary, the mid-Cenomanian and the uppermost Cenomanian, while consideration of the complete faunal succession provides information necessary for the monitoring of global environmental change in the Cretaceous.

Detailed foraminiferal research, much of which remains unpublished in theses, has recently been summarized by Hart *et al.* (1989) and this provides a substantial data base for the Cretaceous that can be used in stratigraphical correlation. Within the UK mid-Upper Cretaceous succession a number of distinct 'events' can be recognized. These 'bioevents' can be placed alongside other sedimentological, geochemical, volcanic, diagenetic, and astronomical(?) events and used in the compilation of a full event stratigraphy. The purpose of this account is to document a variety of bioevents of varying importance from the Cretaceous succession. The data used in this assessment of each bioevent's importance are also presented. The examples chosen are described in ascending order of importance and are ascribed to a variety of causes. In every case the interpretation of the event relies on a very high degree of stratigraphic resolution. In some cases this resolution is on a scale measured in centimetres rather than metres.

The zonal boundary

Every first and last appearance of an identified taxon is a bioevent and as such can be used in the identification of zonal boundaries. Not all taxa are, however, used in this way and the occurrences of some species pass almost un-noticed except in the most detailed research. Some first and last appearances are more noteworthy because they have taken on a greater significance in local, national or international biostratigraphy. Without this wider data base all such bioevents would have the same significance. In the Upper Albian and Lower Cenomanian succession of the Folkestone (Kent) area a number of bio-

events can be identified (Fig. 1), some of which have been used by Carter & Hart (1977) in the recognition of a number of zones (Carter & Hart 1977, zones 5–9). As in all biostratigraphic practice these zones are based on clearly defined ranges which are then used to define Taxon Range Zones, Concurrent Range Zones, Interval Zones, etc. This degree of resolution, which, in the case of Fig. 1 is defined in sections occupying only some 30 m of strata, can only be obtained when there is sufficient density of sampling. The resolution in 'high resolution stratigraphy' is, therefore, directly controlled by the sampling interval. Even without knowledge of other successions, some of these boundaries appear to be more significant as they are based on more than one taxon. This can be seen in the case of the top and bottom of Zone 6A and, to a lesser extent, the 5/6l zonal boundary.

It is, however, by correlation that the importance of some of these boundaries can be assessed and Fig. 2 shows how borehole successions in both the Folkestone–Dover area and Sussex can be compared. Three of the original zonal boundaries can be seen to mark regionally significant unconformities. They are: (i) the base of Zone 7, (ii) the base of Zone 6A and (iii) the base of Zone 6.

Even within the short distance from Folkestone (East Wear Road) to Dover (Borehole P. 000) there are quite significant changes in the stratigraphic succession.

By comparison with standard planktonic foraminiferal zonations it will be seen that the first appearance of *Rotalipora appeninica* (Renz) is not a reliable datum (Robaszynski & Caron 1979) and must be due to water depth or the northward migration of the taxon. Some local bioevents are, therefore, seen to diminish in

From HAILWOOD, E. A. & KIDD, R. B. (eds), *High Resolution Stratigraphy*
Geological Society Special Publication, No. 70, pp. 227–240.

227

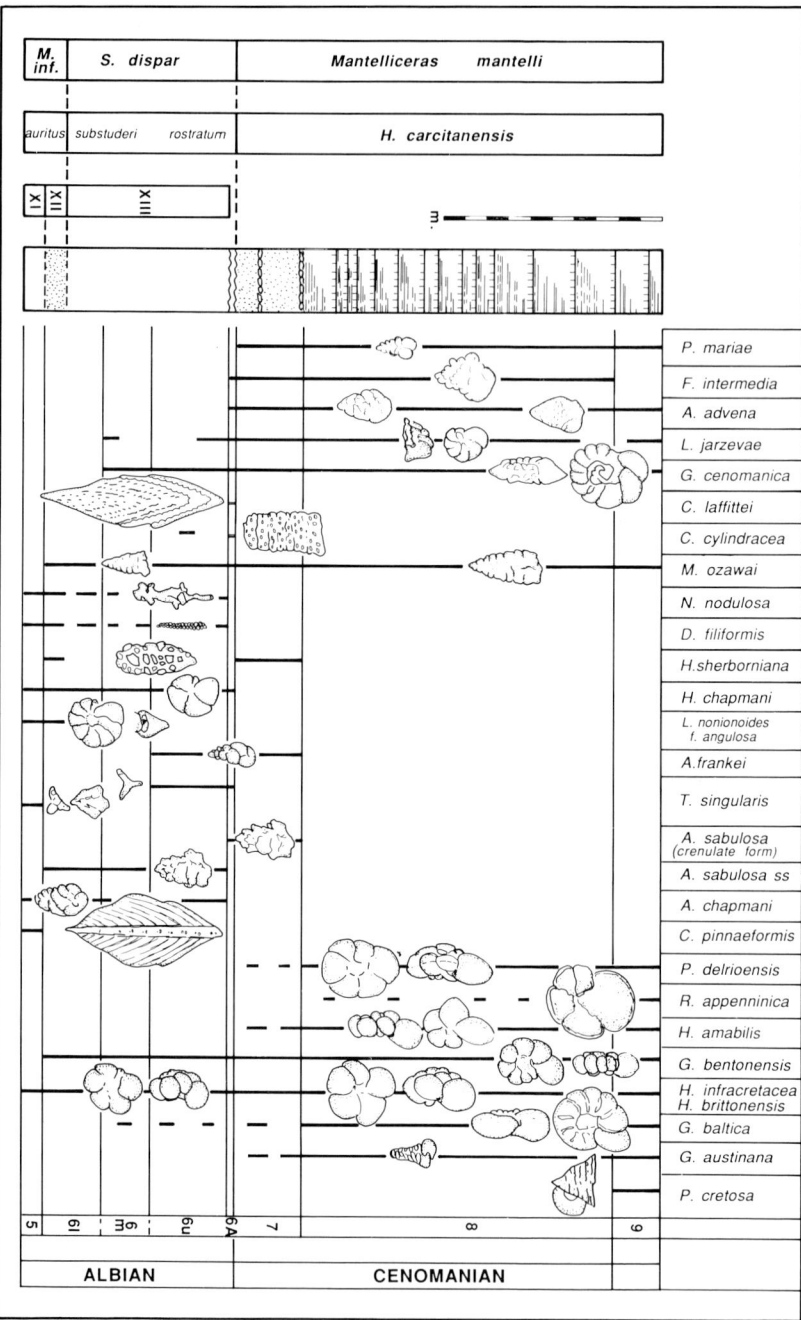

Fig. 1. Foraminiferal distribution across the Albian/Cenomanian boundary. The Glauconitic Marl is indicated at the base of the Cenomanian. In the Folkestone/Dover area this coincides with the benthonic foraminiferal zone 7 of the Carter & Hart (1977) scheme. Drawings of the foraminifers are not all to the same scale. The foraminiferal zonal scheme in the right hand column is that proposed in Carter & Hart (1977). That volume also provides all the taxonomic data for the species represented in the diagram.

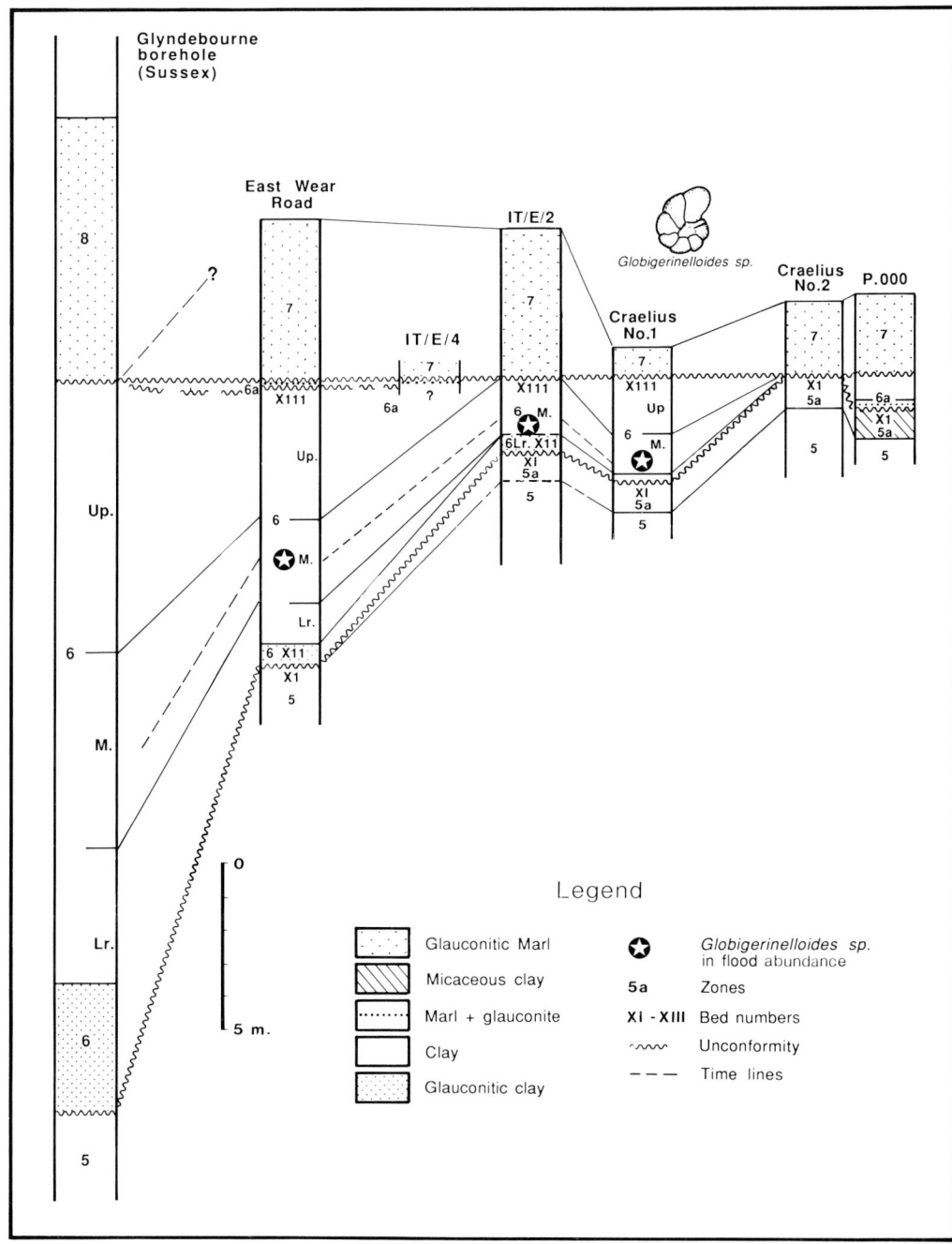

Fig. 2. Correlation of the uppermost Albian and basal Cenomanian across SE England. The Glyndebourne borehole is located some 90 km SSW of Folkestone, while the remaining surface and borehole sections are all in the Folkestone–Dover area (in an approximately SW–NE direction). It is important to recognise the great reduction in thickness of the upper part of the Gault Clay in the Dover area and its replacement by Zone 6A out across the Channel. Note that the Glauconitic Marl of the Glyndebourne borehole is well up within Zone 8 and does not coincide with Zone 7.

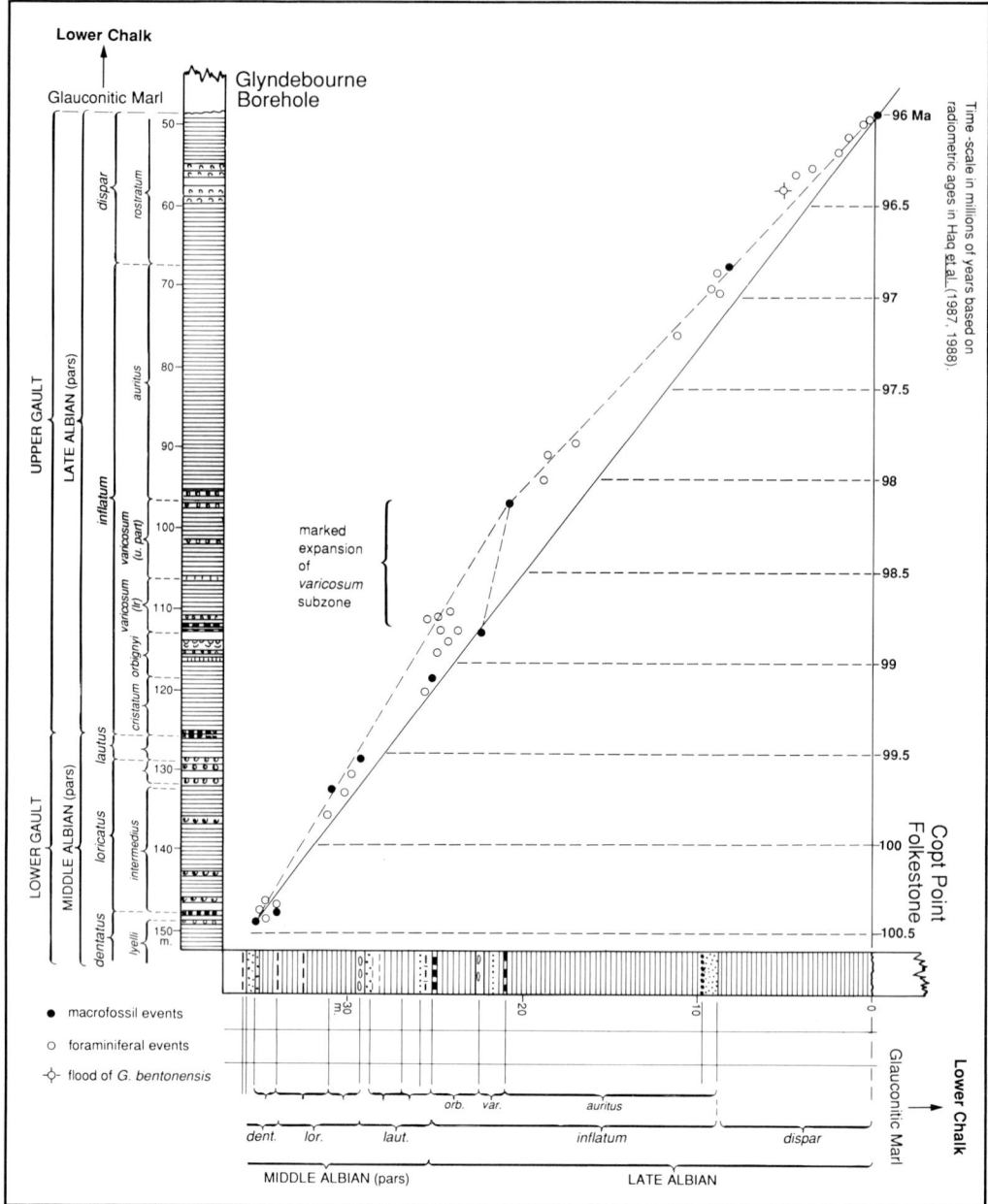

Fig. 3. Graphical correlation of the Copt Point (Folkestone) succession and that recorded in the Glyndebourne borehole. *Some* macrofaunal and foraminiferal first and last appearances are indicated, as is the flood abundance of *Globigerinelloides bentonensis*. Macrofaunal evidence suggests that the *Varicosum* Subzone is expanded at Glyndebourne and alternatives to a straight-line correlation can be developed using increased data. A crude time-scale can also be attached to the Figure.

stratigraphic value when considered on a re-gional/international scale. The flood abundance of *Globigerinelloides bentonensis* (Morrow) in zone 6 (middle), which is a level within the limits of its total range, is also a quite significant local bioevent. This increase in numbers enables this taxon to dominate the 125–63 μm size fraction at this level. The average size of the specimens also increases from a normal figure of 0.20 mm to an average of 0.40 mm at this level. This 'event', which is of relatively short duration, must be due to some palaeoceanographical event and co-incides with the unusual record (Magniez-Jannin 1981) of *Planomalina buxtorfi* (Gandolfi) in such a northerly succession. It would often be danger-ous to use this sort of ecologically-driven bio-event for regional correlation but, in the area covered in Fig. 2, it is probably a near-synchro-nous occurrence. It does seem to be a quite widespread and significant feature in the Upper Albian succession as it can be used over a considerable area of the North Sea Basin as a surrogate basal Cenomanian marker (King 1989).

Within the Gault Clay succession it is possible to use the first and last appearance of every taxon (including those not used in the zonal definitions) to correlate successions. Graphical correlation allows for a precise recognition of faunal events in two successions which can then be used to generate a standard reference section. Figure 3 shows the way in which it is possible to correlate the Copt Point (Folkestone) reference section with that of the BGS Glyndebourne borehole (Sussex). The line of correlation is almost linear, although more detailed analysis allows for the recognition of non-sequences and other erosion surfaces (e.g. *Cristatum* Zone in the Folkestone succession). If one accepts, for the present purpose, the radiometric ages used by Haq *et al.* (1987, 1988) for the top and bottom of the succession, it is possible to calibrate the straight line against time and provide a first order estimate of fine-scale time resolution. At the present time the weakest point in the exer-cise is the radiometric information.

The major non-sequence

In the mid-Cenomanian there were major changes in the microfaunal and macrofaunal assemblages (Kennedy 1969; Carter & Hart 1977; Hart & Bailey 1979; Hart 1980). At the same level in the sections (Fig. 4) there is a significant increase in the number of keeled taxa, especially *Rotalipora cushmani* (Morrow). This has been taken (see references above) to in-dicate a significant change in water depth across southern England and can also be identified in Denmark (Hart 1979; Christensen1984; Packer *et al.* 1989), Colorado (Eicher 1969), and the Pacific Ocean (Sliter 1976). In SE England, and especially across the English Channel (Fig. 5), it can be shown that the water-depth change is associated with a significant hiatus and in mid-Channel it shows up as a significant unconform-

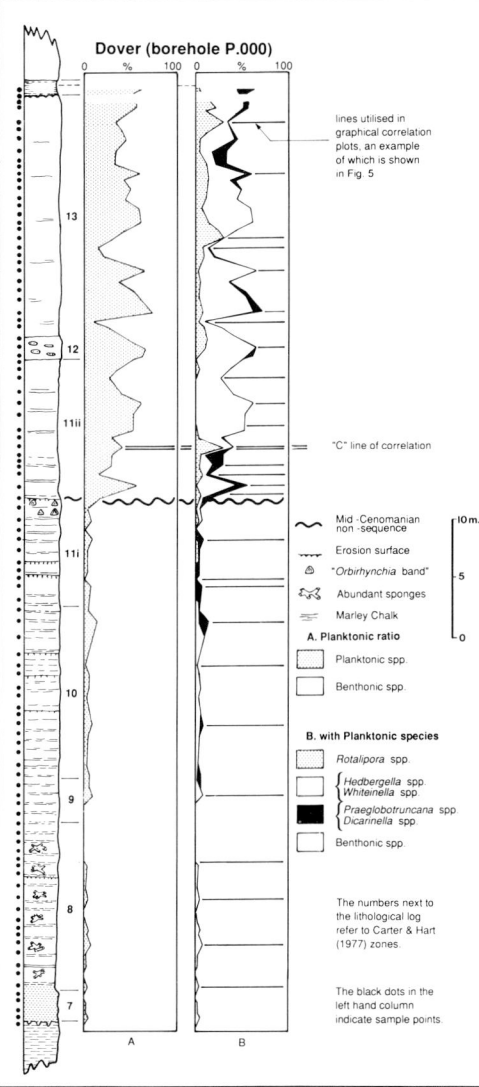

Fig. 4. The Cenomanian succession at Dover showing a schematic lithological log, sample locations (closed circles), benthonic foraminiferal zonation (Carter & Hart 1977), planktonic: benthonic ratio and the development of the 'lines' used in the graphical correlation across the English Channel.

ity (labelled M-C N-S in Fig. 5). The same is true in SW England where the non-sequence be-comes the base of the chalk succession in Dorset and Somerset (Kennedy 1970; Carter & Hart 1977).

This important mid-Cenomanian bioevent is now taken as the boundary between the *Rotali-pora reicheli* (Mornod) Zone and the *R. cush-mani* Zone. In the event stratigraphy models

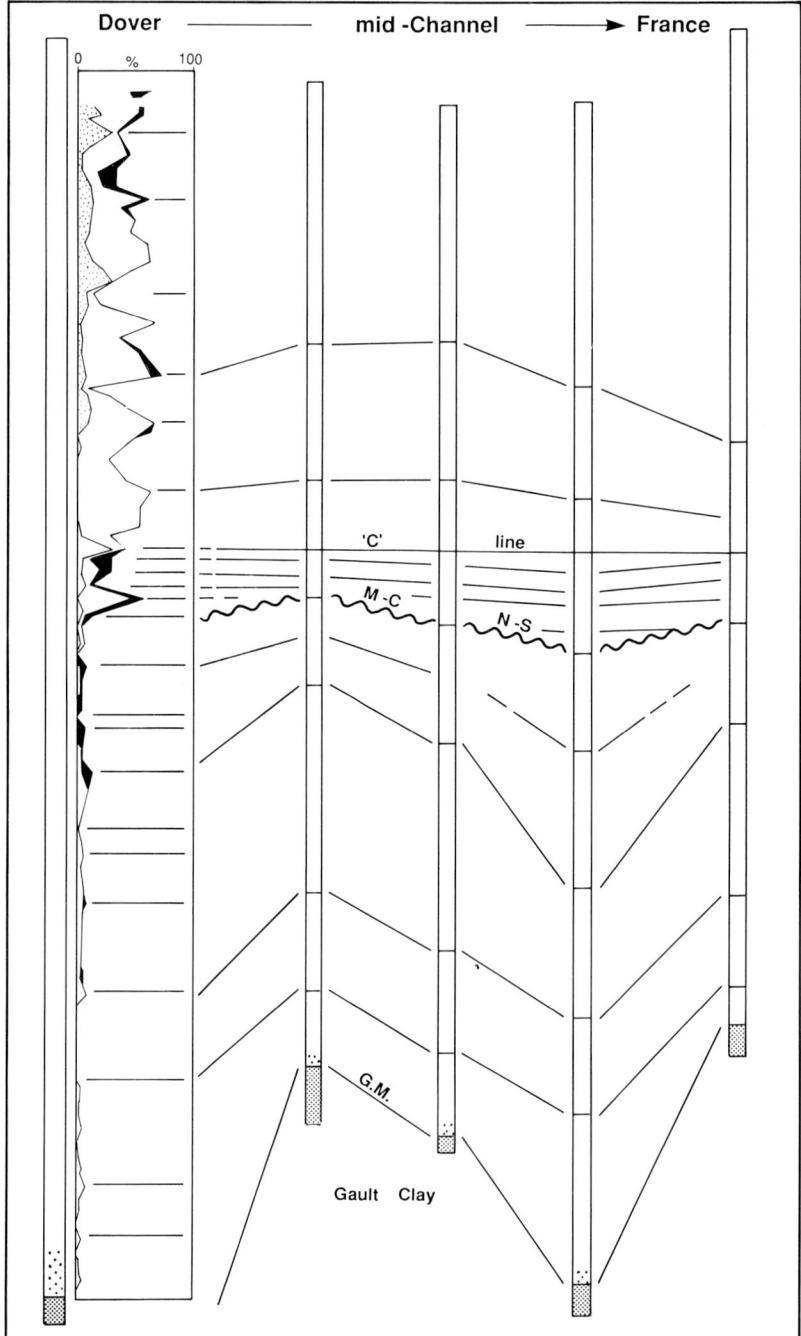

Fig. 5. Correlation of the Dover succession with four unspecified boreholes. The one on the right hand side of the diagram is close to the French coast. The "C" line is indicated, as is the mid-Cenomanian non-sequence. The Glauconitic Marl is indicated immediately above the Gault Clay. (N.B. The correlation is drawn using the 'C' line as a datum for convenience only).

proposed by Hart (1980, 1990a) this is a major feature, yet it is not highlighted to the same extent by Haq *et al.* (1987, 1988).

Major extinction event

In the latest Cenomanian succession there is a major extinction event (Raup & Sepkoski 1984; Hut *et al.* 1987; Jarvis *et al.* 1988; Hart & Leary 1991) which coincides with a $\delta^{13}C$ isotope excursion and evidence of geochemical anomalies (Orth *et al.* 1988). Jarvis *et al.* (1988) have documented the changes in the microflora and microfauna at this level, producing essentially the same results as those being obtained by various workers using the Rock Canyon Anticline succession near Pueblo, Colorado (Eicher & Worstell 1970; Elder 1985; Leckie 1985; Eicher & Diner 1985; Watkins 1985; Kauffman 1986).

In an attempt at understanding these late Cenomanian successions the data from Fig. 6 have been re-plotted as a simple sine curve (Fig. 7) which might, in theory, reflect the rise and fall of the isotopic response in the highly buffered, oceanic environment. Some of the samples from the marls have been plotted distinctively as they fall outside the curve. These are thought to reflect the type of variation reported by Leary *et al.* (1989), in Milankovitch controlled chalk/marl rhythms (Robinson 1986; Gale 1989). This graphical technique leads to the suggestion that the thickness of Beds 6–8 (Plenus Marl Formation) as well as the lower part of the Melborn Rock Beds should be expanded with respect to time.

These beds are clearly condensed when seen in the field, and this must be due, at least in part, to the dramatic reduction in calcareous nannofossil diversity and abundance (Jarvis *et al.* 1988). Against this theoretical curve have been plotted the principal extinction levels for the benthonic foraminifers, planktonic foraminifers, ostracods, dinoflagellate cysts and calcareous nannofossils. This produces the classic 'stepwise mass extinction' pattern (cf. Hut *et al.* 1987 and Fig. 7). It is also noticeable that the disappearance and subsequent re-appearance of the keeled (deeper-water?) taxa occurs at $\delta^{13}C$ value of 4.1% PDB. This stepped disappearance of the various taxonomic groups was used by Jarvis *et al.* (1988) to argue that the pattern was the biological response to a progressive rise (and fall) of an expanded oxygen minimum zone in the water column, although more recently Corfield *et al.* (1990) have disputed this.

Micropalaeontological data, the $\delta^{13}C$ isotope curve, geochemical data and macrofossil information, allow for the accurate correlation of the Cenomanian/Turonian boundary successions despite there being several different versions of the boundary in common currency. The re-appearance of the various taxonomic groups after the event is in an order that *implies* the recolonization of the water column in a way more suggestive of a response to a palaeoceanographic change than the recovery from a sudden shock. Recently Vogt (1989) has suggested a volcanogenic process for the late Cenomanian event. This would be in line with a terrestrial source (Orth, pers. comm.) for the iridium at this level.

Global bioevents

The late Cenomanian event is cited by Raup & Sepkoski (1984) as one of their major extinction events. Similar post-Jurassic events are identified in the mid-Aptian, end-Maastrichtian, late Eocene and mid-Miocene. They follow the 26 Ma periodicity and, following the Alvarez *et al.* (1980) scenario for the K/T boundary, are attributed (by some workers) to extra-terrestrial impact; other authors are more sceptical. Planktonic foraminifers show major changes in diversity throughout the post-Aptian succession and while some of these do seem to follow the Raup and Sepkoski pattern, the results are equivocal. Recently Hart (1990b, fig. 3.2) has replotted the diversity data in terms of the percentage of new taxa in each zone. This produces major 'lows' at the predicted levels followed by periods of major recolonization of the water column. This would appear to support the presence of a 26 Ma periodicity (of reduced levels of speciation, not extinction) although it also highlights a problem. If the late Cenomanian iridium anomalies described above are terrestrial in origin, how can this be a part of a periodic sequence? Periodicity must surely imply the same cause each time, unless one dubiously invokes a periodic effect produced by different mechanisms. It is suggested that all these events are produced by the same or similar causes and that we must still keep an open mind as to the mechanism. It is also apparent (Hart 1990b, fig. 3.2) that between the 26 Ma 'events' there are significant but less important bioevents. Summation of the available biostratigraphic data allows the identification of faunal changes within particular foraminiferal groups. Figure 8 shows the change in diversity and new appearances of various benthonic taxa.

The data for the genus *Gavelinella* and the combined *Arenobulimina/Ataxophragmium/Orbignyna* histogram show remarkably similar

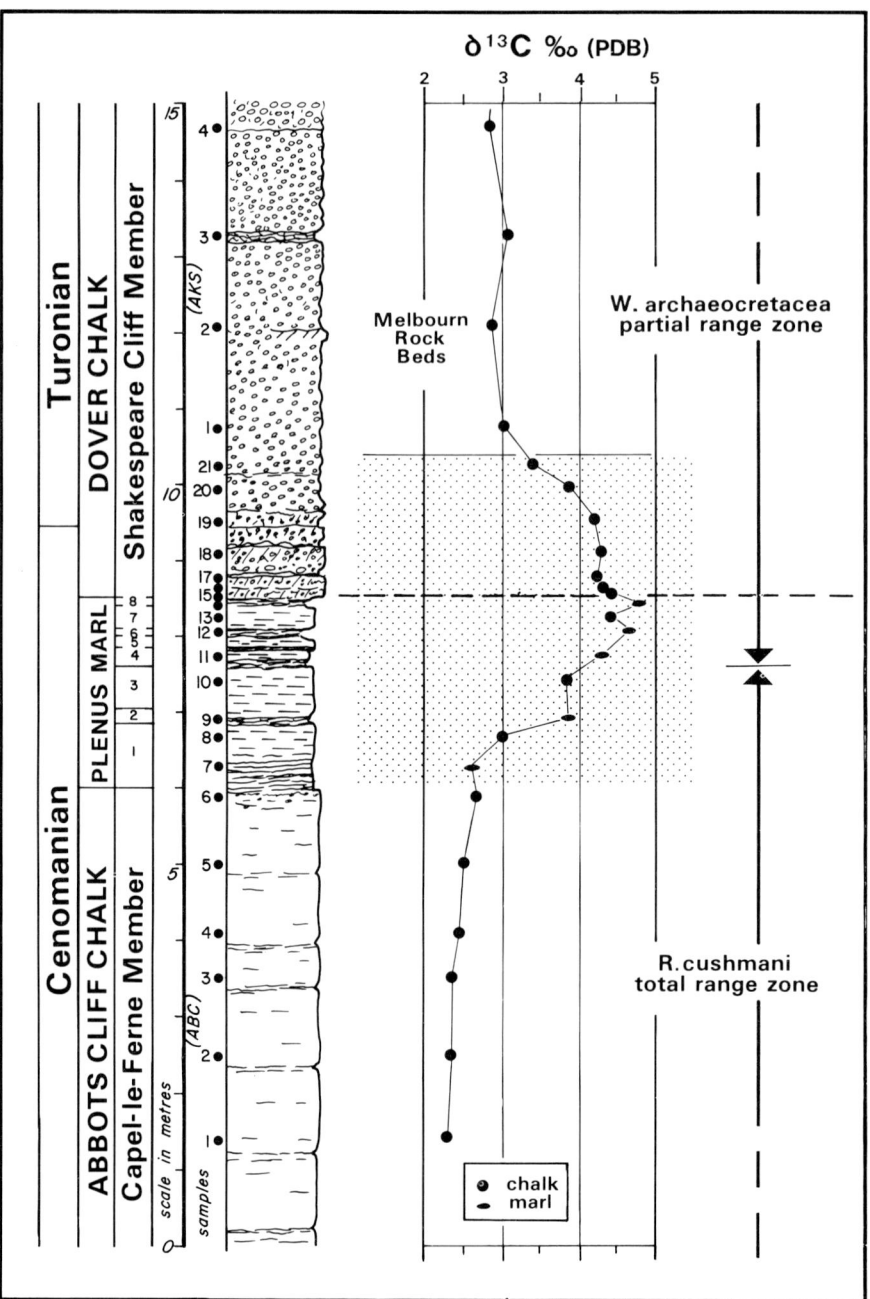

Fig. 6. Carbon isotope curve for the succession at Dover (after Jarvis *et al.* 1988).

patterns despite the two groupings representing very different taxa. In particular it is noticeable how closely Fig. 8b mirrors the eustatic curve of Hancock (1989). Both Fig. 8a and Fig. 8b have been drawn at a very simple level. The histo-

grams could be constructed using each foraminiferal zone (or subzone). This would provide a much higher resolution of the environmental changes in operation during the late Cretaceous. It is noticeable how major event '2' appears in

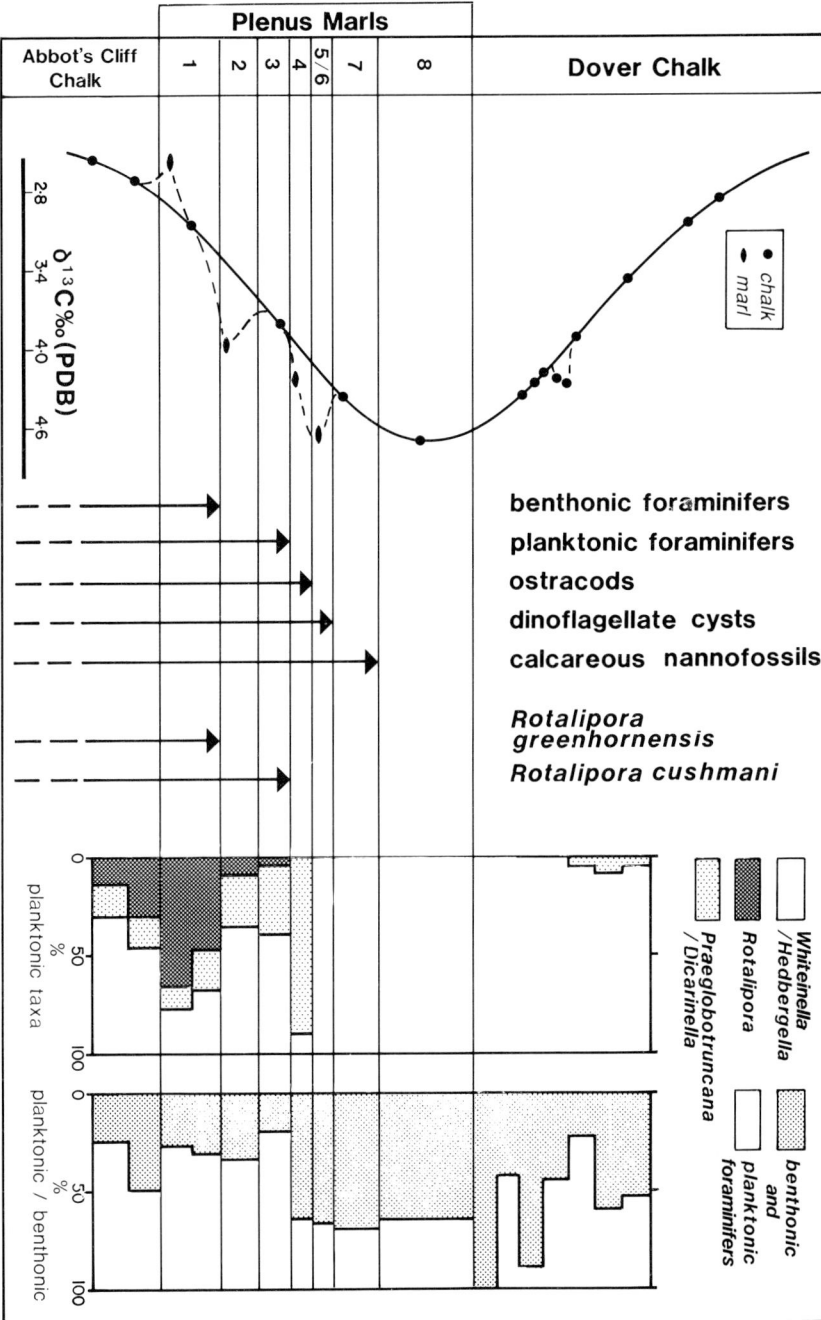

Fig. 7. Generated sine curve based on the carbon isotope data in Fig. 6. Also indicated are the major extinction levels and the foraminiferal distribution data. The lack of keeled planktonic foraminifers within the interval adjacent to the 'carbon isotope spike' is thought to be due to the removal of their preferred niche by the expanded oxygen minimum zone. The benthonic foraminifers are also ecologically reduced at this level (see Jarvis *et al*. 1988 for details). There is no evidence of a shallowing at this level that would provide an alternative explanation for this model.

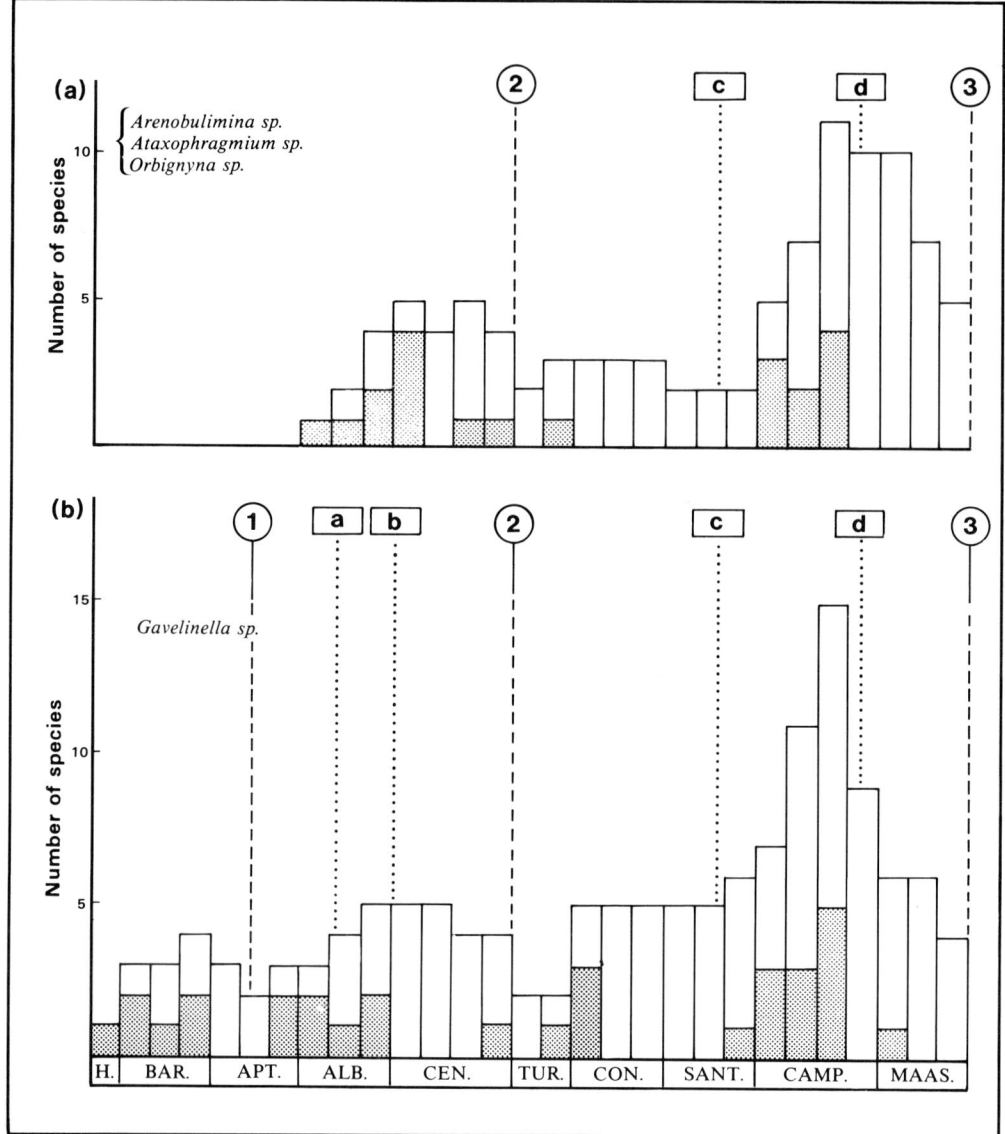

Fig. 8. Distribution of two generic groupings in the mid-Late Cretaceous succession. Newly appearing taxa are represented by the shaded boxes.

both histograms and how minor events 'c' and 'd' also appear to be synchronous. Events 'a' and 'b' are not identifiable in Fig. 8a but this is probably due to the lack of species at this level. It is interesting to note how the overall diversity of both groups falls progressively from a high in the Campanian. The whole of the Maastrichtian interval is one of decline in many foraminiferal groups, a fact that hints at evolutionary stagnation well in advance of a bolide impact(s).

If one uses the data presented in Jarvis *et al.* (1988) and Hart & Duane (1989) it is possible to develop a model for the fluctuations of an oxygen minimum zone in the area immediately west of the UK (Fig. 9). Within DSDP boreholes (Leg 80 and others) there are various levels of dark mudrocks, horizons of weak or strong dissolution, together with evidence of turbiditic flow and faunal mixing. The same, or closely related, events can also be seen in the suc-

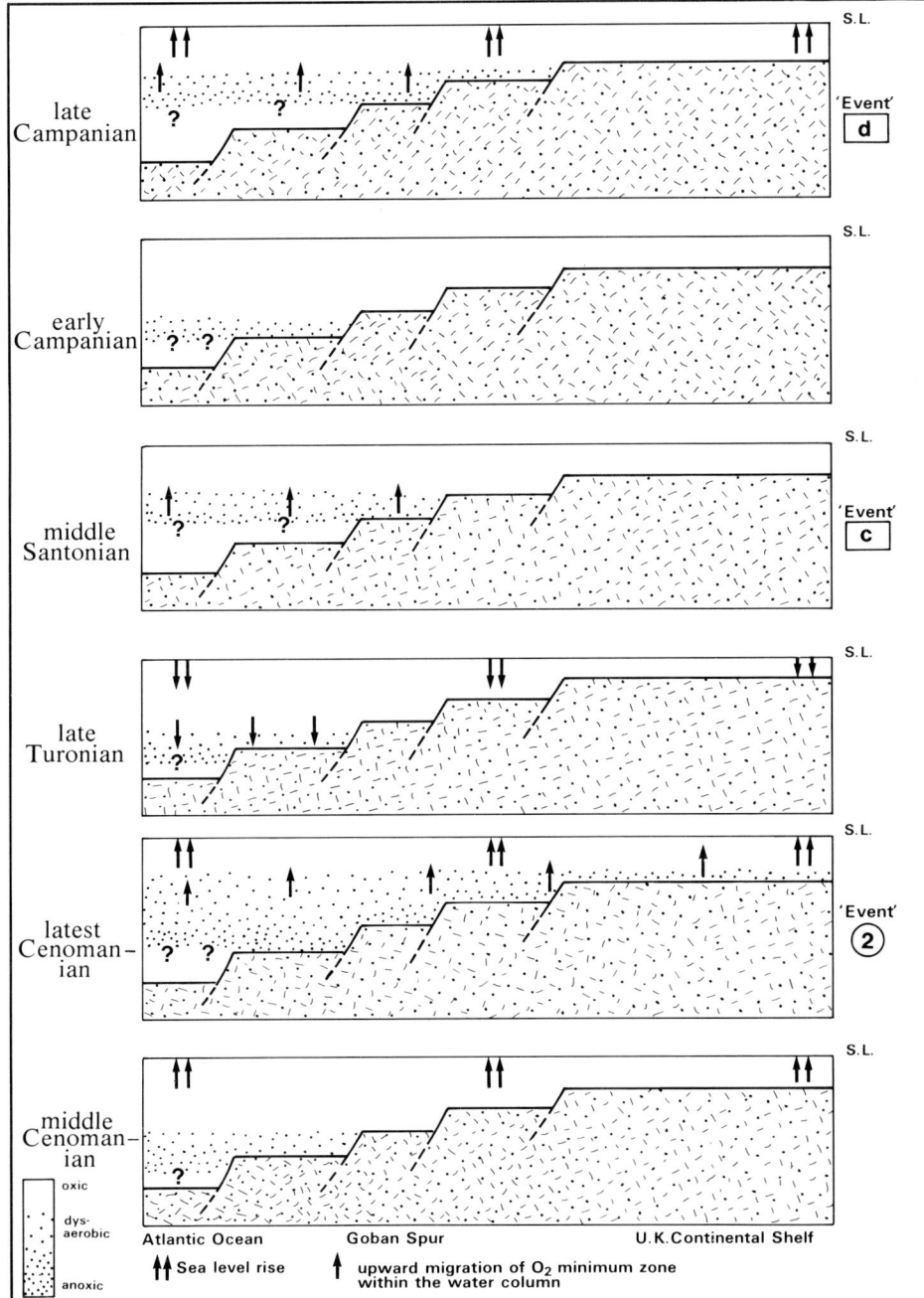

Fig. 9. Eustatic model for the edge of the UK continental shelf modelled on the Goban Spur situation (but not exclusively). The postulated movement and expansion of an oxygen minimum zone is indicated. In all cases it is very difficult to predict if there was a lower level to the minimum zone in the area under discussion.

cessions on the Atlantic margin of the northeastern USA (Olsson & Nyong 1984; Nyong & Olsson 1984; Hart 1987).

Precise correlation of the successions with those in southern England show how deeper-water (?) planktonic foraminifers are more

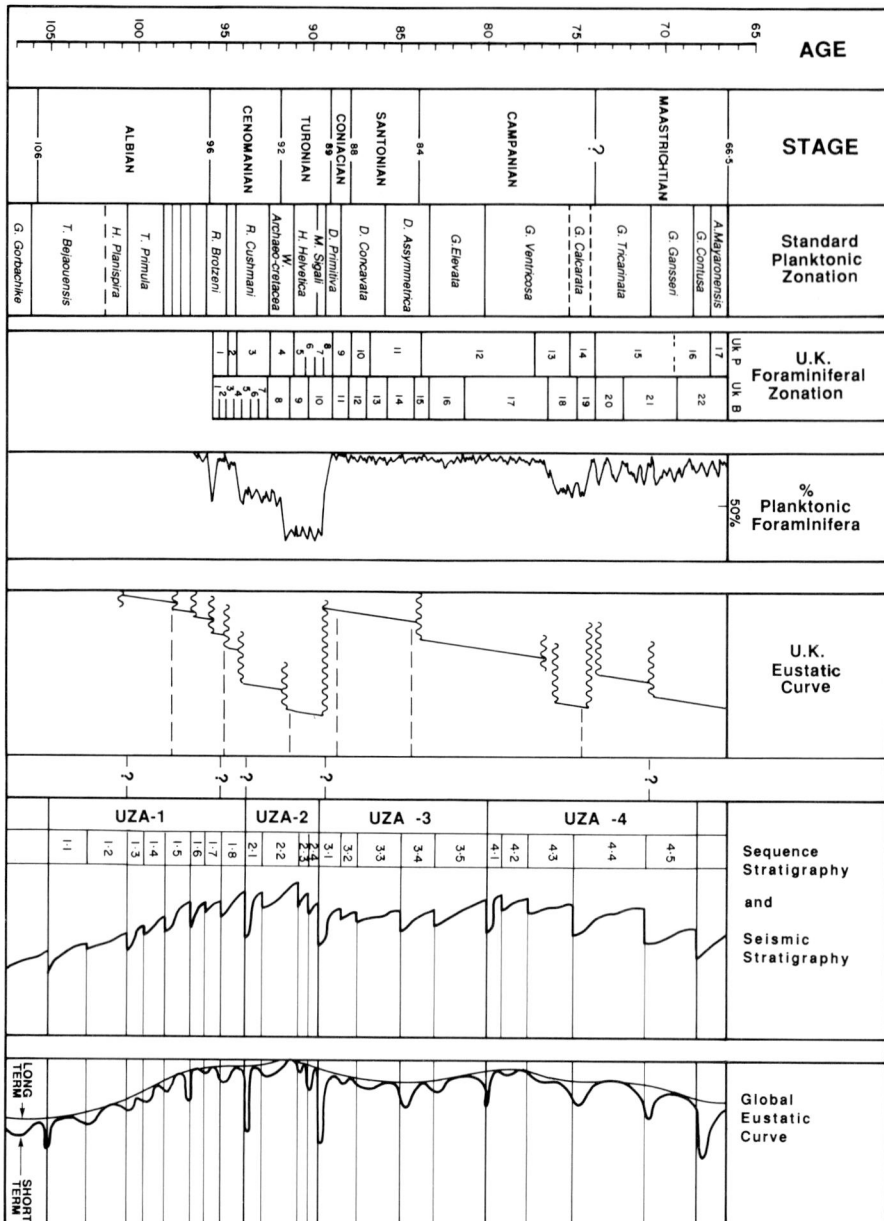

Fig. 10. Compilation of the mid-Late Cretaceous succession using data from Haq *et al.* (1987, 1988) for the time scale, standard international planktonic zonation using planktonic foraminifers, sequence stratigraphy and global eustatic curve. The UKB (benthonic foraminifers) and UKP (planktonic foraminifers) zonations are based on the definitions of Hart *et al.* (1989). The planktonic : benthonic ratio is a composite succession based on sections in the Isle of Wight, Kent, Norfolk and off-shore boreholes. Widespread correlation of sections in the UK has established that the curve plotted here is quite characteristic of Southern England. The total number of samples used in this plot is over 750, with the planktonic : benthonic ratio being based on counts of 500 specimens in the 500–250 μm grain size fraction of every sample (i.e. over 375 000 specimens). Using the other size fractions to obtain data for the palaeoecological interpretation raises the total data for this diagram to a minimum of 1 500 000 individuals.

commonly encountered in this area, thereby allowing a *direct* comparison between the internationally used planktonic foraminiferal zonation and the NW European benthonic foraminiferal zonation.

Summary

Bioevents range in importance from the simple first or last appearance of a single taxon to the global bioevent, exemplified by the late Cenomanian and end-Maaastrichtian extinctions. The relative importance of individual bioevents can only be assessed by accurate correlation on a local, regional or international scale. The UK mid-Upper Cretaceous succession (Fig. 10) can now be sub-divided by many faunal and floral groups, including planktonic and benthonic foraminifers. The zones initially defined by Carter & Hart (1977) in the Cenomanian have now been re-defined and included in a wider zonation; the UKP.1–17 and UKB.1–22 schemes of Hart *et al.* (1989). These zones can be tied into known macrofaunal boundaries (with re-sampling to allow for future movements) and/or stage boundaries.

In chalk successions in the centre of the NW European continental shelf, on-lap, off-lap and maximum flooding surfaces are difficult to determine, but changes in water depth are detectable, both in the numbers of planktonic foraminifera and also the distribution of the various morphotypes. Bioevents and non-sequences can be identified that do indicate changes large enough to leave a signal in more marginal environments (e.g. Hart 1990a; Simmons *et al.* 1991). The distribution of the foraminifera indicated in Fig. 10 and the data base on which it is based indicates a very high degree of stratigraphic resolution. The fact that this data has been used in major, very expensive, engineering projects (Thames Barrier, Channel Tunnel) indicates just how reliable a correlation can now be achieved.

The author acknowledges the financial assistance of Polytechnic South West and the NAB/PCFC Research Funding Initiative. J. Abraham prepared the final versions of the figures and the various typescripts were processed by M. Luscott-Evans.

References

ALVAREZ, L. W., ALVAREZ, W., ASARO, F. & MICHEL, H. V. 1980. Extraterrestrial cause for the Cretaceous-Tertiary extinction. *Science*, **208**, 1095–1108.

CARTER, D. J. & HART, M. B. 1977. Aspects of mid-Cretaceous stratigraphical micropalaeon-

tology. *Bulletin of the British Museum, Natural History (Geology), London*, **29**, 1–135.

CHRISTENSEN, W. K. 1984. The Albian to Maastrichtian of Southern Sweden and Bornholm, Denmark: a review. *Cretaceous Research*, **5**, 313–328.

CORFIELD, R. M., HALL, M. A. & BRASIER, M. D. 1990. Stable isotope evidence for foraminiferal habitats during the development of the Cenomanian/Turonian oceanic anoxic event. *Geology*, **18**, 175–178.

EICHER, D. L. 1969. Palaeobathymetry of the Cretaceous Greenhorn Sea in Eastern Colorado. *Bulletin of the American Association of Petroleum Geologists*, **53**, 1075–1090.

—— & DINER, R. 1985. Foraminifera as indicators of water mass in the Cretaceous Greenhorn Sea, Western Interior. *In*: PRATT, L. M., KAUFFMAN, E. G. & ZELT, F. B.(eds) *Fine-grained Deposits and Biofacies of the Cretaceous Western Interior Seaway; Evidence of Cyclic Sedimentary Processes*. Society of Economic Paleontologists and Mineralogists, 60–71.

—— & WORSTELL, P. 1970. Cenomanian and Turonian Foraminifera from the Great Plains, United States. *Micropaleontology*, **16**, 269–324.

ELDER, W. P. 1985. Biotic patterns across the Cenomanian-Turonian extinction boundary near Pueblo, Colorado. *In*: PRATT, L. M., KAUFFMAN, E. G. & ZELT, F. B. (eds) *Fine-grained Deposits and Biofacies of the Cretaceous Western Interior Seaway: Evidence of Cyclic Sedimentary Processes*, Society of Economic Paleontologists and Mineralogists, 157–169.

GALE, A. S. 1989. A Milankovitch scale for Cenomanian time. *Terra Nova*, **1**, 420–425.

HANCOCK, J. M. 1976. The petrology of the Chalk. *Proceedings of the Geologists' Association*, **86**, 499–535 (for 1975).

—— 1989. Sea-level changes in the British region during the Late Cretaceous. *Proceedings of the Geologists' Association*, **100**, 565–594.

HAQ, B. U., HARDENBOL, J. & VAIL, P. R. 1987. Geochronology of fluctuating sea-levels since the Triassic (250 million years ago to the present). *Science*, **235**, 1156–1167.

——, —— & —— 1988. Mesozoic and Cenozoic chronostratigraphy and cycles of sea level change. *In*: WILGUS, C. K., HASTINGS, B. S., KENDALL, G. ST. C., POSAMENTIER, H. W., ROSS, C. A. & VAN WAGONER, J. C. (eds) *Sea level changes: an integrated approach*. Special Publication of the Society of Economic Paleontologists and Mineralogists, **42**, 71–108.

HART, M. B. 1979. Biostratigraphy and palaeozoogeography of planktonic Foraminiferida from the Cenomanian of Bornholm, Denmark. *Newsletters in Stratigraphy*, **8**, 83–96.

—— 1980. The recognition of mid-Cretaceous sea-level changes by means of Foraminifera. *Cretaceous Research*, **1**, 289–297.

—— 1987. Cretaceous foraminifers from Deep Sea Drilling Project Site 612, Northwest Atlantic Ocean. *In*: POAG, C. W., WATTS, A. B. *et al.*

Initial Reports of the Deep Sea Drilling Project, **95**. US Government Printing Office, 245–252.

—— 1987. Orbitally induced cycles in the Chalk Facies of the United Kingdom. *Cretaceous Research,* **8**, 335–348.

—— 1990*a*. Cretaceous sea level changes and global eustatic curves; evidence from SW England. *Proceedings of the Ussher Society,* **7**, 268–272.

—— 1990*b*. Major evolutionary radiations of the planktonic Foraminiferida. *In*: TAYLOR, P. D. & LARWOOD, G. P. (eds) *Major Evolutionary Radiations*. The Systematics Association Special Volume, **42**, 59–72.

—— & BAILEY, H. W. 1979. The distribution of planktonic Foraminiferida in the mid-Cretaceous of NW Europe. *Aspekte der Kreide Europas, IUGS, Series A,* **6**, 527–542.

——, ——, CRITTENDEN, S., FLETCHER, B. N., PRICE, R. J. & SWIECICKI, A. 1989. Cretaceous. *In*: JENKINS, D. G. & MURRAY, J. W. (eds) *Stratigraphical Atlas of Fossil Foraminifera*. Ellis Horwood, Chichester, 273–371.

—— & DUANE, A. 1989. Late Cretaceous development of the Atlantic Continental Margin off south-west England. *Proceedings of the Ussher Society,* **7**, 165–167.

—— & LEARY, P. N. 1991. Stepwise mass extinctions: the case of the Late Cenomanian event. *Terra Nova,* **3**, 142–147.

HUT, P., ALVAREZ, W., ELDER, W. P., HANSEN, T., KAUFFMAN, E. G., KELLER, G., SHOEMAKER, E. M. & WEISSMAN, P. R. 1987. Comet showers as a cause of mass extinctions. *Nature,* **329**, 118–126.

JARVIS, I., CARSON, G. A., COOPER, M. K. E., HART, M. B., LEARY, P. N., TOCHER, B. A., HORNE, D. & ROSENFELD, A. 1988. Microfossil assemblages and the Cenomanian-Turonian (late Cretaceous) oceanic anoxic events. *Cretaceous Research,* **9**, 3–103.

KAUFFMAN, E. G. 1986. High resolution event stratigraphy; regional and global Cretaceous bioevents. *In: Lecture Notes in Earth Sciences,* **8**, Springer Verlag, 279–335.

KENNEDY, W. J. 1969. The correlation of the Lower Chalk of south-east England. *Proceedings of the Geologists' Association,* **80**, 459–560.

—— 1970. A correlation of the Uppermost Albian and the Cenomanian of south-west England. *Proceedings of the Geologists' Association,* **81**, 613–677.

KING, C. 1989. Cenozoic of the North Sea. *In*: JENKINS, D. G. & MURRAY, J. W. (eds) *Stratigraphical Atlas of Fossil Foraminifera*. Ellis Horwood, Chichester, 418–489.

LEARY, P. N., COTTLE, R. A. & DITCHFIELD, P. 1989. Milankovitch control of foraminiferal assemblages from the Cenomanian of southern England. *Terra Nova,* **1**, 416–419.

LECKIE, R. M. 1985. Foraminifera of the Cenomanian-Turonian boundary interval, Greenhorn Formation, Rock Canyon Anticline, Pueblo, Colorado. *In*: PRATT, L. M., KAUFFMAN, E. G. & ZELT, F. B. (eds) *Fine-grained Deposits and Biofacies of the Cretaceous Western Interior Seaway: Evidence*

of Cyclic Sedimentary Processes, Society of Economic Paleontologists and Mineralogists, 139–150.

MAGNIEZ-JANNIN, F. 1981. Decouverte de *Planomalina buxtorfi* (Gandolfi) et d'autres foraminiferes planctoniques inattendus dans l'Albien Superieur d'Abbots Cliff (Kent, Angleterre): consequences paleogeographiques et biostratigraphiques. *Geobios,* **14**, 91–97.

NYONG, E. E. & OLSSON, R. K. 1984. A paleoslope model of Campanian to lower Maestrichtian foraminifera in the North American basin and adjacent continental margin. *Marine Micropalaeontology,* **8**, 437–478.

OLSSON, R. K. & NYONG, E. E. 1984. A paleoslope model for Campanian-Lower Maestrichtian foraminifera of New Jersey and Delaware. *Journal of Foraminiferal Research,* **14**, 50–68.

ORTH, C. J., ATTREP, M., MAO, X. Y., KAUFFMAN, E. G. & ELDER, W. P. 1988. Iridium abundance maxima in the Upper Cenomanian extinction interval. *Geophysical Research Letters,* **15**, 346–349.

PACKER, S., HART, M. B., TOCHER, B. A. & BRALEY, S. 1989. Upper Cretaceous microbiostratigraphy of Bornholm, Denmark. *In*: BATTEN, D. J. & KEEN, M. C. (eds) *Northwest European Micropalaeontology and Palynology*. Ellis Horwood, Chichester, 236–247.

RAUP, D. M. & SEPKOSKI, J. J. 1984. Periodicity of extinctions in the geologic past. *Proceedings of the National Academy of Science,* **81**, 801–805.

ROBASZYNSKI, F. & CARON, M. 1979. *Atlas of Mid-Cretaceous Planktonic Foraminiferida (Boreal Sea and Tethys)* Parts 1 and 2. Cahiers de Micropaleontologie. Published in both French and English.

ROBINSON, N. D. 1986. Lithostratigraphy of the Chalk Group of the North Downs, southeast England. *Proceedings of the Geologists' Association,* **97**, 141–170.

SIMMONS, M. D., WILLIAMS, C. L. & HART, M. B., in press. Sea-level changes across the Albian-Cenomanian boundary in south-west England. *Proceedings of the Ussher Society*.

SLITER, W. V. 1976. Cretaceous foraminifers from the southwestern Atlantic Ocean, Leg 36, Deep Sea Drilling Project. *In*: BARKER, P. F., DALZIEL, I. W. D. *et al*. *Initial Reports of the Deep Sea Drilling Project,* **36**, US Government Printing Service, Washington, 519–573.

VOGT, P. R. 1989. Volcanogenic upwelling of anoxic nutrient-rich water: a possible factor in carbonate bank/reef demise and benthic faunal extinctions? *Geological Society of America Bulletin,* **101**, 1225–1245.

WATKINS, D. K. 1985. Biostratigraphy and paleoecology of calcareous nannofossils in the Greenhorn marine cycle. *In*: PRATT, L. M., KAUFFMAN, E. G. & ZELT, F. B. (eds) *Fine-grained Deposits and Biofacies of the Cretaceous Western Interior Seaway: Evidence of Cyclic Sedimentary Processes,* Society of Economic Paleontologists and Mineralogists, 151–156.

Transatlantic correlations in the Campanian–Maastrichtian stages by eustatic changes of sea-level

JAKE M. HANCOCK

Department of Geology, Imperial College, University of London, Prince Consort Road, London SW7 2BP, UK

Abstract: There is good evidence that high sea-levels peaked simultaneously during the Late Albian to Turonian in the USA and northwest Europe. Therefore, it is to be expected that the principal peaks during the Campanian–Maastrichtian, now well dated in northwest Europe, should be detectable in the USA. Of the four main peaks in Europe, no. 3, low in the Zone of *Belemnitella langei*, is the most prominent, and can be recognized with confidence in the Western Interior, New Jersey, Alabama–Mississippi and probably in Texas. The other peaks can be identified with various degrees of confidence, but the sea did not persist long enough in the Western Interior for the last peak, early in the Late Maastrichtian, to be developed there. The results change the generally assumed position of the Campanian–Maastrichtian boundary on the foraminiferal scale: the base of the Maastrichtian on the belemnite scale is a considerable distance above the top of the Zone of *Globotruncanita calcarata*, possibly as high as the base of the Zone of *Gansserina gansseri* in north temperate regions. On the ammonite scale in the western interior of the USA the boundary is probably at the base of the Zone of *Baculites baculus* at 70.7 Ma, but could be at the base of the Zone of *B. eliasi*.

Refined biostratigraphy based on ammonites is still unequalled for inter-continental correlation, but inevitably the better ammonite zonations are based on detailed collecting of local successions. Problems sometimes arise when attempts are made to correlate between successions that are geographically far apart. Well known are the difficulties of relating tethyan and boreal ammonite-successions; but there are also less well publicised examples, such as correlating the Campanian-Maastrichtian of the western interior of the USA with other regions, even within the boreal realm. Further potential mistakes can occur when zonations based on other taxa, such as foraminifera, develop different standards for stage-boundaries. All too frequently, two or more standards are assumed to be identical without ever having been actually tested, e.g. Rawson *et al.* (1978) assumed that the Zone of *Globotruncana calcarata* marked the top of the Campanian and that the base of the Zone of *G. calciformis* equalled the base of the Maastrichtian, defined by the base of the Zone of *Belemnella lanceolata* s.s., as recommended by Birkelund *et al.* (1984).

There are still believers and non-believers in eustatic changes of sea-level, e.g. Posamentier *et al.* 1988 v. Burton *et al.* 1987. Whether there were actually world-wide eustatic changes does not matter for this paper: only that they should be coincident in the United States and northwest Europe. Earlier work indicated considerable

agreement between sea-level changes recorded on the two continents during the Albian and Late Cretaceous (Hancock 1975; Hancock & Kauffman 1979). Nevertheless the accuracy of correlations across the Atlantic was relatively crude ten years ago. Improvements in ammonite correlation for the Upper Albian to basal Coniacian allow a more refined test of eustasy between the USA and northwest Europe (Hancock *et al.* in press). During these ages the following major transgressive peaks and regressive troughs were coincident on the separate plates involved.

(1) *Late Albian transgressive peak*. In England this was in the early part of the Subzone of *Callihoplites auritus* (Hancock 1990). In Texas, the peak was at the top of the Zone of *Mortoniceras equidistans* (Young 1986), which probably correlates with the base of the *auritus* Subzone (Hancock *et al.* in press).

(2) *Early Mid-Turonian transgressive peak*. In northwest Europe this was very early in the Zone of *Collignoniceras woollgari*; in the western interior of the USA it was near the top of the underlying Zone of *Mammites nodosoides*; in terms of the 10 ammonite zones that can be recognised in the Turonian of the USA, this is not a significant difference (Hancock & Kauffman 1989).

(3) *Early Late Turonian regressive trough*. Although long lasting, this was most marked at the beginning of the Zone of *Subprionocyclus*

From HAILWOOD, E. A. & KIDD, R. B. (eds), *High Resolution Stratigraphy*
Geological Society Special Publication, No. 70, pp. 241–256.

241

neptuni in England and northern France; this correlates with the Zone of *Prionocyclus macombi* in the Western Interior, which is where the same regression is recorded.

There is one of the five transgressive peaks that Erle Kauffman and I recognized in 1979 that can no longer be sustained as coincident with current correlations: the Coniacian peak. It is now known that in England there was a prolonged sea-level low during the early Coniacian Zone of *Micraster cortestudinarium*, and the Coniacian peak, relatively weak, was about a third of the way up in the Zone of *Inoceramus involutus* (Hancock 1990). In the Western Interior the peak was in the Zone of *Inoceramus deformis* (Kauffman, pers. comm.). Although there is still uncertainty on the correlation between the inoceramid successions in the United States and Europe (Matsumoto & Noda 1986), there is little doubt that the *deformis* Zone is distinctly earlier than the *involutus* Zone (see Schulz *et al.* 1984 for the relative positions of these zones in Germany). I would no longer maintain the Coniacian as an example of a major eustatic peak. What has usually been identified is probably the rebound of sea-level after the very deep Late Turonian trough.

In Europe it is the troughs of regression which can be dated accurately in chalk successions. Ideally, one would make correlations with the USA on the basis of these eustatic lows. In practice, it is easier to use peaks of transgression because the relative intensity of eustatic lows is more difficult to measure in chalk successions (Hancock 1990). There is the further difficulty that more oscillations in local sea-levels have been recorded in parts of the USA than in the Chalk of Britain (compare the oscillations in the Claggett and Gammon Shales in Fig. 2 with those in the Early Campanian in Fig. 1). If there is a tectonic element in small changes in the western interior of the USA, we can assume (possibly erroneously!) that the maximum local high correponds to a recognizable eustatic high in Europe. If local highs are distinguishable by different heights in the western interior, the local lows are often not; note in Fig. 2 the equal easterly limit of the three regressive sandstones in the Gammon Shale.

Note on stage boundaries

The 1983 Copenhagen meeting on Cretaceous stage boundaries (Birkelund *et al.* 1984) recognized that a variety of definitions were in use. Until the International Commission on Stratigraphy has ruled on these boundaries, each author needs to state which definition he is using.

Santonian–Campanian boundary

Birkelund *et al.* (1984) listed nine different biostratigraphic standards. The definition used here is the appearance of *Gonioteuthis granulataquadrata* (Stolley). At Lägerdorf, in north Germany, this appearance coincides with the disappearance of *Marsupites testudinarius* (Schlotheim); this is a more useful standard because *Marsupites* has a much wider geographical distribution. Ammonites are rare at these levels in northern Europe, but in tethyan regions the appearance of *Submortoniceras* has been used (e.g. Young 1963). However, Young himself notes (p. 31) that *Marsupites* has been recorded at the top of his Zone of *Submortoniceras tequesquitense*. This means that the *tequesquitense* Zone is Santonian in the European sense (Hancock 1991). Lillegraven (1991) has come to similar conclusions from independent arguments, and has also shown that the boundary lies within the foraminiferal Zone of *Globotruncana concavata*, not at its summit as has been assumed by most previous authors. This means that the Burditt and Pflugerville Chalk of central Texas are Campanian, not Santonian.

Campanian–Maastrichtian boundary

Six possible biostratigraphic markers were mentioned by Birkelund *et al.* (1984). Most of these have now been placed in stratigraphic order by Burnett *et al.* (in press).

The recommended definition was the appearance of the belemnite *Belemnella lanceolata* (Schlotheim) at Kronsmoor in north Germany. *Belemnella* is geographically limited to the European boreal realm. In tethyan regions the commonest standard has been the top of the Zone of *Globotruncanita calcarata* (e.g. Thompson *et al.* 1991). This is older than the belemnite standard. One of the results of this paper is to place the disappearance of *G. calcarata* at the eustatic peak (No. 3) that lies low within the Zone of *Belemnitella langei* as used in the British Isles. Substantial parts of the succession in the Gulf Coast and New Jersey, previously believed to be Maastrichtian, are now shown to be Campanian in the north European sense.

Campanian–Maastrichtian sea-level peaks and troughs in northwest Europe

By using nodular chalks and hardgrounds to date

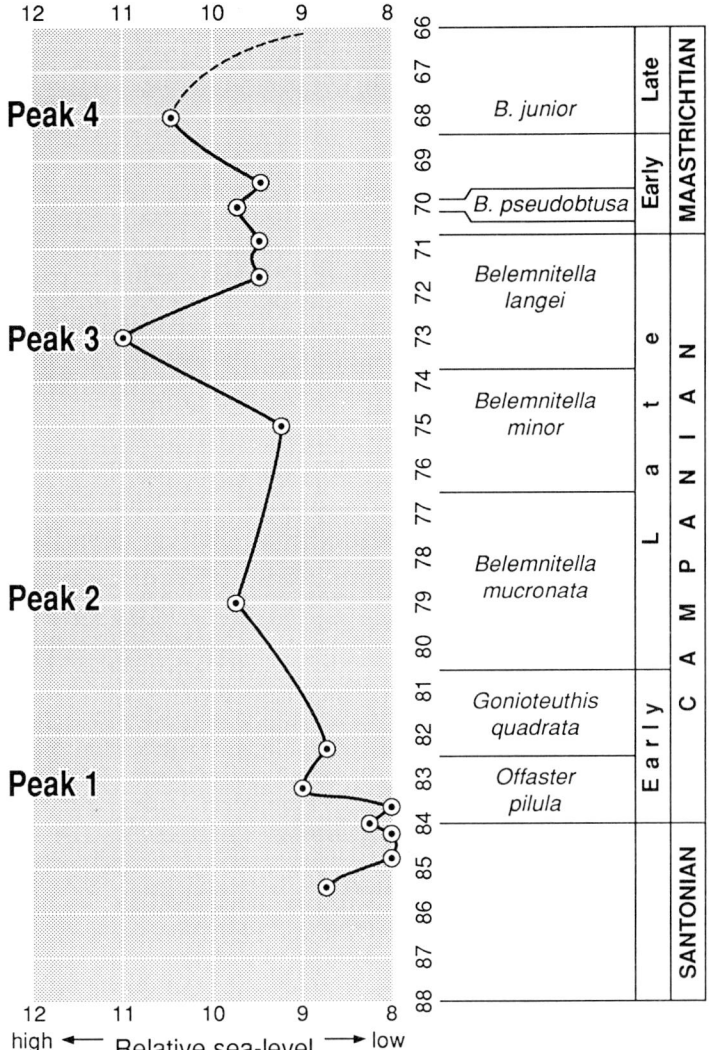

Fig. 1. Campanian–Maastrichtian transgressive peaks and regressive troughs in northwest Europe; modified from fig. 11 in Hancock (1990).

regressive troughs, and the mid-point between pairs of these to date transgressive peaks, it has been possible to elucidate the more accurate pattern of sea-level changes through the Late Cretaceous in north-west Europe (Hancock 1990). Through the Campanian–Maastrichtian there are four major transgressive peaks (Fig. 1).

Peak no. 1 lies in the Subzone of *Echinocorys truncata* in the Zone of *Offaster pilula* of the Chalk in southern England. This is low in the Campanian but not at the base. The 3–4 m of the *E. truncata* Subzone correlates with the 22 m of the *lingua/quadrata* Zone at Lägerdorf in north-west Germany (Ernst 1963). This zone itself

forms the basal subzone of the Zone of *Gonioteuthis quadrata* as used in Germany, which extends much lower than the base of the Zone of *G. quadrata* as used in England (Peake & Hancock 1970, p. 339C; Christensen 1991). This peak also corresponds to the lower part of the partial range Zone of *Gavellinella clementiana* at Lägerdorf (Schönfeld 1988), but unfortunately this benthic foraminiferan appears much higher in the succession in southern England (Bailey *et al.* 1983).

Peak no. 2 is relatively difficult to date accurately because the two defining troughs, between which it lies, are so far apart: the lower trough is at the base of the Zone of *Gonioteuthis quadrata* sensu anglico; the upper trough is represented by the Catton Sponge Beds in Norfolk, high in the Zone of *Belemnitella minor* but not at the summit (Wood 1988). This places peak no. 2 very low in the Zone of *Belemnitella mucronata* s.s. as used in the U.K.; near the base of the Larry Bane Chalk in County Antrim. In Norfolk this lies in Pre-Weybourne[3] of Wood (1988) with *Echinocorys conica* and *E. subglobosa* Goldfuss and *Cardiotaxis heberti*; the middle part of the Basal Mucronata Chalk of Peake & Hancock (1970). In north Germany this peak corresponds to the upper part of the Zone of *Echinocorys conica* and *Belemnitella senior* (= *B. mucronata* s.s., Christensen 1986), from which *Pachydiscus pseudostobaei* Moberg has been recorded (Ernst 1963); possibly at the appearance of *Trachyscaphites spiniger* in Germany (Schmid & Ernst 1975); the lower part of the Zone of *Neancyloceras phaleratum* of Blaszkiewicz (1980) in Poland (see Christensen 1990, fig. 7).

Peak no. 3 is prominent, relatively sharp, and well defined in the lithostratigraphy as being in the middle of the Beeston Chalk of Norfolk (and low in the Portrush Chalk in County Antrim). With the zonation used by Peake & Hancock (1970), Hancock (1990) and in standard works in USSR (e.g. Moskveen 1986–87) this horizon lies low in the Zone of *Belemnitella langei*. However, the detailed study by Wood (1988) shows that the Beeston Chalk lies below the Zone of *B. langei* as used in Germany: Peak no. 3 at Lägerdorf lies near the top of the Zone of *Nostoceras polyplocum* of Schulz (1978), with *Cardiaster cordiformis*; the middle of the *miliaris/incrassata* benthic foraminiferal partial range zone (Schönfeld 1988). Christensen (1990) correlates the *polyplocum* Zone of northwest Germany with the zone of the same name in Poland. Unfortunately, this index species has both a long total range and an erratic occurrence: the lists of Blaszkiewicz (1980) suggests that the correlative of the upper part of the *polyplocum* Zone at Lägerdorf lies in the Zone of *Didymoceras donezianum* in Poland; the belemnite records of Kongiel (1962) are enigmatic.

There are various minor peaks and troughs shortly before and after the boundary between the Campanian and Maastrichtian stages. The most important for the subsequent discussion is the little peak very early in the Maastrichtian that corresponds to the Zone of *Belemnella pseudobtusa*.

Peak no. 4 cannot be seen in England or at Lägerdorf, but in Limburg corresponds with the Craie Grise (= Vijlen Member) which lies low in the Zone of *Belemnitella junior*, i.e. very early in the Late Maastrichtian. The Vijlen Member has yielded *Acanthoscaphites tridens* (Kner) and *Hoploscaphites tenuistriatus* (Kner) (van der Tuuk in Robaszynski et al. 1985), both of which are better known from low in the Lower Maastrichtian (Kennedy & Summesberger 1987). The records of Birkelund (1982) suggests that this peak may correspond with the top of the range of *H. tenuistriatus*. The Vijlen Member also equals the Zone of *Echinocorys* gr. *limburgicus* Lambert (Meijer 1965) and benthic foraminiferal Zones C and D of Hofker (1958). On the microbrachiopod zonation of Surlyk (1984) the peak lies in the middle of the *semiglobularis–humboldtii* Zone.

Montana

Although it is possible to detect, and to some extent to measure, changes in Cretaceous sea-levels throughout the US western interior, viz. the Rocky Mountain foredeep and the cover of the Interior Platform, from Montana in the north to New Mexico in the south, Montana seems to be the best region to study the relationship between facies and sea-levels from surface outcrops during the Santonian to Maastrichtian. The major Cretaceous increase in plate-movements in western North America in general (Engebretson et al. 1985; Avé Lallemant & Oldow 1988), and in the northern US Rockies in particular (Heller et al. 1986), occurred somewhere around the Albian, but the Santonian seems to have marked the start of even faster movements (Globerman et al. in press). This has made it difficult to disentangle tectonics from sea-level changes during the latest Cretaceous, e.g. in Colorado (Molenaar 1983), although this is easier for northwest Colorado than southern Wyoming (Kiteley 1983) or northern Wyoming (Gill & Cobban 1973, fig. 10).

In southern Montana the Sevier orogenic belt seems to have been concentrated around the Boulder Batholith (Peterson 1988), intruded over 80–74 Ma (Tilling et al. 1968). The Crazy Mountain Basin, between Helena and Livingston, in Montana was probably close enough to the Boulder uplift to be affected by it during the latest Santonian and earliest Campanian, but eastwards and later in the succession, e.g. the Campanian around the Porcupine Dome, it was sufficiently independent of earth movements to record a clear picture of sea-level changes from the facies succession (Gill & Cobban 1973, fig. 9; Peterson 1988, fig. 16).

Of the four main peaks recognized in Europe, the greatest in the Montana seaway is unmistakably that represented by the Bearpaw Shale, which therefore corresponds with peak no. 3 in Europe, as already recognized by Hancock & Kauffman (1989). The other peaks can therefore be related to this.

In a region of rapid sedimentation, and the seaway was receiving 40–200 m sediment per million years at this time (Gill & Cobban 1973, fig. 21), regression sets in during the slow-down of the sea-level rise before the actual eustatic peak of transgression (Posamentier et al. 1988, fig. 12; Posamentier & Vail 1988, p. 126). The more off-shore the facies, the less this effect. For the big no. 3 peak, the retardation was probably about one ammonite zone, so that the most westerly feather limit of the Bearpaw Shale is in the Zones of Baculities compressus and B. cuneatus, and the eustatic peak is located around the top of the Zone of Baculites cuneatus.

Peak no. 1. Of the four major peaks in the Campanian–Maastrichtian, this is the most difficult to recognise with certainty in Montana because of the numerous rather small oscillations through the Late Santonian and the early Early Campanian. It is notable that in the sea-level curve of Weimer (1984, fig. 11; 1988, fig. 6), his Niobrara peak is the broadest of the sea-level highs, embracing both the mid-Late Santonian and the early Early Campanian (no. 1) peaks. In the section between Mosby and

the Porcupine Dome there are three peaks within the Gammon Shale. The strongest is the earliest, which is a westerly tongue of shale into the Virgelle Sandstone Member and belongs to the Zone of *Scaphites hippocrepis* I. Further work may show that this peak is slightly higher in the western interior succession because it seems to lie at least one ammonite zone higher in Mississippi and Texas.

This is the weakest sea-level correlation in the Campanian between Europe and the western interior of the USA. It leaves the strong regressive feature of the Eagle Sandstone more lengthy: it embraces approximately all zones from *Desmoscaphites bassleri* in the top Santonian to perhaps the base of the Zone of *Baculites obtusus*. At present, sea-levels cannot give guidance on the position of the boundary between the Lower and Upper Campanian in Montana.

Peak no. 2 is a little above the middle of the Claggett Shale, around the top of the Zone *Baculites mclearni*. The Claggett Shales are about 70–115 m thick in the main seaway and extend west of longitude 111°. This peak is thus both prominent and well dated. Gill & Cobban (1973) noted that the Claggett Shale transgression started aburptly, which is a common feature of initial transgressions (Hancock & Kauffman 1979).

Above the Claggett Shale in west-central

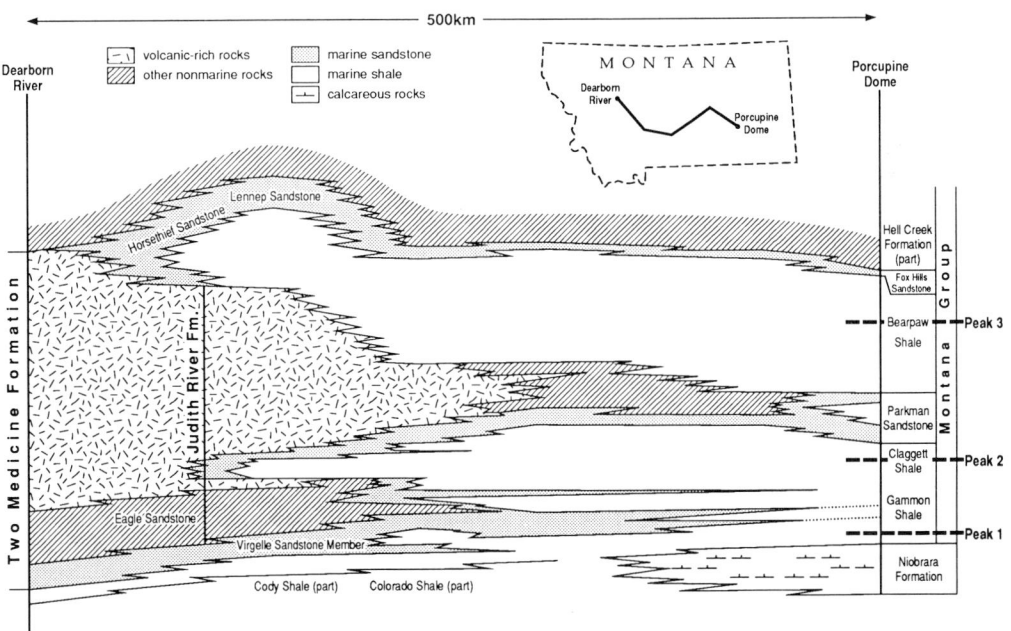

Fig. 2. Lithostratigraphical succession in the Campanian and lower Maastrichtian of central Montana to show positions of transgressive peaks; based on Gill & Cobban (1973, fig. 9) with much simplification.

Montana there are sandstones which rapidly become rich in volcanic debris, the upper half of the Judith River Formation, after which the regressive phase is named. More centrally in the basin, around the Porcupine Dome, the Judith River Formation has passed into trough-bedded feldspathic sands known as the Parkman Sandstone. There are a number of small transgressive tongues of shale within the Sandstone which have not been recognised in Europe, but the centre of the Parkman regressive trough corresponds to the Zone of *Baculites gregoryensis*.

Peak no. 3, the great peak, is a little above the middle of the Bearpaw Shale, approximately at the boundary between the Zones of *Baculites cuneatus* and *B. reesidei*.

The Bearpaw Shale begins abruptly on the Parkman Sandstone, the top of which contains phosphatic pebbles in places (Gill *et al.* 1972). The main mass of the Shale is a slipper-clay facies (Hancock 1975), with several levels of limonite stained, metre-scale, concretions of septarian limestones; bentonites are common and much of the shale contains a high proportion of smectite. Although the top is not as abrupt as the base, the change from silty shale up to the ripple bedded ferruginous sands of the overlying Fox Hills Sandstone occurs over less than 5 m. The Bearpaw Shales extends only about 40–50 km further west than the Claggett Shale, but with a thickness of some 300–330 m it is a much more prominent formation in the Montana Group.

South of Montana it is more difficult to date this third peak. Even eastern Wyoming is not sufficiently far from the tectonically affected margin of the seaway to give a reliable picture: in the Red Bird section the peak must lie within the 'lower unnamed shale member' of Gill & Cobban (1966, fig. 12) in the Pierre Shale, but the latter authors were unable to detect the Zone of *B. cuneatus*. The unconformity at the base of the Teapot Sandstone Member possibly reaches the south-east flank of the Powder River basin (Gill & Cobban 1973, fig. 10).

Peak no. 4 is probably unrepresented in the United States western interior because regional uplift of the Laramide movements removed the sea completely from the region.

There is a further transgression in the record, T_{10} of Kauffman (1977), although this did not reach Montana. In eastern Wyoming it occurs within the 'upper unnamed shale member' of the Pierre Shale, but as mentioned earlier, there seems to be a tectonic distortion of dating peaks and troughs in Wyoming. However, further east in the seaway the peak is marked by the Mobridge Limestone in South Dakota, which belongs to the Zone of *Baculites clinolobatus*. Unless the top half dozen ammonite zones in the marine succession lasted much longer than earlier zones, the *clinolobatus* Zone, only 3–4 zones above the top of the Campanian, can only represent about 1.5 million years after the start of the Maastrichtian; a conclusion confirmed by a K/Ar date of the underlying *B. grandis* Zone at 70.1 ± 0.7 Ma (Obradovich 1988). This *clinolobatus* peak can hardly represent Peak no. 4, early in the Late Maastrichtian. However, there is a minor peak in Europe very early in the Maastrichtian, corresponding to the Zone of *Belemnella pseudobtusa* (Hancock 1990), which could easily be the equivalent of the *clinolobatus* peak. The absence of the 4th peak and the scale of the Maastrichtian ammonite zones suggest that there is no marine Upper Maastrichtian in the western interior of the USA. Similar conclusions were drawn for Canada by Williams & Stelck (1975).

Atlantic coastal plan: New Jersey and Delaware

Geological sections and subsidence studies show that the Cretaceous structure of this region forms a succession of overlapping sediments resting on a basement of metasediments of the Piedmont Province. Watts (1982) and Steckler *et al.* (1988) have argued that the pattern has a tectonic control by downward flexuring, a viewpoint supported by the upward continuation of hundreds of metres of Cenozoic sediments which also have a regional dip towards the Atlantic coast. Superimposed on the simple seaward dip is a series of shallow basins, e.g. the Raritan embayment, interpreted by Brown *et al.* (1972) to be a wrench-fault tectonic framework controlled by lateral compressive stress. However, the details of the higher Cretaceous form a classic example of a facies succession dominated by oscillations of sea-level.

The surface outcrops, with a typical thickness of some 130 m, show a rhythmic alternation between near shore sands and clays, variable in both thickness and lithology, alternating with more uniform glauconite-rich sands. Each rhythm is sufficiently similar to others for confusions of identification to have occurred in the past (Owens *et al.* 1970; Owens & Gohn 1985; Olsson *et al.* 1988). Even now, some of the Merchantville Formation seems to belong to sequence 3 of Owens & Gohn, some to their sequence 4; but the Merchantville is probably diachronous (Petters 1976).

Each glauconite rich formation represents a

marine transgression, but these formations are almost certainly strongly condensed and it is doubtful if the eustatic peaks correspond to the middle of each unit. With the exception of the Tinton Greensand, each greensand is underlain by a disconformity and often contains phosphatic pebbles at the base.

Comparison with transgressive greensands in Europe shows that the top of the most condensed part of the succession (itself just above a disconformity or succession of disconformities) corresponds to the time of regional transgression as postulated by van Wagoner *et al.* (1988), e.g. the transgression at the base of the Aptian Zone of *Parahoplites nutfieldiensis* in England corresponds to the maximum flooding surface of Haq *et al.* (1988); see Hancock (1990) and Hesselbo *et al.* (1990). The eustatic high occurs a little later (see Posamentier *et al.* 1988, fig. 12); in this English case, higher in the *nutfieldiensis* Zone. On the Atlantic coastal margin the eustatic high will be immediately after the time of any basal condensed fauna, below the middle of the unit.

This interpretation disagrees with the analysis of Nyong & Olsson (1984), Olsson & Nyong (1984) and Olsson (1988) who have used a palaeoslope model based on bathymetry measured by benthic foraminifera. The effect is to displace the eustatic peaks to correspond with the shallower water sands and clays, although they recognise that the glauconite-rich units are transgressive. My impression is they have under-estimated the degree of condensation in the transgressive greensands. In this coastal succession, most of Cretaceous time will be represented by glauconitic sediment or no sediment at all; the nearer shore sands and clays are a minor representative of the time involved.

Excluding the base of the Hornerstown Formation, which may just extend down into the Cretaceous according to Olsson (1988), but not according to Owens *et al.* (1970), there are the following glauconitic units from top to bottom: Tinton Greensand, Navesink Greensand (possibly the most prominent), upper part of the Marshalltown Formation, upper part of the Merchantville Formation. These could contain the four eustatic peaks but the Merchantville is often dated as Santonian, and although Olsson (1988) puts the upper Merchantville into a Campanian Zone of *G. elevata*, he also considers that the upper Merchantville represents the second transgression in the succession.

There is no easy way to compare the magnitude of the peaks and one is on safer ground to consider the succession from the top downwards because there is little disagreement on the position of the Cretaceous-Palaeogene boundary. From this approach we get:

Peak 4: a little above the base of the Tinton Greensand
Peak 3: a little above the base of the Navesink Greensand
Peak 2: a little above the base of the upper Marshalltown Fm.
Peak 1: a little above the base of the upper Merchantville Fm.

The base of the Maastrichtian would be somewhere low in the Navesink Greensand. The Mendryk collection of ammonites from the basal part of the Navesink Greensand, described by Cobban (1974), must be very high Campanian. The Late Campanian dating of *Belemnitella americana* (Morton) by Jeletzky (1962) is confirmed. On Olsson's zonation by planktic foraminifera, the base of the Maastrichtian would be around the base of the zone of *Globotruncanita stuarti*. If there is a lower Merchantville eustatic peak, it could be the equivalent of that in the mid Late Santonian.

Most of Olsson's foraminiferal zonal scale gives dates to the New Jersey formations that make them younger than they would be on a macro-faunal correlation: it is possible that not enough allowance has been made for re-working and burrowing in glauconitic facies.

Northern margin of the Gulf of Mexico: Mississippi–Alabama

Cretaceous sediments lap against the Precambrian and Palaeozoic rocks of the southern limits of the Appalachians. The simple pattern seen in surface outcrops is slightly more complicated subsurface, e.g. the thickest Campanian–Maastrichtian (more than 300 m) is found in central Mississippi–Alabama rather than closer to the Gulf coast (Cook & Bally 1975, p. 232). Moreover, small-scale maps fail to bring out that along the strike the facies show deeper water, more open-sea conditions, to the west. Thus the Demopolis Chalk in central Alabama becomes sandy in Montgomery County and passes laterally into the Cusseta Sand in Bullock and Barbour Counties in eastern Alabama (Monroe 1941; Skotnicki & King 1989).

There are near-shore marine sediments in Tennessee but no preserved shoreline for the Campanian–Maastrichtian. Nevertheless, the outcropping sediments are so close to the massif that its slightest vertical movements would have upset the pattern of eustatic movements. Fortunately the Appalachian fold-belt seems to have been passively stable during the latest Cretaceous. Peak 2 is feeble and uncertain, but peaks 3 and 4 are expressed strongly. Major regressive troughs are indicated by disconformities.

Peak no. 1 is represented by the Arcola Limestone at the top of the Mooreville Formation. Most of the Mooreville Formation formed in a shelf mud environment, and is transgressive compared with the underlying Tombigbee Sand (E. E. Russell, pers. comm.; fig. 4). The Arcola Limestone varies from a pair of nodular beds to stratified marly chalk. Although rich in calcispheres, it is essentially a chalk which is a good indicator of a transgressive high in a clastic succession. The position of Peak no. 1 confirms the view of Dowsett (1989) that the whole of the Tombigbee Sandstone is Santonian because this peak is very low in the Campanian in Europe.

Even lower, the transgressive Eutaw Formation overlaps the McShan Formation on to Tuscaloosa Gravels, eventually stepping on to Dinantian and Ordovician rocks in Tennessee (Russell & Keady 1983); this could represent the mid-Santonian peak in the middle of the European Zone of *Uintacrinus socialis*. The line of phosphatic nodules recorded 7 m above the bench of Tombigbee Sandstone at Plymouth Bluff, Mississippi (Stephenson & Monroe 1940) may represent the basal Campanian regressive trough in the Zone of *Echinocorys tectiformis* in England (Hancock 1990).

Peak no. 2 is difficult to place.

The top surface of the Arcola Limestone is bored and possibly glauconitized; the limestone is partly broken up into pebbles. In an otherwise clastic succession, this suggests a considerable break at this horizon. Only 1.8 m above the Arcola Limestone at Tibbee Creek (Clay County, Mississippi) there is another disconformity with baculitids in a white phosphate. Such disconformities generally represent a regressive trough, but in a laminated marl and without a concentration of dark phosphatic nodules (as would be found in the English Gault), it could equally represent the break in sedimentation at a transgressive high.

Unpublished work on ammonites by W. A. Cobban and W. J. Kennedy support the idea that the upper disconformity is much younger than the Arcola Limestone. According to Smith & Mancini (1983) and Dowsett (1989) the Arcola Limestone belongs to the foraminiferal Zone of *Globotruncanita elevata*. Both the Arcola Limestone and the bottom 2.2 m of the Demopolis Formation contain *Bukryaster hayi* (Bukry) at the top of its range, and these beds belong to the lower part of the nannoflora Zone of *Calculites ovalis* (Zone 19a) (Smith & Mancini 1983; Dowsett 1989).

Whilst Peak no. 2 must lie somewhere low in the Demopolis Formation, it cannot be fixed satisfactorily without more detailed logs.

Peak no. 3. Above the Tombigbee Sand in eastern Mississippi and western Alabama there are some 300 m of marly and chalky sediments which at one time were known as the Selma Chalk, but which today are subdivided into the

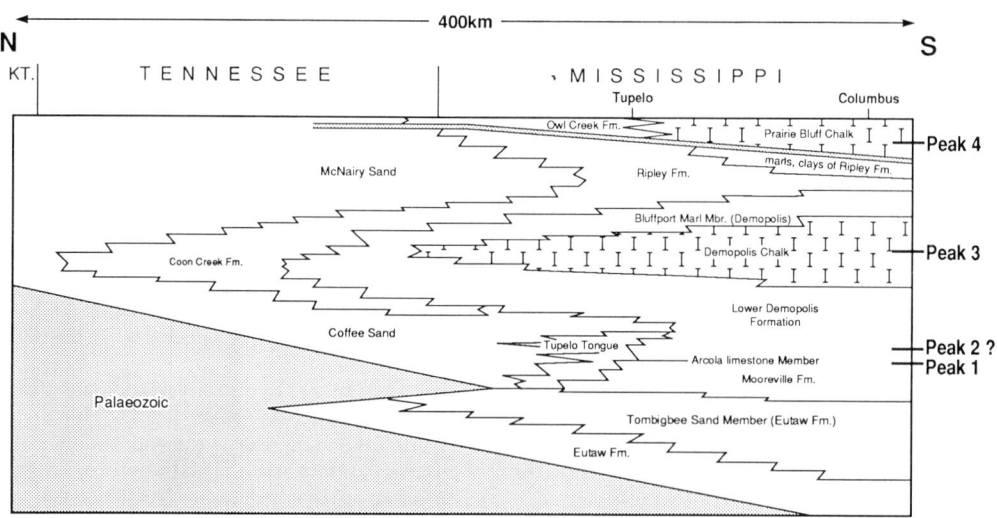

Fig. 3. Lithostratigraphical succession in the Campanian–Maastrichtian on the southwestern flank of the Appalachian massif to show positions of transgressive peaks; based on Russell (1986, fig. 3), with modifications.

Mooreville Formation (with the Arcola Limestone Member at its summit) below and the Demopolis Formation above, with the Bluffport Marl Member at its summit. Within the upper part of the Demopolis Formation there is a 50 m unit with 72–90% $CaCO_3$ content (Russell & Keady, 1983). Much of it is a chalk-marl facies but its centre has the lowest clay content and is without obvious rhythms; this is the closest of all the Alabama-Mississippi succession to ordinary European chalk, complete with occasional incipient hardgrounds. The maximum development and purest Demopolis Chalk represents Peak no. 3. This would appear to be not far above the base of the section on Alexander Schoolhouse Road north-east of Starkville in Oktibbeha County, Mississippi (Russell 1986, fig. 9). This level is assigned by Smith & Mancinci (1983, p. 24) to the upper part of the nannoplankton Zone of *Tetralithus trifidus* (Zone 22) as modified by Perch-Nielsen (1979). Taylor & Russell (1986) record *Globotruncanita calcarata* (Cushman) from 83 to 93 m above the base of the Demopolis Formation, i.e. the extinction of *G. calcarata* is below or at Peak no. 3 in Mississippi, since 93 m above the base falls in the middle of the Demopolis Chalk.

Even in the Demopolis Chalk of Mississippi there are vertical variations in the pelagic element, and in central Alabama there is an alternation between chalk-marl and a micaceous argillaceous silty sand, as the Chalk passes eastwards into Cusseta Sand, a pattern that Skotnicki & King (1989) interpret, correctly in my opinion, as three minor eustatic cycles. This detail has not yet been recognized in Europe.

Peak no. 4 is represented by the Prairie Bluff Chalk. The middle of this formation is in the nannoplankton Zone of *Lithraphidites quadratus* and the foraminiferal Subzone of *Globotruncana gansseri*. The absence of the Subzone of *Abathomphalus mayaroensis* has already been used as evidence that the upper Upper Maastrichtian is missing in this region (Smith & Mancini 1983; Donovan *et al.* 1988). This break is confirmed by the sea-level correlation, because the Clayton Formation, of Palaeocene age, rests directly on the Prairie Bluff Chalk in Alabama.

Texas–Arkansas

Texas forms a region in which it is difficult to date many of the changes of sea-level, which is surprising when one remembers how much research has been published on its Cretaceous stratigraphy, including a survey of transgressions on the San Marcos Platform in central Texas by Young (1986). It is true that near-shore facies which might indicate actual shoreline-positions are rare or difficult to date, but there are chalks in the succession, and further north, in Colorado and Kansas, developments of pelagic chalk can be used as markers of deeper water.

Fortunately, during the Late Albian, central Texas was stable and probably permanently submerged, as shown by the nearly complete successions in the Edwards and Georgetown Limestone Groups along the outcrop north of the Platform (Lozo 1959), and the Stewart City Reef Trend parallel to the outcrop but subsurface downdip (Rose 1972). Hence the date of the Late Albian transgressive peak, already referred to, is in the upper part of the Zone of *Mortoniceras equidistans*, represented by the upper part of the Duck Creek Formation. The difficulty of using such near shore areas for timing sea-level peaks and troughs is shown by the absence on the San Marcos Platform of sediments representing the early Turonian peak close to the *nodosoides–woollgari* zonal boundary. Whatever sediments were deposited, were removed again during the late Turonian regression.

Young (1986) showed that there was an inundation of the San Marcos platform represented by the Dessau Chalk which belongs to the Zone of *Submortoniceras tequesquitense*. Young placed this low in the Campanian which would suggest that it represents Peak no. 1. In fact the upper part of the Dessau Chalk contains *Marsupites* (Marks 1952; Young 1963); in European terms the Dessau Chalk is Upper Santonian, not Campanian (Hancock 1991). The transgressive peak in the Dessau Chalk is well below the top of the formation and corresponds to the European peak in the middle of the Zone of *Uintacrinus socialis*, below the *Marsupites* Zone. Peak No. 1 is higher in the Texan succession.

A more complete succession, in which one should be able to see the sedimentary effects of the Campanian–Maastrichtian transgressive–regressive history, lies in the far northeast of Texas and adjacent areas of Arkansas. There are four chalk formations known in this tract; Gober Chalk, Pecan Gap Chalk, Annona Chalk and Saratoga Chalk. Unfortunately, the very reason for a more complete succession is that there was contemporary downwarping in this East Texas Embayment (Oetking & Feray 1963). This tectonism has complicated the transgressive–regressive history.

Peak no. 1 is represented by the tongue of Gober Chalk in Fannin and Lamar Counties. This Chalk was assigned by Young (1963) to his Zone of *Delawarella delawarensis*, but as I have suggested (Hancock 1991), and has been confirmed by Cobban & Kennedy (1992), this oft-quoted index species is inexact. Thompson *et al.* (1978) put the Gober Chalk into the foraminiferal 'Zonule' of *Pseudotextularia elegans*, apparently now re-named the Zone of *Rugoglobigerina tradinghousensis* (Thompson *et al.* 1991). At present, the eustatic correlation with the western interior Zone of *Scaphites hippocrepis* 1 is the most exact.

Peaks Nos 2 and 3. About 100 m above the Gober Chalk in Fannin County lies the Pecan Gap Chalk, which represents Peak No. 2. Eastwards from Fannin County the Pecan Gap Chalk overlaps the Wolfe City Formation, to rest directly on the Ozan Formation, which itself passes laterally into Annona Chalk in central Red River County (Barnes 1966). The Annona Chalk rests directly on the Brownstown Marl Formation, which itself underlies the Gober Chalk (see also Thompson *et al.* 1991). Whether the maximum thickness of the Annona Chalk is *c.* 140 m (Barnes 1966) or not more than 50–70 m (Thompson *et al.* 1991), it clearly embraces both the Pecan Gap Chalk (and hence Peak No. 2) and older formations; it is impossible to say if it includes an equivalent of the Gober Chalk, but this seems possible. It also extends upwards into Peak No. 3.

Thus the Pecan Gap – Annona Chalk is known as Pecan Gap Chalk more than about 3 km southwest of Clarksville in Red River County, and Annona Chalk to the east and northeast. In spite of a gap in the outcrop south of Dallas (and according to Thompson *et al.* 1991 in western Red River County), this is the most widespread Chalk of Texas, for more than 400 km to the southwest of Fannin County there is still a broad outcrop mapped as Pecan Gap Chalk south of Austin (Barnes 1974); and there is a collapsed block of this Chalk in a graben of Dessau Chalk in Comal County, over the central part of the San Marcos Platform (Young 1986). Eastwards the Annona Chalk thins, the Ozan Formation re-appears beneath it (disconformably; Bottjer 1986), and disappears north of Columbus, Hempstead County, Arkansas, some 200 km east of Fannin county.

The base of the Pecan Gap Chalk near Austin contains a number of disconformities (Frizzel & Anderson 1950) representing an uncertain but probably only slight break. Ammonites from these basal beds are correlated by Cobban & Kennedy (in press) with the Zone of *Baculites*

asperiformis in the Western Interior, one zone higher than Peak No. 2 there.

There is some justification for using different names for this widespread Chalk formation in the main part of Texas from that in Arkansas and the far northeast of Texas. Whilst the downward limit of the Annona Chalk in Texas is uncertain, the upper limit certainly goes higher than the main mass of the Pecan Gap Chalk. The top of the Pecan Gap Chalk in Fannin County is unconformable but the summit of the Annona Chalk passes up into the overlying Marlbrook Marl (Rouse 1944). It probably reaches Peak No. 3 in Texas.

Further east in Arkansas the base of the Annona Chalk is a condensed glauconitic wackestone with phosphatised fossils (Bottjer 1986), including ammonites older than the main mass of the Chalk but equivalent to the Zones of *Baculites gregoryensis* to *Didymoceras stevensoni* in the Western Interior (W.J. Kennedy & W.A. Cobban, pers. comm.). Bottjer (1985, 1986) has shown that the Arkansas Annona Chalk includes beds with more than 90% $CaCO_3$, thus approaching the European whitechalk facies, as well as the more usual American chalk-marl facies. Water depths may have reached 125 m, deeper than the chalks below and above, fitting with the idea that this Annona Chalk reaches up to Peak No. 3, as already recognised by Bottjer (1986). But the mere presence of a chalk facies that connects part or all the succession between Peaks 2 and 3 must be the result of contemporaneous downwarping. When the Annona Chalk finally disappears near Columbus in Arkansas it does so rapidly; in the Okay quarry, northwest of Saratoga, it has become a chalk-marl facies with the clay beds thicker than the limestone beds.

According to Thompson *et al.* (1978) the upper part of the type Annona Chalk and the Annona Chalk in Arkansas belong to the foraminiferal Zone of *Globotruncanita calcarata*. Thompson *et al.* (1991) indicate that both the type Pecan Gap Chalk and type Annona Chalk extend above the extinction level of *G. calcarata*.

As the Annona Chalk disappears laterally northeastwards in Arkansas, the Saratoga Chalk comes in above the local representative of the Marlbrook Marl. The base of the Saratoga Chalk yields numerous *Nostoceas hyatti* (W.J. Kennedy & W.A. Cobban, pers. comm.), which correlates with the upper part of the Zone of *Belemnitella langei* in Europe. Therefore the Saratoga Chalk represents the tail end of Peak No. 3, rather than the centre. Since more recent collecting shows that *N. hyatti* extends some

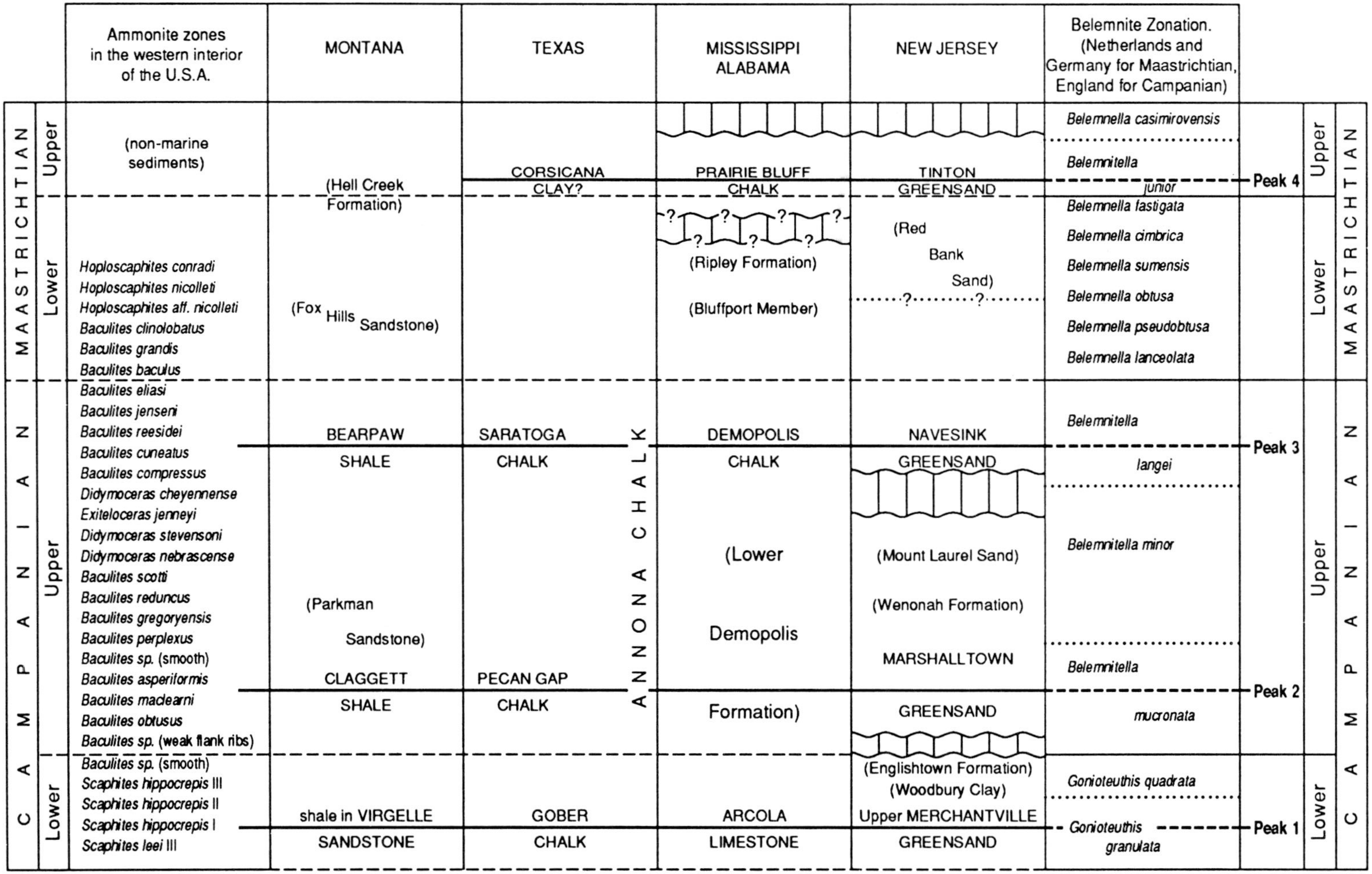
Ammonite zones in the western interior of the U.S.A.
MONTANA
TEXAS
MISSISSIPPI ALABAMA
NEW JERSEY
Belemnite Zonation. (Netherlands and Germany for Maastrichtian, England for Campanian)
MAASTRICHTIAN
CAMPANIAN
Upper
Lower
(non-marine sediments)
Hoploscaphites conradi
Hoploscaphites nicolleti
Hoploscaphites aff. nicolleti
Baculites clinolobatus
Baculites grandis
Baculites baculus
Baculites eliasi
Baculites jenseni
Baculites reesidei
Baculites cuneatus
Baculites compressus
Didymoceras cheyennense
Exiteloceras jenneyi
Didymoceras stevensoni
Didymoceras nebrascense
Baculites scotti
Baculites reduncus
Baculites gregoryensis
Baculites perplexus
Baculites sp. (smooth)
Baculites asperiformis
Baculites maclearni
Baculites obtusus
Baculites sp. (weak flank ribs)
Baculites sp. (smooth)
Scaphites hippocrepis III
Scaphites hippocrepis II
Scaphites hippocrepis I
Scaphites leei III
(Hell Creek Formation)
(Fox Hills Sandstone)
BEARPAW SHALE
(Parkman Sandstone)
CLAGGETT SHALE
shale in VIRGELLE SANDSTONE
CORSICANA CLAY?
SARATOGA CHALK
ANNONA CHALK
PECAN GAP CHALK
GOBER CHALK
PRAIRIE BLUFF CHALK
(Ripley Formation)
(Bluffport Member)
DEMOPOLIS CHALK
(Lower Demopolis Formation)
ARCOLA LIMESTONE
TINTON GREENSAND
(Red Bank Sand)
NAVESINK GREENSAND
(Mount Laurel Sand)
(Wenonah Formation)
MARSHALLTOWN GREENSAND
(Englishtown Formation)
(Woodbury Clay)
Upper MERCHANTVILLE GREENSAND
Belemnella casimirovensis
Belemnitella junior
Belemnella fastigata
Belemnella cimbrica
Belemnella sumensis
Belemnella obtusa
Belemnella pseudobtusa
Belemnella lanceolata
Belemnitella langei
Belemnitella minor
Belemnitella mucronata
Gonioteuthis quadrata
Gonioteuthis granulata
Peak 4
Peak 3
Peak 2
Peak 1

Fig. 4. Suggested correlations at peaks of transgression in the Campanian–Maastrichtian between northwest Europe and the USA.

metres up in the Saratoga Chalk, it is doubtful if the formation extends above the Campanian.

Pessagno (1969) assigned the Saratoga Chalk to the Subzone of *Rugotruncana subcircumnodifer*; it is possibly high in the Zone of *Globotruncana aegyptica* of Caron (1985). Nannofossil Sub-zones CC22C and CC23A are both represented, which confirms a top Campanian age (Burnett *et al*. in press).

Peak no. 4 has not been recognized with certainty. Thompson *et al*. (1991, fig. 23) seem to indicate the Kemp Clay as a transgressive formation in northeast Texas. The known tectonic interference there makes it more probable that Young (1986) is correct in placing a peak in the Corsicana Clay which extends on to the San Marcos Platform. However, the confusion in the use of the names Kemp and Corsicana is such that they may be talking about the same transgression. Of interest, is the existence locally of up to 3 m of 'almost pure chalk' immediately on top of the Nacatoch Sand Formation in Falls to Franklin Counties (north-central Texas) (Crawford in Sellards *et al*. 1933, p. 487). This might represent the *Baculites clinolobatus* peak of the Western Interior.

Texas is a warning that even in a region with chalks, the most convenient indicator of a high sea-level, the interpretation may be difficult without any biostratigraphic check. The absence of a chalk-facies in central Texas above Peak no. 2, in a region with substantial older chalk (the Austin Chalk), implies a northeastward tilt in Texas very late in the Campanian. The existence of more than one peak in the Annona Chalk of Red River County shows that for a time this was a basinal centre in the East Texas Embayment. The Saratoga Chalk represents a brief return of chalk in Arkansas during Peak no. 3 outside the Embayment.

Conclusions

(1) Transgressive peaks (eustatic highs of sea-level) of the Campanian–Maastrichtian, dated in northwest Europe in a pelagic chalk facies, can be recognised, with various degrees of certainty, in the western interior, southern and eastern regions of the USA.

(2) The great transgressive peak, a little above the base of the Zone of *Belemnitella langei* as used in England (Peak no. 3), occurred near the top of the foraminiferal Zone of *Globotruncanita calcarata*. Contrary to common belief, e.g. Pessagno (1969), Caron (1985), the *G. calcarata* Zone does not mark the top of the Campanian; it is not possible at present to quantify how much higher lies the base of the

Maastrichtian, but it must be of the order of 10–20% of the whole Campanian. The illusion that the *calcarata* Zone marked the summit of the Campanian seems to have originated with Dalbiez (1960) who equated the range of *G. calcarata* with the ammonite Zone of *Bostrychoceras polyplocum*, which, in France at the time, was regarded as the top zone of the Campanian. This viewpoint was then accepted by Pessagno (1969) in his very detailed study of the Gulf Coast region.

(3) The Campanian–Maastrichtian boundary in the western interior of the USA was no lower than the junction between the Zones of *Baculites jenseni* and *B. eliasi*, and is more likely to be at the base of the Zone of *Baculites baculus*. If one uses the scale of Olsson (1988) in the eastern United States the boundary would be around the base of the Zone of *Globotruncanita stuarti*. It is more likely to be at or a little below the base of the Zone of *Gansserina gansseri* (see conclusion 6).

(4) The sea had withdrawn from the western interior of the USA, or at least in the area represented by outcropping Cretaceous, before the start of the Late Maastrichtian.

(5) The relative sea-level near the end of the Early Cretaceous compared with that during the Campanian–Maastrichtian was the reverse of that offered by the Exxon curve (Haq *et al*. 1988). In all the American regions studied for this paper the Campanian sea-levels were higher and their marine sediments more extensive than those of the Upper Albian. Similar relative levels are found in northwest Europe (Hancock 1990). In the Exxon curve the sea-levels of the Late Albian are about the same as those of the 'mid-Campanian'.

(6) The detailed peaks and troughs of sea-level during the Campanian–Maastrichtian shown in the Exxon curve are not easily related to those recognized in this paper. Some of this may be due to the mish-mash of biostratigraphy used in the Exxon table for this part of the stratigraphic column. It is possible that the relations will turn out to be as follows: Peak no. 1 could be Exxon downlap-surface 83.75 (but this is too old in comparison with Peak no. 1 in England); the Exxon peak just after downlap-surface 82 is too young; indeed, it appears to correspond to a regressive trough in the English succession.

Peak no. 2 is possibly a combination of Exxon downlap-surfaces 79.5, 78 and 76, but no-one of these is the same age. Peak no. 3 could be the high after downlap-surface 73.5. Peak no. 4 could be the high after downlap-surface 69.5. If these last two have been correctly equated, the base of the Maastrichtian would be at the base of the foraminiferal Zone of *Gansserina gansseri* in the Exxon table (which lies within the top part of the Zone of *Globotruncana aeyptiaca* of tropical areas; see Caron 1985, fig. 5); and the base of nannoplankton Zone of *Arkhangelskiella cymbiformis*; the latter agrees with Perch-Nielsen's observations (1979) on the chalk succession in Denmark. For more detailed discussion of the Campanian-Maastrichtian boundary, see Burnett *et al.* (in press).

I am indebted to W. A. Cobban and W. J. Kennedy for keeping me up to date with their unpublished research. In the interpretation of the succession in Mississippi, I have been saved from several mistakes by E. Russell. Field work in the western interior of the USA, partly paid for by the Natural Environment Research Council, was expertly guided by W. A. Cobban and E. Kauffman. Field work in New Jersey, Alabama and Mississippi would have been impossible without the help of N. Sohl. B. and C. Sperandio gave me hospitality and lent me a car in Texas. R. Parish did almost all the driving and worked as an unpaid field assistant. I was met with friendliness by landowners everywhere except in Wyoming.

References

Avé Lallemant, H. G. & Oldow, J. S. 1988. Early Mesozoic southward migration of Cordilleran transpressional terranes. *Tectonics*, **7**, 1057–1075.

Bailey, H. W., Gale, A. S., Mortimore, R. N., Swiecicki, A. & Wood, C. J. 1983. The Coniacian-Maastrichtian stages of the United Kingdom, with particular reference to southern England. *Newsletters on Stratigraphy*, **12**, 29–42.

Barnes, V. E. 1966. *Geologic atlas of Texas, Texarkana sheet*. University of Texas, Bureau of Economic Geology.

—— 1974. *Geologic atlas of Texas, Austin sheet*. University of Texas, Bureau of Economic Geology.

Birkelund, T. 1982. Maastrichtian ammonites from Hemmoor, Neiderelbe (NW-Germany). *In*: Schmid, F. (ed.) *Die Maastricht-Stufe in NW-Deutschland*, **2**, 13–33, Geologisches Jahrbuch, (A) 61.

——, Hancock, J. M., Hart, M. B., Rawson, P. F., Remane, J., Robaszynski, F., Schmid, F. & Surlyk, F. 1984. Cretaceous stage boundaries – proposals. *Bulletin of the Geological Society of Denmark*, **33**, 3–20.

Blaszkiewicz, A. 1980. *Campanian and Maastrichtian ammonites of the middle Vistula River valley, Poland: a stratigraphic–paleontological study*. Prace Instytutu Geologicznego, 92.

Bottjer, D. J. 1985. Trace fossils and paleoenvironments of two Arkansas Upper Cretaceous discontinuity surfaces. *Journal of Paleontology*, **59**, 282–298.

—— 1986. Campanian-Maastrichtian Chalks of southwestern Arkansas: petrology, paleoenvironments and comparison with other North American and European Chalks. *Cretaceous Research*, **7**, 161–196.

Brown, P. M., Miller, J. A. & Swain, F. M. 1972. *Structural and stratigraphic framework, and spatial distribution of permeability of the Atlantic coastal plain, North Carolina to New York. Professional Papers of the U.S. Geological Survey*, 796.

Burnett, J. A., Hancock, J. M., Kennedy, W. J. & Lord, A. R. in press. Macrofossil, planktonic foraminiferal and nannofossil zonation at the Campanian-Maastrichtian boundary. *Newsletters on Stratigraphy*.

Burton, R., Kendall, C. G. St. C. & Lerche, I. 1987. Out of our depth: on the impossibility of fathoming eustasy from the stratigraphic record. *Earth-Science Reviews*, **24**, 237–277.

Caron, M. 1985. Cretaceous planktic foraminifera. *In*: Bolli, H. M., Saunders, J. B. & Perch-Nielsen, K. (eds) *Plankton stratigraphy*. Cambridge University Press, 17–86.

Christensen, W. K. 1986. *Upper Cretaceous belemnites from the Vomb Trough in Scania, Sweden*. Sveriges Geologiska Undersökning, **Ca 57**.

—— 1990. Upper Cretaceous belemnite stratigraphy of Europe. *Cretaceous Research*, **11**, 371–386.

—— 1991. Belemnites from the Coniacian to Lower Campanian Chalks of Norfolk and southern England. *Palaeontology*, **34**, 695–749.

Cobban, W. A. 1974. *Ammonites from the Navesink Formation at Atlantic Highlands, New Jersey*. Professional Papers of the US Geological Survey, **845**.

—— & Kennedy, W. J. 1992. Campanian *Trachyscaphites spiniger* ammonite fauna in north-east Texas. *Palaeontology*, **35**, 63–93.

—— & —— in press. *Middle Campanian (Upper Cretaceous) ammonites from the Pecan Gap Chalk in central and northeastern Texas*. Bulletin of the US Geological Survey.

Cook, T. D. & Bally, A. W. (eds) 1975. *Stratigraphic atlas of North and Central America*. Princeton University Press, Princeton.

Dalbiez, F. 1960 (mis-dated 1959) Corrélations et résolutions. *Comptes rendus du Congrès des Sociétés Savantes de Paris et des Départements tenu à Dijon en 1959. Section des Sciences, Sous-section de Géologie*: Colloque sur le Crétacé supérieur française, 857–867.

Donovan, A. D., Baum, G. R., Blechschmidt, G. L., Loutit, T. S., Pflum, C. E. & Vail, P. R. 1988. Sequence stratigraphic setting of the Cretaceous–Tertiary boundary in central Alabama. *In*: Wilgus, C. K., Posamentier, H., Ross, C. A. & Kendall, G. St. C. (eds) *Sea-level changes: an integrated approach*. Special Publications of the Society of Economic Paleontologists and Mineralogists, 42, 299–307.

Dowsett, H. J. 1989. *Documentation of the Santonian–Campanian and Austinian–Tayloran stage boundaries in Mississippi using calcareous micro-*

fossils. Bulletin of the United States Geological Survey, 1884.

ENGEBRETSON, D. C., COX, A. & GORDON, R. G. 1985. *Relative motions between oceanic and continental plates in the Pacific basin.* Special Papers of the Geological Society of America, **206**.

ERNST, G. 1963. Stratigraphische und gesteinchemische untersuchungen im Santon und Campan von Lägerdorf (SW-Holstein). *Mitteilungen aus dem Geologischen Staatsinstitut in Hamburg,* **32**, 71–127.

FRIZZEL, D. L. & ANDERSON, I. J. 1950. Diastems in the Pecan Gap Chalk of Travis County, Texas. *Journal of Sedimentary Petrology,* **20**, 55–59.

GILL, J. R. & COBBAN, W. A. 1966. *The Red Bird section of the Upper Cretaceous Pierre Shale in Wyoming.* Professional Papers of the United States Geological Survey, **393-A**.

—— & —— 1973. *Stratigraphy and geologic history of the Montana Group and equivalent rocks, Montana, Wyoming, and North and South Dakota.* Professional Papers of the United States Geological Survey, **776**.

——, —— & SCHULTZ, L. G. 1972. Correlation, ammonite zonation, and a reference section for the Montana Group, central Montana. *Montana Geological Society, 21st Annual Field Conference,* 91–97.

GLOBERMAN, G. R., IRVING, E. & MARQUIS, G. in press. Paleolatitude of cratonic North America, 120–80 Ma, and displacements in the western Cordillera: paleomagnetic evidence. *In*: CALDWELL, W. G. E & KAUFFMAN, E. G. (eds) *Evolution of the Western Interior foreland basin.* Special Papers of the Geological Association of Canada.

HANCOCK, J. M. 1975. The sequence of facies in northern Europe compared with that in the western interior, *In*: CALDWELL, W. G. E (ed.) *The Cretaceous system in the Western Interior of North America.* Special Papers of the Geological Association of Canada, **13**, 83–118.

—— 1990. Sea-level changes in the British region during the Late Cretaceous. *Proceedings of the Geologists' Association,* **100** (for 1989), 565–594.

—— 1991. Ammonite scales for the Cretaceous system. *Cretaceous Research,* **12**, 259–291.

—— & KAUFFMAN, E. G. 1979. The great transgressions of the Late Cretaceous. *Journal of the Geological Society, London,* **136**, 175–186.

—— & —— 1989. Use of eustatic changes of sea level to fix Campanian-Maastrichtian boundary in Western Interior of USA. *International Geological Congress 28 (Washington, D.C.),* Abstracts, **2**, 23.

——, KENNEDY, W. J. & COBBAN, W. A. in press. A correlation of the Upper Albian to basal Coniacian sequences of north-west Europe, Texas and the United States Western Interior, *In*: CALDWELL, W. G. E. & KAUFFMAN, E. G. (eds) *Evolution of the Western Interior Foreland Basin.* Special Papers of the Geological Association of Canada.

HAQ, B. U., HARDENBOL, J. & VAIL, P. R. 1988. Mesozoic and Cenozoic chronostratigraphy and cycles of sea-level change. *In*: WILGUS, C. K., P-OSAMENTIER, H., ROSS, C. A. & KENDALL, C. G. ST. C. (eds) *Sea-level changes: an integrated approach.* Special Publications of the Society of Economic Paleontologists and Mineralogists, 42, 71–108.

HELLER, P. L., BOWDLER, S. S., CHAMBERS, H. P., COOGAN, J. C., HAGEN, E. S., SHUSTER, M. W. & WINSLOW, N. S. 1986. Time of initial thrusting in the Sevier orogenic belt, Idaho-Wyoming and Utah. *Geology,* **14**, 388–391.

HESSELBO, S. P., COE, A. L. & JENKYNS, H. C. 1990. Recognition and documentation of depositional sequences from outcrop: an example from the Aptian and Albian on the eastern margin of the Wessex Basin. *Journal of the Geological Society, London,* **147**, 549–559.

HOFKER, J. 1958. Foraminifera from the Cretaceous of Limburg, Netherlands, 38. The gliding change in Bolivinoides during time. *Natuurhistorisch Maandblad, Maastricht,* **47**, 145–159.

JELETZKY, J. A. 1962. Cretaceous belemnites of New Jersey. *In*: RICHARDS, H. G. *et al. The Cretaceous fossils of New Jersey.* Bulletin of the Geological Survey of New Jersey, Paleontology Series, **61**, 139–161.

KAUFFMAN, E. G. 1977. Geological and biological overview: western interior Cretaceous basin. *The Mountain Geologist,* **14**, 75–99.

KENNEDY, W. J. & SUMMESBERGER, H. 1987. Lower Maastrichtian ammonites from Nagoryany (Ukrainian SSR). *Beiträge zur Paläontologie von Österreich,* **13**, 25–78.

KITELEY, L. W. 1983. Paleogeography and eustatic-tectonic model of Late Campanian (Cretaceous) sedimentation, southwestern Wyoming and northwestern Colorado. *In*: REYNOLDS, M. W. & DOLLY, E. D. (eds) *Mesozoic paleogeography of the west-central United States. Rocky Mountain Paleogeography Symposium,* **2**. Rocky Mountain section of the Society of Economic Paleontologists and Mineralogists, 273–302.

KONGIEL, R. 1962. On belemnites from Maastrichtian, Campanian and Santonian sediments in the Middle Vistula valley. *Prace Muzeum Ziemi,* **5**, 3–148, pls. 1–21.

LILLEGRAVEN, J. A. 1991. Stratigraphic placement of the Santonian-Campanian boundary (Upper Cretaceous) in the North American Gulf Coastal Plain and Western Interior, with implications to global geochronology. *Cretaceous Research,* **12**, 115–36.

LOZO, F. E. 1959. Stratigraphic relations of the Edwards Limestone and associated formations in north-central Texas. *In*: LOZO, F. E. *et al.* (eds) *Symposium on Edwards Limestone in central Texas,* University of Texas Publication, **5905**, 1–19.

MARKS, E. 1952. Occurrence of Santonian crinoid in western Gulf Region. *American Journal of Science,* **250**, 226–227.

MATSUMOTO, T. & NODA, M. 1986. Middle Cretaceous inoceramid stratigraphy, *In*: REYMENT, R. A. &

BENGTSON, P. (eds) *Events of the Mid-Cretaceous. Physics and Chemistry of the Earth*, **16**, 145–153.

MEIJER, M. 1965. The stratigraphical distribution of echinoids in the Chalk and Tuffaceous Chalk in the neighbourhood of Maastricht (Netherlands). *Mededelingen van de Gelogische Stichting, Maastricht*, **17**, 21–25.

MOLENAAR, C. M. 1983. Major depositional cycles and regional correlations of Upper Cretaceous rocks, southern Colorado plateau and adjacent areas. *In*: REYNOLDS, M. W. & DOLLY, E. D. (eds) *Mesozoic paleogeography of the west-central United States. Rocky Mountain Paleogeography Symposium 2*. Rocky Mountain section of the Society of Economic Paleontologists and Mineralogists, 201–224.

MONROE, W. H. 1941. *Notes on deposits of Selma and Ripley age in Alabama*. Bulletin of the Geological Survey of Alabama, **48**.

MOSKVEEN, M. M. 1986–87. Cretaceous system. *In*: *Stratigraphy of the USSR*. Moscow, Academy of Science of the USSR. (in Russian).

NYONG, E. E. & OLSSON, R. K. 1984. A paleoslope model of Campanian to lower Maestrichtian foraminifera in the North American Basin and adjacent continental margin. *Marine Micropaleontology*, **8**, 437–477.

OBRADOVICH, J. D. 1988. A different perspective on glauconite as a chronometer for geologic time scale studies. *Paleoceanography*, **3**, 757–770.

OETKING, P. F. & FERAY, D. E. 1963. *Geological highway map of Texas*. Dallas Geological Society.

OLSSON, R. K. 1988. Foraminiferal modeling of sea-level change in the Late Cretaceous of New Jersey. *In*: WILGUS, C. K., POSAMENTIER, H., ROSS, C. A. & KENDALL, C. G. ST. C. (eds) *Sea-level changes: an integrated approach*. Special Publications of the Society of Economic Paleontologists and Mineralogists, 42, 289–297.

—— & NYONG, E. E. 1984. A paleoslope model for Campanian – lower Maestrichtian foraminifera of New Jersey and Delaware. *Journal of Foraminiferal Research*, **14**, 50–68.

——, GIBSON, T. G., HANSEN, H. J. & OWENS, J. P. 1988. Geology of the northern Atlantic coastal plain: Long Island to Virginia. *In*: SHERIDAN, M. S. & GROW, J. A. (eds) *The Atlantic Continental Margin: U.S.* The Geology of North America, **I-2**. Geological Society of America. 87–105.

OWENS, J. P. & GOHN, G. S. 1985. Depositional history of the Cretaceous series in the U.S. Atlantic coastal plain: Stratigraphy, paleoenvironments, and tectonic controls of sedimentation. *In*: POAG, C. W. (ed.) *Geological evolution of the United States Atlantic Margin*. Van Nostrand Reinhold, New York, 25–86.

——, MINARD, J. P., SOHL, N. F. & MELLO, J. F. 1970. *Stratigraphy of the outcropping post-Magothy Upper Cretaceous formations in southern New Jersey and northern Delmarva Peninsula, Delaware and Maryland*. Professional Papers of the US Geological Survey, **674**.

PEAKE, N. B. & HANCOCK, J. M. 1970. The Upper Cretaceous of Norfolk. *In*: LARWOOD, G. P. &

FUNNELL, B. M. (eds) *The geology of Norfolk*. Geological Society of Norfolk, Norwich, 293–339.

PERCH-NIELSEN, K. P. 1979. Calcareous nannofossils from the Cretaceous between the North Sea and the Mediterranean. *In*: WIEDMANN, J. (ed.) *Aspekte der Kreide Europas*. International Union of Geological Sciences, Series A, no. 6, 223–272.

PESSAGNO, E. A. JR. 1969. *Upper Cretaceous stratigraphy of the Western Gulf Coast area of Mexico, Texas and Arkansas*. Memoirs of the Geological Society of America, Ill.

PETERSON, J. A. 1988. Phanreozoic stratigraphy of the northern Rocky Mountain region. *In*: SLOSS, L. L. (ed.) *Sedimentary Cover – North American Craton: U.S.* The Geology of North America, **D-2**. Geological Society of America, 83–107.

PETTERS, S. W. 1976. Upper Cretaceous subsurface stratigraphy of Atlantic coastal plain of New Jersey. *Bulletin of the American Association of Petroleum Geologists*, **60**, 87–107.

POSAMENTIER, H. W., JERVEY, M. T. & VAIL, P. R. 1988. Eustatic controls on clastic deposition I– conceptual framework. *In*: WILGUS, C. K., POSAMENTIER, H., ROSS, C. A. & KENDALL, C. G. ST. C. (eds) *Sea-level changes: an integrated approach*. Special Publications of the Society of Economic Paleontologists and Mineralogists, **42**, 109–124.

—— & VAIL, P. R. 1988. Eustatic controls on clastic deposition II – sequence and systems tract models. *In*: WILGUS, C. K., POSAMENTIER, H., ROSS, C. A. & KENDALL, C. G. ST. C. (eds) *Sea-level changes: an integrated approach*. Special Publications of the Society of Economic Paleontologists and Mineralogists, **42**, 125–154.

RAWSON, P. F., CURRY, D., DILLEY, F. C., HANCOCK, J. M., KENNEDY, W. J., NEALE, J. W., WOOD, C. J. & WORSSAM, B. C. 1978. *A correlation of Cretaceous rocks in the British Isles*. Special Reports of the Geological Society of London, **9**.

ROBASZYNSKI, F., BLESS, M. J. M., FELDER, P. J., FOUCHIER, J.-C., LEGOUX, O., MANIVIT, H., MEESSEN, J. P. M. T. & VAN DER TUUK, L. A. 1985. The Campanian-Maastrichtian boundary in the chalky facies close to the type-Maastrichtian area. *Bulletin des Centres de Recherches Exploration-Production Elf-Aquitaine*, **9**, 1–113.

ROSE, P. R. 1972. *Edwards Group, surface and subsurface, central Texas*. Bureau of Economic Geology, University of Texas, Report of Investigations, **74**.

ROUSE, J. T. 1944. Correlation of the Pecan Gap, Wolfe City and Annona Formations in east Texas. *Bulletin of the American Association of Petroleum Geologists*, **28**, 522–530.

RUSSELL, E. E. 1986. Shelf marls and chalks in the marine section of the Upper Cretaceous: Mississippi. *In*: NEATHERY, T. L. (ed.) *Southeastern section of the Geological Society of America*. Geological Society of America, Centennial Field Guide 6, 387–392.

—— & KEADY, D. M. 1983. Notes on Upper Cretaceous lithostratigraphy of the eastern Mississippi embayment. *In*: RUSSELL, E. E. *et al. Upper Cretaceous lithostratigraphy and biostratigraphy in northeast Mississippi, southwest Tennessee and northwest Alabama, shelf chalks and coastal clastics.* Society of Economic Paleontologists and Mineralogists, Spring Field Trip, April 1983, 1–15.

SCHMID, F. & ERNST, G. 1975. Ammoniten aus dem Campan der Lehrter Westmulde und ihre stratigraphische bedeutung. 1. Teil: *Scaphites, Bostrychoceras und Hoplitoplacenticeras. Bericht der Naturhistorischen Gesellschaft zu Hannover,* **119**, 315–359.

SCHÖNFELD, J. 1988. *Zur stratigraphie und ökologie benthischer foraminiferen im schreibkreiderichtprofil von Lägerdorf/Holstein.* Dissertation der Mathematisch-Naturwissenschaftlichen Fakultät der Christian-Albrechts-Universität zu Kiel.

SCHULZ, M.-G. 1978. Zur litho- und biostratigraphie des Obercampan-Untermaastricht von Lägerdorf und Kronsmoor (SW-Holstein). *Newsletters on Stratigraphy,* **7**, 73–89.

——, ERNST, G., ERNST, H. & SCHMID, F. 1984. Coniacian to Maastrichtian stage boundaries in the standard section for the Upper Cretaceous white chalk of NW Germany (Lägerdorf-Kronsmoor-Hemmoor): definitions and proposals. *Bulletin of the Geological Society of Denmark,* **33**, 203–215.

SELLARDS, E. H., ADKINS, W. S. & PLUMMER, F. B. 1933 (mis-dated 1932). *The geology of Texas: 1. Stratigraphy.* University of Texas Bulletin, **3232**.

SKOTNICKI, M. C. & KING, D. T. JR 1989. Stratigraphic revision and facies analysis of the Upper Cretaceous Cusseta Sand, coastal plain of Alabama. *South-Eastern Geology,* **29**, 235–253.

SMITH, C. C. & MANCINI, E. A. 1983. Calcareous nannofossil and planktonic foraminifera biostratigraphy. *In*: RUSSELL, E. E. *et al. Upper Cretaceous lithostratigraphy and biostratigraphy in northeast Mississippi, southwest Tennessee and northwest Alabama, shelf chalks and coastal clastics.* Society of Economic Paleontologists and Mineralogists, Spring Field Trip, April 1983, 16–28.

STECKLER, M. S., WATTS, A. B. & THORNE, J. A. 1988. Subsidence and basin modeling at the U.S. Atlantic passive margin. *In*: SHERIDAN, R. E. & GROW, J. A. (eds) *The Atlantic Continental Margin: U.S.* The Geology of North America, **I-2** Geological Society of America, 399–416.

STEPHENSON, L. W. & MONROE, W. H. 1940. *The Upper Cretaceous deposits.* Mississippi State Geological Survey Bulletin, **40**.

SURLYK, F. 1984. The Maastrichtian stage in NW Europe, and its brachiopod zonation. *Bulletin of the Geological Society of Denmark,* **33**, 217–223.

TAYLOR, R. H. & RUSSELL, E. A. 1986. The *Globotruncana calcarata* zone (latest Campanian) in the

Demopolis Formation, Lowndes County, Mississippi. *Abstracts of Papers of the Southeastern and South-Central Sections of the Geological Society of America,* 1986, 268.

THOMPSON, L. B., PERCIVAL, S. F. & PATRICELLI, J. A. 1978. Stratigraphic relationships of the Annona Chalk and Gober Chalk (Upper Campanian) type localities in northeast Texas and southwest Arkansas. *Transactions of the Gulf Coast Association of Geological Societies,* **28**, 665–671.

——, HEINE, C. J., PERCIVAL, S. F. JR & SELZNICK, M. R. 1991. Stratigraphy and micropaleontology of the Campanian shelf in northeast Texas. *Micropaleontology Special Publication,* **5**, 1–147.

TILLING, R., KLEPPER, M. R. & OBRADOVICH, J. D. 1968. K-Ar ages and time span of emplacement of the Boulder batholith, Montana. *American Journal of Science,* **266**, 671–689.

WAGONER, J. C. VAN, POSAMENTIER, H. W., MITCHUM, R. M., VAIL, P. R., SARG, J. F., LOUTIT, T. S. & HARDENBOL, J. 1988. An overview of the fundamentals of sequence stratigraphy and key definitions. *In*: WILGUS C. K., POSAMENTIER, H., ROSS, C. A. & KENDALL, C. G. ST. C. (eds) *Sea-level changes: an integrated approach.* Special Publications of the Society of Economic Paleontologists and Mineralogists, **42**, 29–45.

WATTS, A. B. 1982. Tectonic subsidence, flexure, and global changes of sea level. *Nature,* **297**, 469–474.

WEIMER, R. J. 1984. Relation of unconformities, tectonics, and sea-level changes, Cretaceous of Western Interior, U.S.A. *In*: SCHLEE, J. S. (ed.) *Interregional unconformities and hydrocarbon accumulation.* Memoirs of the American Association of Petroleum Geologists, **36**, 7–35.

—— 1988. Record of relative sea-level changes, Cretaceous of western interior, USA. *In*: WILGUS, C. K., HASTINGS, B. S., POSAMENTIER, H., WAGONER, J. VAN, ROSS, C. A. & KENDALL, C. G. ST. C. (eds) *Sea-level changes: an integrated approach.* Special Publication of the Society of Economic Paleontologists and Mineralogists, **42**, 285–288.

WILLIAMS, G. C. & STELCK, C. R. 1975. Speculations on the Cretaceous palaeogeography of North America. *In*: CALDWELL, W. G. E. (ed.) *The Cretaceous system in the Western Interior of North America.* Special Papers of the Geological Association of Canada, **13**, 1–20.

WOOD, C. J. 1988. The stratigraphy of the Chalk of Norwich. *Bulletin of the Geological Society of Norfolk,* **38**, 3–120.

YOUNG, K. 1963. Upper Cretaceous ammonites from the Gulf Coast of the United States. *University of Texas Publications,* **6304**, 1–373, 82 pls.

—— 1986. Cretaceous, marine inundations of the San Marcos Platform, Texas. *Cretaceous Research,* **7**, 117–140.

High resolution biostratigraphy

JOHN C. W. COPE

Department of Geology, University of Wales, PO Box 914, Cardiff CF1 3YE, UK

Abstract: The degree of stratigraphical resolution obtainable from biostratigraphical data provides, for most parts of the Phanerozoic, a finer discrimination than that obtained by other means. The full potential of biostratigraphy has not yet been realised in many parts of the geological column and much time and effort, in many cases including primary taxonomic work, is needed, in a field which is currently considered unfashionable. Examples are selected from the Phanerozoic to illustrate the degree of resolution currently achievable.

At a meeting entitled 'High resolution stratigraphy' I felt that someone should 'fly the flag' for biostratigraphy; for not only is this subject considered unfashionable to the extent that primary biostratigraphical studies involving detailed bed-by-bed collecting from a series of successions nowadays seldom attracts would-be palaeontological research students, but even when the enthusiastic student has been found, such topics seem to be a low priority for funding by the research councils. Some of the prevailing attutides seem to persist because many otherwise well-informed individuals believe that the limits of resolution in biostratigraphy were reached decades ago and that there is nothing more to do. Likewise, there is the feeling that modern 'black box' approaches offer far more precision and reliability and thus are to be preferred and encouraged. My task here is, therefore, to demonstrate that biostratigraphy is alive and well, that considerable advances are still being made and that, for most parts of the Phanerozoic, biostratigraphy offers far higher resolution than that obtainable by any other method.

One of the best ways to illustrate what is achievable is to cite examples of what degree of biostratigraphical resolution is obtainable in various parts of the Phanerozoic. It should perhaps be noted that the paper on this subject was read at the second of the two meetings convened under the title, as this second meeting was concerned primarily with pre-Tertiary stratigraphy. I raise this point because the Tertiary, in ideal sections, is the one interval for which biostratigraphy cannot compete well with the new techniques of magnetostratigraphy, high-precision radiometric dating etc. However, even here it plays a fundamental role in stratigraphical appraisal, especially in hydrocarbon exploration. In a recent review of stratigraphical resolution in the Cenozoic, Miller (1990) states that 100 000 year resolution is achievable 'in special cases'; in pelagic sections *c.* 500 000 years, and for the middle part of the Oligocene *c.* 2 000 000 years.

In the Cenozoic, there are no good macro-palaeontological guide fossils and biostratigraphy is provided virtually exclusively by microfossils. Some of these provide good biostratigraphical control, but it is increasingly being demonstrated that some groups of microorganisms, often used biostratigraphically by oil companies (for obvious reasons), are more frequency more facies-dependent than some members of the macrofauna. Reservations about their use in the Jurassic (Torrens *in* Cope 1980*a*) have been confirmed in several places since then, a recent example being the demonstration of the facies dependence of foraminiferids and particularly ostracodes in the Upper Jurassic (Wignall 1990). Another example may be quoted from the Lower Carboniferous; here the validity of the foraminiferid zonation is, on occasion, limited by the ease of re-working of the fossils and their incorporation into younger sediments. This problem has proved particularly acute in oil company work, where zones are often documented by first (top) down-hole occurrence. Thus, in the Craven Basin of Northern England, confusion resulting from the re-working of foraminiferids from the basin margin has been resolved in part by the application of macropalaeontological criteria; the combination of ammonoids and trilobites provides a higher degree of biostratigraphical resolution and obviates problems caused by re-working of microfossils, which, in any event, do not yet provide such high resolution (N.J. Riley 1990 and pers. comm.). Microfossils come into their own, however, when moving from the basinal to the shelf carbonate facies; here it becomes clear that the macrofaunas of ammonoids (and to a lesser extent trilobites) are restricted to the basinal facies: the foraminiferids provide the means to correlate between the two facies. Examination

From HAILWOOD, E. A. & KIDD, R. B. (eds), *High Resolution Stratigraphy*
Geological Society Special Publication, No. 70, pp. 257–265.

257

of the ranges in the Craven Basin of the various fossil groups, macrofossil and microfossil (Riley 1990), demonstrates that the greatest resolution is achievable when the biostratigraphical data from macrofossils is combined with that from microfossils. In this way, far greater resolution is obtainable than from any group considered in isolation. This fact has not been fully appreciated by all of those whose prime aim is biostratigraphical resolution; many workers believe, often mistakenly, that their particular specialist fossil group provides unassailable data. It is not out of place here to suggest that anyone (and this includes commercial undertakings) seeking the highest resolution would be well advised to use macropalaeontological data too, whenever they may be available.

Just what can be achieved with macrofossils? I will now look at some examples to demonstrate what degree of resolution is achievable in various parts of the Phanerozoic. To provide the data, I have used available radiometric dates for the base and top of each system (or part of a system) and divided the duration thus obtained by the number of biostratigraphical subdivisions available for that period. A possible alternative approach to obtaining such data was outlined by Schindel (1980) who calculated the ranges of sediment thickness accumulating under differing sedimentary environments; he was able to demonstrate that very fine scale collecting techniques would be needed to resolve fine details of speciation. Although Schindel's methods could provide an independent assessment of the time interval during which particular biozones occurred, the averaging out of the time intervals of a whole series of biozones or sub-biozones over the duration of a geological Period must give a good idea of the *average* duration of a biozone, even if no particular biozone actually had that duration.

The Cambrian

For the Cambrian, there is a large range of degrees of resolution obtainable. Holland (1978) mentions that Lower Cambrian trilobite zones may well represent intervals of several million years each. Whether such a statement would be realistic today is debatable. At the time of publication of Holland's 'Guide to Stratigraphical Procedure' there seemed fairly general agreement that the base of the Cambrian was around 570 Ma. Shortly after this, however, a series of conflicting dates were published and Cowie & Johnson (*in* Snelling 1985) concluded that a range of dates from 530–600 Ma was possible. More recent dates suggest that the

Precambrian/Cambrian boundary could perhaps be younger than *c.* 555 Ma., whilst latest Chinese dates suggest that the boundary may be as young as 523 Ma (M.D. Brasier pers. comm.). As the dates for the Cambrian/Ordovician boundary seem agreed at around 510 Ma (e.g. Snelling 1985; Harland *et al.* 1989) the length of the Cambrian Period is decreasing rapidly; thus the biostratigraphical resolution obtainable is, effectively, increasing.

For the late Cambrian (Merioneth Series) there is a succession of faunas of olenid trilobites, well-documented in Scandinavia and widely applicable over what is now Northwest Europe and the Canadian maritime provinces. Work on the Scandinavian faunas was summarized by Martinson (1974) who listed 32 subzones based on olenids (Fig. 1). How long an interval

ZONES	SUBZONES
Acerocare	Acerocare ecorne
	Westergaardia
	Peltura costata
	Peltura transiens
Peltura scarabaeoides	Peltura megalops
	Parabolina lobata
	Ctenopyge linnarssoni
	Ctenopyge bisulcata
Peltura minor	Ctenopyge affinis
	Ctenopyge tumida
	Ctenopyge spectabilis
	Ctenopyge similis
Protopeltura praecursor	Ctenopyge flagellifera
	Ctenopyge postcurrens
	Leptoplastus neglectus
	Protopeltura holtedahli*
	Protopeltura broeggeri*
Leptoplastus	Leptoplastus stenotus
	Leptoplastus angustatus
	Leptoplastus ovatus
	Leptoplastus crassicornis
	Leptoplastus raphidophorus
	Leptoplastus paucisegmentatus
Parabolina spinulosa	Parabolina spinulosa
	Parabolina brevispina
Homagnostus obesus	Olenus scanicus
	Olenus dentatus
	Olenus attenuatus
	Olenus wahlenbergi
	Olenus truncatus
	Olenus gibbosus
Agnostus pisiformis	

Fig. 1. Upper Cambrian trilobite zones and subzones (modified after Martinsson 1974). * Relative position uncertain.

of time is represented by the Merioneth Series is difficult to estimate. Snelling (*in* Snelling 1985) allocated 20 Ma to the whole of the post-Atdabanian Cambrian, with no suggestion as to how this time should be divided. Cowie & Johnson (*in* Snelling 1985) merely note that the Geological Society of America's time-scale (Palmer 1983) gave a Mid/Late Cambrian boundary at 523 Ma, a figure which coincides with the latest Chinese data for the base of the Cambrian (see above)! By taking the figure given by Harland *et al.* (1989) of 7 Ma (but again based on a 570 Ma boundary date) the resolution per subzone is 225 000 years. If the Merioneth Series proves to be shorter, which appears quite probable, resolution will be even greater. It should seem clear to the reader, however, that resolution of this order, for late Cambrian times, is vastly superior to that currently obtainable by any other means.

The Silurian

In the Silurian, graptolites give the greatest biostratigraphical resolution yet possible. From one of the most recently published zonal

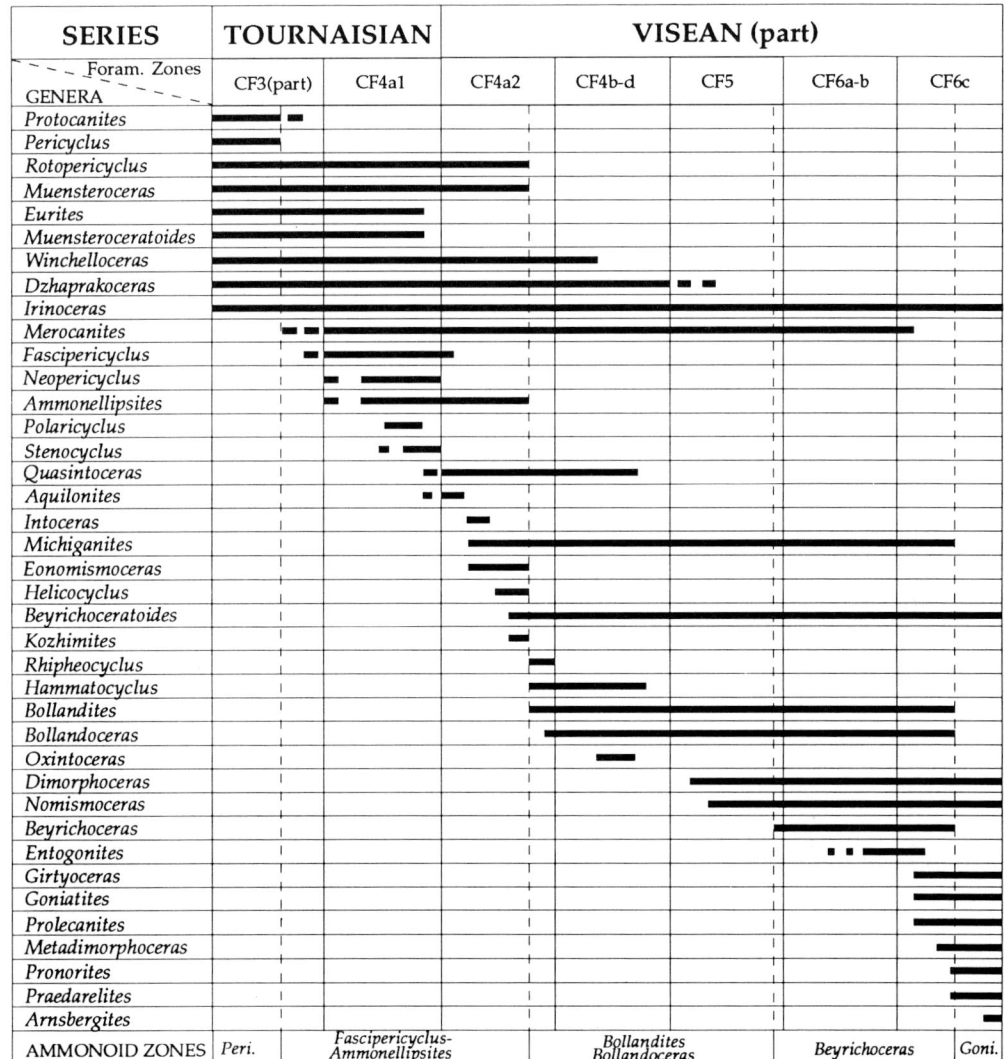

Fig. 2. Global ranges of ammonoid genera in the mid Dinantian compared with foraminiferid zonations. (After Riley 1990).

schemes for the Period (Rickards 1989) graptolite zonation averaged out against the duration of the Silurian Period is *c.* 600 000 years. These zones are largely cosmopolitan in applicability. Locally, considerably finer resolution is possible and Rickards (1989) suggests that within a decade, resolution of the order of 100 000 to 200 000 years will be possible. Work on the *turriculatus* Biozone of the Llandovery by Loydell (this volume) shows that this may already be achieved at some horizons, at least locally. With the Silurian, progress depends upon further taxonomic work and careful collecting.

The Carboniferous

For the Carboniferous, economic considerations have played an important part in the development and use of biostratigraphical subdivision.

STAGE	GENUS-ZONE	CHRONO-ZONE	HORIZON	INDEX
YEADONIAN	CANCELLOCERAS	G1b	*Cancelloceras cumbriense*	G1b1
		G1a	*Cancelloceras cancellatum*	G1a1
MARSDENIAN		R2c	*Verneulites sigma*	R2c2
			Bilinguites superbilinguis	R2c1
	BILINGUITES	R2b	*Bilinguites metabilinguis*	R2b5
			Bilinguites eometabilinguis	R2b4
			Bilinguites bilinguis	R2b3
			Bilinguites bilinguis	R2b2
			Bilinguites bilinguis	R2b1
		R2a	*Bilinguites gracilis*	R2a1
KINDERSCOUTIAN	RETICULOCERAS	R1c	*Reticuloceras coreticulatum*	R1c4
			Reticuloceras reticulatum	R1c3
			Reticuloceras reticulatum	R1c2
			Reticuloceras reticulatum	R1c1
		R1b	*Reticuloceras stubblefieldi*	R1b3
			Reticuloceras nodosum	R1b2
			Reticuloceras eoreticulatum	R1b1
		R1a	*Reticuloceras dubium*	R1a5
			Reticuloceras todmordenense	R1a4
			Reticuloceras subreticulatum	R1a3
			Reticuloceras circumplicatile	R1a2
			Hodsonites magistrorus	R1a1
ALPORTIAN		H2c	*Homoceratoides prereticulatus*	H2c2
			Vallites eostriolatus	H2c1
		H2b	*Homoceras undulatum*	H2b1
	HOMOCERAS	H2a	*Hudsonoceras proteum*	H2a1
CHOKIERIAN		H1b	*Isohomoceras* sp. nov.	H1b2
			Homoceras beyrichianum	H1b1
		H1a	*Isohomoceras subglobosum*	H1a3
			Isohomoceras subglobosum	H1a2
			Isohomoceras subglobosum	H1a1
ARNSBERGIAN	EUMORPHOCERAS	E2c	*Nuculoceras nuculum*	E2c4
			Nuculoceras nuculum	E2c3
			Nuculoceras nuculum	E2c2
			Nuculoceras stellarum	E2c1
		E2b	*Cravenoceratoides nititoides*	E2b3
			Cravenoceratoides nitidus	E2b2
			Cravenoceratoides edalense	E2b1
		E2a	*Eumorphoceras yatesae*	E2a3
			Eumorphoceras bisulcatum	E2a2
			Cravenoceras cowlingense	E2a1
PENDLEIAN		E1c	*Cravenoceras malhamense*	E1c1
		E1b	*Tumulites pseudobilinguis*	E1b2
			Cravenoceras brandoni	E1b1
		E1a	*Cravenoceras leion*	E1a1

Fig. 3. Biostratigraphy and chronostratigraphy of the Namurian Series in Britain. (after Riley MS).

In the Dinantian, the limitations of Vaughan's classic subdivision into a series of assemblage zones have provided a stimulus to more recent work on groups like conodonts (Rhodes *et al.* 1969) and foraminiferids (Conil *et al.* 1980; Fewtrell *et al.* 1981). In North America a scheme of ammonoid zones for their Mississippian ties in well with recent work on the Craven Basin of Northern England (Riley 1990). When used in conjunction with trilobites, conodonts and foraminiferids, the ranges of ammonoid genera are proving to have considerable biostratigraphical resolving power (see above). The global ranges of the ammonoids in the mid-Dinantian are shown in Fig. 2, but note that the diagram shows the ranges of genera. When further work is completed, species described for the first time, and specific ranges established, the degree of resolution obtainable will clearly be further enhanced. At the moment, the scale of resolution available is somewhat better than 2 million years.

Greater resolution is already obtainable in the succeeding late Dinantian and Namurian. The latest scheme of Namurian chronozones compiled by N.J. Riley (reproduced here in Fig. 3) shows 19 chronozones, divisible into 45 horizons. With a duration for the Namurian of 15 Ma each horizon has an average duration of 333 000 years.

The Jurassic

It was in the Jurassic rocks of the Bath area that William Smith first demonstrated that fossils could be used to identify formations. His 'Strata identified by organised fossils' (1815) includes plates of the characteristic macrofossils for each of the formations he listed. Following Smith, work continued apace on the dating of rocks by their fossil content. One of the next major steps forward was the introduction of the zone by Albert Oppel (1856–58). Oppel divided up the Jurassic into 33 zones, each identified by an index fossil. For some two thirds of his zones, Oppel selected an ammonite species as index. Oppel's concept of the zone differed from its present biostratigraphical use (for a summary of current biostratigraphical usage the reader is referred to the new Geological Society guide to stratigraphical procedure (Whittaker *et al.* 1991)) but Oppel's zonal scheme for the Jurassic provided the basis of our modern zonation. Since then, primary Jurassic zones have increased in number and all the index species are ammonites. At the time of production of the Jurassic Correlation Charts (Cope 1980*a*, *b*) 74

zones and 144 subzones were used. Since then the number of subzones has increased, and is probably now about 160. Radiometric dating of the Jurassic boundaries is not precise, but the figures in the Geological Society volume (Snelling 1985) suggest that the Jurassic Period lasted 70 Ma. By a simple calculation we can see that the average subzonal duration was *c.* 437 ka. Accepting the radiometric dates of Harland *et al.* (1989), however, gives a duration for the Jurassic of 62 Ma, and thus an average subzonal figure of *c.* 388 ka. The danger is the assumption that all subzones were of equal duration, which quite clearly they were not, but the figures give a good idea of what is achievable in terms of resolution.

Can we do better than this? One of the classic works on biostratigraphy in the Jurassic is the work of Brinkmann (1929) on the Oxford Clay of the Peterborough region. He measured some 3000 mature ammonites, centimetre by centimetre, through 13 m of the clays and was able to recognize 27 separate ammonite horizons through this interval. Brinkmann's work was a landmark in studies of evolution and has attracted some recent re-evaluations (Raup & Crick 1981, 1982) but this has not affected the biostratigraphical conclusions. 23 of Brinkmann's horizons belong to four of the currently recognized subzones of the Callovian Stage. A recent revision of the Subzones of the early Callovian (Callomon *et al.* 1989) has resulted in 17 subzones now being recognised in the Callovian Stage. Thus, if the Callovian Stage lasted 6 Ma (Hallam *et al. in* Snelling 1985); 17 subzones of the Callovian will average *c.* 353 000 years each; 4 subzones (= 23 horizons) thus approximate to 1 412 000 years; 1 horizon thus approximates to <61 500 years.

If, however, we take an alternative dating for the Callovian, that of Harland *et al.* (1989), its duration was only 4.2 Ma. On that basis the duration of a single ammonite horizon would be <43 000 years. Such horizons as those introduced by Brinkmann can be traced over variable distances; some of them can be recognized widely over southern Britain (Callomon 1984).

One of the earliest detailed demonstrations of the high degree of biostratigraphical resolution obtainable by the use of Jurassic ammonites was made nearly a century ago by S.S. Buckman. In a classic paper on the Inferior Oolite of the North Dorset – South Somerset region (1893) he demonstrated that a very fine resolution was possible. He also introduced the term *hemera* for the time representing the acme of development of an individual species; he went on to produce a series of hemeral tables for the whole of the

Jurassic, published in 'Type Ammonites' (1909–30). Because some of Buckman's later work included theoretical hemeral successions (based in some cases on depositional orders deduced from his examination of ammonite matrices) his work came in for widespread criticism. Recently, Parsons has re-investigated Buckman's work on the southern British Inferior Oolite ammonite faunas (summarized by Parsons, *in* Cope *et al.* 1980*b*) and shown that his work on the Inferior Oolite is reproducible and his biostratigraphy perfectly valid. More recently, Callomon & Chandler (1990) have, by re-excavating some of Buckman's long-defunct localities and meticulously collecting over the Dorset–South Somerset region, shown that it is possible to recognize even finer divisions than Buckman had done. They use the term *horizon* as the smallest recognizable faunal division, and suggest that it is much the same in practice as Buckman's hemera, but bears no implication concerning the acme of a species (Callomon & Chandler 1990, p. 89). The horizons recognized vary from place to place, but this reflects only the condensed nature of the Aalenian–Bajocian succession of southern Britain; their succession of 44 horizons appears widely applicable across Europe. The resolution thus obtainable in the Aalenian and Bajocian, based on a duration of 11 Ma, for these stages (Hallam *et al. in* Snelling 1985) is 250 000 years on average. Using an alternative duration of 10 Ma for these stages (Snelling *in* Snelling 1985) gives *c.* 227 000 years on average.

Why should there be a difference in the resolution obtainable in the Oxford Clay and that in the Inferior Oolite? In the former case we are dealing with a clay/mudstone succession which appears to have been deposited with little interruption (but see Brinkmann 1929 and Raup & Crick 1981, 1982); in the latter we are dealing with a very thin and condensed facies where no two exposures, even in close proximity to each other, show the same succession (Fig. 4). In the case of the Oxford Clay of Peterborough, 13 m of sediment accumulated in *c.* 1 660 000 years (27 horizons), giving a mean rate of *c.* 0.008 mm per year. On the Dorset coast (by no means the thinnest section) 3.8 m of Inferior Oolite (excluding the topmost Lower Bathonian part of the Formation) accumulated in *c.* 11 Ma, i.e. at a mean rate of, 0.00035 mm per year. (Note, however, as pointed out by Callomon & Chandler (1990), the Inferior Oolite of the area includes episodes of rapid deposition separated by major non sequences). The conclusion is, however, that the sediment accumulation (/preservation) rate in the Oxford Clay was, on

HORIZONS		SUBZONES	ZONES
Bj-28	*Parkinsonia bomfordi*	Bomfordi	Parkinsoni
Bj-27	*Strigoceras truelli*	Truelli	
Bj-26	*Parkinsonia acris*	Acris	Garantiana
Bj-25	*Garantiana tetragona*	Tetragona	
Bj-24	*Garantiana garantiana*	Garantiana	
Bj-23	*Leptosphinctes davidsoni*	Baculata	Subfurcatum
Bj-22	*Caumontisphinctes polygyralis*	Polygyralis	
Bj-21	*Caumontisphinctes aplous*	Banksi	
Bj-20	*Teloceras banksi*		
Bj-19	*Teloceras coronatum*	Blagdeni	Humphriesianum
Bj-18	*Teloceras blagdeni*		
Bj-17	*Stephanoceras blagdeniforme*		
Bj-16	*Stephanoceras gibbosum*	Humphriesianum	
Bj-15	*Stephanoceras humphriesianum*		
Bj-14	*Poecilomorphus cycloides*	Cycloides	
Bj-13	*Stephanoceras umbilicum*		
Bj-12	*Stephanoceras rhytum*	Sauzei	Sauzei
Bj-11	*Otoites sauzei*		
Bj-10	*Witchellia laeviuscula*	Laeviuscula	
Bj-9	*Witchellia ruber*		Laeviuscula
Bj-8	*Shirbuirnia trigonalis*	Trigonalis	
Bj-7	*Witchellia connata*		
Bj-6	*Sonninia 'ovalis'*		
Bj-5	*Witchellia romanoides*	Ovalis	Ovalis
Bj-4	*Bradfordia inclusa*		
Bj-3	*Hyperlioceras subsectum*		
Bj-2b	*Hyperlioceras rudidiscites*	Discites	Discites
Bj-2a	*Hyperlioceras walkeri*		
Bj-1	*Hyperlioceras politum*		
Aa-16	*Euhoploceras acanthodes*	Formosum	
Aa-15	*Graphoceras formosum*		Concavum
Aa-14	*Graphoceras concavum*	Concavum	
Aa-13	*Graphoceras cavatum*		
Aa-12	*Brasilia decipiens*	Gigantea	
Aa-11	*Brasilia gigantea*		
Aa-10	*Brasilia bradfordensis similis*		Bradfordensis
Aa-9	*Brasilia bradfordensis baylii*	Bradfordensis	
Aa-8	*Brasilia bradfordensis subcornuta*		
Aa-7	*Ludwigia murchisonae*	Murchisonae	
Aa-6	*Ludwigia patellaria*		
Aa-5	*Ludwigia obtusiformis*	Obtusiformis	Murchisonae
Aa-4	*Ancolioceras opalinoides*	Haugi	
Aa-3	*Leioceras bifidatum*		
Aa-2	*Leioceras lineatum*	Scissum	
Aa-1	*Leioceras opalinum*	Opalinum	

Fig. 4. Division of the Aalenian and Bajocian Stages showing ammonite horizons and their relationship to the current zonal/subzonal scheme. After Callomon & Chandler (1990 and subsequent Addenda & Corrigenda distributed by the authors December 1990). Note that Callomon & Chandler deliberately drew their horizons in discrete boxes, not contiguously as shown here. Modifications of their published scheme have already been made, particularly around Bj2 and 3 and with the introduction of a scheme for the Upper Bajocian (these are included in the Addenda and Corrigenda to the published paper).

average, some 20 times that of the Inferior Oolite of Dorset. There would appear to be little doubt that, if the Inferior Oolite succession were not so condensed, finer subdivision would be possible although even further refined taxonomy might be required to recognize it.

There are other clear cases in the Jurassic where horizons can be recognized. A good example is the Lower Jurassic, particularly the Hettangian to Lower Pliensbachian interval; here it seems clear that a scheme of horizons with a resolution about three to four times better than the present subzonal scheme (Dean *et al.* 1961, with later modifications in Cope 1980*a*) could be applied. This would give a resolution of *c.* 150 000 years for a threefold improvement or *c.* 114 000 for a fourfold improvement. Similar figures are probably applicable to at least some of the other parts of the Jurassic, particularly in the clay facies, but there is scope for a great deal more work. Perhaps some oil companies could be persuaded to part with redundant cores in a good cause?

It will doubtless not have escaped the reader's attention that all the examples of biostratigraphical resolution quoted on the pages above are examples where the index fossils are primarily nektonic or planktonic. Clearly, their relative independence of facies means that they are more widely-dispersed in sediments of that particular age, and thus inherently of more biostratigraphical value. This does not always mean, however, that groups such as ammonoids are invariably the best biostratigraphical indices. As mentioned above, ammonoids are very rare in many Dinantian platform carbonates, probably due to preservational factors. On the other hand, ammonoid faunas such as those of the Namurian of the USSR are extraordinarily diverse, representing post-mortem accumulations in tempestites within platform carbonate sequences (N.J. Riley pers. comm.).

Provincialism within ammonoid faunas presents problems too. Here, examples include some Carboniferous forms, although others, both genera and species, show global distributions. Thus, in the Dinantian, *Goniatites globistriatus* is known from Europe and Australia, whilst *G. granosus* occurs in North America as well. Even Jurassic ammonites are not immune from provincialism, a problem which becomes particularly acute in the latter part of the Upper Jurassic where Tethyan and Boreal forms have nothing in common and where even the Boreal Realm exhibited provincialism between Russia and Poland on the one hand and Britain and Greenland on the other. Although these were all nektonic forms, physical or eco-

logical barriers clearly sometimes prevented wide faunal distributions.

In other parts of the geological column, it is not possible to use nektonic or pelagic organisms; in particular when dealing with predominantly non-marine sediments. Such an example is in the British Westphalian, where marine bands become progressively rarer, and above the top of the Langsettian are infrequent. Stratigraphy of the Coal Measures grew up on an informal basis, founded on the correlation of coal seams from colliery to colliery within each coalfield. Kidston's work on the macrofloras (1923–25) provided some biostratigraphical control, but the first major advance came with the recognition that the non-marine bivalves had biostratigraphical potential. The classic work of Trueman & Weir (1946–58) has been refined, particularly by Calver (1956) and currently provides the best available resolution. Thus, whilst companies working on Westphalian strata in offshore areas currently rely largely upon miospores for their biostratigraphy, resolution is often not of a high order. The Coal Measures macrofauna of non-marine bivalves, however, permits greater biostratigraphical resolution and in the best cases will allow individual coal seams to be correlated. Such resolution probably implies that we are dealing with a scale similar to that of a glacio-eustatic cycle (N.J. Riley pers. comm.).

Conclusions

It is, I hope, now clear that biostratigraphy can provide the best resolving power yet obtainable for much of the Phanerozoic. There is much yet to be done and in some cases progress must depend upon further preliminary taxonomic work – as for instance on many of the Ordovician graptolites (see Rushton 1990).

Taxonomy, that much despised part of the palaeontologist's armoury, is absolutely fundamental to biostratigraphy. Without detailed taxonomy, biostratigraphy would still be in the dark ages. Hallam (1975) claimed that because Jurassic ammonite workers recognized many more species of ammonites than workers on Jurassic bivalves, Jurassic ammonites had been excessively split taxonomically. Without this 'excessive splitting', however, there would be a much less refined biostratigraphy. The fact that Buckman's taxonomy and biostratigraphy on the Inferior Oolite ammonites (often cited as an example of over-splitting) cannot only be reproduced by modern work, but can be improved upon, demonstrates quite clearly that there is much yet to do, in many parts of the Phanerozoic to produce the ultimate in biostratigraphical resolution.

For most parts of the Phanerozoic, nektonic or planktonic organisms provide the best biostratigraphy. It is frequently the case that different groups of fossils used in conjunction can often provide the greatest resolution. This is clearly the case when biozones based on separate fossil groups have different boundaries; use can be made of this non-coincidence to provide greater resolution. In some cases this might mean that if micropalaeontologists and macropalaeontologists pooled their biostratigraphical knowledge, greater resolution could be achieved.

I am particularly grateful to N. J. Riley, who has not only provided me with information on Carboniferous biostratigraphy, but has also undertaken a critical appraisal of the initial draft of the manuscript. I also thank T. Sykes for assistance with drawing-up the diagrams on an Apple-Mac computer.

References

BRINKMANN, R. 1929. Statistisch-biostratigraphische Untersuchungen an mitteljurassischen Ammoniten über Artebegriff und Stammesentwicklung. *Abhandlungen der Gesellschaft der Wissenschaft, Göttingen, mathematische und physische Klasse.* N.F., **13**, 1–249.

BUCKMAN, S. S. 1893. The Bajocian of the Sherborne district: its relation to subjacent and superjacent strata. *Quarterly Journal of the Geological Society of London*, **49**, 479–522.

—— 1909–30. *Type Ammonites*, 1–7. London & Thame.

CALLOMON, J. H. 1984. The measurement of geological time. *Proceedings of the Royal Institution of Great Britain*, **56**, 65–99.

—— & CHANDLER, R. B. 1990. A review of the ammonite horizons of the Aalenian – Lower Bajocian stages in the Middle Jurassic of southern England. *Memorie Descrittive della Carta Geologica d'Italia*, **40**, 85–112. (Addenda and Corrigenda issued by authors December 1990).

——, DIETL, G. & PAGE, K. N. 1989. On ammonite faunal horizons and Standard Zonations of the Lower Callovian Stage in Europe. *In*: ROCHA, R. B. & SOARES, A. F. (eds) *2nd International Symposium on Jurassic Stratigraphy*, Lisboa, **1**, 359–376.

CALVER, M. A. 1956. Die stratigraphische Verbreitung der nichtmarinen Muscheln der penninischen Kohlenfelden Englands. *Zeitschrift der Deutschen geologische Gesellschaft*, **107**, 26–39.

CONIL, R., LONGERSTAEY, P. J. & RAMSBOTTOM, W. H. C. 1980. Matériaux pour l'étude micropaléontologique du Dinantien de Grande Bretagne. *Mémoires de l'Institut géologique de l'Université de Louvain*, **30**, 1–187.

COPE, J. C. W. 1980a. *A Correlation of Jurassic rocks in the British Isles. 1. Introduction and Lower Jurassic.* Special Report of the Geological Society, London, **14**.

—— (ed.) 1980b. *A correlation of Jurassic rocks in the British Isles. 2. Middle and Upper Jurassic.* Special Report of the Geological Society, London, **15**.

DEAN, W. T., DONOVAN, D. T. & HOWARTH, M. K. 1961. The Liassic Ammonite Zones and Subzones of the North-West European Province. *Bulletin of the British Museum (Natural History) Geology*, **4**, 435–505, pls. 63–75.

FEWTRELL, M. D., RAMSBOTTOM, W. H. C. & STRANK, A. R. E. 1981. Carboniferous *In*: JENKINS, D. G. & MURRAY, J. W. (eds) *Stratigraphical atlas of fossil Foraminifera*. British Micropalaeontological Society Series, Ellis Horwood, Chichester, 15–69.

HALLAM, A. 1975. Evolutionary size increase and longevity in Jurassic bivalves and ammonites. *Nature*, **258**, 493–6.

HARLAND, W. B., ARMSTRONG, R. L., COX, A. V., CRAIG, L. E., SMITH, A. G. & SMITH, D. G. 1989. *A geologic time scale 1989*. Cambridge University Press.

HOLLAND, C. H. (ed.) 1978. *A guide to stratigraphical procedure*. Special Report of the Geological Society, London, **11**.

KIDSTON, R. 1923–25. *Fossil plants of the Carboniferous rocks of Great Britain*. Memoir of the Geological Survey of Great Britain.

LOYDELL, D. K. 1993. Worldwide correlation of Telychian (Upper Llandovery) strata using graptolites. *This volume*.

MARTINSSON, A. 1974. The Cambrian of Norden. *In*: HOLLAND, C. H. (ed.) *Cambrian of the British Isles, Norden, and Spitsbergen*. Wiley, 185–283.

MILLER, K. G. 1990. Recent advances in Cenozoic marine stratigraphic resolution. *Palaios*, **5**, 301–2.

OPPEL, 'A. 1856–8. *Die Juraformation Englands, Frankreichs und des Südwestlichen Deutschlands*. Ebner & Seubert, Stuttgart.

PALMER, A. R. (compiler) 1983. The decade of North American geology 1983 Geologic Time Scale. *Geology*, **11**, 503–4.

RAUP, D. M. & CRICK, R. E. 1981. Evolution of single characters in the Jurassic ammonite *Kosmoceras*. *Paleobiology*, **7**, 200–215.

—— & —— 1982. *Kosmoceras*: evolutionary jumps and sedimentary breaks. *Paleobiology*, **9**, 90–100.

RHODES, F. H. T., AUSTIN, R. & DRUCE, E. C. 1969. Conodont biostratigraphy of the Avonian rocks. *Bulletin of the British Museum (Natural History) Geology*, Supplement 3.

RICKARDS, R. B. 1989. Exploitation of graptolite cladogenesis in Silurian stratigraphy. *In*: HOLLAND, C. H. & BASSETT, M. G. (eds) *The global standard for the Silurian System*. National Museum of Wales, Geological Series, **9**, 267–273.

RILEY, N. J. 1990. Stratigraphy of the Worston Shale Group (Dinantian) Craven Basin, north-west England. *Proceedings of the Yorkshire Geological Society*, **48**, 163–187.

—— 1991. A global review of mid Dinantian ammonoid biostratigraphy. *Courier Forschungsinstitut Senckenberg*, **130**, 133–144.

RUSHTON, A. W. A. 1990. Ordovician graptolite biostratigraphy in the Welsh Basin. *Journal of the Geological Society, London*, **147**, 611–614.

SCHINDEL, D. E. 1980. Microstratigraphic sampling and the limits of paleontologic resolution. *Paleobiology*, **6**, 408–426.

SMITH, W. 1816. *Strata identified by organized fossils, containing prints on coloured paper of the most characteristic specimen in each stratum*. E. Williams, London.

SNELLING, N. J. (ed.) 1985. *The chronology of the geological record*. Geological Society, London, Memoir, **10**.

TRUEMAN, A. E. & WEIR, J. 1946–58. *A monograph of British Carboniferous non-marine Lamellibranchia*. Monograph of the Palaeontographical Society, London.

WHITTAKER, A., COPE, J. C. W. *et al.* 1991. A guide to stratigraphical procedure. *Journal of the Geological Society, London*, **148**, 813–824.

WIGNALL, P. B. 1990. Ostracod and Foraminifera micropalaeoecology and its bearing on biostratigraphy; a case study from the Kimmeridgian (Late Jurassic) of NW Europe. *Palaios*, **5**, 219–226.

Devonian goniatite biostratigraphy and timing of facies movements in the Frasnian of eastern North America

MICHAEL R. HOUSE[1] & WILLIAM T. KIRCHGASSER[2]

[1] Department of Geology, The University, Southampton SO9 5NH, UK
[2] Department of Geology, State University of New York, Potsdam, New York 13676, USA

Abstract: The conjunction of detailed goniatite and conodont biostratigraphy has enabled correlation of the Frasnian and adjacent strata of New York with the international standards. A regional goniatite zonation is proposed and as part of the documentation a new Frasnian goniatite assigned to the Triainoceratidae is described as *Wellsites tynani* n.gen. n. sp. The zonation has enabled documentation of the timing and extent of facies movements associated with the Catskill Delta and more easterly deposits in New York State. The broad pattern is one of westward progradation of shallower, more clastic facies. Pulses of deeper water sedimentation, represented by black shale facies indicating anoxic/hypoxic conditions, punctuate the succession and give marker levels. Typically, these black shales initiate shallowing upward cycles. The major pulses are represented by the Tully, Geneseo, Renwick, Genundewa, Middlesex, Rhinestreet, Pipe Creek and Dunkirk units. The cycles enable estimates to be made of deepening against a subsiding shelf; the Rhinestreet Shale represents the greatest. Evidence is given on the regional extent of some of these. International comparison suggests some of these cycles represent global, rather than local events. Microcycles at the Milankovitch Band level also show thin black shales suggesting that climatic control has a bearing on facies developments. General comment is made on the factors controlling local and eustatic changes.

The late Devonian of eastern North America (Fig. 1) is dominated by a terrestrial clastic sequence to the east which has been termed the Catskill Delta. Westward, towards Lake Erie and Ohio, marine rocks enter and units thin towards near continuous black shale sequences of the Ohio, Cleveland, Indiana, New Albany and Antrim shales. Black shales tongue eastward into the clastic facies at several levels and this has formed the basis for the Group terminology generally adopted (Fig. 2; Rickard 1975; Kirchgasser et al. 1985) but there is a general westward progradation of clastic facies so that these marker levels become rather rarer higher in the sequence.

Devonian rocks are well exposed in the Appalachians, but it is the New York sequence that is especially famous for its gently dipping and undisturbed successions that are superbly exposed in innumerable brooks, gullies, creeks and gorges. Of outstanding interest is the Genesee Gorge with its famous pelagic 'Portage Fauna', named from a locality where a portage crossing was possible. More clastic marine facies to the east yielded the benthonic 'Chemung Fauna'. Few areas of the world present such a wealth of opportunity for detailed biostratigraphic and facies analysis. But the enormous variety, the diachroneity of facies, and the difficulties of establishing equivalence, led to a bewildering confusion on local correlation, which in central and western New York has only recently been resolved.

Goniatites were described from this area at the time of the earliest surveys in New York by Vanuxem (1842) and especially James Hall (1843, 1879). However, for the late Devonian it was the 'Naples Fauna' monograph of John Clarke (1899, 1904) that properly documented the presence of the *Manticoceras* faunas and their equivalence to the rocks becoming known in Europe as the Frasnian. Goniatites occur in the three main facies areas, of Naples, Portage and Chemung. A revisionary description of these faunas was given by Miller (1938) but this did not clarify the biostratigraphy.

With a classic paper, *Chemung is Portage*, Chadwick (1935) crystallized recognition of the facies problem for the Frasnian in New York and showed that the facies groups were to a large extent contemporaneous. In a series of papers, some linked to a review of uraniferous black shales, very detailed correlation studies were made in western and central New York by Cooper & Williams (1935), Pepper & de Witt (1950, 1951), Colton & de Witt (1958), de Witt & Colton (1959, 1978) and Sutton (1963). The overall facies correlations were surveyed in masterly reviews by Rickard (1964, 1975) who suggested correlations with the international

From HAILWOOD, E. A. & KIDD, R. B. (eds), *High Resolution Stratigraphy*
Geological Society Special Publication, No. 70, pp. 267–292.

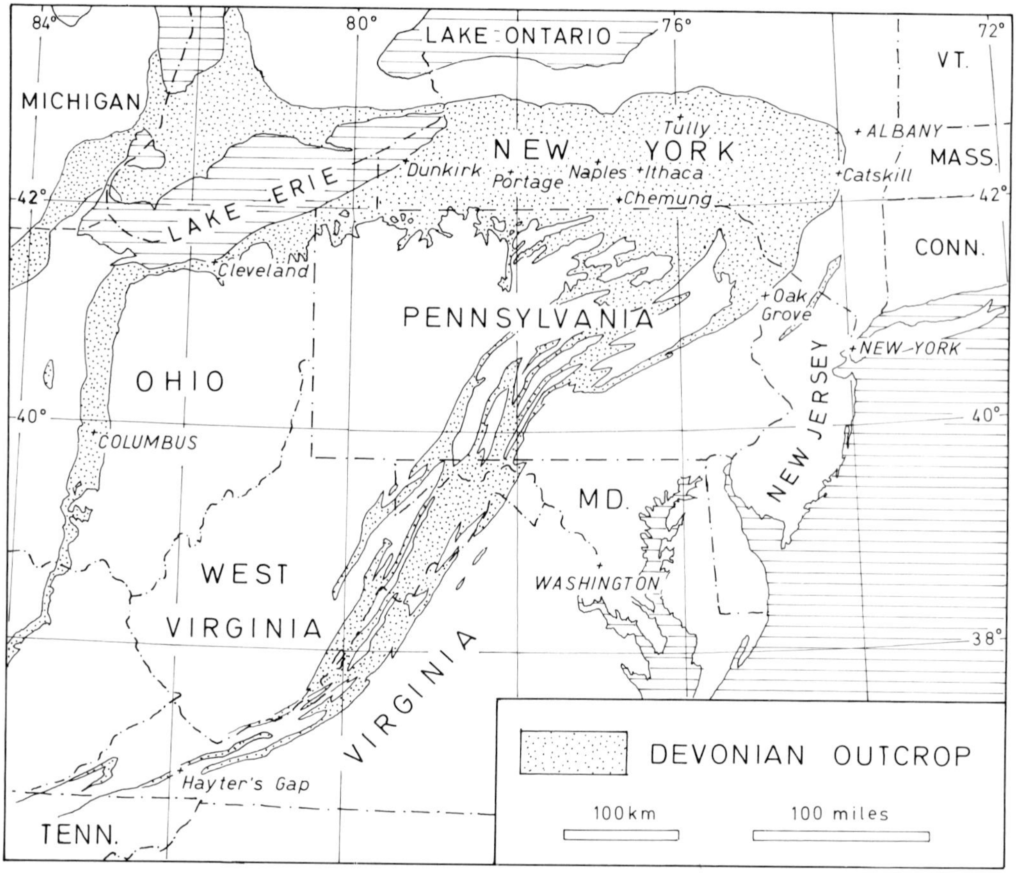

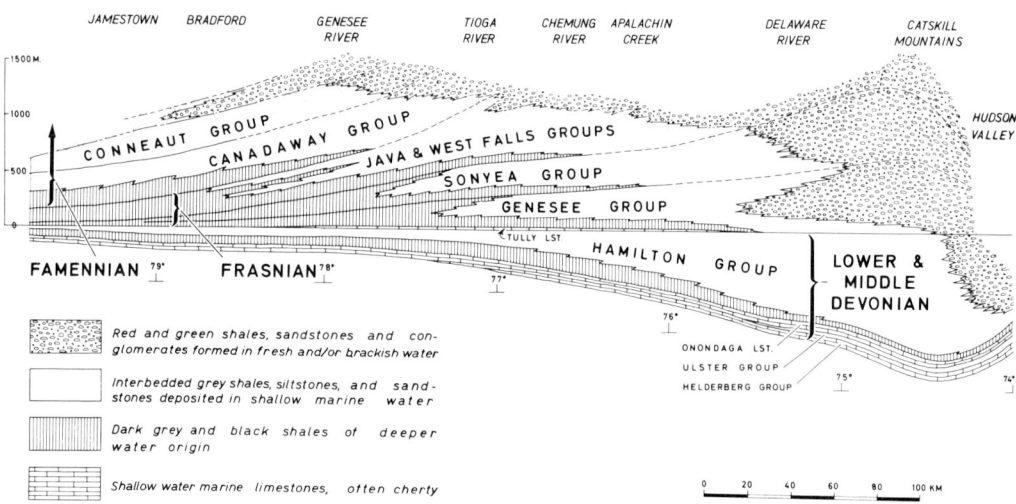

Fig. 1. Map showing the distribution of Devonian rocks in eastern North America.

Fig. 2. Facies transect of the New York State Devonian from the Hudson Valley east to Lake Erie drawn to illustrate the main sedimentary sequences (modified from House 1975).

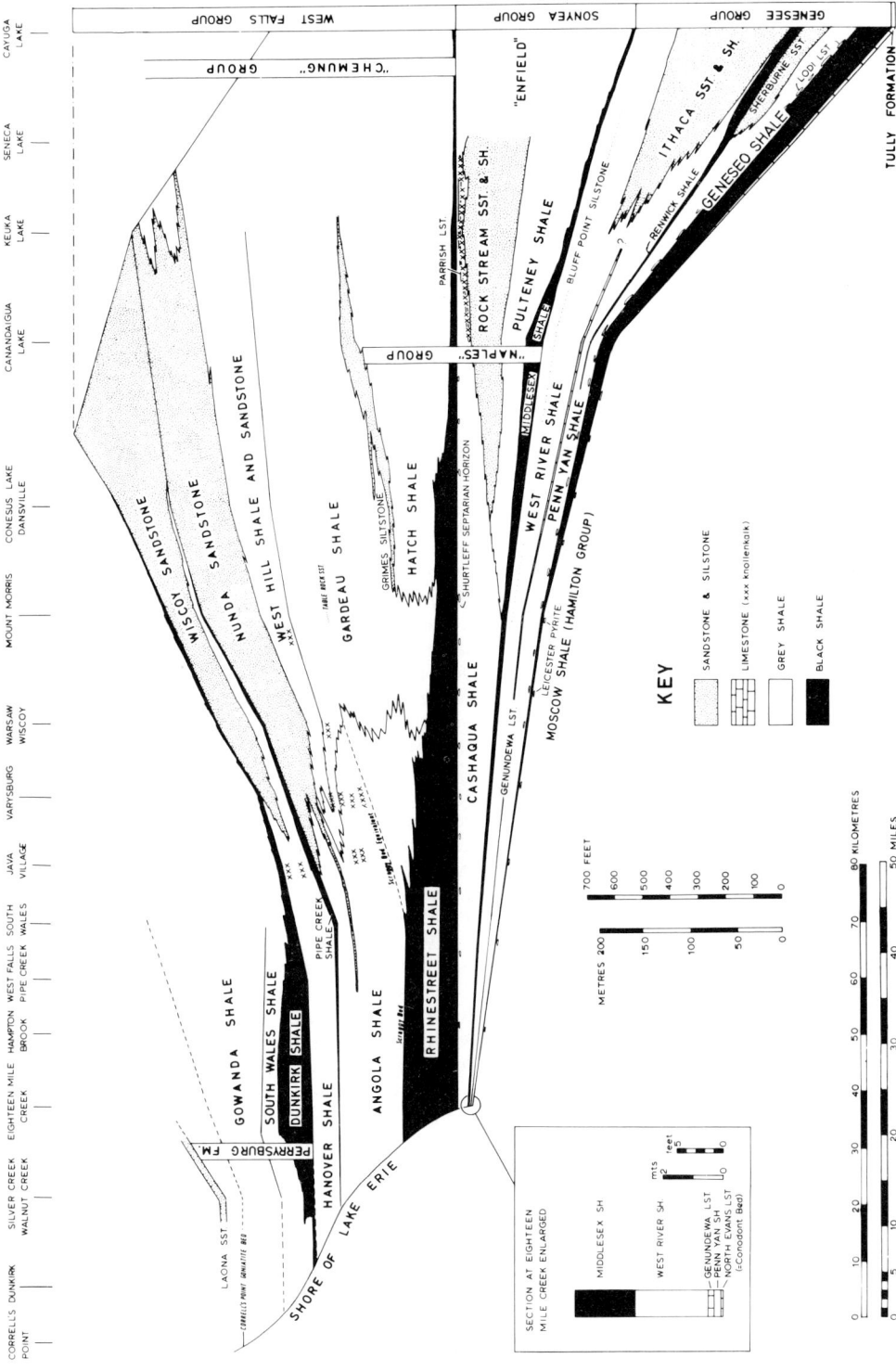

Fig. 3. Diagram of the Frasnian facies developments of New York State in a transect from the finger Lakes area westward to Lake Erie shore (modified from House 1983 and 1985).

conodont and ammonoid zonations. Aspects of the history of progress are given by Wells (1963), Barrell (1913, 1914), Chadwick (1935) and Kirchgasser (1985). A summary of the conclusions of published work with additions from our own field work is given on Fig. 3.

A revision of New York goniatite type material and an attempt to elucidate the Frasnian biostratigraphy was made by House (1962, 1968). This was later extended for outcrops down the Appalachians (House 1978) and provisional zones and a numbered sequence was established. A programme of more systematic collection was commenced and this led to contributions, especially on the early Frasnian (Kirchgasser 1975, 1985). A review of the goniatite stratigraphy as a whole was given in an unpublished work (Kirchgasser & House 1981). This project is the basis for a monographic work in progress by the present authors but the biostratigraphic conclusions are summarized here. Meanwhile, joint work with others in different parts of the world have led to the recognition of a high resolution goniatite biostratigraphy for the Frasnian (Becker et al. this volume) and it is within the new framework that the results for eastern North America are presented.

Documentation and analysis of facies shifts were given for the Devonian by Rickard (1964, 1975). House (1983, 1985) documented facies changes for New York against the improved biostratigraphic scale and made the first attempt to elucidate global eustatic events which might be represented in New York. A similar attempt, concentrating on conodont evidence was given by Johnson et al. (1985, 1986).

Frasnian goniatite zonation

The German standard for Frasnian goniatite zonation was established by Wedekind (1913, 1917), using sections in the Rhenish Schiefergebirge. He also established a subdivision for the whole of the Upper Devonian using Roman numbers I-VI (or do I-VI using the German code letters for the Upper Devonian now replaced by 'to'). Oberdevon I, or the Manticoceras Stufe, comprised the Adorfian or Frasnian in the German orthochronology and this was divided into four divisions Iα-Iδ.

The faunal break at the base of Iα has long been recognized as a major extinction event for goniatites. Genera in the underlying Maenioceras Stufe that do not survive include *Maenioceras*, *Afromaenioceras*, *Holzapfeloceras*, *Wedekindella*, *Cabrieroceras*, *Sobolewia*, *Agoniatites* and *Sellagoniatites*. This break has

been termed the Taghanic Event (House 1985) and corresponds to a major international transgressive pulse of the Middle *varcus* conodont Zone. This is termed the Taghanic Onlap in North America (Johnson 1970) and the international extent of it was documented by House (1975). Formerly, the transgression and faunal break associated with it defined the base of the Upper Devonian (virtually all contributors to Oswald 1968, see House 1982) but the Commission on Stratigraphy has accepted a new definition for the Middle/Upper Devonian boundary at a much higher level, at the base of the Lower *asymmetricus* Zone (Ziegler 1962; Klapper et al. 1987; Klapper & Johnson 1990; Ziegler & Sandberg 1990; Kirchgasser 1992). Subsequently, there has been dispute on the definition and taxonomy of the conodont taxa involved (Sandberg et al. 1988a; Klapper 1988; Johnson 1989).

The Zone of *Pharciceras lunulicosta*, or Iα was excluded from the *Manticoceras* Stufe when first proposed (Wedekind 1913) and this has been formalized by naming part of Wedekind's Iα the *Pharciceras* Stufe (House 1985). In the *Pharciceras* Stufe multilobed pharciceratids enter and diversify but much has still to be learnt about the time sequence of the forms involved. New genera include *Pharciceras*, *Synpharciceras*, *Neopharciceras*, *Petteroceras*, *Epitornoceras* and others. A start at unravelling the biostratigraphy has been made in the Montagne Noire (House et al. 1985) and Morocco (Bensaïd et al. 1985) and about five successive faunas are recognized but not yet formally named. Only the earliest, that of *Pharciceras amplexum*, occurs in New York and this species corresponds closely with the German type species of *Pharciceras*, *Pharc. tridens*. Iβ started the Manticoceras Stufe for Wedekind (1913) and it was marked by the entry of the gephuroceratid genus *Manticoceras*. The base of the Manticoceras Stufe is now taken to correspond with the new definition of the Middle/Upper Devonian boundary, which is marked by the rise to importance of the Gephuroceratidae initiated by *Ponticeras*. It should be borne in mind that the Martenberg (Adorf) succession used by Wedekind is very condensed: the total Frasnian occupies only about 4 metres; this is to be compared with the New York sequence we are using as a standard, which is up to 1.2 km thick (Fig. 3). Consequently, a considerable crudity is to be expected in the Martenberg data despite its revision (House & Ziegler 1977). House (1962) took the occurrence of *Manticoceras nodifer* in the Genundewa of New York to mark the entry of Wedekind's Zone of *Mant. nodulosum*, but the

Fig. 4. Photographs of some Frasnian goniatite markers from the New York State Devonian. (**A**) *Probeloceras lutheri* (Clarke), NYSM. 3648, the Holotype of *Gephyroceras holzapfeli* Clarke (1899) said to have come from the Cashaqua Shale of Eighteen Mile Creek, Erie County × 4.125. (**B**) *Mesobeloceras iynx* (Clarke), NYSM 3586, a cotype, probably from the top Cashaqua Shale or earliest Rhinestreet Shale at Naples, Yates County, × 2.85. (**C** and **D**) *Prochorites alveolatus* Glenister, NYSM. 4063, a specimen figured by Clarke (1899 pl. 7, fig. 4) as *Prob. lutheri*, probably from the Shurtleff Septarian Horizon, Cashaqua Shale, near Honeoye Lake, Ontario County, × 3. (**E** and **F**) *Carinoceras oxy* (Clarke), NYSM. 12036, a specimen from the Point Breeze Goniatite Bed, Angola Shale in Big Sister Creek (Loc. 73/6), Erie County, × 1.5. (**G** and **H**) *Manticoceras simulator* (Hall), NYSM. 3797, the Holotype of the type species of *Manticoceras, Goniatites simulator,* from Ithaca, Tompkins County, × 1.09. (**I** and **J**) *Sphaeromanticoceras rhynchostomum* (Clarke) NYSM. 12132 (6680I), collected by L. V. Rickard from probable upper Rhinestreet Shale equivalents in Relyea Creek (Loc. 60), Wyoming County, × 1.5. (**K**) *Schindewolfoceras chemungense* (Vanuxem), NYSM. 4073, the Holotype, probably from the Cayuta Shale, near Owego, Tioga County (Loc. 69), × 1.5.

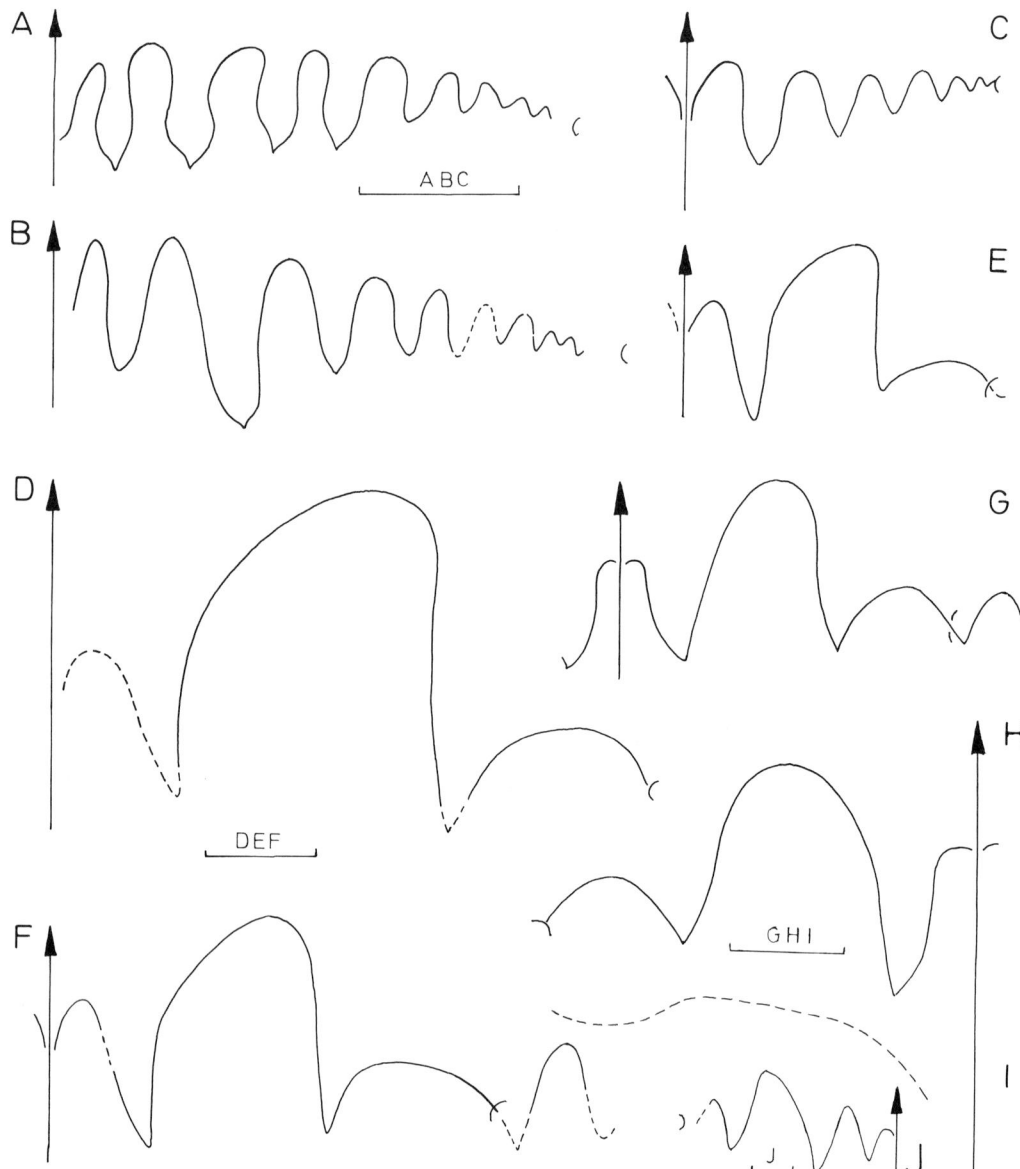

Fig. 5. Suture diagrams of Frasnian goniatites. (**A**) *Wellsites tynani* n.gen. n.sp., SUI 42318, based on the Holotype from the Moreland Shale (early Rhinestreet Shale) near Elmira, Chemung County. B, *Wellsites williamsi* (Wells), CUMP 40023, based on the Holotype from just above the Moreland Shale on Bald Mountain, northeast of White Church, Tioga County. Scale for (**A**) & (**B**) 20 mm. (**C**) *Schindewolfoceras chemungense* (Vanuxem), NYSM. 4073, the Holotype from near Owego, Tioga County, horizon uncertain. Scale for (**C**), 10 mm. (**D–F**) *Sphaeromanticoceras rickardi* n.sp. (**D**) NYSM. 12032, at 137 mm diameter, collected by L. V. Rickard, from the Hanover Shale of Glade Creek, Strykersville, Wyoming County. (**E**) IG 4916, at 165 mm diameter, from 2.7 km east-northeast of Roly, (Sautour Sheet), chemin de Viller's en Fagne, Belgium. (**F**) IG 4449, the Holotype at 100 mm diameter (reversed) figured by Matern (1931*b*, text-fig. 1a,b) from 560 m south-southeast of Romerée, chemin de Matagne-la-Petite (Surice Sheet). Scale for (**D–F**) 20 mm. (**G–I**) *Sphaeromant.* aff. *rickardi* n.sp. collected by G. Kloc and in his collections, from about 1.5 m above the base of the Hanover Shale. (**I**) growth line of specimen figured in (**G**). Scale for (**G–I**), 20 mm. (**J**) *Mesobeloceras naplesense* (Clarke), suture of the Holotype, NYSM 4072, at 30 mm diameter; scale 2 mm.

true *nodulosum* is probably much younger. Zonal usage has been much complicated by past German practice of including complex-sutured forms such as *Triainoceras* and *Koenenites* in Iα although they were not recorded at Martenberg by Wedekind. We now know that such forms range much higher than the *Pharciceras* faunas: in condensed successions detailed discrimination was not always made. Becker *et al.* (this volume, fig. 8) show that the true level of Wedekind's Iβ fauna corresponds to the late Rhinestreet and early Angola levels of New York. The earlier Frasnian faunas are not well documented at Martenberg nor anywhere else in the Rhenish Slate Mountains.

Matern (1931*a*) argued for the combination of Wedekind's Iβ and Iγ (as Iβ(γ)) but House (1962) pointed out that this was premature. The fauna of the true Iγ is distinctive, and includes, in modern terms, species of the genera *Maternoceras*, *Serramanticoceras*, *Trimanticoceras*, *Neomanticoceras* and '*Crickites*'.

Wedekind considered that his terminal Zone of *Crickites holzapfeli* was preceded by a level with *Manticoceras adorfense*. Elements of the German *holzapfeli* Zone were recognized in North America by House (1962), but the discovery by L.V. Rickard of a Hanover Shale goniatite identical with specimens described by Matern (1931*b*) from the Schistes de Matagne as *Crickites holzapfeli* gave confirmatory evidence (House 1968). However, here we reassign Matern's Belgian material to *Sphaeromant. rickardi* n.sp.

Systematic palaeontology

Most of the species referred to in the subsequent text have been figured relatively recently (Miller, 1938; House 1962, 1965, 1978; Kirchasser 1975) and we have a monograph on the whole fauna in preparation. It is, however, necessary to include descriptions of new forms that are important for our regional zonation.

Family Triainoceratidae

New genus *Wellsites*

Type species. Wellsites tynani n.gen. n.sp.
Derivation of name. The genus is named in honour of Prof. J.W. Wells who found so many goniatites in central New York and who fully appreciated their significance.
Diagnosis. An oxyconic genus of triaionoceratid without ribbing and with involute conch, with a single division of the primary ventral lobe, seven umbilical lobes and with all sutural elements rather deep.

New species *Welsites tynani* n.sp.

Holotype. The University of Iowa SUI 42318. Fig. 5A.
Derivation of name. The species is named after Dr M.C. Tynan who, as a student at Iowa, collected the specimen we designate as holotype and drew our attention to it.
Description. Only the holotype is available, which is a large crushed specimen showing a phragmocone, probably oxyconic, of 190 mm diameter. With body chamber, it exceeded 250 mm diameter. The suture has in excess of ten adventitious, lateral and umbilical lobes of which the deepest is the second lateral; the umbilical lobes decrease in size towards the umbilicus. The inner whorls are not visible. The suture of the holotype and type species is illustrated here (Fig. 5A) and it shows a diagnostic suture typical of complex pharciceratids, with swollen lobes pinched at their apices and rounded saddles. For this reason, it is assigned to the Triainoceratidae although the typical ribbing of the family is not seen.
Horizon and locality. The holotype is from spoil in a road cut for US Route 17, 0.16 km east of the crossing of Route 17 and Watercure Road, Elmira at an altitude of about 910 feet. The horizon is the Moreland Shale, the basal tongue of the Rhinestreet Shale. Both the Millport and Dunn Hill Shales crop out in East Church Street Quarries just above the type locality.

Species *Wellsites williamsi* (Wells 1956).

Holotype. Cornell University CUMP 40023. Fig. 5B.
Remarks. This closed umbilicate, large and probably oxyconic form with smooth outer whorls was at that time reasonably referred by Wells to *Beloceras*. But, as now understood, it has fewer adventitious lobes than *Beloceras* and does not have the open umbilicus of *Mesobeloceras*. Although poorly preserved, it seems best placed in *Wellsites* with which it shares the pointed lobes and rounded saddles, although the amplitude of the lobes is greater and they are relatively narrower than in *Wells. tynani*. The rounded saddles and pinched lobes show it should not be placed in the Beloceratidae as recently done by Yatskov (1990).
Horizon and locality. Wells's specimen came from Bald Mountain, 11.3 km southeast of Ithaca, New York, beside a track 3.2 km southeast of Brooktondale and 2.7 km northeast of White Church, Tioga County, at an elevation of about 1770 feet. It was recorded as coming from the lower part of the Cayuta Shale about 150 feet

above the base of the *Thiemella* (?) *danbyi* Zone, that is, slightly above the Moreland Shale.

Family. Gephuroceratidae.
Genus. *Sphaeromanticoceras* Clausen 1971.

Type species. Manticoceras affine by original designation of Clausen 1971, p. 197.
Emended diagnosis: Small to very large involute Gephuroceratidae with rotund cross section converging convexly to a rounded venter. Suture with mid-lateral saddle very high, symmetrical to asymmetrical with steep or overhanging dorsad face. Growth lines bioconvex in known small forms and sub-biconvex to sub-convex in very large specimens. Type species with a distinctive very fine ventral wrinkle layer.

New species *Sphaeromanticoceras rickardi* n.sp.

Holotype. Musée National d'Histoire National, Bruxelles, IG 4449, Fig. 5F (figured by Matern 1931*b* text-fig. 1a,b).
Derivation of name. In honour of Dr L.V. Rickard who has done so much to further New York Devonian studies.
Diagnosis. Giant *Sphaeromanticoceras* characterized by very asymmetric lateral saddle with near vertical to overhanging dorsad face (Fig. 5D) and with subconvex growth lines.
Horizon and locality. The holotype is from 560 m south-southeast of Romerée , chemin de Matagne-la-Petite (Surice Sheet). A conspecific specimen from Roly is also illustrated (Fig. 5E).

Remarks. The group of giant gephuroceratids in the late Frasnian, which include the largest members of the family, poses a number of taxonomic questions. Those from New York have been assigned to *Crickites* (House 1978) on the basis of their similarity to forms from the Schistes de Matagne of Belgium, which Matern (1931*b*) assigned to *Crickites holzapfeli* (Wedekind 1913, p. 72, pl. 7, figs 5, 6). Indeed Matern named one of these the Neotype of *Cr. holzapfeli*, but, as with others of his 'neotypes', this was illegal since original paratypes survive. Additionally, in 1958, a specimen labelled 'Original zu Wedekind', from Wildungen, and in the black lithology of the Kellwasserkalk, was seen in Göttingen; and it appears to be the basis for one of his illustrations (Wedekind 1913, fig. 14b), and with a diameter of rather over 60 mm it corresponds to the second specimen for which he gives dimensions. The possibility that the other two figured specimens have survived cannot be excluded. Typical *Crickites holzapfeli* are not

only much smaller but also consistently less inflated than *Sphaeromant. rickardi*.

A large specimen (New York State Museum 12032) was found by L.V. Rickard in Glade Creek, Strykersville (Loc. 91/7 on Fig. 10) at a level 8 m below the top of the Hanover Shale (Fig. 5D). This was named by House (1968) *Crickites* aff. *holzapfeli* using Matern's literature but that was before our new information and revision. Close relatives of *Sphaeromant. rickardi* are also known from the Kellwasser levels of Morocco (purchased specimen in MRH collection) and in the Montagne Noire (Bergeron 1899, pl. xxii).

Species *Sphaeromanticoceras* aff. *rickardi* House & Kirchgasser, Fig. 5G–I.

Remarks. G. Kloc, now of the University of Rochester, has kindly provided details of specimens from the Hanover Shale about 1.52 above the top of the Pipe Creek Shale which differ from *Sphaeromant. rickardi* in showing a less steep dorsad face to the lateral saddle. He has drawn our attention to the similarity of these forms to *Sphaeromant. lindneri* (Glenister) but that species has biconvex growth lines and a much less asymmetrical lateral saddle.

Regional New York goniatite zones

The subdivision of the Frasnian that has been adopted heretofore has been the scheme proposed by Wedekind (1913, 1917), in which the Manticoceras Stufe was divided into four divisions which, in German practice, are referred to as α, to β, to γ and to δ. It has been known for a long while that more detailed subdivision is possible using goniatites (House 1962, 1979) but this has not been formally attempted. In an accompanying paper in this volume, Becker with the present authors (this volume), propose a scheme numbered from A to L based on the entry points of distinctive genera. That scheme is used here as a framework for discussion of the regional goniatite zones for eastern North America (numbered following House 1978 and Kirchgasser & House 1981 but modified slightly). The zonal divisions are shown on Fig. 6 and the faunas identified and their levels are shown on Fig. 7.

A: *Ponticeras* Division

15: Zone of *Ponticeras perlatum* (Hall)
This zone is almost monospecific in New York. The base is best defined in Hubbard's Quarry where *Pont. perlatum* enters immediately below

KLAPPER ZONES	NEW YORK			GENERIC MARKERS	
	DUNKIRK				
13	HANOVER	Sphaeromanticoceras rickardi	24	CRICKITES	L
		Manticoceras cataphractum	23		
	PIPE CREEK			ARCHOCERAS	K
12	ANGOLA	Sphaeromant. rhynchostomum	22	NEOMANTICOCERAS	J
		Carinoceras oxy			
11		Playfordites tripartitus		PLAYFORDITES	I
	RHINESTREET	Schindewolfoceras chemungense	21	BELOCERAS	H
		Wellsites tynani			
7		(Mesobeloceras iynx)		MESOBELOCERAS	G
6	CASHAQUA	Prochorites alveolatus	20	PROCHORITES	F
5		Probeloceras lutheri	19	PROBELOCERAS	E
4	MIDDLESEX	Sandbergeroceras syngonum	18	SANDBERGEROCERAS	D
	WEST RIVER	Koenenites aff. lamellosus	17	TIMANITES	C
3	GENUNDEWA	Manticoceras contractum			
2	PENN YAN	Koenenites styliophilus	16	KOENENITES	B
1		Ponticeras perlatum	15	PONTICERAS	A
norrisi	LODI	Epitornoceras peracutum	14		
disparilis	GENESEO				

Fig. 6. Stratigraphical column of the New York State Devonian showing the new goniatite biostratigraphic terminology used in the text. Numbered conodont zones are from the Montagne Noire zonation of Klapper (1989) and Klapper & Johnson (1990). Numbered goniatite zones in New York are after House (1978) and Kirchgasser & House (1981). Generic goniatite markers and lettered divisions are after Becker *et al.* (this volume).

the Lodi Limestone (House 1962). The Lodi Limestone yields the best preserved material (Kirchgasser 1975). The conodont *norrisi* fauna in the Lodi Limestone represents the latest Givetian (Middle Devonian) on current definitions (Klapper *et al.* 1987) and the base of the Frasnian and Upper Devonian is taken where the Lower *asymmetrics* Zone occurs, at the junction of the Lodi Limestone and the overlying Penn Yan Shale and Sandstone (Kirchgasser 1991). Thus the *perlatum* Zone commences fractionally below the basal Frasnian. *Epitornoceras* ranges up from the zone below and is

known 12.2 m below the Lodi Limestone in Mill Creek (Lodi Glen), immediately below the Lodi Limestone, with the zone fossil at Hubbard's Quarry, and rarely in the Penn Yale Shale. Some solid specimens from the Lodi Limestone are known. The genus thus appears to occur in the upper Givetian and lower Frasnian. Famennian records of *Epitornoceras* refer to a later homeomorphic derivative of the *Tornoceras* stock as pointed out by House (1963). *Pont.* cf. *regale* occurs in the Ithaca Formation at Cayuga Lake at a level equivalent to the lower Penn Yan below the Crosby Sandstone (Kirchgasser 1985).

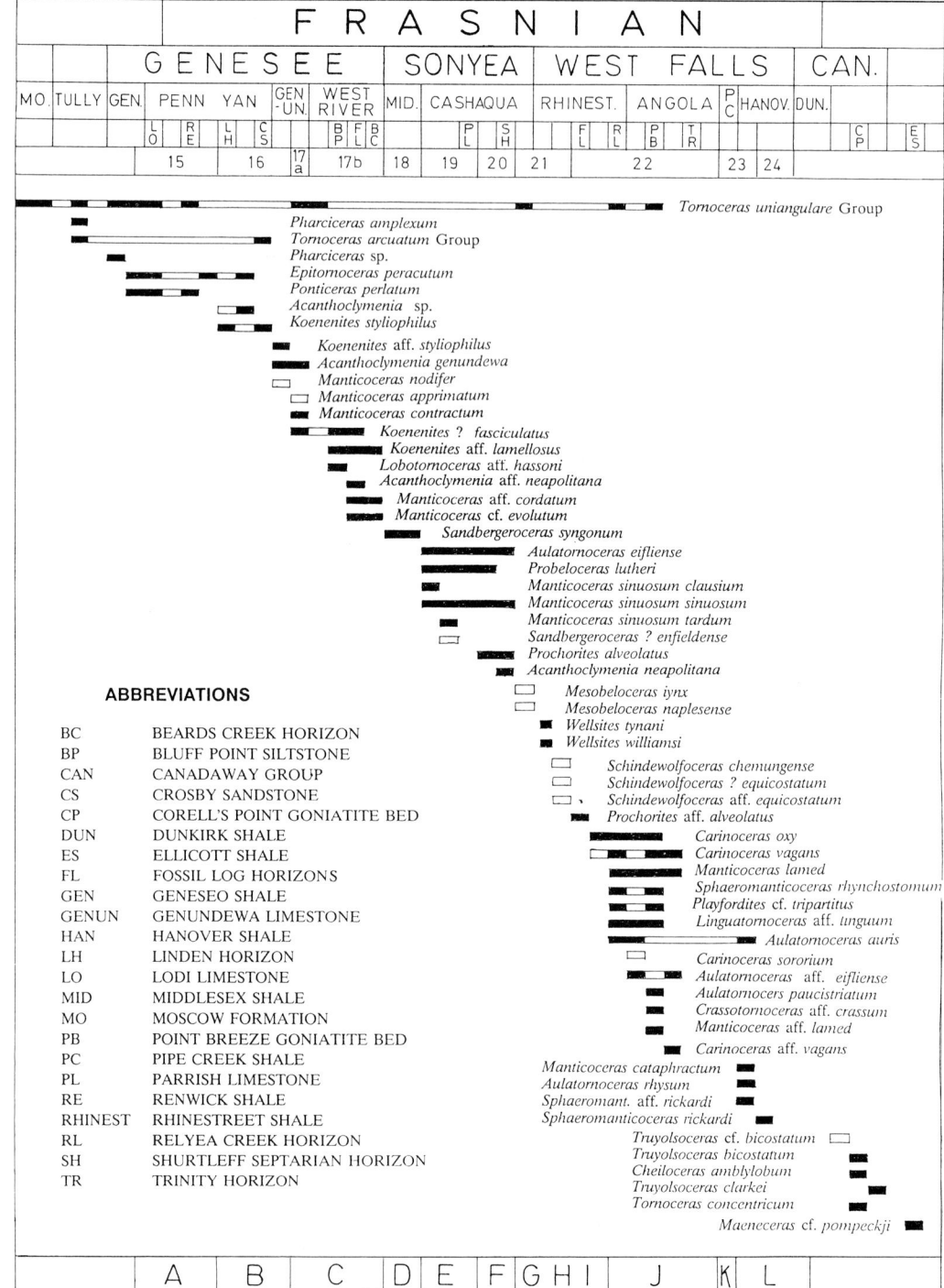

Fig. 7. Stratigraphical column of the Frasnian of New York State Devonian showing the ranges of distinctive goniatite taxa.

B: *Koenenites* Division

16: Zone of *Koenenites styliophylus* (Clarke)
Koenenites enters in the upper Penn Yan Shale at the Linden Goniatite Horizon (Kirchgasser & House 1981) at Linden, New York and is found eastward to Canandaigua Lake. It occurs in the Crosby Sandstone, which is probably slightly younger, around Keuka Lake where *Tornoceras* cf. *arcuatum* House has been recorded. No ponticeratids are known to overlap with the range of *Koenenites* in New York. Forms that were formerly named *Probeloceras* occur, such as *Acanthoclymenia* sp. from the Linden Horizon in the west and at Keuka Lake just beneath the Crosby Sandstone. The richest fauna of this zone is represented in the lower part of the Genundewa Limestone near Bethany Center where *Koen.* aff. *styliophylus* (Clarke), *Acantho. genundewa* (Clarke) and *Tornoceras uniangulare compressum* (Clarke) are found. *Koen. ? fasciculatus* (Clarke) occurs at the type area of the Genundewa Limestone at Canandaigua Lake.

C: *Timanites* Division

Both *Manticoceras* and *Timanites* enter about here but the Australian evidence shows *Timanites* is slightly earlier. Although *Timanites* has not yet been found in eastern North America the name is prefered for the division between because it is a short-lived genus whereas *Manticoceras* continues until the close of the Frasnian.

17a: Zone of *Manticoceras contractum* (Clarke)
Manticoceras enters in the upper Genundewa Limestone at Geneseo, New York, and subsequently the genus is not seen until the upper West River Shale. Together with *Mant. contractum*, two other species were recorded by Clarke (1898) '*Mant*'. *nodifer* (Clarke) and *Mant. apprimatum* (Clarke). *Acantho.* cf. *genundewa* occurs. This is the level which was taken to indicate the base of Iβ (House 1962). It would be expected that this entry of *Manticoceras* represents a transgressive pulse following the sedimentation lull of the Genundewa Limestone proper and hence transitional to the West River Shale. There is a gap in the record in the succeeding lower West River. The type species for *Manticoceras*, *Mant. simulator* (Hall), came from the *Leiorhynchus* Beds of the Naples Formation at or near Ithaca, New York (Fig. 4G,H). The precise horizon is not known but it is expected to be in the equivalents of the West River Shale, and probably high rather than low in the unit.

17b: Zone of *Koenenites* aff. *lamellosus* (G. & F. Sandberger)
The name form occurs immediately below the Bluff Point Silstone and is close to the specimen described from the Montagne Noire under the name *Hoeninghausia* aff. *archiaci* Gürich (House *et al.* 1985, text-fig. 8D,E, pl. 3, figs 10–12) but we have now decided that it is better to keep the genus *Hoeninghausia* for undoubtedly oxyconic forms. The upper West River Shale from the Bluff Point Siltstone to the top of the unit has horizons which yield koenenitids of variable quality of preservation and size, hence the assignment is correspondingly difficult. They are referred to *Koen* aff. *lamellosus* sp. D. In this interval *Acanth.* aff. *neapolitana* and also *Mant.* cf. *cordatum* (G. & F. Sandberger) occur and there is an evolute form which has been referred to as '*Archoceras*' but which is likely to be *Mant.* cf. *evolutum*.

D: *Sandbergeroceras* Division

18: Zone of *Sandbergeroceras syngonum* (Clarke)
The zonal species is the only form known from the Middlesex Shale and it has been widely reported from New York and the Appalachians. *Sandb. syngonum* was described from Snyder's Gully, Canandaigua Lake. This entry of multilobed and ribbed forms, assigned to the Triainoceratidae, and presumably derived from the Pharciceratidae is curious and would seem to correspond with a deepening event and facies shift which introduces exotic elements.

E: *Probeloceras* Division

19: Zone of *Probeloceras lutheri* (Clarke)
Typical faunas of the Zone of *Manticoceras cordatum* enter with the Cashaqua Shale in New York although *Mant.* aff. *cordatum* occurs earlier in the upper West River Shale. In the lower Cashaqua Shale *Prob. lutheri* (Clarke) enters and *Mant. sinuosum* (Hall) occurs with the subspecies *tardum* (Clarke) and a subsp. nov. The relations of *Mant. cordatum* (G. & F. Sandberger 1850) to *Mant. sinuosum* (Hall 1843) have never been satisfactorily dealt with taxonomically. They are extremely close to each other, but *Mant. sinuosum* has priority, yet *Mant. cordatum* is the name used historically and internationally. The slight differences between the type specimens of the two forms (which differ also in size) do not appear higher than of subspecific rank.

F: *Prochorites* Division

20: Zone of *Prochorites alveolatus* (Glenister)
In the upper Cashaqua Shale a new fauna appears, characterised by *Proch. alveolatus* (Glenister), which we now regard as a senior synonym of *Proch. strix* (Kirchgasser 1975). In this interval occurs the celebrated barytic fauna of the Shurtleff Septarian Horizon, known best from Livonia, Livingston County, which yielded (House 1961) the type material of *Acantho. neapolitana* (Clarke) but this form had already appeared earlier.

G: *Mesobeloceras* Division

21a: Zone of *Mesobeloceras iynx* (Clarke)
The fauna of this zone is not known in place but only from museum collections. It may be latest Cashaqua Shale in age or early in the Rhinestreet Shale. The faunas of the Rhinestreet Shale are poorly known, apart from the uppermost part. This black shale is the thickest of the Frasnian black shales in New York and a considerable period of time is thought to be represented by it. Clarke (1899) described pyritic forms now assigned as *Meso. iynx*. This species was formerly assigned to *Eobeloceras* but the types of that genus show it to be quite unrelated to *Mesobeloceras*. The type material of *Meso. iynx* has been restudied, and specimens have the lithology of the Cashaqua Shale, but we have not located the horizon yielding it in the type area near Naples nor in the field anywhere in the sections to the west. Nevertheless the Cashaqua has been extensively searched and the fauna may come from appropriate facies in the early Rhinestreet but we draw attention to Clarke's account which suggests it is a Cashaqua Shale form. A similar situation applies to *Mesobeloceras naplesense* (Clarke), similarly from the Naples area, which we have also not found ourselves. The holotype of *Meso. naplesense*, New York State Museum 4072 (Fig. 5J), has been made the type species of a new genus *Naplesites* by Yatskov (1990). Unfortunately he seems to have based the characters on old inaccurate published figures rather than on an examination of the holotype, which shows a suture on the lower flanks close to *Meso. iynx* but with one or two less umbilical lobes. As with *Meso. iynx* it is close to *Meso. anguisellatum* (Chao) which Yatskov has made the type species of another new genus, *Chaoceras*. The Beloceratidae show a progressive increase in adventitious and umbilical lobes and there seems little point in assigning new species or genera on the basis of every additional lobe, especially when ontogen-etic and stratigraphic details are undescribed. We regard *Naplesites* and *Chaoceras* as junior subjective synonyms of *Mesobeloceras*.

H: *Beloceras* Division

We do not believe the true *Beloceras* is represented in New York and we here assign to a new genus forms which have been refered to *Beloceras*.

21b: Zone of *Wellsites tynani* House & Kirchgasser n.gen. n.sp.
In the main development of the Rhinestreet Shale towards the east, especially south and southeast of Ithaca, are several localities where complex-sutured goniatites are found. Their occurrence well to the east, approaching the more clastic facies of the Catskill Delta, suggests they result from particularly successful transgressions within the major transgression represented by the Rhinestreet Shale and correspond to the triainoceratids in the Middlesex Shale in representing unusual deeper-water forms. The Rhinestreet examples raise problems of their exact time equivalence with the western succession.

Perhaps the earliest is the group represented by the new genus *Wellsites*. A new species from Elmira is in the collections of the University of Iowa (SUI 42418) and is named here *Wellsites tynani* House & Kirchgasser n.gen. n.sp. To the same genus is referred the specimen named by Wells (1956) as *Beloceras williamsi* from Bald Mountain southeast of Brooktondale at a level probably just above the Moreland Shale. *Wells. williamsi* differs from *Wells. tynani* in having lobes of greater amplitude.

21c: Zone of *Schindewolfoceras chemungense* (Vanuxem)
This level marks the incoming of forms with ribbed early whorls characteristic of the Triainoceratidae. The original specimen (Fig. 4K) of *Schindewolfoceras chemungense* (Vanuxem 1842) was not precisely located but, when it was refigured, Hall (1879, p. 69) gave the locality as Chemung near Owego, New York. Other specimens, referred to the genus with doubt, are *Schind.? equicostatum* (Hall) said to come from Athens, Pennsylvania and later said to come from a boulder there. The best localized specimen is *Schind.?* aff. *equicostatum* from a level between the Moreland Shale and Roricks Glen Shale tongues of the Rhinestreet which Sutton and McGhee (1985) indicate to be still rather low in the Rhinestreet but above the levels bearing *Wellsites*.

I: *Playfordites* Division

There is a thick interval, mostly of black shale, within the Rhinestreet Formation that has yielded neither goniatites nor conodonts and hence there is no evidence on where this division should begin. *Playfordites* cf. *tripartitus* occurs in the upper Rhinestreet (NYSM 12074) at Johnson Creek (Fig. 9, Loc. 58/1) about 5.5 m below the Scraggy Bed level. This, and some associated fauna, suggest correlation with the Lower *cordatum* Zone (to Iβ) of Europe but *Playfordites* ranges higher and this may be a late record.

J: *Neomanticoceras* Division

The name genus for the division has not yet been identified in North America. Already in the upper few metres of the Rhinestreet occurs the oxyconic genus *Carinoceras* which is most abundant in the Upper *cordatum* Zone to Iγ (of the older terminology) and Division J but may occur earlier. A search is required of the New York oxyconic forms to see whether any develop the extra division in the ventral saddle which is characteristic of *Neomanticoceras*; no evidence has yet been found. *Carinoceras* is abundant at this level but is known rarely below the division. The distinction between the *Playfordites* and *Neomanticoceras* divisions is therefore at present uncertain in New York and will remain so until more faunal levels are discovered. Comparison with Germany is not helpful since other forms of the Upper *cordatum* Zone, such as *Trimanticoceras cinctum*, *Maternoceras sandbergeri* and relatives, *Timanoceras* and '*Crickites*' *expectatus* (House & Ziegler 1977) have also not been found. In Germany and the Canning Basin *Playfordites* does not range as high as it appears to in New York or, alternatively, only the upper part of the Angola Shale may fall into Division J and carinoceratid faunas are unusually well developed in Division I.

22: Zone of *Sphaeromanticoceras rhynchostomum* (Clarke)

Towards the top of the Rhinestreet Shale near Lake Erie small-scale rhythms with calcareous levels or concretions occur and in this facies, and in equivalents in the Gardeau Shale and Sandstone farther east, a suite of genera appear for the first time. These include *Sphaeromanticoceras* (Fig. 4I, J), *Carinoceras* (Fig. 4E, F), *Playfordites*, *Linguatornoceras* and the group of *Aulatornoceras auris* which is represented by *Aulat. rhysum*. The most productive level still within the Rhinestreet Shale is at Relyea Creek (Fig. 9, Loc. 60) near Warsaw, Wyoming

County, a locality known to Clarke (1899) and possibly Luther (1903, p. 1012) under the name of Gibsons Glen. Near Lake Erie the base of the Angola Shale is taken at the Scraggy Bed (Luther 1903, p. 1023) and a concretionary level six rhythms (about 13 m) above is a particularly rich level which we have termed the Point Breeze Goniatite Bed (Kirchgasser & House 1981) after a locality on Lake Erie shore (Fig. 8, Loc. 72). Goniatites include *Mant. cordatum*, *Mant.* aff. *lamed*, *Carinoceras oxy* (Fig. 4E, F), *Carin. vagans*, *Sphaeromant. rhynchostomum*, *Playfordites* cf. *tripartitus*, *Tornoceras*, *Linguatornoceras*, *Aulatornoceras* and *Crassotornoceras*. *Playfordites* is not known from the Upper Anglo Shale where collecting is poor.

K: *Archoceras* Division

No fauna is known from the Pipe Creek Shale which we consider to represent the Lower Kellwasserkalk level of Europe. Also, in New York there are no records of *Archoceras*, a genus which in Europe becomes opportunistically abundant after the Lower Kellwasser Event (Feist 1990) although *Archoceras wabashense* is known from Indiana (House 1962). Detailed study has been undertaken of the early Hanover faunas by G. Kloc, and we are indebted to him for some information, but unfortunately his results have not been published. The distinction between Divisions K and L in New York is not clear since the name genera are not known; the assignment to the divisions of zones 23 and 24 is therefore tentative.

23: Zone of *Manticoceras cataphractum* (Clarke)

The name species is common in the lower part of the Hanover Shale; it is a distinctive small and rather evolute, periodically constricted manticoceratid which is not unlike *Archoceras varicosum* but has a subdivided ventral lobe. First records in Silver Creek are about 1.8 m above the Pipe Creek Shale and they continue to over 10 m higher. A specimen collected by G. Kloc, 1.52 m above the top of the Pipe Creek Shale on Walnut Creek, is named *Sphaeromant.* aff. *rickardi* n.sp. It was identified earlier by House (1968) and Kirchgasser & House (1981) as *Crickites holzapfeli* but is reassigned for reasons given earlier.

L: *Crickites* Division

24: Zone of *Sphaeromanticoceras rickardi* n.sp.

This is the highest goniatite level so far determined in the New York Devonian and remarks

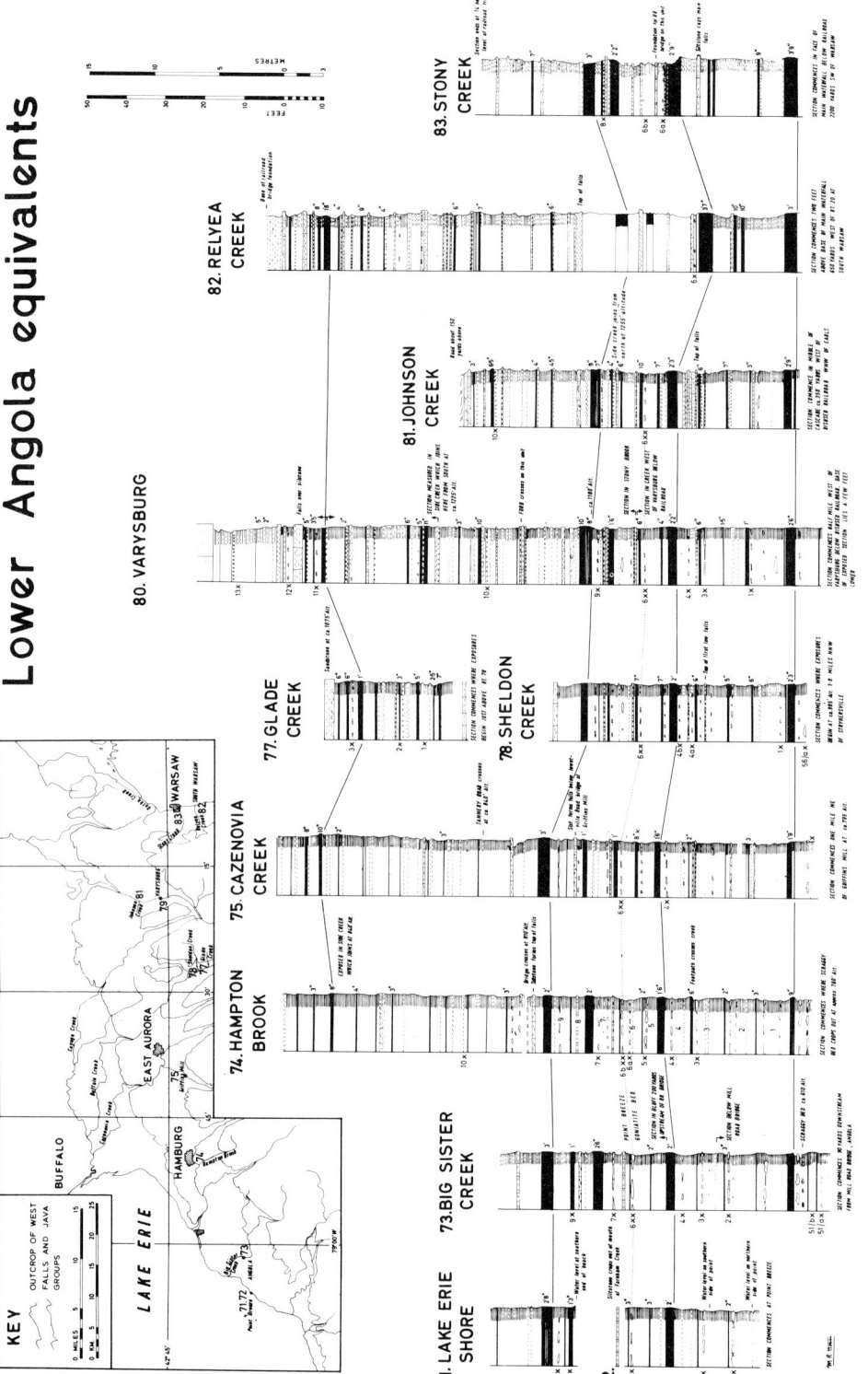

Fig. 8. Sections in the lower Angola Shale equivalents between Lake Erie Shore and Stony Creek showing goniatite-producing horizons and locality numbering.

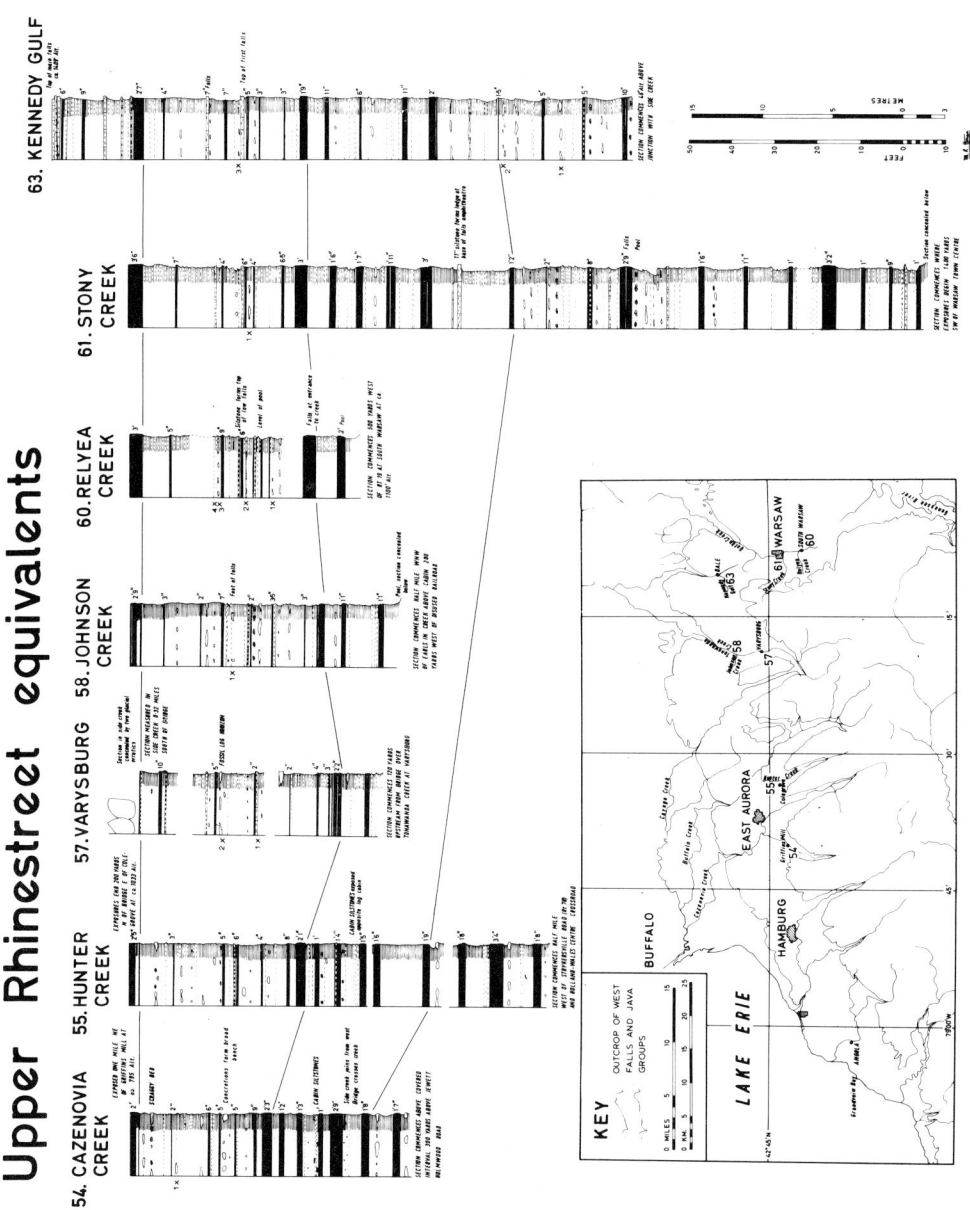

Fig. 9. Sections in upper Rhinestreet Shale equivalents between Cazenovia Creek and Kennedy Gulf showing some goniatite-producing levels and locality numbering.

Hanover equivalents

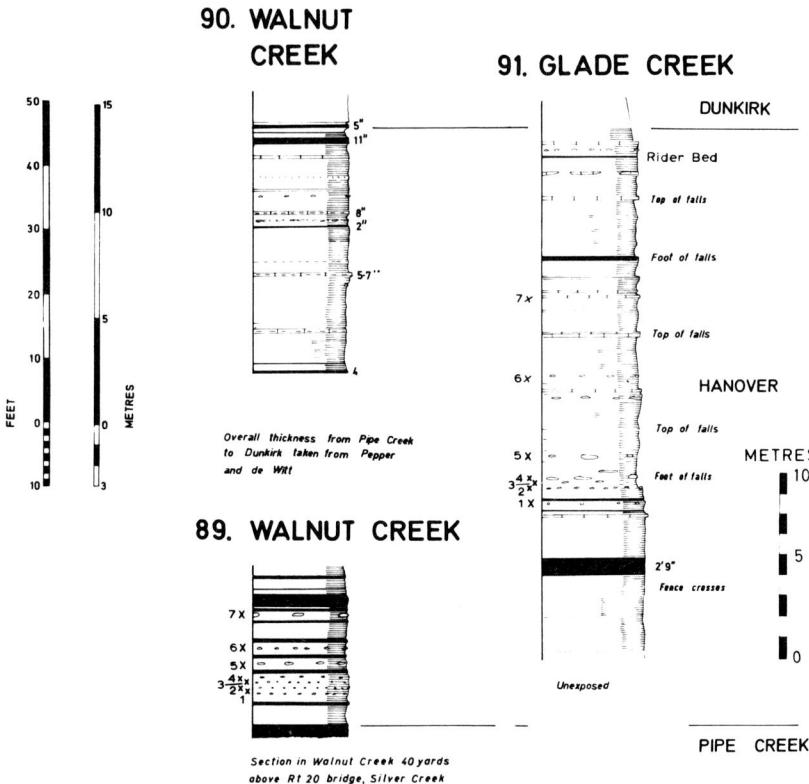

Fig. 10. Sections in the Hanover Shale of eastern New York showing goniatite-producing levels and locality numbering.

of this form are given earlier. The horizon yielding the figured specimen is in Glade Creek (Fig. 10, Loc. 91, Bed 7). Possible anaptychi occur in the Rider Bed, a black shale a metre or so below the top of the Hanover Shale. Homoctenids occur profusely in the first of three stippled units shown on the Java section (number 7) of Pepper & de Witt (1951) 66 feet above the base of the Dunkirk Shale and this was taken to suggest the level was Frasnian, since homectenids become extremely rare after the end Frasnian and only Schindler (1990) has recorded single specimens above the Upper Kellwasser Kalk. Prof. Klapper has informed us that he has a *triangularis* Zone fauna from the Dunkirk Shale at Point Gratiot which would indicate early Famennian but that level may be higher in the Dunkirk Shale. However, Baird & Lash (1990, fig. 5) suggest considerable westward

thinning with non-sequences in the early Dunkirk Shale so the stage boundary has still to be exactly placed but the lower black band (shown on their fig. 5) may represent the Upper Kellwasser level. No goniatites are known higher until those of the Gowanda Shale (House 1962) which now would be assigned to the *subpartitum* Subzone of IIα (Becker 1986, 1990).

Timing and nature of facies movements in New York

The object of the discussion on the biostratigraphical scale has been to provide a framework capable of international correlation against which documentation of Frasnian facies shifts elsewhere in the world can be compared. The New York facies change situation is summarized in Fig. 11 which will be commented on now.

Tully Limestone. The Tully Limestone has been considered to be the most extensive of the Devonian transgressive levels. The *Hypothyridina* faunas of the Tully extend further than any marine level eastwards into the Catskill Delta (Cooper & Williams 1935). Later, its extensive correlatives westwards towards the North American transcontinental arch were demonstrated (Cooper *et al.* 1942; Cooper 1968; Johnson & Friedman 1969). An international transgression at this level was formerly taken to mark the basal Frasnian, but the Subcommission on Devonian Stratigraphy has recommended a higher level (Klapper *et al.* 1987) and this has been ratified by the IUGS. The Tully Limestone (Heckel 1973) is the last significant limestone development in the New York Devonian and this bears witness to a short period of sediment starvation. Below the Tully is a non-sequence which, in New York, is increasingly developed westwards, and is recorded by pyritic lag deposits, bone beds, and condensed sequences (Baird & Brett 1986*a, b*, Baird *et al.* 1988) which indicate an overstepping base to the Tully Limestone westward. The dating of this event in goniatite terms is basal Pharciceras Stufe, *amplexum* Zone, and the event is probably best assigned as Middle *varcus* Zone in conodont terms. It is difficult to quantify the sea level change represented, but it is likely to have been many tens of metres. The rare reefs at Borodino indicate that elsewhere in central New York water depth was below the photic zone.

Geneseo Shale. Although there is interleaving of the basal black shales of the Geneseo Shale with the Filmore Glen Member of the Tully Limestone (Heckel 1973) the facies change is quite sudden. This black shale tongue is so widespread, and the lack of early quartz clastics so marked that a significant deepening is indicated as well as a spread of anoxic waters. The occurrence of late pharciceratids shows that this deepening still falls within the Pharciceras Stufe and the diachronous Leicester Pyrite lag deposits at the base indicate the transgression begins in the *hermanni-cristatus* Zone (Huddle 1981). This may indicate a more extensive transgression than that of the Tully but there is no faunal evidence for this as far as we are aware. However, successful later transgressions may so easily remove the evidence of more extensive earlier transgressions. Thicknesses of the Genesee Group as a whole in the east reach 490 m (Rickard 1975), of which the Geneseo Shale equivalents might total 100 m but this figure represents fill of a basin combining general subsidence with sealevel change. Therefore,

the deepening is unlikely to have exceeded 50 m. Already in the Geneseo Shale west of Fall Brook microrhythms are evidenced by concretionary levels within grey shale intervals.

Lodi Limestone. There are several calcareous levels in the upper Geneseo Shale but the Lodi Limestone, at the base of the Penn Yan Shale, is the most useful marker; another lower level is the Fir Tree Limestone (Baird & Brett 1986*a,b*; Baird *et al.* 1988). The *perlatum* Zone commences below the Lodi at Hubbard Quarry with the entry of *Ponticeras*. The conodont *Skeletognathus norrisi* fauna occurs at the contact of the Lodi Limestone and Geneseo Shale and this level corresponds to the *norrisi* Zone (= lowermost *asymmetrica* Zone. The base of the Upper Devonian, and conodont Zone 1 of Klapper (1989) is taken at the black shales immediately overlying the Lodi Limestone in the Honeoye and Canandaigua Lake area (Kirchgasser *in* Woodrow *et al.* 1989). No more than a slight diminution of sediment supply is inferred for the limestone.

Penn Yan Shale. Coarser clastics gradually enter above the Lodi Limestone and culminate in the Sherburne Siltstones. The Renwick Shale is a black shale tongue initating another small rhythm with the early Ithaca Siltstones and Shales representing shallower deposits which to the west are represented by the upper Penn Yan Shale. *Koenenites* enters at the Linden Goniatite Horizon which marks the base of the *styliophilus* Zone and conodont Zone 2 of Klapper (1989). There are thus two rhythms represented in the Penn Yan Shale above the Lodi Limestone.

Genundewa Limestone. This last limestone marker in the New York Devonian apparently represents a period of sediment starvation. However, at this time abundant styliolinids were accumulated, and these are a pelagic indicator. Several faunal horizons are represented. In the lower Genundewa Limestone the *styliophilus* Zone and conodont Zone 2 of Klapper (1989) terminate. In the upper Genundewa Limestone *Manticoceras* enters and the *contractum* Zone and Zone 3 of Klapper (1989) begins. The same general facies is represented in Michigan with the Squaw Bay Limestone and this suggests that a wide geographical area was affected and oxic pelagic conditions are indicated by the planktonic *Styliolina*. In goniatite terms, several levels are represented within the Genundewa interval. *Timanites* is to be expected but its horizon may correspond to a winnowed non-sequence level or inappropriate environmental conditions. In

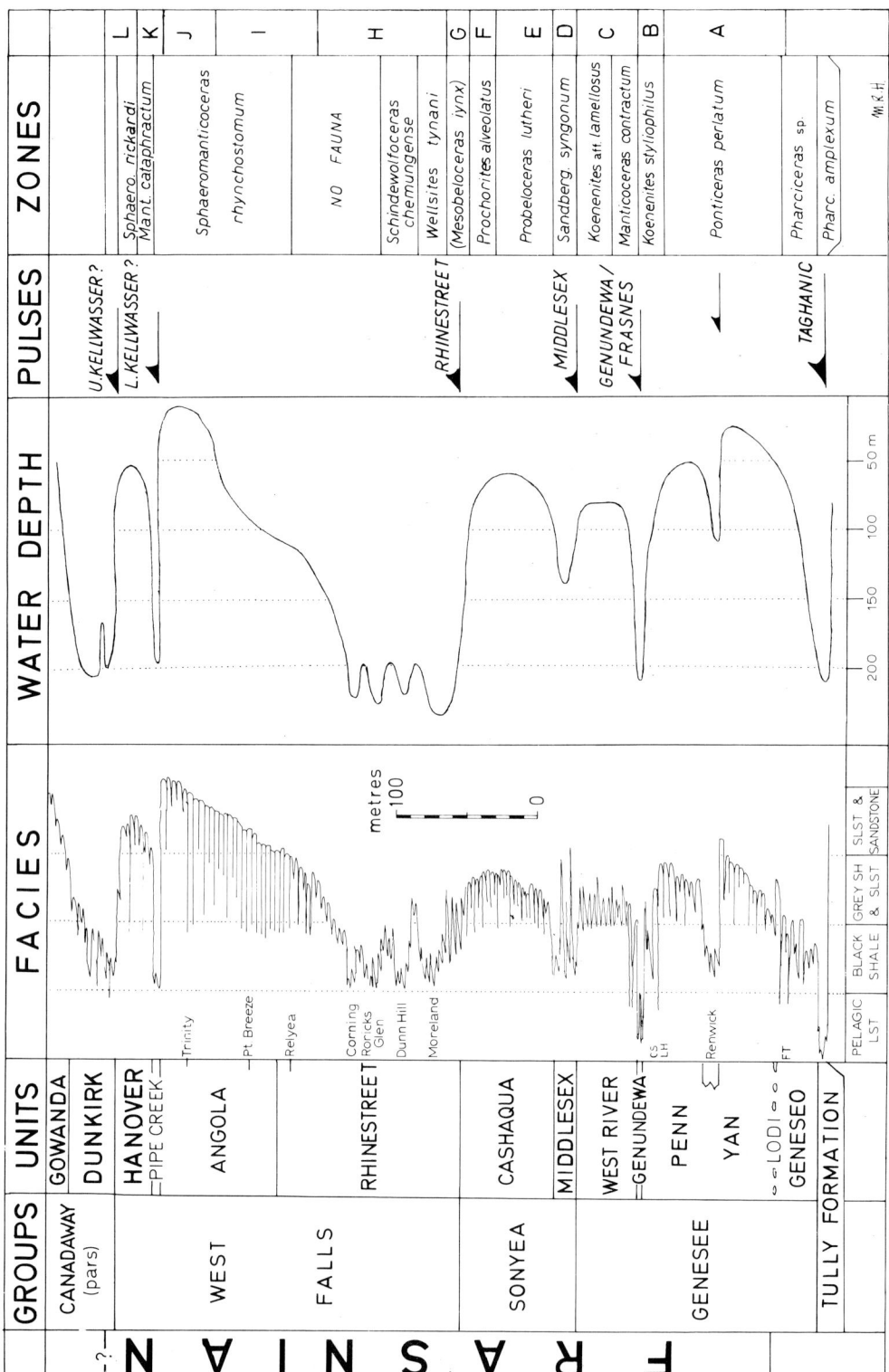

Fig. 11. Diagram indicating the main facies movements recognized in the New York Frasnian, their correlation with the international biostratigraphic divisions and the sea-water depths and transgressive pulses recognized. See explanation to Fig. 7 for abbreviations used.

central New York the level is within the Ithaca, but exactly where is uncertain. University Quarry, Ithaca, has *Ponticeras cf. regale* and would be expected to correlate with the Penn Yan Shale before the entry of *Koenenites* (Kirchgasser 1985). *Koenenites* occurs above the power plant cascade in the Fall Creek section at Ithaca which could represent the Linden, Crosby or Genundewa levels. We therefore regard it as a transgressive level, but unlike the other pulses considered (apart from the Tully Limestone), black shales are not associated with it. Is this because the transgression was so great that even the clay fraction of land-derived material was not available? Or were there chemical aspects of the water tiering that meant anoxia was not involved? The latter point is supported by the presence of benthos in the limestone. Water depth was not sufficient to transgress the CCD although some dissolution of the upper side of goniatite shells is known. Across much of western New York there is a black shale underneath the Genundewa Limestone and styliolinid limestones occur both in the upper Penn Yan Shales and in the lower West River Shales. Thus, it would appear that a greater deepening is represented by the limestone phase when the sediment source is so flooded that starvation results. Reworking and winnowing on the sea floor would explain several of the characteristic Genundewa features; condensation, non-sequences, pyrite lag deposits and irregular bedding (Baird *et al.* 1989). A deepening of the same order as with the Tully Limestone is envisaged.

West River Shales. More stable sedimentation characterises the West River Shales and Milankovitch Band rhythmicity is indicated by some two dozen grey/black shale couplets which are well shown on the correlation chart for the Keuka Lake – Canandaigua Lake region produced by de Witt & Colton (1978). The West River Shale is contemporary with the progradation of the Ithaca siltstone and shale facies farther east. In the upper West River Shale the Bluff Point Siltstone (de Witt & Colton 1978) is a siltstone with convolute bedding which forms an excellent marker west of Seneca Lake westward to near Lake Erie. It is at around this level that *Koenenites* aff. *lamellosus* Zone and conodont Zone 4 of Klapper (1989) begin. If we are correct in attributing the Genundewa to high sea level, then the West River represents shallowing and infill from the clastic wedge of the Ithaca Siltstones and Shales. However, the consistancy of lithology suggests that sea level rise kept pace

with sea floor sedimentation, indicating a rise of only a few tens of metres.

Middlesex Shale. This black shale unit initiates the Sonyea division. It is siltier and thinner than either of the other two major black shales in the Frasnian, the Geneseo Shale below or the Rhinestreet Shale above. It may be siltier because the Pulteney Shale and Rock Stream Siltstone and Shale give evidence of a significant progradational shift. A curious goniatite fauna of triainoceratids enters, giving the *syngonum* Zone; there is no conodont date. We think this represents a major transgressive pulse. There is a specimen of *Triainoceras* aff. *costatum* (d'Archiac & de Verneuil) in the collections of the University of Iowa collected by J. Ragan from a roadcut on Highway 19 near Big Spring Missouri. This is a marker species for this level in North Africa and its occurrence in the Mid West suggests this level was transgressive in North America also.

Cashaqua Shale. The greenish-grey soft shales and mudrocks of the Cashaqua Shale represent a facies not seen earlier. The *lutheri* Zone and conodont Zone 5 commence here. The *alveolatus* Zone and conodont Zone 6 commence at the level of the Shurtleff Septarian Horizon. The lowermost and uppermost parts of the Cashaqua Shale grade into the black shales of the Middlesex Shale and Rhinestreet Shale, giving a large-scale ABCBA cycle (Fig. 12A; House 1983, p. 401). This is interpreted as a symmetrical shallowing cycle. In the central part (C) microcycles are indicated by periodic concretionary horizons and there are few of the thin black shales found in the lowermost and uppermost parts (B). Eastward, the Cashaqua Shale thins as a result of the Rock Stream Siltstone wedge and an anomalous facies of 'cephalopodenkalk', 'kramenzelstein', or 'knollenkalk' is developed as the Parrish Limestone on the sea floor rise (Fig. 3).

Rhinestreet Shale. This is the thickest of the black shale units in the New York Devonian and is only exceeded by the thicknesses of the Millboro Shale southward in the Appalachians. It represents a significant extension of black shale facies into the Catskill Delta (Fig. 2) and Sutton (1963) has demonstrated how particular tongues are especially pervasive (Rickard 1975; Sevon & Woodrow 1985). Whilst a considerable time is represented, the facies is not suitable for a good goniatite record. As with the Middlesex Shale, triainoceratid types return and we regard this as indicating a significant deepening. The

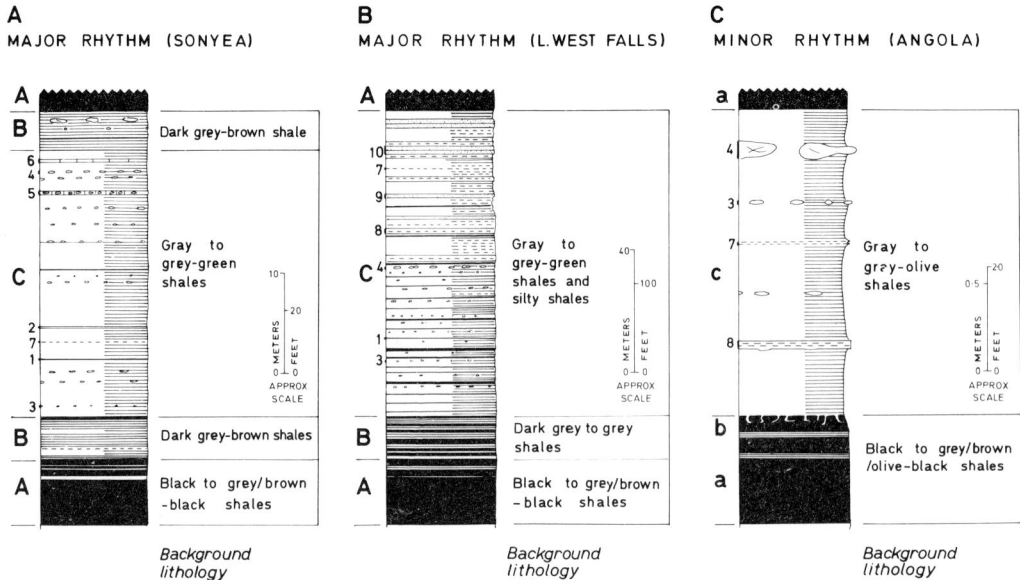

Fig. 12. Illustration of the pattern of sedimentary cycles shown in the New York Frasnian. (**A**) The symmetrical rhythm of the Sonyea Group. (**B**) The asymmetrical rhythm of the Lower West Falls Group. (**C**) A small-scale rhythms from the Angola Shale of the West Falls Group. All scales approximate. (Key: 1, black shale horizons; 2, kerogen-rich shale; 3, pyritic or calcareous nodules; 4, small to large septarian nodules, sometimes barytic; 5, nodular carbonate unit; 6, carbonate unit; 7, siltstone seam; 8, turbidite unit; 9, siltstone or sandstone unit; 10, massive sandstone unit). Based on House (1983).

lower Rhinestreet has conodont Zone 7 of Klapper (1989) and Zone 11 occurs near the top. Good microcycles are developed in the lower Rhinestreet as black/grey shale alternations with rare concretions (as in Buck Run Creek, Mt Morris) and in the upper Rhinestreet Shale as rhythms of thin black shale and thicker grey shale couplets with the grey shale units grading into those of the Angola Shale by developing concretionary horizons in their upper part (Fig. 9). A deepening in excess of 100 m seems required and whilst there is no faunal evidence that the black shale facies extends over the transcontinental arch it seems probable that the marine transgression did so and we regard it as a major transgressive pulse.

Angola Shale. The lower Angola Shale in the type area comprises a succession of small-scale rhythms (Figs 8, 12). The base is drawn at a marked black shale above the Scraggy Bed of irregular concretionary mudstone (Fig. 9, Loc. 54). The third rhythm carries a thin turbidite which we have traced for some 80 km. The calcareous concretions of the sixth rhythm is the Point Breeze Goniatite Bed which carries a fauna of considerable diversity and excellence.

Faunas are much sparser at higher levels as more clastic facies enter, culminating in the Nunda Sandstone equivalents.

Pipe Creek Shale. This thin black shale marker bed indicates a sudden return to anoxic conditions and the boundaries above and below the sharp. Eastward it tongues between the clastic Nunda and Wiscoy Sandstones (Fig. 3) in a way that suggests a sharp transgressive pulse as well as a rise in the pyconcline.

Hanover Shale. The return to grey shales in the early Hanover is interrupted by several black shale horizons (Fig. 10) and there are some, but fewer, in the late Hanover, suggesting a symmetrical cycle.

Dunkirk Shale. This represents a significant transgressive pulse and spread of anoxic conditions. Detailed sequences were given by Pepper & de Witt (1951) and they show a Hume Shale pulse higher in the section and below the Gowanda Shale which yields the first *Cheiloceras* (House 1962) but these levels are within the Famennian.

Milankovitch band cycles

The New York Frasnian is permeated by small-scale rhythmicity and an attempt has been made to show this in a generalized way on a facies shift diagram (Fig. 11). Details are illustrated for parts of the Rhinestreet and Angola Shale (Figs 8 & 9) where the microrhythms may be several metres thick. Comment was made earlier (House 1983, 1985) on how the facies elements of microrhythms matched in many ways those of the major sedimentary rhythms and may suggest a climatic control. An example is given here for the Angola Shale (Fig. 12C) showing how an initiating anoxic black shale gives way to grey shale, with burrowing indicating the return to sea-floor oxycity. Higher levels generally become more calcareous as indicated by nodule formation, and shallowing is suggested by occasional benthos. It is not at present possible to give a satisfactory estimate of total numbers of rhythms for the Frasnian as a whole because in the major black shales rhythmicity is either not seen, or when well shown, as at many localities in the Rhinestreet Shale, the fluctuations are of the order of several to a metre, witnessing to the reduced sedimentation at these times and also to the fact that the large black shales probably represent very great periods of time. In these circumstances a simplistic view of different orders of cycle magnitude is inappropriate.

Sea level and facies shifts

On the grand scale, as has often been pointed out, the New York Devonian consists of sedimentation groups or sequences usually starting with black shales, followed by grey shales and concluded with deltaic sandstones, and rarely limestones (Fig. 2). This is the basis of the Group terminology of Rickard (1964, 1975). These are major Klüpfelian shallowing-upwards rhythms. In detail, however, the succession in punctuated by rhythms at smaller scales. Baird *et al.* (1989) have stressed how these may be delimited by discontinuities especially in the west of the State. Also, it has long been recognized that the whole Devonian sequence is essentially prograding westward (Fig. 2) and this is accompanied by a westward shift of the black shale basins (Ettensohn 1985*a*, p. 46). The total Frasnian thickness (low Penn Yan to basal Canadaway Group) in the east is in the order of 1500 m and taking a duration for the Frasnian of about 7 million years (Palmer 1983), this gives a rate for an average of combined subsidence and sealevel rise of 2 mm per annum, although this could be halved using other estimates of Frasnian duration.

We show in Fig. 11 a preliminary estimate of facies and water depth changes for the Frasnian in New York State for a transect moving with the regional progradation from Seneca Lake for the lowest beds to Lake Erie shore for the uppermost part. Figure 13 attempts a larger-scale interpretation of sea level changes. It differs in several respects from that produced by Johnson *et al.* (1985, 1986). Their model essentially suggests a series of sudden and progressively deepening events with subsequent stability, or stasis, terminated by a sudden regression initiating the next transgressive pulse and culminating in the Upper Kellwasser Event, taken to be the maximum transgression (developing a thesis put forward by House 1975). Firstly, and on the broad scale, we do not consider this model consistent with the field evidence. Successive transgressions often remove evidence of early transgressions yet, for North America, there is evidence that several of the earlier pulses were as extensive or more so than the later ones. Probably the Rhinestreet interval represents the greatest onlap. We see the Taghanic, Genundewa, Middlesex, Rhinestreet, Pipe Creek and Dunkirk as all being significant and these have been widely traced in the subsurface south to Tennessee (Woodrow *et al.* 1989). We give some evidence to suggest these are international rather than local (Becker *et al.* this volume), confirming suggestions of House (1983, 1985) and Johnson *et al.* (1985). Secondly, we differ on the nature of the pulses. Some are not sudden and sharp transgressions, as, for example the Cashaqua/Rhinestreet transition (Fig. 12A), and others, whilst showing a sharp lithological change, indicate merely that there is a switch of pycnocline level which, of necessity, changes lithology sharply, although no great depth change need be involved. Thirdly, with regard to the stasis interval, sediment availability, infill rate and regional subsidence rate are factors that cannot easily be disentangled from global sea-level but we consider most interpretations of the actual cause of eustatic change do not easily lend themselves to this model. Finally there are differences in timing of the events, but these result at least in part from the fact that conodont correlation of the New York Frasnian succession in 1985 was somewhat vague.

In New York, several factors combine to control facies developments. From time to time these have had particular proponents and grouping the proposals systematically here is not to suggest the proponents mentioned would not

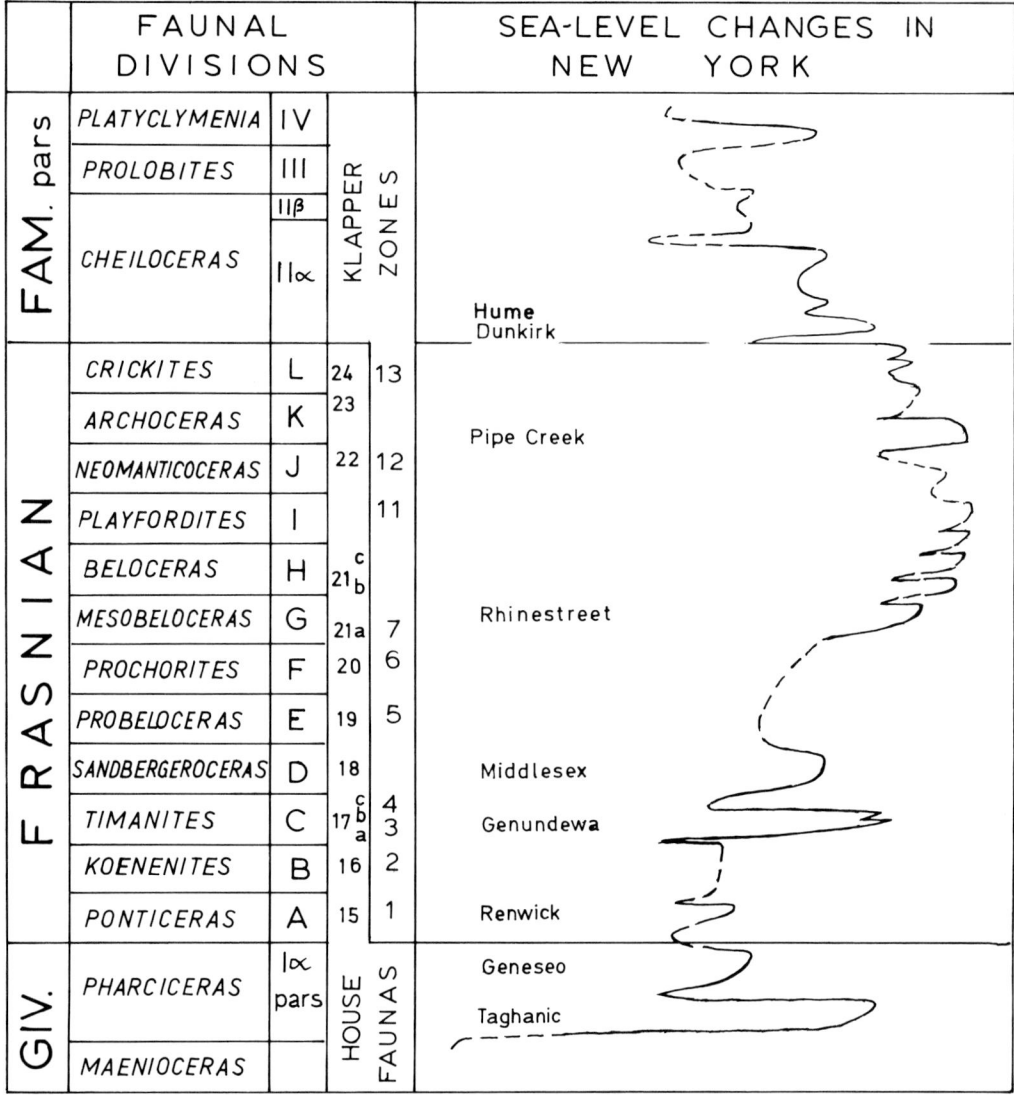

Fig. 13. Diagram suggesting eustatic facies movements which affected clastic rhythmic sequences in the Devonian of New York State and showing correlation with goniatite zones (Becker 1990; Becker *et al.* this volume) and conodont zones (information of G. Klapper and W.T.K.).

exclude the role of other factors. Indeed Woodrow (1985) and Dennison (1985) analyse causations in more detail than is possible here. Views on major factors controlling sedimentation may be categorized as follows.

Orogenic and tectonic pulses. The cause of the terrigenous input from the Catskill Delta after the Eifelian is said to result from the Acadian orogeny and an increasing effect of mountain

areas established east of the Appalachians (Barrell 1913, 1914; Rickard 1981). Some authors relate this to collision between Avalonia and the Appalachian area, others directly with Gondwanaland, the fracturing of which subsequently gave rise to the Avalonia terrace. An oblique convergence is invoked because palaeomagnetic evidence suggests dextral movement between the continental masses. Tectonic depophases resulted (Ettensohn 1985*a*) which may have produced the major rhythms but eustatic changes are also possible.

Major palaeogeographic changes and delta switching. The dominating control of the Catskill Delta itself has led to hypotheses of migrating delta fronts to explain clastic wedges, either on the very large scale (Woodrow 1985), or on the scale of smaller rhythms (Allen & Friend 1968). Such hypotheses are often expressed in a way to imply these are not related to tectonic factors, but that is presumably not necessarily meant.

Climatic factors. Palaeomagnetic evidence suggested a southward migration of the Appalachian area during the Devonian and the migration of climatic belts and high rainfall regimes were invoked to explain increased clastic deposition in the later Devonian (Woodrow *et al.* 1973; Heckel & Witzke 1979). In novel developments of climatic control, but related to depophases, are the models of Ettensohn (1985*b*; Ettensohn & Barron 1981) which relate the main black shale tongues of the New York Frasnian to climatic responses, following episodic tectonism, which control periods of sediment starvation and pycnocline migration.

Orbital control. The importance of orbital control for Milankovitch band rhythms in the New York Devonian seems increasingly conceded. But recently larger-scale cycles related to low frequency orbital cycles have been suggested to be important in explaining periodic sedimentation events (van Tassell 1987) and evolutionary and sedimentary events (Bayer & McGhee 1986). The climatic importance of such events has been stressed by Copper (1986) and others. The possibility that some eight Devonian extinction events, many associated with anoxia were periodic was raised by House (1985), but shelved for lack of precise dating to show the large cycles were of equal duration, and that is also the present situation. We do not see evidence for periodic cometary impact (Sandberg *et al.* 1988*b*).

Eustatic sea-level changes. Such effects as these will only be elucidated by global comparisons. The international nature of the Taghanic onlap and later the possibility that major New York depophases were international was pointed out by House (1975, 1983, 1985) and detailed conodont evidence supported the matching of several Frasnian deepening phases internationally (Johnson *et al.* 1985, 1986). Wignall (1991) has suggested a model linking anoxia with transgression. Such models are not at variance with some of the earlier hypotheses because it is plate tectonic activity, tectonism, and global climatic change that will be the main providers of global sea level changes. However, the rapidity and degree of hypoxic transgressions is still only poorly understood because usual mechanisms for eustatic change operate slowly or are related to glacial melting, for which there is no Frasnian evidence.

Conclusions

A review of the high resolution goniatite biostratigraphy of the Frasnian of eastern North America enables more precise dating of the distinctive coarse meso and micro-rhythmic sedimentary sequences that are so distinctive a feature of the successions. Evidence is given to suggest that the black shale horizons correspond to true deepening events and international comparisons suggest these are eustatic but more detailed discussion appears elsewhere (Becker *et al.* this volume). Acadian tectonic phases may be important internationally in modifying sea level but hypotheses invoking large scale climatic, and orbitally-controlled climatic changes must await improvements in the time scale.

Throughout this work J. W. Wells has been a major source of encouragement and information which is gratefully acknowledged. The advice of L. V. Rickard and his colleagues at Albany and the logistic support of the New York State Geological Survey is much appreciated as are comments on the typescript by G. Klapper, E. B. Selwood and R. T. Becker who kindly prepared Fig. 6. We acknowledge information made available by G. Kloc. The work was commenced under a grant to M.R.H. from the NERC, and has been continued by W.T.K. with support over many years from the State University of New York at Potsdam.

References

ALLEN, J. R. L. & FRIEND, P. F. 1968. Deposition of the Catskill facies, Appalachian region, with notes on some other Old Red Sandstone basins. *Geological Society of America, Special Papers,* **106**, 21–74.

BAIRD, G. C. & BRETT, C. E. 1986*a*. Erosion of an aerobic seafloor: significance of reworked pyrite deposits from the Devonian of New York State. *Palaeogeography, Palaeoclimatology, Palaeoecology,* **57**, 157–193.

—— & —— 1986*b*. Submarine erosion on the dysaerobic sea floor: Middle Devonian corrasional disconformities in the Cayuga Valley region. *New York State Geological Association, 58th Annual Meeting. Field Trip Guidebook,* 23–80.

—— & LASH, G. G. 1990. Devonian strata and

paleoenvironments; Chautauqua County Region: New York State. *Field Trip Guidebook, New York State Geological Association, September 1990*, A1–30.

——, BRETT, C. E. & KIRCHGASSER, W. T. 1988. Genesis of black shale–roofed discontinuities in the Devonian Genesee Formation, Western New York State. *Bulletin of the Canadian Society of Petroleum Geologists*, **14**, 357–375.

BARRELL, J. 1913–1914. The Upper Devonian delta of the Appalachian geosyncline: Parts I–III. *American Journal of Science*, **36**, 429–472; **37**, 87–109, 225–253.

BAYER, U. & McGHEE, G. R. 1986. Cyclic patterns in the Paleozoic and Mesozoic: implications for time scale calibrations. *Paleoceanography*, **1**, 383–402.

BECKER, R. T. 1986. Ammonoid evolution before, during and after the "Kellwasser-event" – review and preliminary new results. *In*: WALLISER, O. H. ed. *Global Bio-Events; a critical approach*. Springer-Verlag, 181–188.

—— 1990. *Stratigraphische Gliederung un Ammonoideen Fauna im Nehdenium (Oberdevon II) von Europe und Nord Africa*. PhD Dissertation, University of Bochum.

——, HOUSE, M. R., KIRCHGASSER, W. T. & PLAYFORD, P. E. 1991. Sedimentary and faunal changes across the Frasnian/Famennian boundary in the Canning Basin of Western Australia. *Historical Biology*, **5**, 183–196.

——, —— & —— 1993. Devonian biostratigraphy and timing of facies movements in the Frasnian of the Canning Basin, Western Australia. *This volume*.

BENSAID, M., BULTYNCK, P., SARTENAER, P., WALLISER, O. H. & ZIEGLER, W. 1985. The Givetian–Frasnian boundary in pre-Sahara Morocco. *Courier Forschungsinstitut Senckenberg*, **75**, 287–300.

BERGERON, J. 1899. Étude des terrains Paléozoiques et de la tectonique de la Montagne Noire. *Bulletin de la Societé Géologique de la France, 3ème Série*, **27**, 617–678.

CHADWICK, G. H. 1935. Chemung is Portage. *Geological Society of America, Bulletin*, **46**, 343–354.

CLARKE, J. M. 1899. The Naples Fauna (fauna with *Manticoceras intumescens*) in Western New York. *New York State Geologist, Annual Report*, **16**, 29–161.

—— 1904. Naples Fauna (fauna with *Manticoceras intumescens*). *New York State Museum, Memoir*, **6**, 31–144.

CLAUSEN, C-D. 1971. Geschichte, Umfang und Evolution der Gephuroceratidae (Ceph.; Oberdevon) in Heutiger Sicht. *Neues Jahrbuch für Geologie und Paläontologie, Abhandlung*, **137**, 175–208.

COLTON, G. W. & DE WITT, JR., W. 1958. *Stratigraphy of the Sonyea formation of Late Devonian age in western and west-central New York*. United States Geological Survey, Oil and Gas Investigations Chart. **OC-54.**

COOPER, G. A. *et al*. 1942. Correlation of the Devonian sedimentary formations of North America. *Geological Society of America, Bulletin*, **53**, 1729–1794.

—— 1968. Age and correlations of the Tully and Cedar Valley Formations in the United States. *In*: OSWALD, D. H. (ed.) *International Symposium on the Devonian System, Calgary, 1967*. **2**. Alberta Society of Petroleum Geologists, Calgary, Alberta, 701–709.

—— & WILLIAMS, J. S. 1935. Tully Formation of New York. *Geological Society of America Bulletin*, **46**, 781–868.

COPPER, P. 1986. Frasnian/Famenian mass extinction and cold water oceans. *Geology*, **14**, 835–839.

DENNISON, J. M. 1985. Catskill Delta shallow marine strata. *Geological Society of America, Special Paper*, **201**, 91–106.

DE WITT JR., W. & COLTON, G. W. 1959. Revised correlations of lower Upper Devonian rocks in western and central New York. *American Association of Petroleum Geologists, Bulletin*, **43**, 2810–2828.

—— & —— 1978. *Physical stratigraphy of the Genesee Formation (Devonian) in western and central New York*. United States Geological Survey, Professional Paper **1032-A**.

ETTENSOHN, F. R. 1985a. The Catskill Delta complex and the Acadian Orogeny: A model. *Geological Society of America, Special Paper*, **201**, 39–49.

—— 1985b. Controls on development of Catskill Delta complex basin facies. *Geological Society of America, Special Paper*, **201**, 65–75.

—— & BARRON, L. S. 1981. Depositional model for the Devonian Mississippian black shales of north America: a paleoclimatic-paleogeographic approach. *In*: ROBERTS, T. G. (ed.) *G.S.A. Cincinnati '81, Field Trip Guidebooks*. **2**. American Geological Institute, 344–361.

FEIST, R. (ed.) 1990. *Guidebook of the Field Meeting Montagne Noire 1990*. Subcommission on Devonian Stratigraphy, IUGS. 1–69.

HALL, J. 1843. *Geology of New York. Part IV, Comprising the survey of the Fourth Geological District*. Carrol & Cook, Albany.

—— 1879. *Descriptions of the Gasteropoda, Pteropoda and Cephalopoda of the Upper Helderberg, Hamilton, Portage and Chemung groups*. New York Geological Survey, Palaeontology of New York, **4** (2).

HECKEL, P. H. 1973. *Nature, origin, and significance of the Tully Limestone*. Geological Society of America, Special Papers, **138**.

—— & WITZKE, B. J. 1979. *Special Papers in Palaeontology*, **23**, 99–123.

HOUSE, M. R. 1961. *Acanthoclymenia* the supposed earliest Devonian clymenid is a *Manticoceras*. *Palaeontology*, **3**, 472–476.

—— 1962. Observations on the ammonoid succession of the North American Devonian. *Journal of Paleontology*, **3**, 247–284.

—— 1963. Evolution observed. *Discovery*, **24**, 12–17.

—— 1965. A study in the Tornoceratidae; the succession of *Tornoceras* and related genera in the north American Devonian. *Transactions of the Royal Society of London*, **B250**, 79–130.

—— 1968. Devonian ammonoid zonation and correlation between North America and Europe. *In*:

OSWALD, D. H. (ed.) *International Symposium on the Devonian System, Calgary, 1967.* **2**. Alberta Society of Petroleum Geologists, Calgary, Alberta, 1061–1068.

—— 1973. Delimitation of the Frasnian. *Acta palaeontologica Polonica*, **23**, 1–14.

—— 1975. Facies and times in Devonian tropical areas. *Proceedings of the Yorkshire Geological Society*, **40**, 233–288.

—— 1978. Devonian ammonoids from the Appalachians and their bearing on international zonation and correlation. *Special Papers in Palaeontology*, **2** v + 70 1–70.

—— 1979. Biostratigraphy of the early Ammonoidea. *Special Papers in Palaeontology*, **23**, 263–280.

—— 1982. The Middle/Upper Devonian Series boundary and decisions of the International Geological Congresses. *Courier Forschungsinstitut Senckenberg*, **55**, 449–462.

—— 1983. Devonian eustatic events. *Proceedings of the Ussher Society*, **5**, 396–405.

—— 1985. Correlation of mid-Palaeozoic ammonoid evolutionary events with global sedimentary perturbations. *Nature*, **313**, 17–22.

—— & ZIEGLER, W. 1977. The goniatite sequence at Adorf, Germany. *Geologica et Palaeontologica*, **11**, 69–108.

——, KIRCHGASSER, W. T., PRICE, J. D. & WADE, G. 1985. Goniatites from the Frasnian (Upper Devonian) and adjacent strata of the Montagne Noire. *Hercynica*, **1**, 1–21.

HUDDLE, J. 1981. *Conodonts from the Genesee Formation in Western New York.* United States Geological Survey Professional Paper, **1032-B**.

JOHNSON, J. G. 1970. Taghanic Onlap and the end of North American Devonian provinciality. *Geological Society of America Bulletin*, **81**, 2077–2106.

—— 1989. Base of the Upper Devonian in the conodont zonation. *Newsletters in Stratigraphy*, **21**, 11–145.

——, KLAPPER, G. & SANDBERG, C. A. 1985. Devonian eustatic fluctuations in Euramerica. *Geological Society of America Bulletin* **96**, 567–587.

——, —— & —— 1986. Late Devonian eustatic cycles around margin of Old Red Sandstone Continent. *Annales de la Société géologique de Belgique.* **103**, 141–147.

JOHNSON, K. G. & FRIEDMAN, G. M. 1969. The Tully Clastic correlatives (Upper Devonian) of New York State: A model for recognition of alluvial dune (?), tidal, nearshore (bar and lagoon), and offshore sedimentary environments in a tectonic delta complex. *Journal of Sedimentary Petrology*, **39**, 451–485.

KIRCHGASSER, W. T. 1975. Revision of *Probeloceras* Clarke, 1898 and related ammonoids from the Upper Devonian of western New York. *Journal of Paleontology*, **49**, 58–90.

—— 1985. Ammonoid horizons in the Upper Devonian Genesee Formation of New York: Legacy of the Genesee, Portage and Chemung. *Geological Society of America, Special Papers*, **201**, 225–235.

—— 1991. Early morphotypes of *Ancyrodella rotundiloba* at the Middle/Upper Devonian boundary, Genesee Formation, west-central New York. *New York State Museum Bulletin*. (in press).

—— & HOUSE, M. R. 1981. Upper Devonian goniatite biostratigraphy. *In*: OLIVER, W. A. & KLAPPER, G. (eds) *Devonian Biostratigraphy of New York, Part 1. Text.* International Union of Geological Sciences, Subcommission on Devonian Stratigraphy, 39–55.

——, OLIVER, JR, W. A. & RICKARD, L. V. 1985. Devonian series boundaries in the Eastern United States. *Courier Forschungsinstitut Senckenberg*, **75**, 233–260.

KLAPPER, G. 1988. Intent and reality in biostratigraphic zonation; a reply to SANDBERG, ZEIGLER & BULTYNCK. *Newsletters in Stratigraphy*, **19**, 179–183.

—— 1989. The Montagne Noire Frasnian (Upper Devonian) conodont succession. *Canadian Society of Petroleum Geologists, Memoir*, **14** (3), 449–478.

—— & JOHNSON, J. G. 1990. Revisions of Middle Devonian conodont zones. *Journal of Paleontology*, **64**, 934–936, 941.

——, FEIST, R. & HOUSE, M. R. 1987. Decision on the boundary for the Middle/Upper Devonian Series Boundary. *Episodes*, **10**, 97–101.

LUTHER, D. D. 1903. Stratigraphy of the Portage Formation between the Genesee Valley and Lake Erie. *New York State Museum, Bulletin*, **69**, 1000–1029.

MATERN, H. 1931*a*. Das Oberdevon der Dill-Mulde. *Abhandlungen der preußischen Geologischen Landesanstalt, N.F.*, **134**, 1–139.

—— 1931*b*. Die Goniatiten-Fauna des Schistes de Matagne in Belgien. *Bulletin Musée Royale d'Histoire Naturelle Belgique*, **7**, 1–15.

MILLER, A. K. 1938. *Devonian ammonoids of America.* Geological Society of North America, Special Papers, **14**.

OSWALD, D. H. (ed.) 1968. *International Symposium on the Devonian System, Calgary 1968.* Alberta Society of Petroleum Geologists.

PALMER, A. R. 1983. The decade of North American Geology 1983 Geologic Time Scale. *Geology*, **11**, 503, 4.

PEPPER, J. F. & DE WITT, JR., W. 1950. *Stratigraphy of the Upper Devonian Wiscoy Sandstone and equivalent Hanover Shale in western and central; New York.* United States Geological Survey, Oil and Gas Investigations. Chart **OC 37**.

—— & —— 1951. *Stratigraphy of the late Devonian Perrysburg Formation in western and central New York. United States Geological Survey, Oil and Gas Investigations. Chart* **OC 45**.

RICKARD, L. V. 1964. *Correlation of the Devonian rocks in New York State.* New York State Museum and Science Series, **4**.

—— 1975. *Correlation of the Silurian and Devonian rocks in New York State.* New York State Museum and Science Service, Map and Chart Series, **24**.

—— 1981. The Devonian System in New York.

In: OLIVER, W. A. & KLAPPER, G. (eds) *Devonian Biostratigraphy of New York, Part 1 Text*. International Union of Geological Sciences, Subcommission on Devonian Stratigraphy, 5–22.

SANDBERG, C. A., ZIEGLER, W. & BULTYNCK, P. 1988a. Middle-Upper Devonian boundary as an example of intent and reality in biostratigraphic zonation. *Newsletters on Stratigraphy*, **18**, 117–121.

——, ——, DREESEN, R. & BUTLER, J. L. 1988b. Late Frasnian mass extinction: conodont event stratigraphy, global changes, and possible causes. *Courier Forschungsinstitut Senckenberg*, **102**, 263–307.

SANDBERGER, G. & SANDBERGER, F. 1850–1856. *Systematische Beschreibung und Abbildung der Versteinerungen des Rheinischen Systems in Nassau*. Wiesbaden.

SCHINDLER, E. 1990. Die Kellwasser-Krise (hohe Frasne Stufe, Ober-devon). *Göttinger Arbeiten zur Geologie und Paläontologie*, **46**, 1–115, 5pls.

SEVON, W. D. & WOODROW, D. L. 1985. Middle and Upper Devonian stratigraphy within the Appalachian basin. *Geological Society of America, Special Paper*. **201**, 1–7.

SUTTON, J. 1963. Correlation of Upper Devonian strata in south-central New York. *Pennsylvania Geological Survey, General Geology Reports*, **G 39**, 87–101.

SUTTON, R. G. & McGHEE, JR, G. R. 1985. The evolution of Frasnian marine "community-types" of south-central New York. *Geological Society of North America, Special Papers*, **201**, 211–224.

VAN TASSELL, J. 1987. Upper Devonian Catskill Delta marine cyclic sedimentation: Brallier, Scherr and Foreknobs Formations of Virginia and West Virginia. *Geological Society of America, Bulletin*, **99**, 414–426.

VANUXEM, L. 1842. *Geology of New York Part III. Comprising the Third Geological District*. White & Fisher, Albany.

WELLS, J. W. 1956. The ammonoids *Koenenites* and *Beloceras* in the Upper Devonian of New York. *Journal of Paleontology*, **30**, 749–751.

—— 1963. Early investigations of the Devonian System in New York, 1656–1836. *Geological Society of America, Special Papers*, **74**, 1–74.

WEDEKIND, R. 1913. Die Goniatitenkalke des unteren Oberdevon von Martenberg bei Adorf. *Sitzungberichte der Gesellschaft Naturforschender Freunde zu Berlin*, **1**, 1–77, 4 pls.

WEDEKIND, R. 1917 (1918). Die Genera der palaeoammonoidea (Goniatiten). *Palaeontographica*, **62**, 85–184, 9 pls.

WIGNALL, P. B. 1991. Model for transgressive black shale? *Geology*, **19**, 167–170.

WOODROW, D. L. 1985. Palaeogeography, paleoclimate, and sedimentary processes of the Late Devonian Catskill Delta. *Geological Society of America, Special Paper*, **201**, 51–63.

——, DENNISON, J. M., ETTENSOHN, F. R., SEVON, W. T. & KIRCHGASSER, W. T. 1989. Middle and Upper Devonian stratigraphy and paleogeography of the central and southern Appalachians and eastern midcontinent, U.S.A. *Canadian Society of Petroleum Geologists, Memoir*, **14** (1), 277–301.

——, FLETCHER, F. W. & AHRNSBRAK, W. F. 1973. Paleogeography and paleoclimate at the deposition sites of the Devonian Catskill delta and Old Red facies. *Geological Society of America Bulletin*, **201**, 3051–3064.

YATSKOV, S. V. 1990. K sistematike semeistva Beloceratidae (Ammonoidea). *Trudy Paleontologscheskogo Instituta*, **243**, 36–51.

ZIEGLER, W. 1962. Taxionomie und Phylogenie oberdevonischer conodonten und ihre stratigraphische Bedeutung. *Abhandlungen des Hessischen Landesamtes für Bodenforschung*, **38**, 1–166, 14 pls.

—— & SANDBERG, C. A. 1990. The Late Devonian standard conodont zonation. *Courier Forschungsinstitut Senckenberg*, **121**, 1–115.

Devonian goniatite biostratigraphy and timing of facies movements in the Frasnian of the Canning Basin, Western Australia

R. THOMAS BECKER[1], MICHAEL R. HOUSE[1] & WILLIAM T. KIRCHGASSER[2]

[1] *Department of Geology, The University, Southampton, SO9 5NH, UK*
[2] *Department of Geology, State University of New York, Potsdam, New York 13676, USA*

Abstract: As part of an international study of Devonian facies movements against cratonic areas an investigation has been undertaken of the timing of facies movements associated with the Frasnian and Famennian reef complexes of the Canning Basin, Western Australia. This led to the discovery of rich goniatite faunas which enable significant improvements to be made to the high resolution ammonoid time scale. A new goniatite classification for the Frasnian is proposed for international use based on the entry of particular genera; this gives 12 divisions. A much more detailed regional zonation is proposed for the Canning Basin. As part of this, a description is given of the following new genera of the Gephuroceratidae: *Gogoceras*, *Playfordites*, *Serramanticoceras* and *Mixomanticoceras*. The genus *Probeloceras* is used in a restricted sense and recognised as the first member of the Beloceratidae; *Acanthoclymenia* is used for forms previously assigned loosely to *Probeloceras*.

The new zonation provides a framework for analysis of facies movements. Occasions of anoxic or hypoxic events are indicated by rich, hematized goniatite faunas. Reef backstepping associated with deepening is dateable by goniatites in associated marginal slope deposits. Several of the international eustatic sea-level changes are recognised, noticeably those correlative with the Genundewa, Middlesex and Rhinestreet deepenings of New York. No evidence has been found for the internationally widespread 'Kellwasser-kalk' facies in the late Frasnian, but a succession of terminal Frasnian sea-level changes is recognized. The 'Irtidium Anomaly' level lies well above the Frasnian/Famennian boundary and hence provides no confirmation of a bolide to explain Frasnian/Famennian boundary extinctions. In the Famennian, evidence for the European Condroz regression and Annulata deepening is presented.

Along the southern border of the Precambrian Kimberley Block in northern Western Australia lies the Lennard Shelf where there is exhumed a Devonian reef complex stretching for over 350 km. It forms the northern margin of the Canning Basin (Fig. 1). The carbonate complex was named the 'Great Devonian Barrier Reef' by Teichert following his initial surveys (Teichert 1941, 1943, 1949). This pristine reef complex, superbly exposed in bare rock karstic plateaux and bush is only weakly folded and faulted. It exhibits three-dimensional aspects of its structure in traversing gorges and has been described in a series of classic papers (Playford & Lowry 1966; Playford 1980, 1981, 1984, Playford *et al.* 1989). It must be the finest Palaeozoic reef complex to be seen anywhere in the world. Playford has shown that Frasnian stromatoporoid–coral reefs of his Pillara Cycle were replaced near the Frasnian/Famennian boundary by cyanobacterian–stromatolite reefs of his Nullara Cycle and he recognized transgressive and regressive changes associated with the changeover.

The possible correlation of Frasnian reef depositional phases of Alberta (Mountjoy 1968; Norris & Uyeno 1983) and Belgium (Tsien 1975; Mouravieff 1975) with major sedimentary cycles in New York State (Rickard 1975; House 1983) seemed likely, and House (1983, 1985) suggested that these resulted from eustatic sea-level movements. This suggestion has found support for Euramerica using conodont data by Johnson *et al.* (1985, 1986). A study of facies movements in the Frasnian reef complexes in relation to the Precambrian Kimberley Block seemed likely to provide another test which might enable the major sea-level movements of the Frasnian and early Famennian to be disentangled from local effects. Although there have been very significant recent studies of the distribution and sedimentology of the reef complexes of the Lennard Shelf, biostratigraphic studies have lagged significantly behind. At the outset it was recognized that detailed biostratigraphic work was required to provide a framework for international correlation and international dating of facies movements. The attraction was the

From HAILWOOD, E. A. & KIDD, R. B. (eds), *High Resolution Stratigraphy*
Geological Society Special Publication, No. 70, pp. 293–321.

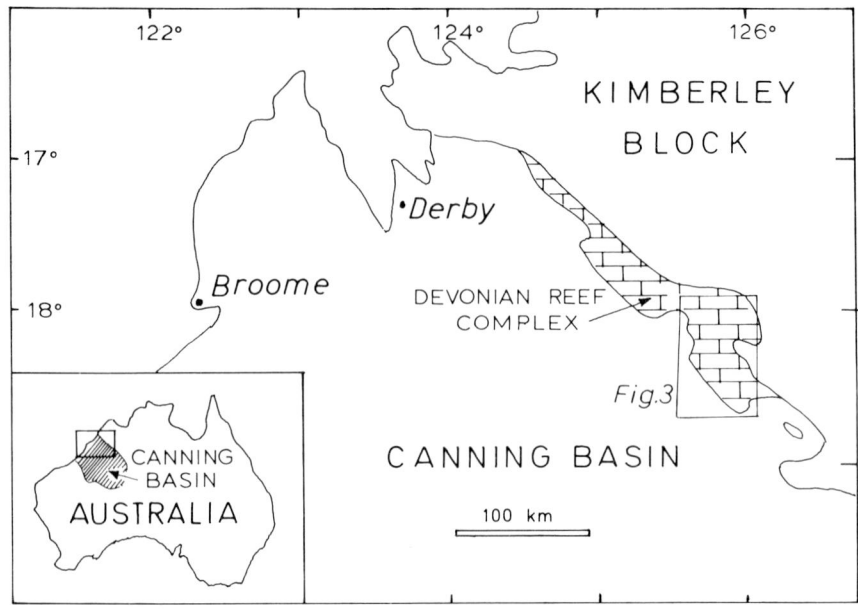

Fig. 1. Map showing the position of the Devonian reef complex of the Canning Basin and the area covered by Fig. 3.

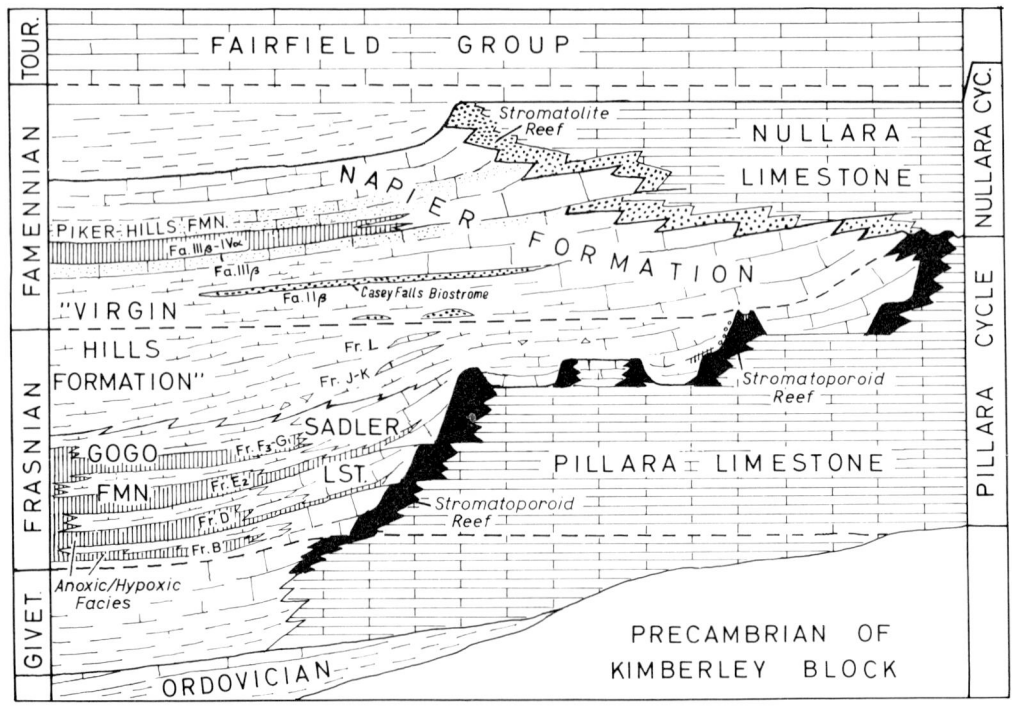

Fig. 2. Facies diagram of the Devonian reef complex on the northern side of the Canning Basin. Based on Playford *et al.* (1989) but showing the anoxic/hypoxic incursions (vertical ruling) in the basin facies and dating the transgressive pulses in the light of the work described in this paper.

knowledge that varied ammonoid faunas were known (Glenister 1958; Petersen 1975), and it was recognized from an earlier visit in 1976 by one of us (M.R.H.) that field work should enable the establishment of an independent goniatite zonation for the area. The discovery of much new material, the careful collection through measured sequences, and the re-examination of museum collections has led to the new regional zonation described here. It has also enabled correlation of some of the main facies movements with the international scales. Figure 2 shows the broad facies relationships of the Devonian reef development in the Canning Basin: it is primarily based on Playford *et al.* (1989) but adds the anoxic/hypoxic levels and some facies shifts recognized in this study. Locality details are given but additionally copies of our section logs, a catalogue of fossils and the mentioned and figured material will be deposited at the Geological Survey of Western Australia (GSWA), Perth, and the Bureau of Mineral Resources, Canberra (Commonwealth Palaeontological Collection, CPI).

Previous work

Frasnian goniatites of Western Australia

Frasnian goniatites were recorded from the Canning Basin by Wade (1938) and Teichert (1941) and, following further work, Teichert (1943, 1949) recognized the goniatites *Manticoceras, Beloceras, Timanites* among others, and provided faunal lists of other groups. It was Glenister (1958) who first gave monographic treatment to the material largely collected by Teichert but also included some of his own. The material, from a wide range of localities, was categorized into the crude stufen and zone system developed early this century in the Rhenish Slate Mountains (Wedekind 1908, 1913) and in Thuringia (Schindewolf 1923). No measured sequences were given; indeed, there is still no detailed measured section available for Frasnian rocks in the Canning Basin. An attempt to clarify relationships between the goniatite localities and the conodont zonation was made by Glenister & Klapper (1966) and this has given helpful information for correlation with the goniatite biostratigraphy established in New York (House 1962, Kirchgasser & House 1981) and the Appalachians (House 1978).

The distribution of Frasnian goniatites in different facies settings of the Devonian reef complexes and inter-reef areas was outlined by Playford & Lowry (1966) who added new records, established the lithostratigraphic framework, and showed the geological setting of many of the goniatite localities that forms the basis for our investigation. No further work has been published subsequently and the question of the biostratigraphical sequence of faunas and the placing of several supposedly endemic species was still unresolved. Also the Australian forms have not been satisfactorily compared with goniatites from North America, Europe, North Africa, Russia and China, largely because of the lack of detailed biostratigraphy; this is particularly true for the cephalopod limestone sequences of Germany. Our two field seasons in the Canning Basin show not only that the Frasnian faunas are among the most diverse anywhere in the world but also that greater stratigraphic discrimination is possible than elsewhere. This paper reports on the newly established local biozonation and gives preliminary comparisons with other areas. Our main localities and their setting are shown on Fig. 3: the numbered localities are those of the Bureau of Mineral Resources programme of R.S. Nicoll and G. Klapper and should be preceded by the index letters (WCB).

International Frasnian goniatite zonation

Many Frasnian goniatite species were described in the last century but a zonation for the Frasnian, or *Manticoceras* Stufe, or Oberdevon (do) I, was not established until the work of Wedekind (1913, 1917) who separated four zones; *Pharciceras lunulicosta* Zone, *Manticoceras nodulosum* Zone, *Manticoceras cordatum* Zone and *Crickites holzapfeli* Zone. These have been also referred to as zones do Iα, Iβ, Iγ and Iδ. Matern (1931a,b) later combined the middle zones as his do Iβ(γ). This scheme was applied to goniatite faunas from Devon and Cornwall (House 1963), North Africa (Petter 1959) and Russia (Bogoslovskiy 1958, 1969) and was found satisfactory, but no new refinement was added. In a revision of the famous Martenberg section, which formed a major basis for Wedekind's zonation, House (*in* House & Ziegler 1977), found good evidence for keeping Iβ and Iγ separate. Clausen (1989) has tried to synthesize new German evidence with work elsewhere. Montesinos (1990) reviewed evidence from northern Spain where tectonic complications hamper precise placing of the goniatite faunas (Montesinos & Henn 1986).

With a revision of the New York sequence House (1962) demonstrated that very much more refinement to a Frasnian zonation was possible, although, in suggesting a time equivalence between *Manticoceras nodifer* (Clarke)

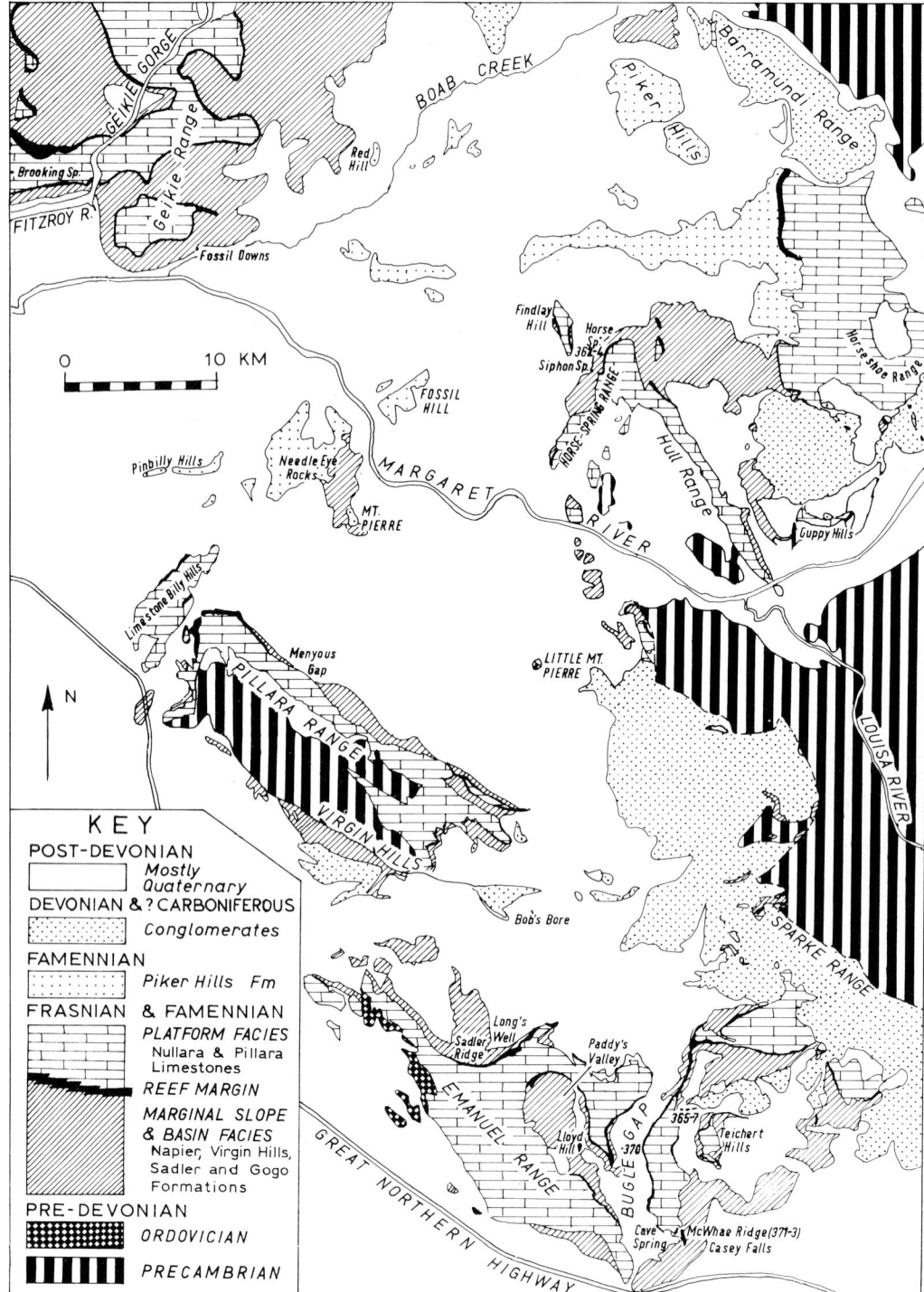

Fig. 3. Map showing the distribution of reef carbonate facies in the area investigated and the localities and WCB locality numbers of the Bureau of Mineral Resources, Canberra refered to in the text. Map based on Playford & Lowry (1966).

and *Manticoceras nodulosum* (Wedekind), he drew the Iβ base too low in New York. More detail of the sequence was later added by House (1978) and Kirchgasser & House (1981) so that, until recently, the New York sequence was the best discriminated goniatite succession in the world: this is mainly because the New York succession is so thick that relative time discrimination is easy whereas in European condensed pelagic limestone sequences it is difficult. For the USSR, through his many and major publications, Bogoslovskiy (1969, 1971; Bogosloviskiy *et al.* 1982) has laid a considerable foundation for synthesis.

Much additional detail has been added for the Montagne Noire as part of the work for the Subcommission on Devonian Stratigraphy (SDS) (House *et al.* 1985). The Global Stratotype Section and Point (GSSP) which was recommended by the SDS and accepted by IUGS for the Givetian/Frasnian boundary and Middle/Upper Devonian boundary is at Puech de la Suque in the Montagne Noire (Klapper *et al.* 1987). The GSSP boundary level places much of Wedekind's do Iα in the Middle Devonian, an unsatisfactory position as far as goniatite workers are concerned (House 1988) and at variance with much international usage. House (1985) separated the part of do Iα now referred to the Middle Devonian as the *Pharciceras* Stufe. Thus the *Manticoceras* Stufe remains equivalent to the newly defined Frasnian. The parts of do Iα remaining in the Upper Devonian include successive faunas with *Koenenites, Timanites* and *Sandbergeroceras* and Becker (1986) has recommended that this interval be termed Lower Frasnian because historically (Wedekind 1913, 1917; House 1978) these faunas were regarded as lower Frasnian. On that basis the Middle Frasnian would correspond to do Iβ and do Iγ. The Upper Frasnian would comprise do Iδ which historically comprises the interval from the Lower to Upper Kellwasser horizons.

Following decisions of the SDS in 1986, the base of the Famennian is taken to coincide with the base of the *Palmatolepis triangularis* conodont zone but the zone has been redefined (Sandberg *et al.* 1988) to exclude early morphotypes named *Pal. praetriangularis* Ziegler & Sandberg which now are taken to be late Frasnian but formerly would have been considered as part of the *triangularis* Zone (Ziegler 1962). Gephuroceratids, mostly assigned to *Manticoceras*, were formerly held to continue into the earliest Famennian (Wedekind 1908; Helms 1959; Schindewolf 1923; Ziegler 1971; House 1973) but this is not now thought to be the case: some are assigned to tornoceratids (Jux & Krath 1974; Becker 1990), others are now regarded as not having been collected in place or not having come from the same conodont levels as claimed (Helms 1959). The Gephuroceratidae became extinct at the end of the Frasnian (Becker & House *in* Feist 1990).

New Genera of Gephuroceratidae

Glenister (1958) has illustrated most of the typical Frasnian goniatites from the Canning Basin. Before description of the new goniatite marker classification and the regional goniatite zones established for the Frasnian of the Canning Basin it is necessary to create some new generic names. Essentially these represent another step in the break-up of the carpet-bag name *Manticoceras* by separating forms with distinctive morphologies and stratigraphical usefulness and drawing attention to some of the many distinct lineages represented in the complex evolution of Frasnian goniatites. The need for this has been obvious for some while but it only becomes appropriate now that forms can be precisely documented stratigraphically and zonally.

New genus: *Gogoceras*

Type-species. Gogoceras nicolli n.sp.
Derivation of name. The genus from the Gogo Formation in which it is found, a name derived from Gogo Station.
Included species. The type species and two unnamed species (Fig. 7)
Diagnosis. Manticoceratids with involute shell and almost closed umbilicus. Suture line as in *Manticoceras* but with a rounded lateral lobe, a very shallow subumbilical saddle and an incipient lobe at the umbilical seam. With convex or biconvex ornament.
Remarks. The genus is unique among Gephuroceratidae in the *Tornoceras*-like shell form, but they show the typical suture and growth lines of the family. The slow increase in coiling is also more characteristic of *Tornoceras* and it is probable that the genus does not reach large size. The suture resembles *Uchtities* more than *Manticoceras* and there is no indication of sharpening of the lateral lobe. The incipient second umbilical lobe is another feature which distinguishes *Gogoceras* from other genera of the family.
Range. Frasnian E_2 to F_2, or *Ponticeras discoidale* to *Probeloceras lutheri* Zones in the Canning Basin.

New species. *Gogoceras nicolli* n.sp. Figs 4H,I, 5A,B

Fig. 4. Photographs of some Frasnian goniatite markers identified in the Canning Basin and illustrations of some new taxa. All are from the Canning Basin unless otherwise stated. (**A** and **B**) *Protimanites pons* (Glenister), GSWA. F5427c, from the Gogo Formation of Sadler Ridge, × 1.5. (**C** and **D**); *Probeloceras lutheri* (Clarke), GSWA F 48518 from Loc. 365/10a, × 1.5. (**E** and **J**); *Manticoceras* sp. nov., CPL. 30654 from Loc. 367, near McIntyre Knolls, × 1.5. (**F** and **G**); *Prochorites alveolatus* (Glenister), GSWA. F 48519, from Loc. 367/10(½), near McIntyre Knolls, × 2.25. (**H** and **I**) *Gogoceras nicolli* n.gen. n.sp., Holotype, GSWA. F 48506, from Loc. 367/10(½), Near McIntyre Knolls, × 2.25. (**K–N**) *Mixomanticoceras exploratum* n.gen., n.sp., (**K** and **L**) a Paratype, GSWA F 48507, × 3.75 (**M** and **N**), the Holotype, GSWA. F 48507, both from Loc. 365/−4.5, near McPhee Knoll, × 3.75. (**O** and **P**) *Playfordites tripartitus* (G. & F. Sandberger), specimen from Martenberg, Germany, profile VI, Bed 12, Univ. Marburg Coll., Mbg. 2594a, figured House & Ziegler 1977, pl. 35 figs 13–15, showing shell covering spiral grooves on internal mould, × 2.25. (**Q**) *Mesobeloceras housei* Montesinos & Henn, GSWA. F 48521, from. Loc. 366/a, × 3.

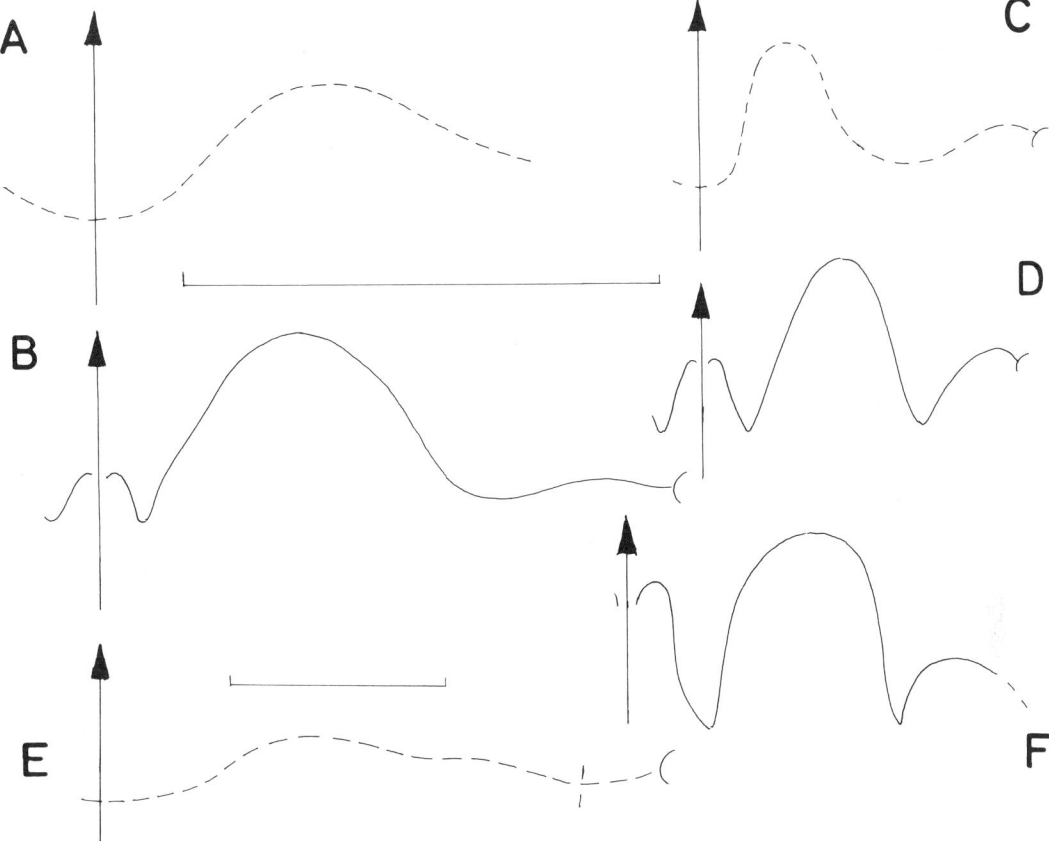

Fig. 5. Suture lines of some marker goniatites recognised in the Frasnian of the Canning Basin. (**A** and **B**) *Gogoceras nicolli* n. gen., n.sp., growth line and suture based on the Holotype, GSWA. F 48506, from Loc. 367/10(½) near McIntyre Knolls, Canning Basin; scale 5 mm. (**C** and **D**) *Mixomanticoceras exploratum* n.gen, n.sp., growth line and suture based on a Paratype, GSWA F 48509, from Loc. 365/−4.5 near McPhee Knoll, Canning Basin; scale as for (**A**). (**E** and **F**) *Sphaeromanticoceras lindneri* (Glenister), growth line based on CPC 35766, and suture based on the Holotype CPC 35765, from the Virgin Hills Formation; scale 5 cm.

Type. Holotype GSWA F 48506, Figs 5A,B, 4H,I.

Type locality. Loc WCB 367, 80 m west of R. S. Nicoll section, Bed1/2, equivalent to 3–5 m above Bed 10, close to McIntyre Knolls.

Derivation of name. In honour of Dr R. S. Nicoll, BMR, Canberra.

Diagnosis of species. Species of *Gogoceras* with almost closed umbilicus and weakly biconvex growth lines.

Comparison. The new collections include two further species of *Gogoceras* from somewhat earlier levels. *Gog.* n.sp. A. has convex growth lines and *Gog.* n.sp. B has a small open umbilicus and a well-developed ventrolateral salient of the growth lines.

Range. Only from the Zone of *Probeloceras lutheri* n.ssp. or F$_2$ in the Canning Basin.

New genus. *Mixomanticoceras*

Type species. *Mixomanticoceras exploratum* n.sp.

Derivation of name. Suggesting the admixture of morphologial features of other genera.

Diagnosis. Gephuroceratid with depressed shell, densely ribbed, with steep umbilical shoulders, flanks with a spiral depression, with flattened venter and marginal furrows. Growth lines strongly biconvex with a narrow salient lying in the marginal furrows and a weak ventral band similar to *Aulatornoceras*. Suture line manticeratid with highly arched and narrow lateral saddle.

Included species. monospecific.

Remarks. The new genus combines features of *Sphaeromanticoceras, Trimanticoceras* and *Aulatornoceras:* some of these features are developed iteratively in later gephuroceratid lineages. The suture line is already more advanced than *Mant. cordatum* or *Mant. lamed* of the same size and from the same bed. In *Sphaeromanticoceras* with similar whorl cross section slight ribbing may be seen in juvenile stages, but there are no ventrolateral furrows or strongly biconvex growth lines. The ventral band suggests the existance of low ventral spines as in *Aulatornoceras* (Becker 1990) or *Kosmoclymenia* (Korn 1979).

New species. *Mixomanticoceras exploratum* n.sp. Figs 4K–N, 5C,D

Type. Holotype GSWA F 48507, Figs 4M,N, Paratype F 48507 4K,L.
Type locality. Near the informally named 'McPhee Knoll', Loc. WCB 365, 4 to 5 m below the base of the section measured by R. S. Nicoll. Frasnian E_1, *Prob. lutheri lutheri* Zone.
Deviation of name. To reflect the fact that the genus explores morphologies not previously adopted in the Gephuroceratidae.
Diagnosis. As for genus.
Range. Frasnian E_1, *Probeloceras lutheri lutheri* Zone.

New genus. *Playfordites*

Type species. Goniatites lamed var. *tripartitus* G. & F. Sandberger 1850.
Derivation of name. In honour of Dr. P. E. Playford, Director, Geological Survey of Western Australia.
Diagnosis. Slender manticoceratid with discoidal shell and spiral thickenings on either side of the venter or on the flanks which are not shown on the external moulds (Fig. 4, O,P). Suture line as in *Manticoceras.*
Included species. Monospecific.
Remarks. The genus *Trimanticoceras* was erected (House *in* House & Zeigler 1977, p.87) for species like the type species *Trimant. cinctum* (Glenister 1958, p. 75, pl. 10, figs 1–3) which have strong furrows on the shell on either side of a narrow raised venter. There are also forms like *Goniatites lamed* var. *tripartitum* G. & F. Sandberger (1850 p. 90 pl. 8, fig. 7) which do not have such deep furrows near the venter and in which the shell form is much more slender and discoidal. Such forms were figured by House & Ziegler (1977, pl. 3, figs 13, 14, 23, 24) at which time significant weight was not placed on the fact that the shell thickenings are purely internal

(House & Ziegler, pl. 3, figs 13, 14). These comprise the new genus. The Canning Basin material is plentiful and distinctive in that there is variety and often asymmetry in the lateral furrows which may be developed and this has led to the manuscript recognition of morphotypes. Similar types are known from Martenberg, Germany. The nature of the thickenings indicates that the genus arose from the *Mant. lamed* stock and specimens sometimes share with *Serramanticoceras* slight crenulations on the venter.
Range. Confined to Frasnian I_2, the *Play. tripartitus* Zone.

New genus. *Serramanticoceras*

Type species. Goniatites serratus Steininger, 1843.
Derivation of name. Referring to the ventral serrations on the internal mould.
Diagnosis. Thinly discoidal shells with flattened flanks and venter rounded, flat, tabular or with a slight median groove and with periodic thickenings of the inside of the shell at the venter giving serrations on the internal mould. Suture as in the *Mant. lamed* Group. Growth lines strongly bioconvex.
Included species. Serramant. serratum serratum (Steininger), *Serramant. serratum obliqueseptatum* (Clausen), *Serramant, serratum* ssp. A (Clausen). (? = *Goniatites dorsicosta* Roemer). Undescribed new forms.
Remarks. Internal shell thickenings are widespread among genera of the Goniatitida, especially in Devonian tornoceratids, cheiloceratids and imitoceratids. Their value for taxonomic separation is limited although there are some distinctive types of constrictions (as in *Prolobites* and *Prionoceras*). Among manticoceratids this morphological feature is limited to certain narrow lineages with high stratigraphic value. Serration of the ventral shell was illustrated by Matern (1931*b*) in species he referred to *Crickites* with convex growth lines.
Range. Serramanticoceras serratum is confined to Frasnian I_1 to J_2, *Playfordites tripartitus* and *Neomanticoceras erraticum* Zones.

Canning Basin Frasnian goniatite zonation

The evidence for the new Frasnian goniatite biozonation established in the Canning Basin will be given here. Very substantial thicknesses of record are available which significantly exceed those known from the classic European sequences. Not all of the succession in the

FRASNIAN GONIATITE LEVELS OF THE CANNING BASIN

	regional zones	other first entries
L2	Manticoceras sp.	
L1	Sphaeromanticoceras lindneri	Manticoceras guppyi
K	Beloceras aff. tenuistriatum	Manticoceras cf. guppyi
J2	Neomanticoceras erraticum	"Crickites" expectatus Trimanticoceras cinctum
J1	Maternoceras retorquatum	"Manticoceras" tuberculatum Sphaeromanticoceras bullatum
I2	Playfordites tripartitus	Trimanticoceras aff. cinctum Linguatornoceras clausum
I1	Serramanticoceras serratum	
H	Beloceras tenuistriatum	
G2	Mesobeloceras thomasi	
G1	Mesobeloceras housei Mesobeloceras n.sp.	Linguatornoceras sp. Mesobeloceras cf. iynx
F3	Sphaeromanticoceras affine	Sphaeromanticoceras cf. orbiculum Maternoceras n.sp. Carinoceras s.str. sp. Aulatornoceras eifliense
F2	Probeloceras lutheri n.ssp.	Gogoceras nicolli n.gen. n.sp. Acanthoclymenia aff. n.sp. Manticoceras ?n.sp. Manticoceras aff. lamed
F1	Prochorites alveolatus	Acanthoclymenia neapolitana Probeloceras aff. lutheri Gogoceras n.sp. B
E3	Manticoceras n.sp.	Ponticeras aff. discoidale
E2	Ponticeras discoidale	Aulatornoceras n.sp. B Gogoceras n.gen. n.sp.A
E1	Probeloceras lutheri lutheri	Acanthoclymenia n.sp. Acanthoclymenia aff. neapolitana Manticoceras lamed Manticoceras cordatum Aulatornoceras n.sp. A Ponticeras sp. Glenister Mixomanticoceras exploratum n.sp.
D	Manticoceras cf. evolutum	Hoeninghausia cf. archiaci
C2	Manticoceras sp.	
C1	Timanites angustus	Tornoceras contractum
B2	Koenenites n.sp.	
B1	Protimanites pons	

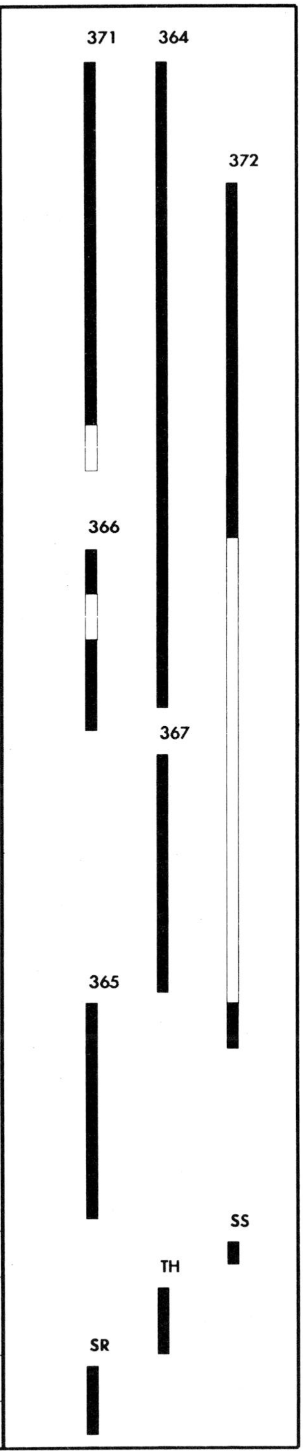

SR = Sadler Ridge   TH = Timanites Hill   SS = Siphon Spring

Fig. 6. Diagram listing the new regional goniatite zones established in the Canning Basin and the range covered by the main sections investigated. Localities 371 and 372 are near McWhae Ridge, Localities 365 to 367 are near McPhee and McIntyre Knolls, Timanites Hill is Locality 370. G. Klapper has conodont evidence which suggests an overlap of parts of sections 365 and 367 without goniatites.

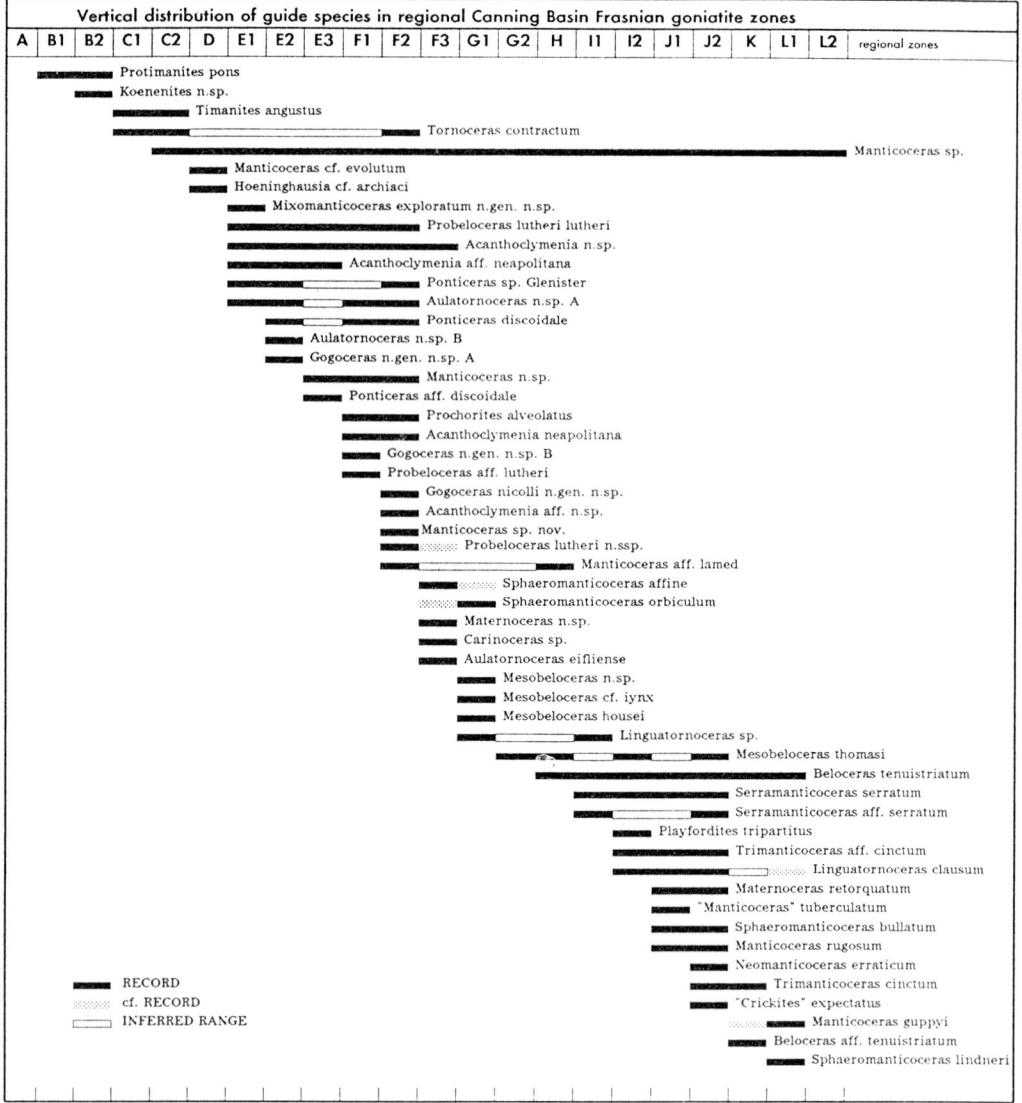

Fig. 7. Table showing the range of important goniatite taxa in relation to the new regional goniatite zones established in the Canning Basin.

Canning Basin is documented for goniatites. It may well be that additional faunal levels will have to be added to the scheme proposed here. The zones are therefore to be regarded as tentative regional zones. Also, for some groups taxonomic work is at an early stage and terminology correspondingly imprecise. The sequence will be described from the oldest and for each biozone the location of critical sections, the characteristic fauna, and taxonomic notes will be given.

We have found no evidence of Givetian goniatites nor of the pharciceratids which, under the new definition, now characterize the latest Middle Devonian. We think the earliest fauna is early in the Frasnian, but goniatite faunas associated with the very earliest Frasnian conodonts, of Zone 1 of Klapper (1989), or the earliest Lower *Asymmetricus* Zone or *Falsiovalis* Zone of Ziegler & Sandberg (1990), have not been identified, that is the *Petteroceras feisti* fauna of the Montagne Noire (House *et al.* 1985) or the *Ponticeras perlatum* fauna of New York (Kirchgasser & House 1981). There is a thick succession, probably in excess of 100 m, below our first goniatite level near Long's Well but unfortunately in our brief reconnaisance, we were unable to locate any goniatite horizons. This would seem to be the most likely succession to search for the pharciceratid and ponticeratid faunas to be expected below the *Protimanites* fauna. Therefore, so far, this *Ponticeras* division is not recognised in the Canning Basin.

The following regional goniatite zones have been recognized in the northern Canning Basin. They are grouped within the framework of the new generic international divisions. Rather detailed locality data are given, since the precise sequences from which the faunas have come have not been adequately described. The new regional zones and ranges of critical taxa are shown on Figs 6 and 7.

B₁: *Protimanites pons* (Glenister)

A collection from Sadler Ridge (GSWA F 5247) made by P. E. Playford in 1958 drew our attention to this fauna. The localities are about 400 m west of Long's Well (1 : 1000 000, Bruten Sheet 4060, Grid Reference 4060–148391). Allochthonous blocks occur to the north and there is a large allochthonous hillock to the southwest. The succession dips southeast and the 35 m section with many goniatite horizons which we have measured is capped by a massive 0.8 m limestone bed which is the thickest seen in the measured section. The typical fauna of the zone was collected from some five levels between 12.0 and 35.0 m below the 0.8 m marker bed. The fauna includes *Protimanites pons* and *Prot. pons* subsp. nov., *Bactrites* cf. *parvus*, *Lobobactrites paucesinuatus*, *Lobo. timanicus*, brevicones, small mostly indeterminate brachiopods, including ambocoeliids, and small gastropods, including naticopsids, pleurotomariaceans and loxonemids. The taxonomic assignment of the *Protimanites pons* group is perhaps debateable. When naming the species, Glenister (1958, p. 77, 78) referred it to *Hoeninghausia* but judging from the type species we now regard that genus as a terminal giant oxyconic genus of the *Koenenites* lineage with only one external umbilical lobe. *Protimanites*, as its name implies, and as Lyaschenko (1956) apparently thought when he proposed it, is ancestral to the true *Timanites* which has more involute inner whorls, and has stabilized the lateral, first and second umbilical lobes with acute lobes and strongly arched intervening saddles.

B₂: *Koenenites* n.sp.

At 12.5 m below the 0.8 m marker bed at Sadler Ridge the first koenenitids appear. It is true that specimens have been found loose lower down, but they are distinctively polished and weathered and we feel confident that they are from hill wash. The new form is *Koenenites* sp. nov., but *Prot. pons* and *Prot. pons* subsp. nov. continue as do the lobobactritids. The new *Koenenites* has a lateral saddle too narrow to match any of the well localized specimens from New York and in this respect it is closer to *Koen. kirchgasseri* House (1978) from Virginia; several slightly different morphotypes seem recognizable.

C₁: *Timanites angustus* Glenister

The famous locality in the Canning Basin for *Timanites* is a small hill in Bugle Gap northeast of Glenister Knolls near the spot height 241 marked on the map (1 : 100.000, Bohemia Sheet 4160, Grid Reference 4160–880323) and called informally 'Timanites Hill'. This is Locality WCB 370. A section of about 7 m can be measured on the northern slopes and the succession is capped by a 15 cm pink limestone brecciated bed. The best collecting level is from 5.6 m above the base of the section and a little below, where *Tim angustus* can be found abundantly in all growth stages. Associated is *Tornoceras contractum* Glenister, a species closely related to *Torn. typum* (G. & F. Sandberger) from Germany that has somewhat higher ventral saddles. *Lobobactrites* sp., brevicones and naticopsids complete the pelagic assemblage. *Tim. angustus* also occurs in a 5 cm bed 5.9 m above the base of the section.

C₂: *Manticoceras* sp.

At about 6.15 m above the base of the section on Timanites Hill, occurs the first specimen referable to *Manticoceras*. *Tim. angustus* continues to occur, and a particularly large specimen (GSWA F 48520) shows a saddle at the venter as figured by Bogoslovskiy (1969, text figure 90i) from the Timan. *Torn. contractum* also occurs.

D: *Manticoceras* cf. *evolutum* Petter

Two faunas are referred here but with some doubt since both are geographically isolated records with no evidence above or below. Neither do we have independant conodont or other evidence to place them with confidence.

The first fauna was noted from two specimens of a distinctively evolute goniatite in a collection

(GSWA F 48522) from 3.2 km south of Siphon Spring (1:100 000, Elma Sheet 4161, near Grid Reference 4161–847812). The suture is that of *Manticoceras* but the shell form is that of *Ponticeras*. The suture is advanced with an acute lateral lobe and the extra umbilical lobe of *Manticoceras*. The closest described species is *Mant. evolutum* Petter (1959, p. 174, pl. 16, fig. 1) from North Africa but there are similarities to *Mant. artum* Bogoslovskiy (1958, p. 114, pl. 6, fig. 3) and *Pont. acutilobatum* Bogoslovskiy (1958, p. 92, pl. 2, fig. 6). The juveniles are similar to *Pont.* aff. *tschernyschewi* sp. juv. (compare with Holzapfel 1989, pl.4, fig. 1, but with the lateral saddle better developed). They show how this group of *Manticoceras* could simply arise from *Ponticeras*. P. E. Playford kindly showed us the locality and there is a circlet of low hills with hematitic faunas low on the slopes of the Sadler Limestone to the east and south all of which yield this form with very little evidence of other goniatites and the low diversity is unusual. The locality is isolated, and we have no means of placing it with respect to the other sections and we position it at this level because of the absence of the distinctive forms of higher or other horizons. Noteworthy in the fauna are the well-preserved hematitic brachiopods, *Spinatrypina, Productella*, rare spiriferids, gastropods and even a hematitic rugose coral. Upslope calcareous or silicified preservation of identical assemblages suggests rolling downslope of benthonic forms into stratified strongly hypoxic to anoxic bottom waters.

The second fauna (GSWA F 5248) is labelled as from three miles southeast of Virgin Bore and collected by P. E. Playford and D. C. Lowry. This includes a specimen approaching an oxyconic form which has a suture similar to a specimen referred to by House *et al.* (1985, p. 9, fig. 8D–E) as *Hoeninghausia* cf. *Hoen. archiaci* Gürich from Bed 24 at La Serre Trench A$_1$ but which now, because it does not have an oxyconic shell, is refered to *Koenenites*. A similar form from New York, high in the West River Shale, and again not oxyconic, is referred to as *Koen. aff. lamellosus*. However we assume that the oxyconic types are later and take them to be early in the *Sandbergeroceras* Division.

E$_1$: *Probeloceras lutheri* (Clarke)

The discovery in the Canning Basin of this marker form of the New York sequence is a notable contribution to international correlation. A series of measured traverses were undertaken by R. S. Nicoll and G. Klapper in the Old Bohemia Valley numbered localities

Loc. 365–367. These form a W–E sequence with Loc. 365 lying west of McPhee Knoll (1:100 000, Bohemia Sheet 4160, Grid Reference 4160–943370) and the traverse continues after an interruption towards McIntyre Knolls. We are indebted to R. W. Brown for introducing us to this area. We consider the goniatite-bearing part of Loc. 366 to lie above both Loc. 365 and 367. Goniatites occur at 4.5 m below the stake which marks the start of the measured Loc. 365 section and contains *Probeloceras lutheri*, *Ponticeras* sp., *Acanthoclymenia* spp., *Mant. lamed* (G. & F. Sandberger), *Mant. cordatum* G. & F. Sandberger), *Aulatornoceras* n. sp. and bactritids and lobobactritids. Restricted to this level is the unique *Mixomanticoceras exploratum* n.gen., n.sp. described herein.

E$_2$: *Ponticeras discoidale* Glenister

This distinctive species, which has a highly arched lateral saddle in contrast to the associated *Acanthoclymenia* species, enters 5.5 m above the base of the measured Loc. 365 section. Already the remarkable *Mixomanticoceras* is missing. Other elements include *Mant.* spp., bactritids, lobobactritids and a new form, *Gogoceras* n.gen. n.sp. A, which has convex growth lines, which occurs 40 m above the base of the section (above Bed 8 of R. S. Nicoll). In the McWhae Ridge area loose hematitic faunas collected from the flats below the trench section (Loc. 372) yielded the typical fauna of this level with *Pont. discoidale*, *Acantho.* n. sp., *Mant.* spp. and bactritids. A second new species of *Aulatornoceras* occurs rarely with this fauna. A similar fauna occurs in the faulted graben at the crest of McWhae Ridge. This allows a precise dating for the termination of Gogo facies in this area and for sediment infilling over the drowned McWhae Ridge reef.

E$_3$: *Manticoceras* n.sp.

The section at Loc. 367 lies just west of Mt Pierre Creek and of the most westerly of the large allochthonous blocks which form McIntyre Knolls. The sequence has been numbered by R. S. Nicoll but it thickens rather sharply immediately to the west where we have another measured section. A laterally compressed manticoceratid with affinities to *Carinoceras menneri* enters (GSWA F 48510) in the lowest part of this section which does not have (at the diameters seen) the oxyconic form with sharp venter to justify placing in *Carinoceras*. Other elements are similar to the underlying fauna. At 8 m above the base of the Loc. 367 section a species transitional between *Pont. discoidale* and *Uchtites* was collected which suggests that *Uchtites* is a descendant of *Ponticeras* rather than being a carinoceratid with a weakly developed umbilical saddle.

F_1: *Prochorites alveolatus* (Glenister)

This distinctive goniatite with a grooved venter we consider identical to *Proch. strix* (Kirchgasser 1975) from New York State. At Loc. 367 the first specimen was encountered at about 12.2–15.8 m above the base of the section and between Beds 6 and 7 of R. S. Nicoll. Immediately below, between 10 and 12.2 m above the base, first appear forms that seem identical with the New York *Acanth. neapolitana*, a form revised by House (1961). Also at this level is a probeloceratid with a subacute, rather than a tabulate venter which we refer to as *Prob.* aff. *lutheri*, and a second new species, here unnamed, of *Gogoceras* (n. sp. B) with an open umbilicus, flattened venter and biconvex growth lines. The rest of the fauna is identical to that of E_3.

F_2: *Probeloceras lutheri* n. ssp.

Also from Loc. 367, faunas from around equivalents of Bed 10, but collected about 100 m west of the Nicoll section, are incredibly rich and well preserved. A characteristic new form appearing is a new subspecies of *Prob. lutheri* with concavities on the upper flanks of the ventral lobe, a sign of the subdivision leading to the advanced beloceratids. Levels 3 m above equivalents to Bed 10 are especially rich in *Probeloceras*. Other typical forms of level F_2 are specimens of *Manticoceras* similar to *Mant. lamed* but with much more evolute inner whorls, and specimens with a slightly flattened venter and sub-trapezoidal cross-section (*Mant.* ? n.sp.). *Gogoceras nicolli* n.gen. n.sp. is characteristic.

F_3: *Sphaeromanticoceras affine* (Steininger)

We think the base of the section at Loc. 366 lies above the faunas of Loc. 367. From 0–1.5 m below the basal bed of the R. S. Nicoll section *Sphaeromanticoceras* enters, and both *Sphaero. affine* (Steininger) and *Sphaero.* cf. *orbiculum* (Beyrich) and *Bactrites schlotheimi* have been identified. *Mant. lamed* continues together with *Acanthoclymenia*. There is also a small specimen (GSWA F 48501) tentatively identified as *Maternoceras* n. sp. which differs from *Mat. calculiforme* in that the lateral saddle is rounded rather

than acute and it has a flattened but not a grooved venter. Of special interest is the first record for Australia of *Aulatornoceras eifliense* but the records of this are rather earlier in New York (Kirchgasser & House 1981) and rather later in the Büdesheim fauna of the Eifel Hills.

The higher part of the Frasnian goniatite sequence in the Canning Basin has been elucidated in an area near Horse Spring (1 : 100 000, Elma Sheet 4161, Grid Reference 4161–858854) which is Locality WCB 364 of R. S. Nicoll, and also near McWhae Ridge. At Horse Spring the Virgin Hills Formation fills in the topography of the earlier reef margin deposits of the Sadler Limestone and the earliest beds of the Virgin Hills Formation overstep in 100 m southwest of Loc. 364 so that younger beds come to lie directly on the Sadler Limestone. The succession continues up into the Famennian, the Frasnian/Famennian boundary being very well exposed (Becker *et al.* 1991).

Beds below Bed 1 of the section at Loc. 364, but 50 m to the north, yielded a poor fauna characteristic of Bed 1 itself but it is included in the interval F_3 because of the entry of the earliest true *Carinoceras* with a sharp venter at 4 cm diameter. Also *Prob.* cf. *lutheri* and a cross-section of *Sphaeromanticoceras* was found.

G_1: *Mesobeloceras housei* Montesinos & Henn

Ledges about 20 m southwest of the Loc. 364 section base have abundant *Probeloceras* and *Mesobeloceras* on an exposed surface. In Bed 2 of section Loc. 364 occur specimens (GSWA F 48502,3) which we refer to *Mesob.* n. sp. with only one or two extra umbilical lobes and are thus transitional from *Prob. lutheri* to *Mesobeloceras*. Also in Bed 2 are widely evolute forms referred to *Mesob.* cf. *iynx* and forms with a small fourth ventral lobe and several umbilical lobes that resemble *Mesob. housei*. At the Loc. 366 section in Bed 0 and up to 16.8 m above, *Mesob. housei* occurs and *Mant. lamed, Mant.* cf. *cordatum, Linguatornoceras* aff. *clausum* (Glenister), orthocones, gastropods and styliolinids were noted. The stratigraphical position of *Mesob. housei* in the Cantabrian Mountains (Montesinos & Henn 1986, p. 74) is not clear because, although a section has been published, the sequence is tectonized.

We believe *Probeloceras lutheri* to be the basal group of the Beloceratidae and that other members of the family arise from *Prob. lutheri* by the addition first of extra umbilical lobes and then of extra adventitious lobes; early forms which are included within *Mesobeloceras* have an open umbilicus but when *Beloceras* is reached not only are there substantially more sutural elements, but the umbilicus closes; the genus *Ceratobeloceras* (House *et al.* 1985), which has ceratitic lobation, is the most complex. We do not accept the origin from *Timanites* preferred by Yatskov (1990) and consider his terminology, in the absence of detailed biostratigraphy, to be premature. In New York the equivalent simple early forms of *Mesobeloceras* are usually named *Eobeloceras iynx* (Clarke) (see Miller 1938, p. 137) and that species already has three outer umbilical lobes. The holotype of the type species of *Eobeloceras, Ammonites multiseptatus* von Buch, was illustrated by Bensaïd (1974) and it is clear that it is not a beloceratid, but a pharciceratid.

G₂: *Mesobeloceras thomasi* Glenister

This form was described by Glenister (1958, p. 83, pl. 8, fig. 2, pl. 10, fig. 6) from the Canning Basin. The species already shows the increase in size which is progressive in the lineage and a somewhat less evolute conch than in the earlier species. It is very close to *Mesob. kayseri* (Holzapfel) from Germany. It occurs in the Horse Spring Loc. 364 in Bed 5 and laterally outcropping beds. In the trench section of R. S. Nicoll, Loc. WCB 372, about 350 m west of McWhae Ridge (1:100 000, Bohemia Sheet 4160, Grid Reference 916260) this type occurs in the basal bed, although the specimen is poorly preserved (GSWA F 48523).

H: *Beloceras tenuistriatum* (d'Archiac & de Verneuil)

Of the several names available for this species we here adopt the oldest, in agreement with Kullmann (1960). This was first located in yellow calcareous siltstones 2–3 m up in the trench of Loc. 372 with *Mant. lamed.* and *Mant.* aff. *lamed.* Thereafter *Bel. tenuistriatum* is common at several levels. At Horse Spring Loc. 364 the species was first encountered in Bed 7 (GSWA F 48504,5) and it ranges almost to the top of the Frasnian there.

I₁: *Serramanticoceras serratum* (Steininger) n.gen.

At 7.6 m above the base of the Loc. 372 trench *Serramant. serratum* enters (BMR 372/24–25e,n) and it is associated with a stock of very slender *Manticoceras* which is so characteristic of the late Frasnian in the Canning Basin and which probably evolves into the large *Mant. guppyi* Glenister.

I₂: *Playfordites tripartitus* (G. & F. Sandberger)

At 12 m above the base of the trench section (Loc. 372) manticoceratids similar to *Mant. lamed* appear but with lateral grooves on the outer flanks which are here named *Playfordites*. The group seems to have diverged from very early *Serramanticoceras* since in some specimens slight transverse internal shell thickenings show over the venter. Forms such as these appear at the Horse Spring section Loc. 364, Bed 24 (GSWA F 48527), and in Loc. 372 between 12.1 and 12.9 m (GSWA F 48512–5, F 48516,7): loose material has been found low in the section at Loc. 371.

J₁: *Maternoceras retorquatum* (Glenister)

Sections have been studied on the north side of a small hillock west of McWhae Ridge and about 180 m southeast of the trench (Loc. 372) discussed above. Loc. 371 is a section of R. S. Nicoll published on by Playford *et al.* (1984) and Nicoll & Playford (1988) and figured in Becker *et al.* (1991). One section (Loc. 371A/ø) we measured about 30 m northeast of Loc 371 where our earliest fauna of this zone is from the basal bed, Bed A (Loc. 371A/ø), which yielded *Maternoceras retorquatum, Sphaer. bullatum, Mant. rugosum,* 'Mant.' *tuberculatum, Mant.* spp., *Trimant.* aff. *cinctum* (which may be *Trimant. retrorsum* (von Buch)), *Bel. tenuistriatum,* and *Linguatornoceras clausum.*

J₂: *Neomanticoceras erraticum* Glenister

We measured another section (Loc. 371B), of 31.5 m thickness from the base of the gully north of the hillock to the *Frutexites* Bed about 20 m west of R. S. Nicoll's section 371. At 9–10 m above the base of this section is a level very rich in goniatites which we refer to as the Lower *Beloceras* Bed in which we have collected *Neomanticoceras erraticum* Glenister, *Trimanticoceras cinctum* (Glenister) *Serramant.* aff. *serratum, Serramant. serratum, Mant. lamed., Mant. cordatum, Mant. rugosum,* 'Crickites' *expectatus, Maternoceras retorquatum, Bel. tenuistriatum, Sphaer. bullatum* and *Linguat. clausum.* Immediately above is a coquina full of scutelluids which we call the Esko Bed after the late Esko Rokylle who first drew it to our attention: the Lower Beloceras Bed is Bed B of Loc. 371A/ø where it is also overlain by the Esko

Bed. The Lower *Beloceras* Bed equivalent, but without such a rich fauna, is about Bed 13 of Loc. 372. Near Horse Spring, about the level of Bed 56 of Loc. 364, but about 300 m southwest, a similar fauna occurs with *Trimant. cinctum, Serramant. serratum, Serramant.* aff. *serratum, Mant. lamed, Linguatornoceras* and *Spinatrypina.*

K: *Beloceras* aff. *tenuistriatum* (d'Archiac & de Verneuil)

At 18.9 m in Loc. 371B there is another goniatite-rich level. *Trimant. cinctum* is still present and there are large gephuroceratids that are close to *Mant. guppyi* Glenister, small *Mant. lamed* and *Bel. tenuistriatum.* Beloceratids are incredibly abundant at 23.0–24.0 m which we refer to as the Upper *Beloceras* Bed and there is a form, *Bel.* aff. *tenuistriatum*, with rather inflated whorls which occurs with typical *Bel. tenuistriatum. Mant.* spp. and harpids are especially abundant here. At Horse Spring Loc. 364, the Upper *Beloceras* Bed is represented by Bed 70 and the lower part of Bed 71, and the same level is represented in the creek below Casey Falls (Becker *et al.* 1991, fig. 1).

L₁: *Sphaeromanticoceras lindneri* (Glenister)

The uppermost level with abundant Frasnian goniatites at Loc. 371B is at 25.0–25.2 m, which yields very large specimens of *Sphaer. lindneri* and *Mant. guppyi* Glenister. Earlier ontogenetic stages are also common following this level up the slope of the drowned reef at McWhae Ridge and *Bel. tenuistriatum* and *Linguatornoceras* also occur. This level with large *Sphaer. lindneri* and *Bel. tenuistriatum* also occurs on the drowned reef crest. *Mant. guppyi*, which is common at this level, was used by Glenister (1958) in a rather wide sense for all compressed manticoceratids, but we restrict the species to the very large terminal Frasnian forms, agreeing with the Holotype, which have a highly arched lateral saddle and a characteristic external lobe. At Horse Spring (Loc. 364) the L₁ level is represented by a gastropod bed (Bed 71) with a mixed pelagic-neritic assemblage unmatched elsewhere in the world. In addition to *Mant. guppyi*, gastropods, trilobites, tabulate corals and atrypids are present which include several last occurrences (Becker *et al.* 1991).

L₂: *Manticoceras* sp.

A few metres to the west of section Loc. 371B, *Manticoceras* sp. was collected at 26 m which is the last known manticoceratid. G. Klapper kindly informs us that his first records of the Lower *triangularis* Zone, and basal Famennian, is just above 26 m in the section. At Horse Spring our highest *Manticoceras* sp. comes from 0.5 m below a number 73 painted on the section (Becker *et al.* 1991): G. Klapper informs us he has Lower *triangularis* Zone conodonts from a higher bed numbered 74 (marked on Becker *et al.* 1991, fig. 1).

International goniatite division of the Frasnian

At present the only formal subdivision of the Frasnian is that of Wedekind (1908, 1917) although there are more detailed schemes for eastern North America (House 1978) and other limited regions. For some years the authors have been discussing between themselves a more appropriate terminology which would take into account the increasing knowledge of the rich evolutionary diversity of Frasnian goniatites. The new Australian succession provides a framework for this to be done, since it is the most detailed known anywhere. We are still working on detailed taxonomy, and when that is completed a fuller and more comprehensive statement will be possible. Meanwhile we offer an outline scheme which, for international purposes, divides the Frasnian into twelve divisions lettered A to L (Figs 6–8). Usually these are named after genera which either appear at the base of the division, or first become abundant at the base. In all but a few cases these genera continue into higher levels and therefore the top of any division is marked by the next defining genus. We do not wish the divisions to be too rigorously construed at present. It is clear that very much more detailed regional zonations are possible and when the relationships between these are better understood it is expected that the divisions proposed here will be refined and better defined. The reason for proceeding at this stage is because it became apparent that the German type sequence at Adorf (House & Ziegler 1977), because of its incomplete goniatite record, is now known to specify only about the upper third of the Frasnian goniatite record as now understood. Hence the Wedekind terminology has to be superceded. A summary of the latest German view was given by Clausen (1989) and we will revise this in detail elsewhere. We base the new system on the classical German zonation of Wedekind (1913), Matern (1931a,b), and House and Ziegler (1977); it incorporates the North America work of House

WEDEKIND ZONES	GENERIC MARKERS		CANNING BASIN	NEW YORK		GERMANY	
I delta **holzapfeli**	CRICKITES s. str.	L2	*Manticoceras sp.*	DUNKIRK	*Sphaeromanticoceras* aff. *rickardi*	*Cricktites holzapfeli*	UKW
		L1	*Sphaeromanticoceras lindneri*	HANOVER	*Sphaeromanticoceras rickardi*		
	ARCHOCERAS	K	*Beloceras* aff. *tenuistriatum* / *Manticoceras* cf. *guppyi*	PIPE CREEK	*Manticoceras cataphractum*	*Manticoceras adorfense* / *Archoceras varicosum*	LKW
I gamma **cordatum**	NEOMANTICOCERAS	J2	*Neomanticoceras erraticum* / *"Crickites" expectatus*	ANGOLA	*Carinoceras oxy*	*Neomanticoceras paradoxum* / *"Crickites" expectatus* / *Carinoceras galeatum*	
		J1	*Maternoceras retorquatum* / *"Manticoceras" tuberculatum*		*Sphaeromanticoceras rhynchostomum* / *Playfordites* cf. *tripartitus*	*Maternoceras sandbergeri* / *"Manticoceras" tuberculatum*	
I beta **nodulosum**	PLAYFORDITES	I2	*Playfordites tripartitus*			*Playfordites tripartitus* / *Maternoceras calculiforme*	
		I1	*Serramanticoceras serratum*				
	BELOCERAS	H	*Beloceras tenuistriatum*	RHINESTREET	*Schindewolfoceras chemungense*		
					Wellsites tynani		
	MESOBELOCERAS	G2	*Mesobeloceras thomasi*				
		G1	*Mesobeloceras housei* / *Mesobeloceras* cf. *iynx*		*(Mesobeloceras iynx)*		
	PROCHORITES	F3	*Sphaeromanticoceras affine*				
		F2	*Probeloceras lutheri* n.ssp.	CASHAQUA	*Prochorites alveolatus*		
		F1	*Prochorites alveolatus*				
	PROBELOCERAS	E3	*Manticoceras* n.sp.				
		E2	*Ponticeras discoidale*				
		E1	*Probeloceras lutheri lutheri*		*Probeloceras lutheri*		
I alpha **lunulicosta**	SANDBERGEROCERAS	D	*Manticoceras* cf. *evolutum*	MIDDLESEX	*Sandbergeroceras syngonum*	*(Sandbergeroc. costatum)*	
	TIMANITES	C2	*Manticoceras sp.*	WEST RIVER	*Koenenites* aff. *lamellosus*	*(Koenenites lamellosus)*	
		C1	*Timanites angustus*	GENUNDEWA	*Manticoceras contractum*		
	KOENENITES	B2	*Koenenites* n.sp.	PENN YAN	*Koenenites styliophilus*		
		B1	*Protimanites pons*				
	PONTICERAS		?	LODI	*Ponticeras perlatum*	*(Ponticeras aequabile)*	
				GENESEO	*Epitornoceras peracutum*		

Fig. 8. Table showing the new goniatite generic marker divisions of the Frasnian and comparing goniatite sequences in the Canning Basin, New York and Germany. L.K.W. and U.K.W. refer to Lower and Upper Kellwasserkalk horizons respectively.

(1962, 1978), Kirchgasser & House (1981), House & Kirchgasser (this volume), the Montagne Noire (House et al. 1985; Becker et al. 1989), unpublished work from North Africa (by M.R.H. and R.T.B.) and published work elsewhere, especially the USSR, by others. This collation has benefited from collaborative work with G. Klapper and use of the conodont scheme he has established in the Montagne Noire (Klapper 1989) and other areas. It represents a considerable advance over an earlier review (House 1979).

A: *Ponticeras* Division

The new Middle-Upper Devonian series boundary places most of the *Pharciceras* faunas (do Iα of Wedekind 1913) in the Middle Devonian. To clarify the Givetian generic division some of these were grouped as the Pharciceras Stufe (House 1985). In the GSSP section at Puech de la Suque (Klapper et al. 1987; House et al. 1985) the top part of the *disparilis* conodont Zone has a rich fauna of pharciceratids with *Ponticeras* and *Epitornoceras* and genera that cross the boundary include *Pharciceras*, *Petteroceras*, and from evidence at Pic de Bissous, *Ponticeras*. The base of this division is marked by the rise to dominance of *Ponticeras* and not its appearance. *Petteroceras* is another characteristic genus but still poorly documented. The advanced *Pett. feisti*, and probably *Acanthoclymenia* (refered to *Probeloceras* in House et al. 1985 but revised herein) are the newcomers at the base of the Upper Devonian, but the first is a straggler and the second not well documented so far. In New York also (House & Kirchgasser this volume) *Ponticeras* appears just before the new series boundary. In the Canning Basin we have yet to locate this fauna.

In the Tafilalt of south Morocco the late Givetian has a thick marker limestone of the upper part of the *disparilis* Zone which is rich in multilobed pharciceratids, *Synphariceras clavilobum*, *Epitornoceras peracutum* and Gen. nov. aff. *Pharc. erraticum*. A reddish hematite-rich level above of the *norrisi* conodont Zone (Lowermost *asymmetricus* Zone or basal part of the *falsiovalis* Zone or Ziegler & Sandberg 1990) is characterized by *Petteroceras errans* and well preserved *Ponticeras* spp.

B: *Koenenites* Division

The evolutionary step from convolute *Ponticeras* to more involute forms with the addition of an outer umbilical lobe as in *Koenenites* marks the base of this division. In New York the earliest *Koenenites* occur in the late Penn Yan (Kirchgasser & House 1981) together with *Acanthoclymenia*. *Koenenites* may overlap with the last pharciceratids in the Montagne Noire (House et al. 1985) but there is uncertainty on the exact collecting levels. The earliest goniatite level in the Canning Basin already has *Protimanites* which in morphological terms would be expected to be a descendant of *Koenenites* and to enter somewhat later. Typically both the Australian and early North American koenenitids of Division B still have rounded outer umbilical lobes. The type species, *Koen. lamellosus* and relatives with an angular U_2 lobe seem characteristic of Division C. House et al. (1985) drew attention to Givetian homeomorphs with rounded lateral lobes in a pharciceratid lineage ('*Koen.*' *juvenocostatus* Bensaïd 1974) and they noted it had a different wrinkle layer ornament.

C: *Timanites* Division

The division is defined by the occurrence of the name genus, *Timanites*, which is confined to it. The type area of the genus is the Domanik Suite of the Petchora Basin, and it is reported in the Tatra (Batanova 1953). It occurs in Western Canada (House & Pedder 1963) and is abundant at Timanites Hill in the Canning Basin. The record at Martenberg, Germany (Kullman & Ziegler 1970) may refer to an earlier pharciceratid homeomorph of the *Pharciceras taouzense* – '*Timanites*' *meridionalis* Group. *Manticoceras* also enters in this division both in New York (Upper Genundewa) and at Timanites Hill (C_2). Higher levels characteristically have more advanced species of *Koenenites* as in the West River Shale of New York (House & Kirchgasser this volume) and Montagne Noire (House et al. 1985). In the Domanik *Komioceras* enters together with *Timanites* (Kushnareva et al. 1978) and this genus was also reported with *Hoeninghausia* in Novaya Zemlya (Cherkesova et al. 1988).

D: *Sandbergeroceras* Division

The ancestors of the evolute, multilobed and ribbed triainoceratids survived the earliest Frasnian extinction which wiped out the smooth earlier pharciceratids. The group is represented in New York by *Sandb.* (?) *syngonum* of the Middlesex Shale. From a similar position between the former conodont Lower and Middle *asymmetricus* Zones in the Tafilalt of Morocco, Bensaïd et al. (1985) reported *Triainoceras costatum* which they refered to *Sandbergeroceras*. These two similar genera are only separated here because of the types of *Sandbergeroceras* (which has page priority), at the diameters known, has an undivided ventral lobe whereas all species refered to *Triainoceras* have a weakly or strongly subdivided ventral lobe. The cono- dont data of Kushnareva *et al.* (1978) needs revision but suggests a similar age for the fauna with *Komioceras* in the Domanik Suite (IIdm) that occurs with the earliest *Manticoceras* above *Timanites. Triain.* aff. *costatum* is reported by House & Kirchgasser (this volume) from Mis- souri. *Triain. gerassimovi* was described by Bogoslovski (1958) from the Altai but the age of neither this species, nor any of the classic German material is precisely known. The triai- noceratids have not been found in the Canning Basin but two faunas are refered to this division.

E: *Probeloceras* Division

The entry in the Canning Basin of *Probeloceras lutheri*, here taken to be the first beloceratid, parallels its entry in the Cashaqua Shale of New York (House & Kirchgasser this volume). No member of the *Koenenites–Hoeninghausia* line- age, *Timanites* or *Komioceras* range this high anywhere in the world. An extinction event seems to be associated with the regression at the close of Division D which would facilitate the subdivision of the Frasnian into substages (cf. Becker 1986). In the Petchora Basin, the rich *Ponticeras* faunas of the Domanik Suite (IIdm$_1$) (Kushnareva *et al.* 1978) correspond well with the ponticeratids of the Canning Basin at this level, although there are differences in size and preservation. Most of the Canning Basin speci- mens are small but these match well the juveniles figured by Holzapfel (1899). The ponticeratids from Division E are clearly differentiated from those near the Middle/Upper Devonian bound- ary in the widespread development of ventro- lateral furrows and tabular venters similar to *Acanthoclymenia*.

F: *Prochorites* Division

The entry of *Probel. lutheri* is followed by *Prochorites alveolatus* both in the Cashaqua Shale of New York and in the Canning Basin. Records of *Probel.* cf. *strix* from the top of the West River Shale (Kirchgasser *et al.* 1985) have to be excluded and they probably refer to *Acanthoclymenia*. In the Timan, poor goniatites from Domanik Suite (IIIdm) do not allow precise correlation but G. Klapper has informed us that the conodont reports, if correct, suggest Zone 5 or 6 of the Montagne Noire (Klapper

1989) and this may be used to refer at least parts of Domanik Suite (IVdm) with its fauna of *Mant. ammon* to Division F. Further support may come from the co-occurrence in Member IVdm of *Uchtites syrjanicus* and *Mant. ammon*, since an intermediate form between *Pont. dis- coidale* and *Uchtites* was recovered from Loc. 367 near McIntyre Knolls immediately below the first *Prochorites*. An assemblage of *Uchtites* with *Manticoceras* was also mentioned from the *asymmetricus* Zone in western Yunan (Kong *et al.* 1985). In the Montagne Noire and Tafilalt, Divisions E and F have not yielded macrofauna.

G: *Mesobeloceras* Division

The Canning Basin bed-by-bed collecting has proved for the first time that there is a smooth and gradual transition between *Probeloceras* and *Mesobeloceras*, the stages defined by the progressive introduction of both adventitious and umbilical lobes. The suture of *Probeloceras* shows by its angular sigmoidal elements that it should be regarded as the first member of the Beloceratidae and we do not accept the deri- vation proposed by Yatskov (1990). Subdivision of Division G in the Canning Basin is possible as a result of the unidirectional evolutionary ad- dition of sutural elements. A similar succession is present in the Tafilalt where very evolute *Mesobel. housei* (a species Yatskov (1990) ass- igns to a new genus *Chaoceras*) were collected at Djebel Erfoud (locality data in Buggisch & Clausen 1972) but the main species is *Mesob. kayseri* which is at a similar stage to *Mesob. thomasi* in Australia. It should be noted that the ranges of *Mesobeloceras* and *Beloceras* overlap considerably. Wedekind (1913) even reported *Mesobeloceras* from the terminal Frasnian Upper Kellwasser Limestone, but this needs confirmation. Bensaïd (1974) showed that *Eobeloceras* is not a beloceratid and many records of *Eobeloceras* in the literature refer to *Mesobeloceras* such as those in central Kazakh- stan, southern Quangsi (Chao 1956) and New York (House & Kirchgasser this volume).

H: *Beloceras* Division

A succession from advanced species of *Meso- beloceras* to *Beloceras tenuistriatum* has been observed both in the Canning Basin and in the Tafilalt (unpublished work of R.T.B. and M.R.H; Bensaïd *et al.* 1985). In the latter area there is also a commonly quarried limestone bed with large *Manticoceras acutiforme*. In the Rhinestreet Shale of New York the giant multi- lobed goniatites refered to the Triainoceratidae

(House & Kirchgasser this volume) are placed in this division. It is the entrance to *Beloceras* which marks the base of this division but the genus is often extremely abundant in higher divisions which are defined by other criteria.

I: *Playfordites* Division

The entrance of descendants of *Mant. lamed* with spiral internal shell thickenings, here named *Playfordites*, defines the base of this division. In the Canning Basin this occurs significantly above the first *Beloceras*. Another characteristic group that enters is *Serramanticoceras* n. gen. In the Rhenish Slate Mountains faunas of this level also have *Mant. nodulosum*, *Mant. inversum* and *Maternoceras calculiforme* (Wedekind 1913; House & Ziegler 1977) and this is Wedekind's original do Iβ fauna. Wedekind found *Bel. tenuistriatum* abundant at this level and this is true also of the Canning Basin. In the Canning Basin and Germany *Playfordites* is not known outside of Division I, but in New York there are records in the upper Rhinestreet and lower Angola Shales (House & Kirchgasser this volume). No faunas of this level have yet been found in the Tafilalt.

J: *Neomanticoceras* Division

This division is characterised by the entry of *Neomanticoceras*, *Timanoceras* and early 'Crickites' and *Carinoceras* are rare abundant here. But all new forms are and caution will be needed in recognizing this interval. This may prove to be best done using species of the lineages of *Maternoceras*, *Sphaeromanticoceras*, *Trimanticoceras*, *Linguatornoceras* and other tornoceratids. Both in Germany, at Martenberg and Seßacker, and in the Canning Basin there is a level with maternoceratids such as *Mat. retorquatum*, *Mat. sandbergeri* and *Mat. calculiforme* (Matern 1931a, trench II, beds 21–20; House & Ziegler, 1977, section VI, beds 10c–10d) with *Mant. tuberculatum* and abundant *Serramant. serratum* at the base. Immediately above are the extremely thin and oxyconic *Neomanticoceras erraticum* or *Carin. galeatum* (Matern 1931a, trench II, bed 19). *Sphaeromant. crassum* (Matern 1931a, trench II, bed 19), 'Crickites' *expectatus* Group and true *Trimanticoceras*, especially *Trim. cinctum* and *Trim. retrorsum*. This is basically the association of the Büdesheim Shales of the Eifel District (Clausen 1969) which yielded *Neomant. paradoxum*. House & Ziegler (1977) called it the upper part of the *cordatum* Zone and compared this with the Lyajolysk Suite of southern Timan

which has *Car. menneri* and *Neomant. ljaschenkoae* together with a diverse *Manticoceras* fauna and *Timanoceras ellipsoidale* (Bogoslovskiy 1969). *Neomanticoceras* is also known from the Dill Syncline of Germany (Buggisch *et al.* 1978) and from Qwangsi, South China (Ruan *et al.* 1985).

K: *Archoceras* Division

If the end of the middle Frasnian is defined by the hypoxic Lower Kellwasser Event then typical evolute gephuroceratid genera such as *Ponticeras*, *Acanthoclymenia*, *Maternoceras* and the *Mant. nodulosum* Group do not range into the upper Frasnian. Oxyconic forms have only rarely been reported (Matern 1931a, trench II, bed 14; Müller 1956, Geipel Quarry, bed 4) and *Serramanticoceras* and possibly *Mesobeloceras* may also disappear with the Lower Kellwasser Event (Becker & House 1990). In his study of the German Martenberg section Wedekind (1913) separated a level with *Mant. adorfense* below the entry of *Crickites holzapfeli* which defined his terminal Frasnian zone (do Iδ). This level was also reported from the Beul section. A coquina with a similar species was encountered between the two Kellwasser Beds in the section at Bine Jebilet in the Tafilalt (by M.R.H. & R.T.B.). However, there are reports of *Mant. adorfense* from earlier levels (Matern 1931a) that require taxonomic investigation. After the extinction of evolute gephuroceratids which has been refered to there is a spread of *Archoceras*; this genus is reported from the Lower Kellwasser Limestone of the Balve area together with *Mant. adorfense* (Paeckelmann 1924), from Seßacker (Matern 1931a, trench II, bed 16), from the lower part of the Matagne Shale of Belgium and just above the Lower Kellwasser Limestone equivalent in the Montagne Noire (Becker *et al.* 1989). Further records are from the Saltern Cove Goniatite Beds of South Devon (House 1963) and the New Albany Shale of Indiana (House 1962). In Australia faunas of the early part of Zone 13 (Klapper 1989) are not very similar.

L: *Crickites* Division

The latest Frasnian Zone of *Crickites holzapfeli* (Wedekind 1913) has been recognised widely in the Rhenish Slate Mountains, Thuringia (Müller 1956), the Montagne Noire (Becker *et al.* 1989), in the Tafilalt (Becker *et al.* 1988) and Novaya Zemlya (faunal lists in Cherkesova *et al.* 1988). House *et al.* (1985) showed that *Crick. holzapfeli* enters between the two Kellwasser Limestones and near the middle and this is substantiated at Seßacker (Matern 1931*a*, trench II, bed 13).

As Glenister (1958) pointed out, taxonomic concepts of *Crickites* are unsatisfactory. The union of all gephuroceratids with convex growth lines in this way is unnatural. A start is made here by restricting *Crickites* to the relatively small forms, corresponding with the holotype of *Crick. holzapfeli*, the type species, which have robustly rounded whorl cross-sections with height and width nearly equal and rarely exceeding 80 mm in diameter and which may be derived from the *Mant. adorfense* Group. Then there is the slender group, probably derived from *Mant. lamed*, which inlcudes '*Crick*'. *expectatus*, '*Crick.*' *sahlgrundensis*, and '*Crick.*' *scheldensis* and this genus group is kept in inverted commas for the present. The oxyconic forms like '*Crick.*' *actus* are probably derived from *Carinoceras* and are best assigned to that genus. This leaves the giant forms so characteristic of Division L and which may reach half a metre in diameter. These are separable into types with convex or biconvex growth lines but the differences may be slight. Often large forms do not show growth lines. They have the shell form of *Sphaeromanticoceras* and are best assigned there. The holotype of *Sphaero. lindneri* has biconvex growth lines (Fig. 5E,F: note that the cross section of CPC 35766 figured by Glenister (1958, text-fig. 4) has a diameter of 209 mm and is not illustrated true scale as indicated). Forms from the early Hanover Shale in New York, and the giants from the Schistes de Matagne described by Matern (1931*b*) have convex growth lines; these have been named *Sphaero. rickardi* by House & Kirchgasser (this volume). All these types became extinct by the end of the Upper Kellwasser Event.

Timing of facies movements in the Canning Basin

The Frasnian reef complexes of the Lennard Shelf have been intensively studied during the last 30 years. For information on the general reef development, carbonate petrology, lithostratigraphy, facies types and their spatial distribution reference should be made to the classic papers by P. E. Playford and coauthors (Playford & Lowry 1966; Playford 1980, 1984; Playford *et al.* 1989; Kerans & Playford 1984; Read 1973*a,b*). The general Frasnian model for the carbonate complex (Fig. 2) shows a basinal facies, the Gogo Formation, passing laterally and up slope into marginal slope and reef talus facies of the Sadler Limestone, to the reef margin (true reef or reef rampart) facies of massive stromatoporoid limestone and then to the back-reef facies of the Pillara Limestone. The later red-coloured infilling sediments are termed the Virgin Hills Formation but regrettably there is no name for the Frasnian part. The Famennian reef margin, in stromatolite facies, is called the Windjana Limestone, but there is no name available for the very different Frasnian stromatoporoid reef margin facies and we would recommend this. Further, we consider the relegation of the reef margin to member status is not in accord with its palaeobiological, sedimentological and framework importance in the reef complex. Furthermore, the reef margin facies has been mapped across the basin in great detail (Playford & Lowry 1966) and this is the usual criterion for assigning formation status.

The high resolution goniatite stratigraphy developed in this work allows precise dating of facies shifts and periods of reef backstepping. It also enables a consideration of international correlation of these. Of these events, we consider that eustatic sea-level changes are the important driving force for some changes in the palaeogeography and shelf configuration and for the development of anoxic/hypoxic facies in the adjacent basin from time to time (Fig. 2). We do not rule out the importance of climatic effects (Copper 1986), but these are still poorly understood. Synsedimentary tectonic activity is, however, well documented and affected discharge of carbonate reef detritus, collapse of reef margins (Playford 1984), or the formation of graben structures with deeper-water sedimentation. Playford & Lowry (1966) and Playford *et al.* (1989) outline the significance of probable fault-controlled fanglomerate complexes, delta and submarine fans in platform areas.

The sections we have examined in detail are limited in area and confined to the eastern part of the Lennard Shelf. We have constructed a simplified facies shift curve on the basis of our sections (Fig. 9) which should be considered as a local interpretation. An attempt at a larger scale should be left to those familiar with a wider region, and subsurface data. Nevertheless, we modify an earlier diagram by Playford *et al.* (1989, p. 197, fig. 18) by eliminating the regional trend, and adding our new dating of movements and the probability that some facies shifts are eustatically driven (Fig. 10).

Facies types

The Gogo Formation is typically a series of dark shales and siltstones which alternate, often rhythmically, with thin detrital limestone beds, certain of which are rich in pelagic dacryocona-rids. The Sadler Ridge, McPhee Knoll (Locs. 365–367) and McWhae Ridge (below Loc. 372) areas produce fine hematitic faunas, resulting from an alteration of primary pyrite to goethite as result of weathering. Nektonic organisms dominate and there are occasional pseudo-planktonic *Buchiola*. Sometimes there are low diversity benthonic faunas with rhynchonellids, and gastropods of the '*Leiorhynchus*' com-munity of Europe and North America which is interpreted as a low-oxygen facies. There is no evidence of bioturbation or any endobenthonic life and this suggests anoxic or strongly hypoxic seafloor conditions. The lack of biodegradation is best emphasized by the occurrence of hema-tised wood in which even cortical tissues are preserved, and in the celebrated Gogo fish fauna. The significance of these anoxic/hypoxic levels of hydrocarbon source-rock facies, is not apparent in the literature. The main Frasnian levels recognised are shown on Fig. 2, as is one Famennian level (corresponding to the Annu-lata Event of House 1985). The small-scale rhythmicity is possibly caused by Milankovitch Band forcing oscillations producing changes in the pycnocline level: these are shown somewhat diagrammatically on Fig. 9.

The Sadler Limestone represents reef talus limestones of light coloured detrital calcare-nites, calcirudites and coarse slump beds which may include giant down-slope transported al-lochthonous blocks. There is a diverse alloch-thonous to parautochthonous, often silicified, neritic, benthonic fauna with brachipods, simple corals, bryozoans and gastropods in which goniatites are normally lacking. Goniatites are also normally absent in the reef margin and back reef facies.

The Frasnian part of the Virgin Hills For-mation embraces both deposits of the marginal slope and inter-reefal shallow basins of the late Frasnian. It includes a variety of lithofacies which mirror, sometimes sharply, changing en-vironmental and depositional conditions. The most important for dating are the goniatite coquinas with lutitic matrix and crinoids, large brevicones, longiconic orthocones, some bi-valves, solitary corals, fish teeth, trilobites, and gastropods, among the macrofauna. The dark bluish coating of goniatites from the Lower *Beloceras* Bed at McWhae Ridge (Loc. 371B) is strikingly similar to the preservation mode of goniatites near the old iron mines of the Rhenish Slate Mountains, especially those from Mar-tenberg. Conodonts belong to the deep-water *Palmatolepis* biofacies. Rich goniatite assem-blages are restricted to distinct beds within

successions of light-grey to reddish, fine-grained reef detritus limestones which grade into more proximal, calcarenitic or even calciruditic beds on the reef slope. Locally, there are very sparite-rich intraformational breccias ('Scheck breccias' of Kerans & Playford 1984) or peloid packstones with '*Globalgae*', possibly former soft-bodied organisms. Reef margin collapse led to the accumulation of coarse slump beds and the displacement of gigantic reef limestone blocks (Playford 1984; Becker *et al.* 1991). Interspersed with coarse-grained reef talus are levels of white limestone or sparite which pinch out downslope (Playford *et al.* 1976). These represent hard-grounds and times of reduced detritus supply and are interpreted (Fig. 9) as regressive inter-vals. They are often associated with short-term high-energy episodes leading to the downslope transport of coarse reef debris. Increased per-centages of icriodids in the stromatolitic layers (Playford *et al.* 1976) supports the assumption of episodic shallowing (Druce 1976) on conodont biofacies models such as Sandberg & Dreesen (1984) or Sandberg *et al.* (1988). Stromatolitic bioherms of any size have only been seen in the Famennian.

Timing of facies movements

The main transgression onto the Kimberley Block is Givetian in age. A correlation with the mid-*varcus* Zone Taghanic onlap was suggested by House (1975) and this has been supported by recent palynological evidence (Grey 1991). We have found no goniatite evidence bearing on the subject since our earliest faunas are already early Frasnian. The Taghanic international trans-gression formerly marking the base of the Upper Devonian (House 1975), which is early *Pharci-ceras* Stufe in age and falls within the *varcus* conodont Zone; some effect is indicated dia-grammatically on Fig. 2. The earliest incursions of anoxic/hypoxic facies are with the B_1 and B_2 faunas which tongue into Sadler Limestone. This interval is terminated by debris flows at the top of the measured section at Sadler Ridge. The light-grey goniatite coquinas with *Timanites* at Timanites Hill (Loc. 370) show that the C_1 and C_2 levels represent well-oxygenated open ma-rine conditions. We regard this as the most offshore or distal facies in the early Frasnian. It corresponds in time with the similar Genundewa pulse in New York (Fig. 10) and the Frasne Event of House (1985). A second anoxic/hypoxic incursion is represented south of Siphon Spring and near Virgin Bore by Division D, in an inferred correlative of the anoxic Middlesex Shale of New York. A third intercalation is

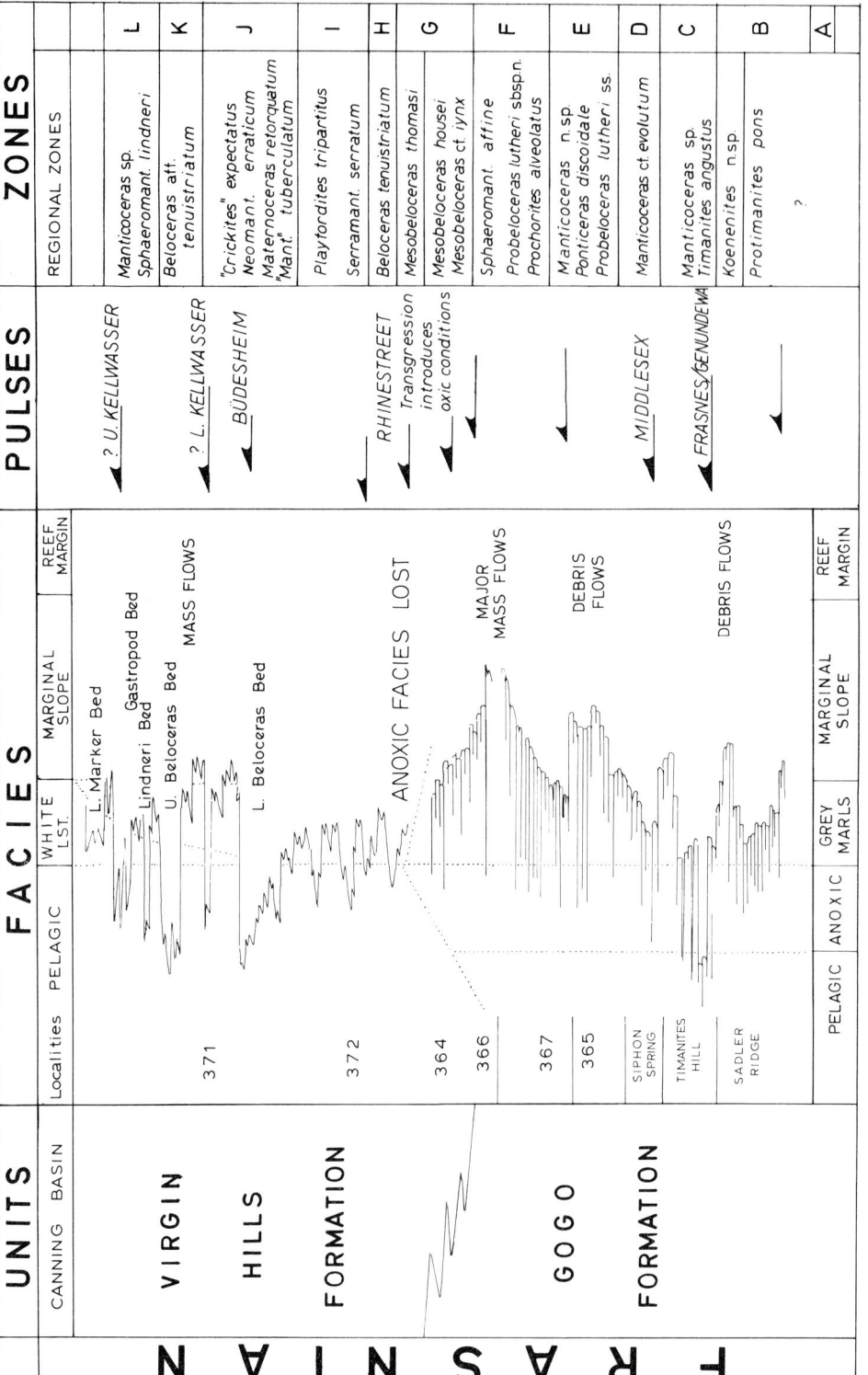

Fig. 9. Diagram simplifying and summarizing local Frasnian facies movements at the localities studies. For details see text.

represented near McPhee Knoll (Loc. 365) by levels E_1 and E_2 which are also terminated by debris flows. The fourth oxygen-poor interval ranges throughout the section at Loc. 367, that is E_3 to F_2, which is again capped by coarse slump beds. The major allochthonous blocks and mass flows around McIntyre Knolls are placed at the base of the section at Loc. 367. The fifth hypoxic interval is between Beds 5–7. This falls in the F_3–G_1 interval which corresponds to the start of the progressive Rhinestreet transgression in New York (Figs 8 and 9). Thereafter we so far have no evidence for later Frasnian anoxia. Following arguments for the causal relation between transgression and the spread of anoxia (Hallam & Bradshaw 1979), as is documented for the Devonian of North America (House 1983; Baird & Brett 1986; Baird et al. 1989), we show these anoxic/hypoxic levels as transgressive on Fig. 9.

In the McWhae Ridge area, as long recognised by Playford, there is a major reef backstepping. This can be dated because faunas from Gogo facies in the plain below trench Loc. 372 of R. S. Nicoll correspond to the third anoxic/hypoxic interval, E_2, near McIntyre Knolls. A similar dating is obtained from the Gogo facies in the graben at the ridge top (Playford 1981). Thus the drowning of the McWhae Ridge reef must have been great enough to allow basinal facies more or less up to the ridge top. Such a significant deepening was already assumed by Playford (1981). A review of the conodonts reported from the steep western slope of McWhae Ridge (Nicoll 1984) and a date kindly given by G. Klapper of his Zone 4, confirm this. Our oldest date here for infilling sediments (base of the trench at Loc. 372) is G_2: a sediment-starved intervening record is to be expected. Sediments later engulfing the old reef come from the north.

At Horse Spring the Virgin Hills Formation also fills in the palaeotopography of the underlying Sadler Limestone. It is also diachronous. The earliest faunas at Loc. 364 belong to F_3 and goniatite facies (with calcareous preservation) is continuous up to Division H, the Zone of Bel. tenuistriatum. This middle Frasnian transgression sees the loss of anoxia and is thus emphasised on Fig. 9. Both at Horse Spring and at McWhae Ridge there are several approximately corresponding fluctuations between goniatite-rich pelagic and more reef-detritic, marginal slope facies. The Lower Beloceras Bed, J_2, represents the peak of this transgressive interval and this can also be traced transgressing over the old reef on the both sides of McWhae Ridge. P. E. Playford kindly showed us a locality near Brooking Spring and Geikie Gorge (Fig. 3) where Beloceras has been washed into platform facies but we cannot determine whether it represents this transgression, J_2, or the later Upper Beloceras Bed Division K transgression.

In the Upper Frasnian of McWhae Ridge (Zone 13 of Klapper) there is a succession of four goniatite horizons separated by detritic limestones and stromatolite layers that wedge out both down and upslope. The Upper Beloceras Bed, Zone K, can be correlated with Beloceras-rich intervals at Horse Spring and Casey Falls (Becker et al. 1991). At Horse Spring the unique gastropod bed must be interpreted as a subsequent shallowing interval which allowed the spread of rich benthonic faunas. The transgressive lindneri Bed, L_1, lies immediately above and at McWhae Ridge overlaps the drowned reef. At Casey Falls (Fig. 3) goniatite beds are separated by stromatolitic hardgrounds with haematite encrustations and crinoid holdfasts. The Lower Marker Bed of Loc. 371 (Becker et al. 1991) belongs to the earliest Famennian. The higher Frutexites Bed (Playford et al. 1976, 1984) is in the crepida conodont Zone and the iridium anomaly associated with it is thus well above the base of the Famennian. The Frasnian–Famennian boundary is not defined by any obvious facies break and the critical interval has no macrofauna.

Eustatic versus Regional Sea Level Changes

The improvement of biostratigraphic data, especially using goniatites and conodonts, has enabled a much more precise analysis of regional facies movements. For the Devonian, House (1983) was the first to correlate the developments in distant sedimentary provinces and to start the separation of eustatic signals from regional effects. Johnson et al. (1985, 1986) carried this further for Euramerica but their actual facies movements were strongly influenced by the New York pattern described by House (1983) which has now been further improved (House & Kirchgasser this volume). Whether the depophase terminology should survive is questionable. It seems better that there should be geographic names and type areas for deepening phases, as House (1985) argued also for Devonian Events, rather than a terminology which presumes correlation. The newly codified goniatite zonation provides an alternative tool (Fig. 10). This is especially important for linking the present work to areas like New York (House & Kirchgasser this volume) where the conodont record at critical levels

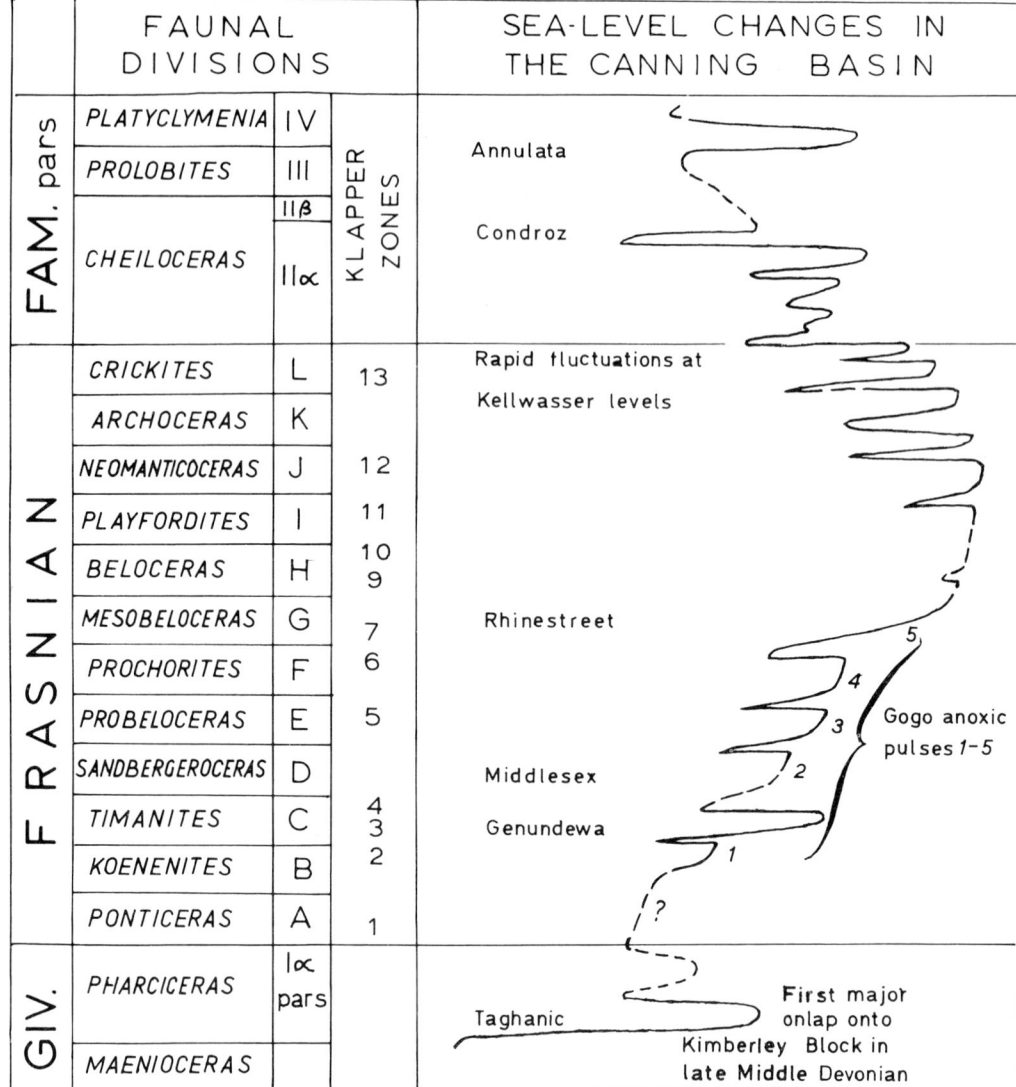

FAUNAL DIVISIONS					SEA-LEVEL CHANGES IN THE CANNING BASIN

Fig. 10. Diagram suggesting eustatic facies movements which affected reef development in the Canning Basin. Trend line taken subtracting the linear regional trend shown by Playford *et al.* (1989) to emphasize deepening phases recognized in this work and showing correlation with goniatite zones (herein and Becker 1990) and conodont zonations (information kindly supplied by G. Klapper). The diagram has benefitted from comments by P. E. Playford.

leaves much to be desired. Talent & Yolkin (1987) drew attention to the particular problems of disentangling transgressive effects in complex reef areas and we would wish to stress that much future work will be needed in other areas of the Lennard Shelf than we have been able to examine. In Fig. 10 the broad trend of deepening is based on Playford *et al.* (1989, fig. 12) with

generalised deepenings obtained by substracting a regional deepening trend, taken as linear, and giving approximate deepenings linked with the events that we recognize.

Johnson *et al.* (1985) dated an international transgression (start of depophase IIb) as the base of the former lowermost *asymmetricus* Zone (now base of the *falsiovalis* Zone of

Ziegler & Sandberg 1990 or *norrisi* Zone of Klapper & Johnson 1990). Yolkin & Talent (1987) suggested the presence of this deepening phase in northwest Australia but this is below any data we have. At present we think the Canning Basin Zone B hypoxic event may be regional. But the deepening pulse of the *Timanites* fauna, Zone C_1, correlates in time with the deepening interval of the Genundewa Limestone of New York and elsewhere in North America, and the spreading of black styliolinid limestones in the Tafilalt of southern Morocco; House (1985) called this the Frasnes Event. The New York Middlesex Shale and the F_{2d} drowning of Belgian reefs was the basis for recognising Depophase IIc. This corresponds well with our second hypoxic interval, Zone D.

The continuing mid-Frasnian sea-level rise within Depophase IIc culminates in the thick black Rhinestreet Shale of New York (Depophase IId). House and Kirchgasser (this volume) demonstrate that this essentially develops from a mid-Cashaqua Shale low. This corresponds well with our deepening in Divisions G to I. The same gradual tendancy is seen in the loss of the oxygen-depleted Gogo facies and the filling of former reef topography following backstepping. The deepening peak of our Division J corresponds with the deposition of the famous Büdesheim goniatite shales of the Eifel Hills (Clausen 1969) and with the wide distribution of goniatite facies in submarine sections in the Rhenish Slate Mountains (House & Ziegler 1977). This is probably the late Frasnian reef backstepping period of Playford *et al.* (1989, p. 192–193. fig. 12).

The New York Pipe Creek Shale is taken to correspond to the Lower Kellwasser Limestone of Germany and North Africa (Buggisch 1972) and is high within Depophase IId: it may correspond to Zone 13 of Klapper (1989) which also embraces the Upper Kellwasser Limestone. The transgressive nature of both black limestones is most obvious by their onlap onto the North African part of Gondwana (Hollard 1961; Becker 1990) and the two levels seem correlated with the Pipe Creek and a late level in the Hanover Shale in New York. We have seen no similar hypoxic intervals in the Canning Basin at this level and it is not clear which pulses were associated with the final emergence and extinction of the stromatorporoid reefs when the Pillara Cycle ended. The date of the earliest large cyanobacterian reefs is also not clear; the small bioherms at Casey Falls (Becker *et al.* 1991) are dated as *verneuili* or Middle *crepida* Zone in age, so there may be a time gap before major cyanobacterian reefs are established. A

regression at Casey Falls is indicated by the massive 4 m thick stomatolite biostrome which is probably late Famennian IIα (*petterae* Subzone) in age and corresponds to the Condroz Event (Becker 1990) and we recognise an anoxic/ hypoxic deepening in early IV which corresponds to the Annulata Event (House 1985). These are indicated on Fig. 2.

Conclusions

(1) A detailed Frasnian goniatite biostratigraphy is described from the Canning Basin which is the finest known anywhere. It at present enables the establishment of 21 regional zones.

(2) Comparison with the global Frasnian goniatite record enables the introduction of 12 international divisions, lettered A–L, each characterized by the appearance or rise to dominance of generic taxa.

(3) The analysis of lateral and vertical facies changes, coupled with the new biostratigraphy, enables a tentative regional facies shift diagram to be produced and, building on the work of P. E. Playford, for improvements in the regional relative sea-level curve. Two major periods of backstepping are recognized, early E_2, and late G-J, in the middle Frasnian. The upper Frasnian is characterized by strong sea-level fluctuations but the correlations with the terminal Frasnian stromatoporoid reef extinction are not yet clear.

(4) Comparison with the depositional intervals of Johnson *et al.* (1985), and our own information from other areas, allows the separation of local fluctuations from eustatic influences. Generally the Canning Basin development reflects sea-level movements elsewhere. The distinctive lithologies of the Kellwasser Events have not been recognised in northwest Australia.

(5) The iridium enriched *Frutexites* Bed postdates the Frasnian–Famennian boundary considerably (Nicoll & Playford 1988) and is later than the terminal Frasnian mass extinctions. We have no evidence which supports a bolide impact at this time.

This work would not have been possible without the unstinted and enthusiastic help, encouragement and information provided by P. E. Playford, Director of the Geological Survey of Western Australia, and logistic support from the Survey. We acknowledge his comments on this paper. We are also indebted to the Director, Bureau of Mineral Resources, Canberra, for logistic support and especially to R. S. Nicoll for advice and R. W. Brown for field assistance. Our field companion G. Klapper has helped us by discussion, comment, and by providing some of his unpublished conodont results. The programme was initiated under

NERC Grant GR3/6958 awarded to M.R.H. The State University of New York at Potsdam, and the NSF, made travel grants to W.T.K.

References

BAIRD, G. C. & BRETT, C. E. 1986. Erosion of an anaerobic seafloor: significance of reworked pyrite deposits from the Devonian of New York State. *Palaeogeography, Palaeoclimatology, Palaeoecology*, **57**, 157–193.

BAIRD, G. C., BRETT, C. E. & KIRCHGASSER, W. T. 1989. Genesis of black shale-roofed discontinuities in the Devonian Genesee formation, Western New York State. *Canadian Society of Petroleum Geologists*, **14** (2), 357–375.

BATANOVA, G. P. 1953. Stratigrafiya Franskikh Otlozhennij v Tatarskoj ASSR. *Doklady Akademii Nauk SSSR*, **89**, 143–146.

BECKER, R. T. 1986. Ammonoid evolution before, during and after the "Kellwasser-event" – review and preliminary new results. *In*: WALLISER, O. H. (ed.) *Global Bio-events. A critical Approach.* Lecture Notes in Earth Sciences, **8**, 181–188.

—— 1987. The Kellwasser-Event and its significance for ammonoid evolution and biostratigraphy. *2nd International Symposium of the Devonian System, Program and Abstracts*, 27.

—— 1990. *Stratigraphische Gliederung und Ammonoideen-Fauna im Nehdenium (Oberdevon II) von Europa und Nord-Afrika.* Dissertation, University of Bochum.

—— & HOUSE, M. R. 1990. *The Montagne Noire goniatite record around the Frasnian/Famennian boundary.* Document submitted to the Subcommission on Devonian Stratigraphy, Frankfurt.

——, ——, KIRCHGASSER, W. T. & PLAYFORD, P. E. 1991. Sedimentary and faunal changes across the Frasnian/Famennian boundary in the Canning Basin of Western Australia. *Historical Biology*, in press.

——, —— & ASHOURI, A-R. 1988. *Potential stratotype section for the Frasnian-Famennian boundary at El Atrous, Tafilalt, Morocco.* Document Submitted to the Subcommission on Devonian Stratigraphy, Rennes.

——, FEIST, R., FLAJS, G., HOUSE, M. R. & KLAPPER, G. 1989. Frasnian-Famennian extinction events in the Devonian at Coumiac, southern France. *Compte Rendu de l'Academie des Sciences, Paris, Serie II*, **309**, 259–266.

BENSAÏD, M. 1974. Etudes sur les Goniatites à la limite du Devonien Moyen et Supérieur, du Sud Marocain. *Notes Service géologique Marocaine*, **36**, 81–140.

——, BULTYNCK, P., SARTENAER, P., WALLISER, O. H. & ZIEGLER, W. 1985. The Givetian-Frasnian boundary in pre-Sahara, Morocco. *Courier Forschungsinstitut Senckenberg*, **75**, 287–300.

BOGOSLOVSKIY, B. I. 1958. *Devonskie Ammonoidei Rudnogo Altaya.* Trudy Paleontologicheskogo Instituta, **64**.

—— 1969. *Devonskie ammonoidei, I. Agoniatity.* Trudy Paleontologicheskogo Instituta, **124**.

—— 1971. *Devonskie ammonoidei, II. Goniatity.* Trudy Paleontologicheskogo Instituta, **127**.

——, POSLAVSKAYA, I. A. & BELYAEV, O.E. 1982. Frasnian ammonoids from Central Kazakhstan. *Paleontolicheskogo Zhurnal*, **16**, 32–37.

BUGGISCH, W. 1972. *Zur Geologie und Geochemie der Kellwasserkalke und ihrer begleitenden Sedimente (Unteres Oberdevon).* Hessisches Landesamt für Bodenforschung, Abhandlungen, **62**.

—— & CLAUSEN, C-D. 1972. Conodonten- und Goniatiten-Faunen aus dem oberen Frasnium und unteren Famennium – Marokkos (Tafilalt, Antiatlas). *Neues Jahrbuch für Geologie und Paläontologie, Abhandlungen*, **141**, 137–167.

BUGGISCH, W., RABIEN, A. & HÜHNER, G. 1978. Biostratigraphie und Faziesvergleich von oberdevonischen Becken- und Schwellen- Profilen E Dillenburg. *Geologisches Jahrbuch Hessen*, **106**, 53–115.

CHAO, K. 1956. Notes on some Devonian ammonoids from southern Kwangsi. *Acta Paleontologica Sinica*, **4**, 101–116.

CHERKESOVA, S. V., SOBOLEV, H. H., SMIRNOVA, M. A. & LAKHOV, V. 1988. Novaye dannye po stratigrafii Novoj Zemli. *Sovetskaya Geologiya*, **1988**, 55–68.

CLAUSEN, C.-D. 1969. Oberdevonische Cephalopoden aus dem Rheinischen Schiefergebirge. II. Gephuroceratidae, Beloceratidae. *Palaeontographica, Abteilung A*, **132**, 95–178.

—— 1989. Die Goniatiten der Bohrung Balve 1 (Sauerland, östliches Rheinische Schiefergebirge). *Fortschritte in der Geologie von Rheinland und Westfalen*, **35**, 31–56.

COPPER, P. 1986. Frasnian/Famennian mass extinction and cold water oceans. *Geology*, **14**, 835–839.

DRUCE, E. C. 1976. *Conodont biostratigraphy of the Upper Devonian reef complexes of the Canning Basin, Western Australia.* Bureau of Mineral Resources, Australia, Bulletin, **158** (2).

FEIST, R. (ed.) 1990. *Guidebook of the Field Meeting Montagne Noire 1990.* Subcommission on Devonian Stratigraphy, IUGS.

GLENISTER, B. F. 1958. Upper Devonian ammonoids from the *Manticoceras* Zone, Fitzroy Basin, Western Australia. *Journal of Paleontology*, **32**, 58–96.

—— & KLAPPER, G. 1966. Upper Devonian conodonts from the Canning Basin, Western Australia. *Journal of Paleontology*, **40**, 777–842.

GREY, K. 1991. A mid-Givetian miospore age for the onset of reef development on the Lennard Shelf, Canning Basin, Western Australia. *Review of Palaeobotany and Palynology*, **68**, 37–48.

HALLAM, A. & BRADSHAW, M. J. 1979. Bituminous shales and oolitic ironstones as indicators of transgressions and regressions. *The Journal of the Geological Society, London*, **136**, 157–164.

HELMS, J. 1959. Conodonten aus dem Saalfelder Oberdevon (Thüringen). *Geologie*, **8**, 634–677.

HOLLARD, H. 1961. Charactère transgressif du Frasnien supérieur dans le Maïder (Marco présaharien). *Comptes Rendu des séances Societé Géologique de la France*, **1961** (2), 41–42.

HOLZAPFEL, E. 1899. *Die Cephalopoden des Domanik in südlichen Timan*. Mémoires du Comité Géologique, **12** (3).

HOUSE, M. R. 1961. *Acanthoclymenia* the supposed earliest Devonian clymenid, is a *Manticoceras*. *Palaeontology*, **3**, 472–476.

—— 1962. Observations on the ammonoid succession of the North American Devonian. *Journal of Paleontology*, **36**, 247–284.

—— 1963. Devonian ammonoid successions and facies in Devon and Cornwall. *Quarterly Journal of the Geological Society of London*, **119**, 1–27.

—— 1973. Delimitation of the Frasnian. *Acta palaeontologica Polonica*, **23**, 1–14.

—— 1975. Facies and times in Devonian tropical areas. *Proceedings of the Yorkshire Geological Society*, **40**, 233–288.

—— 1978. Devonian ammonoids from the Appalachians and their bearing on international zonation and correlation. *Special Papers in Palaeontology*, **21**, 1–70.

—— 1979. Biostratigraphy of the early Ammonoidea. *Special Papers in Palaeontology*, **23**, 263–280.

—— 1983. Devonian eustatic events. *Proceedings of the Ussher Society*, **5**, 396–405.

—— 1985. Correlation of mid-Palaeozoic ammonoid evolutionary events with global sedimentary perturbations. *Nature*, **313**, 17–22.

—— 1988. International definition of Devonian System boundaries. *Proceedings of the Ussher Society*, **7**, 41–46.

HOUSE M. R. & KIRCHGASSER, W. T. 1993. Devonian goniatite biostratigraphy and timing of facies movements in the Frasnian of eastern North America. *This volume*.

—— & PEDDER, A. E. H. 1963. Devonian goniatites and stratigraphical correlations in Western Canada. *Palaeontology*, **6**, 491–539.

—— & ZIEGLER, W. 1977. The goniatite and conodont sequences in the early Upper Devonian at Adorf, Germany. *Geologica et Palaeontologica*, **11**, 69–108.

——, KIRCHGASSER, W. T., PRICE, J. D. & WADE, G. 1985. Goniatites from Frasnian (Upper Devonian) and adjacent strata in the Montagne Noire. *Hercynica*, **1**, 1–21.

JOHNSON, J. G., KLAPPER, G. & SANDBERG, C. A. 1985. Devonian eustatic fluctuations in Euramerica. *Geological Society of America, Bulletin*, **96**, 567–587.

——, —— & —— 1986. Late Devonian eustatic cycles around margin of Old Red Sandstone Continent. *Annales de la Société géologique de Belgique*, **103**, 141–147.

JUX, U. & KRATH, J. 1974. Die Fauna aus dem Mittleren Oberdevon (Nehden-Stufe) des südwestlichen Bergischen Landes (Rheinisches Schiefergebirge). *Palaeontographica, Abteilung A*, **147**, 115–168.

KERANS, C. & PLAYFORD, P. E. 1984. Scheck breccias from Devonian reef complexes of the Canning Basin, Western Australia. *American Association of Petroleum Geologists Bulletin*, **68**, 495.

KIRCHGASSER, W. T. 1975. Revision of *Probeloceras*

Clarke 1898 and related ammonoids from the Upper Devonian of western New York. *Journal of Paleontology*, **49**, 58–90.

—— & HOUSE, M. R. 1981. Upper Devonian biostratigraphy. *In*: OLIVER, JR, W. A. & KLAPPER, G. (eds) *Devonian Biostratigraphy of New York*. Subcommission on Devonian Stratigraphy, IUGS. Washington, 39–55.

——, OLIVER, JR, W. A. & RICKARD, L. C. 1985. Devonian series boundaries in the Eastern United States. *Courier Forschungsinstitut Senckenberg*, **75**, 233–260.

KLAPPER, G. 1989. The Montagne Noire Frasnian (Upper Devonian) conodont succession. *Canadian Society of Petroleum Geologists*, **14** (3), 449–468.

—— & JOHNSON, J. G. 1990. Revisions of Middle Devonian conodont zones. *Journal of Paleontology*, **64**, 934–936.

——, FEIST, R. & HOUSE, M. R. 1987. Decision on the boundary stratotype for the Middle/Upper Devonian series boundary. *Episodes*, **10**, 97–101.

KONG, L., ZHOUG, D. & ZHOU, T. 1985. Devonian of Ninglang and Yangsheng areas of western Yunan. *Journal of Stratigraphy*, **9**, 136–141.

KORN, D. 1979. Mediandornen bei *Kosmoclymenia* SCHINDEWOLF (Ammonoidea, Cephalopoda). *Neues Jahrbuch für Geologie und Paläontologie, Monatshefte*, **1979** (7), 399–405.

KULLMANN, J. 1960. Die Ammonoidea des Devon im Kantabrischen Gebirge (Nordspanien). *Abhandlungen der mathematisch-naturwissenschaftlichen Klasse, Akademie der Wissenschaften, Mainz*, **7**, 1–105.

—— & ZIEGLER, W. 1970. Conodonten und Goniatiten von der Grenze Mittel-/Oberdevon aus dem Profil am Martenberg (Ostrand des Rheinischen Schiefergebirge). *Geologica et Palaeontologica*, **4**, 73–85.

KUSHNAREVA, T. I., KHALAMBADZHA, & BUSYGINA, Y. N. 1978. Biostratigraficheskaya zonalost' domanikovoy svity v razreze stratstipa. *Sovetskaya Geologiya*, **1978**, 60–71.

LYASCHENKO, G. P. 1956. Goniatity osnovaniya Franskogo yarusa Timana. *Izvestiya Akademiya Nauk SSSR, Seriya Geologiya*, **5**, 87–92.

MATERN, H. 1931a Das Oberdevon der Dill-Mulde. *Abhandlungen der preußischen Geologischen Landesanstalt, N.F.* **134**, 1–139.

MATERN, H. 1931b. Die Goniatiten-Fauna der Schistes de Matagne in Belgien. *Bulletin Musée Royale d'Histoire Naturelle Belgique*, **7**, 1–15.

MILLER, A. K. 1938. *Devonian ammonoids of America*. Geological Society of America Special Papers, **14**.

MONTESINOS, J. R. 1990. Las Biozonas de Ammonoideos del Devonico (Emsiense inferior – Famenniense inferior): Critica al sistema de classificación zonal. *Revista Española de Paleontologia*, **5**, 3–17.

—— & HENN, A. H. 1986. La fauna de *Pharciceras* (Ammonoidea) de la Formación Cardaño (Dominio Palentino, Cordillera Cantabrica, NO de España). *Trabajos de Geologia*, **16**, 61–76.

MOURAVIEFF, A. N. 1982. Conodont stratigraphic scheme of the Frasnian of the Ardennes. *In*: BIGEY, F. (ed.) *Papers on the Frasnian-Givetian boundary*. Geological Survey of Belgium, 101–118.

MOUNTJOY, E. W. 1968. Factors governing the development of the Frasnian Miette and Ancient Wall reef complexes (banks and biostromes), Alberta. *In*: OSWALD, D. H. (ed.) *International Symposium on the Devonian System*, 2. Alberta Society of Petroleum Geologists, 387–408.

MÜLLER, K. 1956. Cephalopodenfauna und Stratigraphie des Oberdevons von Schleiz und Zeulenroda in Thüringia. *Beihefte zum Geologischen Jahrbuch*, **20**, 1–93.

NICOLL, R. S. 1984. Conodont distribution in the marginal-slope facies of the Upper Devonian reef complex, Canning Basin, Western Australia. *Geological Society of America, Special Paper*, **196**, 127–141.

—— & PLAYFORD, P. E. 1988. Upper Devonian iridium anomaly and the Frasnian-Famennian boundaryin the Canning Basin, Western Australia. *Geological Society of Australia, Abstracts*, **21**, 296.

NORRIS, A. W. & UYENO, T. T. 1981. Stratigraphy and paleontology of the lowermost Upper Devonian Slave Point Formation on Lake Claire and the lower Upper Devonian Waterways Formation on Birch River, northeastern Alberta. *Geological Survey of Canada Bulletin*, **334**, 1–65.

PAECKELMANN, W. 1924. Das Devon und Karbon der Umgebung von Balve in Werstfalen. *Jahrbuch Preußischen Geologischen Landesanstelt*, **41**, 51–97.

PETERSEN, M. S. 1975. Upper Devonian (Famennian) ammonoids from the Canning Basin, Western Australia. *Journal of Paleontology, Memoir*, **8**, 1–55, 7 pls.

PETTER, G. 1959. *Goniatites dévoniennes du Sahara*. Publications du Sérvice de la Carte Géologique de l'Algérie, N.S., Paléontologie, Mémoires, **2**.

PLAYFORD, P. E. 1980. Devonian "Great Barrier Reef" of Canning Basin, Western Australia. *American Association of Petroleum Geologists Bulletin*, **64**, 814–840.

—— 1981. *Devonian reef complexes of the Canning Basin, Western Australia*. Geological Society of Australia. Fifth Australian Geological Convention, Field Excursion Guidebook.

—— 1984. Platform-margin and marginal-slope relationships in Devonian reef complexes of the Canning Basin. *In*: PURCELL, P. G. (ed.) *The Canning Basin, W. A.*, Proceedings of the Geological Society of Australia and Petroleum Exploration Society of Australia, Symposium Perth 1984, 189–214.

—— & LOWRY, D. C. 1966. Devonian reef complexes of the Canning Basin, Western Australia. *Geological Survey of Western Australia Bulletin*, **118**, 1–150.

—— COCKBAIN, A. E., DRUCE, E. C. & WRAY, J. L. 1976. Devonian stromatolites from the Canning Basin, Western Australia. *In*: WALTER, M. R.

(ed.) *Developments in Sedimentology*. Elsevier Scientific Publishing Company, Amsterdam, 543–563.

——, McLAREN, D. J., ORTH, C. J., GILMORE, J. S. & GOODFELLOW, W. D. 1984. Iridium anomaly in the Upper Devonian of the Canning Basin, Western Australia. *Science*, **226**, 437–439.

——, HURLEY, N. F., KERANS, G. & MIDDLETON, M. F. 1989. Reefal development, Devonian of the Canning Basin, Western Australia. *Society of Economic Paleontologists and Mineralogists, Special Publication*, **44**, 187–202.

READ, J. F. 1973*a*. Carbonate cycles, Pillara Formation (Devonian), Canning Basin, Western Australia. *Bulletin of Canadian Petroleum Geology*, **21**, 38–51.

—— 1973*b*. Paleo-environments and paleogeography, Pillara Formation (Devonian), Western Australia. *Bulletin of Canadian Petroleum Geology*, **21**, 344–194.

RICKARD, L. V. 1975. *Correlation of the Silurian and Devonian rocks in New York State*. New York State Museum and Science Service, Map and Chart Series, **24**.

RUAN, Y.-P., WANG, S.-Q., MU, D.-S., WU, Y., HE, J.-H. & LIANG, J.-D. 1985. New observations on the Upper Devonian Zhaisha section of Luzhai, Guangxi. *Journal of Stratigraphy*, **9**, 262–269.

SANDBERG, C. A. & DREESEN, R. 1984. Late Devonian icriodontid biofacies models and alternate shallow-water conodont zonation. *Geological Society of America, Special Papers*, **196**, 143–178.

——, ZIEGLER, W., DREESEN, R. & BUTLER, J. L. 1988. Late Frasnian mass extinction: conodont event stratigraphy, global changes, and possible causes. *Courier Forschungsinstitut Senckenberg*, **102**, 263–307.

SANDBERGER, G. & SANDBERGER, F. 1850–1856. *Systematische Beschreibung und Abbildung der Versteinerungen des Rheinischen Systems in Nassau*. Wiesbaden.

SCHINDEWOLF, O. H. 1923. Beiträge zur Kenntnis der Paläozoicums in Oberfranken, Ostthüringen und dem Sächsischen Vogtlande. I. Stratigraphie und Ammoneenfauna des Oberdevons von Hof a. S. *Neues Jahrbuch Mineralogie, Geologie und Paläontologie*, **49**, 250–358, 393–509.

STEININGER, J. 1849. *Die Versteinerungen des Übergangs-Gebirges der Eifel*. Trier.

TALENT, J. A. & YOLKIN, E. A. 1987. Transgression-regression patterns for the Devonian of Australia and Southern West Siberia. *Courier Forschungsinstitut Senckenberg*. **92**, 235–249.

TEICHERT, C. 1941. Upper Devonian goniatite successions of Western Australia. *American Journal of Science*, **239**, 148–153.

—— 1943. The Devonian of Western Australia, a preliminary review. *American Journal of Science*, **241**, 69–94, 167–184.

—— 1949. Observations on stratigraphy and palaeontology of Devonian western portion of Kimberley Division, Western Australia. *Bureau of Mineral Resources, Australia, Geology & Geophysics Reports*, **2**, 1–65.

TSIEN, H. H. 1975. Introduction to the Devonian reef development in Belgium. *In*: CONIL, R. (ed.) *Second international Symposium on Fossil Corals and Reefs, Guidebook Excursion C (Nord de la France et Belgique)*. Service Géologique Belgique, 3–43.

WADE, A. 1938. The geological succession in the West Kimberley District of Western Australia. *Australian and New Zealand Association for the Advancement of Science, Report*, **23**, 93–96.

WEDEKIND, R. 1908. Die Cephalopodenfauna des höheren Oberdevon am Enkeberge. *Neues Jahrbuch Mineralogie, Geologie und Paläontologie*, **26**, 565–634.

—— 1913. Die Goniatitenkalke des unteren Oberdevon von Martenberg bei Adorf. *Sitzungberichte der Gesellschaft Naturforschender Freunde zu Berlin*, **1**, 1–77, 4 pls.

WEDEKIND, R. 1917 (1918). Die Genera der Palaeoammonoidea (Goniatiten). *Palaeontographica*, **62**, 85–184, 9 pls.

YATSKOV, S. V. 1990. K sistematike semeistva Beloceratidae (Ammonoidea). *Trudy Paleontologicheskogo Instituta*, **243**, 36–51.

ZIEGLER, W. 1962. Taxionomie und Phylogenie oberdevonischer Conodonten und ihre stratigraphische Bedeutung. *Abhandlungen des Hessischen Landesamtes für Bodenforschung*. **38**, 1–166, 14 pls.

ZIEGLER, W. 1971. Conodont stratigraphy of the European Devonian. *Geological Society of America, Memoir*, **127**, 227–284.

—— & SANDBERG, C. A. 1990. The Late Devonian standard conodont zonation. *Courier Forschungsinstitut Senckenberg*, **121**, 1–115.

Worldwide correlation of Telychian (Upper Llandovery) strata using graptolites

DAVID K. LOYDELL

Institute of Earth Studies, University College of Wales, Aberystwyth, Dyfed SY23 3DB, UK

Abstract: The formal boundaries to the Telychian Stage have been stated to correlate with the bases of the *Monograptus turriculatus* Biozone at the bottom and the *Cyrtograptus centrifugus* Biozone at the top. The evidence for this, however, is far from conclusive.

The Telychian has been divided, mainly using monograptids, into eight or more biozones, although fewer have been recognized in Britain due to the general paucity of recent studies of the graptolite faunas of this age and to the existence of 'barren beds' high within most British Telychian sections. Each biozone has a duration of only a few hundred thousand years, but most are, at present, difficult to recognize in the absence of the index species or where this species' stratigraphical range extends beyond the limits of its biozone.

Intercontinental correlation of Telychian strata using graptolites is hampered to a limited extent by provincialism and water mass specificity, and considerably by taxonomic problems. A thorough taxonomic revision of published species would provide a firm foundation for future biostratigraphical subdivision and correlation of the Telychian Stage. Knowledge of Telychian graptolite biostratigraphy would also be enhanced by restudies of two classic sections, the Trannon river section in Wales, and the Bajiaokou section, Ziyang, China.

The graptoloids offer the potential for some of the finest resolution biostratigraphical subdivision and correlation of strata of any fossil group at any time in the stratigraphical column. Being rapidly evolving, planktic, in many cases (particularly in the Silurian) cosmopolitan, to a certain degree facies independent, and, in theory at least, readily identifiable to specific level, the graptoloids possess the majority of the features required of the ideal 'zonal fossil'.

This paper outlines the current state of Telychian (Upper Llandovery) graptolite biostratigraphy, the degree of precision possible in international and intercontinental correlation, and highlights those areas in which future research might profitably be focused.

The formal boundaries to the Telychian Stage

The boundary stratotype for the base of the Telychian Stage is within the Wormwood Formation in an old quarry on the west side of the Cefn Cerig road [NGR SN 7743 3232] in the type Llandovery area. This point is considered (Bassett 1985) to correlate with the base of the *Monograptus turriculatus* Biozone (although no graptolites were collected from this locality) on the basis of the discovery of *Monograptus runcinatus* Lapworth, 1876 (not Tullberg as stated by Cocks *et al.* 1984 and Cocks 1989) from near the

base of the Cerig Formation (which overlies the Wormwood Fm) at a locality 10 km to the northeast of the stratotype. This was the only graptolite identifiable to specific level collected from the Telychian of the type Llandovery area (Cocks *et al.* 1984). The specimen may be lost, as frequent requests to examine it have met with no success. The presence of *M. runcinatus* (sometimes assigned to *Diversograptus* Manck, 1923) indicates the lower part of the *turriculatus* s.l. Biozone (Bjerreskov 1975; Loydell in press *a, c*).

The boundary stratotype for the base of the Wenlock Series (i.e. the top of the Telychian Stage) is at [SO 5688 9839] in Hughley Brook, near Apedale, Shropshire. This level is coincident with the base of the Buildwas Formation and is stated by Martinsson *et al.* (1981) to correlate with the base of the *Cyrtograptus centrifugus* Biozone. Again, graptolites do not occur at the stratotype. Indeed, the only evidence cited for the existence of the *centrifugus* Biozone in the type Wenlock area was the presence of *Pristiograptus watneyae* Rickards, 1965 from 18.3 m above the base of the Buildwas Fm in the old brickpit [SO 4858 9042] near Ticklerton Brook, over 11 km SW of the stratotype (Bassett *et al.* 1975). *P. watneyae* was previously known from only two definite specimens from the Howgill Fells (Rickards 1965). It is portrayed as confined to the middle of the *centrifugus* Biozone by Rickards (1976).

From Hailwood, E. A. & Kidd, R. B. (eds), *High Resolution Stratigraphy*
Geological Society Special Publication, No. 70, pp. 323–340.

323

SERIES	STAGE		RICKARDS (1989) 'This is not a formally agreed standard of zones, but is the one most widely in use in international correlation.' †	BOHEMIA (BOUČEK 1953)	subzone/band	WALES (various authors) see text	ARCTIC CANADA (MELCHIN 1989)	BORNHOLM (BJERRESKOV 1975)
WE.	Sh.		centrifugus	centrifugus		centrifugus		centrifugus
LLANDOVERY	Telychian			insectus				
				grandis	prosbosciformis		sakmaricus	
				spiralis	geinitzi	spiralis		lapworthi
					anguinus			spiralis
					parapriodon			
			crenulata	crenulata	curvus	crenulata	griestoniensis	
			griestoniensis	griestoniensis		griestoniensis		griestoniensis
						sartorius		
			crispus	crispus		crispus	crispus	crispus
			turriculatus	turriculatus	runcinatus	turriculatus s.l.	turriculatus	turriculatus
					turriculatus			
				linnaei	hispanicus		minor	
					palmeus			
	Aer.		sedgwickii	sedgwickii	rastrum	halli		
						sedgwickii		

Fig. 1. Telychian graptolite biozonal schemes. This is not an exhaustive survey, but includes most, if not all, biozones currently in use. Correlation of biozones is, of necessity, approximate. Note that the *runcinatus* Subzone of Bouček (1953) is based on the occurrence of *Streptograptus runcinatus* sensu Bouček & Přibyl 1942a, a species which is very different from the true *D. runcinatus* (Lapworth, 1876), typical of the lower part of the *turriculatus* Biozone. Loydell (1991a) erects a new species for *S. runcinatus* sensu Bouček & Přibyl. The *sartorius* Biozone has not been formally erected. A more extensive comparison of Telychian biozonal schemes is provided by Fu & Song (1986), whose correlations are broadly similar. WE., Wenlock; Aer., Aeronian; Sh., Sheinwoodian.

Thus, on graptolite biostratigraphical grounds the 'golden spikes' (Holland 1986) defining the bases of the Telychian Stage and the Wenlock series have but little to commend them, particularly when one considers 'that stages are, anyway, exceedingly difficult to define without graptolites, once away from the type section' (Rickards 1989; see also Temple 1988).

Despite the limitations of the stratotypes, the interval discussed herein will range from the base of the *turriculatus* s.l. Biozone to that of the *centrifugus* Biozone, on the understanding that this interval may not be coincident with the Telychian as formally defined. With the exception of Fu & Song (1986) and Huo & Shu (1986), who place the base of the Wenlock at the base of the *Cyrtograptus sakmaricus* Biozone, a similar procedure has been followed by graptolite workers worldwide. Figure 1 correlates approximately some of the different biozonal schemes used around the world. The biozones themselves

are discussed below. Usage of the *turriculatus* Biozone herein is based on a broad definition of the biozone (i.e. the base is at the first appearance of *Monograptus turriculatus* (Barrande, 1850) sensu lato), rather than the more restricted Bohemian usage (Bouček 1953; base of the biozone is at the first appearance of *M. turriculatus* sensu stricto). This is necessary because there is, at present, no agreed basis for distinction between *M. turriculatus* s.s. and its ancestor, which is often referred to the subspecies *minor* Bouček, 1932 (e.g. by Melchin 1989; see Fig. 1). These two taxa comprise *M. turriculatus* s.l. The top of the *turriculatus* Biozone is marked by the first appearance of *M. crispus* Lapworth, 1876.

Telychian graptoloids

The Llandovery as a whole saw the rise, in both diversity and abundance, of the uniserial monograptids and the decline, but not total demise, of the biserial diplograptids. The majority of Telychian sequences are, therefore, dominated by assemblages of monograptids. Locally, however, particularly in the early Telychian, diplograptid diversity is high, the result of a major evolutionary radiation of retiolites and petalolithids.

Diplograptids

Diplograptid classification has recently undergone something of a revolution (Kearsley 1985; Fortey & Cooper 1986; Legrand 1987; Mitchell 1987) with thecal morphology (the basis of the 'Treatise' classification of Bulman 1970) now considered as of secondary importance to early astogeny. Nine patterns of 'primordial astogeny', designated 'A'–'I', have been recognized, with only the simplest, 'G'–'I', represented in the Silurian (Mitchell 1987). Melchin (1987) has recognized a number of additional astogenetic patterns in Silurian taxa and has significantly emended some of Mitchell's (1987) generic diagnoses.

The diplograptids of the Telychian comprise the retiolites and members of the genera *Petalolithus*, *Glyptograptus* and *Metaclimacograptus*, together with a small number of species the generic assignation of which is problematical (e.g. '*Orthograptus insectiformis*' of Paris *et al.* 1980).

The retiolites. These are united in their possession of an ancora which, in the majority of taxa, extends distally to ensleeve the thecate portion of the rhabdosome. In all cases the rhabdosome is composed of a meshwork of lists (Bates 1987). Only one species, *Pseudoretiolites perlatus* (Nicholson, 1868) is known from the Aeronian (Middle Llandovery), but in the lower part of the lowest Telychian *Monograptus turriculatus* Biozone as many as ten retiolite species are known, probably derived from *Psr. perlatus* in most instances, but in the case of, for example, *Retiolites* sp. of Hutt *et al.* (1970), which lacks a reticulum, having evolved perhaps independently from an ancorate 'normal' diplograptid.

Existing retiolite genera are poorly defined as they were erected largely on the basis of flattened material. Detailed structure has only relatively recently begun to be understood, with the advent of scanning electron microscope study of specimens chemically isolated from limestones (e.g. Bates & Kirk 1978, 1986; Lenz & Melchin 1987a, b).

The stratigraphical ranges of most species are imprecisely known. Others, for example *Psr. perlatus*, have long ranges (of several biozones duration). In some lithologies the delicate retiolite meshworks are difficult to see and specimens are often only collected by chance when on a slab together with another more obvious fossil. These factors have combined to limit the biostratigraphical use of the retiolites. Only *Stomatograptus grandis* (Suess, 1851), a broad species with a dense reticulum, has had its name put to a biozone, that of the upper Telychian in Bohemia (Bouček 1931). *Retioclimacis typica* NIGP, 1974, assigned by Ni (1978) to the Archiretiolitinae Bulman, 1955, has been considered in China to be a 'species characteristic of a particular level, but not assigned to a zone' (Chen 1984a). Chen (1984a) correlates this 'level' with the *turriculatus* Biozone. However, *R. typica* is a junior synonym of *Metaclimacograptus undulatus* (Kurck, 1882), a nonretiolite which is much more typical of the late Aeronian (Loydell 1991a).

Bouček & Münch (1944) described the majority of Telychian retiolites then known. Most species erected subsequently are junior synonyms of taxa described therein. However, Obut (1949), Hutt *et al.* (1970) Sennikov (1976), Lenz (1988) and Ge (1990) have described additional species which are valid, although some of these have been left in open nomenclature.

Petalolithus Suess, 1851. Between 15 and 20 species are found in the early part of *turriculatus* Biozone. There seems to be some palaeoecological control on distribution, however, and the *Pe. palmeus* and *Pe. hispanicus* subzones used successfully in Bohemia for the lowest Telychian (Bouček 1953) have proved difficult to apply elsewhere (see below).

The subsequent decline in diversity of the petalolithids was marked. In the upper part of the *turriculatus* Biozone only three species are known worldwide, *Pe. tenuis* (Barrande, 1850), *Pe. posterus* Schauer, 1971 (a homonym, see Loydell 1991a) and *Pe. altissimus* (Elles & Wood, 1908). In the *crispus* Biozone *posterus* and *altissimus* were joined by *Pe. wilsoni* (Hutt, 1974) and a few as yet unpublished species (Howe 1982). None of these survived into the *spiralis* Biozone.

Bouček and Přibyl (1941) describe a number of Telychian petalolithids. New Chinese species appear in Chen (1984b) and Ge (1990). Other works describing valid new species include those of Hundt (1957) and Paškevičius (1979).

Glyptograptus Lapworth, 1873. Earliest Telychian, lower *turriculatus* Biozone faunas include glyptograptid species surviving from the Aeronian (see e.g. Packham 1962) together with a number of new elements, most of which are relatively broad forms. All of these were extinct long before the end of the *turriculatus* Biozone. The only later glyptograptids recorded are *G. ultimus* Chen, 1984*b*, from the middle of the *turriculatus* Biozone and *G. nebula* Toghill & Strachan, 1970 from the *griestoniensis* Biozone. Other than the new early *turriculatus* Biozone species, such as *G. auritus* Bjerroskov, 1975 and *G. fastigatus* Haberfelner, 1931a, the glyptograptids have proved of very little use in Telychian biostratigraphy.

Metaclimacograptus Bulman & Rickards, 1968. Only three species of this genus are known from the Telychian. Two, *Me. undulatus*, from the lower part of the *turriculatus* Biozone and *Me.* sp. (Paris *et al.* 1980), from the upper *turriculatus* and *crispus* biozones, are very rare. The other, *Me. scalaris* (Hisinger, 1837) is locally common (e.g. in Wales) in the middle part of the *turriculatus* Biozone (Loydell 1991*a*). As Rickards (1972) has noted, *Me. scalaris* has 'not been as successfully identified as the literature would suggest.' The overall stratigraphical range of this species is still unclear.

Monograptids

The monograptids differ from the diplograptids both in overall rhabdosome construction and, more fundamentally, in the mode of development of the initial bud from the metasicula (Bulman 1970, p. V72).

Although some Telychian species may be assigned with confidence to particular genera, the thecal morphology of others is not yet known well enough for the genus *Monograptus* Geinitz, 1852 to lose its long-standing portmanteau status (Bulman & Rickards 1970). Problems of generic classification are not as great as in the Rhuddanian, however, where in many species thecal morphology changes markedly along the length of the rhabdosome. This phenomenon is less prevalent in the Telychian, although it does occur in some monoclimacids (Rickards 1968) and spiral monograptids (Melchin & Lenz 1986). It seems likely that the vast majority of Telychian monograptids will fairly soon be assigned to well-defined genera.

Rastrites Barrande, 1850. Although this genus is polyphyletic (Sudbury 1958), Štorch & Loydell (1992) consider that all Telychian species were derived, directly or indirectly, from the true rastritid *R. hybridus* Lapworth, 1876.

Rastritid species diversity is high in the lower part of the *turriculatus* Biozone, a number of the species concerned having first appeared in the late Aeronian, however. Only two species, *R. spengillensis* Rickards, 1970 and *R. distans* Lapworth, 1876, occur in the *crispus* Biozone, with only the latter possibly surviving into the *griestoniensis* Biozone (P. Štorch, pers. comm.).

The delicate nature of the rastritid rhabdosome has resulted in the majority of rhabdosomes having suffered fragmentation. This may make identification difficult, especially given the considerable degree of intraspectific variation recognised in some taxa (Štorch & Loydell 1992), and also makes individual species abundance difficult to assess.

R. linnaei Barrande, 1850 and *R. maximus* Carruthers, 1867 have both been used as biozonal or subzonal indices for the lowest Telychian (lower *turriculatus* Biozone). The *linnaei* Biozone is discussed below (see also Loydell in press *b* for a discussion of the *R. maximus* Biozone).

The most important taxonomic works on Telychian rastritids are those of Schauer (1967) and Štorch & Loydell (1992). New species still remain to be erected for some of the rastritids illustrated in Strachan (1972) and Chen (1984*b*).

Monoclimacis Frech, 1897. Of all the established Telychian monograptid genera the monoclimacids probably pose the greatest difficulties in species identification. Important diagnostic features include the rate of increase of dorsoventral width and the maximum attained, and the nature and number of early hooked or lappeted thecae. All of these require examination of complete specimens or long fragments. Identifications based on short fragments should be treated with considerable caution.

In the *turriculatus* and *crispus* biozones monoclimacid diversity and abundance are very low. Only *Mcl. inchoatus* Přibyl, 1943 is known from the lower part of the *turriculatus* Biozone, from only a few localities in central Europe. *Mcl. galaensis* (Lapworth, 1876) is common in strata of high *turriculatus*-low *crispus* Biozone age in a number of European sections (e.g. Bjerreskov 1975; Cocks & Toghill 1973; Burgess *et al.* 1970). Slender forms, *Mcl. griestoniensis* (Nicol, 1850) and *Mcl. minuta* Přibyl, 1940*a* (which may be conspecific, see Bjerreskov 1975), are often common elements of a *griestoniensis* Biozone fauna, but here again diversity is low, with only two other species known from this biozone.

More robust forms predominate in the remainder of the Telychian.

Telychian monoclimacids from central Europe are described in Přibyl (1940a). Bjerroskov (1975) describes a number of important Scandinavian authored species. Examination of these two references, together with Elles & Wood (1901–18), Přibyl (1943), Bjerreskov (1981) and Fu & Song (1986) provides details of most described Telychian monoclimacid species. The majority of these would benefit greatly from modern redescription.

Three species, *Mcl. griestoniensis*, *Mcl. crenulata* (sensu Elles & Wood, 1911) and *Mcl. geinitzi* (Bouček, 1932) have been used as biozonal indices.

Pristiograptus Jaekel, 1889. With the exception of Přibyl (1940b) and Bjerreskov (1975, 1981), few authors have attempted to 'split' the Telychian pristiograptids. The simple tubular thecae, together with the straight or only very gently curved rhabdosomes, characteristic of almost all representatives of the genus, seem at first sight to offer little to distinguish between species. However, features such as rate of increase in dorso-ventral width, angle of thecal inclination and amount of thecal overlap vary significantly between taxa, and several biostratigraphically useful species have been recognized.

Species diversity is probably greatest in the *turriculatus* Biozone, but, because of the paucity of studies, is difficult to assess for later in the Telychian, although it would appear never to have been very high. One species, *P. renaudi* (Philippot, 1950), has been used as a subzonal index in the *turriculatus* Biozone of Wales (Loydell 1991a).

Streptograptus Yin, 1937. This genus, recently redefined by Loydell (1990a), is one of the most characteristic of the Telychian. Although its earliest representatives, like those of most Telychian genera, appeared earlier in the Llandovery, it was not until the *turriculatus* Biozone that streptograptids became both diverse and abundant. At least ten species occur in the lower to middle part of the *turriculatus* Biozone (compared with one, *S. ansulosus* (Törnquist, 1892) in the latest Aeronian). Two rhabdosome shapes are particularly common in Telychian species: ventrally curved into a hook (e.g. *S. exiguus* Lapworth, 1876), and doubly curved (dorsally proximally, ventrally distally, e.g. *S. pseudobecki* Bouček & Přibyl, 1942a). Post-*turriculatus* Biozone diversity declines, but not dramatically and the genus survives into the Wenlock. Bouček & Přibyl (1942a), Bjerreskov (1975),

Chen (1984b) and Loydell (1991a) describe the majority of Telychian taxa.

Three streptograptid species have been used as a biozonal or subzonal indices, by Chen (1984b), Loydell (1991a) and Bouček (1953).

The thecal morphology of a number of other Telychian monograptids, e.g. *Monograptus crispus* Lapworth, 1876 and *M. nodifer* Törnquist, 1881, suggests that they are related to the streptograptids. Both of these species have considerable biostratigraphical value, although this has been largely masked for the latter as it has frequently been confused with *S. plumosus* (Baily, 1871; see Loydell 1990a).

Stimulograptus Přibyl & Štorch, 1983. This genus comprises few Telychian species of which *St. halli* (Barrande, 1850), a survivor from the Aeronian, *St. utilis* Loydell (1991a) and *St. clintonensis* (sensu Hall, 1852) are the most important biostratigraphically.

St. halli is a relatively common element of low to middle *turriculatus* Biozone faunas worldwide. *St.* utilis, a subzonal index fossil for the middle of the *turriculatus* Biozone in Wales (Loydell 1991a), has not been recorded outside Wales. *St. clintonensis* (sensu Hall, 1852) is probably more widely geographically distributed than the literature suggests. Its precise range is not known, but appears confined to the *crispus* and (lower?) *griestoniensis* biozones.

Monograptus Geinitz, 1852 sensu stricto. The long, usually straight and robust, rhabdosomes of taxa with simply hooked thecae are found throughout the Telychian, and are abundant from the middle part of the *turriculatus* Biozone onwards. Most species appear to be stratigraphically long-ranging, *M. prosciformis* Bouček, 1931, indicative of a level high in the Telychian, being an exception. Much taxonomic work remains to be done, particularly to define precisely (if possible) the species and subspecies erected by Perner (1897, 1899), Kirste (1919) and Obut & Sobolevskaya (1966). Many may prove to be junior synonyms of *M. priodon* (Bronn, 1835). *M.* aff. *becki* sensu Bjerreskov, 1975, *M. marri* Perner, 1897 and *M. parapriodon* Bouček, 1931 are other geographically widespread Telychian species.

Spirally curved monograptids. Přibyl (1945) united a large number of strongly dorsally or spirally curved monograptids in the genus *Spirograptus* Gurich, 1908. Mu (1955) was the first of many to recognize this as an unnatural grouping, as several different thecal types are included within Přibyl's concept of the genus. Přibyl's *Spirograptus* included species with hooked and spinose thecae (e.g. *Monograptus turriculatus* (Barrande, 1850), type species of the genus *Spirograptus*), species with thecae exhibiting torsion of their axes, but not significantly laterally expanded (e.g. *M. planus* (Barrande, 1850)) and others which exhibit torsion of thecal axes only distally in the rhabdosome, but have laterally expanded apertures throughout (e.g. *M. spiralis* (Geinitz, 1842)).

Spirograptus, if redefined to include only species with hooked and spinose thecae, will include only *M. turriculatus*, its smaller ancestor (often referred to *M. turriculatus minor* Bouček, 1932), and *M. andrewsi* Sherwin, 1974. In thecal form these are similar to *Stimulograptus* and Melchin & Lenz (1986) have suggested that *M. turriculatus* evolved from *St. sedgwickii* (Portlock, 1843). *M. turriculatus* sensu lato has been recognised widely and its biozone has been used successfully worldwide. However, distinguishing *turriculatus* s.s. from related species may be difficult and the exact ranges of the different taxa are still imprecisely known, although it is clear that only *M. turriculatus* s.s. survives into the *crispus* Biozone.

The torsion of the rhabdosome axis often present in spiral monograptids results in a variety of thecal views being presented to the observer. In imperfectly preserved material, these may prove difficult to interpret and the thecal morphology of many taxa is poorly known. Frequently, the thecae exhibit torsion of their axes, but are otherwise simple (Bjerreskov 1975).

Thecal morphology is known (from chemically isolated material) in only two other Telychian species, *Monograptus contortus* Perner, 1897 (illustrated by Loydell & Zhao 1990 as *M.* aff. *M. spiralis*) and *M. spiralis* (see Lenz & Melchin 1989). Both have laterally expanded apertures and, other than in the most proximal thecae, exhibit torsion of the thecal axis. The stratigraphical range of *M. contortus* is late Aeronian–middle *turriculatus* Biozone. *M. spiralis* does not appear until the late Telychian *M. spiralis* Biozone. Thus, further species with this thecal morphology may be identified from intervening stratigraphical levels in the future.

A different and highly distinctive spiral rhabdosome form is that of *Monograptus veles* (Richter, 1871), characterized by tight ventral curvature. Howe (1982) has suggested that this species may be related to the *M. priodon* group. *M. drepanoformis* Toghill & Strachan, 1970 and *M. sinicus* (NIGP, 1974), are also ventrally curved throughout much of their length.

Other monograptids. The remainder of the Telychian monograptids remain to be assigned or allied satisfactorily to a genus or species group as their thecal morphology is poorly known. Included in this category are a number of slender forms, such as *M. pergracilis* (Bouček, 1931).

Diversograptus Manck, 1923. Although discussed in some detail by Rickards (1973) this genus continues to pose problems. Rickards (1973) noted that the majority of species assigned to the genus in the past are simply normal monograptid rhabdosomes regenerated after damage, and accepted as valid diversograptids only *D. runcinatus* (Lapworth, 1876) and *D. ramosus* Manck, 1923. The former has simply hooked thecae (Strachan 1952; Bjerreskov 1975) whilst the latter's thecae Rickards (1973) noted to be of varying appearance, *priodon*-like, globosiform or beak-like. Rickards (1973) synonymised *D. rectus* Manck, 1923 and *D. bohemicus* Bouček, 1933 with *D. ramosus* and indicated a remarkably long stratigraphical range for *D. ramosus* (Rickards 1976), of from '*Diplograptus*' *magnus* Biozone (early Aeronian) to *crenulata* Biozone (late Telychian), considerably longer than any other Llandovery graptoloid. Rickards' (1973, 1976) concept of *D. ramosus* probably embraces a number of species. *D. rectus* appears to have hooked thecae and may be related to *D. runcinatus*. The thecae of *D. ramosus* exhibit torsion of the thecal axis. *D. bohemicus* is a junior synonym of *D. ramosus*, but not of *D. rectus*.

It appears that the species with hooked thecae are confined to the early Telychian (in the case of *D. runcinatus* to the lower part of the *turriculatus* Biozone) and that those with twisted thecae occur only in the latter part of the stage. The two groups are probably not phylogenetically related.

Sinodiversograptus Mu & Chen, 1962. This genus is undoubtedly related to the early diversograptids, with *D. runcinatus* a likely ancestor. *Si. lientanensis* (Mu, 1948), its only representative, is known only from the lower part of the *turriculatus* Biozone (Loydell 1990*b*).

Cyrtograptus Carruthers, 1867. Although primarily a Wenlock genus, a number of species have been identified from the late Telychian. The majority are illustrated in Bouček (1933), Fu & Song (1986), Lenz (1978*a*, 1988) and Melchin (1989). Three species, *C. lapworthi* Tullberg, 1882, *C. laqueus* Jackson & Etherington, 1969 and *C. sakmaricus* Koren' 1968 have been used as biozonal indices. There does, however, appear to be some confusion in the literature with regard to the identification of certain taxa, a situation which should be rectified by the taxonomic revision of many cyrtograptid species currently being undertaken by Bjerreskov (pers. comm.).

Chemically isolated specimens of three Telychian cyrtograptids have been illustrated by Lenz & Melchin (1989) who also discuss the possible polyphyletic origins of the genus.

Barrandeograptus Bouček, 1933. Bjerreskov (1975) has described the type material and additional specimens of *B. pulchellus* (Tullberg, 1882) from the *lapworthi* Biozone of Sweden and Bornholm. This species, possibly the only Llandovery representative of the genus (but see Lenz 1982), has only rarely been reported from outside of Scandinavia.

Two important graptolitic sections

Two Telychian sections in particular would benefit from re-examination. The Trannon section in central Wales (Wood 1906) includes the type localities of both the *Monoclimacis griestoniensis* and *Mcl. crenulata* biozones. Although the *M. turriculatus* and *M. crispus* Biozone strata here have suffered tectonically, from the *griestoniensis* Biozone through to the early Wenlock is largely continuous, although not graptolitic throughout. The thickness of the section (several hundred metres), and the afforestation of the surrounding area constitute practical difficulties to research.

The section described by Fu (1986) and Fu & Song (1986) at Ziyang, Shaanxi Province is the most important discovered in China and appears to contain the most complete sequence of Telychian graptolite biozones in the world. That it and Fu's collections (housed in the Xi'an Institute of Geology and Mineral Resources) were not re-examined in any detail in the 'Transhemisphere Telychian' project (see Holland 1988) is astonishing, particularly considering that the aim of the project was 'to test the ultimate limits of precision in correlation by the use of biostratigraphy' (Holland 1988). A major opportunity has been missed.

Factors affecting the precision of correlation

Dispersal rate

The base of any geographically widespread biozone must be diachronous. The degree of diachroneity will depend upon the rapidity of dispersal of the biozonal index species from its locus of origin. Slow dispersal rates would seriously undermine confidence in biostratigraphical correlation but are highly improbable in a planktic group such as the graptoloids, especially when one considers the oceanic circulation patterns proposed for the Telychian (see Melchin 1989).

Palaeobiogeography

The taxonomic problems discussed above and below make analysis of graptolite species distribution difficult. However, recent works (Melchin 1989; Loydell 1990b; Rickards *et al.* 1990) are challenging earlier assertions of a 'lack of plankton provincialism in the Silurian' (Rickards 1976).

High levels of provincialism would create major difficulties in correlation. This does not appear to be the case, but the absence of biozonal index species from some areas is a cause for concern. At present it is difficult to say whether this is due to collection failure, to the existence of unsuitable lithologies for graptolite preservation at the horizons concerned, or to provincialism. For example, Melchin (1989) noted that 'in the uppermost Llandovery, a distinct province is again established in northern North America, Siberia, China, and to a lesser extent central Europe, defined by the diverse cyrtograptid fauna with *C. sakmaricus*, *C.* ex gr. *lapworthi*, and others, as well as a diverse retiolitid fauna.' This observation must be treated with considerable caution for the Telychian graptolites of Gondwanaland are virtually unstudied, and in sections in Eastern Avalonia and Baltica (e.g. Trannon (Wood 1906); Swindale Beck in the Cross Fell Inlier (Burgess *et al.* 1970); in the Howgill Fells (Wilson 1954); Rickards 1970); Stockdale Beck in the English Lake District (Hutt 1974); on Bornholm (Bjerreskov 1975)) the uppermost Telychian is characterised by red beds, oolitic limestones and other lithologies unlikely to yield graptolites in any numbers, if at all. The absence of *C. sakmaricus* from the more continuous sections of central Europe may, though, indicate genuine provincialism.

Palaeoautecology

That graptolites exhibited depth stratification was suggested by Berry (1962) and was quite widely accepted (e.g. Cisne & Chandlee 1982; Bates & Kirk 1984). Depth stratification alone cannot, however, explain the absence or extreme rarity of 'shallow water' taxa from deep basinal deposits. The concept of water mass specificity (Berry 1974, 1977; Watkins & Berry 1977; Finney 1984, 1986) explains more satisfactorily observed local graptolite distribution patterns.

Much work remains to be done on water mass specificity in Telychian graptolites, particularly at the species level. At the generic level, during the lower part of the *turriculatus* Biozone, petalolithids are common in the Prague Basin, the water depth in which has been inferred not to have exceeded 200 m (Štorch & Pašava 1989), and on the Yangtze Platform (Chen 1984b), but are rare or absent in deeper water sequences such as those of the Welsh Basin (Loydell 1991a), the Howgill Fells (Rickards 1970) and Washington Land, western North Greenland (Bjerreskov 1981). A similar, but less pronounced, preference for shallower water sequences appears to be shown by the rastritids and retiolites, but this is more difficult to analyse, as in some lithologies both may be overlooked when collecting. These observations run contrary to those of Lenz & Chen (1985) who compared the Peel River section, in the Richardson Trough (deep water), with the Blackstone River section ('depths significantly less'), both of which are in northern Canada. However, as Lenz & Chen note, the Peel River section was sampled more intensively, and this, rather than difference in water depth, might easily account for the higher diversity of petalolithids, retiolites and rastritids noted here. With the exception of *Pe. intermedius* Bouček & Přibyl, 1942b (which was found to be 'very abundant' in the Blackstone River section), the species of the above genera listed by Lenz & Chen (1985) were, almost without exception, rare or uncommon. This suggests, as does Lenz's (1978b) palaeogeographical reconstruction of northwestern Canada, that conditions were not sufficiently dissimilar in the two areas for lateral differentiation of the graptolite fauna by water mass specificity to have occurred to any great extent.

Taxonomic problems

The correlative problems posed by provincialism and water mass specificity pale into insignificance when compared with those of 'the most underappreciated of all sciences' (Gould 1985), taxonomy. These problems, all of them serious, fall into a number of categories.

Synonymy. Real age equivalence of strata in different areas may be masked if the contained taxa, although identical, are known by different names in these areas. Instances of synonymy abound. An example is illustrated in Fig. 2. These are all figures of the same species, *Stomatograptus longus* Obut, 1949, yet only in the USSR and rarely in China (Chen 1984a) is this specific name used. To add a further element of confusion to the story, it should be noted that *Stomatograptus grandis girvanensis* Bjerreskov, 1975 is probably a different species to Cocks & Toghill's (1973) *nomen nudum* and that Thomas's (1960) figures of *Stomatograptus australis* (M'Coy, 1875) suggest that his concept of the species was different to that of M'Coy (1875). The true *Retiolites australis* appears, from M'Coy's description and illustrations, to be a senior synonym of *Retiolites angustidens* Elles & Wood, 1908). This degree of chaos is by no means unique (see Loydell 1990a).

'Lumping'. This occurs where several different species (whose stratigraphical ranges are almost certainly going to be different) are 'lumped' under one specific name. As an example, Melchin (1989) illustrates as *Monoclimacis griestoniensis* (Nicol, 1850) a specimen which is considerably broader than Nicol's species and may be referred to *Mcl.* cf. *griestoniensis* (sensu Elles & Wood, 1911). From his description of the species (Melchin 1987) it seems almost certain that *Mcl. griestoniensis* sensu stricto is also present in his collections. The precise stratigraphical ranges of these two taxa remain unknown (although both Wilson (1954) and Zalasiewicz (1990) have observed that *Mcl.* of *griestoniensis* appears before *Mcl griestoniensis*) and will remain so if they are lumped in this way.

The Elles & Wood problem. The 'Monograph of British Graptolites' (Elles & Wood 1901–18) has been perhaps the single most influential and widely used publication on graptolites. It was for many years, and in some places (e.g. much of China, Chen pers. comm.) still is, the standard guide to graptolite identification. However, although the figures are, in general, excellent (better than most produced today), this work has been the cause of severe taxonomic problems, resulting either from the grouping of several taxa under the same specific name or simply the misidentification of species initially described from European sequences. Strachan (1971) has partly rectified the situation by reassigning several of Elles & Wood's specimens to different species, but much remains to be done.

The Hundt problem. Rudolf Hundt published more than 50 palaeontological papers (see refer-

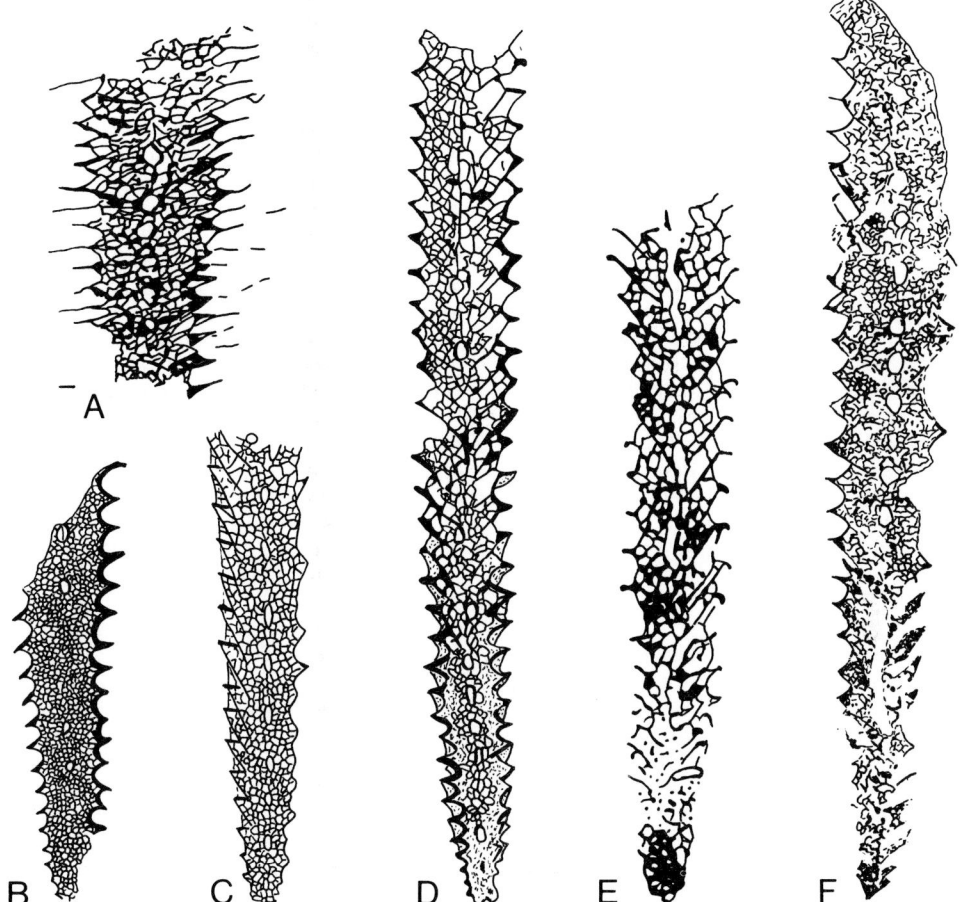

Fig. 2. *Stomatograptus longus* Obut, 1949 and its junior synonyms. (**A**) *Sinostomatograptus? occidentalis* Hutt, 1974, line indicates tectonic lineation – dorso-ventral width and thecal count greatly increased by lateral compression. Re-examination of holotype suggests that the 'spines' may be lineations on the bedding surface rather than part of the rhabdosome. This species name is used in Great Britain (e.g. Rickards 1976; Rickards *et al.* 1977). (**B** and **C**) *Stomatograptus longus*, (C) is at an oblique angle. Species name used in the USSR (e.g. Obut & Sobolevskaya 1966) and, to a much lesser extent, in China (Chen 1984*a*). (**D**) *Stomatograptus girvanensis* Bjerreskov, 1975. Species name used in Denmark (Bjerreskov 1975) and in Canada (Melchin 1989). (**E**) *Stomatograptus australis* (M'Coy, 1875). Species name used by Thomas (1960) in Australia. (**F**) *Stomatograptus shiqianensis* NIGP, 1974, a possible synonym, and (not illustrated) *Stomatograptus asiaticus* Wang, 1978 are species names used in China. All × 5. (A) After Hutt (1974); (B) and (C) after Obut & Sobolevskaya (1966); (D) after Melchin (1989); (E) after Thomas (1960); (F) after NIGP (1974).

ence list in Hundt 1965), many of which are concerned with graptolites and describe new species. Müller & Schauer (1969) observe 'Auch hat vor allem R. Hundt trotz des ihm zur Vefügung stehenden ausgeziechneten Materials durch fehlerafte Arbeitsweise und weitgehende Disziplinlosigkeit bei der Aufstellung neuer

Arten und Gattungen sowie durch phantasievolle Betrachtung den Fortscritt auf diesem Gebiet gehemmt.' This sounds like a recommendation for Hundt's work to be ignored, and indeed Schauer (1971) erects a new subspecies *Petalolithus ovatus scopaecularus* for an existing species *Demicystograptus regius* Hundt, 1957 and includes Hundt's references in his synonymy list. Whatever objections there may be to Hundt's work, it cannot be so discounted without recourse to the International Commission on Zoological Nomenclature, and the policy of Jones & Rickards (1967) of redescribing and redefining Hundt's taxa is to be preferred unless (or until) application is successfully made to the ICZN that any other procedure should be followed.

Lack of redescription of Telychian graptoloid taxa. Standards of illustration and description have varied markedly between graptolite workers. As a result, whilst some species are well-defined, others are difficult or impossible to identify on the basis of original descriptions. It is not surprising that Rickards (1965) erected a new species, *Monograptus danbyi*, which has streptograptid thecae, for the already described *Monograptus kodymi* Bouček, 1931, which in the original illustration is shown to have simply hooked thecae (although the thecae are described as 'coiled into a large tubercle' by Bouček & Přibyl 1951).

Of particular value to Telychian graptolite workers worldwide would be modern redescriptions of the taxa described in the works of Perner (1897, 1899), Bouček (1931), Kirste (1919), Obut (1949), Haberfelner (1931a, b), Hemmann (1931), Přibyl (1940a, b), Ruedemann (1947) and Mu *et al.* (1962). Several species described in more recent, Chinese works (e.g. Wang 1978; Fu & Song 1986) would benefit from similar treatment, particularly those illustrated only by poorly preserved or poorly photographed material. Universal usage of two theca repeat distance (2TRD, Howe 1983) measured standardly at th2, th5, th10, th15, th20, th50 and th100 in these redescriptions would allow more direct comparison of thecal spacing in different taxa than does the measurement of number of thecae in 10 mm, which although considerably less precise is more widely used at present.

In some cases historical collections are lost, for example those of Chang & Sun (1947) and Mu (1948), whilst Huo's (1957) retiolites were destroyed during the 'Cultural Revolution'. In such cases, where the nature of a species is ambiguous, descriptions will have to be based on new collections of topotype material.

Telychian graptolite biozones

Details of the type localities of several Telychian graptolite biozones are given by Curtis (in Whittard 1961) and Rickards (1976) and are not repeated here. Schemes used in different parts

of the world differ more than earlier in the Llandovery, partly as a result of the problems outlined above. Some of these schemes are illustrated in Fig. 1. Almost all the Telychian graptolite biozones are poorly defined, and thus difficult to recognise if the index species is absent from a collection or if the index species has a long range, extending beyond the limits of its biozone.

The British subdivision of the Telychian into four biozones is often described as the standard scale (Cocks *et al.* 1984; Melchin 1989) although 'this is not a formally agreed standard of zones' (Rickards 1989). Its limitations, particularly for the late Telychian, are outlined below.

In many parts of the world the *turriculatus* Biozone has been divided into a lower, *Rastrites linnaei* or *Spirograptus minor* Biozone and an upper *Spirograptus turriculatus* Biozone. Jones & Pugh (1916) and Loydell (1991a) have found that *R. linnaei* Barrande, 1850 appears in the late Aeronian *Stimulograptus halli* Biozone, and thus the *linnaei* Biozone, were it to encompass the full range of the index species, would span the Aeronian–Telychian boundary. Finer subdivision of the *turriculatus* Biozone as a whole has been achieved by Bouček (1953) in Bohemia and by Loydell (1991a) who has recognised six subzones in Wales.

The *turriculatus* Biozone has been widely recognized, partly as a result of its long duration compared with most other Llandovery graptolite biozones (Lenz 1978b; Loydell in press). The *crispus* and *griestoniensis* biozones, almost certainly of shorter duration, have been encountered less frequently. *M. turriculatus* (Barrande, 1850) s.s. ranges high into the *crispus* Biozone (Zalasiewicz 1990; Schauer 1971) and its presence alone cannot be taken as a definite indication of the *turriculatus* Biozone.

Zalasiewicz (1990) has produced a bipartite subdivision of the *crispus* Biozone, which seems to be applicable also on Bornholm (Bjerreskov 1975), and indicates the existence of an interval after the disappearance of *M. crispus* and before the appearance of *Mcl. griestoniensis* characterised by the presence of *M. pragensis* (Přibyl, 1943) sensu lato, *Streptograptus sartorius* (Törnquist, 1881) and, in the upper part, by *Mcl.* cf. *griestoniensis* (sensu Elles & Wood). This interval is present also in the Howgill Fells (Wilson 1954) and may in the future need to be accorded biozonal status (it is referred to informally as the *sartorius* Biozone in Fig. 1).

Monograptus crispus has been recorded in China by Mu *et al.* (1962) and by Fu & Song (1986). Although the *crispus* Biozone is almost certainly represented in Ziyang (Fu & Song

1986), the illustrations of the index species seem referable to *S. exiguus* sensu lato rather than to *M. crispus*. The material from the Qilian Mountains referred by Mu *et al.* (1962) to *M. crispus* is identical to the specimens illustrated by them as *Streptograptus nanshenensis*. The associated fauna suggests a horizon within the *griestoniensis* or *crenulata* Biozone.

The *griestoniensis* Biozone is probably the least well-defined of the Telychian biozones, largely because its type locality, Trannon, has not been recently studied (see above), and because the species associated with *Mcl. griestoniensis* at Grieston Quarry (see Toghill & Strachan 1970) are either long-ranging (retiolites, *M. priodon*, *M. veles*), rare (*M. drepanoformis*), or regularly misidentified (*P. nudus* (Lapworth, 1880), *M. spiraloides* Přibyl, 1945).

Differences between biozonal schemes are most pronounced in post-*griestoniensis* Biozone times. In Britain only the *crenulata* Biozone is recognised (Rickards 1989), whilst elsewhere two, three, four or even five biozones occur between the *griestoniensis* and *centrifugus* biozones. In some cases these have been further subdivided. The reason for this difference has been alluded to above, in that in most British sections studied there is a significant thickness of largely or entirely non-graptolitiferous strata between beds yielding *Mcl. crenulata* (sensu Elles & Wood, 1911) and those yielding *Cyrtograptus centrifugus* Bouček, 1931. Should a standard scheme of Telychian graptolite biozones ever be agreed, Britain would be, on present evidenmce, an unsuitable area for its erection. In Bohemia, where the late Telychian-early Wenlock is, in some sections, represented by an unbroken sequence of graptolitic shales (P. Štorch, pers. comm.), *Mcl. crenulata* (sensu Elles & Wood) disappears long before the end of the Telychian (Bouček 1953). Here, the *Mcl. crenulata* Biozone is succeeded by the *Monograptus spiralis* Biozone and then the *Stomatograptus grandis* Biozone. Melchin's (1989) and Lenz's (1982) suggestions that the *crenulata* Biozone of Britain correlates with these higher biozones (and with other post-*crenulata* biozones recognised elsewhere) are thus incorrect. Fu & Song (1986, table 6) give a more realistic interpretation of the correlation of different Telychian biozonal schemes.

Recent work (Loydell *in* Cave & Dixon in press) has now identified the *spiralis* Biozone in Wales, and a re-examination of Cocks & Toghill's (1973) collections from the Girvan area, Scotland, has indicated the presence of this biozone here also.

A *spiralis* Biozone is widely recognized, and was probably the longest in duration of any of the Llandovery graptolite biozones. For the remainder of the Telychian there is greater variation than before in the biozonal schemes used. In Bohemia the *Stomatograptus grandis* Biozone encompasses this interval, whilst on Bornholm (Bjerreskov 1975), although *Sto. grandis* is present, post-*spiralis* Biozone graptolitic strata are assigned to the *Cyrtograptus lapworthi* Biozone. This cyrtograptid is preferred as a biozonal index by Bjerreskov (1975) because of the difficulty, on the basis of existing descriptions, in distinguishing *Sto. grandis* from similar stomatograptids (see Bouček & Münch 1944) which have different ranges (e.g. *Sto. grandis imperfectus* occurs in the *spiralis* Biozone). In Ziyang, Fu & Song (1986) raised Bouček's *Monoclimacis geinitzi* Subzone to full Biozonal status, with the overlying *C. lapworthi* biozone being succeeded by the *Cyrtograptus sakmaricus* Biozone, assigned by them to the Wenlock. The Ziyang section is the only section in which the *sakmaricus* Biozone has been shown to succeed the *lapworthi* Biozone. In northern Canada Lenz's (1988) range chart has *C. sakmaricus* appearing before *C. lapworthi*, whilst Melchin (1989) has the *sakmaricus* Biozone directly succeeding the *griestoniensis* Biozone. Melchin's (1989) concept of the *griestoniensis* Biozone would appear to be considerably broader than that of other authors (with it encompassing also the *crenulata* and *spiralis* biozones). That the *sakmaricus* Biozone post-dates the *spiralis* Biozone seems clear, but how it correlates precisely with late Telychian biozones elsewhere is difficult to assess on present evidence.

One consequence of designating the base of the *C. centrifugus* Biozone as the base of the Wenlock is the relegation, from the Wenlock to the Telychian, of the *Cyrtograptus insectus* Biozone, recognized in Bohemia (Bouček 1953; Štorch pers. comm.) and in China (e.g. Fu & Song 1986). The stratigraphical ranges of *C. insectus* and *C. centrifugus* do not overlap either in China or in Bohemia. Rickards (1967), however, reports *C. insectus* from the *centrifugus* Biozone of the Howgill Fells. Two of Rickards' (1967) specimens are illustrated in Rickards (1963). Neither seems referable to *C. insectus* and it is possible that both are in fact specimens of *C. centrifugus* (similar to those illustrated by Bouček 1933, pl. 3, figs 1, 4).

The inclusion of the *insectus* Biozone brings the total number of Telychian graptolite biozones to somewhere between seven and ten depending on the scheme in use. Given current estimates of the duration of the Silurian (Holland

1989), it might be reasonable to estimate the duration of the Telychian as of the order of 5 million years. When one considers that many of the biozones have been subdivided, the poten-tial level of resolution of biostratigraphical correlation possible using graptolites may be seen to be very impressive indeed.

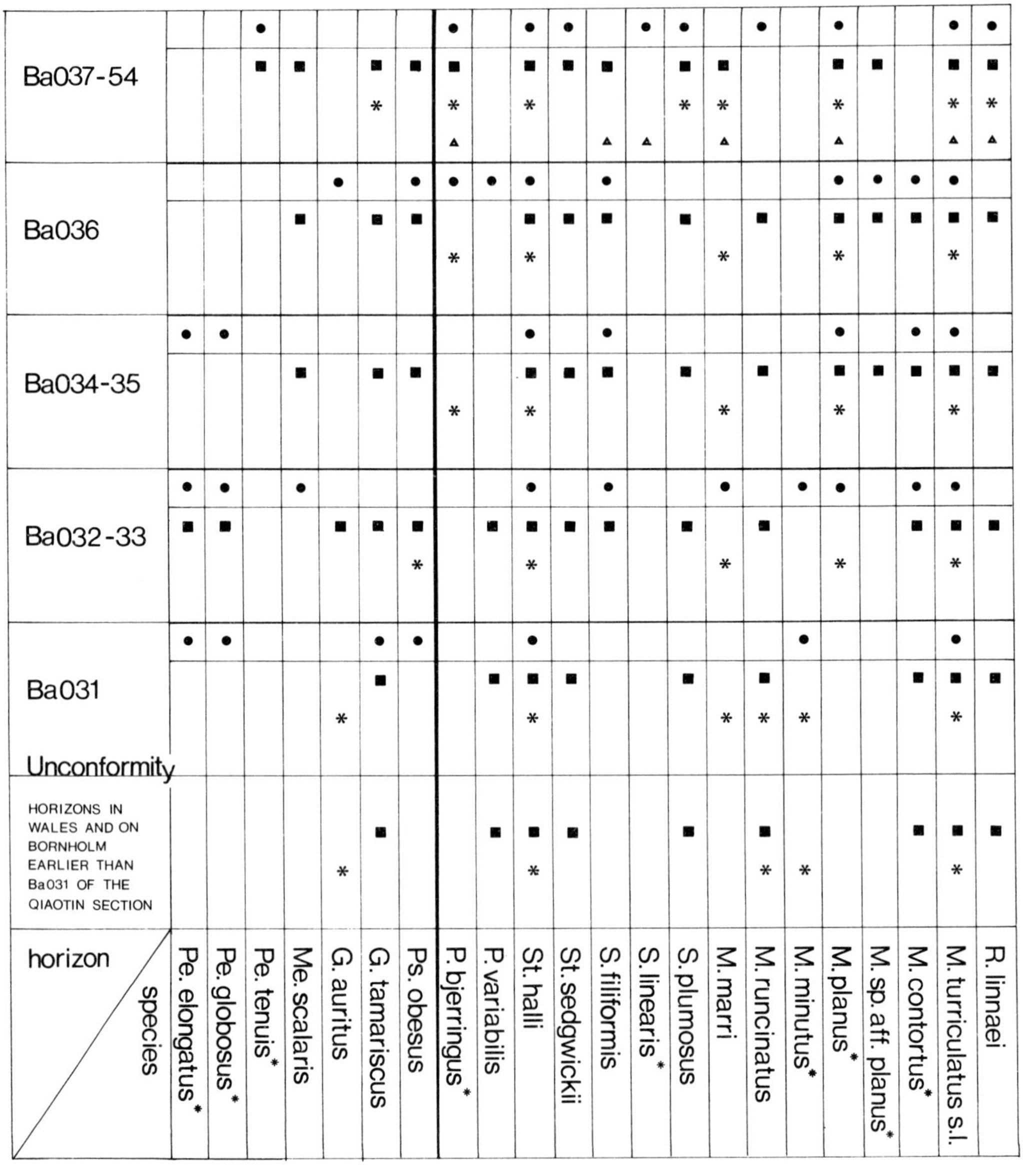

Fig. 3. Comparison of ranges of graptolites from the Qiaotin section, Nanjiang, Sichuan China (circles) with those of the same species in Wales (squares) and on Bornholm (stars). Those species present in Linnarsson's (1881) collection from the middle part of the *turriculatus* Biozone of Klubbudden, Sweden are also indicated (triangles). Further details of the section are given in Chen (1984*b*). Species of value in biostratigraphical correlation of the Qiaotin section are asterisked. Correlation of Qiaotin with Wales, Bornholm and Klubbudden is, of necessity, somewhat approximate. Data is from a re-examination of Chen's (1984*b*) and Linnarsson's (1881) collections, and from Bjerreskov (1975) and Loydell (1991*a*).

An example of intercontinental correlation

Figure 3 compares the ranges of species encountered in a restudy of Chen's (1984*b*) graptolites from the Qiaotin section, Nanjiang, Sichuan, China with those of the same species in Wales (Loydell 1991*a*) and on Bornholm (Bjerreskov 1975). Linnarsson's (1881) species records, from the middle of the *turriculatus* Biozone of Sweden are also plotted (the material having been re-examined).

In comparing a shelf fauna (Qiaotin) with basinal faunas (Wales, Bornholm) several thousand kilometres distant (both now and in the Telychian), one might expect differences in composition and difficulties in correlation. Indeed, of the 37 species present in the Qiaotin section (which may be referred in its entirety to the early to middle parts of the *turriculatus* Biozone), 15 have not been recorded in the Telychian of Bornholm, Wales or Klubbudden, Sweden. Of these, nine are retiolites, petalolithids or rastritids and their absence may be explained by palaeoecological factors (see above) rather than palaeobiogeographical factors, particularly given that several of the species have been recorded from Bohemia. Of the remainder, three are new species, two have recorded only from the Qiaotin section (*G. ultimus* and *G. nanjiangensis*) and the other is *Sinodiversograptus lientanensis*, a species considered by Loydell (1990*b*) to have had a limited geographical distribution.

Of the remaining 22 species, those marked with an asterisk on Fig. 3 show a stratigraphical restriction to their range which is considered significant. On the basis of these, it is possible to recognize three distinct assemblages. The lowest of these (Ba031–Ba033) is characterized by *M. minutus* (Chen, 1984*b*). This species is present at a similar level on Bornholm, in Bohemia, and also in Canada where it has been identified as *Monograptus decipiens valens* by Lenz (1982, Fig. 23G) and as *Monograptus decipiens* n. ssp. by Melchin (1989)). The stratigraphical importance of *M. minutus* (Fig. 4) has not previously been appreciated.

The second of the assemblages (Ba034–Ba036) is less well defined. It lacks *M. minutus* which has been replaced by *M. planus* (Barrande, 1850). *M. contortus* Perner, 1897 makes its last appearance whilst *P. bjerringus* (Bjerreskov, 1975) makes its first. A similar situation is seen in Wales and on Bornholm. The upper assemblage Chen (1984*b*) referred to the *Streptograptus exiguus* Biozone. This species is not present here, however, and horizons Ba037–Ba054 may be assigned to the middle part of the

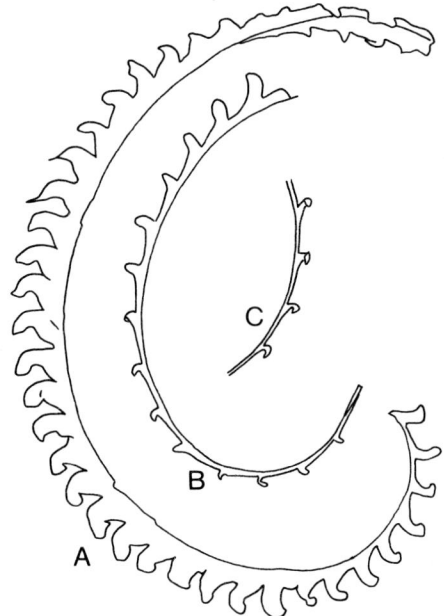

Fig. 4. *Monograptus minutus* (Chen, 1984*b*). (**A**) Mesial fragment (after Melchin 1989, originally illustrated as *Monograptus decipiens* n. ssp.). (**B** and **C**) Proximal fragments (after Chen 1984*b*; (**B**) originally illustrated as *Oktavites planus*). Note the twisting of the thecal apertures (to the reverse side of the rhabdosome). The specific name, applied originally to the proximal end, is something of a misnomer for the rhabdosome as a whole. This species is characteristic of horizons very low in the *turriculatus* s.l. Biozone. All figures × 5.

turriculatus Biozone. The assemblage is characterised by the cooccurrence of *Pe. tenuis* (Barrande, 1850), *P. bjerringus*, *S. linearis* (Chen, 1984*b*), *S. plumosus* (Baily, 1871) and *M. planus*. This assemblage compares well with those at similar levels in Wales and Sweden and on Bornholm.

That intervals of approximately one-sixth of a biozone duration may be correlated over such a distance and between such different palaeoenvironments gives some indication of what might be achieved if one were to attempt to correlate the graptolite assemblages of similar palaeoenvironments (e.g. Bohemia and the Yangtze Platform).

Conclusions

Graptolies offer the most precise means of correlation within the Telychian Stage, currently and in the forseeable future. The Telychian has been divided on the basis of its contained sequence of graptolite faunas into a number of biozones, most of only a few hundred thousand years' duration, and several of these have been further divided into subzones. However, several of the biozones are poorly defined at present and may be difficult to recognise in the absence of the biozonal index species or where the range of an index species extends beyond the limits of its biozone.

Concentration on the taxonomic revision of Telychian species would facilitate a major advance in Telychian biostratigraphy and correlation and hence of our knowledge of the Silurian world. Such taxonomic revision may not be considered a fashionable pursuit at present, but would be of permanent value, and would allow graptolite biostratigraphers to be more confident that the precision which they often claim is something more than illusory.

P. Štorch, Chen Xu and J. Zalasiewicz are thanked for discussion. R. B. Rickards and R. Banger assisted greatly in my search for elusive graptolite literature. M. G. Bassett made Linnarsson's graptolites available for study. Xu Yang helped greatly in preparing Chinese material for my examination. The Royal Society financed one visit to China (1990), whilst the University of Wales supported an earlier visit to China (1988) and one to Czechoslavakia (1990). This support is gratefully acknowledged.

References

BAILY, W. H. 1871. Palaeontological remarks. In: TRAILL, W. A. & EGAN, F. W. Explanatory Memoir to accompany Sheets 49, 50 and Part of 61 of the Maps of the Geological Survey of Ireland including the Country around Downpatrick, and the shores of Dundrum Bay and Strangford Lough, County of Down. Alexander Thom, Dublin and London, 22–23.

BARRANDE, J. 1850. Graptolites de Bohême. Prague.

BASSETT, M. G. 1985. Towards a 'Common Language' in stratigraphy. Episodes, 8, 87–92.

——, RICKARDS, R. B. & WARREN, P. T. 1975. The type Wenlock Series. Report of the Institute of Geological Sciences, 75/13.

BATES, D. E. B. 1987. The construction of graptolite rhabdosomes in the light of ultrastructural studies. Indian Journal of Geology, 59, 1–28.

—— & KIRK, N. H. 1978. Contrasting modes of construction of retiolite-type rhabdosomes. Acta Palaeontologica Polonica, 23, 427–448, pls 1–17.

—— & —— 1984. Autecology of Silurian graptoloids. Special Papers in Palaeontology, 32, 121–139.

—— & —— 1986. The mode of secretion of graptolite periderm, in normal and retiolite graptolites. In: RICKARDS, R. B. & HUGHES, C. P. (ed.) Palaeo-

ecology and Biostratigraphy of Graptolites. Geological Society, London, Special Publications, 20, 221–236.

BERRY, W. B. N. 1962. Graptolite occurrence and ecology. Journal of Paleontology, 36, 285–293.

—— 1974. Types of Early Paleozoic faunal replacements in North America: their relationship to environmental change. Journal of Geology, 82, 371–382.

—— 1977. Ecology and age of graptolites from graywackes in eastern New York. Journal of Paleontology, 51, 1102–1107.

BJERRESKOV, M. 1975. Llandoverian and Wenlockian graptolites from Bornholm. Fossils and Strata, 8, 1–94, pls 1–13.

—— 1981. Silurian graptolites from Washington Land, western N. Greenland. Bulletin Grønlands Geologiske Undersøgelse, 142.

BOUČEK, B. 1931. Předběžná zpráva o některých nových druzích graptolitů z českého gotlandienu. Communication préliminaire sur quelques nouvelles espèces de Graptolites provenant du Gothlandien de la Bohême. Věstník Státního Geologického ústavu Republiky Československé, 7, 293–313.

—— 1932. Predbézna zpráva o nekterych novych druzích graptolitu z ceského gotlandienu. (Cást II.) Preliminary report on some new species of Graptolites from the Gothlandian of Bohemia. Vestník Státního Geologického ústavu Republiky Ceskoslovenské, 8, 150–155.

—— 1933. Monographie der obersilurischen Graptoliten aus der Familie Cyrtograptidae. Práce Geologicko-Paleontologického ústavu Karlovy university v Praze zar. 1933 (č. 1.).

—— 1953. Biostratigraphy, development and correlation of the Želkovice and Motol Beds of the Silurian of Bohemia. Sborník ústredního ústavu Geologického, 20, 473–484, pl. 1.

—— & MÜNCH, A. 1944. Die Retioliten des Mitteleuropäischen Llandovery und der unteren Wenlock. Mitteilungen der Tschechischen Akademie der Wissenschaften, 53, (41), 1–54, pls 1–3.

——, & PŘIBYL, A. 1941. Über die Gattung Petalolithus Suess aus dem böhmischen Silur. Mitteilungen der Tschechischen Akademie der Wissenschaften, 51, 1–17, pls 1–3.

—— —— 1942a. Über böhmische Monograpten aus der Untergattung Streptograptus Yin. Mitteilungen der Tschechischen Akademie der Wissenschaften, 52, 1–23, pls 1–3.

—— —— 1942b. Über Petalolithen aus der Gruppe P. folium (His.) und über Cephalograptus Hopk. Mitteilungen der Tschechischen Akademie der Wissenschaften, 52, (31), 1–22, pl. 1.

—— & —— 1951. On some slender species of the genus Monograptus Geinitz, especially of the subgenera Mediograptus and Globosograptus. Bulletin International de l'Académie Tchèque des Sciences, 13, 1–32, pls 1–3.

BRONN, H. G. 1935. Lethaea Geognostica 1. E. Schweizerbart, Stuttgart.

BULMAN, O. M. B. 1955. Treatise on Invertebrate Paleontology. Part V, Graptolithina. Geological

Society of America and University of Kansas Press.

—— 1970. *Treatise on Invertebrate Paleontology. Part V, Graptolithina, with sections on Enteropneusta and Pterobranchia*. Geological Society of America and University of Kansas Press.

—— & RICKARDS, R. B. 1968. Some new diplograptids from the Llandovery of Britain and Scandinavia. *Palaeontology*, **11**, 1–15.

—— & —— 1970. Classification of the graptolite family Monograptidae Lapworth, 1873. *In*: BULMAN, O. M. B. *Treatise on Invertebrate Paleontology. Part V, Graptolithina, with sections on Enteropneusta and Pterobranchia*. Geological Society of America and University of Kansas Press, V149–157.

BURGESS, I. C., RICKARDS, R. B. & STRACHAN, I. 1970. The Silurian strata of the Cross Fell area. *Bulletin of the Geological Survey of Great Britain*, **32**, 167–182.

CARRUTHERS, W. 1867. Graptolites: their structure and systematic position. *Intellectual Observer*, **11**, (4), No. 64, 283–292, pl. 1; (5), No. 65, 365–374, pl. 2.

CAVE, R. & DIXON, R. J. In press. The Ordovician and Silurian strata of the Welshpool area. *In*: BASSETT, M. G. & WOODCOCK, N. H. (eds) *Geological Excursions in Powys, Wales*. National Museum of Wales.

CHANG, H. C. & SUN, Y. C. 1947. New graptolite faunas from Lientan, Kwangtung. *Contributions from the Geological Institute, National University of Peking*, **29**, 9–17.

CHEN, X. 1984*a* The Silurian graptolite zonation of China. *Canadian Journal of Earth Sciences*, **21**, 241–257.

—— 1984*b*. [*Silurian graptolites from Southern Shaanxi and Northern Sichuan with special reference to classification of Monograptidae.*] Palaeontologica Sinica, New Series B, **166**, No. 20. [in Chinese with English summary].

CISNE, J. L. & CHANDLEE, G. O. 1982. Taconic foreland basin graptolites: age zonation, depth zonation, and use in ecostratigraphic correlation. *Lethaia*, **15**, 343–363.

COCKS, L. R. M. 1989. The Llandovery Series in the Llandovery area. *In*: HOLLAND, C. H. & BASSETT, M. G. (eds) *A Global Standard for the Silurian System*. National Museum of Wales, Cardiff, Geological Series, **9**, 36–50.

—— & TOGHILL, P. 1973. The biostratigraphy of the Silurian rocks of the Girvan District, Scotland. *Journal of the Geological Society, London*, **129**, 209–243, pls 1–3.

——, WOODCOCK, N. H., RICKARDS, R. B., TEMPLE, J. T. & LANE, P. D. 1984. The Llandovery Series of the Type Area. *Bulletin of the British Museum (Natural History), Geology series*, **38**, 131–182.

ELLES, G. L. & WOOD, E. M. R. 1901–18. *A Monograph of British Graptolites*. Palaeontographical Society Monograph.

FINNEY, S. C. 1984. Biogeography of Ordovician graptolites in the southern Appalachians. *In*: BRUTON, D. L. (ed.) *Aspects of the Ordovician*

System. Palaeontological Contributions of the University of Oslo, 167–176.

—— 1986. Graptolite biofacies and correlation of eustatic, subsidence, and tectonic events in the Middle to Upper Ordovician of North America. *Palaios*, **1**, 435–461.

FORTEY, R. A. & COOPER, R. A. 1986. A phylogenetic classification of the graptoloids. *Palaeontology*, **29**, 631–654.

FRECH, F. 1897. *Lethaea Geognostica, 1, Theil., Lethaea Palaeozoica, 1*, **11**, *Graptolithen*. Schweizerbach, Stuttgart, 544–684.

FU, L. P. 1986. Graptolite zones of upper Ordovician to middle Silurian age in a continuous section at Ziyang, Shaanxi, China. *In*: HUGHES, C. P. & RICKARDS, R. B. (eds). *Palaeoecology and Biostratigraphy of Graptolites*. Geological Society, London, Special Publications, **20**.

—— & SONG, L. S. 1986. [*Stratigraphy and paleontology of Silurian in Ziyang Region (Transitional Belt).*] Bulletin of the Xi'an Institute of Geology and Mineral Resources, Chinese Academy of Geological Sciences, **14** [in Chinese].

GE, M. Y. 1990. [*Silurian graptolites from Chengkou, Sichuan.*] Palaeontologica Sinica, New Series B, **179**, No. 26 [in Chinese with English summary].

GEINITZ, H. B. 1842. Über die Graptolithen. *Neues Jahrbuch für Mineralogie, Geografie, Geologie und Petrefak.-Kunde*, 697–710, pl. 10.

—— 1852. *Die Versteinerungen der Grauwackenformation in Sachsen und den angrenzenden Länder-Abtheilungen (Die Graptolithen)*. Engelmann, Leipzig.

GOULD, S. J. 1985. *The flamingo's smile*. Norton, New York.

GÜRICH, G. 1908. *Leitfossilien (Kambrium bis Silur)*. Lief 1, Berlin.

HABERFELNER, E. 1931*a*. Graptolithen aus dem Obersilur der Karnischen Alpen. I. Teil: Hochwipfel, Nordseite. Sitzungsberichte. *Akademie der Wissenschaften in Wien*, (1), **140**, (1–2), 89–168, pls 1–3.

—— 1931*b*. Eine Revision der Graptolithen der Sierra Morena (Spanien). Abhandlungen der Senckenbergischen Naturforschenden Gesellschaft, **43**, 19–66, pl. 1.

HALL, J. 1852. *Palaeontology of New York. Volume II, containing descriptions of the organic remains of the Lower Middle Division of the New-York System*. Van Benthuysen, Albany, New York.

HEMMANN, M. 1931. Die Graptolithen der Zone 16 vom Grobsdorfer Berg bei Ronneburg. *Beiträge zur Geologie von Thüringen*, **3**, 120–122, pl. 5.

HISINGER, W. 1837. *Lethaea Suecica seu Petrifacta Suecica, Supplementum 1.*, Stockholm.

HOLLAND, C. H. 1986. Does the golden spike still glitter? *Journal of the Geological Society, London*, **143**, 3–21.

—— 1988. Transhemisphere Telychian: a biostratigraphical experiment. *Lethaia*, **21**, 188.

—— 1989. Classification, *In*: HOLLAND, C. H. & BASSETT, M. G. (eds) *A Global Standard for the Silurian System*. National Museum of Wales, Cardiff. Geological Series, **9**, 23–26.

Howe, M. P. A. 1982. *Upper Llandovery graptolites and stratigraphy of the Northern Oslo region*. PhD Thesis, University of Cambridge.

—— 1983. Measurement of thecal spacing in graptolites. *Geological Magazine*, **120**, 635–638.

Hundt, R. 1957. Schwebeblasen bei Graptolithen (Diplograptidae) ein Beitrag zu ihrer Lebenweise. *200 Jahre Naturkundesmuseum, Heidecksburg*, 79–95.

—— 1965. *Aus der Welt der Graptolithen*. Commerzia Verlag und Druckerei Seidel & Co., Berlin/Bonn.

Huo, S. C. 1957. Some Silurian graptolites of the family Retiolitidae from Liangshan, Hanchung. *Acta Palaeontologica Sinica*, **5**, 513–522, pls 1–3.

—— & Shu, D. G. 1986. The Silurian graptolite-bearing strata in China. *In*: Hughes, C. P. & Rickards, R. B. (eds) *Palaeoecology and Biostratigraphy of Graptolites*. Geological Society, London, Special Publications, **20**, 173–179.

Hutt, J. E. 1974–75. *The Llandovery graptolites of the English Lake District*. Palaeontographical Society Monograph.

——, Rickards, R. B. & Skevington, D. 1970. Isolated Silurian graptolites from the Bollerup and Klubbudden · stages of Dalarna, Sweden. *Geologica et Palaeontologica*, **4**, 1–23.

Jackson, D. E. & Etherington, J. R. 1969. New Silurian cyrtograptid graptolites from northwestern Canada and northern Greenland. *Journal of Paleontology*, **43**, 1114–1121, pls 129, 130.

Jaekel, O. 1889. über des Alter des sogen. Graptolithengesteins mit besonderer Berücksichtigung der in demselben enhaltenen Graptolithen. *Zeitschrift der Deutschen Geologischen Gesellschaft*, **41**, 653–690, pls 28, 29.

Jones, O. T. & Pugh, W. J. 1916. The geology of the district around Machynlleth and the Llyfnant Valley. *Quarterly Journal of the Geological Society of London*, **71**, 343–385, pl. 27.

Jones, W. D. V. & Rickards, R. B. 1967. *Diploraptus penna* Hopkinson, 1869, and its bearing on vesicular structures. *Paläontologische Zeitschrift*, **41**, 173–185.

Kearsley, A. T. 1985. A new phylogeny of diplograptine graptoloids, and their classification based on proximal and thecal construction. *Newsletter of the Graptolite Working Group of the International Palaeontological Association*, **6**, 8–22.

Kirste, E. 1919. Die Graptolithen des Altenburger Ostkreises. *Mitteilungen Osterlände*, **16**, 60–222, pls 1–3.

Koren', T. N. 1968. [Novye Rannesiluriiskie graptolity yuzhnogo Urala.] *Paleontologicheskii Zhurnal*, **4**, 101–103 [in Russian].

Kurck, C. 1882. Några nya Graptolitarter från Skåne. *Geologiska Föreningens i Stockholm Förhandlingar*, **6**, 294–304, pl. 14.

Lapworth, C. 1873. Notes on the British graptolites and their allies. 1. On an improved classification of the Rhabdophora. *Geological Magazine*, **1**, (10), 500–504, 555–560, Table 1.

—— 1876. On Scottish Monograptidae. *Geological Magazine*, **2**, (3) 308–321, 350–360, 499–507, 544–552, pl. 10–13, 20.

—— 1880. On new British graptolites. *Annals and Magazine of Natural History*, **5**, (5) 149–177, pls 4, 5.

Legrand, P. 1987. Modo de desarrollo del Suborden Diplograptina (Graptolithina) en el Ordovício superior y en el Silúrico. Implicaciones taxonómicas. *Revista Española de Paleontologia*, **2**, 59–64.

Lenz, A. C. 1978a. Llandoverian and Wenlockian *Cyrtograptus*, and some other Wenlockian graptolites from northern and Arctic Canada. *Géobios*, **11**, 623–653.

—— 1978b. Llandoverian graptolite zonation in the northern Canadian Cordillera. *Acta Paleontologica Polonica*, **24**, 137–153.

—— 1982. Llandoverian graptolites of the Northern Canadian Cordillera: *Petalograptus, Cephalograptus, Rhaphidograptus, Dimorphograptus*, Retiolitidae, and Monograptidae. *Life Sciences Contributions, Royal Ontario Museum*, **130**, 1–154.

—— 1988. Upper Llandovery and Wenlock graptolites from Prairie Creek, southern Mackenzie Mountains, Northwest Territories. *Canadian Journal of Earth Sciences*, **25**, 1955–1971.

—— & Chen, X. 1985. Graptolite distribution and lithofacies: some case histories. *Journal of Paleontology*, **59**, 636–642.

—— & Melchin, M. J. 1987a. Silurian retiolitids from the Cape Phillips Formation, Arctic Islands, Canada. *Bulletin of the Geological Society of Denmark*, **35**, 161–170.

—— & —— 1987b Peridermal and interthecal tissue in Silurian retiolitid graptolites: with examples from Sweden and Arctic Canada. *Lethaia*, **20**, 353–359.

—— & —— 1989. *Monograptus spiralis* and its phylogenetic relationship to early cyrtograptids. *Journal of Paleontology*, **63**, 341–348.

Linnarsson, G. 1881. Graptolitskiffrar med *Monograptus turriculatus* vid Klubbudden nära Motala. *Geologiska Föreningens i Stockholm Föhandlingar*, **5**, 503–526, pls 22, 23.

Loydell, D. K. 1990a. On the graptolites described by Baily (1871) from the Silurian of Northern Ireland and the genus *Streptograptus* Yin. *Palaeontology*, **33**, 937–943.

—— 1990b. *Sinodiversograptus* – its occurrence in Australia and northern Canada. *Journal of Paleontology*, **64**, 847–849.

—— 1991a. The biostratigraphy and formational relationships of the upper Aeronian and lower Telychian (Llandovery, Silurian) formations of western mid-Wales. *Geological Journal*, **26**, 209–244.

—— 1991b. Dob's Linn – the type locality of the Telychian (Upper Llandovery) *Rastrites maximus* Biozone? *Newsletters on Stratigraphy*, **25**, 155–161.

—— In press *Upper Aeronian and lower Telychian (Llandovery) graptolites from western mid-Wales*. Palaeontographical Society Monograph.

—— & Zhao, Y. H. 1990. Studies of some monograptids using the scanning electron microscope. *Acta Palaeontologica Sinica*, **29**, 331–336, pls 1–3.

Manck, E. 1923. Untersilurische Graptolithenarten

der Zone 10 des Obersilurs, ferner *Diversograptus* gen. nov. sowie einige neue Arten anderer Gattungen. *Natur, Leipzig*, **14**, 282–289.

MARTINSSON, A., BASSETT, M. G. & HOLLAND, C. H. 1981. Ratification of standard chronostratigraphical divisions and stratotypes for the Silurian System. *Lethaia*, **14**, 168.

M'COY, F. 1875. Prodromus of the palaeontology of Victoria. Dec. II. *Geological Survey of Victoria Melbourne*, 29–37, pl. 20.

MELCHIN, M. J. 1987. *Late Ordovician and early Silurian graptolites, Cape Phillips Formation, Canadian Arctic Archipeligo.* PhD Thesis, The University of Western Ontario, London, Ontario, Canada.

—— 1989. Llandovery graptolite biostratigraphy and paleobiogeography, Cape Phillips Formation, Canadian Arctic Islands. *Canadian Journal of Earth Sciences*, **26**, 1726–1746.

—— & LENZ, A. C. 1986. Uncompressed specimens of *Monograptus turriculatus* (Barrande, 1850), from Cornwallis Island, Arctic Canada. *Canadian Journal of Earth Sciences*, **23**, 579–582.

MITCHELL, C. E. 1987. Evolution and phylogenetic classification of the Diplograptacea. *Palaeontology*, **30**, 353–405.

MU, A. T. 1948. Silurian succession and graptolite-fauna of Lientan. Bulletin of the Geological Society of China, 28, 207–231, pls 1–3.

—— 1955. On *Spirograptus* Gurich. *Acta Palaeontologica Sinica*, **3**, 1–10.

—— & CHEN, X. 1962. *Sinodiversograptus multibrachiatus* gen. et sp. nov. and its developmental stages. *Acta Palaeontologica Sinica*, **10**, 143–154, pls 1, 2.

——, LI, J. J., GE, M. Y. & YIN, J. X. 1962. [*Graptolites from the Qilian Mountains.*] Geology of Qilian Shan, **4** [in Chinese].

MÜLLER, A. H. & SCHAUER, M. 1969. Über Schwebeeinrichtungen bei Diplograptidae (Graptolithina) aus dem Silur. *Freiberger Forschungshefte*, C, **245**, 5–26.

NANJING INSTITUTE OF GEOLOGY AND PALAEONTOLOGY, ACADEMIA SINICA (NIGP). 1974. [*Handbook of stratigraphy and palaeontology of southwest China*]. Science Press, Beijing [in Chinese].

NI, Y. N. 1978. [Lower Silurian Graptolites from Yichang, Western Hubei.] *Acta Palaeontologica Sinica*, **17**, 387–416, pls 1–4 [in Chinese].

NICHOLSON, H. A. 1868. On the graptolites of the Coniston Flags; with notes on the British species of the genus *Graptolites*. *Quarterly Journal of the Geological Society of London*, **24**, 521–545, pls 19, 20.

NICOL, J. 1850. Observations on the Silurian strata of the south east of Scotland. *Quarterly Journal of the Geological Society of London*, **6**, 53–65.

OBUT, A. M. 1949. [*Polevoi atlas rukovodyaashchich graptolitov verchnego silura Kirgizskoy SSR.*] Publishing House of the Kirgiz Branch of the Academy of Sciences of the USSR, Frunze [in Russian].

—— & SOBOLEVSKAYA, R. F. 1966. [Graptolitit rannego silura i Kazachstane.] *Akademii Nauk SSSR, Sibirskoe Otdelenie, Institut Geologii, Geofiziki, Ministerstvo Geologii SSSR, Nauchno-issledovatelskie Institut Geologii Arktiki*, 1–56, pls 18– [in Russian].

PACKHAM, G. H. 1962. Some diplograptids from the British Lower Silurian. *Palaeontology*, **5**, 498–526, pls 71, 72.

PARIS, F., RICKARDS, B. & SKEVINGTON, D. 1980. Les assemblages de graptolites du Llandovery dans le synclinorium du Ménez-Bélair (Massif Armoricain). *Géobios*, **13**, 153–171.

PAŠKEVIČIUS, J. 1979. [*Biostratigraphy and graptolites of the Lithuanian Silurian.*] Vilnius Mokslas Publishers [in Russian].

PERNER, J. 1897. Études sur les Graptolites de Bohême. *IIIième Partie. Monographie des Graptolites de L'Étage E. Section a.* R. Gerhard, Prague. 1–25, pls 9–13.

—— 1899. Études sur les Graptolites de Bohême. *IIIième Partie. Monographie des Graptolites de L'Étage E. Section b.* R. Gerhard, Prague. 1–24, pls 14–17.

PHILIPPOT, A. 1950. *Les Graptolites du Massif Armoricain. Étude stratigraphique et paléontologique.* Memoires de la Société Géologique et Minéralogique de Bretagne, **8**.

PORTLOCK, J. E. 1843. *Report on the geology of the county of Londonderry, and of parts of Tyrone and Fermanagh.* Milliken; Hodges & Smith; Longman, Brown, Green & Longmans, Dublin and London.

PŘIBYL, A. 1940*a*. Revise českych graptolitů rodu *Monoclimacis* Frech. Rozpravy II. *Třídy České Akademie*, **50**, no. 23, 1–19, pls 1–3.

—— 1940*b*. O. Českých zastupcích mongraptidů ze skupiny *Pristiograptus nudus*. Rozpravy II. *Třídy České Akademie*, **50**, no. 16, 1–14, pls 1, 2.

—— 1943. O několika nových graptolitech z českého a německého siluru. *Vestnik Královské České Společnosti Nauk, Třída Matematicko-Přírodovědecká*, **6**, 1–16, pls 1, 2.

—— 1945. The Middle-European monograptids of the genus *Spirograptus* Gürich. *Bulletin International de l'Académie Tchèque des Sciences*, **45**, 185–231, pls 1–11.

—— & ŠTORCH, P. 1983. Monograptus (Stimulograptus) subgen. n. (Graptolites) from the Lower Silurian of Bohemia. *Vestník ústředního ústavu Geologického*, **58**, 221–226, pls 1, 2.

RICHTER, R. 1871. Aus dem thüringischen Schiefergebirge. *Zeitschrift der Deutschen Geologischen Gesellschaft*, **23**, 231–256, pl. 5.

RICKARDS, R. B. 1963. *The Silurian strata of the Howgill Fells.* PhD Thesis, University of Hull.

—— 1965. New Silurian graptolites from the Howgill Fells. (Northern England). *Palaeontology*, **8**, 247–271, pls 29–31.

—— 1967. The Wenlock and Ludlow succession in the Howgill Fells (north-west Yorkshire and Westmoreland). *Quarterly Journal of the Geological Society of London*, **123**, 215–249, pl. 12.

—— 1968. The thecal structure of *Monoclimacis? galaensis*. *Lethaia*, **1**, 303–309.

—— 1970. *The Llandovery (Silurian) graptolites of the*

Howgill Fells, Northern England. Palaeontographical Society Monograph.

—— 1972. *Climacograptus scalaris* (Hisinger) and the subgenus *Glyptograptus (Pseudoglyptograptus)*. *Geologiska Föreningens i Stockholm Förhandlingar*, **94**, 271–280.

—— 1973. Bipolar monograptids and the Silurian genus *Diversograptus* Manck. *Paläontologische Zeitschrift*, **47**, 175–187.

—— 1976. The sequence of Silurian graptolite zones in the British Isles. *Geological Journal*, **11**, 153–188.

—— 1989. Exploitation of graptoloid cladogenesis in Silurian stratigraphy. *In*: HOLLAND, C. H. & BASSETT, M. G. (eds) *A Global Standard for the Silurian System*. National Museum of Wales, Cardiff. Geological Series, **9**, 267–274.

——, HUTT, J. E. & BERRY, W. B. N. 1977. Evolution of the Silurian and Devonian Graptoloids. *Bulletin of the British Museum (Natural History) Geology series*, **28**, 1–120, pls 1–6.

——, RIGBY, S. & HARRIS, J. H. 1990. Graptoloid biogeography: recent progress, future hopes. *In*: McKERROW, W. S. & SCOTESE, C. R. (eds). *Palaeozoic Palaeogeography and Biogeography*. Geological Society, London, Memoirs, **12**, 139–145.

RUEDEMANN, R. 1947. *Graptolites of North America*. Memoir of the Geological Society of America, **19**.

SCHAUER, M. 1967. Biostratigraphie und Taxionomie von *Rastrites* (Pterobranchiata, Graptolithina) aus dem anstehenden Silur Ostthüringens und des Vogtlandes. *Freiberger Forschungshefte*, (C), **213**, 171–199.

—— 1971. Biostratigraphie und Taxionomie der Graptolithen des tieferen Silurs unter besonderer Berücksichtigung der tektonischen Deformation. *Freiberger Forschungschefte*, (C), **373**, Paläontologie, 1–185.

SENNIKOV, N. V. 1976. [*Graptolity i stratigrafiya nizhnego silura Gornogo Altaya*.] Nauka, Moscow [in Russian].

SHERWIN, L. 1974. Llandovery graptolites from the Forbes District, New South Wales. *In*: RICKARDS, R. B., JACKSON, D. E. & HUGHES, C. P. (eds) *Graptolite studies in honour of O. M. B. Bulman*. Special Papers in Palaeontology, **13**, 149–175, pls 10–12.

ŠTORCH, P. & LOYDELL, D. K. 1992. Graptolites of the *Rastrites linnaei* Group from the European Llandovery (lower Silurian). *Neues Jahrbuch für Geologie und Paläontologie*, **184**, 63–86.

—— & PAŠAVA, J. 1989. Stratigraphy, chemistry and origin of the Lower Silurian black graptolitic shales of the Prague Basin (Barrandian, Bohemia). *Vestník ústředního ústavu Geologického*, **64**, 143–162.

STRACHAN, I. 1952. On the development of *Diversograptus* Manck. *Geological Magazine*, **89**, 365–368.

—— 1971. *A synoptic supplement to "A monograph of British graptolites by Miss G. L. Elles and Miss E. M. R. Wood"*. Palaeontographical Society Monograph.

—— 1972. Distribution of *Rastrites* of the *Linnaei* group. Proceedings of the IPU, XXIII International Geological Congress, 431–436.

SUDBURY, M. 1958. Triangulate monograptids from the *Monograptus gregarius* Zone (Lower Llandovery) of the Rheidol Gorge (Cardiganshire). *Philosophical Transactions of the Royal Society of London*, **B241**, 485–554, pls 19–23.

SUESS, E. 1851. Über böhmische Graptolithen. *Naturwissenschaftliche Abhandlungen*, **4**, (4), 87–134, pls 7–9.

TEMPLE, J. T. 1988. Biostratigraphical correlation and the stages of the Llandovery. *Journal of the Geological Society, London*, **145**, 875–879.

THOMAS, D. E. 1960. The zonal distribution of Australian graptolites. *Journal & Proceedings of the Royal Society of New South Wales*, **94**, 1–58.

TOGHILL, P. & STRACHAN, I. 1970. The graptolite fauna of Grieston Quarry, near Innerleithen, Peeblesshire. *Palaeontology*, **13**, 511–521, pls 103–105.

TÖRNQUIST, S. L. 1881. Om några graptolitarter från Dalarne. *Geologiska Föreningens i Stockholm Förhandlingar*, **5**, 434–445, pl. 17.

—— 1892. Undersökningar öfver Siljansområdets graptoliter. II. *Geologiska Föreningens i Stockholm Förhandlingar*, **28**, 1–47, pls 1–3.

TULLBERG, S. A. 1882. On the graptolites described by Hisinger and the older Swedish authors. *Bihang Till KoniigaSvenska Vetenskaps-Akademiens Handlingar (Stockholm)*, **6**, (13), 1–23, pls 1–3.

WANG, X. F. 1978. [A restudy on the graptolites and the age of the Wentoushan Formation from Liantan, Kwangtung.] *Acta Geologica Sinica*, **4**, 303–317, pls 1–4 [in Chinese].

WATKINS, R. & BERRY, W. B. N. 1977. Ecology of a Late Silurian fauna of graptolites and associated organisms. *Lethaia*, **10**, 267–286.

WHITTARD, W. F. 1961. *Lexique Stratigraphique International, 1, Europe, fascicule 3a, Angleterre, Pays de Galles, Écosse, fascicule 3aV, Silurien*. Centre National de la Recherche Scientifique.

WILSON, D. W. R. 1954. *The stratigraphy and palaeontology of the Valentian rocks of Cautley (Yorks W. R.)*. PhD Thesis, University of Birmingham.

WOOD, E. M. R. 1906. The Tarannon Series of Tarannon. *Quarterly Journal of the Geological Society of London*, **62**, 644–701, 2 pls.

YIN, T. H. 1937. Brief description of the Ordovician and Silurian fossils from Shihtien. *Bulletin of the Geological Society of China*, **16**, 281–302, pls 1, 2.

ZALASIEWICZ, J. 1990. Silurian graptolite biostratigraphy in the Welsh Basin. *Journal of the Geological Society, London*, **147**, 619–622.

Towards a carbon isotope stratigraphy of the Cambrian System: potential of the Great Basin succession

M. D. BRASIER

Department of Earth Sciences, University of Oxford, Parks Road, Oxford, OX1 3PR, UK

Abstract: The Cambrian spans a time of major biosphere change, during which the major invertebrate groups appeared and radiated. Recent discoveries of major carbon isotope excursions in carbonate rocks around the world have stimulated a working group on 'Late Precambrian–Cambrian Event Stratigraphy' to explore their potential for correlation (IGCP Project 303). Bioevents, sequence boundaries and trace element markers may also be used for calibration. A major impediment to this objective has been the endemicity and diachroneity of many polymeroid trilobites and early skeletal microfossil taxa associated with shallow water carbonates. The use of deeper-water carbonates, bearing more widely-distributed agnostoids and polymeroids should, however, provide an opportunity to test independently the correlative value of carbon isotope excursions on a global scale. Studies on the Cambrian of the Great Basin, western USA, show that such facies can bear similar $\delta^{13}C$ signatures to contemporaneous shallow-water carbonates with more endemic trilobites. Hence 'agnostoid facies' do have the potential to provide a global and integrated carbon isotope and fossil stratigraphy. A test case is provided by a major positive carbon excursion in the late Dresbachian (or Steptoean) Stage, coincident with shoaling and emergence at the top of the Sauk II sequence across the Laurentian craton. If global, this distinctive marker should be traceable into other Cambrian carbonate successions since it lies between two of the most widely correlatable biohorizons: above the last occurrence of *Glyptagnostus reticulatus* and below the first appearance of *Irvingella* spp.

This paper examines the potential of carbon isotope stratigraphy for the international correlation of the Cambrian System, which lasted from about 540 to 500 Ma BP (e.g. Odin *et al.* 1983; Compston *et al.* 1990). Invertebrate groups made their first appearance and radiated through this interval, although fossils tend to be relatively scarce and sporadic. Trilobites provide the traditional means of correlation in Cambrian strata, having reached the acme of their diversity at that time (Robison 1988). Trilobite assemblages, however, tend to be markedly endemic with relatively few polymeroid taxa crossing the oceanic waters between major cratons (Palmer 1973, 1974; Cowie 1971). Only the mesopelagic? agnostoids were sufficiently pandemic to provide global markers and these were largely restricted to facies along the continental margins or otherwise open to oceanic influence (Robison 1972; Robison *et al.* 1977). International correlation of Cambrian rocks has therefore proceeded at a slow pace and the system is still without internationally agreed subdivisions of series or stage rank.

Trilobites are absent from the earliest Cambrian and older strata. International correlation of Tommotian and Nemakit–Daldynian rocks has therefore been attempted by means of 'small shelly fossils', which include a variety of micromolluscs, cap-shaped shells, tubular shells, sclerites of multi-element skeletons and cono-dont-like spines (Rozanov *et al.* 1969; Matthews & Missarzhevsky 1975). For many years it was assumed that sub-trilobitic assemblages of 'Tommotian type' were of Tommotian age but this has proved too simplistic. Some 'Tommotian' small shelly fossil taxa range high into equivalents of the succeeding Atdabanian and Botomian stages (e.g. Landing *et al.* 1980) while the first trilobites did not appear until middle to later parts of the Atdabanian stage in Avalonia and eastern Gondwana (e.g. Brasier 1989*a*).

Over the last few years, attempts have been made to improve the resolution of small shelly fossils by identifying evolutionary successions of taxa (e.g. Missarzhevsky 1982, 1983; Brasier 1989*a*). This approach allows the recognition of at least five successive assemblages through the latest Precambrian-early Cambrian interval (e.g. Brasier 1989*a*). A further check on the veracity of successive assemblages is provided by a study of first appearance datum (FAD) and last occurrence datum (LOD) of earliest skeletal fossil taxa. Where the FADs appear in the same order in two or more successions then diachroneity of taxa is unproven and the datum points can be used as a guide to provide points of correlation (*see* Scott 1985 for discussion of this technique). Both English and Siberian boundary sections, for example, share nine successive points of 'homotaxial' correlation, involving fifteen widely distributed taxa (Brasier 1986,

From HAILWOOD, E. A. & KIDD, R. B. (eds), *High Resolution Stratigraphy*
Geological Society Special Publication, No. 70, pp. 341–350.

1989*b*). No less than 32 taxa can be used to correlate the boundary interval along the Asiatic margins of Gondwana from China to Iran, with about seven points of homotaxial correlation (Brasier 1989*c*; Brasier *et al*. 1990). In these Gondwana sections, correlations can be calibrated against the position of sequence boundaries and time-specific facies such as phosphorites and sponge cherts.

In all known cases, however, there were strong ecological and taphonomic controls on the first appearances of skeletal fossils. Each FAD therefore records a local to regional migration, a shift in the environment or a change in preservational conditions. In Avalonia, for example, the first appearance of diverse assemblages of shelly fossils was brought about by the spread of shallow water carbonate facies (e.g. Landing & Benus 1988). In China and India, ephemeral phosphogenic conditions favoured the preservation of poorly biomineralized taxa, whose first appearance is therefore likely to be diachronous (Brasier 1990*a*; Brasier *et al*. 1990). Early skeletal fossils may therefore provide a broad framework of assemblages, and FADs may highlight useful bioevents for local to regional correlation but neither early skeletal fossils nor polymeroid trilobites can yet provide a high resolution stratigraphy of the Cambrian on an international scale.

The problems of Cambrian chronostratigraphy may be transformed, however, by the discovery of major carbon isotope excursions in carbonate rocks of Late Precambrian and Cambrian age (e.g. Tucker 1986; Magaritz *et al*. 1986; Brasier *et al*. 1990). A working group has recently been established on 'Late Precambrian and Cambrian Event Stratigraphy' to explore the potential of such non-biostratigraphic techniques for regional to global correlation (IGCP Project 303). This aims to develop an integrated and refined biostratigraphy, chemostratigraphy, even stratigraphy and palaeomagnetic scale for the latest terminal Proterozoic and Cambrian Systems. It will involve the use of FADs and LODs as discussed above, together with analysis of sequence boundaries (e.g. Palmer 1981*a*, *b*) and sea-level changes (e.g. Bond *et al*. 1988), trace element markers (e.g. Hsu *et al*. 1985; Xu *et al*. 1989), palaeomagnetic reversals (e.g. Kirschvink & Rozanov 1984), strontium isotope stratigraphy (e.g. Donnelly *et al*. 1990) and, of course, carbon isotope stratigraphy.

Carbon isotopes

This technique makes use of the natural variation between stable isotopes of carbon. The lighter ^{12}C isotope is preferentially taken up by primary producers during photosynthesis, leaving the hydrosphere–atmosphere relatively depleted in ^{12}C and enriched in ^{13}C. Variations in the level of primary productivity (owing to ocean surface fertility) or in the level of burial of organic matter (e.g. owing to anoxia or sedimentation rates) therefore lead to temporal fluctuations in the ratio of these two isotopes. Temporal and spatial fluctuations can then be reconstructed by mass spectrometry of carbon preserved within carbonate rocks or organic matter.

The $\delta^{13}C$ of carbon in carbonate rocks derives from a variety of pathways operating on different scales. On the local scale, light carbon is released during the decomposition of organic matter in sediment and soils (especially from methanogenic and sulphate-reducing bacteria) so that vadose cements, burial cements and ferroan calcite or dolomite concretions can yield relatively negative values of $\delta^{13}C$ relative to marine carbonate (e.g. Hudson 1977; Tucker & Wright 1990). Variation is also likely between regions of high productivity and carbon burial (e.g. at lower latitudes, in areas of upwelling or nutrient run-off, and in restricted basins) and regions of lower productivity and carbon burial. On the cratonic to global scale, these fluctuations will be superimposed upon temporal variations in the global budget brought about by changes in the area of the highly productive shelf, of anoxic water masses, igneous activity and climate (e.g. Broecker 1982; Shackleton & Pisias 1985; Schlanger *et al*. 1987).

The main challenge to workers within IGCP Project 303 will be to separate the components of local, regional and global signals within the $\delta^{13}C$ profiles of each succession. It is suggested here that such studies focus on carbonate cratons which meet a variety of criteria. The Laurentian cration is cited below as an example.

(1) Carbonate rocks should range through most of the Cambrian. This will allow analysis of long term trends.

(2) Carbonate facies should range from shallow platform to deep sea apron deposits (e.g. Cook & Taylor 1975; Taylor & Palmer 1981). This will allow for analysis of onshore-offshore trends.

(3) Sedimentology and sequence stratigraphy should be reasonably well constrained. On the Laurentian craton, for example, Cambrian strata are packaged into Grand Cycles that provide markers across much of the craton (e.g. Aitken 1966, 1981; Palmer 1981*a*; Bond *et al*. 1988; James *et al*. 1989). This allows for consideration of the effects of changing sea level.

(4) A relatively constant (especially low) latitudinal position should be maintained throughout most of the Cambrian, as was the case for the Laurentian craton (e.g. Scotese & McKerrow 1990; Mckerrow *et al*. 1992). This should reduce the amount of 'noise' due to drifting through climatic zones.

(5) Different margins should provide contrasting sedimentological histories, as with the eastern and western margins of Laurentia (e.g. Holland 1971; Palmer 1981*a*). This should assist understanding of regional and latitudinal effects upon stable isotope stratigraphy.

Below is given an account of preliminary results from the Cambrian of the Great Basin, western United States. Here, the relatively complete Cambrian succession allows comparison between onshore carbonates with endemic polymeroids and offshore carbonates with more pandemic polymeroids and agnostoids.

Cambrian of the Great Basin

Cambrian rocks are well-exposed in the Great Basin of the western United States, from the White-Inyo Mountains of California, through Nevada and Utah to Idaho (Palmer 1971; Taylor & Palmer 1981). Three main sedimentary belts can be distinguished. An 'inner detrital belt' in the east of the area comprises light-coloured shales, siltstones and sandstones with some carbonates. A 'middle carbonate belt' of shallow-water limestones and dolomites predominated across the centre of the region. An 'outer detrital belt' of dark coloured shales and thin-bedded calcareous siltstones and limestones lay to the west. These belts tended to move eastwards with transgression of the craton (Fig. 1). It is therefore possible to study the chemostratigraphy of a range of carbonate environments, from onshore to offshore.

To this picture must be added the phenomenon of Grand Cycles (Aitken 1966, 1981) in which a lower unit, mainly of clastics, is succeeded by an upper unit, predominantly of carbonates. The top of the latter tends to mark an abrupt change in depositional regime. Twelve such boundaries have been traced between Grand Cycles along the Rocky Mountains (Palmer 1981*a*), of which Grand Cycles A, B and C have been named from the Lower Cambrian (Fritz 1975) and the remaining nine are given corresponding letters in Fig. 1. In the Middle Cambrian, several Grand Cycle tops can be traced through the Arctic into the Appalachians and western Newfoundland, suggesting eustatic controls on sea level (Bond *et al*. 1988; James *et al*. 1989).

Maximum sea level on the Laurentian craton was probably reached during the *Cedaria* and *Aphelaspis* Zones (Bond *et al*. 1988). This was followed by major regression and a widespread hiatus between the Dresbachian and Franconian Stages (i.e the Steptoean Stage of Ludvigsen & Westrop 1985, spanning the *Aphelaspis* to lower *Elvinia* Zones). This break can be traced across a wide area, as follows: in the eastern part of the Great Basin, in the Rocky Mountains from Montana to Arizona, in the mountains of northwestern and southwestern Canada, in the midcontinent region of Texas, Missouri and the upper Mississippi Valley, in western Newfoundland and in northwestern Greenland (Palmer 1981*b*; Westrop 1986). Only in the outer shelf sequences of the Great Basin (sampled here) and in British Columbia was there a continuous depositional record across this interval. This major regression marks the boundary between the so-called Sauk II and Sauk III sequences and its wide extent suggests a eustatic event (Palmer 1981*b*; James *et al*. 1989).

A similar regression and hiatus (the 'Hawke Bay Event') intervened between the Lower and Middle Cambrian sequences of Sauk I and Sauk II (Palmer 1981*b*; Palmer & James 1980). This can be traced from northern British Columbia, through Alaska and northeast Greenland and down through the Appalachians, but it is not present in the Great Basin, where deposition appears to have been continuous (Palmer 1981*b*). A comparable break can be traced in the platform succession of Comley, England (Brasier 1989*a*) and on the Baltic Platform (Bergström 1981) while widespread regression and evaporites developed at about this interval in Siberia (Zhuravleva *et al*. 1990), North China, Tarim, South China (Xiang 1981) and Australia (Shergold & Brasier 1986). A eustatic fall in sea level is also likely for this event.

Carbon isotopes of the Great Basin

Stable isotopes of carbon and oxygen have been obtained from whole-rock analyses of 34 selected samples of carbonate (Fig. 1), using the VG Isogas mass spectrometer in the Oxford Age Labs.

A plot of carbon against oxygen shows little of the covariance that might be expected from meteoric diagenesis (Fig. 2). The wide spread of $\delta^{18}O$ results does, however, indicate diagenetic modification of oxygen isotopes. For example, dolomitic samples display heavier values than calcitic samples (Fig. 2) which may be due to ^{18}O enrichment of evaporitic brines or to fractionation during the precipitation of dolomite

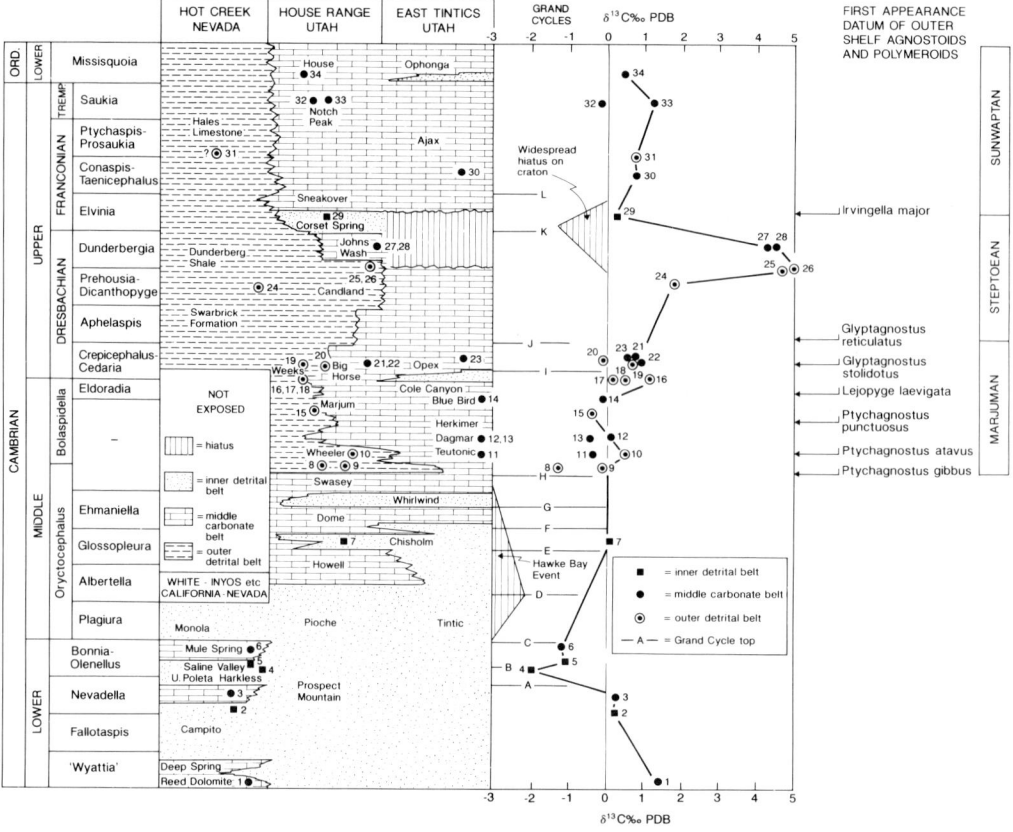

Fig. 1. Cambrian succession of the Great Basin, showing carbon isotope sample horizons and results (see text for discussion). Lithostratigraphy and biostratigraphy essentially after Taylor & Palmer (1981). Markers A to L shown indicate the tops of Grand Cycle sequences described by Fritz (1975) and Palmer (1981a). The widespread hiatuses marked by the 'Hawke Bay Event' and the un-named late Dresbachian or Steptoean event (shown by vertical ruling) separate the major sequences of Sauk, I, II and III (Palmer 1981b). Further details of the sample localities are given in the Appendix. At right are shown first appearance datum points of agnostoids and polymeroids known from outer shelf facies of the Laurentian craton (including the Great Basin) and useful for global correlation. These are calibrated against the stage names of Ludvigsen & Westrop (1985).

relative to calcite (e.g. Tucker & Wright, 1990, p. 384). The more negative values of $\delta^{18}O$ (e.g. from −7.0 to −17.0) suggest thermal fractionation of oxygen isotopes during burial. This phenomenon indicates that $\delta^{18}O$ signals are less pristine than those from the Siberian Platform (e.g. Magaritz et al. 1986) and may limit the scope of isotope studies in the Great Basin.

Carbon isotopes in carbonate rocks appear less affected by dolomitization and burial diagenesis. None of the carbonates studied from the Great Basin proved ferroan or pyritic (i.e. there is little evidence for methanogenic or sulphate-reducing diagenesis), and they plot close to the

range of modern marine carbonates and dolomites (e.g. Hudson 1977; Tucker & Wright 1990). Even so, a cluster of samples (25 to 28) appear anomalous, with carbon isotope values raised by about 4‰ (Fig. 2). As will be discussed below, these provide evidence for a major carbon isotope excursion in the Late Cambrian.

The late Dresbachian (or Steptoean) excursion

When examined in stratigraphic context (Fig. 1), the anomalous, positive $\delta^{13}C$ excursion noted above is seen to lie within the late Dresbachian

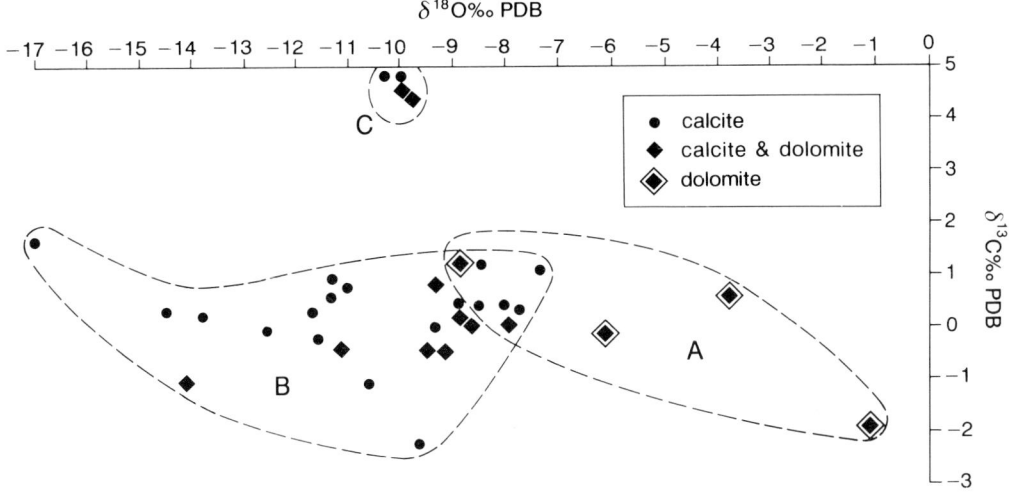

Fig. 2. Carbon-oxygen isotopic compositions of samples from the Cambrian of the Great Basin (see Fig. 1 for stratigraphic setting and Appendix for details). (**A**) Cluster of dolomitic samples. (**B**) Cluster of majority of calcitic samples. (**C**) Cluster of calcitic samples 25–28, associated with the late Dresbachian carbon isotope excursion.

(or Steptoean) Stage, though the shape of the excursion is still poorly constrained. These results were obtained from analyses of two contrasting lithofacies: shelly wackestones of the upper Candland Shale and oolitic grainstones of the Johns Wash Limestone, House Range Utah. The Candland Shale reflects conditions on the outer flank of the middle carbonate belt/inner margin of the outer detrital belt. The rock contains abundant trilobite remains, phosphatic brachiopods and an unusual abundance of phosphatic peloids. It is interesting to note that the diversity of both trilobites and brachiopods reached a peak in the Great Basin during the late *Dunderbergia* Zone (Rowell & Brady 1976).

The overlying Johns Wash Limestone records the seaward progradation of oolite shoals and hypersaline lagoons, whose absence of fauna may relate to extremes of temperature and salinity (Rowell & Brady 1976). The succeeding Corset Spring Shales mark an influx of clastics from the inner detrital belt, coincident with a rapid decline in the diversity of trilobites and brachiopods prior to extinctions at the end of the 'Pterocephaliid Biomere' (Rowell & Brady 1976).

Further material should be collected to study whether a similar excursion took place during the 'Hawke Bay Event' across the Lower–Middle Cambrian boundary.

Onshore–offshore trends in carbon isotopes

Global chronostratigraphy based on carbon isotope fluctuations should, ideally, be based upon the surface water signal, such as recorded by certain Cenozoic planktonic foraminifera (e.g. Williams *et al.* 1988). It might seem sufficient, therefore, to confine Cambrian analyses to the most shallow-water carbonates or biominerals. However, analysis of onshore-offshore trends in carbon isotopes is desirable for a variety of reasons. Knowledge about onshore-offshore gradients (or better still, surface to bottom water gradients) is needed to characterize the patterns of surface-water productivity implicit in carbon isotope shifts (e.g. Shackleton & Pisias 1985). Equally important is the fact that offshore deposits contain more pandemic faunas such as agnosoid arthropods, allowing more independent testing of correlations between cratons.

Several factors must make one wary, however, of carbon isotope stratigraphy in offshore strata of Cambrian age. The supply of biological, planktonic carbonate at this time was negligible and calcareous sediments must have derived their carbonate from three other major sources: cements, autochthonous biominerals of benthic invertebrates, or allochthonous carbonate grains (both biogenic and inorganic).

Cement-dominated carbonates are typical of

the deeper-water, clastic-dominated facies of the Cambrian in southeastern Newfoundland (Brasier et al. 1992). Their relatively high $\delta^{13}C$ and ferroan calcite composition suggest formation during the release of respiratory CO_2 from bacterial degradation of organic matter in the sediment. As such, the pore water or lower watermass signal strongly overprints the regional to global, surface-water signal, and they are of limited value for anything but local to regional basin analysis. A similar problem may attend the negative $\delta^{13}C$ of dark, ferroan dolomites of the Qiongzhusi Formation in South China, which appear to have been affected by the anoxic degradation of organic matter (Brasier et al. 1990). The widespread occurrence of black shales and phosphorites on Cambrian shelves (e.g. Cook & Shergold 1986) confirms that anoxia was widespread and its local effects must therefore be eliminated in studies that go beyond the confines of local watermass history.

Biomineral detritus of autochthonous invertebrates dominates the carbonates of deeper waters in some non-ferroan calcite nodules of southeastern Newfoundland and England (Brasier et al. 1992). Here the carbonate appears to have formed from carbonate skeletal fragments (mainly trilobites) which may be diagenetically altered. Stable isotopic values may therefore approach both those of invertebrate shells and shallow water limestones but care must be taken to assess the variable input from microbial and burial cements. This may be possible by comparing the $\delta^{13}C$ of biominerals and matrix; in Avalonian samples, for example, the matrix is typically lighter than the shells, suggesting that biominerals were cemented by lighter, diagenetic cements.

Allochthonous-dominated carbonates of deeper waters may originate from grains and muds transported from adjacent shallow water platforms by storms and turbidity currents, like modern periplatform ooze. The carbon isotopes of such carbonates should, in theory, be close to those of their shallow water precursors, modified perhaps by microbial or burial cements.

Carbonates of the Great Basin examined here show little sedimentological evidence for anoxia in bottom waters. Allochthonous sediment, and to a lesser extent autochthonous or pelagic trilobite biomineral detritus, are suggested to have provided the principal source of carbonate to the outer detrital belt and ocean-based margins (e.g Cook et al. in Taylor & Palmer 1981). This interpretation is supported by the similarity of $\delta^{13}C$ signatures from middle carbonate and outer detrital belt facies (fig. 1). It implies that deeper-water facies, with more widely distrib-

uted agnostoids and polymeroids, can carry the signature of shallow-water carbonates and, by implication, the surface water $\delta^{13}C$ signal. Hence suitable 'agnostoid facies' may yet provide a standard scale for international correlation of Cambrian strata, analogous to the deep-sea record and planktonic biostratigraphy of Mesozoic and younger strata.

Discussion

Positive carbon isotope excursions have been found at several levels in late Proterozoic and early Cambrian carbonate rocks (e.g. Brasier et al. 1990). The latter authors noted the coincidence between the Dahai carbon isotope maximum near the Precambrian–Cambrian boundary and an interval of shoaling, evaporite formation, phosphorite deposition and emergence across much of southern and central Asia. A broad coincidence between patterns of phosphate biomineralization, phosphogenesis and carbon isotope fluctuations suggests that major changes in sea level may have brought about changes in nutrient supply and changes in primary productivity (Brasier 1990b; Brasier 1991).

This study indicates the presence of another such excursion, in the late Dresbachian of the Great Basin. It also appears to have coincided with an interval of shoreline progradation and hiatus over much of the Laurentian craton. Offshore strata contain conspicuous amounts of phosphatic pellets and inarticulate brachiopods. Thus a similar explanation may apply: major still-stand or regression associated with increased nutrient flux and mixing, leading to raised levels of surface-water primary productivity and carbon isotope maxima.

Whatever the interpretation of such perturbations, the potential for onshore-offshore correlation of carbon isotope excursions looks promising. A better understanding of both carbon and nutrient cycles may also arise from such studies.

Conclusion

The following general conclusions can be made from this study.

(1) Biostratigraphic and chronostratigraphic correlation of the Cambrian System is hampered by strong ecological and taphonomic controls on the appearance and preservation of faunas and by restricted geographic distributions of many taxa.

(2) Carbon isotope stratigraphy may provide a means for higher resolution stratigraphy of the Cambrian System on a regional to global scale.

Before this is possible, however, the effects of local and regional variations in $\delta^{13}C$ must be analysed.

(3) Researches should therefore focus upon large carbonate platforms with good biostratigraphic and sedimentological control, enabling comparison between the carbon isotopic records of contrasting facies (e.g. middle carbonate and outer detrital belts) and different regions (e.g. Great Basin, Canadian Rocky Mountains, northeastern Greenland, Appalachians). Similar contrasts should be studied on the Baltic, Yangtze, Siberian, Australian and Afro-Arabian cratons.

(4) Preliminary results from the Great Basin Cambrian indicate that $\delta^{18}O$ signatures appear to have been reset by burial diagenesis but $\delta^{13}C$ signatures were little affected. Most importantly, carbonates of deeper-water facies preserve similar $\delta^{13}C$ signals to those of contemporaneous deposits in shallow waters. This may be because the carbonates were primarily allochthonous (grains, biominerals) transported from adjacent shallow water carbonate systems by storms and turbidity currents.

(5) Such deeper-water carbonates contain a more complete record of pandemic agnostoid and polymeroid zones. These have the potential to test the global biostratigraphic calibration of stable isotope excursions in a manner more precise and objective than can currently be demonstrated using early Cambrian trilobites and shelly fossils.

(6) A major carbon isotope excursion has been discovered in the late Dresbachian (i.e. Steptoean) Stage, coincident with a suggested eustatic fall in sea level. This compares in scale and setting with excursions close to the Precambrian–Cambrian boundary and provides a test case, in trilobitic strata, for correlation beyond the Laurentian craton. Further work should aim to test its correlation onto the eastern margin of Laurentia (e.g. Appalachians) and into Scandinavia, China and Queensland, Australia.

The material from the Great Basin was collected under the guidance of A. R. Palmer (GSA, Boulder, Colorado), D. Robison (University of Kansas), H. Cook and M. Taylor (USGS, Denver). The stable isotopes were processed by R. Corfield and J. Cartlidge, and their assistance is gratefully acknowledged. A. R. Palmer, D. Robison, A. Rushton and M. Tucker are thanked for their comments on the manuscript. This paper is a contribution to IGCP Project 303 on Precambrian–Cambrian Event Stratigraphy.

Appendix: Great Basin sample data

(1) Top of the Reed Dolomite Formation, with *Wyattia* sp., from Schulman Grove, White-Inyo Mountains, California (1981 excursion, stop 2.3; sample RD).

(2) Upper Montenegro Member of the Campito Formation, with "*Ethmophyllum*" sp., from Deep Spring Valley, White-Inyo Mountains, California (1981, stop 2.9; sample MM).

(3) Lower Poleta Formation, Stewart's Mill, Esmeralda County, Nevada (1981, stop 3.3; sample 3.3).

(4) Upper Harkless Formation, with *Salterella* sp., from Cucomungo Canyon, California (1981, stop 3.2; sample UHF).

(5) Echo Shale Member of Carrara Formation, with *Bristolia* sp., from Titanothere Canyon, California (1981, stop 1.2; sample ESM). Correlated with upper Saline Valley Formation (Taylor & Palmer 1981).

(6) Gold Ace Limestone Member of Carrara Formation, from Titanothere Canyon, California (1981, stop 1.2; sample GALM). Correlated with Mule Spring Limestone (Taylor & Palmer 1981).

(7) Spence Shale, Twomile Canyon, Malad Range, Idaho (1981, stop 9A.1; sample 9A.1).

(8) Lower Wheeler Shale, *Oryctocephalus* Zone, Marjum Canyon, House Range, Utah (1979, sample SL1).

(9) Lower Wheeler Shale, with *Peronopsis* spp. and *Ptychagnostus* spp., Drum Mountains, Utah (1981, stop 6A.1; sample WSa).

(10) As (9) with silicified *Peronopsis* sp. (sample WSb).

(11) Teutonic Limestone, East Tintic, Utah (1979, sample TL).

(12) Microbial carbonate from Dagmar Dolomite, East Tintic, Utah (1979, sample DLA).

(13) Fragmental algal carbonate from Dagmar Dolomite, East Tintic, Utah (1979, sample DLB).

(14) Bluebird Dolomite, East Tintic, Utah (1979, sample BD).

(15) Marjum Limestone, Marjum Canyon, Utah (1979, sample MLF).

(16) Weeks Limestone, with phosphatic brachiopods and *Lejopyge calva*, from Marjum Canyon, House Range, Utah (1981, stop 5.4; sample WKL).

(17) As (16) (sample WKa, from micritic matrix of following sample).

(18) As (16) (sample WKb, from calcitic skeleton of *Lejopyge calva*).

(19) Weeks Limestone with *Tricrepicephalus* sp., from *Cedaria* Zone of Marjum Pass, House Range, Utah (1981, stop 5.4; sample CZ).

(20) Basal Big Horse Limestone Member of Orr Formation, with trilobite hash facies, Little

Horse Canyon, House Range, Utah (1979, sample BHM).

(21) Big Horse Limestone Member, oncolitic grainstone facies, from Little Horse Canyon, House Range, Utah (1979, sample BH).

(22) As (21), thrombolite facies with *Epiphyton* sp. and *Renalcis sp.* (1979, sample ER).

(23) Opex Formation, with oolitic dolomite facies, East Tintic, Utah (1979, sample OF).

(24) Basal Dunderberg Shale, Tybo Canyon, Hot Creek Range, Nevada (1981, stop 4.1; sample 4.1).

(25) Candland Shale, Orr Ridge, House Range, Utah (1979, sample CS1).

(26) As (25) (sample CS2).

(27) Johns Wash Limestone, oolitic grainstone facies, Orr Ridge, House Range, Utah (1979, sample JWL1).

(28) As (27) (sample JWL2).

(29) Corset Spring Member, Orr Formation, Orr Ridge, House Range Utah (1979, sample EZ).

(30) Ajax Fomation, dolomitic facies, East Tintic, Utah (1979, sample AL). Level within formation uncertain.

(31) Hales Limestone, Tybo Canyon, Hot Creek Range Nevada (1981, stop 4.2; sample 4.2).

(32) Whipple Cave Formation, Sawmill Canyon, Egan Range, Nevada (1981, stop 4.3; sample 4.3).

(33) Uppermost Cambrian of Notch Peak Formation, Lava Dam, House Range, Utah (1981, stop 6B.2; sample UC).

(34) Lower most Ordovician of Notch Peak Formation, ditto (sample 6.1).

References

AITKEN, J. D. 1966. Middle Cambrian to Middle Ordovician cyclic sedimentation, southern Rocky Mountains of Alberta. *Bulletin of Canadian Petroleum Geology*, **14**, 405–441.

—— 1981. Generalizations about Grand Cycles. *Open-File Report of the United States Department of the Interior Geological Survey*, **81–743**, 8–14.

BERGSTRÖM, J. 1981. Lower Cambrian shelly faunas and biostratigraphy in Scandinavia. *Open-File Report of the United States Department of the Interior Geological Survey*, **81–743**, 22–25.

BOND, G. C., KOMINZ, M. A. & GROTZINGER, J. P. 1988. Cambro-Ordovician eustasy: evidence from geophysical modelling of subsidence in Cordilleran and Appalachian passive margins. *In*: KLEINSPEHN, K. L. & PAOLA, C. (eds) *New Perspectives in Basin Analysis*. Springer-Verlag, New York, 129–160.

BRASIER, M. D. 1986. The succession of small shelly fossils (especially conoidal microfossils) from English Precambrian-Cambrian boundary beds. *Geological Magazine*, **123**, 327–356.

—— 1989a. Towards a biostratigraphy of the earliest skeletal biotas. *In*: COWIE, J. W. & BRASIER, M. D. (eds) *The Precambrian–Cambrian Boundary*. Oxford Monographs in Geology and Geophysics, **12**. Clarendon Press, Oxford, 117–165.

—— 1989b. Sections in England and their correlation. *In*: COWIE, J. W. & BRASIER, M. D. (eds) *The Precambrian–Cambrian Boundary*. Oxford Monographs in Geology and Geophysics, **12**. Clarendon Press, Oxford, 82–104.

—— 1989c. China and the Palaeotethyan belt (India, Pakistan, Iran, Kazakhstan and Mongolia). *In*: COWIE, J. W. & BRASIER, M. D. (eds) *The Precambrian-Cambrian Boundary*. Oxford Monographs in Geology and Geophysics, **12**. Clarendon Press, Oxford, 40–74.

—— 1990a. Phosphogenic events and skeletal preservation across the Precambrian-Cambrian boundary interval. *In*: NOTHOLT, A. J.G. & JARVIS, I. (eds) *Phosphorite Research and Development*. Geological Society, London, Special Publication, **52**, 289–303.

—— 1990b. Nutrients in the early Cambrian. *Nature*, **347**, 521–522.

—— 1992 Nutrient flux and the evolutionary explosion across the Precambrian–Cambrian boundary interval. *Historical Biology*, **5**, 85–93.

——, ANDERSON, M. M. & CORFIELD, R. M. 1992. Oxygen- and carbon-isotope stratigraphy of early Cambrian carbonates in southeastern Newfoundland and England *Geological Magazine*, **129**, 265–279.

——, MAGARITZ, M., CORFIELD, R., LUO, H., WU, X., OUYANG, L., JIANG, Z., HAMDI, B., HE, T. & FRASER, A. G. 1990. The carbon- and oxygen-isotope record of the Precambrian-Cambrian boundary interval in China and Iran and their correlation. *Geological Magazine*, **127**, 319–332.

BROECKER, W. S. 1982. Ocean chemistry during glacial time. *Geochimica et Cosmochimica Acta*, **46**, 1689–1705.

COMPSTON, W., WILLIAMS, I. S., KIRSCHVINK, J. & ZHANG, Z. 1990. Zircon-U-Pb ages relevant to the Cambrian numerical time scale. *Geological Society of Australia, ICOG 7 Abstracts*, 27.

COOK, H. E. & TAYLOR, M. E. 1975. Early Paleozoic continental margin sedimentation, trilobite biofacies and the thermocline, western United States. *Geology*, **3**, 559–562.

COOK, P. J. & SHERGOLD, J. H. (eds) 1986. *Proterozoic and Cambrian Phosphorites*. Cambridge University Press, Cambridge.

COWIE, J. W. 1971. Lower Cambrian faunal provinces. *Geological Journal Special Issue*, **4**, 31–46.

DONNELLY, T. H., SHERGOLD, J. H. & SOUTHGATE, P. N. & BARNES, C. J. 1990. Events leading to global phosphogenesis around the Proterozoic/Cambrian transition. *In*: NOTHOLT, A. J. G. & JARVIS, I. (eds) *Phosphorite Research and Development*. Geological Society, London, Special Publications, **52**, 273–287.

FRITZ, W. H. 1975. Broad correlations of some lower

and middle Cambrian strata in the North American Cordillera. *Paper of the Geological Survey of Canada,* **75–1**, part A, 533–540.

HOLLAND, C. H. (ed.) 1971. *Cambrian of the New World.* Wiley-Interscience, London.

HSÜ, K. J., OBERHANSLI, H., GAO, J. Y., SUN, S., CHEN, H. & KRAHENBÜHL, U. 1985. 'Strangelove' ocean before the Cambrian explosion. *Nature,* **316**, 809–811.

HUDSON, J. D. 1977. Stable isotopes and limestone lithification. *Journal of the Geological Society, London,* **133**, 637–660.

JAMES, N. P., STEVENS, R. K., BARNES, C. R. & KNIGHT, I. 1989. Evolution of a lower Paleozoic contintental-margin carbonate platform, north-western Canadian Appalachians. *SEPM Special Publication,* **44**, 123–146.

KIRSCHVINK, J. L. & ROZANOV, A. YU. 1984. Magnetostratigraphy of Lower Cambrian strata from the Siberian Platform: palaeomagnetic pole and preliminary time-scale. *Geological Magazine,* **121**, 189–203.

LANDING, E., NOWLAND, G. S. & FLETCHER, T. P. 1980. A microfauna associated with early Cambrian trilobites of the *Callavia* Zone, northern Antigonish highlands, Nova Scotia. *Canadian Journal of Earth Sciences,* **17**, 400–18.

LANDING, E. & BENUS, A. 1988. Stratigraphy of the Bonavista Group, Southeastern Newfoundland: Growth Faults and the distribution of Subtrilobitic Lower Cambrian. *Bulletin of the New York State Museum,* **463**, 59–71.

LUDVIGSEN, R. & WESTROP, S. 1985. Three new Upper Cambrian Stages for North America. *Geology,* **13**, 139–43.

MAGARITZ, M., HOLSER, W. T. & KIRSCHVINK, J. L. 1986. Carbon-isotope events across the Precambrian/Cambrian boundary on the Siberian Platform. *Nature,* **320**, 258–259.

MATTHEWS, S. C. & MISSARZHEVSKY, V. V. 1975. Small shelly fossils of late Precambrian and early Cambrian age: a review of recent work. *Journal of the Geological Society, London,* **131**, 289–304.

MISSARZHEVSKY, V. V. 1982. Subdivision and correlation of the Precambrian-Cambrian boundary using some groups of the oldest skeletal organisms. *Byulleten' Moskovskogo Obschestva Ispytatelei Prirody, Otdelenie Geologii,* **57**, 52–67. (In Russian).

McKERROW, W. S., SCOTESE, C. R. & BRASIER, M. D. 1992. Early Cambrian continental reconstructions. *Journal of the Geological Society, London,* **149**, 599–606.

MISSARZHEVSKY, V. V. 1983. Stratigraphy of oldest Phanerozoic deposits of Anabar Massif. *Soviet Geology,* **9**, 62–73. (In Russian).

ODIN, G. S. *et al.* 1983. Numerical dating of the Precambrian-Cambrian boundary. *Nature,* **301**, 21–23.

PALMER, A. R. 1971. The Cambrian of the Great Basin and adjacent areas, western United States. *In:* HOLLAND, C. H. (ed.) *Cambrian of the New World.* Wiley-Interscience, London, 1–78.

—— 1973. Cambrian trilobites. *In:* HALLAM, A. (ed.) *Atlas of Palaeobiogeography,* Elsevier, New York, 3–11.

—— 1974. Search for the Cambrian World. *American Scientist,* **62**, 216–224.

—— 1981*a*. On the correlatability of Grand Cycle tops. *United States Department of the Interior Geological Survey, Open-File Report,* 81–743, 160–162.

—— 1981*b*. Subdivision of the Sauk sequence. *United States Department of the Interior Geological Survey, Open-File Report,* **81–743**, 160–162.

—— & JAMES, N. P. 1980. The Hawke Bay Event: a circum-Iapetus regression near the Lower-Middle Cambrian boundary. *In:* WONES, D. R. (ed.) *The Caldeonides in the USA. Proceedings of IGCP Project 27 on the Caledonide Orogen.* Memoir of the Department of Geological Sciences, Virginia Polytechnic Institute and State University, **2**, 15–18.

ROBISON, R. A. 1972. Mode of life of agnostid trilobites. *24th International Geological Congress, Montreal,* **7**, 20–24.

—— 1988. Superclass Trilobitomorpha. *In:* BOARDMAN, R. S., CHEETHAM, A. H. & ROWELL, A. J. (eds) *Fossil Invertebrates.* Blackwell Scientific Publications, Palo Alto, 221–240.

——, ROSOVA, A. V., ROWELL, A. J. & FLETCHER, T. P. 1977. Cambrian boundaries and divisions. *Lethaia,* **10**, 257–262.

ROWELL, A. J. & BRADY, M. J. 1976. Brachiopods and biomeres. *Brigham Young University Geology Studies,* **23**, 165–180.

ROZANOV, A. YU *et al.* 1969. The Tommotian Stage and the Cambrian Lower Boundary Problem. *Trudy Geol. Institut Nauka, Moscow.* 206. (In Russian; English translation, US Department of the Interior, 1981).

SCHLANGER, S. O., ARTHUR, M. A., JENKYNS, H. C. & SCHOLLE, P. A. 1987. The Cenomanian-Turonian Oceanic Anoxic Event, I. Stratigraphy and distribution of organic carbon-rich beds and the marine $\delta^{13}C$ excursion. *In:* BROOKS, J. & FLEET, A. (eds) *Marine Petroleum Source Rocks.* Geological Society of London Special Publication, **26**, 371–399.

SCOTESE, C. R. & McKERROW, W. S. 1990. Revised world maps and introduction. *In:* McKERROW, W. S. & SCOTESE, C. R. (eds). *Palaeozoic Paleogeography and Biogeography.* Geological Society of London, Memoirs, **12**, 1–21.

SCOTT, G. 1985. Homotaxy and the biostratigraphical theory. *Palaeontology,* **28**, 777–782.

SHACKLETON, N. J. & PISIAS, N. G. 1985. Atmospheric carbon dioxide, orbital forcing and climate. *American Geophysical Union, Geophysics Monograph,* **32**, 303–317.

SHERGOLD, J. H. & BRASIER, M. D. 1986. Biochronology of Proterozoic- Cambrian phosphorites. *In:* COOK, P. J. & SHERGOLD, J. H. (eds) *Proterozoic and Cambrian Phosphorites.* Cambridge University Press, 295–326.

TAYLOR, M. F. & PALMER, A. R. (eds) 1981. *Cambrian stratigraphy and paleontology of the Great Basin and vicinity, western United States.* Guide Book for Field Trip 1, Second International Symposium on the Cambrian System, Denver, Colorado.

TUCKER, M. E. 1986. Carbon isotope excursions in the Late Precambrian-Cambrian boundary beds, Anti-Atlas, Morocco. *Nature*, **319**, 48–50

—— & WRIGHT, V. P. 1990. *Carbonate Sedimentology*. Blackwell Scientific Publications, Oxford.

WESTROP, S. R. 1986. *Trilobites of the Upper Cambrian Sunwaptan Stage, southern Canadian Rocky Mountains, Alberta*. Palaeontographica Canadiana, **3**.

WILLIAMS, D. F., LERCHE, I. & FULL, W. E. 1988. *Isotope Chronostratigraphy. Theory and Methods*. Academic Press Geology Series, San Diego.

XIANG, L. (ed.) 1981. *The Stratigraphy of China. Volume 4. The Cambrian System*. Geological Publishing House, Beijing. (In Chinese).

XU, D.-Y., YANG, Z., SUN., Y.-Y., HE, J.-W., ZHANG, Q.-W. & CHAI, Z.-F. 1989. *Astrogeological Events in China*. Scottish Academic Press, Edinburgh and Geological Publishing House, Beijing.

ZHURAVLEVA, I. T., REPINA, L. N. & ROZANOV, A. YU. 1990. *Stage subdivision of the lower Cambrian*. Abstracts of the Third International Symposium of the Cambrian System, Novosibirsk, 178–179.

Index

abyssal sedimentation rates 20
Acarinina wilcoxensis berggreni 131
accumulation rate determination 19–20, 46–8
acme zones, Quaternary 146–7
Africa, North West 41–2
Ailsa Craig 30
Alabama Cretaceous stratigraphy 247–9, 251
Alborian Sea 142
alcohols as biomarkers 51–3
alkanediols as biomarkers 52, 53
alkanes as biomarkers 51–3
alkenones as biomarkers 52, 53
aluminium in marls 213–14
ammonoid zones
    Carboniferous 261
    Jurassic 10, 262
Anglo-Paris Basin marl stratigraphy 214–22
Angola Shale 286, 287
Annona Chalk 249, 250
Antrim Plateau 30
apatite 173, 176
Aquilapollenites 14
$^{40}Ar/^{39}Ar$ dating 179, 181
Archoceras spp. 279
Arcola Limestone 248, 251
Arkansas Cretaceous stratigraphy 249–52
Arran granite 30
Atlantic Ocean
    Cenozoic sediments 127
    coastal stratigraphy 246–7
    foraminifera distribution 127–9
    nannofossil distribution 142
    Quaternary stratigraphy 147–9
    tephrostratigraphy 176
Australia, Western
    facies studies
        transgression 313–15
        transgression causes 315–17
        types 312–13
    geological setting 293, 294
    goniatite stratigraphy
        early studies 295
        new species 297–300
        recent studies 300–7

Barandeograptus spp. 327
Barton Beds 22
Bearpaw Shale 245, 251
Beeston Chalk 244
Beggars Knoll 220
Beinn à Ghraig granophyre 30
Beinn an Dubhaich granite 30
Belgium 112, 114–16
Beloceras spp. 278, 306, 307, 310–11
bentonite, 169, 171
biogeochemical markers *see* biomarkers
biomarkers
    analytical methods 54–8
    introduction 51–3
    numerical analysis 58–9

sampling 53–4
techniques discussed 59–60
biostratigraphy
    macrofossil 257–8
        Cretaceous 199, 205
        Jurassic 261–3
        Devonian
            goniatite regional studies 274–82, 295–300
            goniatite systematics 273–4
            goniatite zones 270–3, 307–12
        Carboniferous 260–1
        Silurian 259–60
            correlation factors 329–32, 335
            Telychian biozones 332–4
            Telychian graptoloids 322–9
            Telychian stage boundaries 323–4
        Cambrian 258–9
    microfossil 3, 257
        see foraminifera; nannofossils; ostracoda;
        radiolaria
Blackstone River Section 330
Blanc, Cap 41–2
Blind Rock dyke 30
Bluffport Marl 249
Bornholm section 334, 335
boundary stratotypes 9, 16–17
Bracklesham Beds 22
Bridgewick Marl 223
British Tertiary Igneous Province magnetic timescale
    29–34
Brownstown Marl Formation 250
bundle 187
Burghclere 220

$\delta^{13}C$ stratigraphy
    Cretaceous 203, 234, 235
    Cambrian
        applications 343–4
        interpretation 345–6
        technique 342–3
Caburn Marl 222, 223
calcium carbonate stratigraphy 3, 140–1
    correlation with magnetic stratigraphy 67, 70, 72
    use in palaeoclimate analysis 88–91
calibration methods 9
Cambrian studies
    biostratigraphy 258–9, 341–42
    $\delta^{13}C$ stratigraphy
        applications 342–3
        interpretation 345–6
        technique 342–3
    lithostratigraphy 343
Campanian
    sea levels 242–4
    stratigraphy 251
Canada 14, 330
Canadaway Group 276, 284
Candland Shale 345
Canning Basin
    facies studies

transgression 312–15
transgression causes 315–17
types 312–13
geological setting 293, 294
goniatite stratigraphy
early studies 295
new species 297–300
recent studies 300–7
carbon isotope stratigraphy
Cretaceous 203, 234, 235
Cambrian
applications 343–4
interpretation 345–6
technique 342–3
carbonates *see* calcium carbonate
Carboniferous studies
biostratigraphy 260–1
cyclicity 188
tephrostratigraphy 171, 173
Caribbean Sea 128
Carlingford Complex 30
Cashaqua Shale 278, 285
Cassigerinelloita amekiensis 131, 134
Castile evaporites 191
Catskill Delta 267, 287–9
Catton Sponge Beds 244
Cerig Formation 323
chalk
biostratigraphy 199
lithology 198
lithostratigraphy 198–9
marl band stratigraphy 214–22
$^{87}Sr/^{86}Sr$ isotope stratigraphy
methods 201–3
results 203–5
summary 206–8
Chalk Rock 218, 224–5
chemical analyses 173–6
Chemung Fauna 267
Chiloguembelitria spp. 131
chronograms 10–15
chronometric scale 9, 16–17
chronostratic scale 9, 16–17
Claggett Shale 245, 251
Clayton Formation 249
CLIMAP 4
climate analysis 88–91, 141–3, 289
clinoptinolite 213, 214
coccolith stratigraphy 148, 150
Coire Vaigneach granite 30
completeness in sediment record 19–20
Conochair granite 30
contamination in cores 87–8
Copt Point 230
cores and coring
core handling problems
contamination 87–8
correlation 73–81
gaps 84
repetition 84–7
core logs 68–72
methods of coring
extended core barrel (XCB) 4
hydraulic piston (HPC) 4

piston 2–3
Corset Spring Shales 345
Corsicana Clay 251, 252
Craven Basin 257
Cretaceous studies
cyclicity 188
foraminiferal stratigraphy
extinction events 233
global bioevents 233–9
non-sequences 231–3
zonal boundaries 227–31
marl correlations 214–22
ostracod stratigraphy 155
$^{87}Sr/^{86}Sr$ stratigraphy
correlations 206
methods 201–3
modelling 206
results 203–5
standardization 195–8
summary 206–8
sea levels
Europe 242–4
USA 244–52
stage boundaries 242
tephrostratigraphy 171, 177
Crickites 279
Cusseta Sand 247
cyclicity
causes 190–1, 287–9
identification 191–2
mathematical analysis 192
Milankovitch type 188–90, 287
nature 187–8
Cypridea 155
Cyrtograptus spp. 328–9

Dachstein Kalk 191
Danian 11–12, 14, 15, 18, 19
Dartry Limestone 191
Deep Sea Drilling Project 4–5
foraminifera studies 127–9
Delaware Cretaceous stratigraphy 246–7
delta switching 289
Demopolis Chalk 247, 248, 251
Denmark 171, 176
density logs 171
Dessau Chalk 249
Devonian goniatite stratigraphy
cyclical sedimentation 287–9
facies indicators 282–6
history of work 267–70
international divisions 307–12
regional studies
Canning Basin
early work 295
new species 297–300
recent work 300–7
New York State 274–82
systematics 273–4
zonation 270–3
diatom stratigraphy 143
dilution cycle 190
dispersal rates 329
dissolution cycle 190

Diversograptus spp. 328
Donegal granite 30
Dresbachian Stage 344–5
DSDP *see* deep sea drilling project
Duck Creek Formation 249
Dunkirk Shale 276, 282, 286

Eagle Sandstone 245
Earnley Formation 101, 119
Eastchurch Gap 108, 109
eccentricity, orbital
   cycles 189
   recognition 191–2
EDIP 58, 62
Edwards Group 249
Egem 115–16
Eigg lava 30
electric logs 171
Emiliania spp. 145, 146
Eocene
   boundary studies 5, 131
   foraminifera 131, 132
   ostracoda 157, 159, 161–2
   regional studies
      Belgium 112–17
      England 106–7
      France 117–18
   sea level 123
   tephrostratigraphy 171, 176
eustasy
   Eocene 123
   Cretaceous
      Europe 237, 238, 241–2, 243–4
      USA 244–52
   Devonian
      Australia 315–17
      USA 289
Eutaw Formation 248
Ewelme 220
extended core barrel corer (XCB) 4
extinction 233

feldspars in tephras 175
first appearance datum (FAD) 3
   Cambrian fauna 341–42
   nannofossil stratigraphy 146
   ostracod stratigraphy 161–5
Fishnish dyke 30
fission track dating 180
flaser marl 211
Fognam Marl 219, 222
foraminifera
   Quaternary
      palaeoclimate analysis 141–2
      zones 143–6
   Tertiary
      lineage zones 135
      stratigraphic ranges 131–3
   Cretaceous
      extinction events 233
      global bioevents 233–9
      non-sequences 231–3
      zonal boundaries 227–31
Fourier spectral analysis 58

Fourier transforms 192
Fox Hills Sandstone 246, 251
France 117–18
Frasnian Stage
   Australia
      facies movements 312–15
      goniatite zones 295, 300–7
      sea level changes 315–17
   International goniatite zones 295–7, 307–12
   USA
      goniatite zones 270–3
      sea level curves 288
Frontier Formation 177

Gachlingen 10
gamma ray logs 171
Gammon Shale 245
gaps in core sequences 84
gas chromatography 54–5
Genersee Gorge 267
Genesee Group 284
Geneseo Shale 276, 283
Genundwa Limestone 276, 283
geomagnetic events *see* magnetic polarity stratigraphy
Georgetown Group 249
Gephyrocapsa spp. 145, 146
Germany 243–4
glacial terminations 60, 61
glaciations and orbital forcing 39
glass phase mineralogy 172, 174
glauconite 213, 214
Globigerina spp. 131, 144
Globoquadrina spp. 144
Globorotalia spp. 129
   G. crassaformis 144
   G. fimbriata 144
   G. hirsuta 144
   G. menardii 143, 144
   G. tumida flexuosa 144
Globorotalia spp.
   G. truncatulinoides
      distribution
         palaeogeography 127–8
         present day 127
      evolution 128–9
      use in datum fixing 127, 143–4
Glynde Marl 214–16, 222, 223
Glyndebourne borehole 229, 230
Glyptograptus spp. 324
Gober Chalk 249, 250, 251
Gogo Formation 281, 312
Gogoceras spp. 297–9
goniatites
   cyclical sediment indicators 287–9
   facies indicators 282–6
   history of studies 267–70
   international zonation 307–12
   regional studies
      Canning Basin
         new species 297–300
         new zones 300–7
      New York State 274–82
   systematics 273–4
   zonation 270–3

Gowanda Shale 282
Grand Cycles 341
graptolites 259–60
    correlation factors 329–32
    Telychian biozones 332–4
    Telychian characters 324
        diplograptids 324–6
        monograptids 326–9
Great Basin
    Cambrian lithologies 343
    Cambrian stratigraphy 343–4
Green River Formation 191
Greenland 328

Hampshire Basin 22
    magnetic polarity stratigraphy 110–11
    lithostratigraphy 101, 102, 103
Hanover Shale 276, 279, 282, 286
Harty borehole 107, 108, 110
Hawke Bay Event 341
hemera 261
hiatus recognition 42–3
hopanes and hopanols as biomarkers 52, 53
Hornerstown Formation 247
Howgill Fells 330
hydraulic piston corer (HPC) 4, 39

Ieper Clay Formation 101, 112, 116–17, 119
illite 214
Indian Ocean
    Cenozoic marine correlations 127
    diatom zones 142
    foraminifera distribution 127
    ODP sites 68
isothermal remanent magnetism (IRM) 105–6
isotope stratigraphy see carbon; oxygen; strontium
Ithaca Formation 275

Jenkinsina triseriata 131, 134
Johns Wash Limestone 345
JOIDES 2
Judith River Formation 246
Jurassic studies 10, 261–3

K–T boundary 12–14
KAr dating 10, 179, 181
kaolinite
    in marls 214
    in tonstein 169, 172, 173
Kellwasser Limestone 317
Kensworth 220
Koenenites spp. 277, 303, 309
Kortemark 115
Kortrijk 114–15

Lägerdorf 243, 244
lake basin sedimentation rates 20
Landen Formation 101
Lang Randen 10
Larry Bane Chalk 244
last appearance datum (LAD) 3
    Cambrian fauna 341–2
    foraminifera 146
    ostracod stratigraphy 161–5

Latimer Marl 219, 222
limonite 214
Lockatong Formation 191
Lodi Limestone 275, 276, 283
London Basin 22, 101, 102–3
London Clay Formation 22, 101, 102
    magnetic polarity stratigraphy 106–8, 119
lumping 328
Lundy Island 30

Maastrichtian 11–12, 14, 15, 18, 19
    sea level 242–4
    stratigraphy 251
macrofossils 257–8
    Cretaceous 199, 205
    Jurassic 261–3
    Carboniferous 260–1
    Silurian 259–60
    Cambrian 258–9
    see also goniatites also graptolites
MACROGEN 58, 62
McShan Formation 248
Madeira Abyssal plain 147, 148
magnetic polarity
    stratigraphy
        methods of analysis 103–6
        results
            Belgium 112–17
            England 106–11
            France 117–18
        results discussed 118–22
    timescale development 23
        applications 11, 29–34
        introduction 9, 27
        mathematics 34–6
        problems 27–8
        testing 28–9
magnetic susceptibility
    defined 66
    methods of measurement 67
    results
        core contamination 87–8
        core correlation 73–81
        core gaps 84
        core logs 68–72
        core repetition 84–7
        tephrostratigraphy 81–3
    results discussed
        core correlation 91–2
        core repetition 92
        use in palaeoclimate analysis 88–91, 92–3
magnetostratigraphy 3
Manticoceras spp. 279, 303–4, 304–5, 307
Marlbrook Marl 250
marls
    correlations 214–22
    formation 211
    geochemistry 212–14
Marshalltown Formation 247, 251
mass spectroscopy
    isotope analysis 39
    molecular stratigraphy 55–8
Massignano 5
Maternoceras spp. 306

Mediterranean Sea tephrostratigraphy 178, 179
Melborn Rock 233
Merchantville Formation 246, 251
Mesobeloceras 278, 305–6, 310
Metaclimacograptus spp. 326
Mexico, Gulf of
    Cenozoic stratigraphy 127–8
    Cretaceous stratigraphy 247–9
microfossils *see* diatoms; foraminifera, nannofossils;
        radiolaria
Middlesex Shale 276, 285
Milankovitch cycles
    mathematical analysis 192
    recognition 23–4, 191–2
    sedimentary evidence 138, 189–90
        Devonian 287, 289
    theory 4, 6, 188–9
mineralogy of tephras 172–3
Miocene
    boundary studies 5
    ostracoda 163–4
Miravetes Formation 191
Mississippi Cretaceous stratigraphy 247–9, 251
Mixomanticoceras spp. 299–300
Mobridge Limestone 246
molecular stratigraphy
    analytical methods 54–8
    introduction 51–3
    numerical analysis 58–9
    sampling 53–4
    techniques discussed 59–60
Monoclimacis spp. 324–5
    M. griestoniensis 336
Monograptus spp. 327
Montana Cretaceous stratigraphy 244–6, 251
Mooreville Formation 248
Mourne Mountains granite 30
Mowry Formation 177
Muck lava 30
Mull granite 30

Nacatoch Sand Formation 252
nannofossils
    Cretaceous 11, 199, 202
    Quaternary 142–3, 146
Naples Fauna 267
natural remanent magnetism (NRM) 103–5
Navesink Greensand 247, 251
Neomanticoceras spp. 279, 306–7, 311
Neogloboquadrina pachyderma 143, 144
New Jersey Cretaceous stratigraphy 246–7, 251
New Pit Marl 216, 223
New York State
    cycles of sedimentation 287–9
    Devonian outcrops 268, 269
    facies distributions 282–6
    goniatite zones 274–82
North Sea
    Eocene palaeogeography 99, 100
    Eocene stratigraphy 101, 102, 176
Norwegian Sea diatom zones 142
Nullara Limestone 293, 294

δ¹⁸O stratigraphy 3–4

Cretaceous 203
Quaternary 137–40
role in sequence slotting 41–2
obliquity, orbital
    cycles 189
    recognition 191–2
Ocean Drilling Program (ODP) 5
    leg 115 magnetic susceptibility measurements
        core contamination 87–8
        core correlation 73–81
        core gaps 84
        core logs 68–72
        core repetition 84–7
        marine tephra 81–3
    site 658 core matching
        accumulation rate 46–8
        data manipulation 43–4
        reference sequence 41–3
        sequence tightness 44–6
        slotting constraints 46
oceanic sediment record, value of 2–3
Oedelem Sands 101, 119
Oldhaven Formation 22, 101, 108
Oligocene
    boundary studies 5
    foraminifera 132, 133
    ostracoda 163
oozes, rate of accumulation of 20
orbital forcing 39, 92–3, 289
    see also Milankovitch cycles
Ordovician studies 176
organic chemical markers *see* biomarkers
ostracoda
    benthonic characters 156
    Cenozoic FAD/LAD data 161–5
    diversity studies 156–9
    orders 160
    use in biostratigraphy 155–6, 160
oxygen isotope stratigraphy 3–4
    Cretaceous 203
    Quaternary 137–40
    role in sequence slotting 41–2
Ozan Formation 250

Pacific Ocean
    Cenozoic sediments 127, 128
    diatom zones 142
    foraminifera distribution 128
    ostracod stratigraphy 157, 158, 159
    tephrostratigraphy 172
Paddy's Point 109, 110
Pakistan 21, 22
palaeoautecology, graptolite 330
palaeobiogeography, graptolite 327
Palaeocene ostracoda 157, 159, 161
palaeoclimate analysis 141–3
    cyclicity 289
    role of magnetic susceptibility 88–91
palaeogeography, North Sea 100
Parkman Sandstone 246, 251
Pecan Gap Chalk 249, 250, 251
Peel River Section 330
Penn Yan Shale 275, 283
Permian studies 191

Petalolithus spp. 325
phenocrysts in tephras 172, 174–5
Pierre Shale 246
Pillara Limestone 293, 294, 312
Pipe Creek Shale 276, 279, 286, 316
piston cores 2–3
platform sedimentation rate 20
Playfordites spp. 279, 300, 306, 311
Plenus Marl Formation 233
Pliocene
    boundary studies 5, 127–8
    ostracoda 164
polarity reversals *see* magnetic polarity stratigraphy
polarity timescale *see under* magnetic polarity
Ponticeras spp. 274–5, 304, 309
Portage Fauna 267
Portrush Chalk 244
Praetenuitella spp. 133, 134, 135
Prague Basin 328
Prairie Bluff Chalk 249, 251
precession, orbital
    cycles 189
    recognition 191-2
principal components analysis 59
Pristiograptus spp. 325
Probeloceras spp. 277, 304, 305, 310
Prochorites spp. 278, 305, 310
production cycles 190
Protimanites spp. 303
Pseudoemiliania spp. 146
Pulleniatina finalis 144
pyroxenes in tephras 175

Qiaotin section 334, 335
Qiongzhusi Formation 346
quartz in tephras 176
Quaternary studies
    base defined 127, 137
    micropalaeontological zones 143–6
    ostracoda correlations 165
    stage identification 138–40

radioisotopic dating 179, 181
radiolaria 143
Rastrites spp. 326
Rb/Sr dating 10, 179, 181
Reading Beds 22
Red Deer Valley 14
redox cycles 190
Reed Marl 219, 222
regressions, Cretaceous 241–2
    Europe 243–4
    USA 244–52
repetition in cores 84–7
resolution in stratigraphy 2
Retiolites spp. 325
    R. angustidens 330
    R. australis 330
Rhinestreet Shale 276, 285–6, 287
Rhum granophyre 30
rhythmical sedimentation
    causes 190–1, 287–9
    identification 191–2
    mathematical analysis 192

Milankovitch type 188–90, 287
    nature 187–8
river channel sedimentation rates 20
Rosselo, Capo 5

Sadler Limestone 313
St Kilda granite 30
Sandbergoceras spp. 277, 309–10
Sandy Braes 30
Saratoga Chalk 249, 251
Scaglia bianca 191
scandium in marls 214, 216, 217, 219
Schindewolfoceras spp. 278
Scisti a Fucoidi 188, 191
scour cycles 190–1
sea level curves
    Eocene 123
    Cretaceous 238, 241–2, 243–4, 244–52
    Devonian 288, 316
sea level and facies distribution
    Australia 315–17
    USA 287–9
sea surface temperature and U^{k}_{37} parameter 60
sedimentation rates 19–20
seismic sections 171–2
Selma Chalk 249
sequence slotting
    data manipulation 43–4
    methodology 40–1
    slotting constraints 46
    slotting tightness 44–6
sequence stratigraphy 5
Serramanticoceras spp. 300, 306
shelf sedimentation rates 20
Sheppey, Isle of 106–9
Shillingstone Hill 222
Silurian biostratigraphy 259–60
    correlation factors 329–32
        intercontinental 335
    Telychian biozones 332–4
    Telychian boundaries 323–4
    Telychian graptoloids 324
        diplograptids 325–6
        monograptids 326–9
Sinodiversograptus spp. 328
Siwalik Group 21, 22
Skye granite 30
Slemish plug 30
Sm/Nd dating 181
smectite 169, 214
Sonyea Group 284
Southerham Marl 222, 223
Southern Ocean diatom zones 142
Spartivento, Capo 5
Sphaeromanticoceras spp. 274, 279, 305, 307
Spirograptus spp, 325–6
Spurious Chalk Rock 221–2, 224–5
^{87}Sr/^{86}Sr stratigraphy *see* strontium
Steptoean Stage 344–5
sterols as biomarkers 22, 53
Stimulograptus spp. 327
Stomatograptus spp. 330, 331
Streptograptus spp. 327
strontium isotope stratigraphy

correlations 206
methods 201–3
modelling 206
results 203–5
standardization 195–8
summary 206–8
Studland Bay 196, 199, 206
suture diagrams 272
Switzerland 10
synonymy 330

Taghanic Event 270
taxonomy, graptolite 330–2
Teapot Sandstone 246
Telychian Stage
biozones 332–4
boundaries 323–4
graptolites 324–9
Tenuitella spp. 133, 134, 135
tephrochronology
applications 178–82
defined 169, 170
tephrostratigraphy
applications 176–8
defined 170
relation to magnetic susceptibility 81–3
techniques
chemical 173–6
physical 170–3
terminations 138
Tertiary studies
boundary recognition 5, 131
foraminifera 131–5
ostracoda 161–5
Texas Cretaceous stratigraphy 249–52
Thanet Beds 22
thermoluminescence dating 180–1
thickness-time relationships 19–20
Timanites spp. 277, 303, 309
Tinton Greensand 247, 251
titanite in tephras 176
titanium in marls 214, 216, 217, 219
titanomagnetite in tephras 175
Tombigbee Sand 248
tonstein 169, 171, 173, 178
trace element analysis
marls 214, 225
tephras 173–4
Trannon section 329
transgressions, Cretaceous 241
Europe 243–4
USA 244–52
Triassic studies 188
trilobite zones 258–9
Trubi Marls 191
Trunch borehole
isotope stratigraphy
methods 201–3
results 203–5
summary 206–8
lithologies 198
location 196
stratigraphy 197, 198–9

Tully Limestone 283
turbidites 43
Madeira Abyssal Plain 147–9
Turonian subsidence 224
Tuscaloosa Gravels 248

U/Pb dating 181
U^k_7 parameter 60
USA
Cretaceous stratigraphy
Atlantic coast 246–7
Mexican Gulf 247–9
Montana 244–6
Texas 249–52
Wyoming 177
Devonian stratigraphy
cyclicity 287–9
facies changes 282–6
goniatite systematics 273–4
goniatite zones 270–3
history of studies 267–70
New York State 274–82
Ordovician stratigraphy 176
Cambrian stratigraphy 343–4

vanadium in marls 214, 216, 217, 219
Varengeville 117–18
Varengeville Formation 101
Virgin Hills Formation 281, 312, 313
Virginia Water Formation 103, 108, 119
Vlierzele Formation 101, 119
volume susceptibility 66, 67
VW Formation 101

Walsh transforms 192
Warden Bay borehole 107, 108, 110
Warden Point 108, 109
Wardrecques 113
Washington Land 330
Wealden deposits 155
Wellsites spp. 273–4, 278
Welsh Basin 328
West Falls Group 284
West River Shales 285
Whitecliff Bay 110–11
whole-core magnetic susceptibility (WCMS) *see* magnetic susceptibility
Wight, Isle of 222
Windjana Limestone 312
wireline logs 171
Wittering Formation 101, 102–3, 110–11, 119
Wolfe City Formation 250
Woolwich Beds 22
Wormwood Formation 323
Wyoming tephrostratigraphy 177

Yangtze Platform 330
Younger Dryas 142
Ypresian Stage 98–102
yttrium in marls 214, 216, 217, 219

zircon in tephras 176
Ziyang section 329